Soil Health Management for Plantation Crops

George V Thomas • V. Krishnakumar
Editors

Soil Health Management for Plantation Crops

Recent Advances and New Paradigms

Springer

Editors
George V Thomas
Former Director
ICAR- Central Plantation Crops Research Institute
Kasaragod, Kerala, India

V. Krishnakumar
Former Principal Scientist and Head
ICAR- Central Plantation Crops Research Institute
Kayamkulam, Kerala, India

ISBN 978-981-97-0091-2 ISBN 978-981-97-0092-9 (eBook)
https://doi.org/10.1007/978-981-97-0092-9

This Springer imprint is published by the registered company Springer Nature Singapore Pte Ltd.
The registered company address is: 152 Beach Road, #21-01/04 Gateway East, Singapore 189721, Singapore

Foreword

Plantation crops, being perennial tree crops with long life span, are important with respect to their capacity to deliver ecosystem services, including regulatory and supportive services, to build up healthy agro-ecosystems. Plantation crops sustain livelihood and socio-economic status of millions of people globally in tropical agro-ecosystem. The production systems of plantation crops are experiencing stress due to environmental factors concerning production base deterioration, dwindling biodiversity resources and decline in soil fertility. There is an urgent need to transform the plantation agriculture systems to make them more productive, environmentally sustainable and resilient while preserving and enhancing the ecological base of farming.

It is now recognized that anthropogenic interventions since the advent of organized agriculture, particularly intensive use of synthetic fertilizers alone as nutrient source and application of chemical pesticides with the objective of increasing production and the unscientific management practices followed, had adverse impact on soil resource base and fertility. In the prevailing agricultural scenario, degradation of soil characterized by decline in quality and decrease in goods and services of ecosystems is considered an important constraint in achieving sustainable agricultural production. The situation calls for new directions in soil health management based on integrated approaches with focus on ecological principles to attain sustainable intensification in plantations. Soil, a dynamic system in which close interplay among abiotic and biotic entities governs several chemical and biogeochemical processes and enzyme activities, needs to be managed intelligently to harness its potential to sustain biological productivity, maintain environmental quality and promote plant health and productivity. To mitigate the climate change, the great potential of plantation crops as carbon sinks needs to be harnessed as they are committed to the same piece of land for long number of years and capable of retaining more carbon per unit land area than other vegetation.

The book provides information on new paradigms of soil health-based pathways for sustaining plantation crops and diversified systems based on perennial crops. This compilation identifies technological interventions to translate science into action to mitigate environmental footprint and climate change. I congratulate the editors for taking the initiative in bringing out this compilation which will be of great value to researchers, students, extension personnel, developmental agencies and policy planners.

Government of India, Department of Agricultural Research and Education (DARE), Indian Council of Agricultural Research (ICAR), Ministry of Agriculture and Farmers Welfare

New Delhi, India
31 October 2023

Himanshu Pathak

Preface

Soil, the critical resource which forms the basis of all terrestrial life, is very vital in food systems in view of the fact that sustainable agricultural production depends on the health, quality and functionality of the soil. Enhancing and sustaining soil health are critical to the health of plants, animals, people, environment and the planet. In the past century, the intensification of agriculture using chemicals and fertilizers, though contributed to several fold increase in production and productivity of crops, had adverse impact on soil quality and environment. Unsustainable agricultural practices, soil erosion, pollution, climate change and dwindling biodiversity resources have led to deterioration of soil health. In the emerging scenario, as the world population grows and global demand for food is escalating, it has become crucial to enhance agricultural production ensuring soil quality and health as well as the ecosystem services of soil.

Plantation crops including palms—coconut, arecanut and oil palm; beverage crops—tea, coffee and cocoa; commercial crops—cashew and rubber and spices—black pepper and small cardamom, in general, are high-value crops with considerable significance in terms of livelihood security, commerce and trade, and they are grown in several countries in tropical humid regions in resource-limited soils with poor physical properties and low native fertility. Their cultivation in high rainfall regions in light-textured soils leads to leaching of nutrients, thereby depleting the soil of precious nutrients. Being perennial in nature, they mine nutrients from very limited volume of soil and hence, nutrients have to be replenished to maintain the fertility of the soil. Soil constraints with respect to very low organic carbon content, multi-nutrient deficiencies, low carbon sequestration, strong acidic or alkaline reactions, salinity and low cation exchange capacity have been reported to affect the plantation crops, necessitating technological interventions with appropriate soil health improvement strategies to achieve sustainability in production. Global climate change has become a reality and plantation crops, committed to the same land for decades, face bigger challenge from climate change effects. Therefore, it is felt necessary to prepare a book by compiling research results from around the world (where various plantation crops are being cultivated) to refocus attention on the dangers of continuing soil degradation as well as on the need to take up appropriate measures for ensuring a positive nutrient balance for various plantation crops.

At the time when soil deterioration has become a serious constraint affecting the agricultural productivity, it was felt necessary to gather the basic information as well as the latest findings on soil health available globally on each of the plantation crops to form the strong basis in our scientific pursuit to address the challenges faced by the crops. Soil health management strategies should aim at improvement in biological, chemical and physical features of soil that are essential to long-term, sustainable agricultural productivity with minimal environmental impact. Future strategies for sustainable agriculture through soil health management need to be based on biodiversity and conservation-based plantation agriculture to enhance sustainable system productivity and ecosystem services. Strategies to enhance carbon sequestration by terrestrial woody biomass and below ground systems assume great significance to mitigate the adverse effect of climate change on plantation production systems. Optimizing beneficial interactions with rhizosphere microorganisms has been proposed to reduce reliance on external inputs, increase pathogen resistance, and alleviate abiotic stresses. Also, further research is needed to elucidate the dynamic interaction between crop plants, the rhizosphere micro-biome and the environment to harness micro-biome to increase crop yield and quality.

This book containing various chapters contributed by experts in various plantation crops elucidates state-of-the-art information, providing new paradigms of soil health-based pathways for sustaining plantation crops and diversified systems. The book covers soil characteristics, soil fertility constraints, issues of soil contamination, indicators of soil health (physico, chemical and biological properties) and soil health assessment, use of soil amendments, impact of climate change on plantation soils, fertility management for healthy soils, soil biodiversity and biological functions and micro-biome and meta-genomic approach. The book also includes technological options to achieve sustainable production encompassing soil health improvement comprising multistrata multispecies cropping systems, integrated farming system, conservation agriculture practices, cover cropping and green manuring, crop residue recycling, bio-fertilizer and bio-stimulant technologies and organic farming systems. The book is also enriched with a special directional chapter focusing soil biological fertility as a new foundation for sustainability in plantation crops. We are sure that this book will be immensely useful to professionals, academicians and other stakeholders interested in sustainable production of plantation crops.

Konni, India — George V. Thomas
Kayamkulam, India — V. Krishnakumar

Acknowledgement

It is our proud privilege to place on record our respectful gratitude to Dr Himanshu Pathak, Secretary, Department of Agricultural Research and Education and Director General, Indian Council of Agricultural Research, Government of India, for his encouragement and guidance by contributing foreword for the book.

We are indeed grateful to the directors of plantation crops research institutes viz. Dr KB Hebbar, ICAR-Central Plantation Crops Research Institute, Kasaragod; Dr R Dinesh, ICAR-Indian Institute of Spices Research, Kozhikode; Dr K Suresh, ICAR-Indian Institute of Oil Palm Research; Dr Dinakara Adiga, ICAR-Directorate of Cashew Research, Puttur; Dr MD Jessy, Rubber Research Institute of India, Kottayam; Dr R Victor J Ilango, Tea Research Institute, Valparai and Dr M Senthil Kumar, Central Coffee Research Institute, Balehonnur, for their valuable support in permitting the scientists to contribute the chapters on the respective crops. Our sincere gratitude to Dr KUK Nampoothiri and Dr Anitha Karun, Former Directors of ICAR-Central Plantation Crops Research Institute, Kasaragod, for their encouragement and support to bring out this book. We thank Shri SR Prabhu, Chief Scientific Officer, TerraBioGen Technologies, Burnaby, British Columbia, Canada, for his sincere help in enriching this book. The editors wish to thank profusely the authors of individual chapters for their painstaking efforts in collecting voluminous information and compiling them within the timeframe.

It is with great pleasure that we thank the publisher, M/s Springer Nature, for the excellent printing and time bound publication of the book.

ICAR-Central Plantation Crops Research Institute
Kasaragod, Kerala, India — George V. Thomas

ICAR-Central Plantation Crops Research Institute
Regional Station, Kayamkulam, Kerala, India — V. Krishnakumar

About the Book

Plantation crops are perennials, cultivated in tropical and sub-tropical agroecosystems in a contiguous area and comprise estate crops which include tea, coffee and rubber and small holders' plantation crops such as coconut, arecanut, oil palm, cashew, cocoa and spices. They are high-value crops with considerable significance in livelihood security, commerce and trade and are grown in ecologically vulnerable regions such as coastal belts, hilly areas and regions with high rainfall and high humidity, which makes maintenance of soil health a major challenge.

Sustainable agricultural production depends on the health, quality and functionality of the soil, the critical resource which forms the basis of all terrestrial life. Enhancing and sustaining soil health are critical to the health of plants, animals, people, environment and the planet. In the emerging scenario, it has become very crucial to enhance agricultural production ensuring soil quality and health as well as the ecosystem services of soil.

This book contains chapters contributed by experts in various plantation crops and elucidates state-of-the-art information, provides new paradigms of soil health-based pathways for sustaining plantation crops and diversified systems. The book covers soil characteristics, soil fertility constraints, issues of soil contamination, indicators of soil health and soil health assessment, use of soil amendments, impact of climate change on plantation soils, fertility management for healthy soils, soil biodiversity and biological functions and micro-biome as well as meta-genomic approach.

The book also includes technological options to achieve sustainable production encompassing soil health improvement comprising multistrata multispecies cropping systems, integrated farming system, conservation agriculture practices, cover cropping and green manuring, crop residue recycling, bio-fertilizer and bio-stimulant technologies and organic farming systems. The book is also enriched with a special directional chapter focusing soil biological fertility as a new foundation for sustainability in plantation crops. The book intends to share the advances made in soil health management for the benefit of researchers, academicians and other stakeholders to address the complex problems.

Contents

Editors and Contributors

About the Editors

George V. Thomas is formerly Director of the Central Plantation Crops Research Institute (CPCRI), Kerala, India, one of the main Horticultural Institutes under the Indian Council of Agricultural Research (ICAR). After completing his Master's and Doctoral degrees in Microbiology from Punjab Agricultural University, Dr Thomas joined Agricultural Research Service of ICAR at CPCRI in 1978 and rose to the position of Director of the Institute. During his tenure as Director (2006–2014), he guided research programmes on plantation crops-coconut; arecanut and cocoa and plantation crop-based cropping systems. Dr Thomas, in his research career spanning 36 years on plantation crops, on basic and applied aspects of soil microbiology and on-farm residue biomass recycling, could highlight the potential of microorganisms and organic farming technologies for management of soil health and fertility leading to sustainable improvement in yields of these crops. His work towards creating wealth from waste biomass of coconut has fetched him the prestigious ICAR Outstanding Team Research Award for the biennium 2005–2006, as team leader, for the work 'Value addition to coconut waste biomass through biological approaches for enhancing soil health, crop productivity and income generation'. He contributed immensely with significant outcomes as Consortia Leader for a National Agricultural Innovation Project—NAIP of ICAR and Country Leader for a project from International Fund for Agricultural Development (IFAD-COGENT).

For his leadership role in crop improvement programme of coconut, Dr Thomas was selected as Chairman of the International Coconut Genetic Resources Network (COGENT) of Biodiversity International in July 2012. Dr Thomas visited China, Philippines, Thailand, Cote d' Ivoire and Sri Lanka, for scientific deliberations in connection with international collaborative projects on plantation crops. He is a Fellow of the Indian Society for Plantation Crops (ISPC), Confederation of Horticulture Associations of India (CHAI), Association of Microbiologists of India (AMI) and Honorary Fellow, Society for Applied Biotechnology (SAB). He was also President of the Indian Society for Plantation Crops (2008–2009) and served as Editor, Journal of Plantation Crops (2001–2006) and recipient of Dr CS Venkataram Memorial Award.

Dr Thomas has to his credit over 200 research publications including research papers in high impact factor international and national journals, reviews, technical bulletins, and book chapters and edited 10 books. As Emeritus Scientist (2016–2020) of Kerala State Council for Science, Technology and Environment, he contributed significantly to the research on 'Bioconversion of residues from major crops of Kerala to value-added organic resource for sustainable farming'. Dr Thomas also served as Director, Council for Food Research and Development of Government of Kerala.

Address for correspondence: Vettimoottil house, Peroorkulam, Vakayar PO, PIN 689698, Kerala, India

Contact No.: 91-9446051847; E mail: georgevthomas52@gmail.com

V. Krishnakumar Born on 6 September 1957, in Kayamkulam, Kerala, India, earned his graduation and post-graduation in Agriculture from the College of Agriculture, Vellayani, Thiruvananthapuram, under the Kerala Agricultural University and doctoral degree in Agronomy from the Tamil Nadu Agricultural University, Coimbatore, India. He started his career as Assistant Professor, KAU during 1982. He has worked both in the State and Central Government service as Assistant Agronomist in the Rubber Research Institute of India (Rubber Board), Kottayam (Ministry of Commerce, Govt. of India); Senior Scientist (Transfer of Technology and Agronomy and Soil Science) in the Indian Cardamom Research Institute (Spices Board), Myladumpara, Idukki, Kerala (Ministry of Commerce, Govt. of India). He joined the ICAR service as Senior Scientist (Agronomy) at ICAR-Central Plantation Crops Research Institute, Kasaragod, during December 2001 and later served as Principal Scientist (Agronomy) and Head, ICAR-CPCRI Regional Station, Kayamkulam (from April 2005 to September 2019).

The main areas of functioning include Undertaking Field Research Programmes on cropping system for resource use efficiency, organic farming in plantation and spice crops; Farmer Community-Oriented Technology Dissemination Programmes; and Administration and Manpower Management Programmes. As a member of the team, he has bagged the prestigious 'Outstanding Interdisciplinary Team Research in Agriculture and Allied Sciences for the Biennium 2005–2006' instituted by the Indian Council of Agricultural Research. He is a recipient of the RL Narasimha Swamy Award and Fellow of the Indian Society for Plantation Crops. He has functioned as the Principal Investigator/Co-Investigator for various externally funded projects of the Dept. of Biotechnology, Dept. of Science and Technology, GOI; Coconut Development Board, GOI; Kerala State Planning Board, National Agricultural Technology Project-NATP and National Agricultural Innovation Project-NAIP, Government of India; and International Fund for Agriculture Development-IFAD/COGENT.

He was Assistant Editor, Journal of Natural Rubber Research of RRII (1987–1989) and Editor, Journal of Plantation Crops

(ISPC) during 2006–2011 and served as a member of the Editorial Board of the Annual Reports of ICAR-CPCRI and AICRP on Plantation Crops for several years. He has edited 14 books and published more than 100 research and technical articles apart from contributing chapters in various books, technical bulletins, popular articles, etc. He currently functions as a member of the Editorial Board of Current Science, a Fortnightly Research Journal, being published by the Current Science Association, Bengaluru, India, in collaboration with the Indian Academy of Sciences.

Address for correspondence: Kaleeckal Madhom, Palace Ward, Krishnapuram (P.O)-690533, Alappuzha Dist., Kerala, India.

Contact No.: 91-9447364877; E mail: dr.krishnavkumar@gmail.com

Contributors

J. D. Adiga ICAR-DCR, Puttur, Karnataka, India

K. S. Anil Kumar ICAR-NBSS & LUP, Regional Centre, Bangalore, Karnataka, India

C. Babou CCRI, Coffee Board of India, Chikkamagaluru, Karnataka, India

S. K. Behera ICAR-IISS, Bhopal, Madhya Pradesh, India

Ravi Bhat ICAR-CPCRI, Kasaragod, Kerala, India

Bhavishya ICAR-CPCRI, Regional Station, Vittal, Karnataka, India

N. K. Binitha College of Agriculture, KAU, Padannakkad, Kerala, India

K. Chandrappa CCRI, Coffee Board of India, Chikkamagaluru, Karnataka, India

N. Chandrashekar CCRI, Coffee Board of India, Chikkamagaluru, Karnataka, India

R. Dinesh ICAR-IISR, Kozhikode, Kerala, India

Murali Gopal ICAR-CPCRI, Kasaragod, Kerala, India

Alka Gupta ICAR-CPCRI, Kasaragod, Kerala, India

S. B. Hareesh CCRI, Coffee Board of India, Chikkamagaluru, Karnataka, India

N. Hariyappa CCRI, Coffee Board of India, Chikkamagaluru, Karnataka, India

S. Induja ICAR-CPCRI, Regional Station, Kayamkulam, Kerala, India

Mercykutty Joseph RRII, Rubber Board, Kottayam, Kerala, India

V. Krishnakumar ICAR-CPCRI, Regional Station, Kayamkulam, Kerala, India

Prafulla Kumari CCRI, Coffee Board of India, Chikkamagaluru, Karnataka, India

T. Raj Kumar UPASI TRF, Regional Centre, Coonoor, Tamil Nadu, India

S. Mangalassery ICAR-DCR, Puttur, Karnataka, India

K. Manorama ICAR-IIOPR, Pedavegi, Andhra Pradesh, India

Jeena Mathew ICAR-CPCRI, Regional Station, Kayamkulam, Kerala, India

M. Murugan CRS, KAU, Pampadumpara, Kerala, India

P. Murugesan UPASI TRF, Regional Centre, Coonoor, Tamil Nadu, India

A. Nadaf CCRI, Coffee Board of India, Chikkamagaluru, Karnataka, India

J. S. Nagaraja CCRI, Coffee Board of India, Chikkamagaluru, Karnataka, India

S. R. Prabhu TerraBioGen Technologies, Burnaby, Canada

U. K. Priya ICAR-CPCRI, Regional Station, Vittal, Karnataka, India

N. Rajib Pati CCRI, Coffee Board of India, Chikkamagaluru, Karnataka, India

V. Selvamani ICAR-CPCRI, Kasaragod, Kerala, India

M. Senthil Kumar CCRI, Coffee Board of India, Chikkamagaluru, Karnataka, India

P. Shivaprasad CCRI, Coffee Board of India, Chikkamagaluru, Karnataka, India

P. R. Sowmya CCRI, Coffee Board of India, Chikkamagaluru, Karnataka, India

V. Srinivasan ICAR-IISR, Kozhikode, Kerala, India

P. Subramanian ICAR-CPCRI, Kasaragod, Kerala, India

S. Sujatha ICAR-IIHR, Bengaluru, Karnataka, India

Surekha ICAR-CPCRI, Kasaragod, Kerala, India

K. Suresh ICAR-IIOPR, Pedavegi, Andhra Pradesh, India

George V. Thomas ICAR-CPCRI, Kasaragod, Kerala, India

Amok Uchoi ICAR-CPCRI, Research Center, Kahikuchi, Assam, India

G. L. Veena ICAR-DCR, Puttur, Karnataka, India

Plantation Crops and Soil Health Management: An Overview

George V. Thomas and V. Krishnakumar

Abstract

Plantation crops are perennials, cultivated in a contiguous area in agroecosystems mostly confined to tropical belt, and comprise estate crops which include tea, coffee, and rubber and small holders' plantation crops, viz., coconut, oil palm, arecanut, cashew, cocoa, and spices. The cultivation of these crops in ecologically vulnerable regions including coastal belts and hilly areas and in locations endowed with high rainfall and high humidity makes the maintenance of soil health a serious challenge. The agroecosystems of plantation crops are confronted with several constraints including multi-nutrient deficiencies, nutrient imbalances, decline in SOM and biological attributes affecting the soil health, and climatic aberrations, which influence the sustainability of production systems. Research programs undertaken in different countries resulted in the development of technologies to improve soil health and to achieve sustainability in production systems of plantation crops. This chapter covers a brief introduction to plantation crops, their production systems, and the strategies for soil health management such as diversification of nutrient sources including inorganic, organic, and bio-fertilizers; conservation agriculture practices; efficient biomass resource recycling practices; agroforestry/cropping/integrated farming systems; and microbiome approaches to achieve sustainable production. Efforts to augment agroecosystem properties through various nature-based technologies would enhance carbon sequestration and soil health in general, finally contributing to sustainability and resilience of this group of crops under adverse weather and edaphic conditions.

G. V. Thomas (✉)
ICAR-Central Plantation Crops Research Institute, Kasaragod, Kerala, India

V. Krishnakumar
ICAR-Central Plantation Crops Research Institute, Regional Station, Kayamkulam, Kerala, India

G. V. Thomas, V. Krishnakumar (eds.), *Soil Health Management for Plantation Crops*, https://doi.org/10.1007/978-981-97-0092-9_1

Keywords

Plantation crops · Cropping systems · Soil health management · Biomass recycling · Carbon sequestration · Soil microbiome · Soil biodiversity · Integrated nutrient management · Agroforestry systems

Abbreviations

AMF	Arbuscular mycorrhizal fungi
CEC	Cation exchange capacity
CSA	Climate-smart agriculture
DRIS	Diagnosis and recommendation integrated system
FYM	Farm yard manure
GHG	Greenhouse gases
GIS	Geographical information system
HDMSCS	High-density multispecies cropping system
INM	Integrated nutrient management
MBC	Microbial biomass carbon
MRT	Mean residence time
PGPR	Plant growth promoting rhizobacteria
PSB	Phosphorus solubilizing bacteria
RS	Remote sensing
SOC	Soil organic carbon
SOM	Soil organic matter

1 Introduction

Plantation crops are perennials, cultivated on an extensive scale in a contiguous area mainly in the tropics, and also in certain areas in subtropics, owned and agricultural operations carried out by individual farmers or companies at the estate level. They are important for livelihood and economy as they meet a wide range of human requirements including food, edible oil, beverages, and other confectionary items, fuel, fiber, and industrial raw materials (Rethinam 2000). A range of plantation crops including coconut, arecanut, oil palm, cashew, cocoa, coffee, tea, rubber, and spices are cultivated in humid tropics and subtropical belts in different parts of the world (Table 1).

Plantation crops are broadly classified as estate crops which include tea, coffee, and rubber and small holder's plantation crops such as coconut, arecanut, oil palm, cashew, cocoa, and spices. They are high-value crops with significant commercial importance and play a vital role in improving economy of many countries through export earnings and employment generation and by reducing poverty of millions of people, more so in the rural sector (Anandaraj 2015). The major socioeconomic features of small holder plantation crop sector include predominance of fragmented,

Table 1 Plantation crops of economic importance and their geographic distribution and economic produce

Common name	Scientific name	Botanical family	Geographic distribution	Economic produce
Coconut	*Cocos nucifera* L.	Arecaceae	Tropical	All parts of the tree including nut for edible oil, natural drink, fiber, food, and other products
Arecanut	*Areca catechu* L.	Arecaceae	Tropical	Nut for chewing
Oil palm	*Elaeis guineensis* Jacq.	Arecaceae	Tropical	Fruit bunches for oil
Cashew	*Anacardium occidentale* L.	Anacardiaceae	Tropical	Culinary nut
Cocoa	*Theobroma cacao* L.	Sterculiaceae	Tropical	Beans for production of chocolate
Tea	*Camellia sinensis* (L.) O Kuntze	Theaceae	Tropical/ subtropical	Processed leaves for beverage preparation
Coffee	*Coffea arabica* L. *C. canephora* Pierre	Rubiaceae	Tropical/ subtropical	Processed beans for beverage preparation
Rubber tree	*Hevea brasiliensis* Muell. Arg	Euphorbiaceae	Tropical	Latex as natural rubber
Spices				
Black pepper	*Piper nigrum* L.	Piperaceae	Tropical	Black pepper berries used as spice
Small cardamom	*Elettaria cardamomum* Maton	Zingiberaceae	Tropical	Dried capsules used as spice

small, and marginal holdings and resource-poor farm environment, often with less marketable and marketed surplus.

Most of the commercial plantations are established after clearing natural forests. Given this fact, maintaining soil fertility throughout the lifespan of these perennial crops is a critical issue to be addressed in the plantation sector. Despite these facts, being perennial crops with long a lifespan, plantation crops are also important in view of their role in ecosystem services through carbon sequestration resulting from the generation of massive belowground and aboveground biomass, continuous lignocellulosic biomass addition through litter fall, and protection of soil from heavy rainfall through massive canopies and their overall impact on the protection of the environment. Sustainable and mature vegetation is the key to a balanced ecosystem, which provides a permanent solution to accelerated erosion problems (Santha 2006). Well-established and sustainably managed plantations help to protect soil health. To be sustainable, a managed land use system should imitate the structure and functioning of natural ecosystems, which are the results of natural selection over long periods (Ewel 1999). Even in agroforestry, loss of nutrients during the harvest, especially when rotations are short, may exceed the rate of replenishment by weathering of minerals and by atmospheric inputs. Furthermore, global warming accelerates soil organic matter (SOM) oxidation, making degradation of nutrient-poor soils faster in the tropics. Consequently, there is a major uncertainty, that is, whether the tropical tree plantations or agroforests could be grown perpetually on the same site without serious risk to their vitality and productivity (Kumar 2008). The only viable option available is to manage these plantation crops more sustainably by following principles of agroecology. In various chapters included in this book, a great deal of information has been provided on sustainable intensification of plantation crop sector.

2 Global Status of Plantation Crops

The global scenario of plantation crops with respect to area and major producing countries is presented in Table 2. Coconut, arecanut, and oil palm are the tropical palms grown as plantation crops at global level.

Coconut (*Cocos nucifera* L.), the environmentally friendly multipurpose palm mostly grown in small holdings in the tropics, is an agriculturally and economically important crop, interwoven with lives and cultures of people in Southeast Asia, the Pacific region, Africa, and some countries in Latin America. Asia and Pacific region accounts for a major share of area occupying 10.67 million ha (87%) and

Table 2 Global status of area and major producing countries of plantation crops

Crop	Area (m ha)	Major producing countries	Reference
Coconut	12.26	Indonesia, Philippines, India, Sri Lanka, Malaysia, Oceania	ICC (2020)
Arecanut	1.57	India, Bangladesh, Myanmar, Indonesia	FAOSTAT (2021)
Oil palm	24.13	Indonesia, Malaysia, Thailand, Colombia, Nigeria, Guatemala, Papua New Guinea, Cote d'Ivoire	USDA (2022)
Cashew	7.10	Cote d'Ivoire, India, Vietnam, Burundi, Philippines, Tanzania	https://numerical.co.in/numerons/collection/
Cocoa	14–15	Cote d'Ivoire, Ghana, Nigeria, Cameroon, Indonesia	Amponsah-Doku et al. (2021)
Coffee	10.0	Brazil, Indonesia, Ethiopia, Cote d'Ivoire, Colombia, Vietnam, Mexico, Uganda, India, and Peru	Jha et al. (2014)
Tea	5.0	China, India, Sri Lanka, Kenya, Malawi, Rwanda, Tanzania, Uganda	Wang et al. (2020)
Rubber	11.0	Thailand, Indonesia, Malaysia, India, Vietnam, Côte d'Ivoire, China	Singh et al. (2021)
Black pepper	1.5	Vietnam, Brazil, Indonesia, India	https://www.ipcnet.org/history-of-pepper/
Small cardamom	70,410 ha for India	Guatemala, India, Indonesia	Spices Board (2021)

production of 10.11 million metric tons of copra equivalent (86%) (ICC 2020). The coconut palm, revered as *Kalpavriksha* or *tree of life*, is important as a source of large number of products meeting daily requirements of people and traded internationally including coconut oil as vegetable oil, coconut water as a natural health drink, virgin coconut oil, desiccated coconut, coconut milk, milk powder, coconut shell powder, activated carbon, coir as the best natural fiber, coir products, and many others.

Arecanut (*Areca catechu* L.), a monocot palm, is an important cash crop, and the economic produce is the fruit known as *betel nut* or *supari* used for masticatory purposes. The arecanut forms the economic backbone to the extent of 10 million people in India and a few other countries. The arecanut products are consumed globally by about 700 million people, particularly among the South, East, and Southeast Asian communities. India is a major producer of arecanut contributing to 57% of the global production, and the remaining 43% is contributed by China, Bangladesh, Myanmar, and other countries (Bedi and Scully 2014; Patil et al. 2022).

Oil palm (*Elaeis guineensis* Jacq.), popularly known as "African oil palm" or "red oil palm," is the world's highest oil yielding crop, with a capacity of 5–10 times higher oil yield ha^{-1} than any other vegetable oil crops. Palm oil produced from the mesocarp of the fruit of oil palms has gained importance not only as an important edible oil traded internationally but also for its commercial applications including biofuel production.

Cashew (*Anacardium occidentale* L.), also referred to as "wonder nut," one of the leading edible nuts traded internationally, was primarily introduced as soil binding crop by Portuguese in India. Grown primarily in less fertile soils under rainfed conditions in terrains including slopes of hills, this perennial horticultural crop has become an important plantation crop due to its potential to provide livelihood security to the growers, empower rural women in the processing sector, create employment opportunities, and generate foreign exchange through exports (Rejani and Yadukumar 2010).

Cocoa, coffee, and tea are the commercial beverage crops grown in plantation scale in several countries. The cocoa tree (*Theobroma cacao* L.), though a native to the lowland rainforests of the Amazon basin, is now grown in 58 countries mostly around the equator. Africa is the world's largest producer, and about 70% of the production comes from four West African countries: Côte d'Ivoire, Ghana, Nigeria, and Cameroon (www.worldcocoafoundation.org/in-the-news/world-cocoa-production/). Cocoa known as the "food of God's" is the backbone of chocolate industry and the second most important organic crop, known for its capabilities to safeguard ecosystems. Considering the insufficient supply of cocoa beans, ten of the largest chocolate multinationals launched a cocoa sustainability scheme named "cocoa action," with an investment of 500 million USD to enhance sustainable cocoa production in West Africa to support cocoa planters and the chocolate industry (van Vliet et al. 2021).

Coffee (*Coffea* spp.), one of the most widely consumed beverages, is grown in more than 80 countries in tropical and subtropical regions of the world. The three primary coffee growing regions of the world, viz., Central and South America,

Africa, and the Middle East and Southeast Asia, are located along the equatorial zone between the Tropic of Cancer and the Tropic of Capricorn, widely known as the "bean belt." Even though coffee crop is cultivated within the "bean belt," the products from different regions have unique regional flavors, making the popular phrase "geography is a flavor" very meaningful. Different factors contribute to the flavors in a particular region, and the distinctiveness of the flavors is attributed to differences in soil chemical properties, weather parameters, sunshine, altitude, and processing method adopted.

Tea (*Camellia sinensis* (L.) O. Kuntze) crop, an evergreen perennial shrub or small tree, is the source of the world's most consumed drink after water. Though native to East Asia, tea crop is cultivated in 52 countries in the world mostly in tropics and subtropics. As an important cash crop, tea contributes to significant export earnings in several developing countries apart from its socioeconomic importance, employment generation, and poverty alleviation.

Rubber tree (*Hevea brasiliensis* Muell. Arg), the perennial tree crop, contributes over 90% of natural rubber in the world. A native of Amazonian river basin, rubber crop is an economically important industrial crop that produces latex, which is converted to various rubber products, used in aviation, automobile, medical, industrial, and consumer sectors. Over 85% of global production is from small holdings in Asia. Growing demand for rubber latex has been the driving force for rapid expansion of area in rubber cultivation. Thailand, Indonesia, Vietnam, and India are the major rubber producers in the world. Rubber tree is grown mostly in monoculture as well as in various diversified systems.

Spices have been used in Indian cooking since at least 2000 BC. Black pepper (*Piper nigrum* L.), known as "the king of spices," is a flowering vine, and its fruit is dried and used as spice in food and also used in medicines. A native of Malabar Coast of India, black pepper spread from India through trade to the rest of the world.

Small cardamom (*Elettaria cardamomum* Maton), known as "the Queen of Spices," is a herbaceous perennial rhizomatous plant native to Western Ghats of South India. Its dried ripe fruit having strong and punchy aroma and taste is used as a spice and is traded as an important spice commodity in international markets.

Clove (*Syzygium aromaticum* (L.) Merr. and Perry), nutmeg (*Myristica fragrans* Houtt.), cinnamon (*Cinnamomum verum* Brecht. & Presl.), allspice (*Pimenta dioica* (L.) Merr., and vanilla (*Vanilla planifolia* Andr.) are the tropical spices grown mainly as mixed crops in coconut and arecanut plantations.

3 Soil and Climatic Requirements of Plantation Crops

Plantation crops are cultivated under humid tropical conditions, and the cultivation of certain plantation crops are also extended to subtropical conditions. This group of crops is grown essentially in a wide range of soils between 28°N and 28°S latitudes of the equator. As these crops and their cropping systems are in ecologically vulnerable regions including coastal belts, hilly areas, and undulating terrains, experiencing high rainfall, soil health management is vital due to poor nutrient status of soil caused by soil erosion, fast oxidation of SOM, and nutrient loss. In view of the very narrow range of climatic requirements for the optimal growth and productivity of plantation crops, the information on ideal climatic and soil conditions is of vital importance due to current scenario of climatic aberrations and deterioration of the production environment. The suitable soil and climatic requirement of different plantation crops are presented in Table 3. Among the plantation crops, the coconut palm grows up to 1000 m above MSL, but most of the cultivation is confined to an altitude of 600 m (Ohler 1999). The highly adaptable nature of coconut palm is evident from the establishment and productivity of the crop under diverse soil

Table 3 Soil and climatic requirements of plantation crops

Crop	Altitude (m)	Temperature (°C)	Rainfall (cm $year^{-1}$)	Soil requirement	Soil pH
Arecanut	Up to 1000[a]	14–36; growth adversely affected <10 and >40	150–500	Laterite, red and sandy loam, alluvial and deltaic	Acidic to neutral
Cashew	Up to 1000	15–35	100–200	Hard degraded laterites, red sandy loam, and coastal sands	5.5–7.5
Cocoa	Up to 900	Maximum annual average of 30–32 and minimum average of 18–21	150–200	Deep soil with sufficient organic matter, nutrients, and a depth of at least 1.5 m	Acidic to neutral
Coconut	Up to 1000	27 ± 5 to 7	100–300	Coarse sand is its natural habitat; deep soils with good physical structure and chemical properties are most ideal. Laterite, coastal sandy, alluvial, red sandy loam suitable	5.5–7.0
Coffee	1000–1500	*C. arabica*: 15–24 *C. canephora:* 24–30	100–250	Light, deep friable, well-drained, loamy, rich in humus and exchangeable bases, especially K	4.5–6.5
Oil palm	Up to 200	22–32	200–300	Loamy or alluvial, well-drained soils, at least 1 m deep; rich in organic matter	4.0–6.0
Rubber	Up to 450	21–35	200–300	A variety of soils, mostly acidic, of the humid tropics. Deep soils from underlying sheet rocks	3.2–6.2

(continued)

Table 3 (continued)

Crop	Altitude (m)	Temperature (°C)	Rainfall (cm year^{-1})	Soil requirement	Soil pH
Tea	700–2500	23–30	250–300	High land well-drained and deep sandy loam to silty loam soils and >2% organic matter	4.0–6.5 (4.5–5.5 optimal)
Spices—black pepper	Up to 1500	Favorable range: 23–32 Ideal: around 28	125–200	Heavy clay to light sandy clays rich in humus with a porous friable nature, high organic matter, and high base saturation with Ca and Mg	Near neutral pH
Small cardamom	600–1200	10–35	150–400	Grows well in forest loamy soils	Acidic, 4.2–6.8

[a] Nuts from higher altitudes are of inferior quality than nuts from lower altitudes

conditions varying in texture, chemical composition, drainage, acidity, and alkaline calcareous nature (Khan et al. 1978).

Cashew crop is flexible in its soil requirement and is cultivated in high and low rainfall regions. The crop shows best growth in deep and well-drained red sandy loam, lateritic, and coastal sands which are slightly acidic pH. For ideal growth, oil palm requires well-drained, deep, loamy alluvial soils rich in organic matter with good water permeability. Rubber plants require highly deep, well-drained porous soils which are moderately acidic. Coffee and tea crops require cooler temperature for ideal growth. The optimum temperature for coffee is around 18 °C, but can withstand temperatures up to 30 °C. Altitude variation also makes a difference in the performance of crops. According to Khan et al. (2002), tea crops prefer high altitude and are followed by coffee. Hill slopes are ideal for black pepper. The general soil characteristics and the range of available nutrients in plantation crop soils are summarized by Tandon and Ranganathan (1988).

4 Soil Degradation as a Major Constraint Affecting Sustainability

In the present crop production scenario, the health of agroecosystems is under severe stress causing a decline in ecosystem services due to sub-optimal management practices followed, use of agrochemicals, and excessive soil disturbance, adversely affecting the productivity and sustainability of crops. At a global level, the land area prone to soil degradation is about 2 billion ha, of which 29.7% (596 million ha) is agricultural land (Lal 2001). The agroecosystems of plantation crops, predominantly occupied by small and marginal holdings and characterized by acidic soils with poor physical characteristics and low native fertility, are confronted with multiple soil health problems. Malhotra et al. (2017) in a review on soil fertility constraints in coconut stated that the production base had multiple soil health problems related to soil acidity, low cation exchange capacity (CEC), and poor nutrient reserve. Soil health in littoral sandy soils was affected by the physicochemical characteristics including high bulk density, poor aggregate stability, poor water holding capacity, high soil temperature, and poor soil fertility status, impeding the productivity of the palm (Reddy and Upadhyay 2002). The symbiotic association of arbuscular mycorrhizal fungi (AMF) with coconut roots was reported to be significantly affected as evidenced by the degree of colonization and intensity of infection in coconut roots in plots which received the recommended doses of fertilizers (Harikumar and Thomas 1991), indicating that chemical fertilization in coconut plantations adversely affect soil properties and soil biology. A study on mapping of soil constraints in Tamil Nadu using remote sensing (RS) and geographical information system (GIS) indicated that 31.5% of total coconut area had soil constraints, particularly related to soil reaction, texture, drainage, and soil depth (Selvamani and Duraisami 2018). Another study on coconut plantations under different management treatments indicated that earthworm community was completely absent in monocropped coconut gardens maintained with tillage and recommended dose of chemical fertilizer inputs, while the density and biomass of earthworms were significantly high in cropping systems with profuse aboveground diversity and maintained with integrated nutrient management (Thomas et al. 2022).

Evidence shows that high rainfall in oil palm plantations leads to soil erosion which subsequently leads to leaching of soil nutrients, decline in organic matter and soil biota, loss of soil structure, and changes in infiltration rates (Lord and Clay 1999). Soil constraints have also been reported in several locations in coffee-growing areas of Wayanad district of Kerala, India (Shivaprasad et al. 2018). Deterioration of soil condition with respect to very low organic carbon content, multi-nutrient deficiencies, and low carbon sequestration have been reported in intensively cultivated cashew plots when compared to that in forest and scrub lands (Srinivasan et al. 2019). Nutrient constraints affecting cashew production in India have been delineated by Mangalassery et al. (2021). Arecanut-growing soils of Karnataka had strong acidic or alkaline reactions and low CEC in many locations as per the recently conducted state-wide survey (Vasundara et al. 2020). Most of the cashew growing soils had soil health constraints including deficiency of nitrogen (N) and micronutrients such as zinc (Zn) and boron (B) and low base status and CEC (Rejani et al. 2013). A study on spices growing soils revealed that soil acidification was alarming and 90% of the soils were found to be slightly to highly acidic warranting immediate intervention through amendments to restore the productivity. Indiscriminate use of phosphorus fertilizers has resulted in increased P concentration in soil, and it interfered with the uptake of other essential nutrients (Dinesh et al. 2014). These reports indicated that soil degradation is a major constraint influencing the sustainable productivity of plantation crops and it necessitates technological interventions to mitigate such problems with appropriate soil health improvement strategies to achieve sustainability in production. A well-managed plantation can also help to reduce soil degradation due to continuous organic matter addition and long-term soil cover, as proved in the case of date palm plantation's influence on controlling and mitigating desertification (Mihi et al. 2019).

5 Soil Health: The Most Important Foundation for Plantation Sustainability

The intensive agricultural practices including the application of chemical fertilizers and pesticides though enabled to enhance agricultural production several fold, but has led to degradation of soil and pollution of environment, thereby adversely impacted sustainability of agricultural production. Now, there is a greater realization that soil health is the most important foundation necessary to achieve sustainability in agriculture. Jarecki and Lal (2003) reported that strategies are to be formulated to manage the crop production systems as agroecosystems based on ecological principles with the objective of enhancing soil organic carbon (SOC) pool, soil life, and soil quality to achieve sustainability in agricultural production and food security.

Soil is now considered as a dynamic living entity and has the most important role in the functioning of terrestrial ecosystems. Lehman et al. (2015b) reported that soil degradation can be addressed only by augmenting soil health through adoption of appropriate agronomic management strategies. Accordingly, Tiedje et al. (2001) opined that soil biology can become at least one, if not the major driver, for soil functions to support crop production. According to Doran et al. (1996), soil health is "the continued capacity of soil to function as a living system, within the ecosystem and land use boundaries, to sustain biological productivity, quality of air and water environment, promote plant, animal and human health." Soil biota are the drivers of biological functions in soil to deliver ecosystem services. The delivery of ecosystem services is made possible by soil biota through the regulation of biogeochemical cycles, making available nutrients to primary producers, enhancing soil structure and fertility, microbial breakdown of pollutants, supply of clean drinking water, reducing the impact of floods and droughts, control of soil erosion, reduction of atmospheric greenhouse gases (GHG), bio-control of pests and pathogens, and supporting agricultural production through biological pathway.

The management strategies suggested by Lal (2015) for mitigation of soil degradation include steps for minimization of soil erosion, creating positive SOC and N budgets, enhancing the activity and species diversity of soil biota and improvement in structural stability and pore geometry. In order to reach the goal of sustainable production, there needs to be fresh thinking on the crop production strategies, with a distinct difference from the approach being followed now and caused the deterioration of the production environment. The new approach should focus on soil health as the central point and the management strategies designed in such a way to realize the potential of soil life and biological processes to build the soil biological fertility with less dependence on external chemical inputs.

Important biological indicators as far as soil quality is concerned are organic matter content, the biological functioning as evidenced by respiration, microbial biomass content, and mineralizable N levels. Activities of soil enzymes indicate the extent of biological functioning of soil as well as the transformation of nutrients and are considered as integrative indicators of soil health. These indicators will be helpful in comparing the soil health under different systems. Organic matter content is an important component of soil which determines not only its quality but also its water holding capacity and susceptibility to degradation (Feller et al. 2001). In addition, SOM performs an important function of carbon sequestration by serving as a source or sink to atmospheric CO_2 (Lal 2009). Soil quality and functionality depend on physical, chemical, biological, and ecological attributes which individually and through interaction develop conducive soil health conditions for growth and nutrition of crops (Lal 2016).

6 Organic Farming in Plantation Crops as a Soil Life-Centered Approach

The ever-increasing demand for safe and healthy food, free from pesticides and other residues, and concerns on environmental pollution due to indiscriminate use of agro-chemicals,are the major driving forces responsible for the interest in alternate systems of farming in modern agriculture. Since many of the plantation crops are consumed either directly (e.g., coconut, cocoa, coffee, tea, spices, cashew, oil palm) or used for industrial purposes (e.g., coconut oil, palm oil, and arecanut), organic production to meet consumer's needs is very much essential. In order to adopt organic farming, it is necessary that such technologies without the use of any inorganic fertilizers and plant protection chemicals are made available to the farmers for adoption in their farmlands.

Varied crop farming practices such as cultural, mechanical, and biological methods can be adopted to achieve sustainable agricultural production by encouraging and enhancing biological cycles within farming system by giving due credit to the soil flora and fauna, plants, and animals. Organic farming has been reported to be a viable strategy for sustainable production of certain horticultural crops and plantation crops (Singh and Thomas 2010). The holistic approach of crop production also provides social and ecological advantages and ensures long-term fertility of soil. In organic agriculture, nutrients are supplied through organic sources which become available through the activity of soil biota, are considered as "biologically derived nutrients," and work on the principle of "feeding the soil to make it living" rather than "feeding the plants."

Organic farming is practiced in 191 countries at the international level, covering an area of almost 76 million ha of land, which represents 1.6% of agricultural land worldwide, with the involvement of 3.4 million farmers. The international trade of organic food products is valued at 125 billion euros during 2021 (Willer et al. 2023). The region-wise land area under organic management in the world during 2021 for some of the perennial plantation crops is presented in Table 4. The crops considered are cocoa, coffee, coconut, and tea. The amenability of coconut palm for organic farming has been reported by Nampoothiri (2001). The prospects of organic cashew production in India have been reported by Chattopadhyay et al. (2015).

Table 4 Region-wise land area (ha) under organic management in the world (2021) (permanent crops)[a]

Region	Cocoa	Coffee	Coconut	Tea/mate, etc.
Africa	287,665	313,325	4069	24,280
Asia	380	73,799	250,263	178,160
Latin America	179,689	466,922	7061	3324
North America	–	115	–	–
Oceania	1935	68,238	48,226	–
Total	469,669	922,399	309,619	205,764

[a] Data tabulated only for a few plantation crops (Source: Willer et al. 2023; FiBL survey 2023)

7 Soil Fertility Management in Plantation Crops

7.1 Nutrient Management

Plantation crops, being perennial in nature, require continuous supply of nutrients, which are mined from limited volume of soil in the root zone of crops, necessitating nutrient additions to maintain fertility level of soils and plant productivity. The annual nutrient removal by different plantation crops has been worked out for a particular yield situation, and it varied from 19.1 to 93.5 kg N ha^{-1}, 8.7 to 25.2 kg P_2O_5 ha^{-1}, and 1.6 to 111.2 kg K_2O ha^{-1} (Khan et al. 2002). Plant productivity is determined by soil fertility and nutrient availability. In plantation ecosystems, due to continuous removal of nutrients from soil through harvesting, fertilizer application is necessary for maintenance of sustained growth and yield of perennial plantation crops. Nutrient management strategies are to be planned considering the native soil fertility, leaf nutrient status, nutrient uptake pattern, and yield level. A comparative assessment of nutrient partitioning in adult coconut palms revealed the uptake pattern of various nutrients to be in the order of $K > N > Ca > S > P > Mg$ (Mathew et al. 2021). It is important to replenish the soil nutrients removed during crop production process and the nutrients lost by leaching and various environmental processes. Based on field experimentation, fertilizer dose for various plantation crops has been recommended (Table 5) to obtain optimum growth and productivity (Rethinam and Venugopal 1994). Most of the plantation crops require higher quantities of K for growth and fruiting.

Application of fertilizers in split doses has been recommended according to the stages of plant growth and nutrient demand. Soil test-based fertilizer application is proved to be an effective strategy to achieve higher productivity and nutrient use efficiency. Sadanandan et al. (2000) worked out diagnosis and recommendation integrated system (DRIS) norms for optimum yield in black pepper.

Fertigation systems, in which the fertilizers in soluble form are applied through the drip irrigation system in different splits, have been standardized for different plantation crops with considerable savings in the quantity of fertilizers as well as higher nutrient use efficiency (Dhanapal et al. 2005). Lignin, lignin sulfonate, and lignin acetate extracted from empty fruit bunches from oil palm served as additive

Table 5 Nutrient recommendation for various plantation crops in India

Sl. no.	Plantation crop	Recommended N, P_2O_5, K_2O
1	Coconut	500:320:1200
2	Arecanut, cocoa, and black pepper	100:40:140
3	Cashew	500:125:125
4	Oil palm	1200:600:1200
5	Coffee	140:90:450
6	Tea	450:90:450
7	Rubber	30:30:30
8	Cardamom	75:75:150

Sl. no. 1–4, the recommended dose is in g $palm^{-1}$ or $plant^{-1}$ $year^{-1}$, and for 5–8, the values are in kg ha^{-1} $year^{-1}$

substance for urea fertilizer, and the urea lignin formation showed slow-release properties (Arianie 2010).

Integrated nutrient management (INM) involving the combined application of inorganic, organic, and bio-sources of nutrients enables to achieve the high productivity in plantation crops in a sustainable manner (Khan et al. 2002). Organic manures are important sources of nutrients for plantation crops but are required in higher quantities. Generation of green manures by growing of cover crops can be sustainable to generate biomass right in the gardens. Foliar application of nutrients, especially micronutrients, wherever needed are being adopted to overcome their deficiency.

7.2 Availability of Crop Residue Biomass in Plantations and Plantation-Based Cropping Systems

Crop residues form the largest harvest in the crop production process and a vital resource that can be recycled and utilized for a number of useful purposes. On the global scale, the availability of crop residue biomass is estimated to be 3.8 billion Mg (Lal 2005). One of the most important applications of crop residues is in the crop production process itself to enhance, maintain, and sustain the soil quality, to derive the essential environmental services, and to develop sustainable agroecosystems (Smil 1999). The multiple beneficial attributes derived from the application/incorporation of crop residues in soil include the reduction of soil erosion, enhancement of nutrient availability, stabilization of soil structure, reduction of bulk density, improvement in water retention, the use as energy source for microbial processes, increase of CEC, and finally enhancement of crop productivity (Lal 2005). The utilization of residue biomass in the crop production process is also important in the agroecosystem functioning as they promote biological functions by supplementation of SOC pool, enhanced proliferation, species diversity and activity of soil fauna, and mitigation of climate change by augmenting C sequestration and above all contributing significantly to global food security through enhancement of soil quality (Lal 2005).

Plantation crops produce large quantities of recyclable biomass which are valuable resources containing plant nutrients. Recycling of these residue biomass using biological agents or by amendments results in the production of value-added organic resource, which has application in crop production as a nutrient source as well as for soil health enhancement. The residues from certain plantation crops (e.g., coconut waste biomass) are lignocellulosic complexes with high lignin content. The natural decomposition of these wastes and the nutrient release are very slow processes due to chemical composition. The organic wastes have applications in crop production directly as mulch or after conversion to composts with the help of earthworms or microbial cultures. The recycled biomass can meet the major portion of nitrogen and a part of other nutrients required by the palms.

The quantity of recyclable biomass available from plantation crops and plantation-based cropping systems are presented in Table 6. The recyclable biomass available from coconut plantations includes mainly the fronds and husk, while bunch waste is available in very less quantities. The total availability of waste biomass from 1.93 million ha of coconut plantation in India has been estimated as 14.36 million tons annually (Biddappa et al. 1996). Waste biomass availability is higher in cropping systems when compared to that in monocrop cultivation. The coconut-based high-density multispecies cropping system (HDMSCS) at ICAR-CPCRI, Kasaragod, which has black pepper, clove, banana, and pineapple as component crops yielded 23.51 t ha^{-1} $year^{-1}$ of recyclable biomass. When the biomass from the HDMSCS is recycled, it contributed as much as 130 kg N ha^{-1}, 8.5 kg P ha^{-1}, and 121 kg K ha^{-1} (Palaniswami et al. 2007).

The extraction of coir fiber from coconut husk results in the generation of coir dust or coir pith, which constitutes 50–70% of the husk, and it accumulates near coir processing factories (Coir Board 2016). The global coconut production annually is 68 billion nuts from an area of 12 million ha (ICC 2020). Based on the report that processing of 10,000 coconut husks results in the production of one ton of coir pith as a by-product (Prabhu and Thomas 2002), it is possible to generate 6.8 million tons of coir pith annually, if all husks obtained from coconut cultivation are used for coir extraction (Jayasekhar and Chandran 2021). In practice, all husks produced are not used for coir extraction, and there is also lack of accurate data on coir pith production as all coir extraction units are not in organized sector. It is now estimated that the global production of coir pith is around 0.7 million tons annually by India and Sri Lanka together constituting 90% of the global production, of this Indian share is 0.5 million tons of coir pith annually (van Doren et al. 2017).

There is increasing interest in utilization of coir pith in a more productive way as a value-added organic resource in agrihorticulture as a suitable medium for growth of plants, organic manure after composting, soil amendment, carrier for biofertilizers, and substitute for peat moss (Prabhu and Thomas 2002). Coir fiber-based products are of value in environmental protection through soil bioengineering measures, to protect the banks of water resources by combining biological systems with engineering principles to restore deteriorated soil masses (Santha 2006). Coconut husk burial technology can help to conserve soil and water in all plantation crops. This technology was effective in reducing the runoff from 37% in control to 22% and soil

Table 6 Recyclable biomass generated in different plantation crops and plantation-based cropping systems

Plantation crop/cropping system	Type of recyclable biomass	Quantity of recyclable biomass (t ha^{-1} $year^{-1}$)	Reference
Coconut	Fronds, bunch waste, husk	14–16	Biddappa et al. (1996)
	Coir pith	0.7 million tons in India and Sri Lanka	van Doren et al. (2017)
Coconut-based HDMSCS[a] (coconut-black pepper-banana-clove-pineapple)	Coconut fronds, bunch waste, husk, leaves, pineapple stover, banana pseudo-stem, leaves	19.1–27.55	Thomas and Palaniswami (2005)
Arecanut	Leaves with leaf sheath, bunch waste, husk	8.7	Bavappa et al. (1986)
Arecanut-based HDMSCS[a] (arecanut-black pepper-cacao-banana-pineapple-clove-coffee)	Leaves with leaf sheath, bunch waste, husk, banana pseudo-stem, leaves, pineapple stover	14.3	Bavappa et al. (1986)
Cocoa	Leaf litter	8.5	Santana et al. (1988)
Oil palm	Fronds, male inflorescences, bunches that fail to develop	15–17	Sunitha and Krishnakumar (2006)
Rubber	Leaf litter	10–13	Onyibe and Gill (1992)
Cashew	Leaf litter	5.0	Usha and Vikraman Nair (2002)
Tea	Pruning	7.5–8.3	Zaman et al. (2011)
Coffee[b]	Solid and liquid from the cherries and shrubs	6.0–6.5	Glover and Beer (1986)

[a] *HDMSCS* high-density multispecies cropping system
[b] Includes the biomass addition from shade trees

loss by 49% compared to that in control in a cashew plantation and resulted in 30% increase in cashew yield (Rejani and Yadukumar 2010).

The arecanut palm yields significant quantities of recyclable biomass in the form of leaves with sheath, husk, and bunch waste. As the arecanut husk constitutes approximately 60–80% of the total weight of the nut, the quantity of husk produced is to the extent 5.5–6 metric t ha^{-1} $year^{-1}$, posing a significant disposal challenge (Vannarath and Thalla 2022). Arecanut husk fibers have high lignin and low ash content (Yuan et al. 2023) and might be of value in producing products useful in soil bioengineering similar to coconut fiber-based products. The high-density multispecies cropping system of arecanut produces 8.76–10.50 t ha^{-1} $year^{-1}$ of waste biomass (Hussain et al. 2008). The waste biomass available in oil palm plantations include pinnae, petiole, rachis, waste inflorescence, empty fruit bunches, and mesocarp waste and shell.

The recyclable biomass from cashew consists of leaf litter, dried leaves, and cashew apples. Cashew leaves are not easily degraded due to high lignin (134 g kg^{-1}) and phenol (48 g kg^{-1}) contents (Isaac and Nair 2005). The availability of litter fall from a shaded cocoa system was between 5 and 10 t ha^{-1} $year^{-1}$ of dry matter (van Vliet and Giller 2017). A study on the availability of plant residue and nutrient cycling in cacao gardens in Brazil revealed the highest quantity of plant residue to the tune of 8500 kg ha^{-1} in plantations without shade (Santana et al. 1988).

Leaf litter and dried biomass from leguminous cover crops are available for recycling in rubber plantations. In India, rubber trees, which are deciduous in nature, shed leaves during December, followed by leaf sprouting and new leaf production in January. This process termed as "wintering" returns significant organic biomass to soil through fallen leaves, petioles, and small twigs. Thus, there is constant yearly natural nutrient cycling through this biomass and is like what happens in natural forest ecosystems. The quantity of litter produced is determined by the age of plantation, clone, planting density, and geographical area. In India, over seven tons of litter fall ha^{-1} of rubber plantation is available annually. Experiments indicated self-sufficiency of phosphorous in a mature rubber plantation through litter-derived P (Jessy et al. 2009). It is known that pectin activates P solubilizing bacteria (PSB) and promotes soil phosphorus solubilization (Mise et al. 2020). It would be interesting to know if the pectin fraction from rubber litter is involved in solubilizing insoluble P in rubber plantation soils, thus contributing to P self-sufficiency.

The recyclable biomass from coffee processing includes coffee pulp, processing effluent, parchment husks, and coffee husks. Wet processing of coffee in the estates results in production of large quantities of pulp or epicarp as by-products. Ten of the major coffee producers on the global scale produce >10 million tons of liquid and solid residues on a yearly basis to be disposed of. The pollution and environmental threat from the residues and the by-products of coffee berry processing units such as defective coffee beans and spent coffee grounds and coffee husks (pulp and mucilage) that constitute around 60% of the wet weight of the fresh fruit have been reported by Echeverria and Nuti (2017). Crop residue biomass in tea gardens includes shade tree litter, lopping, tea litter, tea pruning, and weeds. The crop residue wastes and by-products from tea production and processing have great potential in ameliorating soil acidity, enhancing soil health, productivity, and quality of tea (Le et al. 2022).

7.3 Recycling Technologies for Lignocellulosic Residue Biomass from Plantation Crops

Several bio-resources-based low-cost technologies have been developed to generate organic resources from residue biomass, which are important as substitute for chemical fertilizers and enable to grow the plantation crops in a sustainable mode.

Recycling of lignin-rich leaf biomass from coconut palms is now possible due to the standardization of a vermicomposting technology using a local epigeic earthworm belonging to *Eudrilus* sp., at ICAR-CPCRI, Kasaragod (Prabhu et al. 1998). The vermicompost obtained from coconut leaf biomass is valued as a source of plant nutrients, growth hormones, soil enzymes, and plant beneficial microbes such as free living diazotrophs, phosphate-solubilizing microbes, fluorescent pseudomonads, and silicate solubilizers (Gopal et al. 2009). It has been proved that co-composting of lignocellulosic leaf waste from date palms with rock phosphate and sheep manure results in the production of a good quality compost, which can be used to promote circular agriculture in low fertile areas (El Janati et al. 2022). Recyclable biomass from arecanut and cocoa gardens were efficiently converted to acceptable organic manure by vermicomposting process using African night crawler (*Eudrilus eugeniae*). The recovery of vermicompost ranged from 74.65% to 87.75% in a composting period of 3 months. Vermicompost had lower C:N ratio; pH; higher content of macronutrients (NPK) and micronutrients such as Cu, Zn, Fe, and Mn; and beneficial microbial communities (Chowdappa et al. 1999). Vermicomposting with indigenous epigeic earthworm—*Perionyx excavatus*—enabled production of quality vermifertilizer from cashew leaf litter admixed with various animal dungs (Parthasarathi et al. 2016). The efficiency of earthworm species was in the following descending order: *E. eugeniae* > *E. fetida* > *P. excavatus* for bioconversion of rubber leaf litters (*Hevea brasiliensis*, var. RRIM 600) to vermicompost (Chaudhuri et al. 2003).

A large-scale composting technology has been reported to convert coir pith to organic resource useful for improving plantation soil (Ghosh et al. 2007). Lignocellulosic crop residue biomass from coconut could be successfully used as substrate for edible mushroom production using basidiomycete fungus *Pleurotus sajor-caju* (Fr.) Singer, and the spent substrate could be utilized as organic manure to improve soil fertility and growth of coconut seedlings (Thomas et al. 1989). Co-composting coir pith with the amendments of poultry manure, lime, and rock phosphate resulted in the production of quality organic manure with physicochemical and biological characteristics for application as a good-quality manure in agriculture (Thomas et al. 2013). A technology has also been developed for conversion of recalcitrant lignin-rich coconut palm residue biomass to biochar for application in soil to improve nutrient availability, aeration, and water filtration (Gopal et al. 2020). There are also reports that coir pith can be a substitute for peat moss for horticultural purposes (van Doren et al. 2017).

7.4 Microbial Inoculants Possessing Functional Attributes to Enhance Growth and Productivity

Function-specific microorganisms are the groups of microorganisms which possess beneficial traits, and the selected efficient strains of these microorganisms are used as inoculants to enhance the growth and productivity of crops. The function-specific microorganisms reported in plantation crops belong to five groups, viz., N_2-fixing bacteria forming associative and rhizospheric relationship, phosphate-solubilizing microbes, potassium-solubilizing bacteria, plant growth-promoting rhizobacteria (PGPR), and arbuscular mycorrhizal fungi (AMF). The function-specific microbes isolated from coconut include species of diazotrophs, *Beijerinckia*, *Azospirillum*, *Burkholderia*, *Herbaspirillum*, and *Azoarcus*; phosphate-solubilizing fungi, *Aspergillus* and *Penicillium*; phosphate-solubilizing bacteria, *Bacillus*, *Pseudomonas*, and *Micrococcus*; potassium-solubilizing bacteria, *Acinetobacter*, *Alcaligenes*, and *Micrococcus*; PGPR, fluorescent pseudomonads, *Bacillus*, *Brevibacillus*, and *Serratia*; and AMF, *Glomus*, *Gigaspora*, *Scutellospora*, *Sclerocystis*, and *Acaulospora* (Thomas et al. 1991; Thomas and Prabhu 2003; George et al. 2012; Rajesh Kumar et al. 2015; Gupta et al. 2016). Whole-genome sequencing of plant growth-promoting bacteria was a significant step toward identifying the genes involved in an array of plant growth-promoting functions (Gupta et al. 2014). Carrier-based biofertilizer formulations have been developed with efficient strains of PGPR, AMF, phosphate solubilizers, and nitrogen-fixing bacteria for application to seedlings and adult plantation crops as low-cost and eco-friendly inputs. A talk-based formulation of PGPR, *Pseudomonas* sp., and a soil-based AM formulation with AMF, *Claroideoglomus etunicatum*, have been commercialized as effective bio-inoculants for coconut seedlings (CPCRI 2014; Thomas et al. 2016). These bioformulations were found effective not only to substitute the inorganic nutrients in crop nutrition but also provide a host of other benefits such as production of growth-promoting substances to stimulate crop growth and impart tolerance to stress conditions.

7.5 Harnessing Symbiotic Nitrogen Fixation by Leguminous Crops

Apart from the efficient nitrogen fixation benefits, the cover legume *Rhizobium* symbiotic interaction bestows plantation crops with a host of other advantages to enhance soil health and crop growth. Nutrient cycling studies in oil palm plantations having cover crops of legumes revealed that the total benefit due to the introduction of legume cover was to tune of 239 kg N ha^{-1} $year^{-1}$ (Agamuthu and Broughton 1985). Application of ^{15}N natural abundance technique has conclusively proved fixation of nitrogen by bacteria associated with oil palm ecotypes (de Carvalho et al. 2008). A simple and easily adoptable agro-technique involving cultivation of leguminous nitrogen-fixing cover crops, *Pueraria phaseoloides* and *Calopogonium mucunoides*, in basins of coconut palms yields 15–20 kg of fresh biomass and 150–200 g of N per basin, and its incorporation brings about significant improvement in soil health properties and enabled substitution of 50% fertilizer nitrogen (Thomas et al. 1989). A spurt in microbial communities and soil respiration were evident following in situ cultivation and incorporation of green manure legumes in coconut basins. Nitrogen fixation studies based on natural abundance method revealed that *Pueraria phaseoloides* fixed N to the tune of 250 kg N ha^{-1} $year^{-1}$ and high level of transfer of fixed nitrogen from the legume to the rubber trees, contributing from 39% to 46% of tree leaf nitrogen (Clermont-Dauphin et al. 2016). Nutrient-rich legume biomass could also trigger formation of humus in the plantation soils (Lal 2003).

7.6 Integrated Nutrient Management to Enhance Soil Health and Productivity

The integrated use of different nutrients, viz., chemical, organic, and bio-fertilizers, is an effective nutrient management strategy to maintain soil fertility, restore soil health, and provide continuous supply of nutrients to crops at optimum levels to sustain the desired crop productivity, maximize the profitability, and protect the environment. Field experiments with adult coconut palms in different locations revealed that chemical fertilizer application to the extent of 50% of the recommended dose of N could be substituted by various organic sources such as green manure legumes (*Calopogonium mucunoides*/*Mimosa invisa*) (Thomas et al. 2001), application of *Gliricidia* lopping (Subramanian et al. 2005), vermicompost (Nath et al. 2012), and composted coir pith (Kalpana et al. 2008), along with the benefits in terms of improvement in physical, chemical, and biological properties of soil. In oil palm, INM contributed to sustainable oil palm cultivation by enhancing the biodiversity of microbial communities in soil toward a balanced ecosystem (Sundram et al. 2019). Integrating the application of a reduced dose of inorganic NPK fertilizers (up to 50% of recommended dose) along with FYM at 1.5 t ha^{-1} and bio-inoculants consortium consisting of *Azospirillum*, *Azotobacter*, PSB, *Pseudomonas fluorescence*, and AMF were the effective combination to provide significantly higher yield in coffee (Nadaf et al. 2019). A comparison of the integrated and fully organic treatments in an HDMSCS experiment in coconut revealed that coconut productivity was on par in integrated and fully organic treatments comprising of biomass recycling, biofertilizers, green manuring, vermiwash, husk burial, and mulching the coconut basins (Maheswarappa et al. 2013). Farming system maintained with integrated nutrient treatments had superior soil biological fertility parameters including the beneficial microbial communities, microbial biomass C, and soil enzymes compared to that in monocropping of coconut with recommended fertilizer treatments (Thomas et al. 2022). Apart from enhanced crop growth and productivity, the benefits accrued due to INM include enhanced fertilizer use efficiency, resistance to diseases, improvement in fertility of soil, increased soil carbon stock, lower levels of emissions of CO_2 and N_2O, and lower water pollution (Sundram et al. 2019).

As constant cycling and uptake of nutrients take place, rubber plantations can be considered as closed ecosystems (Watson 1989). George and Joseph (2011) were of the view that fertilizer application can be skipped at least for a short period in well-maintained rubber plantations in favorable terrain, when adequate manuring was done during the immature phase and leguminous cover crop was established.

8 Plantation Crop-Based Cropping/Farming Systems as Agroecosystem-Driven Approach

One of the great advantages in plantation farming is the potential of most of the plantation crops to accommodate a range of compatible crops in the interspaces to develop diversified cropping systems reaching new spatiotemporal dimensions in land use. Integrated farming systems involve raising of fodder crops and compatible subsidiary crops in the interspaces of the palms and integration of livestock and other enterprises to increase the productivity and net returns from unit area of plantations. These systems are characterized by high and stable productivity and high returns per unit of area, time, labor, and energy inputs and are endowed with a strong agroecological base. Many of these traditional cropping systems mimic nature and create well-established agroecosystems with minimum dependency on external inputs to sustain production. It is reported that the small-scale diversified systems involving plantation crops, traditionally followed by tropical small farmers, are dependent on local resources and have a range of crops in various combinations and provided reasonable productivity and stability and high return per unit of labor and energy (Altieri and Nicholls 2004). The agroecological approach followed in these complex poly-cultures and agroforestry systems provided benefits such as stability, enhanced belowground biodiversity, efficient nutrient cycling, less incidence of pests and diseases, and higher economic returns.

Several highly productive and economically viable cropping system models of intercropping, mixed cropping, multi-storied cropping, and agroforestry systems have been developed (Nair 1977, 2017; Nelliat et al. 1974), wherein the agroecology principles and soil health benefits form the basis for stability and sustainability. Soil health benefits are well established in coconut-based high-density multispecies cropping systems at ICAR-CPCRI, Kasaragod, Kerala, India, which was started in 1983 with 17 crops grown in high density to maximize production and economic returns per unit area but later on was stabilized with black pepper, clove, banana, and pineapple (Bavappa et al. 1986). Mixed farming system involving cultivation of forage grasses as feed for dairy cows which are maintained in the system, along with integration of subsidiary enterprises like pisciculture, poultry, goat rearing, rabbitry, and sericulture along with recycling of wastes back to the system as a source of nutrients for coconut palm, was found to be a typical example of utilization of soil-plant-animal relations in a manmade agroecosystem (Maheswarappa et al. 1998). The coconut farmers in India, especially in the homestead gardens of Kerala state are accustomed to grow a variety of crops in the interspaces of coconut palms to meet the food, fodder, fuel, wood, and timber requirements in addition to generating supplementary income (Krishnakumar and Reddy 2007). Mohan Kumar and Kunhamu (2022) reported that homestead farming systems in Kerala with coconut as its nucleus has characteristics of a healthy agroecosystem to promote sustainable production and to deliver a range of ecosystem services.

8.1 Soil Biological Fertility Responses to Enhancement of Aboveground Biodiversity in a System Mode

In perennial plantation crop-based cropping systems involving annuals and perennials and their combinations, complementary and synergistic interactions were evident among the component crops in the system due to the biotic, structural, and microclimatic complexity in the multispecies cropping systems. Several experiments in plantation-based cropping systems in several locations by various investigators revealed that increased aboveground diversity enhanced belowground diversity which contributed to enhanced soil biological activities, nutrient transformations, arthropod diversity, and carbon sequestration and contributed to sustainability of the system (Thomas et al. 2022). The stimulatory effects of living roots to encourage higher microbial activity and organic matter decomposition were reported (Cheng and Coleman 1990), and a variety of crops with varied root exudation patterns and litter chemistry can mediate transformation of plantation soils to achieve the new frontier of soil science—soils teeming with life (Tiedje et al. 2001).

The diverse annual and perennial crops in multistoried and mixed farming systems in coconut supported higher load of plant beneficial microbial community (Bopaiah and Shetty 1991a, b), function-specific microbes such as associative and asymbiotic diazotrophs, mineral phosphate solubilizers and PGPR (Nair and Subba Rao 1977; Thomas and Prabhu 2003; Thomas et al. 1991), and arbuscular mycorrhizal species abundance and richness (Rajesh Kumar et al. 2015) in the belowground rhizospheric region of main crop and the component crops. Crop diversity and management practices significantly influenced the amount of soil microbial biomass (Bopaiah and Shetty 1991a). Enzymes, which catalyze nutrient transformations and degradation of biomass, are indicators of soil biological activity and have relationship with microbial proliferation and changes in soil management. Activities of phosphatase and urease were significantly higher in farming system when compared to that in coconut monocrop (Bopaiah and Shetty 1991a). Microbial metabolic quotient, considered as a dynamic indicator of soil health, also responded to enhancement of aboveground biodiversity and management practices particularly recycling of residue biomass. A study on the long-term (12 years) effects of leguminous cover crops, viz., *Atylosia scarabaeoides*, *Centrosema pubescens*, *Calopogonium mucunoides*, and *Pueraria phaseoloides*, on important soil biochemical and biological properties revealed that microbial biomass-C, microbial biomass-N, and microbial biomass-P, net N mineralization rates, CO_2 evolution, metabolic quotient (qCO_2), and the activities of L-asparaginase, L-glutaminase, and β-glucosaminidase enzymes were significantly higher in the cover cropping site compared to the levels in the control site (Dinesh et al. 2006).

8.2 Agroforestry-Based Plantation Production Systems

Multiple cropping systems serve as drivers for providing multiple ecosystem services and enable reliance on biotic interactions to reduce chemical inputs (Gaba et al. 2015). Introducing trees or other woody perennials into monoculture coffee plantation will generate benefits for both the coffee plantations and the farmers. When trees are introduced, it enables restoration of soil health as the deep root network of trees can retrieve nutrients from deeper levels of soil; and the N fixation capabilities of trees add N into the soil ecosystem, thus enabling improvement and sustenance of soil fertility. The trees function as windbreaks, protect soil from erosion and overexposure to changing weather conditions, lower soil and air temperatures, and provide congenial microclimate for coffee plants. Therefore, an agroforestry-based coffee production system leads to increased biodiversity, increases water availability, and restores soil health and productivity and mitigation of climate change (Gomes et al. 2020).

Shade trees in tea plantation also provide various beneficial effects as in the case of coffee. They mitigate climate change impacts and provide ecosystem services as well. During tea cultivation, diverse agroforestry is an important and established intercropping measure. Various agroforestry intercropping systems in tea plantations provided benefits in terms of nutrient dynamics, microbial diversity and abundance, and ecological services. The tea-agroforestry systems also sequester significant quantities of CO_2 (Zhang et al. 2021).

There is a welcome trend in some coffee growing countries to convert monocrop gardens to agroforestry systems to derive the benefits of ecosystem (Muschler and Bonnemann 1997). Agroforestry or polyculture is reported to be the best management strategy to improve soil health, quality of environment, and ecosystem functions compared to monocropping of rubber (Chen et al. 2017; Singh et al. 2021). The agroforestry systems in rubber were superior with respect to improvement in SOC and nitrogen pools, reduced carbon and nitrogen losses, and improvements in soil aggregation. Jessy et al. (2017) suggested having diverse rubber-based agroforestry systems for small holdings in India to improve the health of ecosystems. Quantification of nutrient stocks and nutrient cycling in cocoa-agroecosystems revealed that large amounts of nutrients are recycled each year, and such nutrient cycling is essential for maintaining sustainable cocoa production systems (Hartemink 2005). High biodiversity has been combined with high yields in tropical agroforests (Clough et al. 2011).

9 Biodiversity and Ecosystem Services

The relevance of soil biodiversity not only for sustainability in agriculture but also for sustenance of life on this planet by delivery of various ecosystem services is well recognized (Bach et al. 2020). As the soils harbor more than 25% of total biodiversity of earth, soil biota-centric approach is most vital to achieve sustainability in agriculture. The multiple roles of biodiversity in agriculture to support the crop production process include nutrient cycling, enhancing soil fertility, carbon sequestration, pest regulation, and pollination. In agriculture, increasing tree cover contributes to the buildup of biodiversity and delivery of ecosystem services (Barrios et al. 2018). In this respect, agroforestry systems practiced in plantation crops, particularly those involving maintenance of shade trees, provide ideal habitat for maintenance of species diversity as well as for conservation of biodiversity (Bhagwat et al. 2008). Among the plantation production systems, the value of cocoa-agroforestry systems to meet the livelihood security of small farmers and to deliver the ecosystem services to sustain the production systems has been well recognized, as reported by Clough et al. (2009).

In coffee agroforestry system, absence of shade trees led to loss of native earthworm populations and resulted in 76% lower soil macro-porosity. The advantages of increased tree cover included 53% increase in tea crop yield and maintenance of 93% of crop pollinators as occurring in the natural forest and, in combination, contributed to 86% lower incidence of coffee berry borer (Barrios et al. 2018). In Peru, the agroforestry management systems, cover cropping, and edaphic factors significantly influenced the soil microbial communities under cocoa, as evidenced by phospholipid fatty acid analysis and terminal restriction fragment length polymorphism analysis (Buyer et al. 2017). The plant mixtures significantly influenced microbial parameters, viz., microbial biomass, bacterial biomass, fungal biomass, and fungi, bacteria ratio, and microbial respiration, when compared to monocrops, as revealed in a global meta-analysis from 106 studies (Chen et al. 2019).

Alley cropping with cacao, black pepper, pineapple, bactris palms, and bamboo was found to be an effective strategy in oil palm for enhancing productivity along with ecological benefits in terms of faunal biodiversity and related ecosystem services (Ashraf et al. 2018). A study on arecanut palm-based cropping systems (ABCS) in farmers gardens in west coast of India indicated higher levels of ecosystem services and energy use efficiency under organic farming practices followed by that in INM (Paramesh et al. 2019). The cacao agroforestry systems in the southern Bahia region of Brazil have beneficial effects as a conservation system for soil fauna including soil and litter faunal communities (da Silva Moço et al. 2009). The beneficial effects were also reflected on non-monitory inputs such as soil fertility status, microbial activity, and nutrient transformations, leading to sustainability in crop production.

10 Microbiome Engineering to Enhance Plant Health and Productivity

Microbial communities form the driving force for biogeochemical cycles in the earth (Falkowski et al. 2008). Soil biology is absolutely critical for healthy and reliant soil (Lehman et al. 2015a). Crop plants as a whole and soils where they are grown are always associated with a variety of microorganisms. The sum of the genome of the host and associated microbiota is termed as hologenome (Zilber-Rosenberg and Rosenberg 2008). Genetic wealth of diverse microbial symbionts can play an important role both in adaptation and evolution of higher organisms.

Increasing evidence points to the possibilities of manipulation of soil microbiome as an effective strategy to enhance nutrient availability and crop production (Chaparro et al. 2012). Plant microbiome engineering, if incorporated into traditional agricultural practices, can improve ecosystem services for growth and resilience (Arif et al. 2020). Application of next-generation sequencing technologies to study coffee-associated microorganisms with a metabarcoding/metagenomic approach provided deeper insight into coffee plant-associated microbial communities establishing the identity of "coffee core microbiota" and their implications not only in the coffee plant fitness but also in the quality of final product (Duong et al. 2020). Investigation of coffee diversity between the coffee growing continents revealed that 11 bacterial genera along with 10 fungal genera were shared all across the world and are ubiquitous across continents and they adapted themselves to various environmental conditions. A total of ten bacterial genera and three species as well as six fungal genera and one species colonized all coffee compartments and therefore exhibited better competitive ability to colonize rhizosphere, tissues, and organs of coffee plant. Plant health and productivity are influenced when microorganisms function in a coordinated manner with the overall microbiome of soil. The root exudates secreted by the plant have an important role in shaping the soil microbiome associated with the plant (Chaparro et al. 2012). Metagenomic analysis during different stages of coconut leaf vermicomposting by Gopal et al. (2017) revealed that bacterial communities involved in lignin and lipid degradation were predominant from initial to mid-phase of vermicomposting, while soil and plant health-improving communities were predominant in the mature vermicompost. In agriculture, a paradigm shift is taking place in the focus from microbe to microbiome to achieve sustainability (Ray et al. 2020). It would be of interest in the coming years to investigate how to enhance the activities of microbiomes of interest through manipulation of SOM, specific fertilizers, and intercrops and cover crops. The guiding principle should be based on the fact that the intensity of land use is shaping the soil microbiome and pathogenic organisms, and therefore, efforts should be made to preserve the natural ecosystems keeping the land disturbance to minimum (Idbella and Bonanomi 2023).

11 Climate Smart Agricultural Practices for Soil Health Improvement

Climate change is affecting crop productivity worldwide, and plantation crops are no exception. It also further accelerates processes of degradation in already-degraded soil ecosystem and environments and has negative impact on food production and food system. It is reported that climate change may pose more problems for a crop like cashew as the crop is cultivated in ecologically sensitive areas such as coastal belts, hilly areas, and areas with high rainfall and humidity (Rupa et al. 2013). Climate change adversely affects soil health by enhancing the degradation of organic matter in the soil; affecting the structure of the soil; and increasing its vulnerability to erosion and other degradation processes. Therefore, new solutions to adapt crops to these environmental changes are needed. The impact of global warming and climate change will be different for each plantation crop. To address the problems caused by climate change in plantation crops, its possible regional impacts are to be investigated. The impact of increasing temperature and precipitation patterns is projected to be different over the various agroecological regions, and hence, the climate change will affect plantation crops differently in various cultivating regions.

Crop-based and agroecological region-based mitigation strategies are to be developed by integrating all the measures available for sustainable productivity. In this context, it will be quite appropriate to adopt climate-smart agriculture (CSA) practices to minimize the adverse effects caused by aberrations in climate impacting crop production. CSA envisages a number of green and resilient practices that can make a change in the production systems in the face of climate change (Qureshi et al. 2022). CSA aims to integrate various issues of climate change into planning and implementation of sustainable agricultural practices. The three important core outcomes expected from CSA are (a) improved productivity by adopting all those practices which improve soil quality and structure. The concept is also that healthy soils are important carbon sinks that hold CO_2 and keep it out of the atmosphere, thus helping fight climate change; (b) enhanced resilience by reducing adverse effects of biotic and abiotic stresses; and (c) reduced emissions from agriculture and identify better management practices to absorb carbon out of the atmosphere (Matteoli et al. 2021). Rupa et al. (2013) reported that in order to reduce the impact of climate change on cashew, strategies such as soil and water conservation measures, mulching, supplemental irrigation, drip irrigation, fertigation, and carbon sequestration are essential. Agriculture contributing up to 29% of the annual anthropogenic GHG emissions is also part of the solution as it can alleviate climate change through SOC capture, plant biomass production, and improved land use management (Vermeulen et al. 2012). The plantation crops represent remarkable carbon pools as they retain much more carbon per unit of land area than other vegetations and show great potential as carbon sinks. Therefore, the best management practices including conservation agriculture practices, viz., mulching, waste recycling, precision application of nutrients, and water, need much emphasis to safeguard the soil health of the plantation crop ecosystem. Plantation crops grown in agroforestry system are sites of significant levels of carbon sequestration. Substantial levels of carbon sequestration have been reported when cocoa is grown in agroforestry system (Middendorp et al. 2018).

As climate change has become a reality, there is a need to have a relook at the soil and land management practices in crop production in relation to mitigation and adaptation to climate change with focus on soil health management to achieve sustainability in agriculture (Lal 2005).

12 Plantation Agroecosystems Management for Efficient Carbon Sequestration

Humus and biomass of living organisms such as plants and animals are carbon pools in soil. The enhancement of carbon sequestration in soil is possible by reducing oxidation of soil carbon, continuously adding organic matter to finally transform to humus and covering the land area with plants which convert CO_2 to plant biomass. Soil is the largest organic carbon pool (up to 2 m of soil depth) accommodating approximately 2400 Gt C, and it is three times higher than the amount of carbon in the atmosphere. Therefore, even small changes in SOC stocks can significantly impact atmospheric CO_2 levels and contribute to climate change (Beillouin et al. 2023). In order to harness maximum C sequestration in a soil ecosystem, the inputs and the outputs should reach steady-state or dynamic equilibrium (Lal et al. 1998). Most of the plantation crops are established after clearing natural forests. A recent study revealed that the site with 10-year arecanut plantation had 35 Mg ha^{-1} less SOC stock than that under the forests (Das et al. 2021). Natural rubber is a fast-growing and high biomass-producing perennial tree crop, and, in general, it attains 50 cm girth in the first 7 years. Rubber trees can stock large amounts of carbon in their standing biomass. The world's rubber plantations can sequester carbon to the tune of 0.0782 PgC $year^{-1}$ (Jacob 2003), and it reduces 2% of the current rate of rise in atmospheric CO_2. GHG emissions from rubber production were low compared to land use change effects on plant and soil C stocks and C accumulation in latex, wood products, and seed oil. C stocks were influenced by rubber clone and management strategies like tapping frequency or planting density (Blagodatsky et al. (2016). Carbon sequestration in coconut-based cropping system under integrated nutrient management has been reported by Shinde et al. (2020). Coconut-based cropping systems developed with perennial fruit trees recorded higher C sequestration levels with the highest total carbon sequestration reported in coconut + jamun (140.06 t ha^{-1}) system, followed by coconut + mango system (138.91 t ha^{-1}) and coconut + garcinia (131.72 t ha^{-1}) system, compared to 98.2 C t ha^{-1} under coconut monocrop (Bhagya et al. 2017). Growing cocoa in agroforestry system would sequester higher carbon than in mixed cropping or monocropping (Abou Rajab et al. 2016; Balasimha and Naresh Kumar 2013; Jacobi et al. 2014). Oil palm plantations act as important carbon sinks by assimilating more CO_2 and producing more biomass per unit area than forests by virtue of its capacity to support high fruit production (Kotowska et al. 2015).

13 Conclusions

Intensification of agriculture with the sole objective of maximizing productivity using chemical inputs resulted in degradation of the environment, biodiversity decline, and diminished ecosystem services. The situation warrants a change in our approach to agriculture minimizing the application of chemicals but relying on practices and inputs which will not harm the environment and prevent the loss of biodiversity (Gaba et al. 2015). Crop biostimulants which include microbial products, humic acids, etc. are predicted to play a leading role in reducing the effects of climatic aberrations on crop production (Del Buono 2021). Maintaining soil health by raising and incorporating leguminous cover crops, application of organic manures and reuse of organic biomass into the crop production system as a regular practice, mulching the crop basin or other areas in the plantation, use of biofertilizers, plant growth promoters, nutrient solubilizers, etc. can help in the buildup of beneficial microbial communities, leading to better availability of nutrients and development of antagonistic populations in the rhizosphere. The integrated nutrient management through the judicious use of customized nutrient formulations, biofertilizers, organic manures, and harnessing nutrient cycling mechanisms that augment nutrient use efficiency can form the best option in soil health management of plantation crops. Equally, clear insight into the effect of chemical inputs on the nature and diversity of the rhizosphere microbiome will be very much useful in formulating effective management strategies for achieving sustainability in soil health and productivity with the best quality products for human consumption.

This book covers various chapters on the importance of each plantation crop prepared by experts from each area of relevance. The topics covered include general soil characteristics and requirements for each crop, indicators of soil health and soil health assessment, soil fertility constraints, use of soil amendments, issues of soil contamination, impact of climate change and plantation soils (environmental impact), fertility management for healthy soils including fertilizer recommendations, soil biodiversity and biological functions, organic farming, microbiome and metagenomic approach, technological options for soil health improvement and their impact, and future strategies for research in soil health management of important plantation crops.

References

Abou Rajab Y, Leuschner C, Barus H (2016) Cacao cultivation under diverse shade tree cover allows high carbon storage and sequestration without yield losses. PLoS One 11(2):1–22. https://doi.org/10.1371/journal.pone.0149949

Agamuthu P, Broughton WJ (1985) Nutrient cycling within the developing oil palm legume ecosystem. Agric Ecosyst Environ 13:111–123. https://doi.org/10.1016/0167-8809(85)90054-4

Altieri MA, Nicholls CI (2004) An agroecological basis for designing diversified cropping systems in the tropics. J Crop Improv 11:81–103. https://doi.org/10.1300/J411v11n01_05

Amponsah-Doku B, Daymond A, Robinson S, Atuah L, Sizmur T (2021) Improving soil health and closing the yield gap of cocoa production in Ghana—a review. Sci Afr:e01075. https://doi.org/10.1016/j.sciaf.2021.e01075

Anandaraj M (2015) Status of plantation crops sector in India. J Plantn Crops 43:1–8

Arianie L (2010) Potential of lignin, lignin sulfonate and lignin acetate from palm empty bunch as an additive substance in urea fertilizer as an effort to reduce the solubility of urea nitrogen. In: Advanced materials research, vol 93. Trans Tech Publications Ltd., pp 409–412. https://doi.org/10.4028/www.scientific.net/AMR.93-94.409

Arif I, Batool M, Schenk PM (2020) Plant microbiome engineering: expected benefits for improved crop growth and resilience. Trends Biotechnol 38:1385–1396. https://doi.org/10.1016/j.tibtech.2020.04.015

Ashraf M, Zulkifl R, Sanusi R, Tohiran KA, Terhem R, Moslim R, Norhisham AR, Adham Ashton-But AA, Azhar B (2018) Alley-cropping system can boost arthropod biodiversity and ecosystem functions in oil palm plantations. Agric Ecosyst Environ 260:19–26

Bach EM, Ramirez KS, Fraser TD, Wall DH (2020) Soil biodiversity integrates solutions for a sustainable future. Sustainability 12(7):2662. https://doi.org/10.3390/su12072662

Balasimha D, Naresh Kumar S (2013) Net primary productivity, carbon sequestration and carbon stocks in areca-cocoa mixed crop system. J Plantn Crops 41(1):8–13

Barrios E, Valencia V, Jonsson M, Brauman A, Hairiah K, Mortimer PE, Okubo S (2018) Contribution of trees to the conservation of biodiversity and ecosystem services in agricultural landscapes. Int J Biodivers Sci Ecosyst Serv Manag 14:1–16. https://doi.org/10.1080/21513732.2017.1399167

Bavappa KVA, Kailasam C, Khader KBA, Biddappa CC, Khan HH, Kasturi Bai KV, Ramadasan A, Sundararaju P, Bopaiah BM, Thomas GV, Misra LP, Balasimha D, Bhat NT, Bhat KS (1986) Coconut and arecanut based high density multispecies cropping systems. J Plantn Crops 14:74–87

Bedi R, Scully C (2014) Tropical oral health. In: Farrar J, Hotez PJ, Junghanss T, Kang G, Laloo D, White NJ (eds) Manson's tropical infectious diseases, 23rd edn. Elsevier, pp 1073–1083. https://doi.org/10.1016/B978-0-7020-5101-2.00090-X

Beillouin D, Corbeels M, Demenois J, Berre D, Boyer A, Fallot A, Feder F, Cardinael R (2023) A global meta-analysis of soil organic carbon in the Anthropocene. Nat Commun 14:3700. https://doi.org/10.1038/s41467-023-39338-z

Bhagwat SA, Willis KJ, Birks HJB, Whittaker RJ (2008) Agroforestry: a refuge for biodiversity? Trends Ecol Evol 23:261–267. https://doi.org/10.1016/j.tree.2008.01.005

Bhagya HP, Maheswarappa HP, Surekha, Bhat R (2017) Carbon sequestration potential in coconut-based cropping systems. Indian J Hort 74(1):1–5. https://doi.org/10.5958/0974-0112.2017.00004.4

Biddappa CC, Upadhyay AK, Hegde MR, Palaniswami C (1996) Organic matter recycling in plantation crops. J Plantn Crops 24:71–85

Blagodatsky S, Xu J, Cadisch G (2016) Carbon balance of rubber (*Hevea brasiliensis*) plantations: a review of uncertainties at plot, landscape and production level. Agric Ecosyst Environ 221:8–19. https://doi.org/10.1016/j.agee.2016.01.025

Bopaiah BM, Shetty HS (1991a) Soil microflora and biological activities in the rhizosphere and root region of coconut based multistoried cropping and monocropping systems. Soil Biol Biochem 23:89–94. https://doi.org/10.1016/0038-0717(91)90167-I

Bopaiah BM, Shetty HS (1991b) Microbiology and fertility in coconut-based mixed farming and coconut monocropping systems. Trop Agric (Trinidad) 68(2):135–138

Buyer JS, Baligar VC, He Z, Arévalo-Gardini E (2017) Soil microbial communities under cacao agroforestry and cover crop systems in Peru. Appl Soil Ecol 120:273–280. https://doi.org/10.1016/j.apsoil.2017.09.009

Chaparro JM, Sheflin AM, Manter DK, Vivanco JM (2012) Manipulating the soil microbiome to increase soil health and plant fertility. Biol Fertil Soils 48:489–499. https://doi.org/10.1007/s00374-012-0691-4

Chattopadhyay N, Hore JK, Bhattacharjya A (2015) Prospect of organic cashew production in India. Acta Hort 1080:337–341. https://doi.org/10.17660/ActaHortic2015.1080.45

Chaudhuri PS, Pal TK, Bhattacharjee G, Dey SK (2003) Rubber leaf litters (*Hevea brasiliensis* var. RRM 600) as vermiculture substrate for epigeic earthworm, *Perionyx excavatus*, *Eudrilus eugeniae* and *Eisenia fetida*. Pedobiologia 47:796–800. https://doi.org/10.1078/0031-4056-00261

Chen C, Liu W, Jiang X, Wu J (2017) Effects of rubber-based agroforestry systems on soil aggregation and associated soil organic carbon: implications for land use. Geoderma 299:13–24. https://doi.org/10.1016/j.geoderma.2017.03.021

Chen C, Chen HYH, Chen X, Huang Z (2019) Meta-analysis shows positive effects of plant diversity on microbial biomass and respiration. Nat Commun 10:1332. https://doi.org/10.1038/s41467-019-09258-y

Cheng W, Coleman DC (1990) Effect of living roots on soil organic matter decomposition. Soil Biol Biochem 22:781–787. https://doi.org/10.1016/0038-0717(90)90157-U
Chowdappa P, Biddappa CC, Sujatha S (1999) Efficient recycling of organic wastes in arecanut (*Areca catechu* L.) and cocoa (*Theobroma cacao* L.) plantations through vermicomposting. Indian J Agric Sci 69:563–566
Clermont-Dauphin C, Suvannang N, Pongwichian P, Chevlan V, Hammecker C, Harmand JM (2016) Dinitrogen fixation by the legume cover crop *Pueraria phaseoloides* and transfer of fixed N to *Hevea brasiliensis*—impact on tree growth and vulnerability to drought. Agric Ecosyst Environ 217:79–88. https://doi.org/10.1016/j.agee.2015.11.002
Clough Y, Faust H, Tscharntke T (2009) Cacao boom and bust: sustainability of agroforests and opportunities for biodiversity conservation. Cons Lett 2:197–205. https://doi.org/10.1111/j.1755-263X.2009.00072.x
Clough Y, Barkmann J, Juhrbandt J (2011) Combining high biodiversity with high yields in tropical agroforests. Proc Natl Acad Sci 108:8311–8316. https://doi.org/10.1073/pnas.1016799108
Coir Board (2016) Coir pith-wealth from waste, a reference. Coir Board, Ministry of MS and ME, Coir House, Kochi, 92 p
CPCRI (2014) Annual report 2013–14. Central Plantation Crops Research Institute, Kasaragod
da Silva Moço MK, da Gama-Rodrigues EF, da Gama-Rodrigues AC, Machado RCR, Baligar VC (2009) Soil and litter fauna of cacao agroforestry systems in Bahia, Brazil. Agrofor Syst 76:127–138. https://doi.org/10.1007/s10457-008-9178-6
Das M, Nath PC, Sileshi GW, Pandey R, Nath AJ, Das AK (2021) Biomass models for estimating carbon storage in Areca palm plantations. Environ Sustain Indic 10:100115. https://doi.org/10.1016/j.indic.2021.100115
de Carvalho AL, Alves BJ, Baldani VL, Reis VM (2008) Application of ^{15}N natural abundance technique for evaluating biological nitrogen fixation in oil palm ecotypes at nursery stage in pot experiments and at mature plantation sites. Plant Soil 302(1):71–78. https://doi.org/10.1007/s11104-007-9456-5
Del Buono D (2021) Can biostimulants be used to mitigate the effect of anthropogenic climate change on agriculture? It is time to respond. Sci Total Environ 751:141763. https://doi.org/10.1016/j.scitotenv.2020.141763
Dhanapal R, Subramaniam P, Mathew AC, Palaniswami C (2005) Drip fertigation to increase the productivity of coconut. Indian Coconut J 31:17–19
Dinesh R, Suryanarayana MA, Chaudhuri SG, Sheeja TE, Shiva KN (2006) Long-term effects of leguminous cover crops on biochemical and biological properties in the organic and mineral layers of soils of a coconut plantation. Eur J Soil Biol 42:147–157. https://doi.org/10.1016/j.ejsobi.2005.12.004
Dinesh R, Srinivasan V, Hamza S, Anandaraj M (2014) Massive phosphorus accumulation in soils: Kerala's continuing conundrum. Curr Sci 106(3):343–344
Doran JW, Sarrantonio M, Liebig MA (1996) Soil health and sustainability. Adv Agron 56:1–54
Duong B, Marraccini P, Maeght JL, Lebrun VPM, Duponnois R (2020) Coffee microbiota and its potential use in sustainable crop management—a review. Front Sustain Food Syst 4:607935. https://doi.org/10.3389/fsufs.2020.607935
Echeverria MC, Nuti M (2017) Valorization of the residues of coffee agro-industry: perspectives and limitations. Open Waste Manag J 10:13–22. https://doi.org/10.2174/1876400201710010013
El Janati M, Robin P, Akkal-Corfini N, Bouaziz A, Sabri A, Chikhaoui M, Thomas Z, Oukarroum A (2022) Composting date palm residues promotes circular agriculture in oases. Biomass Conv Bioref 13:14859. https://doi.org/10.1007/s13399-022-03387-z
Ewel JJ (1999) Natural systems as models for the design of sustainable systems of land use. Agrofor Syst 45:1–21
Falkowski PG, Fenchel T, Delong EF (2008) The microbial engines that drive Earth's biogeochemical cycles. Science 320:1034–1039. https://doi.org/10.1126/science.1153213
FAOSTAT (2021) Food and Agricultural Organization, Rome
Feller C, Albrecht A, Blanchart E, Cabidoche YM et al (2001) Soil organic carbon sequestration in tropical areas. General considerations and analysis of some edaphic determinants

for Lesser Antilles soils. In: Martius C, Tiessen H, Vlek PLG (eds) Managing organic matter in tropical soils: scope and limitations. Springer, Netherlands, pp 19–31. https://doi.org/10.1007/978-94-017-2172-1_3

Gaba S, Lescourret F, Boudsocq S, Enjalbert J, Hinsinger P, Journet EP, Navas ML, Wery J, Louarn G, Malézieux E, Pelzer E, Prudent M, Ozier-Lafontaine H (2015) Multiple cropping systems as drivers for providing multiple ecosystem services: from concepts to design. Agron Sust Dev 35:607–623. https://doi.org/10.1007/s13593-014-0272-z

George S, Joseph P (2011) Natural rubber plantation: a nutritionally self-sustaining ecosystem. Nat Rubber Res 24(2):197–202

George P, Gupta A, Gopal M, Thomas GV (2012) Screening and *in vitro* evaluation of phosphate solubilising bacteria from rhizosphere and roots of coconut palms (*Cocos nucifera* L.) growing in different states of India. J Plantn Crops 40(1):61–64

Ghosh PK, Sarma US, Ravindranath AD, Radhakrishnan S, Ghosh P (2007) A novel method for accelerated composting of coir pith. Energy Fuel 21(2):822–827. https://doi.org/10.1021/ef060513c

Glover N, Beer J (1986) Nutrient cycling in two traditional Central American agroforestry systems. Agrofor Syst 4:77–87. https://doi.org/10.1007/BF00141542

Gomes LC, Bianchi FJJA, Cardoso IM, Fernandes RBA, Fernandes Filho EIF, Schulte RPO (2020) Agroforestry systems can mitigate the impacts of climate change on coffee production: a spatially explicit assessment in Brazil. Agric Ecosyst Environ 294:106858. https://doi.org/10.1016/j.agee.2020.106858

Gopal M, Gupta A, Sunil E, Thomas GV (2009) Amplification of plant beneficial microbial communities during conversion of coconut leaf substrate to vermicompost by *Eudrilus* sp. Curr Microbiol 59:15–20. https://doi.org/10.1007/s00284-009-9388-9

Gopal M, Bhute SS, Gupta A, Prabhu SR, Thomas GV, Whitman WB, Jangid K (2017) Changes in structure and function of bacterial communities during coconut leaf vermicomposting. Antonie Van Leeuwenhoek 110:1339–1355. https://doi.org/10.1007/s10482-017-0894-7

Gopal M, Gupta A, Hameed KS, Sathyaseelan N, Rajeela THK, Thomas GV (2020) Biochars produced from coconut palm biomass residues can aid regenerative agriculture by improving soil properties and plant yield in humid tropics. Biochar 2:211–226. https://doi.org/10.1007/s42773-020-00043-5

Gupta A, Gopal M, Thomas GV, Manikandan V, Gajewski J, Thomas GV (2014) Whole genome sequencing and analysis of plant growth promoting bacteria isolated from the rhizosphere of plantation crops coconut, cocoa and arecanut. PLoS One 9(8):e104259. https://doi.org/10.1371/journal.pone.0104259

Gupta A, Gopal M, Ravindran D, Thomas GV (2016) Prevalence of potassium solubilizing bacteria in the rhizosphere of coconut palms (*Cocos nucifera* L.) growing in different soil types. In: Chowdappa P et al (eds) Abstracts of third international symposium on coconut research and development. ICAR-CPCRI, Kasaragod, p 178

Harikumar VS, Thomas GV (1991) Effect of fertilizers and irrigation on vesicular—arbuscular mycorrhizal association in coconut. Philipp J Coconut Stud XVI:20–24

Hartemink AE (2005) Nutrient stocks, nutrient cycling, and soil changes in cocoa ecosystems: a review. Adv Agron 86:227–253. https://doi.org/10.1016/S0065-2113(05)86005-5

Hussain M, Ray AK, Maheswarappa HP, Krishnakumar V, Bhat R, Subramanian P, Thomas GV (2008) Recycling of organic biomass from arecanut based high density multi-species cropping system models under Assam condition. J Plantn Crops 36(1):53–57

ICC (2020) Coconut statistical yearbook 2020. International Coconut Community, Jakarta

Idbella M, Bonanomi G (2023) Uncovering the dark side of agriculture: how land use intensity shapes soil microbiome and increases potential plant pathogens. Appl Soil Ecol 192:105090. https://doi.org/10.1016/j.apsoil.2023.105090

Isaac SR, Nair MA (2005) Biodegradation of leaf litter in the warm humid tropics of Kerala, India. Soil Boil Biochem 37:1656–1664. https://doi.org/10.1016/j.soilbio.2005.02.002

Jacob J (2003) Carbon sequestration potential of natural rubber plantations. In: Proceedings of IRRDB symposium on challenges for natural rubber in globalization, 15–17 Sept 2003, Chiang Mai, Thailand

Jacobi J, Andres C, Schneider M, Pillco M, Calizaya P, Rist S (2014) Carbon stocks, tree diversity, and the role of organic certification in different cocoa production systems in Alto Beni, Bolivia. Agrofor Syst 88:1117–1132. https://doi.org/10.1007/s10457-013-9643-8

Jarecki MK, Lal R (2003) Crop management for soil carbon sequestration. Crit Rev Plant Sci 22(6):471–502. https://doi.org/10.1080/713608318

Jayasekhar S, Chandran KP (2021) World economic importance. In: Rajesh MK, Ramesh SV, Perera L, Kole C (eds) The coconut genome. Springer, pp 1–12. https://doi.org/10.1007/978-3-030-76649-8_1

Jessy MD, Nair AS, Bai MM, Rajendran P, Punnoose KI (2009) Self–sustainability of phosphorous cycle in rubber (*Hevea brasiliensis*) plantations: annual recycling through litter and removal through latex. J Plantn Crops 37(3):177–184

Jessy MD, Joseph P, George S (2017) Possibilities of diverse rubber based agroforestry systems for smallholdings in India. Agrofor Syst 91:515–526. https://doi.org/10.1007/s10457-016-9953-8

Jha S, Bacon CM, Philpott SM, Ernesto Méndez V, Läderach P, Rice RA (2014) Shade coffee: update on a disappearing refuge for biodiversity. BioSci 64(5):416–428. https://doi.org/10.1093/biosci/biu038

Kalpana M, Gautam B, Srinivasulu B, Rao DVR, Arulraj S, Jayabose C (2008) Impact of integrated nutrient management on nut yield and quality of coconut under coastal ecosystem. J Plantn Crops 36(3):249–253

Khan HH, Sankaranarayan MP, Narayanan BB (1978) Characteristics of coconut soils of India. 1. Morphology, some physic chemical characteristics and taxonomy. In: Nelliat EV et al (eds) Proceedings of PLACROSYM I. Indian Society of Plantation Crops, Kasaragod, pp 54–79

Khan HH, Upadhyay AK, Palaniswami C (2002) Integrated nutrient management in plantation crops. In: Rethinam P, Khan HH, Reddy VM, Mandal PK, Suresh K (eds) Plantation crops research and development in the new millennium. Coconut Development Board, Kochi, pp 9–22

Kotowska MM, Leuschner C, Triadiati T (2015) Quantifying above- and belowground biomass carbon loss with forest conversion in tropical lowlands of Sumatra (Indonesia). Glob Chang Biol 21:3620–3634. https://doi.org/10.1111/gcb.12979

Krishnakumar V, Reddy DVS (2007) Homestead farms in Northern Kerala—their basic characteristics and socio-economic status. J Plantn Crops 35(2):106–110

Kumar BM (2008) Litter dynamics in plantation and agroforestry systems of the tropics-a review of observations and methods. In: Batish DR, Kohli RK, Jose S, Singh HP (eds) Ecological basis of agroforestry. CRC Press, Boca Raton, FL, pp 181–216

Lal R (2001) Soil degradation by erosion. Land Degrad Dev 12(6):519–539. https://doi.org/10.1002/ldr.472

Lal R (2003) Global potential of soil carbon sequestration to mitigate the greenhouse effect. Crit Rev Plant Sci 22:151–184. https://doi.org/10.1080/713610854

Lal R (2005) World crop residues production and implications of its use as a biofuel. Environ Int 31(4):575–584. https://doi.org/10.1016/j.envint.2004.09.005

Lal R (2009) Challenges and opportunities in soil organic matter research. Eur J Soil Sci 60:158–169. https://doi.org/10.1111/j.1365-2389.2008.01114.x

Lal R (2015) Restoring soil quality to mitigate soil degradation. Sustainability 7:5875–5895. https://doi.org/10.3390/su7055875

Lal R (2016) Soil health and carbon management. Food Energy Secur 5(4):212–222. https://doi.org/10.1002/fes3.96

Lal R, Kimble JM, Follett RF, Cole CV (1998) Potential of U.S. cropland to sequester carbon and mitigate the greenhouse effect. Ann Arbor Press, an imprint of Sleeping Bear Press, Chelsea, MI

Le VS, Herrmann L, Hudek L, Nguyen TB, Baru L, Lesuer D (2022) How application of agricultural waste can enhance soil health in soils acidified by tea cultivation: a review. Environ Chem Lett 20:813–839. https://doi.org/10.1007/s10311-021-01313-9

Lehman RM, Acosta-Martinez V, Buyer JS, Cambardella CA, Collins HP, Ducey TF, Halvorson JJ, Jin VL, Johnson JM, Kremer RJ, Lundgren JG (2015a) Soil biology for resilient, healthy soil. J Soil Water Conserv 70(1):12A–18A. https://doi.org/10.2489/jswc.70.1.12A

Lehman RM, Cambardella CA, Stott DE, Acosta-Martinez V, Manter DK, Buyer JS, Maul JE, Smith JL, Collins HP, Halvorson JJ, Kremer RJ, Lundgren JG, Ducey TF, Jin VL, Karlen DL

(2015b) Understanding and enhancing soil biological health: the solution for reversing soil degradation. Sustainability 7:988–1027. https://doi.org/10.3390/su7010988
Lord S, Clay L (1999) Environmental impacts of oil palm—practical considerations in defining sustainability for impacts on the air, land and water. New Britain Palm Oil Ltd. Centre, Dami Oil Palm Research Station, 38 p
Maheswarappa HP, Hegde MR, Dhanapal R, Biddappa CC (1998) Mixed farming in coconut gardens; its impact on soil physical and chemical properties, coconut nutrition and yield. J Plantn Crops 26:139–143
Maheswarappa HP, Dhanapal R, Subramaniam P, Palaniswami C (2013) Evaluation of coconut based high density multispecies cropping system under organic and integrated nutrient management. J Plantn Crops 41(2):130–135
Malhotra S, Maheswarappa HP, Selvamani V, Chowdappa C (2017) Diagnosis and management of soil fertility constraints in coconut (*Cocos nucifera*): a review. Indian J Agric Sci 87(6):711–726
Mangalassery S, Nayak MG, Susan PP, Rupa TR, Behera SK, Srinivasan V (2021) Delineating the nutrient constraints and developing nutrient norms for cashew (*Anacardium occidentale* L.) in coastal India. J Plant Nutr 44:2627–2639. https://doi.org/10.1080/01904167.2021.1921198
Mathew J, Haris AA, Bhat R, Krishnakumar V, Muralidharan K, John KS, Surendran U (2021) A comparative assessment of nutrient partitioning in healthy and root (wilt) disease affected coconut palms grown in an Entisol of humid tropical Kerala. Trees 35:621–635. https://doi.org/10.1007/s00468-020-02064-w
Matteoli F, Schnetzer J, Jacobs H (2021) Climate-smart agriculture (CSA): an integrated approach for climate change management in the agriculture sector. In: Luetz JM, Ayal D (eds) Handbook of climate change management. Springer, Cham. https://doi.org/10.1007/978-3-030-57281-5_148
Middendorp R, Vanacker V, Lambin E (2018) Impacts of shaded agroforestry management on carbon sequestration, biodiversity and farmers income in cocoa production landscapes. Landsc Ecol 33:1953–1974. https://doi.org/10.1007/s10980-018-0714-0
Mihi A, Tarai N, Chenchouni H (2019) Can palm date plantations and oasification be used as a proxy to fight sustainably against desertification and sand encroachment in hot drylands? Ecol Indic 105:365–375. https://doi.org/10.1016/j.ecolind.2017.11.027
Mise K, Koyama Y, Matsumoto A, Fujita K, Kunito T, Senoo K, Otsuka S (2020) Pectin drives microbial phosphorus solubilization in soil: evidence from isolation-based and community-scale approaches. Eur J Soil Biol 97:103169. https://doi.org/10.1016/j.ejsobi.2020.103169
Mohan Kumar B, Kunhamu TK (2022) Nature-based solutions in agriculture: a review of the coconut (*Cocos nucifera* L.)-based farming systems in Kerala, "the Land of Coconut Trees". Nat Based Solutions 2:100012. https://doi.org/10.1016/j.nbsj.2022.100012
Muschler RG, Bonnemann A (1997) Potentials and limitations of agroforestry for changing land-use in the tropics: experiences from Central America. Forest Ecol Manag 91:61–67. https://doi.org/10.1016/S0378-1127(96)03887-X
Nadaf SA, Hariyappa N, Chandrappa K, Manonamani GK, Hareesh SB, Raghuramulu Y (2019) Integrated use of organic and inorganic fertilizers in coffee—a key under the climatic resilience. In: Paper presented in the 23rd plantation crops symposium (PLACROSYM XXIII), 6–8 Mar 2019. CCRI, Chikkamagaluru, p 84
Nair PKR (1977) Multispecies crop combinations with tree crops for enhanced productivity in the tropics. Gartenbauwissenschqft 42:145–150
Nair PKR (2017) Managed multi-strata tree + crop systems: an agroecological marvel. Front Environ Sci 5:88. https://doi.org/10.3389/fenvs.2017.00088
Nair SK, Subba Rao NS (1977) Microbiology of the root region of coconut and cocoa under mixed cropping. Plant Soil 46:511–519
Nampoothiri KUK (2001) Organic farming-its relevance to plantation crops. J Plantn Crops 29(1):1–9
Nath JC, Arulraj S, Maheswarappa HP (2012) Integrated nutrient management in COD × WCT hybrid coconut under alluvial clay-loam soil of Assam. J Plantn Crops 40(2):105–110
Nelliat EV, Bavappa KVA, Nair PKR (1974) Multi-storied cropping—a new dimension in multiple cropping for plantations. World Crops 26(6):262–266

Ohler JG (1999) Modern coconut management: palm cultivation and products. Intermediate Technology Publications, 458 p
Onyibe HI, Gill LS (1992) Litter production and disappearance in three clones of rubber. J Plantn Crops 20(1):22–31
Palaniswami C, Thomas GV, Dhanapal R, Subramanian P, Maheswarappa HP, Upadhyay (2007) Integrated nutrient management in coconut based cropping system. Technical bulletin no. 49. CPCRI, Kasaragod, 24 p
Paramesh V, Arunachalam V, Nath AJ (2019) Enhancing ecosystem services and energy use efficiency under organic and conventional nutrient management system to a sustainable arecanut based cropping system. Energy 187:1–10. https://doi.org/10.1016/j.energy.2019.115902
Parthasarathi K, Balamurugan M, Prashija KV, Jayanthi L, Ameer Basha S (2016) Potential of *Perionyx excavatus* (Perrier) in lignocellulosic solid waste management and quality vermifertilizer production for soil health. Int J Recycl Org Waste Agric 5:65–86. https://doi.org/10.1007/s40093-016-0118-6
Patil B, Hegde V, Sridhara S, Narayanaswamy H, Naik MK, Patil KK, Rajashekara H, Mishra AK (2022) Exploring the impact of climatic variables on arecanut fruit rot epidemic by understanding the disease dynamics in relation to space and time. J Fungi 8(745):745. https://doi.org/10.3390/jof8070745
Prabhu SR, Thomas GV (2002) Biological conversion of coir pith into a value-added organic resource and its application in agri-horticulture: current status, prospects and perspective. J Plantn Crops 30(1):1–7
Prabhu SR, Subramaniam P, Biddappa CC, Bopaiah BM (1998) Prospects for improving coconut productivity through vermiculture technology. Indian Coconut J 29:79–84
Qureshi MRNM, Almuflih AS, Sharma J, Tyagi M, Singh S, Almakayeel N (2022) Assessment of the climate-smart agriculture interventions towards the avenues of sustainable production—consumption. Sustainability 14:8410. https://doi.org/10.3390/su14148410
Rajesh Kumar PP, Thomas GV, Gupta A, Gopal M (2015) Diversity, richness and degree of colonization of arbuscular mycorrhizal fungi in coconut cultivated along with intercrops in high productive zone of Kerala, India. Symbiosis 65:125–141. https://doi.org/10.1007/s13199-015-0326-2
Ray P, Lakshmanan V, Labbé JL, Craven KD (2020) Microbe to microbiome: a paradigm shift in the application of microorganisms for sustainable agriculture. Front Microbiol 11:622926. https://doi.org/10.3389/fmicb.2020.622926
Reddy DVS, Upadhyay AK (2002) Impact of integrated nutrient management of the mineral nutrition and yield of WCT coconut in littoral sandy soil at Kasaragod. In: Sreedharan K, Vinod Kumar PK, Jayarama, Chulaki BM (eds) Proceedings of PLACROSYM-XV. Indian Society of Plantation Crops, Kasaragod, pp 274–282
Rejani R, Yadukumar N (2010) Soil and water conservation techniques in cashew grown along steep hill slopes. Sci Hort 126(3):371–378. https://doi.org/10.1016/j.scienta.2010.07.032
Rejani R, Rupa TR, Nayak MG (2013) Suitability of cashew growing areas in India—an appraisal using GIS. J Agrometeorol 15:123–128
Rethinam P (2000) Future of plantation crops in the millennium. In: Rethinam P, Khan HH, Reddy VM, Mandal PK, Suresh K (eds) Plantation crops research and development in the new millennium. Coconut Development Board, Kochi, pp 33–37
Rethinam P, Venugopal K (1994) Cropping systems in plantation crops. In: Chadha KL, Rethinam P (eds) Advances in horticulture-plantation and spice crops. Malhotra Publishing House, New Delhi, pp 605–620
Rupa TR, Rejani R, Bhat MG (2013) Impact of climate change on cashew and adaptation strategies. In: Singh HP, Rao NKS, Shivasankara KS (eds) Climate resilient horticulture: adaptation and mitigation strategies. Springer Nature, pp 189–198. https://doi.org/10.1007/978-81-322-0974-4_17
Sadanandan AK, Hamza S, Bhargava BS, Ragupathi HB (2000) Diagnosis and recommendation integrated system (DRIS) norms for black pepper (*Piper nigrum* L.) growing soils of South India. In: Muraleedharan N, Raj Kumar R (eds) Recent advances in plantation crops research. Allied Publishers Ltd, New Delhi, pp 203–220

Santana MBM, Cabala Roland P, Serodio MH, Rosand PC (1988) Nutrient cycling in cocoa agroecosystems. In: Proceedings of international cocoa research conference, Santo Domingo, Dominican Republic, pp 233–237

Santha CR (2006) Coir products for soil engineering. J Soil Water Conserv 61(3):88A–93A

Selvamani V, Duraisami VP (2018) Mapping soil constraints for coconut using RS and GIS for the major coconut growing region of Tamil Nadu. J Plantn Crops 46(3):190–195. https://doi.org/10.25081/jpc.2018.v46.i3.20181

Shinde VV, Maheswarappa HP, Ghavale SL, Wankhade SM, Haldankar PM (2020) Productivity and carbon sequestration potential of coconut based cropping system as influenced by integrated nutrient management practices. J Plantn Crops 48(2):103–110. https://doi.org/10.25081/jpc.2020.v48.i2.6368

Shivaprasad P, Hanyappa N, Chandrasekhar N, Manonmani GK, Haresh SB, Nadaj SA, Jagadeesan M, Venkatesh DH, Sujatha K (2018) Studies on soil fertility status of coffee growing districts of Wayanad district. J Plantn Crops 46(3):180–189

Singh HP, Thomas GV (eds) (2010) Organic horticulture—principles, practices and technologies. Westville Publishing House, 440 p

Singh AK, Liu W, Zakari S, Wu J, Yang B, Jiang XJ, Zhu X, Zou X, Zhang W, Chen C, Singh R (2021) A global review of rubber plantations: impacts on ecosystem functions, mitigations, future directions, and policies for sustainable cultivation. Sci Total Environ 796:148948. https://doi.org/10.1016/j.scitotenv.2021.148948

Smil V (1999) Crop residues: agriculture's largest harvest: crop residues incorporate more than half of the world's agricultural phytomass. Biosci 49(4):299–308. https://doi.org/10.2307/1313613

Spices Board (2021). https://www.indianspices.com/

Srinivasan R, Natarajan A, Anil Kumar AS, Lalitha M (2019) Carbon stocks in major cashew growing soils of coastal Karnataka, India. J Plantn Crops 47(1):55–61

Subramanian P, Dhanapal R, Sanil P, Palaniswami C, Sairam CV, Maheswarappa HP (2005) Glyricidia (*Gliricidia sepium*) as green manure in improving soil fertility and productivity of coconut. J Plantn Crops 33(3):179–183

Sundram S, Angel LPL, Sirajuddin SA (2019) Integrated balanced fertilizer management in soil health rejuvenation for a sustainable oil palm cultivation: a review. J Oil Palm Res 31(3):348–363

Sunitha S, Krishnakumar T (2006) Biomass production and potential nutrient contribution from oil palm at felling. J Plantn Crops 34(3):309–311

Tandon HLS, Ranganathan V (1988) In: Tandon HLS (ed) Fertilizer management in plantation crops—a guide book. Fertilizer Development and Consultation Organization, New Delhi, pp 26–60

Thomas GV, Palaniswami C (2005) Effect of different components of farming system and on farm recycling on coconut crop and soil fertility. In: Joshi GD, Rangwala AD, Nagwekar DD, Haldankar PM (eds) Proceedings of national seminar on more remunerative coconut based farming systems. Dr Balasaheb Sawant Konkan Krishi Vidyapeeth, Dapoli, pp 49–61

Thomas GV, Prabhu SR (2003) Association of diazotrophic and plant growth promoting rhizobacteria with coconut palm (*Cocos nucifera* L.). In: Proceedings of sixth international PGPR workshop, 5–10 Oct 2003, Kozhikode, India, pp 20–26

Thomas GV, Prabhu SR, Reeny MZ, Bopaiah BM (1989) Evaluation of lignocellulosic biomass from coconut palm as substrate for cultivation of *Pleurotus sajor caju* (Fr) Singer. World J Microbiol Biotechnol 14:879–882. https://doi.org/10.1023/A:1008881124903

Thomas GV, Iyer R, Bopaiah BM (1991) Beneficial microbes in the nutrition of coconut. J Plantn Crops 19:127–138

Thomas GV, Biddappa CC, Prabhu SR (2001) Evaluation of N_2 fixing cover legumes as green manures for nitrogen substitution in coconut palm. Trop Agric (Trinidad) 78(1):13–18

Thomas GV, Palaniswami C, Prabhu SR, Gopal M, Gupta A (2013) Co-composting of coconut coir pith with solid poultry manure. Curr Sci 104(2):245–250

Thomas GV, Gopal M, Gupta A, Subramaniam P, Prabhu SR (2016) Bio-resources based biological soil fertility management for sustainable coconut production. In: Chowdappa P et al (eds) Abstracts of third international symposium on coconut research and development. ICAR-Central Plantation Crops Research Institute, Kasaragod

Thomas GV, Subramanian P, Gopal M, Gupta A, Prabhu SR (2022) Unraveling the potential of belowground and aboveground biodiversity for sustainable management of the health of plantation crop soils in coastal agroecosystem. In: Lama T, Burman D, Mandal UK, Sarangi SK, Sen H (eds) Transforming coastal zone for sustainable food and income security. Springer, Cham. https://doi.org/10.1007/978-3-030-95618-9_48

Tiedje JM, Cho JC, Murray A, Treves D, Xia B, Zhou J (2001) Soil teeming with life: new frontiers for soil science. In: Rees RM, Ball BC, Campbell CD, Watson CA (eds) Sustainable management of soil organic matter. CAB International, Wallingford, pp 393–412

USDA (2022) Oil seeds and products update. Available at: https://apps.fas.usda.gov/newgain-api/api/Report/DownloadReportByFileName?fileName=Oilseeds%20and%20Products%20Update_Jakarta_Indonesia_ID2022-0021

Usha KE, Vikraman Nair R (2002) Recycling potential of litter in cashew. In: Rethinam P, Khan HH, Reddy VM (eds) Plantation crops research and development in the new millennium. Coconut Development Board, Kochi, pp 348–350

van Doren J, Van den Hurk K, Broekmeulen D (2017) Can coir pith replace peatmoss for horticultural purposes in the future? Acta Hort 1266:43–48. https://doi.org/10.17660/ActaHortic.2019.1266.7

van Vliet JA, Giller KE (2017) Mineral nutrition of cocoa: a review. Adv Agron 141:185–270. https://doi.org/10.1016/bs.agron.2016.10.017

van Vliet JA, Slingerland MA, Waarts YR, Giller KE (2021) A living income for cocoa producers in Côte d'Ivoire and Ghana. Front Sustain Food Syst 5:732831. https://doi.org/10.3389/fsufs.2021.732831

Vannarath A, Thalla AK (2022) Effects of chemical pretreatments on material solubilization of *Areca catechu* L. husk: digestion, biodegradability, and kinetic studies for biogas yield. J Environ Manag 316:15322. https://doi.org/10.1016/j.jenvman.2022.115322

Vasundara R, Prakash NB, Anil Kumar KS, Hegde R (2020) Characterisation and classification of arecanut growing soils of Karnataka. J Plantn Crops 48(2):91–102

Vermeulen SJ, Campbell BM, Ingram JS (2012) Climate change and food systems. Ann Rev Environ Resour 37(1):195–222. https://doi.org/10.1146/annurev-environ-020411-130608

Wang Y, Yao Z, Pan Z, Wang R, Yan G, Liu C, Su Y, Zheng X, Butterbach-Bahl K (2020) Tea-planted soils as global hotspots for N_2O emissions from croplands. Environ Res Lett 15(10):104018. https://doi.org/10.1088/1748-9326/aba5b2

Watson G (1989) Nutrient addition. In: Webster CC, Baulkwill WJ (eds) Rubber, Tropical agricultural series. Longman Scientific and Technical, London, p 348

Willer H, Schlatter B, Trávníček J (2023) The world of organic agriculture. statistics and emerging trends 2023. Research Institute of Organic Agriculture FiBL, Frick, and IFOAM—Organics International, Bonn, 358 p

Yuan J, Zhang H, Zhao H, Ren H, Zhai H (2023) Study on dissociation and chemical structural characteristics of arecanut husk. Molecules 28(3):1513. https://doi.org/10.3390/molecules28031513

Zaman Q, Sarwar S, Ahmad F, Hamid FS (2011) Effect of nitrogenous fertilizer on the growth and yield of tea (*Camellia sinensis* L.) pruned in curved vs. flat shape. J Agric Res 49:477–482

Zhang G, Chu X, Zhu H, Zou D, Li L, Du L (2021) The response of soil nutrients and microbial community structures in long term tea plantations and diverse intercropping systems. Sustainability 13(14). https://doi.org/10.3390/su13147799

Zilber-Rosenberg I, Rosenberg E (2008) Role of microorganisms in the evolution of animals and plants: the hologenome theory of evolution. FEMS Microbiol Rev 32(5):723–735. https://doi.org/10.1111/j.1574-6976.2008.00123.x

George V. Thomas is a former Director of the Central Plantation Crops Research Institute, Kasaragod of Indian Council for Agricultural Research, Government of India. He has a research and research management career spanning 36 years in plantation crops particularly on coconut. He made significant contributions in soil microbiology, soil fertility, plant-microbe interactions, biomass recycling, and bioformulation development in plantation crops. He was Chairman, International Coconut Genetic Resources Network; President, Indian Society for Plantation Crops; Editor, Journal of Plantation Crops; and Fellow of four Professional Societies. He is a recipient of ICAR-Outstanding Team Research Award in Agriculture and Allied Sectors, as Team Leader for 2005–2006. He also served as Emeritus Scientist of Kerala State Council for Science, Technology and Environment and Director of Council for Food Research and Development, Konni, Kerala, India. Dr. Thomas has over 200 research publications and edited 10 books.

V. Krishnakumar completed his M.Sc. (Agri) in Agronomy from the College of Agriculture, Vellayani, Thiruvananthapuram of the KAU and Ph.D. in Agronomy from the Tamil Nadu Agricultural University, Coimbatore, India, and served various state and central government research institutions. He had worked as the Principal Investigator, Co-investigator of various National and International Projects. He superannuated as Principal Scientist (Agronomy) and Head, ICAR-Central Plantation Crops Research Institute, Regional Station, Kayamkulam, Kerala, India. As an agronomist, he has undertaken various field research programs, farmer community-oriented technology dissemination activities, administration, and manpower management programs. He is a recipient of the ICAR Award for Outstanding Interdisciplinary Team Research in Agriculture and Allied Sciences for the biennium 2005–2006. In his career of 37 years, he has published more than 100 research articles both in the National and International journals; edited 10 books and contributed chapters; and edited and published technical bulletins, training manuals, and other compilations. He served as Editor, Journal of Plantation Crops, and currently is a member of the Editorial Board of Current Science.

Coconut (*Cocos nucifera* L.)

P. Subramanian, Alka Gupta, Murali Gopal, V. Selvamani, Jeena Mathew, Surekha, and S. Indhuja

Abstract

Coconut (*Cocos nucifera* L.), the perennial tree crop, has a productive lifespan of six to seven decades. During its life cycle, soil plays a crucial role in sustaining its productivity, by providing the necessary physico-chemical and biological soil quality attributes for its sustenance. Soil health is the foundation for sustainable coconut production. The various soil health attributes need to be assessed and understood for designing the suitable location-specific management practices for sustainable production. In this regard, the constraints for production are to be assessed and monitored. Soil pollution as well as climate change also aggravate the already existing soil fertility constraints hindering the palm productivity. Technological options to improve the productivity while maintaining soil health, viz. efficient organic recycling techniques, integrated nutrient management, conservation agriculture, cropping system, cover cropping, and production of location-specific beneficial microbial consortia and nutrient application based on soil test, were reported. The fertility management strategies encompassing both the integrated nutrient management and organic management depending on the socioeconomic as well the ecological scenario need to be utilized for realizing the optimum production potential. Being 'Kalpa Vriksha' for human kind, residue recycling through composting techniques is of utmost importance. Approaches such as DRIS, digitization of soil fertility map at micro level, leaf

P. Subramanian (✉) · A. Gupta · M. Gopal · V. Selvamani · Surekha
ICAR-Central Plantation Crops Research Institute, Kasaragod, Kerala, India
e-mail: Subramanian.p@icar.gov.in; murali.gopal@icar.gov.in; selvamani.v@icar.gov.in; surekha.r@icar.gov.in

J. Mathew · S. Indhuja
ICAR-Central Plantation Crops Research Institute, Regional Station, Kayamkulam, Kerala, India
e-mail: jeena.mathew@icar.gov.in; indhuja.s@icar.gov.in

G. V. Thomas, V. Krishnakumar (eds.), *Soil Health Management for Plantation Crops*, https://doi.org/10.1007/978-981-97-0092-9_2

analysis-based nutrition, and conservation agricultural practices are to be given focus in this regard. Microbial solubilizers for particular nutrients such as phosphorus, potassium, zinc, and silicon are to be identified for formulating location-specific soil health management strategies in coconut-based cropping system.

Keywords

Coconut · Soil health indicators · Soil fertility · Climate change · Cropping system · Biomass recycling · Microbial resources · Organic farming systems

Abbreviations

$	US dollar
%	Percentage
@	At the rate
<	Less than
>	More than
µg	Microgram
ACC-1	Aminocyclopropane-1-carboxylate
ADP	Adenosine diphosphate
AICRP	All India Co-ordinated Research Project
Al	Aluminium
AMP	Adenosine monophosphate
APM	Adult palm mixture
ATP	Adenosine triphosphate
B	Boron
Ba	Barium
Bi	Bismuth
C	Carbon
C:N	Carbon/nitrogen ratio
Ca	Calcium
CaO	Calcium oxide
Cd	Cadmium
CD	Critical difference
CEC	Cation exchange capacity
CFE	Chloroform fumigation-extraction
CFI	Chloroform fumigation-incubation
CFU	Colony-forming unit
CH_4	Methane
CILY	Côte d'Ivoire Lethal Yellowing
cm	Centimetre
cm^2	Square centimetre
cm^3	Cubic centimetre
cmol	Centimol

CO_2	Carbon dioxide
COD	Chowghat Orange Dwarf
Cr	Chromium
Cu	Copper
CUE	Carbon use efficiency
DHA	Docosahexaenoic acid
DRIS	Diagnosis and Recommendation Integrated System
EAN	Enzyme activity number
Ec	Electrical conductivity
Etc.	So on
Fe	Iron
FYM	Farm yard manure
g	Gram
h	Hour
H_2O_2	Hydrogen peroxide
ha	Hectare
IAA	Indole acetic acid
INM	Integrated nutrient management
K	Potassium
K_2O	Potassium oxide
KCl	Potassium chloride
kg	Kilogram
L	Litre
Mg	Magnesium
mg	Milligram
MGD	Malayan Green Dwarf
MgO	Magnesium oxide
mm	Millimetre
MNF	Multi-nutrient fertilizer
Mo	Molybdenum
N	Nitrogen
Na	Sodium
NaCl	Common salt
NGS	Next-generation sequencing
NMR	Nitrogen mineralization rate
NO_3	Nitrate
OM	Organic matter
P	Phosphorus
P_2O_5	Phosphorus pentoxide
Pb	Lead
PCR	Polymerase chain reaction
PGPR	Plant growth-promoting rhizobacteria
pH	Power of hydrogen
PLFA	Phospholipid fatty acid
ppm	Parts per million

qCO_2	Microbial metabolic quotient
rDNA	Recombinant deoxyribonucleic acid
RNA	Ribonucleic acid
S	Sulphur
Si	Silicon
SMB	Soil microbial biomass
SMBC	Soil microbial biomass carbon
SOC	Soil organic carbon
SOM	Soil organic matter
SQI	Soil quality index
t	Tonne
TPF	Thiamine phyrophosphate
VAM	Vesicular-arbuscular mycorrhiza
WCT	West Coast Tall
WHO	World Health Organization
yr	Year
Zn	Zinc

1 Introduction

Soil, as a vital natural resource, forms the basis of our planet's existence. It is not merely an inert blend of sand, silt, and clay, but a living, dynamic entity that functions synergistically. The well-being of all life on earth depends on the condition of our soil. The richness of soil is innate and depends on the geological composition of the parent rock from which it derives, while its productivity is primarily determined by human activity. Various factors are currently jeopardizing soil health, including population growth, deforestation, shifting land use patterns, intensive farming, the widespread adoption of high-yielding crop varieties, increased fertilizer and pesticide usage, irrigation practices, the transition to fossil fuels, and modernized soil management approaches which collectively lead to degradation of soil health posing threats to sustainable agriculture (Katyal et al. 2016).

The definition of soil health described by the Natural Resources Conservation Service of United States Department of Agriculture is 'the continued capacity of soil to function as a vital living ecosystem that sustains plants, animals, and humans'. The Food and Agriculture Organization similarly describes soil health as the soil's ability to function as an ecosystem within specific land use boundaries, promoting the productivity of plants and animals, enhancing water and air quality, and nurturing the well-being of both flora and fauna (www.fao.org/agriculture/crops/thematic.../soil...of-soil/healthy-soil/en/). Maintaining optimal soil health is essential to ensure adequate food grain production and sustain ecosystem services. This is particularly crucial because the earth's land surface, covering 30% (14.8 Bn ha), is relatively small compared to its oceans (http://www.planetpals.com/planet2.html). Only about 11% (~1.6 Bn ha) of land is dedicated to food cultivation (FAOSTAT 2014); however, it accounts for 98% of the world's food production, with oceans contributing only 2% (http://core.ecu.edu/geology/woods/SEARESOURCES.htm). Thus, preserving soil productivity is of paramount importance, as it underpins food and biomass production, supporting heterotrophic life.

Coconut (*Cocos nucifera* L.), committed to the cultivating land for nearly a century, demands meticulous soil health management to ensure its sustained growth and production. Coconut ecosystem is located either in undulating terrain with mild to steep slopes, rendering them vulnerable to soil erosion and nutrient loss due to runoff, or in high rainfall areas susceptible to water logging or soil desiccation during dry spells. In both the scenarios, preserving soil health is of great importance for sustainable coconut production. Rather than relying solely on chemical fertilizers, a balanced approach with appropriate inputs at optimal levels can significantly enhance soil health. To ensure long-term coconut yields, proactive soil health management strategies are essential.

2 Importance of Coconut

The coconut palm, often referred as the 'tree of life', offers a multitude of essential resources, including food, beverages, oil, medicinal products, fibre, timber, thatch, mats, fuel, and domestic implements. This versatile plant is a prominent commercial crop cultivated on both large plantations and by smallholders. This palm carries substantial agricultural significance throughout the Asian and Pacific regions, flourishing in more than 90 countries, encompassing around 12.26 million ha. In the 2020–2021 period, it yielded a total of 65.67 Bn coconuts, with an average productivity of 5357 nuts ha^{-1} (CDB 2023). The primary coconut-producing countries, including the Philippines, Indonesia, India, Sri Lanka, Thailand, Malaysia, and Papua New Guinea, collectively contribute to approximately 80% of the world's coconut cultivation area. The coconut palm plays a crucial role in shaping rural economies and supporting the livelihoods of approximately 20 million individuals worldwide. In 2021, coconuts (fresh or dried) were ranked as the 1474th most traded commodity globally, with a combined trade value of $1.92 Bn. Notably, between 2020 and 2021, exports of fresh or dried coconuts experienced a substantial growth of 32.6%, increasing from $1.45 to $1.92 Bn. The global trade in coconuts (fresh or dried) represents only a small fraction, accounting for 0.0091% of the total world trade. In India alone, the livelihoods of 12 million people are intricately linked to the coconut industry through cultivation, processing, marketing, and related trade activities (Indian Trade Portal 2023).

3 General Soil Characteristics and Requirements

Soil stands as the paramount natural resource for sustainable coconut cultivation. It provides requisite anchorage, supplies water and vital minerals, hosts microflora, and facilitates nutrient transformations. It has been observed that the physical and chemical characteristics of the soil profoundly impact the palm's establishment and subsequent development. The initial stages of palm establishment profoundly influence growth rates and phenological development. The care and management being imparted to the young palm in its juvenile stage is very important as it will be reflected in its further growth stages. Given coconut's perennial nature, meticulous attention to site selection and soil characteristics is imperative for successful plantation. The soil's attributes at the chosen coconut plantation site significantly influence crop establishment and long-term sustainability.

3.1 Soil Types Suitable for Coconut Cultivation

Coconut exhibits remarkable adaptability to a diverse range of soil textures, from coastal soils to heavy clay soil. It thrives in gravelly sands, alluvial soils, laterite, peaty soils, volcanic soils, marine, and coral soils. This adaptability encompasses poorly drained clayey soils to sandy soils having good drainage as well as and the peaty soils which are strongly acidic to calcareous soils which are highly basic (Khan and Krishnakumar 2018; Thomas et al. 2018). Soil texture, a fundamental attribute, influences coconut palm establishment, root development, water and nutrient uptake, nutrient reserves, aeration, and microbial activity. It is crucial to avoid underlying sub-surface hard pans that impede root growth and development. Likewise, soils excessively rich in organic matter need to be avoided as they may fail to provide adequate anchorage. Soils with sufficient cation exchange capacity (CEC) are essential for growth and development. Heavy clay soils hinder initial establishment and limit subsequent growth due to restricted root development. Light sandy soils, while permitting unhindered root growth, suffer from poor nutrient and water-holding capacities and low cation exchange capacity. Loamy soils stand as the most favourable choice for coconut cultivation, boasting unhindered drainage, reduced susceptibility to waterlogging, favourable cation exchange capacity, and inherent fertility. In light sandy soils, conditioning through the addition of clay or 'kayal silt' can enhance nutrient and water-holding capacity. Conversely, heavy clay or waterlogged soils can be made suitable by layering with sand to improve texture and moisture retention.

The chemical composition, influenced by soil texture, significantly impacts coconut palm establishment and growth. Higher levels of iron and aluminium oxides and hydroxides in red and laterite soils can lead to phosphorus fixation, reducing its availability for crop uptake. Managing hard pans and mineral fixation are critical from the time of planting onwards. In hard laterite soil, incorporating 1 kg common salt (NaCl) can alleviate soil compaction, facilitating the establishment of juvenile palms. Gravelly soils, with their increased pore spaces, encourage root spread, enabling greater exploration of water and nutrients. However, their inherent low fertility requires supplementation with organic manures and fertilizers.

Child (1974) noted that coconut-growing regions worldwide feature a spectrum of soils, ranging from 97% sandy to 70% clay. Coconut palm growth tends to thrive on lower slopes, particularly in coastal areas with efficient moisture drainage, allowing for extended moisture retention. Cosico and Fernandez (1983) emphasized the significant impact of factors such as drainage, soil depth, and soil series on coconut productivity. Their research highlighted that the highest yields occur in well-drained areas with 80 cm soil depth and coarse to medium soil texture. Subsoil compaction was found to be a more influential factor than topsoil compaction, with a negative correlation between nut yield and subsoil bulk density. Felizardo (1972) observed that in compacted soils, ploughing and cultivation operations could enhance nut yield.

3.2 Water Table

A shallow water table is detrimental to coconut palm growth. In such areas, heavy rainfall can lead to waterlogging, impeding root growth, aeration, and causing palm health decline, including foliar yellowing and root decay, making palms susceptible to pathogens. In light sandy soils, planting depth must be adjusted to prevent root zone waterlogging, which inhibits water and nutrient uptake and harms palm health. In areas with a shallow water table, planting on raised bunds or mounds with moderately deep pits helps to prevent water stagnation. However, poorly drained low-lying areas are unsuitable for coconut cultivation.

3.3 Soil Depth

Soil depth plays a pivotal role in establishing coconut palms. To support the extended lifespan of these perennial crops, favourable growth conditions are imperative. Maheswarappa et al. (2000) found that approximately 81.3% of coconut palm roots are concentrated within the 0–90 cm depth, with 18.6% in the surface layer (0–30 cm) and 62.7% in the sub-surface layer (31–90 cm). Radioisotope studies by Anil Kumar and Wahid (1988) confirmed that around 80% of roots are within a 200 cm radius. This root distribution underscores the necessity of a minimum depth of 100 cm for successful establishment. Fremond (1964) recommended an ideal depth of 80–100 cm for coconut cultivation, emphasizing that subsoil hard pans, even with fertile surface soil hinder root proliferation. A crucial factor for optimal coconut soil is effective drainage. Coconut palms are sensitive to waterlogging, which can lead to chemical imbalances, root decay, and foliar issues due to poor aeration. Prolonged submergence for young palms, exceeding 48 h, can result in palm loss. Therefore, adequate drainage systems are essential in poorly drained conditions.

3.4 Soil pH

pH is a critical parameter influencing the physico-chemical and biological properties of soil. The ideal soil pH for coconut cultivation ranges from 6.4 to 7.0. Child (1974) noted that pH values exceeding 7.5 or falling below 5.4 are less conducive to growth. Cosico and Fernandez (1983) observed coconut palms to thrive across a wide pH spectrum, from 3.6 to 8.7, with optimal performance near pH 8.0 in coarse-textured coastal flatlands. Jemziya (2018) reported pH values ranging from 7.82 ± 0.04 to 8.12 ± 0.06 for surface and subsoil in Batticaloa, Sri Lanka. Srinivasan et al. (2020) found that red gravelly soils in Karnataka, India, with substantial Fe, Al, and Mg content and an acidic to neutral pH could enhance nut setting in coconut. These findings emphasize the importance of soil pH management for successful coconut cultivation across various regions.

3.5 Electrical Conductivity

Electrical conductivity (EC) measures salt concentration and indicates soil salinity. EC is influenced by factors including soil texture, moisture percent, salinity, porosity, CEC, and temperature. Coconut, considered a semi-halophyte, demonstrates adaptability to various growth conditions (Purseglove 1975). Sankaranarayanan et al. (1958) observed coconut palms tolerating salinity up to 0.6%. This resilience to high salt levels was evident in post-tsunami situations in Indonesia and other countries where coconut palms suffered only physical damage (Marohn et al. 2012), unlike other tree crops.

During the rainy season, waterlogging can affect saline and saline-sodic soils due to drainage issues, heavy clay content, and reduced hydraulic conductivity (Yu et al. 2012). Soil salinity can enhance nutrient uptake in coconut seedlings (Remison and Iremiren 1990). Coconut is moderately salt tolerant as indicated by Ferreira et al. (2007) and Marinho et al. (2006). Arya (2020) even found that dwarf coconut varieties like Malayan Green Dwarf (MGD) could tolerate up to 10% substitution with sea water. These findings highlight coconut's ability to adapt to saline conditions.

3.6 Available Mineral Nutrients

Soil fertility is an index of the availability of essential mineral nutrients. Potassium (K) is needed in the highest proportion for coconut cultivation among all the essential nutrients. However, in tropical acidic soils, deficiency of key nutrients such as K, Ca, and Mg is common. Improving the buffering capacity of soil and addition of nutrient sources along with the improvement of organic matter content is an effective strategy to combat mineral nutrient deficiencies. Boron deficiency is particularly critical among micronutrients. Coconut palms are perennial, and they exhaustively extract major, secondary, and micronutrients from the soil. Being the 'Kalpa Vriksha', each and every part of the palm finds its application in the daily life of human kind. As each component of the palm contains valuable nutrients, a substantial portion of the nutrients extracted from the soil is eventually lost through these various palm parts. The total nutrient uptake considering the nutrient composition in different palm parts and the biomass content is presented in Table 1.

Table 1 Total nutrient uptake by apparently healthy coconut palms

	N	P	K	Ca	Mg	S	Mn	Cu	Zn	B
Category	(g palm^{-1})						(mg palm^{-1})			
Apparently healthy	889.0	109.4	1075.0	389.7	71.6	229.7	1784.0	569.0	2304.0	321.6

Source: Mathew et al. (2021a)

Therefore, it is crucial to assess the soil fertility status of coconut-growing areas to ensure adequate nutrient supply and sustainable cultivation practices.

Scheewe (2022) observed that enhancing soil health in coconut farms involves practices like organic matter recycling, responsible herbicide use, incorporating deep-rooting plant species into cropping systems, and implementing mulching. Tennakoon (2004) noted the varying significance of soil nutrients during different crop growth stages. In Indonesia, Irawan et al. (2022) found that soil quality index and land suitability are positively correlated for coconut cultivation. Thuvasan et al. (2019) reported significant correlations between coconut palm yield and major nutrients such as nitrogen and potassium as well as micronutrients such as iron, manganese, zinc, and copper. These findings underscore the importance of soil management and nutrient balance for successful coconut production.

4 Soil Biological Properties

Soil biological property deals with the structural and functional aspects of the organic portion comprising of living organisms and decaying/decomposing debris. Organic matter complex is primarily composed of carbon, oxygen, and hydrogen (92%) and other elements that form the nutritional components. The soil biological component's quality and quantity dimension includes soil biota, soil organic matter (SOM), and metabolites including enzymes, which have telling effect on the soil physico-chemical properties thereby exerting a substantial influence on soil health.

Coconut, the epitome of tropical coastlines and islands, is largely adapted for cultivation in more than 90 countries. Being a tall, sturdy, and perennial crop, with almost all parts economically useful, coconut is an exhaustive crop in terms of nutrients and water. Coconut palms under natural conditions lowered major nutrient status (2.7–11.5 times) and elevated water stress (15%) in the soil which resulted in the alteration of the faunal communities and their interaction (Young et al. 2017). However, coconut is amenable and adaptable for flexible and diverse cropping/farming system models because of its unique growth pattern and canopy. This enables integration of sustainable technologies in coconut farming that would ensure socioeconomic and ecological benefits (Thomas et al. 2018). Such integrated systems are prospective for resource sharing and resource conservation, resulting in enhanced soil organic matter, nutrients status, carbon capture, and carbon storage in coconut soils (Kumar and Kunhamu 2022).

4.1 Soil Organic Matter and Soil Organic Carbon

Soil organic matter includes the organic constituents of soil at various stages of decomposition (<2 mm in size) and soil organisms. It is the key indicator of soil health, and its turnover plays a crucial role in soil ecosystem. SOM is critical in stabilization of soil structure and in retention/release of moisture and nutrients. Soil organic carbon (SOC) comprises the carbon (C) stock present in SOM which is roughly 55–60% (FAO and ITPS 2015).

Roupsard et al. (2008a) estimated that coconut palm, being a monocot, allocates much (>85%) of its net primary productivity into perishable structures like fruits, leaves, peduncles, and fine roots that turn into litter and be respired by the ecosystem or contributing to the build-up of SOM, assuming that the mode of management of the huge annual litter production would have a crucial impact on SOM build-up. Fertile areas of Vanuatu (an island nation in the South Pacific Ocean) with coconut-based small holders' cropping system and large coconut plantations measured 12% of SOM in topsoil (0–20 cm) and 5% in subsoil (20–30 cm) with an average value of C:N ratio 10 (Lamanda et al. 2004). Rapid decomposition of SOM in cultivated soils of tropics and semi-arid regions affect the soil functions. Significant loss of SOM (47.5%) was noticed in the surface layers of conventional coconut gardens to the native forest in the Ultisols in north east region of Brazil, whereas the integrated coconut gardens showed an increase in SOM of approximately 37% (Guimaraes et al. 2013). Findings of Atapattu et al. (2017) provided preliminary evidence on the influence of short-rotation forestry-based agroforestry on the soil properties where the highest SOM was observed in *Gliricidia sepium* planted plots (1.86%) followed by *Acacia mangium* (1.69%). Sandy soils where coconut was predominantly grown lacked adequate organic matter, and hence farmyard manure, compost, and green leaves were generally recommended to maintain organic matter and nutrient status of the soil. The local deficit in organic manures could be overcome by adopting sustainable agro-management practices like in situ cultivation of green manure crops (Thomas et al. 2018). The main coconut-producing state of Guerrerro in Mexico suffered problems of low fertility wherein the total average of SOM varied from 1.29% to 0.71%, all below the reference medium value of 1.6–3.5%, and the same deficiency trend was shown by the foliar N contents (Ramirez-Silva et al. 2021).

Varghese et al. (1978) reported that the organic carbon content and the soil fertility improved considerably under the mixed crop and were reflected in the enhanced yield of coconut. Coconut with cocoa in double hedge, pineapple, and black pepper using coconut as standard recorded SOC of 0.7% and 0.58% in the top (0–15 cm) and subsoil (15–30 cm), respectively, during the experimental period against the pre-experimental record of 0.2–0.3%. Studies on substitution of N fertilizer with glyricidia as green manure in coconut under coastal littoral sandy soil of Kerala (India) established higher level of SOC (0.56%), available N, and soil moisture content as compared to application of inorganic fertilizers alone (Subramanian et al. 2005a). High-density multispecies cropping systems (HDMCS) involving coconut, clove, banana, and pineapple produced around 17–22.5 t ha^{-1} $year^{-1}$ of biomass and if efficiently utilized and recycled into the coconut production system would contribute immensely to soil health by enriching SOM, nutrients, and microbial biomass (Palaniswami et al. 2010). From a detailed study on assessment of the fertility and soil-related constraints in coconut production in major coconut-growing soils of Kerala, India, Nair et al. (2018) observed that the SOC content was low for both surface and sub-surface soils of Central Kerala, Onattukara, and coastal sandy plains. The highest SOC (2.09%) was recorded in surface soils of Northern Kerala followed by Central and Eastern Palakkad (1.2%) and Southern Kerala (0.93%). In the major coconut-growing areas of southern Karnataka (India) with high rainfall, soils were strongly acidic and rich in organic carbon, whereas the other locations of sub-humid to semi-arid were neutral to moderately alkaline with medium SOC content (central dry zone recorded maximum of 0.49%, whereas Coastal zone-Western Ghats foot hills recorded a maximum of 1.49%) (Avinash et al. 2019). Technologies like coconut leaf vermicomposting (Gopal et al. 2017; Prabhu and Thomas 1998), urea-free coir pith composting (Thomas et al. 2013), and pyrolysis techniques (Gopal et al. 2020; Suman and Gautam 2017) for the conversion of lignin-rich recalcitrant residues of coconut palms like coconut leaf, coir pith, and husk would provide viable technological options for quicker recycling and use as soil amendment to improve soil health.

4.2 Soil Microbiota

Coconut, being a plantation crop, is amenable for flexible cropping systems/patterns and enriches/conserves soil microorganisms and other higher trophic organisms. Microbial biomass is considered as the most significant among the labile pools of organic matter. Above-ground vegetation, agro-management practices, and the innate soil quality strongly influence the below-ground diversity, abundance, and activities of soil microflora and fauna. Pioneering studies in India on the microbial status of the coconut soils indicated that coconut under mixed cropping with crops like fodder/legumes (Potty 1977) and cocoa (Nair and Subba Rao 1977) stimulated microbial activities including that of N fixers and P solubilizers. Bopaiah and Shetty (1991) recorded higher microbial biomass (181.98/219.59 μg C g^{-1} soil) in mixed farming than in monocropping system (151.06/187.50 μg C g^{-1} soil) in the coconut/napier grass system. Introduction of annual and perennial crops in coconut plantations induced a shift in biology, in turn proving beneficial to the system. Augmentation of biological activities in the soil during the maximum vegetative phase of fodder legumes in coconut mixed cropping was reported (Thomas 1987; Thomas et al. 2010). Coconut-based multiple cropping system was found to be more productive in terms of soil organisms' density and diversity owing to high SOC, SOM, litter, root biomass, etc. (Attapattu et al. 2017).

Associative symbiotic bacteria, *Azospirillum* were reported from different coconut-based farming systems. *A. lipoferum*, *A. brasilense*, and *A. amazonense* were a few species isolated from the coconut (Ghai and Thomas 1989). Merilyn and Thomas (1992) reported *Beijerinckia indica* as a predominant N fixer in acidic soils of coconut. Free living N fixer, *Azotobacter* and *Azoarcus*, were also reported.

Bacillus, *Burkholderia*, *Enterobacter*, *Pseudomonas*, *Ralstonia*, and *Streptomyces* were the common bacterial genera found in coconut system (Morales-Lizcano et al. 2017). Root exudates released by the coconut and component crops in cropping system proliferated beneficial microorganisms such as *Azospirillum*, *Bacillus*, *Beijerinckia*, *Pseudomonas*, *Azoarcus*, *Arthrobacter*, etc. (Gupta et al. 2010), and many of them were also found to be effective bioinoculants for the better establishment of coconut nursery seedlings. *Pseudomonas*, *Enterobacter*, *Bacillus*, *Micrococcus*, *Corynebacterium*, *Alcaligenes*, and *Klebsiella* were a few P-solubilizing bacterial genera encountered in coconut plantation soils. K-solubilizing bacteria belonging to genera *Acinetobacter*, *Alcaligenes*, and *Micrococcus* were also found in coconut rhizosphere soils (Gupta et al. 2016b; Pandey and Gupta 2020; Thomas and Shantaram 1986).

Rhizospheric soil samples of healthy palms in root (wilt) disease tract had a good representation of plant beneficial bacteria with a higher population of N fixers, phosphate, and silicate solubilizers in comparison with diseased palms (Gopal et al. 2005). Earthworms are also known to increase the population of beneficial microbes such as fluorescent pseudomonads, N fixers, and nutrient solubilizers (Gopal et al. 2012). Bacteria belonging to *Pseudomonads* and *Bacillus* spp. with multiple plant growth-promoting traits were reported in coconut rhizosphere sample following strategic screening approaches (George et al. 2018). Pandey and Gupta (2020) isolated rhizospheric bacteria belonging to *Pseudomonas*, *Bacillus*, *Azospirillum*, *Azotobacter*, *Escherichia*, *Paenibacillus*, *Burkholderia*, and *Klebsiella* exhibiting ACC deaminase activity and other plant growth-promoting traits from coconut grown in Lakshadweep islands of India of which two isolates *P. putida* and *Bacillus paramycoides* had shown the ability of ameliorating the negative impact of salinity stress when experimented with French beans. *Pseudomonas*, *Atlantibacter*, *Enterobacter*, *Klebsiella*, and *Acinetobacter* having multiple mineral nutrient (P, K, Zn, and Si)-solubilizing traits were isolated from field-resistant palms in root (wilt) disease-affected tract (Indhuja et al. 2021). Indole acetic acid-producing *Escherichia* sp. (Nair and Subba Rao 1977), *Enterobacter* sp., *Serratia* sp. (George et al. 2013), *Klebsiella*, and *Pseudomonas* sp. (Indhuja et al. 2021) were isolated from the root zone of coconut soils.

Like bacteria, fungi were also found to be more pronounced in the root zone soils of coconut compared to the interspace soils. The mixed farming system (coconut/napier grass) significantly increased vesicular-arbuscular mycorrhizal (VAM) colonization when compared to monocropping (Bopaiah 1988a). However, tree species such as *Eucalyptus*, *Leucaena*, *Casuarina*, and *Ailanthus* in coconut-based agroforestry system had deleterious effect on soil microflora (CPCRI 1988). Gibberellin (like-substance)-producing *Aspergillus flavus* and *A. fumigatus* (Nair and Subba Rao 1977), P-solubilizing *A. niger* and *Penicillium citrinum* (Bopaiah 1988a), antagonistic *Trichoderma* spp. (Johnson et al. 2017; Iyer et al. 2018), *Candida*, and *Neodeightonia* (Morales-Lizcano et al. 2017) were reported from the root zones of coconut soils.

Coconut cropping system was found to harbour diverse AM fungi of genera, viz. *Glomus*, *Gigaspora*, *Sclerocystis*, and *Acaulospora*. Spores representing *Glomaceae*,

Claoideoglomeraceae, *Diversisporaceae*, *Acaulosporaceae*, and *Gigasporaceae* families were detected in coconut-based cropping system (Ambili et al. 2012; Rajeshkumar et al. 2015). Coconut garden enriched with organic biomass hosts a variety of macrofungi such as *Agaricus*, *Termitomyces*, *Macrocybe*, *Trametes*, *Leucocoprinus*, *Volvariella*, *Polyporus*, etc., establishing saprophytic/symbiotic/parasitic associations in coconut plantations (Gupta et al. 2018).

4.3 Soil Enzymes and Metabolites

Organic compounds in the soil include metabolites and enzymes produced by the plants and microbes/other soil organisms and are formed and released during litter/manurial decomposition. These include root exudates (sugars, amino acids, organic acids, phenolic compounds, and growth regulators), microbial metabolites (growth hormones, siderophores, organic acids, exopolysaccharides, and antimicrobial compounds), enzymes (phosphatase, dehydrogenase, hydrolases, peroxidases, etc.), and products of organic matter decomposition (humic and non-humic substances, carbohydrates, carboxylic acids, and hydrocarbons).

Soil enzyme activity is considered as an indicator of microbial activity and variability in microbial communities. Enzymes in soil such as phosphatase, β-glucosidase, protease, amylase, and catalase, that are released mainly by microorganisms and plants, adhere to and form complexes with soil particles and are involved in decomposition of organic matter, degradation of xenobiotics, and nutrient cycling, thus indicative of the biological capacity of the soil. Considerable increase in dehydrogenase, phosphatase, and urease activity was observed due to the application of green manures such as *Pueraria phaseoloides*, *Mimosa invisa*, and *Calopogonium mucunoides* in coconut basins (Thomas and Shantaram 1984). Variable trends of urease, dehydrogenase, and phosphatase activities were observed in the root zone and rhizosphere regions of different crops in coconut-based multistoreyed cropping and coconut monocropping system (Bopaiah and Shetty 1991). Reduction in soil dehydrogenase and phosphatase activities were noticed in rhizosphere soils of coconut seedlings grown in elevated CO_2 and temperature conditions in controlled field facility. Under elevated temperature, increase in soil urease and invertase activities and decline in amylase activity were observed (Sunoj et al. 2015). Influence of elevated temperature and CO_2 on soil enzymes suggested the need of better soil management strategies in coconut plantations in this changing climatic scenario. Rajeela et al. (2017) observed cross-compatible nature of PGPR strains of coconut rhizosphere with vegetable intercrops as indicated by the increased soil enzyme activities (phosphatase, dehydrogenase, and urease). In the mid-laterite soil regions of Kerala, India, the highest values of dehydrogenase (462.52 μg of TPF released g^{-1} soil h^{-1}), acid phosphatase (80.35 μg of *p*-nitrophenol released g^{-1} soil h^{-1}), and catalase (3.9% H_2O_2 hydrolysed g^{-1} soil h^{-1}) were noticed in coconut + banana under organic system (Usharani and Aparna 2019). Improvements in soil enzyme activities such as DHA (4–47%), urease (2–27%), β-glucosidase (7–49%), acid phosphatase (3–73%), and alkaline phosphatase (4–71%) were noticed in different waste recycling regimes of integrated nutrient management (INM) schedule (Kumar et al. 2022).

Coconut with profuse root system containing more than 4000–7000 roots exudes large quantities of sugars, phenols and other organic compounds that directly influence the population and structure of microbial communities. Studies on biochemical characterization of the root exudates of coconut palm by Bopaiah et al. (1987) detected the presence of sugars such as raffinose, lactose, glucose, fructose, and amino acids like serine/glycine, glutamine, alanine, phenyl alanine, and isoleucine/leucine. Succinic acid was found in the organic acid fraction, and phenol was measured as 11.6 $\mu g\ g^{-1}$ of root. The highest total sugars, phenols, and K were found in coconut canopy leachate when compared to the other crops of coconut cropping system which would ultimately reach the soil. Experiments with coconut leachate suggested that microflora of black pepper was able to overcome the inhibitory allelopathic effect of coconut (Gopal and Gupta 2005). Studies indicated that the microflora of bulk soils were capable of utilizing the leachates. Diazotrophs were found sensitive to both root and leaf leachates, but the leachates were found harmless to P solubilizers and most of the other PGPRs. Gopal et al. (2010c) suggested that the high amount of sugars would have helped microbes to offset the phenol effect, whereas the leaf leachate known to have 4-hydroxyl benzoic acid (an antibacterial compound) supported the suppressive nature. Persistence and biological effect of allelochemicals were found to be influenced by microorganisms through biotransformation of biomolecules. Plant residue mulches sourcing allelochemicals and the microbial activity might result in allelopathic interference as evidenced in soil sickness as a resultant of high quantity of crop residues.

Coconut leaf vermicompost is considered a good soil organic amendment in supplying growth-promoting hormones such as cytokinins and gibberellins and in enhancing humic acid content in the soil (Gopal et al. 2010a). Organic acids released by nutrient-solubilizing microbes act as chelators of cations and compete for nutrient adsorption sites in soils. Low molecular weight organic acids such as tartaric acid, glycolic acid, oxalic acid, gluconic acid, etc. are released into the soil by microorganisms enabling the release of fixed K from K bearing minerals. In a study on systemic screening strategies for identification of elite PGPR from coconut rhizosphere, George et al. (2018) had screened more than 480 rhizospheric bacteria for beneficial/plant growth-promoting traits. The isolates were found to secrete chitinases (35%), antagonistic metabolites (87–95%), and indole acetic acid (85%). More than half of the bacterial isolates produced organic acids aiding in P solubilization. While analysing the diversity of ACC deaminase-producing bacteria from the rhizospheric bacteria of coconut palms grown in Lakshadweep islands, Pandey and Gupta (2020) observed several of these isolates secreting IAA (10–80 $\mu g\ mL^{-1}$), ammonia, and siderophores which are the key metabolites involved in the improvement of root architecture, disease suppression, and nutrient acquisition. PGPR metabolites such as phytohormones, extracellular polysaccharides, and volatile metabolites enhanced the antioxidant defence system and induced transcriptional regulation of stress response genes thereby modifying the host physiological responses against biotic/abiotic stress (Bouremani et al. 2023).

Gopal et al. (2022) opined that the autochthonous nutrients from the coconut biomass recycled by the soil microbial communities could be a reason for sustaining high coconut productivity in Lakshadweep islands. In a review on coconut cultivation in Sri Lanka, Dissanayaka et al. (2023) suggested mixed coconut agroforestry as a pathway to achieve regional and global sustainability by balancing the ecosystem functions by increasing species richness, enhancing soil health attributes, and opening up new carbon sequestering pathways. Soil food web, an integrative framework invigorating plant-soil-microbial-faunal interactions, is an essential component of healthy agro-ecosystem, and right biology is requisite for nutrient recycling, soil fertility, and health. Ecologically amenable cropping system models, agro-residue recycling, and bioresource management strategies to ensure beneficial crop interactions would enable the coconut plantation to be productive and sustainable.

5 Indicators of Soil Health and Their Assessment for Sustainable Coconut Cultivation

Successful cultivation of coconut palms relies to a great extent on the soil health attributes. Exploring nutrients and water from a limited soil volume by a tree crop having a lifespan of more than eight decades is indeed an indication of its magnitude of influence on the palm productivity. It has been documented that long-term cultivation of coconut palms could deteriorate the soil quality due to the non-adoption of recommended management practices finally leading to the declined production (Atapattu et al. 2017b; Herath et al. 2016). Nair et al. (2018) assessed the soil fertility indicators of the different agro-ecological regions of the Kerala, India. The soil-related constraints affecting coconut production in these localities of the state were outlined, and they suggested that liming and application of deficient nutrients could ensure the satisfactory yield from coconut palms. The important physico-chemical and biological indicators of soil health for coconut production are discussed below.

5.1 Physical Indicators

Soil physical properties were significantly related to coconut yield (Arachchi et al. 2000). The major physical indicators of soil health include bulk density, soil compaction, soil texture, depth of soil, porosity, and available water content. As defined by Dexter (2004), bulk density is the mass of soil solids per unit volume of the soil. It is influenced by soil composition as well as by the management practices such as compaction by machinery tilling and by cropping (Dexter et al. 2008). It is also related to the SOM; higher the organic matter content, lower will be the bulk density. High soil compaction limited the available water and aeration capacity of soils which can result in the reduced activity of coconut roots (Arachchi et al. 2000).

Bruce (1956) found that the bulk density of soils of poor yielding coconut palms is higher than good yielding palms. The bulk density of ideal soil for cultivation is 1.33–1.55 g cc^{-1} (Reichert et al. 2009). In red-yellow podzolic soils, inclusion of cover crops such as *Pueraria* and *Brachiaria* in coconut gardens enhanced the aggregate stability and reduced the bulk density (Herath et al. 2016). Davis (1968) reported that the number of respiratory organs for coconut roots decreases with increase in depth of soil. Soil compaction resulting in increased bulk density is presented as a major constraint limiting water absorption and gas exchange process in coconut (Arachchi et al. 2000). It was suggested that inclusion of N_2-fixing leguminous crops, tree species inter-planted in coconut gardens, and provision of drip irrigation systems could improve these physical constraints (Arachchi and Liyanage 1997). Compaction of subsoil than topsoil adversely affected coconut yields (Cosico and Fernandez 1983).

The relative proportion of sand, silt, and clay particles is indicated as soil texture. Owing to the deep root system, coconut palms are particularly adapted to well-drained sandy soils (Anderson 1967). Soil textures of coconut-growing tracts which are demarked as suitable for cultivation range from coarse sand containing 97% sand to heavy soils with 70% clay (Child 1974). Issaka et al. (2012) found that gravelly or stony soils and exceedingly porous soils are not suitable for coconut cultivation.

5.2 Chemical Indicators

The major soil chemical indicators affecting coconut production include soil reaction, electrical conductivity, available nutrients, and percent base saturation. Considering the land suitability for coconut cultivation, it was outlined that highly suitable soils should have >35% base saturation, >1.5% SOC, and EC of 0.8 dSm^{-1} and the sum of basic cations should be >2.8 cmol kg^{-1} (Sys et al. 1993).

Coconut palms could thrive on soil pH ranging from 5.0 to near neutral reaction (Fremond 1964). However, the most ideal pH is 6.4–7.0. Generally tropical soils are red and laterite soils with pH range of 4.2–5.8 (Malhotra et al. 2017). When the pH values go below 5.2, aluminum ions come to soil solution (Kochian 1995). Subsoil aluminium upon hydrolysis could reduce the pH apart from creating direct injury to the roots. Root growth is negatively related to aluminum content in the soil solution (Vizcayno et al. 2001). Mathew et al. (2022) found subsoil acidity is predominant in the tropical laterite soils under coconut cultivation than the sandy soils. Soil pH also influences the soil biological process, microbial growth, and proliferation and modulates the bio-geo-chemical cycle.

EC signifies the salt concentration in the soil. Though coconut palms are semi halophytic in nature, higher salt concentration as indicated by higher EC values could pose salt injury to the roots directly. It also could hinder the nutrient and water uptake from the soil. A salt tolerance level of 0.6% by coconut was recorded by Sankaranarayanan et al. (1958). Inadequate irrigation, poor quality of irrigation water, and drainage problems can cause soil salinization (Silva et al. 2010). Wahid et al. (1974) reported negative correlation between the root CEC of coconut and yield. They found that the cationic nutrient uptake was dependent on the nutrient ratios in soil. Significantly higher correlation was found between K/Na, K/(Ca + Mg) and K/Mg in soil and their corresponding ratios in soil. Significant positive correlation existed between potassium content in the soil and leaf with the yield of coconut palms. Hence, among the macronutrients, potassium is considered as the most critical element for coconut nutrition. However, Tennakoon (2004) found that for young trees, the order of nutrient requirement is N > P > K > Mg, while the adult bearing palms follows the order K > Mg > N > P. Mathew et al. (2021a) while studying the nutrient uptake pattern of healthy palms of 25 years age through destructive sampling found that the total nutrient uptake followed the order K > N > Ca > S > P > Mg. Foliar yellowing of the outer whorls and reduction in the number of female flowers are the main features of nitrogen deficiency (Sobral 1998). Yield of coconut palms is directly related to the addition of potassium (Sreelatha 1993). More than 200% increase in yield was observed with potassium addition over no potassium treatment. Among the micronutrients, boron was found to be the most critically essential for pollen development, nut setting, and enhanced nut yield.

5.3 Biological Indicators

Biological indicators of soil include the measurable properties that define the biotic components in soil which could be used as metrics in determining soil functionality over a wide range of ecological conditions (Bhaduri et al. 2022). A diverse set of biological indicators are available to indicate the multifunctional properties of soil ecosystem and can be grouped according to the different soil ecosystem parameters. Based on the techniques/approaches, they are broadly classified as physiological, metabolic, functional, and molecular. Microbial characteristics of soil such as biomass, diversity, and activity are increasingly being evaluated as sensitive indicators as they are related to soil and plant quality and ecosystem sustainability. Biological indicators help to detect changes in soil quality providing signs of soil improvement or degradation and thus can give an integrated measure of soil health. A few biological indicators that are commonly used to assess the soil quality of coconut plantations and their applications are discussed below.

5.3.1 Soil Microbial Biomass

Soil microbial biomass (SMB) is a measure of the mass of the living component (bacteria, fungi, protozoa, etc.) of soil organic matter. A change in the microbial biomass that responds quickly to differences in soil management is considered as an early and sensitive indicator of soil changes like total soil organic carbon. It is commonly detected using chloroform fumigation-extraction (CFE) method (Vance et al. 1987) or chloroform fumigation incubation (CFI) method (Jenkinson and Powlson 1976).

Bopaiah (1988b) reported higher microbial biomass carbon (187.8 g C g^{-1}) soil in the root regions of coconut under mixed cropping as compared to coconut monocrop (112 C g^{-1} soil). Long-term experiment on the effects of leguminous cover crops grown in coconut plantations by Dinesh et al. (2006) recorded marked enhancement in the SOM and microbial activity as indicated by the significant increase (two- to threefold higher in the organic and mineral layers of the cover cropped location) in SMB-carbon (C_{MIC}), SMB-nitrogen (N_{MIC}), and SMB-phosphorus (P_{MIC}). C_{MIC}, N_{MIC}, and P_{MIC} levels were found significantly higher in the organic layers (5.5 g kg^{-1}, 642, and 199 mg kg^{-1}, respectively) compared to their respective levels in the mineral layers (0.36 g kg^{-1}, 31, and 13 mg kg^{-1}, respectively). The ratios of various microbial variables, viz. C_{MIC}:organic C, C_{MIC}:N_{MIC}, and C_{MIC}:P_{MIC}, reflected the availability of substrates to soil microbiota. While evaluating the impact of a 15-year-old *Pueraria* cover crop under a coconut plantation, C_{MIC} was found to increase 41% under the CC compared to the control. This ex situ experiment showed P application increasing C_{MIC}, which along with other parameters had indicated fertility enhancement in coconut plantations of humid tropics for further recommendations (Pandey and Begum 2010). In a study on working out soil quality index (SQI) of coconut soils by Irawan et al. (2022) in Pacitan district of Indonesia, C-biomass of microbes were found to be in a very low range of 0.22–0.18 which showed a positive correlation with SOC.

5.3.2 Soil Microbial Respiration

Soil respiration is a measure of the carbon dioxide (CO_2) released from soil as a result of decomposition of SOM and plant litter by soil microbes and through plant roots and soil fauna. It indicates the level of microbial activity, SOM, plant litter, and decomposition and may also signify soil conditions that limit biological activity and decomposition. Measured by simple field methods or more sophisticated field and laboratory methods, the rate of CO_2 release is expressed as CO_2-C lbs acre^{-1} day^{-1} (or kg ha^{-1} day^{-1}) (Parkin et al. 1996).

Significantly increased basal respiration rate in fertilizer and manure-added treatments in a coconut-growing sandy regosol provided an insight on the effects of fertilization on soil microbial activity (Nirukshan et al. 2016). The study on role of soil organisms and functions on wet and intermediate zones of different coconut-based cropping systems indicated increased soil microbial respiration rate in comparison to barren land and was maximum (>700 μg g^{-1} day^{-1}) with cocoa intercropping followed by banana and mango intercropped coconut plantations (Atapattu et al. 2017a). While assessing the potential of using coconut-based agroforestry systems to improve soil quality of marginal coconut lands in Sri Lanka, Atapattu et al. (2017b) observed the highest microbial respiration (CO_2 μg g^{-1} soil day^{-1}) of 128.5 in *Acacia auriculiformis* plots, followed by *Macaranga peltata* (102), while the lowest activity was recorded in the control plots (65.56). Senarathne and Udumann (2019) had recorded higher soil microbial activity (indicated by soil respiration) in *Anacardium occidentale* intercropped system which was 22% higher than sole coconut system. In a study by de Carvalho Junior et al. (2019), cumulative soil microbial respiration indicated the suitability of higher rates of application of coconut biochar for soil carbon dynamics, contributing to a positive carbon priming effect. Coconut-glyricidia mixed cropping systems improved soil nutrients in dry and wet regions of Sri Lanka wherein microbial respiration was found to be influenced by the site characteristics, with wet site showing higher values of microbial respiration (33%) (Raveendra et al. 2021).

5.3.3 Microbial Metabolic Quotient (qCO_2)

It is the measure of microbial C use efficiency (CUE) and is an important soil health indicator that can be used as a tool for ecosystem monitoring and assessment in remediation programmes due to its low-cost measurement (Ashraf et al. 2022). It is the amount of carbon respired per unit of microbial biomass carbon, which is a more sensitive indicator of soil microbial reaction to cropping systems. Dinesh et al. (2004) examined the effect of change in land use on microbial indices by comparing forest soils with plantation soils including coconut. The lower metabolic quotient (mg CO_2-C g^{-1} biomass C day^{-1}) of 16.6 ± 2.3 observed in plantation soils compared to the forest soils (33.5 ± 8.3) indicated the lower levels of readily degradable organic matter with longer turnover times of the microbial biomass in plantation soils. Long-term (12 years) leguminous cover cropping in coconut plantation indicated higher qCO_2 (mg CO_2-C g^{-1} biomass C day^{-1}) of 42–58.2 than the control (15.1), reflecting the positive effect of cover crops on soil properties (Dinesh et al. 2006). Topsoils subjected to various treatments with crushed green coconut shell (differing in the time period of residue retention) in the dwarf coconut palm plantation recorded higher metabolic quotient than the control without residues indicating the possibility of use for application between the rows of the coconut palm tree plantation (De Paiva et al. 2022).

5.3.4 Nitrogen Mineralization

N mineralization is a microbial process in which nitrogen-rich organic matter is converted to inorganic forms of nitrogen that is influenced both by the quantity and quality of soil organic matter. The most important enzyme classes involved in nitrogen mineralization are deaminases, *O*-glycosidases, and acetyl hydrolases (Mengel 1996). N mineralization is an important process in determining soil fertility and is a potential indicator to comprehend the soil's response to biological change (Stamatiadis et al. 1999). In situ N-mineralization was determined as the difference between extractable N in incubated and initial soil samples.

Study on seasonal dynamics of mineral N pools and N-mineralization in soils under home garden trees (coconut, clove, and nutmeg) in South Andaman (India) by Pandey et al. (2007) indicated that the N mineralization rate (NMR) was the highest during the post-wet season and the lowest during the dry season. The annual mineralization was found higher in nutmeg (561 mg kg^{-1} $year^{-1}$) soil than the coconut trees (393 mg kg^{-1} $year^{-1}$). While investigating the impact of a 15 year old *Pueraria* cover crop under a coconut plantation, with and without phosphorus (P) application, on the soil NMR, Pandey and Begum (2010) observed that the amount of mineralized N was greater under the cover crop and cover crop with P treatments by 39% and 73%, respectively, than the control.

5.3.5 Other Indicators of Soil Microbial Biomass

The use of other biochemical indicators like soil ergosterol and adenylates (adenosine tri-(ATP), mono-(AMP), di-(ADP) phosphate) were limited in case of coconut soils. The membrane lipid ergosterol is found almost exclusively in fungi, and hence it is a reliable and relatively inexpensive biochemical indicator of fungal biomass in soil ecosystems, whereas ATP and the sum of adenylates are closely related to C_{MIC}. Cover cropping (long-term) coconut plantations with legumes was found to increase ergosterol to C_{MIC} and decreased the ATP-to-C_{MIC} ratio at all depths which indicated the dominance of fungal community in such soils. Further conversion of ergosterol content to fungal biomass pointed to the major constitution of fugal biomass in C_{MIC} (Dinesh et al. 2009). Plantations such as coconut, rubber, etc. in mangrove forest had reduced soil ergosterol content (90%) (Dinesh and Chaudhuri 2013) indicating reduced level of fungal biomass.

5.3.6 Soil Enzyme Activity

Soil enzyme activity described as 'biological fingerprints' are used as indicators in evaluation of soil quality, as they are sensitive, integrative, and measurable. They are involved in soil nutrient cycle through formation, transformation, and decomposition of organic matter and are related to soil physico-chemical characters, microbial community structure, vegetation, disturbance, and succession (Caldwell 2005). Dehydrogenases (DHA) that occur in all living cells were one of the most important and widely used as an indicator of overall soil microbial activity. Phosphatases (involved in P mineralization), β-glucosidases, α-amylases (C mineralization), proteases, urease (N mineralization), and catalase (aerobic microbial activity) were a few commonly analysed enzyme activities in coconut soils to indicate the associated functions.

Green manure cultivation and incorporation in coconut basins had considerably increased soil enzyme activities like dehydrogenase, phosphatase, and urease indicating increased biochemical activities (Thomas and Shantaram 1984). Bopaiah and Shetty (1991) observed decrease in microflora and enzyme activities with the increasing soil depth and recorded greater urease, dehydrogenase, and phosphatase enzyme activities in the rhizosphere soil than in the root zone soil of coconut. While investigating the long-term effect of leguminous intercrops in coconut plantations, Dinesh et al. (2009) observed that the organic layers of the cover cropped site exhibited a fourfold increase in L-asparaginase and L-glutaminase activities and a tenfold increase in β-glucosaminidase activity compared to the mineral layer with a positive correlation to organic C, net N mineralized, and other soil parameters.

Usharani and Aparna (2019) estimated dehydrogenase, acid phosphatase, glucosidase, protease, catalase, and amylase in five coconut-based cropping systems in mid-laterite soils of Kerala (AEU 9), to work out biological fertility index in terms of enzyme activity number (EAN). Coconut with banana intercropping under organic management had recorded maximum EAN (94.01) indicating it as a sustainable intercropping model in this zone. Recycling coconut residue biomass through different INM modules (25%, 50%, and 100% recycling) had improved soil enzyme activities such as dehydrogenase (4–47%), urease (7–49%), β-glucosidases (3–73%), and phosphatases (4–71%) (Kumar et al. 2022).

5.3.7 Microbial Population, Communities, and Diversity

Microbial analyses of coconut plantation soils, root zone, and rhizosphere soils were commonly carried out to determine the population count of general microorganisms (bacteria, fungi, and actinomycetes) and function-specific microbiota (fluorescent pseudomonads, phosphate solubilizers, and free-living N fixers). Serial dilution and plating on general and specific media and most probable number (MPN) techniques were commonly followed. Generally higher proliferation of bacteria, N fixers, P solubilizes, nitrifiers, etc., were recorded in inter-/mixed cropped coconut plantations indicating better soil characteristics (Potty and Jayasankar 1976; Sahasranaman et al. 1983). Spores of vesicular-arbuscular mycorrhizal (VAM) fungi were reported in all the crops (8–420 spores 100 g^{-1} soil) of coconut-based multi-storeyed cropping system (Ramesh 1982). In situ cultivation and incorporation of green manure legumes in coconut basins improved the microbial activity as indicated by the increase in populations of bacteria, fungi, actinomycetes, asymbiotic nitrogen fixing, and phosphate solubilizing bacteria and endomycorrhizal chlamydospores in root zone soil (Thomas 1987).

The diversity of culturable bacterial and fungal communities associated with coconut palms infected and non-infected by the Côte d'Ivoire lethal yellowing (CILY) phytoplasma was done using a high-throughput 'dilution-to-extinction' cultivation method coupled with PCR and sequencing with primers that amplified both the 16S ribosomal RNA and intergenic transcribed spacer (ITS) genes (Morales-Lizcano et al. 2017). Microbial enumeration and isolation studies in coconut palms in several parts of southern India under different soil and ecological conditions clearly indicated the preponderance of ammonifying bacteria and antagonistic bacteria in the rhizosphere soils of coconut. Species level diversity of fluorescent pseudomonads and *Bacillus* spp. with plant beneficial traits were confirmed based on 16S rDNA gene sequence analysis (George et al. 2012, 2018). The diversity of cultivable ACC deaminase-producing bacteria associated with rhizosphere of coconut tree grown in Lakshadweep islands of India, analysed by PCR-RFLP (Restriction Fragment Length Polymorphism) method using three endonucleases *AluI*, *MspI*, and *HaeIII*, led to the grouping of these isolates into six clusters at 80% similarity index (Pandey and Gupta 2020). George et al. (2022) assessed diversity of coconut rhizosphere bacteria by carbon use pattern with BIOLOG™ and RFLP.

Investigation on the degree of Arbuscular Mycorrhizal (AM) association and its diversity in the coconut palm cultivated in mixed cropping system by Rajeshkumar et al. (2015) indicated high level of AM richness (40 species belonging to 10 genera) in coconut rhizosphere. Lara-Pérez et al. (2020) assessed the seasonal shifts in the diversity and community structure of arbuscular mycorrhizal fungi (AMF) associated with green dwarf coconut roots by next-generation sequencing (NGS) using partial sequences of the 18S rDNA gene and by spore isolation and morphological identification. Nirukshan et al. (2022) followed phospholipid fatty acid (PLFA) fingerprinting analyses to study the effect of biochar under two different moisture conditions on the abundance of major microbial groups in the root zone and hyphal zone of coconut in pot experiments. They observed a shift toward a more fungal-composed microbial community with the addition of BC-Gly (biochar from glyricidia) under the dry condition, whereas in the hyphal zone, biochar treatment had a significant effect on the concentration of AMF-related PLFAs 16:1ω5 and 20:5.

To conclude, soil health indicators encompassing its physical, chemical, and biological attributes play a crucial role on sustaining the growth and productivity of coconut palms. These three should be acting in tandem for facilitating the favourable establishment of the palm. Moreover, ideal physical conditions in the soil facilitate the nutrient use efficiency as well as the growth and proliferation of microbes in the rhizosphere. Soil chemical attributes such as soil reaction and the availability of nutrients have got direct influence on the indicators such as hydraulic conductivity as well as activity of microbes which are pertinent only at a specific pH in the soil. Soil microbes play their role on decomposition of SOM as well as an imminent link in the biogeochemical cycle.

6 Application of Metagenomics and Microbiome Approaches

Cutting-edge developments in sequencing technologies and high-resolution bioinformatic programmes have ushered in new frontiers in microbiology: metagenomics and microbiome. They have played an important role in unravelling the complete structure and function of microbial community available in any niche by extracting and sequencing their nucleotides and thus giving rise to metagenomics which has led to microbiome domain. Microbiome is the diversity of microbiota and its genome available in a well-defined habitat with distinct physico-chemical properties.

6.1 Whole-Genome Sequencing of Coconut PGPR

In the year 2009, ICAR-CPCRI had signed the first MoU of the institute for a public-private partnership with SciGenom, Kochi, to undertake whole-genome sequencing of promising PGPR isolated from coconut and other mandate crops. The analysis was done adopting shotgun multiplexed sequencing using the 454-platform based on pyrosequencing concept (Nyren et al. 1993) which later was developed into a method for sequencing-by-synthesis (Ronaghi et al. 1998). The PGPR, CPCRI-1, isolated from coconut rhizosphere had a genome size of 4.7 Mb with 56% GC content (Fig. 1). The CPCRI-1 genome encoded for 4056 protein-coding genes and

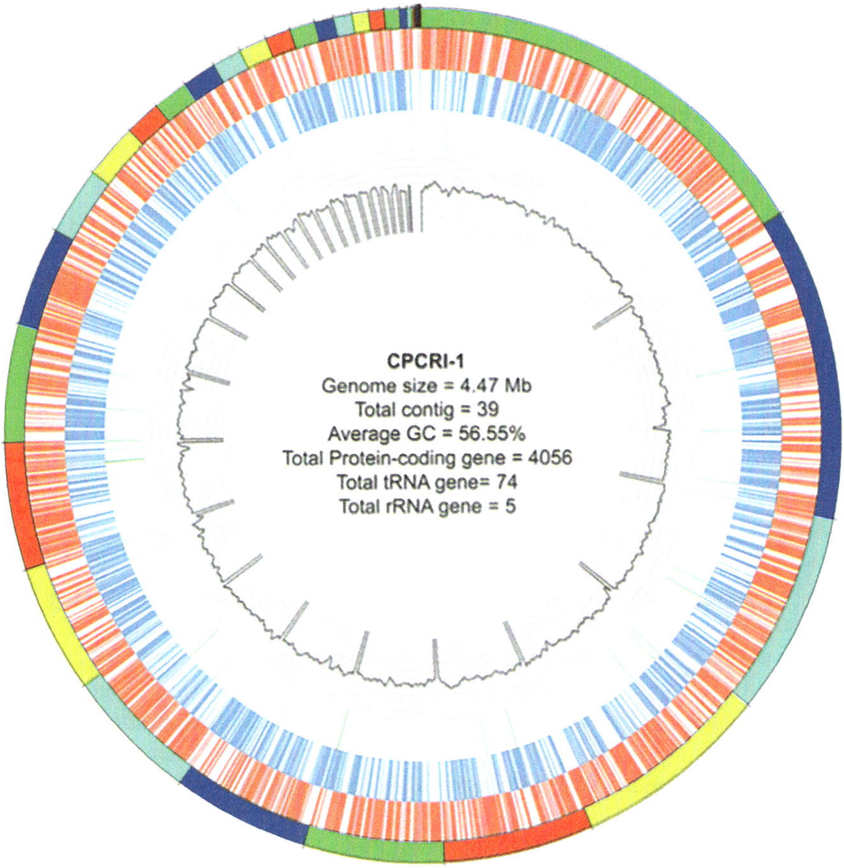

Fig. 1 The whole genome details of PGPR isolated from coconut rhizosphere. (Source: Gupta et al. 2014)

the phylogenetic analysis identified it related to *Enterobacteriaceae* family possessing plant growth-promoting functions such as phosphate solubilization, production of siderophores, acetoin, butanediol, 1-aminocyclopropane-1-carboxylate (ACC) deaminase, chitinase, phenazine, 4-hydroxybenzoate, trehalose, and quorum-sensing molecules. Additionally, for the first time, this whole-genome analysis indicated the presence of genes involved in synthesis of hydrogen sulphide (H_2S), which earlier had been proposed to promote plant growth (Gupta et al. 2014).

6.2 Metagenomic Studies in Coconut Leaf Vermicomposting

Research in metagenomics using next-generation sequencing platform was initiated by ICAR-CPCRI in 2010, in collaboration with University of Georgia, Athens, USA, and National Centre for Cell Sciences, Pune. By employing the 454-sequencing platform based on pyrosequencing principle, a next-generation protocol (Nyren et al. 1993; Ronaghi et al. 1998), microbiome diversity, particularly bacteriome, fluctuations during recycling of recalcitrant coconut leaves rich in lignin to vermicompost with help of CPCRI earthworm isolate, *Eudrilus* sp. was evaluated. Sampling for study was done from two separate composting tanks at four different times: at beginning (15th day) during the pre-decomposition phase when earthworms are not yet added, at 45th day when vermicomposting initiation starts by earthworms, at 75th day when vermicomposting was in active mid-phase, and 105th day when the coconut leaf vermicompost got fully matured. Sampling for the vermicompost and earthworms is explained in Fig. 2. Each tank was filled with 400–500 kg composting substrates composed of weathered coconut leaves + cow dung + *Eudrilus* sp. maintained at 40–50% moisture level (Gopal et al. 2009; Prabhu et al. 1998).

Overall, it was observed that the bacterial diversity increased until the mid-phase of vermicomposting (75th day) and then decreased at the fully composted stage (105th day). On the other hand, β diversity analysis clearly showed that at different stages, different species of bacteria was found in abundance. Certain groups of bacteria like *Acidobacteria*, *Actinobacteria*, *Chloroflexi*, *Gemmatimonadetes*, *Nitrospirae*, *Planctomycetes*, *TM7*, and *WS3* increased in abundance until compost was about 50–70% done, while *Bacteroidetes* and *Proteobacteria* decreased concomitantly. In contrast, *Firmicutes* increased throughout the entire vermicomposting production phases until the compost maturity was attained (Fig. 3).

Using advanced bioinformatic programme, PICRUSt, the functions each bacterial community was performing at different phase of the coconut leaf vermicomposting were determined (Fig. 4).

Predominance of lignin and lipid breakdown bacterial communities were observed during the initial to mid-phase of the vermicomposting, and bacterial communities with ability to improve soil and plant health were preponderant in the mature coconut leaf vermicompost (Gopal et al. 2017). The results from the metagenomic studies mirrored the observations in culture-based microbial analysis of the coconut leaf vermicomposting process (Gopal et al. 2009).

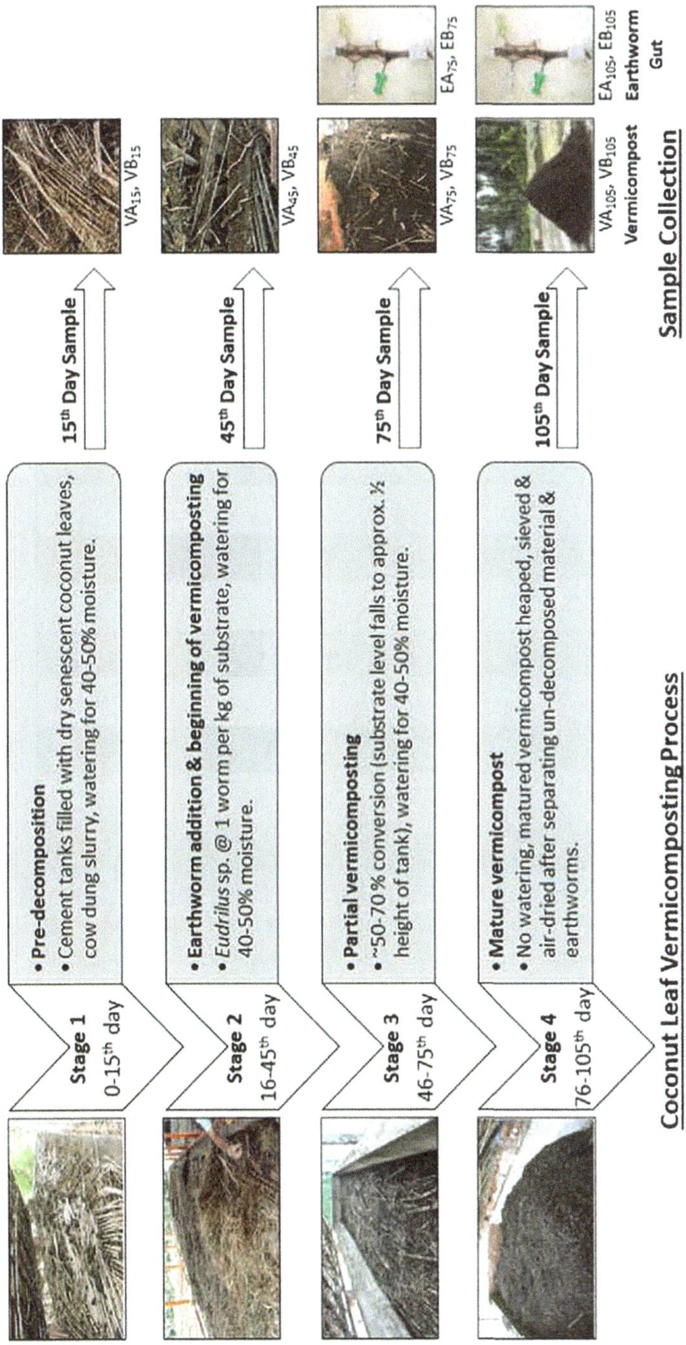

Fig. 2 Stages of sampling during coconut leaf vermicomposting for metagenomic analysis. (Source: Gopal et al. 2017)

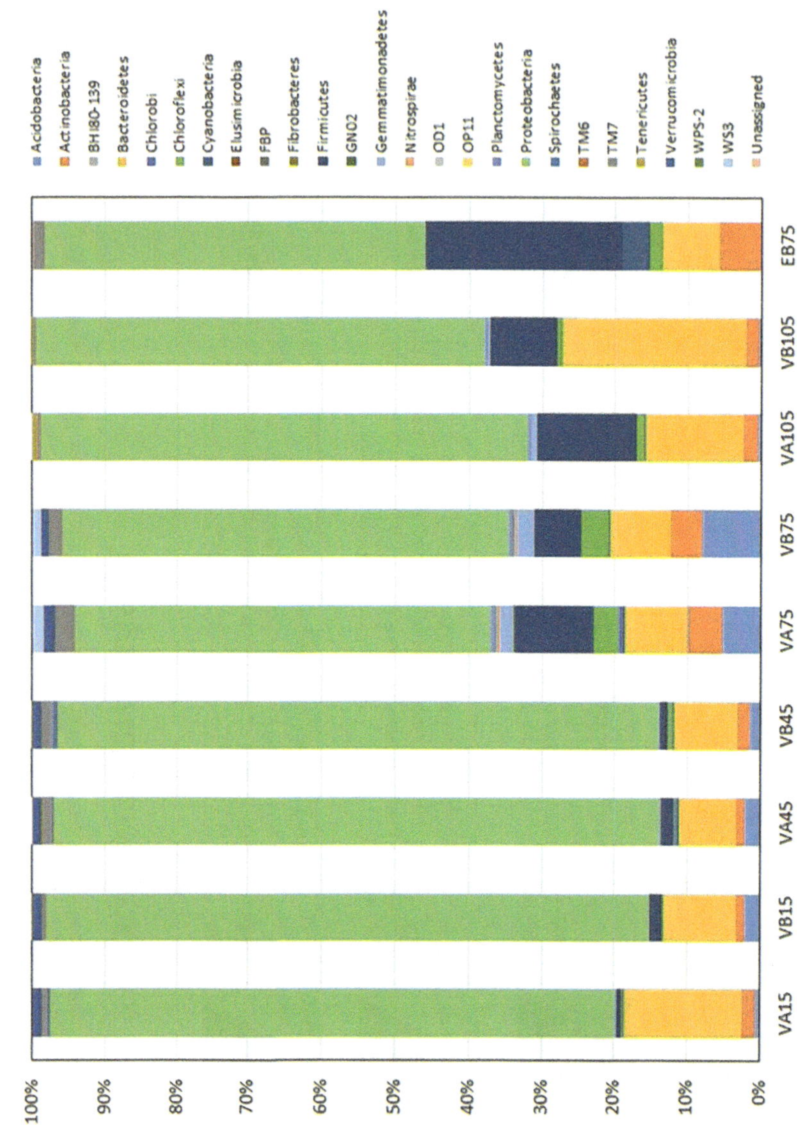

Fig. 3 Changes in bacterial phyla during coconut leaf vermicomposting and in earthworm gut. (Source: Gopal et al. 2017)

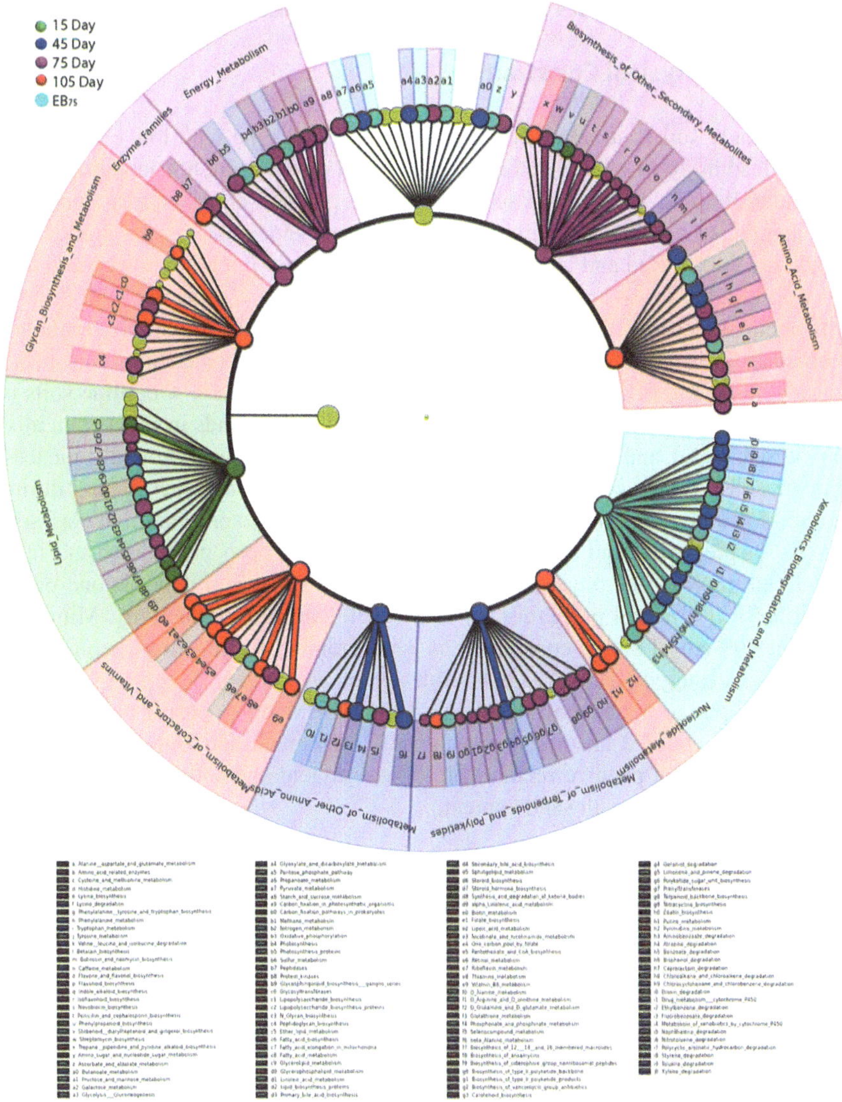

Fig. 4 Functional imputation of bacterial metagenome in coconut leaf vermicompost and earthworm gut. (Source: Gopal et al. 2017)

7 Soil Constraints and Their Management

7.1 Soil Physical Constraints

Coastal sandy soil is structurally a single-grained soil type with low clay and silt content leading to low water retention power and prone to high leaching losses in high rainfall areas. Sandy soils suffer with poor physico-chemical characters like poor aggregate stability and high soil temperature and bulk density (Reddy and Upadhyay 2002). Red and laterite soils are poor in water retention power, and they are well-drained loose and porous soils. Alluvial soils are generally well drained with adequate level of fine and coarse sand. In addition, they are highly fertile and deep soils with minimum level of production constraints for coconut and hence considered as one of the best soil types for coconut cultivation. Coralline soils of Lakshadweep (Arabian Sea) and Andaman and Nicobar Islands in India are a mixture of rock, coral, and sand; almost entirely they are calcareous. Generally, the least suitable soil type for coconut cultivation is clay soils due to waterlogging during monsoon and cracking during dry weather, and especially soil with high clay subsoil is not preferred, and leguminous cover cropping, deep ploughing, and husk burial can improve the physical properties (Child 1974). Soil depth is one of the important soil physical properties for sustainable coconut productivity. Minimum depth of 100 cm of soil is considered as a requirement for planting coconut for a good physical support as well as to provide a good anchorage. Presence of bed rock or hard soil pan or permanent water table within this depth is posing physical constraint and considered unsuitable for coconut. Selvamani and Duraisami (2018b) reported significant positive relationship between yield and soil depth, while relationship was negative with sand content.

7.2 Soil Fertility-Related Constraints

Soil fertility-related constraints are the most important yield limiting constraints and can be managed effectively compared to the other soil constraints. Coconut is cultivated in the tropical condition, and the major soil types found under coconut are laterite, red, sandy, alluvium, clayey, peaty, volcanic, and coralline soils.

One of the major soil types of the soils under coconut cultivation is red and laterite soils. These are highly leached soils with acidic pH and low exchangeable capacity. Therefore, these soils are poor in nutrient retention capacity and also have excess quantity of Al, Fe, and Mn. Among the major nutrients, contents of K and Mg are low followed by N and Ca. High solubility of micronutrients observed in general in these tropical acid soils leads to the availability of Fe, Mn, Zn, and Cu, and they are generally not limiting. Coastal sandy soils of the coastal belts of the coconut-growing countries are the second major soils under coconut and are very poor in water and nutrient retention capacity and also highly prone to leaching losses of the applied nutrients and observed to be deficient in organic matter and all essential nutrients. The third major soil type under coconut cultivation is alluvial soils. Loamy soils of river and estuarine alluvial deposits of India and Sri Lanka, limestone-derived alluvial soils of the Philippines (Cooke 1936), and Indonesia's marine alluvial soils are highly fertile; deep and well-drained soils are also found to be best suitable soils for coconut. Coralline soils of Lakshadweep and some parts of Andaman and Nicobar Islands of India are calcareous and are poor in nutrient content except P, Ca, and Mg, but the partially saline underground water supplies sufficient nutrient leading to excellent coconut production (Newton 1967). In some parts of Alleppey and Kottayam districts of Kerala in India, peaty or kari soils are found with strong acidity (pH 3.0–4.5), high organic matter, and total nitrogen content with satisfactory level of K, but they are often deficient in P and Ca. Coconut can perform well in these soils with appropriate soil reclamation and management practices.

7.3 Management Strategies for Soil Fertility Constraints

Red and lateritic soils found under coconut are highly leached soils formed under humid tropical conditions having very low pH ranging from 4.2 to 5.8 with $A1_3^+$ as the dominant cation. High Al and Fe content observed in these soil types are reverting the soluble phosphates into insoluble ones. Applying liming material according to the lime requirement of the soil can improve soil pH. In case of acid sulphate soils, magnesium silicate is used to inactivate Al and to manage soil acidity.

Excess salinity may create problems to coconut production in low-lying areas where frequent inundation of sea water or seasonal submergence with brackish water is observed. In these regions, providing suitable drainage facilities and reducing sea water inundation with check bunds can improve the coconut production. Coconut cultivated in saline or alkaline areas with high sodium, carbonates, and bicarbonates can be managed with the application of acid-forming chemical amendments such as S or gypsum. Application of equal amounts of soil-sand mixture in the planting pits with more organic matter also can reduce the salt problems.

Soil series map of coconut land cover of the major coconut-growing region (Coimbatore and Tiruppur) of Tamil Nadu state of India revealed that soil reaction-based constraints are found to be the major widespread soil constraint for coconut (Selvamani and Duraisami 2018a). Strong soil acidity and high Al content in the soil solution are the soil constraints severely affecting the coconut cultivation in the soils of Onattukara sandy plain and in the Northern and Central Kerala, India (Nair et al. 2018).

Nitrogen was observed to be low in the sandy soils, and its leaching is found to be more in the sandy soils followed by sandy loam and red soils. Green leaf manuring with glyricidia for 50% N substitution along with 50% of N and full dose of recommended P and K through chemical fertilizers increased the coconut yield (Subramanian et al. 2005a). Nath et al. (2012) recorded the highest nut yield of COD × WCT hybrid coconut with the substitution of 50% N through vermicompost and the remaining N requirement met through chemical fertilizers along with 500 g P_2O_5 and 2000 g K_2O palm^{-1} year^{-1}.

Khan et al. (1983) recommended to regulate application of P fertilizers by monitoring the build-up of soil available P and suggested to skip P application when the level reaches 20 ppm in coconut-growing soils. In highly acidic soils, Ca content is low, and application of Ca containing P fertilizers and also liming can supply the required Ca and also improve the soil pH. But in the coral and calcareous soils, calcium is very high and even antagonistic for the uptake of other nutrients. In such soils, organic manure amendments can improve the soil fertility. Coral and black soils are often found to have low Zn and Cu content. In the humid tropics, B deficiency has been reported in most of the laterite, red, alluvial, and coastal sandy soils. Around 65% of the soils under coconut land cover were low in organic carbon status in the major coconut-growing districts (Coimbatore and Tiruppur) of Tamil Nadu, India. Moreover, available N, Zn, Cu, and B were also deficient in 65%, 89%, 62%, and 5% area, respectively (Selvamani and Duraisami 2014a). Leaf nutrient status revealed that N, P, K, and Mg were deficient in 95.99%, 86.19%, 1.42%, and 0.81% area of the coconut land cover. Leaf content of Fe, Mn, Zn, Cu, and B was also observed to be deficient in 0.64%, 0.84%, 0.54%, 2.62%, and 10.72%, respectively, of the coconut land cover (Selvamani and Duraisami 2014b). The primary yield-limiting nutrient deficiencies identified for the coconut cultivation in the two major coconut-growing districts (Coimbatore and Tiruppur) of Tamil Nadu were N, Mg, K, B, and P as observed from the leaf nutrient status, and the stepwise multiple regression analysis revealed that K, N, and Mg are the most important nutrients influencing coconut yield (Selvamani and Duraisami 2018b).

Soil fertility-related constraints identified from the coconut cultivating regions of Northern Kerala, Central Kerala, Central and Eastern Palakkad, Southern Kerala, Onattukara sandy plain, and coastal sandy plain of the state of Kerala were evaluated. Available boron deficiency was observed in Central and Eastern Palakkad, Central Kerala, Northern Kerala, Onattukara sandy plain, and coastal sandy plains. The major coconut-growing acid soils of Kerala showed the deficiencies of available K, Mg, Zn, and Cu. Phosphorus deficiency is also observed in few parts of Northern and Central Kerala and in the coastal sandy plains. Laterite soils of Central Kerala, sandy soils of Onattukara, and Coastal Plains are found to be low in Ca (Nair et al. 2018).

The critical range of boron in soil and leaf for coconut in the tropical acidic sandy soils of Kerala was found to be 0.48 ppm in soil and 13.27 mg kg^{-1} in leaf. This study also suggested the application of 160 g of borax $palm^{-1}$ (i.e. 16 g of boron $palm^{-1}$) in four split doses as the optimal dose for soil application (Mathew et al. 2018a).

Binitha et al. (2017) evaluated the soil fertility status of the coconut-based cropping systems of Kasaragod district, Kerala, India, and observed that magnesium, boron, and zinc are highly deficient, and among them, Mg and B deficiencies are found to be in a most alarming status in majority of the soils. Moreover, pH of the 54% of the soil samples were categorized as strongly acidic; 38%, 6%, and 2% were moderately, extremely, and slightly acidic, respectively; and they recommended application of dolomite, magnesium sulphate, borax, and zinc sulphate for this region. Coconut soils of Onattukara sandy plain region of Kerala, India, was found to be low in the organic carbon status, P, and K in 74.2%, 0.6%, and 82.6%. Secondary nutrients Ca and Mg were deficient in 96.8% area and S deficient in 8.6% area. Among the micronutrients Zn, Cu, and B were found to be deficient in 90.3%, 73.5%, and 93.1% area, respectively (Mini and Usha 2018). Soils of coconut in the different agro-climatic zones of Karnataka, India, showed available N and K fertility status to be low to medium in surface and sub-surface soils, and available P status is also found to be low in most of the soils.

In Brazil, a study using DRIS revealed that lack of K was the most common deficiencies observed in the municipality of Moju, Pará State, for the hybrid coconut cultivation. The order of nutrient limitation identified was K > P > Ca > Fe > N > O > B > Zn > Cu > Mn > Mg (Saldanha et al. 2017). In this region, the nitrogen fertilization did not influence yield, but P and K application increased the number of coconuts per palm, and Mg application increased the production only in the presence of K (Lins et al. 2021). Coconut yield response to N fertilization was observed to be varying in Brazil, and it was found to be positive in the Brazilian Northeast region (Silva et al. 2006), whereas it was a negative response in São Paulo State (Teixeira et al. 2005). The nutrient deficiency and response of coconut to fertilizer application were found to be region specific.

In the Philippines, positive coconut yield response was observed for the supply of N, Cl, and S through inorganic fertilizers (Magat et al. 1977; Margate et al. 1979). The 8-year-old experiments on integrated soil fertility management conducted in different agro-climatic conditions for the hybrid coconut cultivation revealed that application of different levels of inorganic fertilizers like ammonium sulphate along with KCl or NaCl increased nut and copra yield. Substitution of organic fertilizers for ammonium sulphate has also increased the yield when applied along with KCl or NaCl, but yield increase was lower than the inorganic fertilizer treatments (Margate et al. 1997).

In Sri Lanka, soil pH was found to be favourable, and organic matter content is low in the major coconut-growing soil series such as Dambakanda, Mawillu, Mampuri, Medagama, Rajakadaluwa, and Wilpattu. Soil available nitrogen was moderate in general, and available P status found to be medium to low. Exchangeable K was found to be low in all the soil series except Wilpattu series where it is medium in status. In all the major coconut-growing soil series, exchangeable Ca and Mg were observed to be in moderate status, and micronutrient status is observed to be lower than the average value of tropical soils (Herath et al. 2007). The alluvial soils of tidal area in Indonesia suffer with N and K deficiency, whereas the peat soils of tidal swampy region are deficient in N, P, K, Zn, B, Mo, and Cu (Danoesastro 1977). In order to improve the coconut cultivation in these regions, the specific drainage and fertility management strategies are required to improve soil aeration and nutrient supply to coconut.

The SOM in both Costa Grande and Costa Chica regions, Guerrero, Mexico, were lower than the medium rate value (1.6–3.5%) at both depths of 0–30 and 30–60 cm. Except for Las Tunas, all other localities of this region showed lower values than the critical level of 20 ppm of nitrate nitrogen (Ramirez-Silva et al. 2021). Sulphur deficiency was also reported from Papua and New Guinea (Southern

1969) and some of the South East Asian and East African countries (Ollagnier and Ochs 1972). Okeri et al. (2007) assessed the nutritional status of soils of coconut cultivation and found low N and K status and recommended their application in some localities of Edo State of Nigeria.

Soil fertility constraints are observed to be specific to soil type and agro-climatic conditions. Moreover, they are also dependent on the management practices followed. Therefore, regular assessment for the identification of soil fertility constraints and imposing site-specific management intervention based as per the constraints are the necessary way forward for augmenting and sustaining coconut productivity and soil fertility.

7.4 Soil Contamination-Related Constraints

The effect of various metals (Pb, Al, Bi, Cd, Cr, Cu, and Ba) on P, K, Ca, and Mg concentration in the leaves of coconut palms was studied by supplying the metals to the roots of coconut palms. The leaf P was reduced by all the metals tested. Al, Cd, and Ba reduced K concentration in the leaves. Leaf Ca concentration was decreased by Al and Pb. Leaf Mg was low in palms supplied with the Al and Cr. Leaf K content was increased by Bi and Cr. Leaf Ca content was increased by Cd, Cr, and Ba. Leaf Mg content was increased by the Pb (Biddappa et al. 1987). Coconut trees grown in the polluted area situated near a factory and irrigated with factory waste-contaminated (Cr and Ni) water in Mysore, India, were evaluated by Somashekar (1984) for heavy metal contamination. The accumulation of Cu and Ni in coconut root was 10.89 and 46.84 mg^{-1}, and in leaf extract, it was 19.70 and 32.32 mg^{-1}. Coconut water was also found to have 12.6 mg^{-1} of Ni, 15.45 mg^{-1} of Cr, and 14.13 mg^{-1} of Cu. Though the direct use of wastewaters from a paper mill for irrigating coconut did not show adverse effect on the growth of palms as they could survive indicating some tolerance to toxic heavy metals, the concentrations of heavy metals (Pb, Ni, Cd, Co, Zn, and Cu) in coconut water, leaves, and roots were higher than the limits suggested by World Health Organization (Fazeli et al. 1991).

Islam et al. (2018) observed that due to the highly contaminated Hazaribagh area in Dhaka, Bangladesh, some of the local coconut water samples of this area contained high concentration of toxic heavy metals (Cr, Ni, Pb, and Cd). Pantipkayee et al. (2018) from the analysis of residues of heavy metals such as Cr, Cd, Pb, Cu, Fe, and Zn in coconut and soil from coconut orchard in Thailand found that the concentration of Cr, Pb, and Cd was higher in 25.80%, 96.77%, and 100%, respectively, of coconut juice (tender coconut water) samples tested than the permissible limit recommended by WHO and Thailand. They opined that the heavy metal residues found in coconut samples and soil might have derived from pesticides and chemical fertilizer use. The results of various studies indicate that though the coconut palm is showing some tolerance to the heavy metal contamination, the coconut products especially food products should be monitored regularly for the heavy metal residue.

8 Impact of Climate Change on Coconut Soils

Climate change profoundly alters agriculture, affecting essential soil processes (organic matter breakdown, nutrient cycling, erosion, microbial activity) and crop productivity, impacting food security and sustainability. Varied rainfall adversely affects coconut seedling growth. Coconut in its early growth stages demands life-saving irrigation in high rainfall and regular irrigation in dry regions. Excess rain or water logging leads to root decay and *Phytophthora palmivora*-induced bud rot. Drought diminishes productivity, drying leaves and roots, hampering nutrient transport, leading to leaf loss, detopping, and mortality. In adult coconut, water scarcity hampers inflorescence and female flower development, causing spadix abortion. Results encompass incomplete inflorescence, delayed spathes opening, dropped fertilized flowers, drooping bunches, reduced setting, poor endosperm growth, and smaller nuts. Recovery takes 2–3 years, affecting yield and income. Consecutive droughts caused significant losses in India (Naresh Kumar 2011; Rajagopal et al. 1996).

9 Fertility Management for Healthy Soils: Fertilizer Recommendations

Coconut palms occupy the land for decades together and hence are to be provided with congenial conditions in terms of management practices so as to attain sustained crop growth and productivity. When a coconut palm enters its flowering phase, both vegetative and reproductive processes occur simultaneously. During this period, a significant amount of nutrients is absorbed from the soil by the palm. However, a substantial portion of these nutrients is lost from the system when nuts and other parts of the palm are harvested. This can lead to soil nutrient depletion. To maintain soil fertility, it is necessary to regularly supplement the soil with fertilizers and manure. Fertilizers provide essential plant nutrients, but only a portion of these nutrients become available to the palm, as some get fixed in the soil and remain inaccessible. Additionally, nutrients can be lost due to leaching and erosion and as gases escaping from the soil. Therefore, it is crucial to add sufficient amounts of fertilizer to compensate for these losses, meet the palm's nutrient requirements, and preserve soil fertility and health. The aim of managing soil fertility for healthy soils is to produce maximum yield per unit area without compromising the soil's well-being. This is necessitated in view of shrunken land availability for agricultural purpose and burgeoning population. In coconut, the cultivable area is stagnating around 12.3 million ha during the period between 2010 and 2020 (CDB 2023), but rising demand for coconut products across the world necessitates area expansion and production. Considering the constrained area available, raising productivity is the available option for coconut.

Understanding nutrient depletion in various ecosystems and across different productivity categories is crucial for developing fertilizer recommendations (Somasiri et al. 1994). Information about the rate and quantity of nutrient depletion from coconut lands with varying yield potentials can also help in devising more systematic approaches to coconut nutrition and soil management (Somasiri et al. 2003). Studies on nutrient depletion have shown that the proportional requirement of NPK for the coconut palm is 2:1:3, with K dominating the nutrient requirement, while P is the least required (Pillai and Davis 1963; Ramadasan and Lal 1966). An annual calculation for 1 ha of 173 palms in sandy loam soil revealed nutrient exhaust values of 65.6 kg of N, 29.7 kg of P_2O_5, 84.5 kg of K_2O, 47.4 kg of CaO, and 20.3 kg of MgO, considering the nutrient content in nuts, fallen leaves, spathes, and stem growth (Pillai and Davis 1963). In Sri Lanka, where typica x typica coconut palms yield an average of 17,380 nuts ha^{-1} $year^{-1}$, the amounts of macronutrients lost through the removal of plant components were 116.79 kg of N, 14.02 kg of P, 245.43 kg of K, 40.47 kg of Ca, and 33.66 kg of Mg ha^{-1} $year^{-1}$. Additionally, the amounts of micronutrients lost were 1.14 kg of Fe, 0.63 kg of Mn, 0.13 kg of Cu, 0.44 kg of Zn, and 0.26 kg of B ha^{-1} $year^{-1}$ (Somasiri et al. 2003).

The total nutrient depletion in coconut-based cropping systems varies significantly. For instance, in full recommended fertilizer doses, it can range from 130.45 kg ha^{-1} of N, 18.29 kg ha^{-1} of P, and 172.64 kg ha^{-1} of K. However, when no fertilizers are provided, the nutrient depletion can be reduced to 97.11 kg ha^{-1} of N, 13.06 kg ha^{-1} of P, and 125.45 kg ha^{-1} of K (Subramanian et al. 2005a). According to Wijebandara et al. (2015), coconut palms remove major nutrients from all plant parts at rates (kg ha^{-1} $year^{-1}$) of 164 for K, 53 for N, 37 for Na, 22 for Ca, 19 for Mg, and 9.85 for P. Besides, micronutrients are also removed which include 0.93 for Fe, 0.53 for Mn, 0.28 for Zn, 0.15 for B, and 0.09 for Cu.

In India, ICAR-CPCRI, in collaboration with the All India Coordinated Research Project on Palms (AICRP), has conducted integrated nutrient management (INM) studies and developed region-specific recommendations. Similar recommendations from a few other coconut-growing countries are also provided in Table 2.

9.1 Fertigation Studies

Fertigation is a highly advanced and efficient approach for applying fertilizers, ensuring their direct application to the crop's root zone throughout the entire growing season. Fertigation optimizes both water and fertilizer use, resulting in increased crop yields while promoting the health of the soil and the environment. Extensive fertigation studies have been carried out at various ICAR-AICRP (All India Coordinated Research Project) centres in India, and the

Table 2 INM recommendation for higher productivity in coconut

Sl. no.	Agro-ecological subregion (ICAR)	Recommended fertilizer level	Soil type	Reference
1	Assam and Bengal plain, hot sub-humid to humid (inclusion of per humid) eco-region (15.2)	500 g N wherein 50% N substituted by vermicompost: 500 g P_2O_5: 2000 g K_2O palm^{-1} year^{-1}	Alluvial soil	Nath et al. (2012)
2	Konkan, Karnataka, and Kerala coastal plain, hot humid to per humid eco-subregion (19.3) (Kasaragod)	50% of recommended N through glyricidia + 500 g N: 500 g P_2O_5: 2000 g K_2O palm^{-1} year^{-1}	Littoral sandy soil	Subramanian et al. (2005b)
3	Konkan, Karnataka, and Kerala coastal plain, hot humid to per humid eco-subregion (19.3) (Kasaragod)	1000 g N: 437 g P_2O_5: 1667 g K_2O palm^{-1} year^{-1} for hybrid palms	Red sandy loam	Reddy et al. (2002)
4	Konkan, Karnataka, and Kerala coastal plain, hot humid to per humid eco-subregion (19.3) (Dakshin Kannada)	50% N through vermicompost (19.2 kg) + 50% NPK combination	Lateritic soil	Maheswarappa et al. (2011)
5	Eastern Ghats and Tamil Nadu uplands and D (8.3) (Veppankulam)	50% N as composted coir pith + 500 g N: 500 g P_2O_5: 2000 g K_2O palm^{-1} year^{-1}	Sandy loam	Mohandas (2012)
6	Eastern coastal plain, hot sub-humid to semi-arid eco-region (12.1, 18.4) (East Godavari)	50% N as composted coir pith + 500 g N: 500 g P_2O_5: 2000 g K_2O palm^{-1} year^{-1}	Coastal alluvial	Kalpana et al. (2008)
7	Tamil Nadu uplands and leeward flanks of south Sahyadris, hot, dry semi-arid eco-subregion (8.1) (Aliyarnagar)	50% N as composted coir pith + 250 g N: 320 g P_2O_5: 1200 g K_2O palm^{-1} year^{-1}	Sandy loam	Venkatasamy (2003)
8	Sri Lanka	4.5 kg of adult palm mixture (APM) and 1.5 kg of dolomite palm^{-1} year^{-1} supplied 87.21 kg N, 17.07 kg P, 188.90 kg K, 104.80 kg Ca, and 28.59 kg Mg ha^{-1} year^{-1}	Madampe series soil	Somasiri et al. (2003)

(continued)

Table 2 (continued)

Sl. no.	Agro-ecological subregion (ICAR)	Recommended fertilizer level	Soil type	Reference
9	The Philippines	Combination of multi-nutrient fertilizer (MNF): N–P–K–Cl–S–B inorganic/mineral fertilizer grade 14–5–20–0.02 (B) with 15.5% Cl and 4.5% S and the organic fertilizers (commercial and/or coconut husks) at different proportions on coconut + lanzones cropping system	Clay soil	Magat (2009)
10	The Philippines	Recommended rates of inorganic fertilizers (ammonium sulphate and NaCl) and combination of organic (goat manure) + NaCl		Secretaria and Margate (2001)

Table 3 Recommended fertigation level/dose for different coconut-growing regions of India

Sl. no.	Agro-ecological subregion (ICAR)	Recommended fertigation level/dose	Soil type	Reference
1	Eastern Ghats and Tamil Nadu uplands and Deccan (Karnataka) plateau, hot semi-arid eco-region (8.2) (Arsikere)	75% NPK through drip irrigation	Red sandy loam	Basavaraju et al. (2014)
2	Konkan, Karnataka, and Kerala coastal plain, hot humid to per humid eco-subregion (19.3) (Kasaragod)	50% NPK through drip irrigation with 50 kg of organic manures	Red sandy loam	Subramanian et al. (2012)
3	Tamil Nadu uplands and leeward flanks of south Sahyadris, hot, dry semi-arid eco-subregion (8.1) (Coimbatore)	100% NPK through drip irrigation and polythene mulching	Sandy clay loam	Jayakumar et al. (2017)
4	Bengal and Assam plains, hot sub-humid (moist) to humid (inclusion of per humid) eco-subregion (15.1) (Mondouri)	75% NPK through drip irrigation	Clay loam	Bandyopadhyay et al. (2019)

recommended fertilizer doses to achieve higher yields in different regions are presented in Table 3.

9.2 Diagnosis and Recommendation Integrated System Approach

The Diagnosis and Recommendation Integrated System (DRIS), introduced by Beaufils (1973), is a method used for diagnosing nutrient balance and deficiencies.

It helps to identify all the nutritional factors that may limit crop production and thus enhances the chances of achieving higher crop yields through improved fertilizer recommendations. DRIS relies on nutrient ratios in a suitable plant part to diagnose nutrient imbalances within the plant. This method involves calculating an index for each nutrient. Saldanha et al. (2017) applied DRIS to coconut and determined optimal leaf content values for hybrid coconut as follows: 20.2 g kg^{-1} for N, 1.5 g kg^{-1} for P, 17.2 g kg^{-1} for K, 4.1 g kg^{-1} for Ca, 1.2 g kg^{-1} for Mg, and 1.3 g kg^{-1} for S to achieve higher yields. Additionally, the optimal values for B, Cu, Fe, Mn, and Zn to enhance yield were found to be 15, 4, 115, 101, and 21 mg kg^{-1}, respectively. These values serve as a valuable guide for effective nutrient management in coconut cultivation.

9.3 Reference Soil Values for OM and Inorganic N Approach

According to Semarnat (2002), critical levels of organic matter percentage and inorganic assimilable nitrogen, specifically N–NO_3 in ppm, can be used as indicators for soil fertility. To achieve optimal productivity, the soil should be maintained at medium to higher levels. Soil test ratings are provided in Table 4 to help guide soil fertility management.

9.4 Leaf Analysis-Based Nutrient Management

Another approach to nutrient management aimed at increasing coconut productivity involves leaf nutrient analysis. This technique directly assesses the nutritional status of coconut palms. It estimates the optimal range for primary and secondary macro- and micronutrients, and accordingly, soil fertility can be managed through the application of organic and inorganic fertilizers, enabling the palm to maintain optimal leaf nutrient content. It is important to note that defining standard nutrient levels with broad applicability can be challenging due to variations in climate, soil conditions, crop management, and other factors. Therefore, it is more appropriate to define optimum nutrient levels for a specific cultivar in a well-defined set of

Table 4 Soil test ratings, soil attributes for OM and N-NO_3

	Soil attributes	
Soil-test rating	OM (%)	N-NO_3 (ppm)
Very low	<0.5	0–10
Low	0.6–1.5	10–20
Medium	1.6–3.5	20–40
High	3.6–6.0	40–60
Very high	>6.0	>60

Table 5 Optimum range of concentration of nutrient elements in frond 14 in mature bearing palms

		Variety		
Sl. no.	Nutrient element as % (dry matter)	Local tall	Dwarf	Hybrid
1	N	1.80–2.00	1.80–2.00	1.80–2.20
2	P	0.12–0.14	0.12	0.12–0.13
3	K	0.80–1.00	0.60–0.80	0.90–1.40
4	Mg	0.20–0.30	0.25	0.20–0.35
5	Ca	0.30–0.50	0.15–0.20	0.3–0.35
6	S	0.13–0.17	–	0.15–0.16
7	Na	0.10–0.20	–	0.15–0.17
8	Cl	0.30–0.6	–	0.45–0.50
	As ppm (dry matter)			
9	Mn	60–115	60	–
10	B	9–11	8	–

Source: Chew (1978) and Magat (1992)

conditions. For example, Magat (1992) worked out optimum nutrient content levels for achieving higher yields, as detailed in Table 5. Based on these findings, the application of organic manures and inorganic fertilizers can be tailored to improve soil fertility and increase yield.

Similarly, the International Institute of Rural Reconstruction (IRHO) established critical nutrient limits for West African Tall palms using the 14th leaf, which included values for N, P, K, Ca, Mg, Fe, and Mn (Thampan 1981), as shown in Table 5. Additionally, the sulphur (S) content in the sixth leaf from the apex of coconut palms was found to be a sensitive indicator for sulphur (De Silva et al. 1985). The critical nutrient concentration range for boron (B) in the third leaf was determined to be 8–10 ppm, and deficiency could be addressed by soil application of sodium tetraborate (Jayasekara and Loganathan 1988).

9.5 Customized Nutrient Mixtures for Coconut

Nutrient requirement varies with the soil type as well as the uptake pattern of the crop. Hence customized nutrient mixtures could play a great role on improving soil health and crop productivity. Mathew et al. (2021b) reported that customized nutrient mixtures such as Kalpa Poshak and Kalpa Vardhini could improve the performance of juvenile and adult coconut palms, respectively, in acid soils.

10 Technological Options to Promote Soil Health and Sustainable Agriculture

10.1 System Approach

System approach is a nature-based solution that provides ecosystem services, helping humans protect, sustainably manage, and restore both natural and modified ecosystems. It effectively addresses issues such as climate change adaptation, disaster risk reduction, and food security in an adaptive manner. A notable example of this is coconut-based cropping/farming system. Coconut's perennial nature with wide spacing, canopy structure, and rooting pattern allows intercropping for temporal diversity and also the other diverse kinds of livestock. This will result in improved portioning and utilization of resources.

Osei-Bonsu et al. (2002) noted increased soil moisture retention in cacao + coconut mixtures in Ghana compared to cacao + *Gliricidia sepium* systems. In India, Nair et al. (1975) and Varghese et al. (1978) reported that coconut + cocoa intercropping enhances phosphate solubilizer activity, elevating soil phosphorus availability while improving the microclimate by reducing evaporation and conserving soil moisture. Similar findings were observed by Ghosh et al. (2017) in coconut and flower crop intercropping. Arachchi and Liyanage (1997) demonstrated enriched SOM, bulk density, aeration, and water content in *Acacia auriculiformis* and glyricidia intercropped plots in Sri Lanka, surpassing sole coconut, *Calliandra calothyrsus*, and *L. leucocephala* intercropped plots. Coconut-based multistrata systems exhibit heightened resilience against accelerated SOM decomposition driven by global warming, particularly when compared to monocrop stands (Nair et al. 2018), thereby reinforcing the sustainability of coconut-based farming systems. The integration of nitrogen-fixing trees, showcasing self-sustaining stand management, expedites nutrient uptake, especially advantageous in low fertility conditions. Profound tree roots intercept leached nutrients, particularly from shallow-rooted crops. The efficacy of mixed-species systems relies on associated tree component proximity (Kumar and Jose 2018) and root spatial arrangement (Van Noordwijk and Cadisch 2002).

Senarathne (2017) demonstrated that cultivating *Gliricidia sepium* and *Tithonia diversifolia* as green manure in coconut plantations enhances SOM, macro-/micronutrients, and biological activity, fostering soil health and fertility. High-density multispecies coconut cropping enhances soil water retention, organic carbon, pH, total N, available P, and K, surpassing coconut monocrops (Bopaiah and Shetty 1991; Liyanage and Dassanayake 1993; Maheswarappa et al. 2005). Cover crops mitigate nutrient leaching, shaping soil structure, water movement, temperature regulation, and reducing runoff and erosion. They amplify soil microbial diversity, likely due to augmented plant field diversity, carbon-rich root exudates, shifting moisture conditions, and diverse functional plant groups (Badri and Vivanco 2009; Herath et al. 2016; Kim et al. 2020; Lange et al. 2014; Nuwarapaksha et al. 2022; Singh et al. 2006; Vukicevich et al. 2016).

In Indonesia, cocoa + coconut agroforestry exhibits minimal global warming, acidification, and eutrophication impact 3.67E+01 kgCO_2-eq, 4.31E−02 kg SO_2-eq, and 2.25E−05 kgPO_4-eq per metric ton of cocoa pod. This agroforestry enhances organic carbon, C/N ratio, and SOM, stimulating beneficial soil microbe activity at cocoa farms (Utomo et al. 2016). The enhancement of microbiological activity through organic matter has been well documented (Khan and Upadhyay 2002; Savithri and Hameed 1994; Srinivasan et al. 2020).

Overall, the systems approach, particularly within coconut-based cropping system, underscores a promising avenue for sustainable agriculture, harmonizing productivity, enhanced soil health, and ecological preservation.

10.2 Conservation Agricultural Practices

Conservation agriculture (CA) refers to a conceptual framework for resource-efficient cultivation of crops, involving the coordinated implementation of strategies that minimize soil disturbance, maintain a continuous soil cover using crop residues or living plants, and implement crop rotation (FAO 2015). The implementation of CA practices has demonstrated a substantial and positive impact on soil quality, thereby fostering a sustainable and productive agriculture. Through the reduction of soil disturbance via techniques such as no- till or reduced tillage, the inherent soil structure remains intact, mitigating erosion and preserving its physical characteristics. The persistent coverage of the soil with crop residues or cover crops acts as a protective shield against adverse weather conditions, thereby retaining moisture and preventing the loss of essential nutrients. Furthermore, these practices enhance soil fertility by curbing the proliferation of pests and diseases. As a result, these collective measures foster a flourishing ecosystem of beneficial microorganisms in the soil, thereby enhancing nutrient recycling and overall soil vigour. In essence, the harmonious integration of CA practices serves to fortify soil health, ultimately propelling the advancement of sustainable and productive agricultural systems.

10.3 Cultivation/Tillage

Coconuts primarily thrive in regions with abundant rainfall, yet they encounter drought during certain part of year due to uneven distribution of rainfall. Soil constraints, particularly physical attributes like inadequate water content, significantly impact coconut root physiology and morphology (Arachchi et al. 2000). Soil water scarcity arises from low precipitation, compounded by reduced water retention and infiltration due to soil compaction, resulting in diminished crop growth and yield (Boone and Veen 1994; Rawitz et al. 1983). Soil compaction also curtails water and nutrient storage, necessitating more fertilizers and raising production costs (Hamza and Anderson 2005). In poorly drained soils, coconut roots tend to proliferate, restricting their distribution to the topsoil layers and leading to the formation of root mats. During dry periods, these surface roots are exposed to heat damage, rendering them inactive for nutrient and water absorption. Excessive water in the root zone adversely affects root activity by limiting gas exchange in and out of the soil. This, in turn, keeps the spring soil temperature low, compelling new roots to grow just below or even on the soil surface. Consequently, this makes the root system more vulnerable to attacks by soil pathogens, contributing to the well-established issue of tree leaning (Fuller 1999).

These challenges can be managed through tillage which enhances water infiltration, retention, and root penetration, addressing drought susceptibility. Moreover, coconut root vigour hinges on soil aeration (Arachchi and Somasiri 1997), prompting surface tillage/minimum tillage in coconut plantations of all regions to augment aeration, especially in less compacted, poorly drained soils. Currently, farmers often overlook essential soil management practices due to a lack of sufficient knowledge about their land characteristics, particularly concerning soil compaction levels and the presence of clay and gravel. As a consequence, these neglected practices may have adverse effects on the soil water status of their lands.

In Sri Lanka, deep ploughing of highly compacted soil (bulk density >1.5 g cm^{-3}, penetrometer resistance >250 N cm^{-2}) during wet and dry periods improved water conservation. In less compacted soil (<1.5 g cm^{-3}, penetrometer resistance <250 N cm^{-2}), deep ploughing did not enhance water retention, especially within root zones. Thus, substituting live grass cover with dead mulch from coconut leaves may boost coconut yield in less compacted soils (Arachchi 2009).

In India, with the ICAR-CPCRI Kasaragod, Kerala, a long-term experimental plot is being maintained since 1919, modified in 1972 to assess the impact of long-term cultural and manurial practices on coconut palm performance and soil properties. Cultivation and proper manuring were linked to higher nitrogen (1.90% in summer, 1.92% in rainy season) and potassium levels (0.4% in summer, 1.0% in rainy season) in leaves. Similar trends were noted for soil nitrogen and potassium content at 0–30 and 30–60 cm depths (Ramanandan and Pillai 1974). Minimum disturbance of the soil by way of forking after inorganic fertilization helped to maintain better physical environment in basins of coconut palm over other tillage and manurial treatments (Joshi et al. 1986). During the 2017–2018 period, studies on potassium fractions in the coconut basin revealed that soil potassium levels from fertilizer-applied treatments exhibited higher exchangeable, lattice, and total potassium compared to no manure treatments. While fertilizer application improved total soil potassium compared to unfertilized soils, there was a decreasing trend in non-exchangeable potassium status. The reduction in lattice potassium in unfertilized plots and a slight decrease in non-exchangeable potassium in fertilized plots highlight the importance of applying potassium based on removal to compensate for soil loss, reducing long-term potassium mining and sustaining soil fertility. In the coconut basin, Ca-P dominated at 0–30 cm (Ca-P > RS-P > Fe-P > A1-P > ES-P), with RS-P prevailing at 30–60 cm (RS-P > Ca-P > Fe-P > Al-P > ES-P). Fe-P was consistently dominant, followed by Ca-P and RS-P. Manured plots exhibited higher inorganic phosphorus fractions at both depths, with a decrease observed at 30–60 cm compared to 0–30 cm.

An untilled soil exhibits compaction, causing surface runoff rather than percolation. Ploughing enhances soil porosity, absorption, and aeration. Organic matter like grass and weeds when incorporated releases nutrients upon decomposition for palm absorption. Monsoon-initiated ploughing, following contours, improves aeration, moisture retention, and organic content. Disc harrowing in summer conserves soil moisture through surface dust. This regimen, augmenting porosity, aeration, and organic enrichment, fosters improved soil structure and humus content, thereby enhancing palm productivity.

10.4 Soil Covering (Mulching)

The practice of retaining permanent soil cover, often through cover crops or crop residues, shields the soil from erosive forces of wind and water while also enhancing soil organic matter content. Surface mulching conditioned the soil temperature and makes soil damper by reducing the evaporation rate and more permeable to water. Mulch reduces rain runoff, consequently minimizing soil erosion and increasing water percolation into the soil. To conserve soil moisture in coconut plantations, consider mulching with organic materials such as coconut leaves (in two to three layers), husks (in two to three layers, 250–300 husks per basin), and coir pith (10 cm thickness, approximately 50 kg palm^{-1}). This practice helps lower soil temperature, reduces evaporation, and fosters optimal conditions for root growth and the proliferation of soil flora and fauna. The recommended time for mulching is before the end of the monsoon, preventing the topsoil from drying up.

Coconut husk burial in layers with the bottom layers facing up and top layer facing down in the trenches (50 cm width × 50 cm depth and convenient length) dug out in the interspace of coconut will also help in soil moisture conservation (Fig. 5) (Mathew et al. 2018b).

Fig. 5 Trenches filled with coconut husk in the coconut interspaces

10.5 Crop Diversification

Crop diversification plays a pivotal role in enhancing soil health. Coconut has a huge potential to accommodate number of different crops in the interspace because of its unique canopy structure, wider spacing, and root habit. The beneficial effect of coconut-based crop diversification on soil health is multifaceted.

1. Crop diversification disrupts pest and disease cycles. Varied susceptibility reduces single-pest dominance, decreasing reliance on pesticides, promoting beneficial organisms, and enhancing soil microbiome vitality.
2. Crop diversification mainly the inclusion of leguminous crops enhances nutrient cycling and soil fertility. Varied nutrient requirements and acquisition mechanisms optimize soil nutrient utilization. Leguminous crops fix atmospheric nitrogen and function as natural fertilizers in diversified systems.
3. It mitigates erosion as varied root structures of different crop types bind soil particles, reducing wind and water erosion risk. Cover crops during fallow periods prevent soil exposure, minimizing rain drop impact and runoff.
4. It increases SOM content by adding varied organic residues, enriching soil carbon which in turn helps in enhancing soil structure, water retention, microbial activity, and carbon sequestration for climate mitigation.

10.6 Cover Cropping and Green Manuring

Cover cropping, also known as green manuring, plays a crucial role in soil quality management within sustainable production systems. The cultivation of leguminous green manure crops in coconut gardens contributes to atmospheric nitrogen fixation, enhances overall soil health by reducing both water and wind erosion, improves soil structural properties such as aggregate stability, and enhances soil hydraulic properties like water infiltration. This practice also boosts SOC, increases the soil microbial population, and reduces nitrate nitrogen leaching, thereby mitigating groundwater pollution.

Furthermore, cover crops act as a protective barrier, shielding the soil from potential pests and pathogen carriers like farm equipment, labour, and other crop materials. They promote the colonization of non-harmful, pathogen-suppressive fungi, leading to effective control of pests and pathogens. Additionally, flowering cover crop species attract predatory and beneficial insects, including pollinators, lacewings, ladybirds, hoverflies, and parasitic wasps, due to the provision of nectar and pollen.

Cover crops provide organic substrates that ensure the optimal growth of beneficial microbes in the field, particularly bacteria, non-pathogenic *Fusarium* species, actinomycetes, and *Streptomyces* communities (Reddy 2017). This holistic approach to soil management contributes to sustainable and enriched ecosystems in coconut gardens. The Sustainable Agriculture Farming Systems Project in Davis, California, conducted research suggesting that legume cover crops play a crucial role in developing key humic acid fractions, distinguishing organic soils from conventionally managed soils and serving as significant indices of soil quality improvement (Mitchell et al. 2000). Green manuring, practiced as a cover crop, is applicable in the basin (1.8 m radius area from the trunk) and in the interspaces of coconut gardens (Thomas and Shantaram 1984).

Thomas et al. (2001b) reported on the growth of *Mimosa invisa* and *C. mucunoides* in coconut palm basins, producing 20.5 and 12.8 kg of above-ground wet biomass and 134.8 and 70.0 g N basin^{-1}, respectively, after 140 days of growth for incorporation. *M. invisa* demonstrated higher N-fixation efficiency, evident in nodule biomass and acetylene reduction activity, suggesting the potential for substituting 50% of fertilizer-N with nitrogen contributed by leguminous crops. Herath et al. (2016) observed that soil from coconut land with *P. phaseoloides* exhibited significantly higher electrical conductivity, organic carbon content (0.66%), number of bacterial colonies (19.76 × 10^6 cfu g^{-1}), microbial activities (1.55 mg CO_2 day^{-1}), aggregate stability, and lower bulk density. This highlights the positive impact of specific leguminous cover crops on soil properties in coconut plantations. Huang et al. (2023) found that the intercropping and green manuring utilization pattern (GMUP) of *Stylosanthes guianensis* in the interspaces of coconut gardens in Hainan Province, China, resulted in significantly higher total nitrogen content in the soil, recording a 58.1% increase compared to areas without green manuring. Thomas and Shantaram (1984) reported that *P. phaseoloides*, *M. invisa*, and *C. mucunoides* were identified as suitable crops for cultivation in coconut basins, with mean yields of 19.43, 17.00, and 14.71 kg basin^{-1}, respectively (Table 6). Senarathne et al. (2015) noted that cover cropping of *P. phaseoloides* had similar effects as herbicides (glyphosate) for controlling weeds and was more effective than mechanical weed

Table 6 Growth and nodulation of green manure crops in coconut basins

	Growth		Nodulation	
Legume species	Fresh weight (kg basin^{-1})	Total N added (g basin^{-1})	Nodule number/ five plants	Nodule dry weight (g/five plants)
Calopogonium mucunoides	14.71	102.61	145	0.485
Macrotyloma axillare	0.95	6.67	58	0.132
Mimosa invisa	17	153.19	125	1.86
Pueraria phaseoloides	19.43	121.29	132	1.128
Leucaena leucocephala	2.95	16.55	0	0
Sesbania aegyptiaca	1.3	9.98	56	0.535
Macroptilium atropurpureum	9.1	66.64	60	0.108
Glycine wightii	2.35	19.2	95	0.205
Stylosanthes guianensis	3.5	12.7	350	0.055
CD at 5%	7.59	53.24	65	2.905

control methods in the Sri Lankan context. Dinesh et al. (2009) reported that cover cropping of *Atylosia scarabaeoides* (L.) Benth., *Centrosema pubescens* Benth., and *P. phaseoloides* (Roxb.) Benth led to significantly higher levels of total nitrogen, organic carbon, dissolved organic carbon, dissolved organic nitrogen, and labile organic nitrogen compared to coconut plantations without cover cropping. Growing leguminous cover crops can enhance microbial biomass in terms of carbon, nitrogen, and phosphorus by 61%, 51%, and 50%, respectively, compared to no cover-cropped coconut plantations.

Ideal cover crops should exhibit quick germination and emergence, tolerate adverse climatic conditions, fix atmospheric nitrogen, absorb nutrients through deep root development, produce substantial biomass rapidly, facilitate easy cultivation, avoid competition with the main crop, show tolerance to insect pests and diseases, possess weed suppression capabilities, and be cost-effective for cultivation (Reddy 2016). The suitability of various green manure crops for diverse soil and climatic conditions across coconut-growing regions is detailed in Table 7.

Table 7 Green manure crops suitable for coconut-growing regions

SI. no.	Common name	Scientific name	Suitable soil pH range	Characteristics
1	Berseem/Egyptian clover	*Trifolium alexandrinum*	6.4–7.1	Annual clover, can grow all soils except very light sandy soils
2	Black gram	*Vigna mungo*	6.5–7.8	Suitable for sandy loam soils
3	Black lentil	*Lens culinaris*	5.5–7.5	Suitable for alluvial and black cotton soils
4	Buffalo bean	*Mucuna aterrima*	5.5–7.5	Woody tropical legume crop
5	Calopogonium	*Calopogonium mucunoides*	4.5–5.0	Vigorous, annual, or short-lived perennial legume, suitable to warm humid tropic and withstand water stagnation
6	Centro	*Centrosema pubescens*	6.1–6.4	Annual or perennial with alternate leaves, suitable for sandy loam soils
7	Cluster bean	*Cyamopsis tetragonoloba*	7.0–8.5	Annual herb legume, can grow medium to light textured soils
8	Cow pea	*Vigna unguiculata*	5.0–6.5	Annul herbaceous legume
9	Daincha	*Sesbania aculeata*	4.5–7.5	Plant is well adapted to variety of soil conditions, ranging from water logged to saline and from sandy to clay soils
10	Desmanthus	*Desmanthus virgatus*	>6.5 (neutral to alkaline)	Woody perennial shrub, suitable clay or loamy soil
11	Field beans	*Vicia faba*	6.5–9.0	It can grow in well-drained clay and silt soils in addition to sandy soils

(continued)

Table 7 (continued)

Sl. no.	Common name	Scientific name	Suitable soil pH range	Characteristics
12	Indigo	*Indigofera tinctoria*	6.0–7.0	Shows drought resistance and is well suited for cultivation in light sandy, medium loamy, and heavy clay soils with proper drainage
13	Lab lab	*Lablab purpureus*	6.5–8.5	Can grow under diverged soils from sandy to clay as long as adequate drainage is maintained
14	Lucerne or alfalfa	*Medicago sativa*	6.2–7.8	Perennial pasture legume, well suited for well-drained soils ranging from sandy to moderately heavy clays with a slightly acid to alkaline pH
15	Manila agathi	*Sesbania rostrata*	5.5–slightly alkaline	Woody, erect, robust, annual or short-lived perennial, can tolerate water logging
16	Mimosa	*Mimosa invisa*	6.0–7.0	Non-thorny, profuse vegetative growth single and cluster nodules on its roots and fixing high amounts of atmospheric nitrogen, is also advised to be grown on unfertile land and interspaces of orchard crops, and can be incorporated easily
17	Peas	*Pisum sativum*	6.0–7.5	Herbaceous annual legume plant, suitable for well-drained loamy soil
18	Pillipesara	*Phaseolus trilobus*	7.0–8.5	Annual, spreading, and herbaceous runner, can grow in variety of soils including black cotton soils, red lateritic soils, and sandy loam soils
19	Pueraria	*Pueraria phaseoloides*	3.5–5.5	A vigorous twining and climbing perennial legume, capable of thriving in a diverse range of soils from sandy loam to clay, although its growth may be limited on clay soils
20	Sithagathi	*Sesbania speciosa*	5.5–8.0	Annual to biennial herb
21	Sun hemp	*Crotalaria juncea*	5.0–7.5	Annual leguminous herb, suitable for loamy sand
22	White clover	*Trifolium pratense*	5.6–7.0	Herbaceous perennial plant, well suited for clay and silt soils, and it thrives successfully on sandy soils as well
23	Wild indigo (Kolingi)	*Tephrosia purpurea*	7.0–9.0	Erect or spreading annual or short-lived perennial herb, well suited for dry, gravelly, rocky, and sandy soils including saline and sodic soil conditions

Table 7 (continued)

Sl. no.	Common name	Scientific name	Suitable soil pH range	Characteristics
24	Glyricidia (*madre cacao in Spanish, kakawate in the Philippines)*	*Glyricidia sepium*	4.5–7.2	Evergreen deciduous and fast-growing medium-sized leguminous tree. Grows well in acidic soils
25	Stylo	*Stylosanthes guianensis*	4.0–8.3	Erect or semi-erect perennial legume that can reach a height of 1–1.5 m
26	Siratro (purple bush-bean)	*Macroptilium atropurpureum*	5.0–8.0	Perennial, climber, or creeper average salinity and shade tolerance
27	Wild pigeon pea creeper-climber	*Cajanus scarabaeoides*	6.2–8.0	Drought tolerant

Dinesh et al. (2009), Dissanayaka et al. (2022), Herath et al. (2016), Solangi et al. (2010), Subramanian et al. (2005b), Subramanian et al. (2018), Thomas and Shantaram (1984)

Various green manure crops can be grown in the inter-spaces as well in gardens where land is available after growing intercrops (Fig. 6a–e). Approximately 15–20 tons of fresh biomass can be incorporated into the soil by such practice from the interspaces of 1 ha coconut garden.

10.7 Alley Cropping of Glyricidia in Coconut

By cultivating glyricidia, a fast-growing perennial leguminous green leaf manure tree crop in coconut plantations along the borders (Fig. 7), it is possible to generate a large quantity of nitrogen-rich biomass. Glyricidia can thrive in littoral sandy soils where other green manure crops struggle. Propagation is done through either vegetative cuttings or seeds, with 1-m-long stem cuttings or 3- to 4-month-old seedlings used for planting in poly bags or raised beds.

Planting is typically done during the southwest or northeast monsoon season at a spacing of 1 m × 1 m. Alternatively, two rows of glyricidia can be planted along the boundary of a coconut garden in a zigzag manner. Plant stem cuttings or seedlings in an upright position in pits sized 30 × 30 × 30 cm. For better establishment, a basal dose of 50 g of rock phosphate per pit may be applied.

Regular pruning, starting 1 year after planting and done at least thrice a year, helps maintain the plants at a height of 1 m. Studies conducted at ICAR-CPCRI indicate that planting three rows of glyricidia in a 1 ha coconut garden with 1 m × 1 m spacing between two rows of coconut and pruning leaves during February, June, and October could yield around 8 tons of biomass. This biomass, upon incorporation and decomposition, fulfils approximately 90%, 25%, and 15% of the requirement of N, P, and K, respectively (Subramanian et al. 2005b). Mixed cropping of glyricidia in the interspaces of coconut, along with the application of prunings from glyricidia, has been shown to result in higher organic matter (22%),

Fig. 6 (**a–e**) Various green manure crops that can be grown in coconut garden

available phosphorus content (20%), and total exchangeable potassium content (69%) (Raveendra et al. 2021). Senarathne (2017) reported that in situ cultivation of two green manure crops (*Gliricidia sepium* and *Tithonia diversifolia*) in coconut plantations, along with the incorporation of their green biomass into the palm

Fig. 7 Growing glyricidia as alley crop in coconut

manure circle soil, increased the organic matter content of the soil. This process ultimately added soil macro- and micronutrients while improving soil physical and chemical properties.

The addition of green manure significantly improved soil parameters, including soil nitrogen (by 75%), organic carbon content (by 86%), and soil microbial activity (by 52%). Soil bulk density was reduced by 20%, promoting enhanced root growth. The improvement in soil moisture content by 92% is particularly crucial for coconuts in withstanding frequent dry spells, especially considering the increasing trend of such dry conditions due to present climatic change situations.

10.8 Bioinoculants

Bioinoculants familiarly known as biofertilizers denote the carrier-based live formulations of beneficial microorganisms with a specified or multiple functional traits beneficial for the growth and health of agricultural crops. They are the key components in organic and integrated farming. Based on the highlighted biological functions, they are categorized as biofertilizers (nitrogen (N) fixers, phosphate (P) solubilizers/nutrient (K, Si, Zn) solubilizers), plant growth promoters (PGP rhizobacteria such as fluorescent pseudomonads, *Bacillus*, etc. with plant growth-promoting ability), and biocontrol agents (bacteria/fungi with antagonistic properties against host pathogens). Root-associated symbiotic arbuscular mycorrhizal fungi (AMF) and free-living *Trichoderma* spp. are also being explored as bioinoculants in a wide range of cultivated crops. Like any other crops, the benefits of microbial functions for coconut growth and productivity had been deeply explored, and several microbial formulations were effectively put into use in coconut cropping systems.

Natural symbiotic legume-rhizobial association as green manure crops (in situ cultivation and incorporation) for supplementing nitrogen (N) like *Pueraria phaseoloides*, *Mimosa invisa*, *Calopogonium mucunoides*, *Vigna unguiculata*, *Desmodium* (100–250 g N coconut basin^{-1}), and others had become a popular and widely adopted technology for combo-benefits of microbial fertilization and organic manuring owing to the deliberate efforts of several researchers (Anithakumari 2008; Thamban et al. 2016; Thomas and Shantaram 1986, 1993; Vijayaraghavan and Ramachandran 1989). Free-living and associative nitrogen fixers such as *Azospirillum*, *Bacillus*, *Burkholderia*, *Azoarcus*, *Herbaspirillum*, etc. isolated from coconut rhizosphere had enhanced growth and vigour of poly bag coconut seedlings by increasing root biomass and branching of secondary roots. *Azospirillum brasilense* and *Beijerinckia indica* were identified as promising N biofertilizers for coconut seedlings (Gupta et al. 2010; Reddy et al. 2007; Thomas et al. 2001a). Ilangamudali and Senarathne (2016) documented an increase in root numbers, volumes, and dry weights of primary, tertiary, and quaternary roots, as well as stem girth at later stages of seedling growth with the application of mycorrhizae-based biofertilizer. The field established seedlings treated with 500 g of AMF exhibited significantly higher production and stem girth. This observation suggests the suitability of AMF biofertilizers for the production of high-quality coconut seedlings and their improved field establishment (Senarathne 2018). George et al. (2018) adopted systematic screening strategies, incorporating in vitro characterization of plant growth-promoting (PGP) traits and seedling bioassays. This approach was employed to identify efficient bioinoculants for utilization in coconut nurseries and plantations.

Suitable carrier materials were an absolute requirement to maintain the population level and viability of live cells/spores of bioinoculants for the purpose of commercialization. The carrier material composing mixture of coir dust, soil, and FYM (1:1:1) was found suitable for production of N biofertilizer of *Beijerinckia indica*. Sterilized vermicompost was utilized to formulate nitrogen-fixing *Azospirillum brasilense* and phosphate-solubilizing *Bacillus* sp. The vermicompost produced from coconut leaves resulted in high-quality biofertilizers containing more than 10^8 cfu of bacteria per gram of the carrier material, meeting BSI standards for biofertilizers. The recommended dosage of biofertilizers for coconut was set at 100 g of carrier-based inoculant per palm, applied twice a year (pre-monsoon and post-monsoon). This application involved mixing with topsoil, followed by the addition of organic amendments such as vermicompost, coir pith compost, farmyard manure, neem cake, green manures, etc.

For additive benefits biofertilizers (N-fixing bacteria like *Beijerinckia indica* and phosphate-solubilizing bacteria) were enriched in organic amendments such as coir pith compost, for application in coconut basins (Gupta et al. 2010). Talc-based bioformulations of PGPR, *Bacillus megaterium* (now *Priestia megaterium*), and a soil-based AMF formulation of *Claroideoglomus etunicatum*, isolated from coconut rhizosphere that increased growth parameters of coconut seedlings, were developed and released by ICAR-CPCRI Kasaragod for the benefit of coconut farmers (Thomas et al. 2016, 2018). Gopal et al. (2019) developed a simple method for mass-multiplying microbial inoculants of bacterial plant growth-promoting rhizobacteria (PGPR), including *Pseudomonas* spp., *Bacillus* spp., free-living nitrogen-fixers like *Beijerinckia* spp., *Azotobacter* spp., phosphate-solubilizing bacteria like *Bacillus megaterium*, and fungal antagonists/entomopathogens such as *Trichoderma* spp. and *Metarhizium anisopliae*. This method is based on a sterilized mixture of rice gruel and mature coconut water, providing an effective and immediate application for coconut and other crops.

PGPR inoculant *Bacillus cereus* (UFRABC 40) was found to improve the performance of Brazilian Green Dwarf coconut palm seedling (improved stem, root growth, and nutrient uptake) under reduced fertilizer regime (Cardoso et al. 2021). Indigenous endophytic plant growth-promoting rhizobacteria belonging to *Bacillus*, possessing abiotic stress-tolerant traits resulted in highest dry weight biomass increase in coconut seedlings indicating their potential as effective bioinoculants (George et al. 2022). Gome-falcon et al. (2023) observed increased survival (1.19–1.24 fold) and increased growth (height, leaf area, and stem diameter) in micropropagated coconut plantlets inoculated with certain native and commercial AMF inoculants.

In addition to their contributions in quality seedling production, bioinoculants were found effective for better field establishment of coconut seedlings and also for the improvement of coconut traits. N_2-fixers and P solubilizers enhancing the productivity and copra yield in coconut had been reported (Gupta et al. 2010). Organic management of coconut-based high-density multispecies cropping system which included biomass recycling, biofertilizers, and green manuring produced similar yield in comparison to inorganic fertilizer regime (Maheswarappa et al. 2013). Bioinoculants such as *Trichoderma* and microbial consortia (*Pseudomonas*, *Bacillus*) along with neem cake were found equally effective as fungicide treatments in inhibiting the growth of coconut basal stem rot pathogen in field trials (Surulirajan et al. 2014). Cross-compatibility of coconut PGPR (*B. megaterium* TSP 16) with vegetable crops and their yield improvement of vegetable intercrops (tomato, chilli) indicated the suitability of a single bioinoculant for the coconut main crop as well as vegetable intercrops in coconut-based cropping system (Rajeela et al. 2017). In a 5-year trial, yield of coconut and intercrops were found to increase with INM of 50% N recycling with vermicompost, biofertilizers (*Azotobacter*), and in situ green manuring (Bhalerao et al. 2021). Despite reduction of chemical fertilizers to half the recommendation, treatments with biofertilizer Myko VAMTM recorded increased coconut yield and copra amount (Crisostomo et al. 2023).

10.9 Recycling of Coconut Biomass

Coconut, the sturdy perennial crop with year-round yield of economic produces, generates huge amounts of crop residues of ligno-cellulosic nature, presenting both challenges and opportunities in crop residue management. Typically, 1 ha of a coconut plantation with bearing palms generates around 8–11 tons of biomass waste, including leaves, spathe, bunch stalk, and husk (Biddappa et al. 1996), and in coconut-based high-density multispecies cropping system, the total biomass removed from the system ranges from 19.1 to 27.6 t ha^{-1} $year^{-1}$ (Subramanian et al. 2005b).

Biomass, when properly managed through biomass recycling, results in sustainable production system with low external inputs. The simplest way of residue management is surface mulching with coconut leaves, husk, and coir pith due to their ligno-cellulosic nature and recalcitrance to microbial decomposition. Biomass mulches available from different component crops of coconut-based cropping system contribute to soil organic pool and also serve as nutrient sources when decomposed after a period of time (Sudha and George 2011). Apart from N and P, higher amount of K is available from coconut husks (around 1 kg of K for 100 husks) (Jothimani 1994). Burying husk in linear trenches (1.5–2.0 m wide × 0.3–0.5 m deep) between rows of palms has proven highly effective for moisture conservation in coconut gardens. As surface mulch, husks can be placed around the base (up to 2 m radius from the base) of the palms (Gupta et al. 2010; Thomas et al. 2018). Coconut residue mulching also had effects on soil quality and resilience. Along with cover crops, coir dust mulching was demonstrated as soil and water conservation technique for semi-arid environments (Montenegro et al. 2020). Benefits of coconut residue mulch on weed suppression and yield augmentation had been noticed in other intercrops also (Maheswarappa et al. 2017; Thankamani et al. 2022).

Composting of coconut crop residues to nutrient-rich and beneficial microbe-enriched organic amendment was widely acknowledged as viable technological option for crop residue management, soil health, and entrepreneurial prospects. Lignin-rich coconut leaves and coir pith were found amenable for composting processes that were mediated by earthworms or microbial cultures. Natural decomposition of coconut leaves normally took 12–18 months to degrade because of their high lignin (39%), cellulose (24%), and phenol (8%) contents. The conversion of coconut leaf litter, mixed with 10% cow-dung into carbon-rich granular vermicompost, was produced within 60–75 days using an indigenous strain of *Eudrilus* earthworm (Prabhu et al. 1998) and was widely disseminated at the grass-root level from ICAR-CPCRI. Year-round recycling of coconut leaves along with biomass from other intercrops was possible with high recovery (70%). In situ vermicomposting following trench or heap method utilizing interspaces of coconut formed an inexpensive and eco-friendly residue recycling model. As a stabilized manure boasting high organic carbon content (17–20%), coconut leaf vermicompost (CLV) contains essential plant nutrients (1.8% N, 0.2% P, 1.6% K), plant growth substances (humic acid 10–13%, IAA, and gibberellins), and beneficial microbial communities (N_2-fixers, phosphate solubilizers, plant growth-promoting *Pseudomonas* and *Bacillus* spp., and cellulose degraders). Thus, it serves as an ideal source of organic manure for all types of crops, including coconut (Gopal et al. 2010a; Thomas et al. 2012).

Coconut leaf vermicompost produced using *Eudrilus* sp. at ICAR-CPCRI was released under the trade name of 'Kalpa Organic Gold'. Vermiwash, a liquid organic fertilizer derived from CLV production, was observed to enhance the productivity of several intercrops (Gopal et al. 2010b, 2012; Thomas et al. 2018). A comparative study of vermicomposting of coconut husk and spent coconut flakes employing

Eudrilus eugeniae was reported by Tahir and Hamid (2012). A soilless medium consisting of coconut leaf vermicompost, coir-pith compost, and PGPR was successfully formulated to nurture healthy seedlings of cocoa and arecanut (https://cpcri.icar.gov.in/page/research_achievements_crop_production/).

Lignin degrading microbial starter cultures enhanced coir pith decomposition (Nagarajan et al. 1985). Co-composting of coir pith with coconut leaves mediated by *Eudrilus* sp. had generated granular vermicompost with CN ratio of 16.7:1 in 2 months (Thomas et al. 2001a). Ghosh et al. (2007) demonstrated accelerated on-site composting of coir pith hillocks using edible oyster mushroom and urea. This process yielded organic manure within just 21 days, characterized by a reduced C/N ratio (20:1) and near neutral pH (6.5). Co-composting of coir pith using poultry manure and other amendments further accelerated the composting process reducing the bioconversion duration to 45 days (Gopal et al. 2016; Thomas et al. 2013), and the urea-free coir pith compost thus produced was released by ICAR-CPCRI under the trade name of 'Kalpa Soil Care' (Thomas et al. 2018). Jayakumar et al. (2022) reported the detoxification of coir pith through refined vermicomposting employing *Eudrilus eugeniae*. This involved amending nitrogenous legume plant, *Gliricidia sepium*, together with cattle dung in various combinations and pre-composting using *Pleurotus sajor-caju* spawn for 21 days.

Utilization of coconut stalks and bunch waste as substrate for oyster mushroom production resulted in 57–70% bioconversion efficiency, and the spent substrate after mushroom cultivation served as quality organic manure (Gupta et al. 2016a; Thomas et al. 1998). The substrates with more lingo-cellulosic contents like banana and coconut residues were proved better for producing oyster mushroom with a higher percentage of total protein (Lindao-Pérez et al. 2022).

Exploring alternative methods for utilizing coconut biomass residues is crucial to address the environmental challenges linked to coconut residue disposal. Coconut biomass residues with high carbon content and unique physico-chemical properties were ideal for producing biochar which has applications in wastewater treatment, soil amendment, carbon sequestration, and others (Ighalo et al. 2023). Gopal et al. (2020) highlighted the potential of pyrolysis for rapidly recycling of highly recalcitrant coconut palm biomass residues into biochars. These biochars can serve as soil amendments, promoting regenerative agriculture by enhancing soil characteristics and crop yield in humid tropics. Additionally, Pandiselvam et al. (2023) developed a portable pyrolysis system specifically for producing biochar from tender coconut.

11 Carbon Sequestration

Escalating greenhouse gases, primarily CO_2, methane (CH_4), and nitrous oxide (N_2O), are contributing to global temperature rise, adversely affecting agriculture. Forest preservation and establishment via afforestation and reforestation are crucial strategies. Notably, coconut plantations hold substantial carbon sequestration potential due to their prolonged lifespan and standing biomass storage capacity, aiding CO_2 mitigation (Lal 2005; Navarro et al. 2008). Simulation by Naresh Kumar et al.

(2008) estimates coconut's dry matter production potential at 52–62 Mg ha^{-1} $year^{-1}$, aligning with Corley (1983) data (51 Mg ha^{-1} $year^{-1}$). However, labile components like fruits and leaves reduce coconut's carbon storage potential compared to forests (Ranasinghe and Silva 2007). Across diverse Indian agro-climates, coconut monocrop's net primary production suggests 8–32 Mg CO_2 ha^{-1} $year^{-1}$ carbon sequestration potential, influenced by factors like cultivar, zone, soil, and management. Over ~60 years, stem carbon stocking approximates 120 Mg CO_2 ha^{-1}, averaging 2.0 Mg CO_2 ha^{-1} $year^{-1}$ (Naresh Kumar et al. 2007; Naresh Kumar et al. 2008). A 20-year-old coconut plantation with optimal inputs shows net ecosystem productivity at ~8 Mg C ha^{-1} $year^{-1}$ (~29 Mg CO_2 ha^{-1} $year^{-1}$), leading to 3–5 Mg C ha^{-1} $year^{-1}$ net sequestration after accounting for copra extraction (Roupsard 2008; Roupsard et al. 2008b). Coastal Tanzania studies unveil substantial biomass, and carbon stocks with mainland areas hold ~10,818,072.1 tons biomass (101,689.9 tons carbon), while Zanzibar has ~2,466,520.5 tons biomass (23,185.29 tons carbon), translating to ~US$406,759.5 (mainland) and ~US$92,741.17 (Zanzibar) in carbon markets (Zella and Lawi 2019).

Ghavale et al. (2020) investigated carbon sequestration across Maharashtra's coconut varieties and hybrids and found that East Coast Tall exhibits the highest biomass and carbon stock (312 kg $plant^{-1}$ and 27.32 t ha^{-1}, respectively). In Tamil Nadu, tall coconut varieties aged 5–25 years sequester 1.32–3.96 t $acre^{-1}$ $year^{-1}$, while dwarf varieties sequester 1.45–2.63 t $acre^{-1}$ $year^{-1}$. Over 15 years, both tall and dwarf types store 1.15 million tons of carbon (Boomiraj et al. 2020). CHN analysis indicates higher carbon (48.1–48.7%) in dried leaves than trunk wood (45.9–46.3%). In West Bengal's Sristi Baganbari, biomass was in the range of 100.02–226.45 t ha^{-1}, while stored carbon was in the range of 47.01–107.60 t ha^{-1} (Arpan Mitra et al. 2018). Coconut palms enhance carbon storage when cultivated with diverse tree species, particularly in mixtures (Manna et al. 2016; Nair et al. 2018). Multi-strata coconut-based systems provide diverse ecosystem services, including soil-crop carbon sequestration for CO_2 mitigation. Nonetheless, a few studies comprehensively assess their carbon sequestration potential. In Thrissur district's humid lowlands, India, Saha et al. (2010) compared carbon content among land use systems, revealing higher potential in mixed stands with diverse tree species than in sole coconut stands. Intercropping boosts palm soil carbon, with the baby corn-gherkin sequence exhibiting the highest soil carbon stock (19.17–20.43 Mg C ha^{-1}). Improved nutrient management enhances soil carbon stock. In the Philippines, converting grasslands to coconut plantations yields 3.4–3.6 t C ha^{-1} $year^{-1}$ carbon savings, while forest conversion leads to 3.6 t C ha^{-1} $year^{-1}$ (Vega 2011).

12 Organic Farming Systems

Organic farming has been primarily fuelled by rising consumer demand for organic produce and a growing interest in soil health management to enhance and sustain soil productivity. The core objective of organic farming is to cultivate crops with minimal or no synthetic chemical inputs. Nutrient management in organic farming revolves around efficiently utilizing natural resources on the farm, involving biomass and animal waste recycling, promoting soil nutrient release through biological agents, and incorporating nutrients through biological nutrient fixers.

Wider spacing of palms, coexistence of vegetative and reproductive phases throughout the year, unbranched single stem, phyllotaxy of the crown, and rooting pattern could allow to accommodate diverse crops with high population in the interspace of coconut garden, and the potential of generation of lot of biomass material in the coconut-based farming system would help in recycling of organic biomass which is the key component of organic farming. All these make coconut palm to be highly amenable for organic farming.

ICAR-CPCRI Kasaragod, India, has developed organic farming package of practices for cultivating coconut which involves recycling of organic biomass generated from the system through growing of green manure crops in the basin, growing of glyricidia as border crop and incorporating vegetative biomass, biofertilizer application (*Azospirillum* and phosphobacterium @ 200 g palm^{-1}), and soil moisture conservations. Adoption of such practices recorded sustainable coconut yield in the range of 90–100 and 120–130 nuts palm^{-1} for the varieties Kerakeralm and Chandra Sankara, respectively. Organic cultivation demonstrated a clear enhancement in both the biological and chemical properties of the soil. The biological analysis revealed a higher microbial load, including bacteria, fungi, actinomycetes, free-living nitrogen fixers, cellulose degraders, phosphate solubilizers, and *Azospirillum*, under organic treatments compared to the control (Subramanian et al. 2016, 2022).

Studies on a coconut-based high-density multispecies cropping system model comprising coconut (planted at a spacing of 8.0 × 8.0 m), black pepper (trailed on each coconut trunk), banana cv. Kadali, robusta, cinnamon (inter-row space of palm), and nutmeg (between four coconut palms) with recycling of biomass (as vermicompost @ 30 kg palm^{-1}) + biofertilizer application (*Azospirillum* and *Phosphobacterium* @ 200 g palm^{-1}) + green manuring (in situ) + vermiwash application (10 L palm^{-1}) + coconut husk burial + mulching with coconut leaves revealed that the efficient recycling of available organic materials can sustain the coconut productivity over a longer period (136–155 nuts palm^{-1}) (CPCRI 2022).

13 Coconut-Based Integrated Farming System

Integrating of crops and livestock, including dairy, poultry, duck rearing, aquaculture, and cultivating shade-tolerant fodder crops in the interspaces of coconut gardens, ensures the diversity needed for organic cultivation. This approach not only generates additional income but also create more employment opportunities, providing relief against the fluctuating price of nuts (Subramanian et al. 2014). Cop residues and fodder serve as a valuable, animal feed, and the manure and litter from the livestock contribute renewable sources of organic matter and plant nutrients fostering a sustainable and integrated system in agriculture. The recycling of crop and livestock wastes plays a crucial role in maintaining a high-degree soil health.

Integrating such practices not only contributes to sustainability but also maximizes the beneficial impact of species diversity on soil fertility. Fodder grasses like hybrid napier and guinea grass when grown under coconut shade can yield about 80–120 tons of fodder ha^{-1} in a year. This quantity is sufficient to sustain 8 to 12 crossbred milch cows generating enough farm yard manure to meet the on-farm organic manure requirement. Additionally, installing a biogas plant for use in the farm house and slurry for manuring coconut and other component crops.

In the mixed farming system, higher microbial biomass carbon and dehydrogenase activity were observed in the root zone of coconut and hybrid napier grass compared to the monocropping system with only coconut. The microbial distribution in both the coconut basin and interspaces revealed significantly elevated populations of bacteria, fungi, and actinomycetes in the mixed farming, particularly under the 100% organic farming treatment (Palaniswami et al. 2008).

The soil health benefits of organic farming were evident in the coconut-based mixed farming system, which involved cultivating fodder grass in the interspaces of coconut palms and integrating animal husbandry with dairy, poultry, and pisciculture components. The system, managed with three treatments (100% organics, 100% inorganic, and 50% organic + 50% inorganic), demonstrated significantly higher population densities of soil microbial groups and function-specific microorganisms in the organic treatment compared to the other approaches. The increase in microbial biomass and activities of soil enzymes, such as phosphatase and dehydrogenase, indicated that soil biological activity was more pronounced under the organic cultivation system than under the inorganic or integrated treatment (Subramanian et al. 2014).

14 Coconut-Based High-Density Multispecies Cropping System (HDMSCS)

In the study of the HDMSCS model, which included coconut, black pepper, pineapple, banana, clove, turmeric, ginger, vegetable crops, sweet corn, and baby corn, different nutrient management strategies were examined. Three treatments, viz. two/third of recommended NPK fertilizer + recycling biomass (through vermicompost); one/third of recommended NPK fertilizer + recycling biomass (through vermicompost) + bio-fertilizer + green manuring + vermiwash; and fully organic with recycling biomass (through vermicompost) + bio-fertilizer + green manuring + vermiwash + husk burial + mulching coconut basin, were evaluated. Coconut yield did not vary significantly among the treatments, suggesting that fully organic management could produce yields similar to inorganically fertilized plots while providing the highest economic return (Maheswarappa et al. 2013).

Research conducted at ICAR-CPCRI Kasaragod, India, revealed that a coconut-based HDMSCS, comprising coconut, black pepper, banana cv. Kadali, robusta, cinnamon, and nutmeg (Fig. 8) maintained under organic management practices (recycling of biomass, biofertilizer application, green manuring, coconut husk burial in the interspaces, and vermiwash application) recorded similar yields with

Fig. 8 Coconut-based HDMSCS

integrated nutrient management practices. This indicates the viability of organic maintenance for the cropping system. The organic system also showed higher net energy, energy use efficiency, energy productivity, and energy profitability, resulting in higher net returns (CPCRI 2022).

The impact of organic nutrition sources in coconut garden with sandy loam soil intercropped with vanilla was evaluated and examined. The application of cow dung slurry (6 t ha^{-1}) resulted in a significantly higher mean fresh yield of vanilla beans, followed closely by the application of vermicompost (5 kg $plant^{-1}$) + biofertilizers (*Bacillus* and *Azospirillum*) and vermiwash, which showed comparable results (Maheswarappa et al. 2016).

According to Subramanian et al. (2016), black pepper maintained organically in the coconut-based HDMSCS recorded a yield of 2.4 kg $vine^{-1}$, similar to that of integrated nutrient management practices. Soil nutrient analysis showed that available K content was significantly lower than in integrated nutrient management practices, indicating a higher rate of K depletion in the absence of inorganic sources. This highlights the need to substitute the required quantity of K through organic resources when adopting organic nutrient management.

15 Future Strategies

Soil health is at risk in majority of the coconut-growing regions globally, emphasizing the need to develop sustainable strategies for soil health improvement. Coconut is grown in diverse agro-climatic conditions, and the factors responsible for declining soil also vary widely depending upon the region. Hence physico-chemical and biological constraints responsible for declining soil health should be identified at micro level. For a given agro-climatic region, digitization and soil mapping of soil health constrains should be undertaken. Based on this, there should be focus on developing smart, resilient technologies as short-term, medium-term, and long-term measures. Any soil management intervention should be site specific with health strategies tailor made to the specific location. Further, soil health card issued for the benefit of farmers in India currently focuses on chemical factors only, and measures should be taken to include physical and biological properties of the soil. Recycling of organic matter is an important practice that would help in improving the soil health. Efforts should be made for effective recycling of organic materials and release of nutrients in a balanced manner. In this regard, efforts should be made to find out effective species of earthworms or microorganisms for quick and quality composting. Emphasis should be given to develop fast-growing, high biomass-yielding green manure crop varieties which are adaptable to different physico-chemical and biological situations. Location-specific PGPR consortium should be developed as bioinoculants with extended shelf life. Strengthening research to improve nutrient use efficiency through nano-fertilizer techniques is crucial for ensuring nutrient supply and reducing losses. Developing soil management techniques by integrating advancements in artificial intelligence, mechanical science for mechanization, nano-science technology for fertilizer preparation and utilization, as well as sensors and communication technologies will empower farmers to leverage the anticipated technological boom in these sectors for coconut farming.

References

Ambili K, Thomas GV, Indu P, Gopal M, Gupta A (2012) Distribution of arbuscular mycorrhizae associated with coconut and arecanut based cropping systems. Agric Res 1:338–345. https://doi.org/10.1007/s40003-012-0036-4

Anderson GD (1967) Increasing coconut yields and income on the sandy soils of the Tanganyika Coast. East Afr Agric Forest J 32(3):310–314

Anil Kumar KS, Wahid PA (1988) Root activity pattern of coconut palm. Oleagineux 43:337–342

Anithakumari P (2008) Clustering coconut farmers—a successful extension approach for enhancing adoption and income from marginal and small holdings. CORD 24(2):29–39

Arachchi LPV (2009) Effect of deep ploughing on the water status of highly and less compacted soils for coconut (*Cocos nucifera* L.) production in Sri Lanka. Soil Tillage Res 103(2):350–355

Arachchi LPV, Liyanage MS (1997) Soil physical conditions and root growth in coconut plantations interplanted with nitrogen fixing trees in Sri Lanka. Agrofor Syst 39:305–318

Arachchi LPV, Somasiri LLW (1997) Assessment of oxygen availability in soils of different drainage classes and its effect on coconut root growth. Proc Sri Lanka Assoc Adv Sci 53(2):159–160

Arachchi LPV, Yaspa PAJ, Mapa RB, Somapala H (2000) Soil physical constraints and their effect on morphological characters of coconut (*Cocos nucifera* L.) roots. CORD 16(1):14–33. https://doi.org/10.37833/cord.v16i01.337

Arya S (2020) Effect of salinity on growth physiological and biochemical processes of coconut seedlings (*Cocos nucifera* L.). Doctoral dissertation. Academy of Climate Change Education and Research, Kerala Agricultural University, Vellanikkara, Thrissur, 140 p

Ashraf MN, Waqas MA, Rahman S (2022) Microbial metabolic quotient is a dynamic indicator of soil health: trends, implications and perspectives (review). Eurasian Soil Sci 55:1794–1803. https://doi.org/10.1134/S1064229322700119

Atapattu AJ, Raveendra SAST, Liyanagedara DS, Piyaratna MGNCK, Herath HMSK (2017a) The role of soil organisms and functions in different coconut based multiple cropping systems. Int J Environ Agric Res 3(2):67–84

Atapattu AJ, Senarathne SH, Raveendra SA, Egodawatte WC, Mensah S (2017b) Effect of short term agroforestry systems on soil quality in marginal coconut lands in Sri Lanka. Agric Res J 54(3):324–328

Avinash RK, Anil Kumar KS, Karthika KS, Kalaiselvi B, Sujatha K (2019) Coconut-growing soils in southern Karnataka: characterization and classification. J Plantn Crops 47(2):96–106

Badri DV, Vivanco JM (2009) Regulation and function of root exudates. Plant Cell Environ 32(6):666–681

Bandyopadhyay A, Ghosh DK, Biswas B, Parameswarappa MH, Timsina J (2019) Fertigation effects on productivity, and soil and plant nutrition of coconut (*Cocos nucifera* L.) in the eastern indo-gangetic plains of South Asia. Int J Fruit Sci 19(1):57–74

Basavaraju TB, Bhagya HP, Prashanth M, Arulraj S, Maheswarappa HP (2014) Effect of fertigation on the productivity of coconut. J Plantn Crops 42(2):198–204

Beaufils ER (1973) Diagnosis and recommendation integrated system (DRIS). University of Natal, Pietermartizburg, South Africa. Soil Sci Bull 1:1–132

Bhaduri D, Sihi D, Bhowmik A, Verma BC, Munda S, Dari B (2022) A review on effective soil health bio-indicators for ecosystem restoration and sustainability. Front Microbiol 13:938481. https://doi.org/10.3389/fmicb.2022.938481

Bhalerao PP, Maheswarappa HP, Sumitha S (2021) Effect of integrated nutrient management on coconut (*Cocos nucifera*) based cropping systems in South Gujarat. Cur Hort 9(2):52–55

Biddappa CC, Khan HH, Joshi OP, Manikandan P (1987) Effect of root feeding of heavy metals on the leaf concentration of P, K, Ca and Mg in coconut (*Cocos nucifera* L.). Plant Soil 101:295–297

Biddappa CC, Upadhyay AK, Hegde MR, Palaniswami C (1996) Organic matter recycling in plantation crops. J Plantn Crops 24:71–85

Binitha NK, Ashique TK, George BS, Suresh PR (2017) Soil fertility evaluation of coconut based cropping systems of Kasaragod. Int J Appl Pure Sci Agric 3(7):91–94. https://doi.org/10.22623/IJAPSA.2017.3075.ADVUQ

Boomiraj K, Jagadeeswaran R, Karthik S, Poornima R, Jothimani S, Sudhagar RJ (2020) Assessing the carbon sequestration potential of coconut plantation in Vellore district of Tamil Nadu, India. Int J Environ Clim Chang 10(12):618–624

Boone FR, Veen BW (1994) Mechanism of crop response to soil compaction. In: Soane BD, Van Ouwerker C (eds) Soil compaction in crop production. Elsevier Science Publishers, New York, pp 237–264

Bopaiah BM (1988a) Microbiological studies in relation to high density multispecies cropping systems in coconut. Ph.D. thesis. University of Mysore, Mysore, 168 p

Bopaiah BM (1988b) Microbiological and biochemical studies in relation to high density multispecies cropping system. CPCRI, Annual report for 1987, pp 30–31

Bopaiah BM, Shetty HS (1991) Soil microflora and biological activities in the rhizospheres and root regions of coconut-based multistoreyed cropping and coconut monocropping systems. Soil Biol Biochem 23(1):89–94

Bopaiah BM, Shetty HS, Nagaraja KV (1987) Biochemical characterization of the root exudates of coconut palm. Curr Sci 56(16):832–833

Bouremani N, Cherif-Silini H, Silini A, Bouket AC, Luptakova L, Alenezi FN, Belbahri L (2023) Plant growth-promoting rhizobacteria (PGPR): a rampart against the adverse effects of drought stress. Water 15(3):418. https://doi.org/10.3390/w15030418

Bruce RC (1956) The relationship among physical properties of soil, distribution of coconut roots and productivity of coconut in the town of Magdalene Laguna. BSA thesis. University of Philippines, College of Agriculture

Caldwell BA (2005) Enzyme activities as a component of soil biodiversity: a review. Pedobiologia (Jena) 49(6):637–644

Cardoso AF, Alves EC, da Costa SDA, de Moraes AJG, da Silva Júnior DD, Lins PMP, da Silva GB (2021) *Bacillus cereus* improves performance of Brazilian green dwarf coconut palms seedlings with reduced chemical fertilization. Front Plant Sci 12:649487

CDB (2023) Coconut statistics. https://coconutboard.gov.in/, 17 Jun 2023

Chew PS (1978) Nutrition of coconuts—a review for formulating guidelines on fertilizer recommendations in Malaysia. Planter 54:115–141

Child R (1974) Coconuts, 2nd edn. Longman Group Ltd., London, 335 p. https://books.google.lk/books?id=2ccfAQAAIAAJ

Cooke FC (1936) The coconut industry of the Philippine Islands. Dept. Agric. S.S. and F.M.S. Bull. (Gen. Series) No. 23

Corley RHV (1983) Potential productivity of tropical perennial crops. Exp Agric 19:217–237

Cosico WC, Fernandez NC (1983) Effect of some land qualities and soil properties on productivity of coconut in the Philippines. Philipp J Crop Sci 8(3):153–158

CPCRI (1988) Microbiological and biochemical studies in relation to high density multispecies cropping system. CPCRI annual report, pp 30–31

CPCRI (2022) Annual report. ICAR—Central Plantation Crops Research Institute, Kasaragod, 104 p

Crisostomo SD, Cruz CDD, Quilloy RB, Reaño CE (2023) Narrowing the yield gap of coconut (*Cocos Nucifera* L.) through integrated nutrient management in the Philippines: an on-farm experiment approach. IOP Conf Series Earth Environ Sci 1235(1):012008

Danoesastro H (1977) Kemungkinanpengembangantanamanpekarangan di daerah pasang surut Kalimantan Selatan dan Kalimantan Tengah. Bulletin test farm P4S UGM no. 1, Yogyakarta, pp 42–69

Davis TA (1968) A study on the respiratory organs of *Cocos nucifera* Linn. Ceylon Coconut Quart 3:116–136

De Carvalho Junior JIT, Gonzaga MIS, de Almeida AQ, Araújo J, Santos LCO (2019) Type and quantity of biochar influenced soil microbial activity and carbon priming effect. Semina Ciências Agrárias 40(4):1405–1416. https://doi.org/10.5433/1679-0359.2019v40n4p1405

De Paiva AB, Taniguchi CAK, Romero RE, Garruti DS, Pagan MC, Weber OB (2022) Chemical and microbiological attributes of sandy soil fertilized with crushed green coconut shell. Rev Cienc Agron 53:e20207778

De Silva MAT, Anthonypillai GM, Mathes DT (1985) The sulphur nutrition of coconut. Cocos 3:22–28

Dexter AR (2004) Soil physical quality: part I. Theory, effects of soil texture, density, and organic matter, and effects on root growth. Geoderma 120(3–4):201–214

Dexter AR, Richard G, Arrouays D, Czyż EA, Jolivet C, Duval O (2008) Complexed organic matter controls soil physical properties. Geoderma 144(3–4):620–627

Dinesh R, Chaudhuri SG (2013) Soil biochemical/microbial indices as ecological indicators of land use change in mangrove forests. Ecol Indic 32:253–258. https://doi.org/10.1016/j.ecolind.2013.03.035

Dinesh R, Chaudhuri SG, Sheeja TE (2004) Soil biochemical and microbial indices in wet tropical forests: effects of deforestation and cultivation. J Plant Nutr Soil Sci 167(1):24–32

Dinesh R, Suryanarayana MA, Chaudhuri SG, Sheeja TE, Shiva KN (2006) Long-term effects of leguminous cover crops on biochemical and biological properties in the organic and mineral layers of soils of a coconut plantation. Eur J Soil Biol 42(3):147–157

Dinesh R, Chaudhuri SG, Sheeja TE, Shiva KN (2009) Soil microbial activity and biomass is stimulated by leguminous cover crops. J Plant Nutr Soil Sci 172:288–296. https://doi.org/10.1002/jpln.200700300

Dissanayaka DMNS, Nuwarapaksha TD, Udumann SS, Dissanayake DKRPL, Atapattu AJ (2022) A sustainable way of increasing productivity of coconut cultivation using cover crops: a review. Circ Agric Sys 2:7. https://doi.org/10.48130/CAS-2022-0007

Dissanayaka DMNS, Dissanayake DKRPL, Udumann SS, Nuwarapaksha TD, Atapattu AJ (2023) Agroforestry—a key tool in the climate-smart agriculture context: a review on coconut cultivation in Sri Lanka. Front Agron 5:1162750. https://doi.org/10.3389/fagro.2023.1162750
FAO (2015) The impact of natural hazards and disasters on agriculture and food security and nutrition. FAO guidelines. Food and Agriculture Organization of the United Nations, Rome, 15 p
FAO and ITPS (2015) Status of the world's soil resources (SWSR)—main report. Food and Agricultural Organization of the United Nations and Intergovernmental Technical Panel on Soils, Rome
FAOSTAT (2014). http://faostat3.fao.org. Food Agricultural Organization, Rome
Fazeli SM, Sathyanarayan S, Satish PN, Muthanna L (1991) Effect of paper mill effluents on accumulation of heavy metals in coconut trees near Nanjangud, Mysore district, Karnataka, India. Environ Geol Water Sci 17:47–50. https://doi.org/10.1007/BF01716073
Felizardo BC (1972) Response of coconut trees to different cultural management practices. In: Proceedings of the second ASEAN soil conference, Jakarta, Indonesia, pp 263–268
Ferreira NM, Gheyi HR, Fernandes PD, Holanda JSD, Blanco FF (2007) Leaf emission, ionic relations and production of coconut irrigated with saline water. Ciência Rural 37:1675–1681
Fremond Y (1964) The contribution of IRHO to the study of mineral nutrition of the coconut palm. In: Proceedings of second session. FAQ tech. wkg. pty. cocon. prod. prot. and process, Colombo, Sri Lanka
Fuller K (1999) Bolstering the soil environment-site preparation. Paper presented at the 42nd annual IDFTA conference, Dec 20–24, Hamilton, Ontario, Canada, pp 26–28
George P, Gupta A, Gopal M, Thomas L, Thomas GV (2012) Screening and in vitro evaluation of phosphate solubilizing bacteria from rhizosphere and roots of coconut palms (*Cocos nucifera* L.) growing in different states of India. J Plantn Crops 40(1):61–64
George P, Gupta A, Gopal M, Thomas L, Thomas GV (2013) Multifarious beneficial traits and plant growth promoting potential of *Serratia marcescens* KiSII and *Enterobacter* sp. RNF 267 isolated from the rhizosphere of coconut palms (*Cocos nucifera* L.). World J Microbiol Biotechnol 29:109–117
George P, Gupta A, Gopal M, Thomas L, Thomas GV (2018) Systematic screening strategies for identifying elite plant growth promoting rhizobacteria for coconut (*Cocos nucifera* L.). Int J Curr Microbiol Appl Sci 7(5):1051–1074
George P, Gupta A, Gopal M, Thomas L, Thomas GV (2022) Indigenous rhizobacteria possessing abiotic stress tolerant traits promote vigorous growth of coconut seedlings via increased nutrient uptake and positive plant–microbe feedback. Proc Indian Natl Sci Acad 88(1):64–79. https://doi.org/10.1007/s43538-022-00067-4
Ghai SK, Thomas GV (1989) Occurrence of *Azospirillum* spp. in coconut-based farming systems. Plant Soil 114:235–241
Ghavale SL, Shinde VV, Wankhede SM, Maheswarappa HP, Haldankar PM (2020) Carbon sequestration and productivity potential of coconut (*Cocos nucifera* L.) hybrids and varieties under coastal eco-system of Maharashtra. Curr J Appl Sci Technol 39(22):30–37
Ghosh PK, Sarma US, Ravindranath AD, Radhakrishnan S, Ghosh P (2007) A novel method for accelerating composting of coir pith. Energy Fuel 21:822–827
Ghosh DK, Bandyopadhyay A, Maheswarappa HP, Sahu PK, Chakrabarthi K, Biswas B (2017) Flowering crops in coconut based cropping system increases the productivity under Indo Gangetic plains of South Asia. Bioscan 12(3):1653–1659
Gome-Falcon N, Saenz-Carbonell LA, Andrade-Torres A, Lara-Perez LA, Narvaez M, Oropeza C (2023) Arbuscular mycorrhizal fungi increase the survival and growth of micropropagated coconut (*Cocos nucifera* L.) plantlets. In Vitro Cell Dev Biol Plant 17:1–2
Gopal M, Gupta A (2005) Microflora of bulk soil from coconut garden and that of rhizosphere of pepper and nutmeg possibly modulate the allelopathic interactions in coconut based cropping system. In: Harper JDI, An M, Wu H, Kent JH (eds) Proceedings of fourth world congress on allelopathy. International Allelopathy Society, pp 287–229

Gopal M, Gupta A, Nair RV (2005) Variations in hosting beneficial plant-associated microorganisms by root (wilt)-diseased and field-tolerant coconut palms of West Coast Tall variety. Curr Sci 89(11):1922–1927

Gopal M, Gupta A, Sunil E, Thomas GV (2009) Amplification of plant beneficial microbial communities during the conversion of coconut leaf substrate to vermicompost by *Eudrilus* sp. Curr Microbiol 59:15–20

Gopal M, Gupta A, Thomas GV (2010a) Opportunity to sustain coconut ecosystem services through recycling of the palm leaf litter as vermicompost: Indian scenario (a technology/research note). CORD 26(2):42–55

Gopal M, Gupta A, Palaniswami C, Dhanapal R, Thomas GV (2010b) Coconut leaf vermiwash: a bio-liquid from coconut leaf vermicompost for improving the crop production capacities of soil. Curr Sci 98:1202–1210

Gopal M, Gupta A, Sunil E, Sunoj J, Thomas GV (2010c) Ecological interaction of soil microorganisms and plant allelochemicals in coconut based cropping ecosystem. In: Thomas GV, Krishanakumar V, Maheswarappa HP, Palaniswami C (eds) Coconut based cropping/farming systems. ICAR-Central Plantation Crops Research Institute, Kasaragod, 231 p

Gopal M, Gupta A, Thomas GV (2012) Vermicompost and vermiwash add beneficial micro flora that enhance soil quality and sustain crop growth. Int J Innov Hort 1(2):93–100

Gopal M, Gupta A, Thomas GV (2016) Production of coir pith compost without adding urea. Indian Coconut J 59(4):29–31

Gopal M, Bhute SS, Gupta A, Prabhu SR, Thomas GV, Whitman WB, Jangid K (2017) Changes in structure and function of bacterial communities during coconut leaf vermicomposting. Antonie Van Leeuwenhoek 110:1339–1355. https://doi.org/10.1007/s10482-017-0894-7

Gopal M, Gupta A, Hameed KS, Chandramohanan R, Thomas GV (2019) A simple, quick and contamination-free method for mass-multiplication of plant-beneficial microbes by small and marginal farmers using coconut water and rice gruel medium. Indian J Agric Sci 89(2):339–343

Gopal M, Gupta A, Shahul Hameed K, Sathyaseelan N, Khadeejath Rajeela TH, Thomas GV (2020) Biochars produced from coconut palm biomass residues can aid regenerative agriculture by improving soil properties and plant yield in humid tropics. Biochar 2:211–226. https://doi.org/10.1007/s42773-020-00043-5

Gopal M, Gupta A, Arunachalam V, Maheswarappa HP, Thomas GV, Jacob PM (2022) Autochthonous nutrient recycling driven by soil microbiota could be sustaining high coconut productivity in Lakshadweep Islands sans external fertilizer application. World J Microbiol Biotechnol 38(11):213. https://doi.org/10.1007/s11274-022-03373-7

Guimaraes DV, Gonzaga MIS, da Silva TO, da Silva TL, da Silva DN, Matias MIS (2013) Soil organic matter pools and carbon fractions in soil under different land uses. Soil Tillage Res 126:177–182. https://doi.org/10.1016/j.still.2012.07.010

Gupta A, Gopal M, Subramanian P, Thomas GV (2010) Bio inputs in coconut based cropping system. In: Thomas GV, Krishnakumar V, Maheswarappa HP, Palaniswami C (eds) Coconut based cropping/farming systems. ICAR-Central Plantation Crops Research Institute, Kasaragod, pp 80–96

Gupta A, Gopal M, Thomas GV, Manikandan V, Gajewski J, Thomas GV (2014) Whole genome sequencing and analysis of plant growth promoting bacteria isolated from the rhizosphere of plantation crops coconut, cocoa and arecanut. PLoS One 9(8):e104259. https://doi.org/10.1371/journal.pone.0104259

Gupta A, Murali G, Thomas GV (2016a) Edible mushroom production technology using coconut residues. Indian Coconut J 59(6):21–23

Gupta A, Gopal M, Ravindran D, Thomas GV (2016b) Prevalence of potassium solubilizing bacteria in the rhizosphere of coconut palms (*Cocos nucifera* L.) growing in different soil types. In: Abstracts of third international symposium on coconut research and development. ICAR-CPCRI, Kasaragod

Gupta A, Indhuja S, Vrinda KB, Gopal M (2018) Diversity of macrofungi in coconut ecosystem. In: Patil S, Kumar AR, Reddy PK et al (eds) Book of abstract, 23rd plantation crops symposium. PLACROSYM XXIII, 6–8 Mar 2019, Chikkamagaluru, Karnataka, p 126

Hamza MA, Anderson WK (2005) Soil compaction in cropping systems: a review of the nature, causes and possible solutions. Soil Tillage Res 82(2):121–145

Herath HMIK, Saumyakumara APA, Tennakoon NA (2007) Soil fertility status of common coconut growing soil series in Sri Lanka. J Soil Sci Soc Sri Lanka 18:35–40

Herath HMIK, Silva HM, Synthya PG, Arachchi LPV (2016) Impact of cover crops on soil quality parameters of coconut (*Cocos nucifera* L.)-grown red yellow podzolic soil. J Food Agric 9(1–2):14–23. https://doi.org/10.4038/jfa.v9i1-2.5203

Huang D, Liu X, Huan H, Liu G, Hu A (2023) Intercropping of *Stylosanthes* green manure could improve the organic nitrogen fractions in a coconut plantation with acid soil. PLoS One 18(3):e0277944. https://doi.org/10.1371/journal.pone.0277944

Ighalo JO, Conradie J, Ohoro CR, Amaku JF, Oyedotun KO, Maxakato NW, Akpomie KG, Okeke ES, Olisah C, Malloum A, Adegoke KA (2023) Biochar from coconut residues: an overview of production, properties, and applications. Ind Crops Prod 5(204):17300

Ilangamudali IMPS, Senarathne SHS (2016) Effectiveness of arbuscular mycorrhizal fungi based biofertilizer on early growth of coconut seedlings. Cocos 22:1–12

Indhuja S, Babu M, Gupta A, Gopal M, Mathew J, Thomas RJ, Krishnakumar V (2021) Screening and characterization of nutrient solubilizing phytobeneficial rhizobacteria from healthy coconut palms in root (wilt) diseased tract of Kerala, India. J Environ Biol 42(3):625–635

Indian Trade Portal (2023). https://www.indiantradeportal.in/, 17 Jun 2023

Irawan S, Antriyandarti E, Laia DH (2022) A study the relationship of soil quality with land suitability for coconut (*Cocos nucifera* L.) development in Kebonagung subdistrict, Pacitan District. IOP Conf Series Earth Environ Sci 1111(1):12–13. https://doi.org/10.1088/1755-1315/1111/1/012013

Islam M, Rahaman A, Afrose A (2018) Assessment of heavy metal concentration in coconut water. Recent Res Sci Technol 10:7–10

Issaka RN, Senayah JK, Andoh-Mensah E, Ennin SA (2012) Assessment of fertility status of soils supporting coconut (*Cocos nucifera*) cultivation in western and central regions of Ghana. West Afr J Appl Ecol 20(1):47–56

Iyer R, Gunasekharan M, Hegde V (2018) Coconut: maladies and remedies. In: The coconut palm (*Cocos nucifera* L.)—research and development perspectives (eds) Nampoothiri KUK, Krishnakumar V, Thampan PK, Nair MA Springer Nature Singapore Pvt. Ltd., 489–518 https://doi.org/10.1007/978-981-13-2754-4_10

Jayakumar M, Janapriya S, Surendran U (2017) Effect of drip fertigation and polythene mulching on growth and productivity of coconut (*Cocos nucifera* L.), water, nutrient use efficiency and economic benefits. Agric Water Manag 182:87–93. https://doi.org/10.1016/j.agwat.2016.12.012

Jayakumar M, Emana AN, Subbaiya R, Ponraj M, Kumar KK, Muthusamy G, Kim W, Karmegam N (2022) Detoxification of coir pith through refined vermicomposting engaging *Eudriluseugeniae*. Chemosphere 291:132675

Jayasekara KS, Loganathan P (1988) Boron deficiency in young coconut (*Cocos nucifera* L.) in Sri Lanka: symptoms and corrective measures. Cocos 6:31–37

Jemziya MBF (2018) Assessment on physical and chemical properties of surface soil and subsoil of coconut cultivated lands located in region of Batticaloa, Sri Lanka. Int J Res 7(1):1–6

Jenkinson DS, Powlson DS (1976) The effects of biocidal treatments on metabolism in soil. I. Fumigation with chloroform. Soil Biol Biochem 8:167–177

Johnson I, Ramjegathesh R, Sheela J, Shoba N, Maheshwarappa HP (2017) Development of microbial consortia for the management of leaf blight disease of coconut. Acta Phytopathol Entomol Hung 52(1):1–14

Joshi OP, Verghese PT, Nelliat EV, Hameed Khan H (1986) Influence of long-term cultural operations on physical and water retention characters of a red sandy loam soil. Int J Trop Agric 4(2):108–115

Jothimani S (1994) Organic farming in coconut. Indian Coconut J 25:48–49

Kalpana M, Gautam B, Srinivasulu B, Rao DVR, Arulraj S, Jayabose C (2008) Impact of integrated nutrient management on nut yield and quality of coconut under coastal ecosystem. J Plantn Crops 36(3):249–253

Katyal JC, Chaudhari SK, Dwivedi BS, Biswas DR, Rattan RK, Majumdar K (2016) Soil health: concept, status and monitoring. Bull Ind Soc Soil Sci 30:1–98

Khan HH, Krishnakumar V (2018) Soil productivity and nutrition. In: Nampoothiri KUK, Krishnakumar V, Thampan PK, Nair MA (eds) The coconut palm (*Cocos nucifera*)—research and development perspectives. Springer, Singapore, pp 323–441. https://doi.org/10.1007/978-981-13-2754-4_10

Khan HH, Upadhyay AK (2002) Integrated nutrient management in coconut based cropping farming system for sustained productivity. In: Strategic agenda to make coconut industry globally competitive. Proceedings of the XXXIX COCOTECH meeting, 1–5 Jul 2002, Pattaya, Thailand

Khan HH, Sankaranarayanan MP, George MV, Narayana KB (1983) Effect of phosphorus skipping on the yield and nutrition of coconut palm (*Cocos nucifera* L.). J Plantn Crops 11(2):129–134

Kim N, Zabaloy MC, Guan K, Villamil MB (2020) Do cover crops benefit soil microbiome? A meta-analysis of current research. Soil Biol Biochem 142:107701. https://doi.org/10.1016/j.soilbio.2019.107701

Kochian LV (1995) Cellular mechanisms of aluminum toxicity and resistance in plants. Annu Rev Plant Biol 46(1):237–260

Kumar BM, Jose S (2018) Phenotypic plasticity of roots in mixed tree species agroforestry systems: review with examples from peninsular India. Agrofor Syst 92:59–69

Kumar BM, Kunhamu TK (2022) Nature-based solutions in agriculture: a review of the coconut (*Cocos nucifera* L.)-based farming systems in Kerala, "the Land of Coconut Trees". Nat Based Solutions 2:100012. https://doi.org/10.1016/j.nbsj.2022.100012

Kumar D, Rakshit R, Rani R, Bharti P, Das A, Kundu M (2022) Integrated nutrient management in coconut (*Cocos nucifera* L.): an assessment of soil chemical and biological parameters under subtropical humid climate. J Soil Sci Plant Nutr 22(2):2695–2706

Lal R (2005) Forest soils and carbon sequestration. For Ecol Manag 220(1–3):242–258

Lamanda N, Malézieux E, Roupsard O, Joffre R, Martin P (2004) Soil organic matter in coconut-based agroforestry systems in Vanuatu: a key to sustainability? In: Book of abstracts, first world congress of agroforestry: working together for sustainable land-use systems, 27 Jun–2 Jul 2004, Orlando, USA. IFAS, Gainesville, p 314

Lange M, Habekost M, Eisenhauer N, Roscher C, Bessler H, Engels C, Gleixner G (2014) Biotic and abiotic properties mediating plant diversity effects on soil microbial communities in an experimental grassland. PLoS One 9(5):e96182

Lara-Pérez LA, Oros-Ortega I, Córdova-Lara I et al (2020) Seasonal shifts of arbuscular mycorrhizal fungi in *Cocos nucifera* roots in Yucatan, Mexico. Mycorrhiza 30:269–283. https://doi.org/10.1007/s00572-020-00944-0

Lindao-Pérez JD, Cedeño AL, Villamar-Torres RO, Santillán AD, Mera-Pérez HA, Jazayeri SM (2022) Production and bromatological analysis of the oyster mushroom (*Pleurotus ostreatus* (jacq. ex fr.) p. kumm.) grown with cocoa, banana, coconut and African palm husk substrates. Acta Agric Sloven 118(1):1–8

Lins PMP, Viegas IDJM, Ferreira EVDO (2021) Nutrition and production of coconut palm cultivated with mineral fertilization in the state of Pará. Rev Bras Frutic 43:e-113. https://doi.org/10.1590/0100-29452021113

Liyanage MS, Dassanayake KB (1993) Experiences in coconut-based farming systems in Sri Lanka. In: Advances in coconut research and development. Oxford and IBH Publishing Co., New Delhi, pp 357–368

Magat SS (1992) Coconut. In: IFA world fertilizer use manual. International Fertilizer Industry Association, Paris, pp 234–244

Magat SS (2009) Integrated soil fertility management (ISFM) on coconut + lanzones (*Lansium domesticum* Corr) agro-ecosystem in Southern Mindanao, Philippines (1993–2007): with emphasis on the multi-nutrient coconut-specific mineral fertilizer. Part II. Influence on Le. CORD 25(2):21–21

Magat SS, Margate RZ, Prudente RL (1977) Utilization of common salt (sodium chloride) as a fertilizer and for the control of leaf spot disease of coconut seedlings. Philippines J Coconut Stud 2(3):39–45

Maheswarappa HP, Subramanian P, Dhanapal R (2000) Root distribution pattern of coconut (*Cocos nucifera* L.) in littoral sandy soil. J Plantn Crops 28(2):164–166

Maheswarappa HP, Kumari PA, Kamalakshiamma PG, Shanavas M (2005) Influence of integrated nutrient management and high density multi-species cropping system on soil properties, plant nutrition and yield in root (wilt) affected coconut palms. CORD 21(2):34. https://doi.org/10.37833/cord.v21i02.406

Maheswarappa HP, Thomas GV, Bhat R, Palaniswami C, Jayasekhar S (2011) Impact of inorganic fertilizer substitutions by vermicomposted coconut leaves on productivity and economics of coconut. J Plantn Crops 39(1):30–34

Maheswarappa HP, Dhanapal R, Subramanian P, Palaniswami C (2013) Evaluation of coconut based high density multi species cropping system under organic and integrated nutrient management. J Plantn Crops 41(2):130–135

Maheswarappa HP, Krishnakumar V, Gupta A, Geetha Kumari A (2016) Influence of organic sources of nutrients on vanilla (*Vanilla planifolia*) as an intercrop in coconut garden. J Plantn Crops 44(2):85–89

Maheswarappa HP, Subramanian P, Kumari AG (2017) Influence of organic nutrient management practices for noni (*Morinda citrifolia*) when grown as an intercrop in coconut garden. Indian J Agric Sci 87(9):1155–1157. https://doi.org/10.56093/ijas.v87i9.74160

Malhotra SK, Maheswarappa HP, Selvamani V, Chowdappa P (2017) Diagnosis and management of soil fertility constraints in coconut (*Cocos nucifera*): a review. Indian J Agric Sci 87:711–726

Manna MC, Singh M, Wanjari RH, Mandal A, Patra AK, Lal R (2016) Soil nutrient management for carbon sequestration. In: Encyclopedia of soil science, 3rd edn. CRC Press, Boca Raton, FL, pp 288–293. https://doi.org/10.1081/E-ESS3-120052914

Margate RZ, Magat SS, Alforja LM, Habana JA (1979) A long-term KCI fertilization study of bearing coconuts in an inland-upland area of Davao. Oleaginueux 34(5):235–240

Margate RZ, Padrones GD, Maravilla JN, Magat SS, Mantiquilla JA, Silva EC, Rivera V (1997) Integrated soil fertility management of hybrid coconut grown in different agro climatic conditions of the Philippines. CORD 13(01):34–34

Marinho HJL, Gheyi HR, Fernandez PD, Holanda JSD, Ferreira Neto M (2006) Cultivation of coconut irrigated with saline waters. Brazil Agric Res 41(8):1277–1284

Marohn C, Distel A, Dercon G, Wahyunto Tomlinson R, Noordwijk MV, Cadisch G (2012) Impacts of soil and groundwater salinization on tree crop performance in post-tsunami Aceh Barat, Indonesia. Natl Hazards Earth Syst Sci 12(9):2879–2891

Mathew AC, Thamban C, Samuel MP (2018a) Efficacy of water conservation measures in coconut plantations to enhance ground water resource and coconut yield in West Coast region. J Plantn Crops 46(1):12–20

Mathew J, Krishnakumar V, Srinivasan V, Bhat R, Namboothiri CG, Haris AA (2018b) Standardization of critical boron level in soil and leaves of coconut palms grown in a tropical Entisol. J Soil Sci Plant Nutr 18(2):376–387

Mathew J, Haris AA, Bhat R, Krishnakumar V, Muralidharan K, John KS, Surendran U (2021a) A comparative assessment of nutrient partitioning in healthy and root (wilt) disease affected coconut palms grown in an Entisol of humid tropical Kerala. Trees 35:621–635. https://doi.org/10.1007/s00468-020-02064-w

Mathew J, Haris AA, Anithakumari P, Bhat R (2021b) Nutrient mixtures—'Kalpa Poshak' and 'Kalpa Vardhini' for enhancing the productivity of coconut palms. In: Proceedings of the international virtual conference on challenges, opportunities and innovations in agriculture, plantations and allied (APA) domains posed by the pandemic, 21 Oct 2021. Indian Institute of Plantation Management, Bengaluru, pp 7–8

Mathew J, Haris AA, Indhuja S, Krishnakumar V, Nair KM, Bhat R, Kumar KA (2022) Effectiveness of site-specific management practices on the amelioration of soil acidity in the coconut growing Entisol and Ultisol of humid tropics. J Soil Sci Plant Nutr 22:1060–1073. https://doi.org/10.1007/s42729-021-00715-6

Mengel K (1996) Turnover of organic nitrogen in soils and its availability to crops. Plant Soil 181:83–93

Merilyn VJ, Thomas GV (1992) Distribution of nitrogen fixing *Beijerinckia* in the rhizosphere of coconut. J Plantn Crops 20(1):149–155

Mini V, Usha M (2018) Spatial variability of soil fertility in a coconut based agro ecological unit in the sandy plains of Kerala, India. Asian J Soil Sci 13(1):58–62

Mitchell, J, Gaskell, M, Smith, R (2000) Soil management and soil quality for organic crops. UC ANR Publication 7249, 5 p. https://doi.org/10.3733/ucanr.7248

Mitra A, Biswas S, Pal N, Pramanick P, Datta U, Biswas P, Mitra A (2018) Biomass and stored carbon in the above ground structures of coconut tree. Int J Basic Appl Res 8(2):60–65

Mohandas S (2012) Effect of NPK fertilizer levels on mineral nutrition and yield of hybrid (tall × dwarf) coconut. Madras Agric J 99(1/3):87–91

Montenegro AAA, Almeida TAB, Lima CAD, Abrantes JRCB, de Lima JLMP (2020) Evaluating mulch cover with coir dust and cover crop with *Palma cactus* as soil and water conservation techniques for semiarid environments: laboratory soil flume study under simulated rainfall. Hydrol 7(3):61. https://doi.org/10.3390/hydrology7030061

Morales-Lizcano NP, Hasan AM, To HS, Lekadou TT, Copeland J, Wang PW, Diallo HA, Konan JL, Yoshioka K, Moeder W, Scott JA, Rosete YA (2017) Microbial diversity in leaves, trunk and rhizosphere of coconut palms (*Cocos nucifera* L.) associated with the coconut lethal yellowing phytoplasma in Grand-Lahou, CtedIvoire. Afr J Biotechnol 16(27):1534–1550

Nagarajan R, Manickam TS, Kothandaraman GV, Ramaswamy K, Palaniswamy G (1985) Manurial value of coir pith. Madras Agric J 72:533–535

Nair SK, Subba Rao NS (1977) Microbiology of the root region of coconut and cacao under mixed cropping. Plant Soil 46:511–519. https://doi.org/10.1007/BF00015910

Nair PKR, Varma R, Nelliat EV, Bavappa KVA (1975) Beneficial effects of crop combination of coconut and cacao. Indian J Agric Sci 45:165–171

Nair KM, Haris A, Mathew J, Srinivasan V, Dinesh R, Hamza H, Subramanian P, Thamban C, Chandran KP, Krishnakumar V, Bhat R, Hegde R, Singh SK (2018) Coconut-growing soils of Kerala: 2. Assessment of fertility and soil related constraints to coconut production. J Plantn Crops 46(2):84–91. https://doi.org/10.25081/jpc.2018.v46.i2.3719

Naresh Kumar S (2011) Climate change and Indian agriculture: current understanding on impacts, adaptation, vulnerability and mitigation. J Plant Biol 37(2):1–16

Naresh Kumar S, Rajagopal V, Thomas TS, Cherian VK, Narayanan MKR, Ananda KS, Nagawekar DD, Hanumanthappa M, Vincent S, Srinivasulu B (2007) Variations in nut yield of coconut and dry spell in different agro-climatic zones of India. Indian J Hort 64(3):309–313

Naresh Kumar S, Kasturi Bai KV, Rajagopal V, Aggarwal PK (2008) Simulating coconut growth, development and yield using InfoCrop-coconut model. Tree Physiol 28:1049–1058

Nath JC, Arulraj S, Maheswarappa HP (2012) Integrated nutrient management in COD × WCT hybrid coconut under alluvial clay-loam soil of Assam. J Plantn Crops 40(2):105–110

Navarro MNV, Jourdan C, Sileye T, Braconnier S, Mialet-Serra I, Saint-André L, Roupsard O (2008) Fruit development, not GPP, drives seasonal variation in NPP in a tropical palm plantation. Tree Phys 28(11):1661–1674

Newton K (1967) In: Pomier M (ed) Coconut research at Rangiroa. South Pacific Commission Tech. Paper No. 153, iv, 15 p

Nirukshan GS, Herath HMIK, Wijebandara DMDI, Dissanayake DMPD (2016) Soil microbial population and activity affected by fertilizer and manure addition in a coconut growing Sandy Regosol. In: Proceedings of the sixth symposium on plantation crop research, plantation agriculture towards national prosperity, pp 2–4

Nirukshan GS, Ranasinghe S, Sleutel S (2022) The effect of biochar on mycorrhizal fungi mediated nutrient uptake by coconut (*Cocos nucifera* L.) seedlings grown on a sandy regosol. Biochar 4:68. https://doi.org/10.1007/s42773-022-00192-9

Nuwarapaksha TD, Udumann SS, Dissanayaka DMNS, Dissanayake DKRPL, Atapattu AJ (2022) Coconut based multiple cropping systems: An analytical review in Sri Lankan coconut cultivations. Circ Agric Syst 2(1):1–7

Nyren P, Pettersson B, Uhlen M (1993) Solid phase DNA minisequencing by an enzymatic luminometric inorganic pyrophosphate detection assay. Anal Biochem 208(1):171–175

Okeri HA, Alonge PO, Udoh JJ (2007) Assessment of nutritional status of soil supporting coconut (*Cocos nucifera*) cultivation in some localities of Edo State of Nigeria. Afr J Biotechnol 6(3):258–262

Ollagnier M, Ochs R (1972) Sulphur deficiencies in the oil palm and coconut. Oleagineux 27(4):193–198

Osei-Bonsu K, Opoku-Ameyaw K, Amoah FM, Oppong FK (2002) Cacao-coconut intercropping in Ghana: agronomic and economic perspectives. Agrofor Syst 55:1–8

Palaniswami C, Thomas GV, Subramanian P, Dhanapal R, Selvamani V (2008) Comparative evaluation of soil health and fertility under organic, inorganic and integrated nutrient management in coconut based mixed farming system. In: Singh HP, Thomas GV (eds) Organic horticulture: principles, practices and technologies. Westeville Publishing House, New Delhi, 231 p

Palaniswami C, Thomas GV, Dhanapal R, Subramanian P, Maheswarappa HP (2010) Nutrient management in coconut based high density multispecies cropping system. In: Thomas GV, Krishanakumar V, Maheswarappa HP, Palaniswami C (eds) Coconut based cropping/farming systems. CPCRI (ICAR), Kasaragod, pp 80–96

Pandey CB, Begum M (2010) The effect of a perennial cover crop on net soil N mineralization and microbial biomass carbon in coconut plantations in the humid tropics. Soil Use Manag 26(2):158–166

Pandey S, Gupta S (2020) Diversity analysis of ACC deaminase producing bacteria associated with rhizosphere of coconut tree (*Cocos nucifera* L.) grown in Lakshadweep islands of India and their ability to promote plant growth under saline conditions. J Biotechnol 324:183–197

Pandey CB, Rai RB, Singh L (2007) Seasonal dynamics of mineral N pools and N-mineralization in soils under home garden trees in South Andaman, India. Agrofor Syst 71:57–66. https://doi.org/10.1007/s10457-007-9073-6

Pandiselvam R, Shaji A, Ramesh SV, Sathyan N, Manikantan MR, Mathew AC (2023) Development, evaluation, and optimization of portable pyrolysis system for the production of biochar from tender coconut husk. Biomass Convers Biorefin 11:1–10

Pantipkayee, Krongkwanpooniem, Nawapornkeatkawinwong (2018) Heavy metal residues in coconut and soil from coconut orchard in Ratchaburi, Thailand. In: Proceedings of research for a 15th international conference, 22–23 Feb 2018, Hamburg, Germany, pp 52–55

Parkin TB, Doran JW, Franco-Vizcaíno E (1996) Field and laboratory tests of soil respiration. In: Doran JW, Jones AJ (eds) Methods for assessing soil quality. Soil Science Society of America, Madison, WI, pp 231–245

Pillai NG, Davis TA (1963) Exhaust of macronutrients by the coconut palm: a preliminary study. Indian Coconut J 16(2):81–87

Potty VP (1977) Rhizosphere microflora of coconut palm with special reference to root (wilt) disease. Doctoral dissertation. University of Kerala, Thiruvananthapuram

Potty VP, Jayasankar NP (1976) Influence of crop mixing of hybrid Napier on the root zone microflora of the coconut palm. In: Nayar NM (ed) Coconut research and development. Wiley Eastern Ltd., New Delhi, pp 300–303

Prabhu SR, Thomas GV (1998) Occurrence of *Azospirillum amazonense* in the root environment of coconut palm. In: 30th annual conf. assn. microbiologists. College of Fisheries, Mangalore

Prabhu SR, Subramanian P, Biddappa CC, Bopaiah BM (1998) Prospects and improving coconut productivity through vermiculture technology. Indian Coconut J 29:79–84

Purseglove JW (1975) Tropical crops—monocotyledons, vol 1 & 2. Longman Ltd., London, 607 p

Rajagopal V, Jacob M, Mathew J, Shivashankar S (1996) Impact of dry spells on the ontogeny of coconut fruits and its relation to yield. Plant Res Dev 3(4):251–255

Rajeela TK, Gopal M, Gupta A, Bhat R, Thomas GV (2017) Cross-compatibility evaluation of plant growth promoting rhizobacteria of coconut and cocoa on yield and rhizosphere properties of vegetable crops. Biocatal Agric Biotechnol 9:67–73

Rajeshkumar PP, Thomas GV, Gupta A, Gopal M (2015) Diversity, richness and degree of colonization of arbuscular mycorrhizal fungi in coconut cultivated along with intercrops in high productive zone of Kerala, India. Symbiosis 65:125–141. https://doi.org/10.1007/s13199-015-0326-2

Ramadasan A, Lal SB (1966) Exhaust of nutrients from coconut garden—factors affecting production. Coconut Bull 20:173–175

Ramanandan PL, Pillai NG (1974) Effect of continuous cultivation and manuring on the leaf nutrient composition and soil nutrient status of coconut palms. J Plantn Crops 2(2):1–3

Ramesh CR (1982) Root infection and population density of VA mycorrhizal fungi in a coconut based multistoreyed cropping system. In: Proc. PLACROSYM-V. ISPC, CPCRI, Kasaragod, pp 548–554

Ramirez-Silva JH, Cortazar-Ríos M, Ramírez-Jaramillo G, Oropeza-Salín CM, Rondón-Rivera DD (2021) Soil organic matter and nitrogen content as related to coconut nutrition in Guerrero, Mexico. OALib J 8(8):1–16. https://doi.org/10.4236/oalib.1107727

Ranasinghe CS, Silva LRS (2007) Photosynthetic assimilation, carbohydrates in vegetative organs and carbon removal in nut producing and sap producing coconut palms. Cocos 18:45–57

Raveendra SAST, Nissanka SP, Somasundaram D, Atapattu AJ, Mensah S (2021) Coconut-glyricidia mixed cropping systems improve soil nutrients in dry and wet regions of Sri Lanka. Agrofor Syst 95:307–319. https://doi.org/10.1007/s10457-020-00587-2

Rawitz E, Morin J, Hoogmoed WB, Margolin M, Etkin H (1983) Tillage practices for soil and water conservation in the semi-arid zone. I. Management of fallow during the rainy season preceding cotton. Soil Tillage Res 3(3):211–231

Reddy PP (2016) Cover/green manure crops. In: Sustainable intensification of crop production. Springer, Singapore, pp 55–67. https://doi.org/10.1007/978-981-10-2702-4_4

Reddy PP (2017) Cover/green manure cropping. In: Agro-ecological approaches to pest management for sustainable agriculture. Springer, Singapore, pp 91–107. https://doi.org/10.1007/978-981-10-4325-3_7

Reddy DVS, Upadhyay AK (2002) Impact of integrated nutrient management on the mineral nutrition and yield of WCT coconut in littoral sandy soil at Kasaragod. In: Proceedings of the 15th plantation crops symposium placrosym XV, Mysore, India. Central Coffee Research Institute, Coffee Research Station, pp 274–281

Reddy DVS, Subramanian P, Gopalasundaram P (2002) Coconut based high density multi-species cropping system under different levels of fertilizers in red sandy loam soils. In: Rethinam P, Khan HH, Reddy VM, Mandai PK, Suresh K (eds) Plantation crops research and development in the new millennium (PLACROSYM XIV). Coconut Development Board, Kochi, pp 106–111

Reddy DS, Kumar SN, Prabhu SR (2007) Evaluation of alternative media to potting mixture for raising coconut seedlings in polybags. J Plantn Crops 29(1):62–65

Reichert JM, Suzuki LEAS, Reinert DJ, Horn R, Håkansson I (2009) Reference bulk density and critical degree-of-compactness for no-till crop production in subtropical highly weathered soils. Soil Tillage Res 102(2):242–254

Remison SU, Iremiren GO (1990) Effect of salinity on the performance of coconut seedlings on two contrasting soils. Cocos 8:33–39

Ronaghi M, Uhlén M, Nyrén P (1998) A sequencing method based on real-time pyrophosphate. Science 281:363–365

Roupsard O (2008) Coconut carbon sequestration, part 2: strategies for the carbon market & simulating potential incomes for coconut CDM projects. CORD 24(1):16–34

Roupsard O, Dauzat J, Nouvellon Y, Deveau A, Feintrenie L, Saint-André L, Bouillet JP (2008a) Cross-validating sun-shade and 3D models of light absorption by a tree-crop canopy. Agric For Meteorol 148(4):549–564

Roupsard O, Lamanda N, Jourdan C, Navarro M, Mialet-Serra I, Dauzat J, Sileye T (2008b) Coconut carbon sequestration, part 1: highlights on carbon cycle in coconut plantations. CORD 24(1):1–14

Saha SK, Nair PKR, Nair VD, Kumar BM (2010) Carbon storage in relation to soil size fractions under some tropical tree-based land use systems. Plant Soil 328:433–446

Sahasranaman KN, Pillai NG, Jayasankar NP, Potty VP, Varkey T, Amma PGK, Radha K (1983) Mixed farming in coconut gardens: economics and its effect on root (wilt) disease. In: Nayar NM (ed) Coconut research and development. Wiley Eastern Ltd., New Delhi, pp 160–165

Saldanha ECM, Silva MLD, Lins P, Pontes M, Farias SCC, Wadt PGS (2017) Nutritional diagnosis in hybrid coconut cultivated in northeastern Brazil through diagnosis and recommendation integrated system (DRIS). Rev Bras Frutic 39(1):1–9

Sankaranarayanan MP, Verghese EJ, Menon KPV (1958) A note on the tolerance of salinity by coconut palm. Indian Coconut J 11:133–139

Savithri P, Hameed KH (1994) Characteristics of coconut coir peat and its utilization in agriculture. J Plantn Crops 22:1–18

Scheewe W (2022) Feeding the soil for climate resilient coconut farms. Manila Bulletin Agriculture, Philippines. https://snrd-asia.org/climate-resilient-coconut-farm-management

Secretaria MI, Margate RZ (2001) Integrated soil fertility management on local coconut hybrid in a farmer-managed coconut-based cropping system. CORD 17(1):34–34

Selvamani V, Duraisami VP (2014a) Identifying and mapping soil fertility related constraints for coconut in Coimbatore and Tiruppur districts of Tamil Nadu state, India. J Plantn Crops 42(3):348–357

Selvamani V, Duraisami VP (2014b) Identifying and mapping leaf nutrient based constraints for coconut productivity in Coimbatore and Tiruppur districts of Tamil Nadu state, India. J Plantn Crops 42(2):163–169

Selvamani V, Duraisami VP (2018a) Mapping soil constraints for coconut using GIS for a major coconut growing region of Tamil Nadu. J Plantn Crops 46(3):190–195

Selvamani V, Duraisami VP (2018b) Evaluating the primary yield limiting nutrient deficiency of coconut (*Cocos nucifera* L.) in a major coconut growing zone of Tamil Nadu. J Plantn Crops 46(2):112–117

Semarnat (2002) NOM-021-RECNAT-2000. Que establece las especificaciones de fertilidad, salinidad y clasificación de suelos. Estudios, muestreo y análisis. Secretaría de Medio Ambiente y Recursos Naturales, México D.F., 75 p

Senarathne SHS (2017) Influence of long term application of green manure on the productivity of coconut cultivated in reddish brown latosolic soils in Sri Lanka. CORD 33(2):25–34

Senarathne SHS (2018) Effect of arbuscular mycorrhizal fungi based biofertilizer on coconut seedlings growth in nursery. CORD 34(2):30–41

Senarathne SH, Udumann SS (2019) Evaluation of coconut based Anacardium occidentale agroforestry system to improve the soil properties of coconut growing lands in wet, intermediate and dry zone of Sri Lanka. CORD 35(1):1–10

Senarathne SHS, Sangakkara UR, Raveendra SAST (2015) Weed biomass and seedling emergence patterns as affected by different ground cover management systems in coconut plantations of Asian humid tropics Sri Lanka. Int J Res Agric Sci 2:245–252

Silva RAD, Cavalcante LF, Holanda JSD, Pereira WE, Moura MFD, Ferreira Neto M (2006) Fruits quality of green dwarf coconut fertirrigation with nitrogen and potassium. Rev Bras Frutic 28(2):310–313

Silva ED, Ribeiro RV, Ferreira-Silva SL, Viégas RA, Silveira JAG (2010) Comparative effects of salinity and water stress on photosynthesis, water relations and growth of *Jatropha curcas* plants. J Arid Environ 74(10):1130–1137

Singh RK, Lama TD, Saikia US, Satapathy KK (2006) Economics of rainwater harvesting and recycling for winter vegetable production in mid hills of Meghalaya. J Agric Eng 43(2):33–36

Sobral LF (1998) Coconut nutrition and fertilization. In: Ferreira JMS, Warwick DRN, Siqueira LA (eds) The coconut culture in Brazil. Aracaju, Embrapa, pp 129–157

Solangi AH, Mal B, Kazmi AR, Iqbal MZ (2010) Preliminary studies on the major characteristic, agronomic feature and nutrient value of *Gliricidia sepium* in coconut plantations of Pakistan. Pak J Bot 42(2):825–832

Somashekar RK (1984) Trace metal status of coconut plants of a polluted are in Mysore. Proc Indian Natl Sci Acad B 50(5):535–538

Somasiri LLW, Nadaraja N, Amarasinghe L, Gunathilake HAJ (1994) Land suitability assessment of coconut growing areas in the coconut triangle. Coconut Research Institute, Lunuwila

Somasiri LLW, Wijebandara DMDI, Panditharatna BDP, Sabaratnam S, Kurundukumbura CPA (2003) Loss of nutrients in a high yielding coconut plantation through removal of plant materials from the field. Cocos 15:12–22
Southern PJ (1969) Sulphur deficiency in coconut. Oleagineux 24:211
Sreelatha AK (1993) Potassium dynamics in Neyyattinkara soil series under coconut cultivation. Doctoral dissertation. Department of Soil Science and Agricultural Chemistry, College of Agriculture, Vellayani, p 145
Srinivasan R, Maddileti N, Kasthuri Thilagam V, Hegde R (2020) Coconut production in red gravelly soils of southern Karnataka state of India. Popular Kheti 8(4):11–12
Stamatiadis S, Doran JW, Kettler T (1999) Field and laboratory evaluation of soil quality changes resulting from injection of liquid sewage sludge. Appl Soil Ecol 12(3):263–272
Subramanian P, Dhanapal R, Sanil P, Palaniswami C, Sairam CV, Maheswarappa HP (2005a) Glyricidia (*Gliricidia sepium*) as green manure in improving soil fertility and productivity of coconut. J Plantn Crops 33(3):179–183
Subramanian P, Reddy DS, Palaniswami C, Gopalasundaram P, Upadhyay AK (2005b) Studies on nutrient export and extent of nutrient recycling in coconut based high density multispecies cropping system. CORD 21(1):34–40
Subramanian P, Dhanapal R, Mathew AC, Palaniswami C, Upadhyaya AK, Naresh Kumar S, Reddy DVS (2012) Effect of fertilizer application through micro-irrigation technique on nutrient availability and coconut productivity. J Plantn Crops 40(3):168–173
Subramanian P, Dhanapal R, Thomas GV, Maheswarappa HP, Thamban C (2014) Coconut based integrated farming system. Technical bulletin. ICAR-CPCRI, Kasaragod, 32 p
Subramanian P, Krishnakumar V, Alka Gupta MG, Surekha R, Selvamani V (2016) Impact of long term organic cultivation practices on soil health and productivity of coconut. In: Abstracts of third international symposium on coconut research and development. ICAR-CPCRI, Kasaragod, p 54
Subramanian P, Thamban C, Hegde V, Hebbar KB, Bhat R, Krishnakumar V, Niral V, Josephrajkumar A (2018) Coconut, technical bulletin no. 133. ICAR-CPCRI, Kasaragod, 56 p
Subramanian P, Alka G, Murali G, Selvamani V, Joseph R, Surekha R, Krishnakumar V, Bhat R, Vinayaka H, Thomas GV (2022) Organic cultivation practices in coconut. In: Vanaja T, Balakrishnan PC, Satheesan KN (eds) Compendium on coconut. RARS, Pilicode, KAU, pp 140–155
Sudha B, George A (2011) Tillage and residue management for organic carbon sequestration in coconut (*Cocos nucifera*)-based cropping systems. Indian J Agron 56(3):223–227
Suman S, Gautam S (2017) Pyrolysis of coconut husk biomass: analysis of its biochar properties. Energy Sources A Recov Util Environ Effect 39(8):761–767. https://doi.org/10.1080/15567036.2016.1263252
Sunoj VS, Kumar SN, Muralikrishna KS, Padmanabhan S (2015) Enzyme activities and nutrient status in coconut (*Cocos nucifera* L.) seedling rhizosphere soil after exposure to elevated CO_2 and temperature. J Indian Soc Soil Sci 63(2):191–199
Surulirajan M, Rajappan K, Satheesh Kumar N, Annadurai K, Jeevan Kumar K, Asokhan M (2014) Management of basal stem rot disease in coconut through bio-inoculants and chemicals. Int J Trop Agric 32(3/4):407–414
Sys C, Van Ranst E, Debaveye J, Beernaert F (1993) Land evaluation. Part III: crop requirements. Agricultural publications no. 7. GADC, Brussels, 191 p
Tahir TA, Hamid FS (2012) Vermicomposting of two types of coconut wastes employing *Eudriluseugeniae*: a comparative study. Int J Recycl Org Waste Agric 1:7
Teixeira LAJ, Bataglia OC, Buzetti S, Furlani Junior E, Isepon JDS (2005) NPK fertilization on dwarf green coconut (*Cocos nucifera* L.)-yield and fruit quality. Rev Bras Frutic 27:120–123
Tennakoon S (2004) Rural livelihood strategies and the five capitals: a comparative study in selected villages of Sri Lanka. In: Proceedings of the 18th European conference on modern South Asian studies, Lund, Sweden, Jul, pp 6–9

Thamban C, Subramanian P, Jayasekhar S, Jaganathan D, Muralidharan K (2016) Group approach for enhancing profitability of small holders through technology integration—reflections from coconut farming. J Plantn Crops 44(3):158–164

Thampan PK (1981) Hand book on coconut palm. Oxford and IBH Publishing Co., Kolkata, pp 81–157

Thankamani CK, Prakash KM, Srinivasan V, Kandiannan K, Jayarajan K (2022) Coconut leaf mulching—a boon for ginger farming. Indian Coconut J LXIV(9):16–18

Thomas GV (1987) Microbial population, enzyme activity and VA-mycorrhiza in the root region of coconut in relation to in situ green manuring. In: Proc PLACROSYM-VI, pp 267–274

Thomas GV, Shantaram MV (1984) *In situ* cultivation and incorporation of green manure legumes in coconut basins: an approach to improve soil fertility and microbial activity. Plant Soil 80:373–380

Thomas GV, Shantaram MV (1986) Solubilization of inorganic phosphates by bacteria from coconut plantation soils. J Plantn Crops 14(1):42–48

Thomas GV, Shantaram MV (1993) Biomass production and nodulation of green manure legumes in coconut basins in laterite soil and their influence on soil fertility. CORD 9(1):34–34

Thomas GV, Prabhu SR, Reeny MZ, Bopaiah BM (1998) Evaluation of lignocellulosic biomass from coconut palm as substrate for cultivation of *Pleurotus sajor-caju* (Fr.) Singer. World J Microbiol Biotechnol 14:879–882

Thomas GV, Prabhu SR, Subramanian P, Iyer R (2001a) Organic farming technologies in coconut. ATIC series publication No. 4. CPCRI, Kasaragod, p 14

Thomas GV, Biddappa CC, Prabhu SR (2001b) Evaluation of N2-fixing cover legumes as green manures for N substitution in coconut (*Cocos nucifera* Linn.) palm. Trop Agric Lond Then Trinidad 78(1):13–18

Thomas GV, Gopal M, Gupta A (2010) Microbial interactions in coconut based cropping/farming systems. In: Coconut based cropping/farming systems, Central Plantation Crops Research Institute (Indian Council of Agricultural Research), Kasaragod, 231 p

Thomas GV, Palaniswami C, Gopal M, Gupta A (2012) Recycling coconut leaf-agro wastes mixture using *Eudrilus* sp. and growth promotion properties of coconut leaf vermicompost. Int J Innov Hort 1(2):113–118

Thomas GV, Palaniswami C, Prabhu SR, Gopal M, Gupta A (2013) Co-composting of coconut coir pith with solid poultry manure. Curr Sci 104(2):245–250

Thomas GV, Gopal M, Gupta A, Subramanian P, Prabhu SR (2016) Bio-resources based biological soil fertility management for sustainable coconut production. In: Chowdappa P et al (eds) Abstracts of third international symposium on coconut research and development. ICAR—Central Plantation Crops Research Institute, Kasaragod

Thomas GV, Krishnakumar V, Dhanapal R, Reddy DVS (2018) Agro-management practices for sustainable coconut production. In: Nampoothiri KUK, Krishnakumar V, Thampan PK, Nair MA (eds) The coconut palm (*Cocos nucifera* L.)—research and development perspectives. Springer Nature Singapore Pvt. Ltd., pp 227–322. https://doi.org/10.1007/978-981-13-2754-4_16

Thuvasan KT, Selvi D, Kannan B (2019) Impact of soil available nutrients on yield of coconut under Dindigul and Erode districts of Tamil Nadu. J Pharmacogn Phytochem 8(6):304–305

Usharani KV, Aparna B (2019) Enzyme activities and biological fertility index of the soils of coconut based cropping systems in mid laterite soils of Kerala (AEU 9). Int J Agric Food Sci 1(4):7–11. https://doi.org/10.33545/2664844X.2019.v1.i4a.18

Utomo B, Prawoto AA, Bonnet S, Bangviwat A, Gheewala SH (2016) Environmental performance of cocoa production from monoculture and agroforestry systems in Indonesia. J Clean Prod 134:583–591

Van Noordwijk M, Cadisch G (2002) Access and excess problems in plant nutrition. In: Progress in plant nutrition: plenary lectures of the XIV international plant nutrition colloquium: food security and sustainability of agro-ecosystems through basic and applied research. Springer, Dordrecht, pp 25–40

Vance ED, Brookes PC, Jenkinson DS (1987) An extraction method for measuring soil microbial biomass C. Soil Biol Biochem 19:703–707

Varghese PT, Nelliat EV, Balakrishnan TK (1978) Beneficial interactions of coconut–cocoa crop combination. In: Nelliat EV, Ranganathan V, Vishveswara S, Potti SN, Zachariah PK (eds) Proceedings of the first symposium on plantation crops. Indian Society for Plantation Crops, pp 383–392

Vega EV (2011) Cleaner production opportunities for improvement of carbon saving in the production of coconut biodiesel. Int J Chem Environ Eng 2(5):356–361

Venkatasamy R (2003) Integrated nutrient management in coconut with composted coir pith. Madras Agric J 90(1–3):54–56

Vijayaraghavan H, Ramachandran TK (1989) Effect of in situ cultivation and incorporation of green manure crops on the yield of coconut. Cocos 7:26–29

Vizcayno C, Garcia-Gonzalez MT, Fernandez-Marcote Y, Santano J (2001) Extractable forms of aluminum as affected by gypsum and lime amendments to an acid soil. Commun Soil Sci Plant Anal 32(13–14):2279–2292

Vukicevich E, Lowery T, Bowen P, Úrbez-Torres JR, Hart M (2016) Cover crops to increase soil microbial diversity and mitigate decline in perennial agriculture: a review. Agron Sustain Dev 36:1–14

Wahid PA, Kamala Devi CB, Pillai NG (1974) Inter-relationships among root CEC, yield and mono-and divalent cations in coconut (*Cocos nucifera* L.). Plant Soil 40:607–617

Wijebandara DMDI, Kurudukubura CPA, Panditharathna DP (2015) Nutrient removal through plant components of low yielding coconut plantation. Cocos 21:1–13

Young HS, Miller-terKuile A, McCauley DJ, Dirzo R (2017) Cascading community and ecosystem consequences of introduced coconut palms (*Cocos nucifera*) in tropical islands. Can J Zool 95(3):139–148

Yu X, Luo N, Yan J, Tang J, Liu S, Jiang Y (2012) Differential growth response and carbohydrate metabolism of global collection of perennial ryegrass accessions to submergence and recovery following de-submergence. J Plant Phys 169(11):1040–1049

Zella AY, Lawi Y (2019) Contribution of coconut trees (*Cocos nucifera*) in biomass and carbon store and its' role in improving livelihood of small scale farmers of coastal areas of Tanzania. Clim Chang 5(20):253–260

P. Subramanian is Principal Scientist (Agronomy) and Head, Division of Crop Production at ICAR-CPCRI, Kasaragod, Kerala, India. He has 30 years of experience in coconut and cocoa production technology. He developed agro-techniques for coconut management under coastal sandy soil, organic coconut cultivation, coconut-based integrated farming system, soil moisture conservation measures, and fertigation. He has published more than 50 research papers. He is recipient of Outstanding ICAR Team Research Award for Agriculture and Allied Sectors in 2005–2006.

Alka Gupta is currently Principal Scientist at ICAR-CPCRI, Kasaragod, India. She has 29 years of experience on biofertilizer technology, biomass recycling, plant growth-promoting rhizobacteria, and arbuscular mycorrhizal fungi associated with coconut, arecanut, and cocoa. She has 200 publications in national/international journals, reviews, and book chapters. She is a recipient of ICAR Award for Outstanding Interdisciplinary Team Research in Agriculture and Allied Sciences as a Team Member.

Murali Gopal is Principal Scientist (Ag. Microbiol.) and Head of the Division of Plant Physiology, Biochemistry and Post-Harvest Technology at ICAR-CPCRI, Kasaragod. He has 28 years of service at CPCRI, focusing on leveraging soil, plant, and food microbiome for achieving one-health goal. He published about 60 research articles in international/national journals. He is recipient of ICAR-Outstanding Team Award for the Biennium 2006–2007.

V. Selvamani is a Senior Scientist with a doctorate degree in soil science and agricultural chemistry. She has 16 years of experience in the field of soil and plant nutrient management of plantation crops. Her areas of specialization include remote sensing and GIS applications and precision farming in coconut.

Jeena Mathew is serving as Senior Scientist (Soil Science) at ICAR-Central Plantation Crops Research Institute, Regional Station, Kayamkulam, Kerala. Her research areas include soil health management, nutrient dynamics, customized nutrient mixtures, and amelioration strategies for surface and subsoil acidity in tropical soils. She has published 32 research papers and 8 book chapters. She developed and popularized the nutrient mixtures, 'Kalpa Poshak' and 'Kalpa Vardhini', for juvenile and adult coconut palms, respectively.

Surekha has a doctorate degree in agronomy and is serving as Scientist at ICAR-Central Plantation Crops Research Institute, Kasaragod, Kerala. She has 11 years research experience in coconut production technology, focusing on coconut-based cropping/farming systems, coastal sandy soil management, organic coconut cultivation, and water and nutrient management.

S. Indhuja is a Scientist at ICAR-Central Plantation Crops Research Institute, Regional Station, Kayamkulam, Kerala. Her areas of research in coconut include plant-microbe interaction, developing microbial inoculants for biotic stress management, and microbial valorization of crop residues.

Oil Palm (*Elaeis guineensis* Jacq)

K. Manorama, S. K. Behera, and K. Suresh

Abstract

Oil palm with its low land foot print for oil production is the most important global crop of present times. The two oils produced from palm tree are used in a variety of ways, viz. cooking, cosmetics, lubrication, sanitizing, and cleansing. Furthermore, the by-products of palm oil industry have numerous utilities such as feed, fuel, water purifier, etc. Besides all these apparent advantages, this crop also faces huge criticism in terms of GHG emissions, loss of biodiversity, and deforestation. To produce oil palm sustainably, Roundtable on Sustainable Palm Oil (RSPO), a voluntary global-level organization, set up the standards for certified palm oil production. These practices include right approach of planting, nutrient and water management, waste recycling, integrated pest management, etc. At present, globally, oil palm cultivation is concentrated in three continents, viz. humid tropical Asia, Africa, and Latin America in 46 countries. Though soil health is not much affected due to oil palm cultivation, improvement in soil condition can be achieved through sustainable management. In this chapter, we discuss suitable soil conditions for oil palm cultivation, how soil health is assessed through suitable indicators, constraints of soil fertility, nutrient management, use of soil amendments, the impact of climate change, soil biodiversity, and technological opportunities available for maintaining soil health and sustaining oil palm production.

K. Manorama (✉)
Agronomy, ICAR-Indian Institute of Oil Palm Research, Pedavegi, Andhra Pradesh, India
e-mail: Manorama.Kamireddy@icar.gov.in

S. K. Behera
Soil Science, ICAR-Indian Institute of Soil Science, Bhopal, Madhya Pradesh, India
e-mail: Sanjib.Behera@icar.gov.in

K. Suresh
Plant Physiology, ICAR-Indian Institute of Oil Palm Research,
Pedavegi, Andhra Pradesh, India
e-mail: Suresh.Kancherla@icar.gov.in

G. V. Thomas, V. Krishnakumar (eds.), *Soil Health Management for Plantation Crops*, https://doi.org/10.1007/978-981-97-0092-9_3

Keywords

Oil palm · Soil health · Fertilizers · Amendments · Mulching · Soil biodiversity · Best management practices

Abbreviations

%	Percent
<	Less than
>	More than
AMF	Arbuscular mycorrhizal fungi
ARDRA	Amplified Ribosomal DNA Restriction Analysis
AUDPC	Area under disease progression curve
BCA	Bio-control agent(s)
BMPs	Best management practices
BSR	Basal stem rot
BVOC	Biogenic volatile compound
C	Carbon
C/N	Carbon/Nitrogen
CARDI	Caribbean Agricultural Research and Development Institute
CEC	Cation exchange capacity
CFU g^{-1}	Colony-forming unit
cm	Centimetre(s)
CMP	Compost
CO_2	Carbon dioxide
CP	Conservation practice(s)
CPO	Crude palm oil
DAP	Di-ammonium-phosphate
DGGE	Denaturing Gradient Gel Electrophoresis
DNA-SIP	DNA-stable isotope probing
DRIS	Diagnosis and Recommendation Integrated System
dS m^{-1}	deciSiemens per metre
EB	Endophytic bacteria
EC	Electrical conductivity
ECO	Eco-mat(s)
EFB	Empty fruit bunches
EFBN	Nitrogen content of EFB
FFB	Fresh fruit bunches
FYM	Farm yard manure
g	Gram(s)
GHG	Greenhouse gases
GJ	Giga joules

ha	Hectare(s)
ICAR	Indian Council of Agricultural Research
IFS	Integrated farming system
IIOPR	Indian Institute of Oil Palm Research
ind. m^{-3}	Individuals per cubic metre
IOPA	Indonesian oil palm association
IOT	Internet of things
IPM	Integrated pest management
IsoMO	Isoprene monooxygenase
kg	Kilogram(s)
LCA	Life cycle analysis
LIG	Lignin
LIG/N	Lignin to N ratio
m	Metre(s)
Mha	Million hectares
MHB	Mycorrhiza helper bacteria
Min N	Mineral N
mm	Millimetre(s)
MPOB	Malaysia palm oil board
MRT	Mean resident time
Mt	Million metric tonnes
NDFA	Nitrogen-derived from atmosphere
OC	Organic carbon
OPF	Oil palm frond(s)
PCoA	Principal co-ordinates analysis
PGP	Plant growth promoters
PGPR	Plant growth-promoting rhizobacteria
PKO	Palm kernel oil
POME	Palm oil mill effluent
PP	Polyphenol
PSB	Phosphate-solubilizing bacteria
rRNA	Ribosomal RNA
RSPO	Roundtable on Sustainable Palm Oil
SHMP	Smallholder management practices
SIL	Silt pit(s)
SISKA	Sistem Integrasi Sapi-Kelapa Sawit or System Integration Cattle–Oil Palm Plantations
SOC	Soil organic carbon
SOM	Soil organic matter
t	Tonne(s)/ton(s)
T-RFLP	Terminal Restriction Fragment Length Polymorphism
USDA	United States Department of Agriculture
WRF	White root fungus

1 Introduction

To date, oil palm is the highest edible oil-producing crop (4–6 t of oil ha^{-1} $year^{-1}$), with an economic life span of 30 years. Though the humid tropics of West Africa are the centre of origin for oil palm, it has spread to even semi-arid tropics of Southeast Asia, and it is a successful crop in many of the East Asian countries, viz. Indonesia, Malaysia, Thailand, and India. It is grown extensively in Southeast Asian countries (Indonesia, Malaysia, and Papua New Guinea), African countries (Nigeria, Ivory Coast, Ghana, Liberia, Sierra Leone, Cameroon, Republic of Congo, and Zaire), and South American countries (Costa Rica, Panama, Columbia, British Guyana, Peru, Ecuador, Venezuela, and Brazil). Globally, it produces 81 million tonnes (Mt) of oil from about 19 million ha (Mha) (Murphy et al. 2021). In 2022, crude palm oil (CPO) production was 46.729 Mt from an area of 12.5 Mha in Indonesia (IPOA 2022; USDA 2022). Similarly, in Malaysia, 18.45 Mt of CPO were produced from 5.67 Mha with an average FFB yield of 15.49 t ha^{-1} during the same year (MPOB 2022). In contrast, the second and third largest vegetable oil crops, soybean and rapeseed, yield combined 84 Mt oil but occupied over 163 Mha of arable land. Oil palm is a perennial monoecious, monocot tree crop (of the family Arecaceae). In oil palm, fresh fruit bunches (FFBs) are harvested from the fourth year onwards after main field planting, and the yield stabilization takes place from the 12th year. The two oils produced from mesocarp and kernels of oil palm, viz. crude palm oil (CPO) and palm kernel oil (PKO), respectively, are in much demand for both consumption and industrial use purposes.

Soil acts as storehouse of nutrients. It is the medium to supply water and nutrients to plants. The capacity of soil to function, within ecosystem and land use boundaries, to sustain biological productivity, maintain environmental quality, and promote plant and animal health represents soil health (Doran and Parkin 1994). Nutrient imbalances, excessive fertilization, soil pollution, and soil loss processes affect soil health adversely. A soil, capable of supporting the production of food and fibre to a level, and with a quality, sufficient to meet human requirements and to continue to sustain those functions that are essential to maintain the quality of life for humans and the conservation of biodiversity, is considered as a healthy agricultural soil (Kibblewhite et al. 2008). Although the decline in soil fertility status is less in plantation crop areas (Hartemink 2005), due to intensive cultivation, widespread differences in chemical properties of soils are observed under plantations. Therefore, maintenance of soil quality and health in perennial crops like oil palm is a major challenge mainly due to intensive land use, higher water and nutrient requirements, availability of large inter-space thereby weed menace, and being grown mostly under tropical conditions with higher soil temperature. However, the generation of huge quantities of plantation bio-waste, sequestering large quantities of organic carbon, subjecting the land to minimal disturbance, etc. contribute to changes in physical properties and bio-geo-chemical characteristics. Crop management practices play an imperative role in soil health management of plantation crops. Understanding the physical, chemical, and biological factors controlling soil fertility and productivity is essential for effective exploitation of yield potential. This is possible from research findings and knowledge generated through the cultivation of crops for a longer period of duration. Demand for edible oil warrants the expansion of oil palm to newer and non-traditional areas. Variability in soil physical and chemical conditions further complicates the issue, as each limitation needs to be tackled differently under varying climatic situations. Better soil health management can be achieved by precisely understanding the crop's requirements for water, nutrients, and climatic factors and providing them appropriately. The increasing population demands higher productivity and thereby mounts pressure on limited natural resources, especially land and water. This leads to the irrational use of inputs that impact soil and environmental health negatively resulting in unsustainable production situations. In the process of growing resource-intensive crops like oil palm, different anthropogenic activities alter all the agricultural soils from their natural state due to huge applications of chemical fertilizers. Soil is a non-renewable natural resource. Therefore, it needs to be managed with utmost care. The risks of soil degradation (physical, chemical, biological, and ecological) can be minimized by improving soil quality (i.e. increasing soil organic carbon (SOC) pool, improving soil structure, enhancing soil fertility) (Lal 2015). Therefore, a balance needs to be struck between the balanced use of inputs and dwindling natural resources to meet the ever-increasing demands for food and nutrition. Better soil health management should aim at maintaining soil productivity without causing damage to its physical, chemical, and mineralogical properties while ensuring them at permissible levels. Therefore, efforts should be made to achieve this by minimizing the negative effects.

2 The Oil Palm

Vegetable oil demand at global level is expected to become double between 2010 (120 Mt) and 2050 (240 Mt) (Stichnothe and Bessou 2017). With just more than 7% of land under oil crops, oil palm meets 30% of the world's edible oil and fat demands. It is regarded as fastest-expanding agricultural crop (Vargas et al. 2015). Crude palm oil is extracted from the fleshy mesocarp of fruits (45–55% oil) and the palm kernel oil, a good source of lauric acid, from kernels of stony seeds. Because of its high oil yield potential, oil palm is believed capable of meeting a large part of future global demand for vegetable oils. Palm oil is nutritive, eco-friendly, environmentally sustainable, and source of bio-fuel and suits well for intercropping. Thus, oil palm is the crop for the future. But, quite often, forests are cut down, and peat lands are drained to provide land for new oil palm plantations in certain locations of Southeast Asia (especially in Indonesia and Malaysia). This irresponsible palm oil production can have severe global environmental and social ramifications as it influences biodiversity and ecological balances. If these issues are addressed by strong policies, palm oil is going to be the oil and fat resource to a great extent, ensuring vegetable oil security at global level.

Palm oil is an important source of vegetable oil and also used in a wide range of products. The theoretical yield potential is reported at 18 t of oil ha^{-1} year^{-1} in oil palm (Hoffmann et al. 2014), which is the highest ever in all the oil-producing crops. Because of longer life cycle, release of greenhouse gases (GHGs) is minimal in oil palm, as land preparation, ploughing, and cleaning are not required every year. Carbon sequestration potential is high, and it is in the range of 29.3 t ha^{-1} $year^{-1}$ of CO_2. In comparison with other oil-producing crops such as soybean, sunflower, and rapeseed, oil palm has the highest land productivity and very low land footprint. Though oil palm lives up to 200 years, economic life is considered to be up to 30 years (Verheye 2010).

E. guineensis Jacq. or African oil palm and *E. oleifera* or American oil palm are the two major species of genus *Elaeis* (Palmae-Cocoideae). American oil palm is generally used in producing hybrids only. There are three distinct fruit forms in African palm, viz. "*Dura*" (with large-sized nuts, thick endocarp, and thin mesocarp), "*Pisifera*" (fruits with naked kernels and no endocarp), and "*Tenera*" (fruits with small nuts, thin endocarp, and thick mesocarp), which is a hybrid between *Dura* and *Pisifera* forms.

Oil palm demands large amounts of nutrients as its dry matter production is comparable with the top producers among C_3 plants. Majority of the required nutrients for sustaining reasonable yields can be supplied only through external fertilizer application. Therefore, plant nutrient management plays a key role in profitable oil palm production. Over the past few decades, extensive work has been carried out on this aspect, and good progress has been made in understanding the same.

3 General Soil Characteristics and Requirements

Oil palm grows on a wide variety of tropical soils, and it comes up well in almost all types of soils. With respect to soils, this crop is not much demanding type. However, loamy or alluvial soils with good drainage are the best suited ones. In Malaysia, most of the oil palm is grown on upland areas where Ultisols and Oxisols occur. Ultisols and Oxisols in Malaysia are considered suitable for oil palm cultivation as oil palm is found to be acid tolerant, and hence, it can still grow even at the soil pH of 4.3 and at high Al concentration (Auxtero and Shamshuddin 1991). In Indonesia, oil palm is grown mostly on spodosols and inceptisols which are widespread in Kalimantan and Sumatra islands (Wigena et al. 2009). In southern Benin, Acrisols (dominant), Plintics and luvisols, Hydromorphic soil, and Vertisols are dominant soils of oil palm cultivation (Kindohounde et al. 2021). Oil palm grows in about 28 states in Nigeria. The states of Lagos, Ondo, Osun, Oyo, Ekiti, Edo, Delta, Rivers, Cross Rivers, Akwa Ibom, Imo, Abia, Anambra, Ebonyi, Abia, and Bayelsa are the states with the highest number of oil palm plantations in Nigeria. The soils of oil palm belt in Ghana include Temang (Lixisols), Akroso (Acrisol), Kokofu (Alisols), Basitia (Acrislos), Firam (Acrisols), and Nkwanta (Acrisols). These soils generally have a good physical structure, but have a low inherent fertility (IPNI 2015) due to high rainfall, strong acidity, low available cations, and low to very low available phosphorus status (Danso et al. 2019). In India, it is grown in different types of soils ranging from red sandy loam/sandy clay loam (Alfisols), alluvial silty loam (Inceptisols/Entisols), heavy black soils (Vertisols), and acidic red laterite soils

(Oxisols). However, the best suited soils are well-drained, deep loamy alluvial soils with good organic matter content and water-holding capacity. In Andhra Pradesh, India, the palm is grown in red sandy, red loamy, red loam with clay base, deltaic alluvium soils (pH 6.5–8.0); in Karnataka, it is grown in red and brown sandy loams and black clayey soils (pH 6.5–8.0). In Tamil Nadu, the palm is grown in river alluvium and red sandy soils, red soils, and red loamy and clay loam soils, while it is grown on laterite soils in Kerala (Behera et al. 2022). Oil palm grows well in soils with pH ranging between 4.3 and 6.5 (Tiemann et al. 2018). From different surveys in India, oil palm is reported to be grown in soil orders of Alfisols, Inceptisols, Entisols, and Vertisols. In these locations, the soil pH, electrical conductivity (EC), and organic carbon (OC) contents were estimated to range between 4.11 and 8.74, 0.01 and 2.54 dS m^{-1}, and 0.10 and 48.4 g kg^{-1}, respectively (Behera et al. 2022). Further, alkaline, highly saline, water-logged, and coastal sandy soils are reported as not suitable for oil palm cultivation.

Oil palm crop growth and FFB yield gets affected in acid sulphate soils with a pH below 4. Acid sulphate soils are unfavourable for oil palm cultivation because they contain significant amounts of (1) free and absorbed sulphate, (2) pale yellow mottles of jarosite along old root channels and on ped surfaces, and (3) low pH (in water), i.e. below 4. Out of these, Al toxicity and excess sulphates are the major constraints to FFB production in oil palm (Pupathy and Paramananthan 2014). Along the west coast of Malaysia, around 0.11 million ha of acid sulphate soils and in Indonesia about 102.80 million ha of acidic soils, are estimated to be present. With proper management, oil palm production with comparable FFB yields is possible even in these soils (Pupathy and Sundian 2020).

Most of the active roots in oil palm are found in the upper 30–40 cm of the soil. The root system is relatively shallow and adventitious type (Suresh et al. 2003; Tinker 1976). Majority of primary roots providing anchorage and support to the tree are located around 60 cm radius from the trunk (Fig. 1). The functional roots responsible for uptake of water and nutrients are located from 60 cm to 3 m radius around the palm trunk.

Soil physical, chemical properties and depth influence the growth and performance of the crop. Physical parameters include structure, texture, parent material, stoniness, and drainage and soil depth. Soil chemical properties that influence oil palm performance are pH, electrical conductivity, soil available macro- and micronutrients, cation exchange capacity (CEC), base saturation, and soil organic carbon. The physical, chemical, and mineralogical properties of tropical soils, which determine the yield levels and type of management in oil palm, depend on parent materials from which the soils are formed. Iron content, structure, and porosity are linked to soil colour; internal drainage is related to texture and structure of the soil. Moisture availability, rooting depth, and vulnerability to damage by winds depend upon effective soil depth (Paramananthan 2000).

Steep slopes affect oil palm growth in many ways. Problems like runoff, erosion, loss of nutrients, and insufficient soil depth causing problems in anchorage are the major constraints in steep slopes. Plain areas with 0–4° slope are suitable, whereas slopes up to 12° can be managed with circular platforms or half-moon terracing, and

Fig. 1 Root distribution in 10-year-old oil palm tree. (Adopted from Suresh et al. 2003. Int. J. Oil Palm 2(3–4): 39–41)

the slopes of 12–20° require bench terracing as a management intervention. However, slopes of >25° are unsuitable for oil palm cultivation. Additionally, elevation of the land affects the lapse rate of temperature. Due to low temperature at higher elevations, plant growth is slow and takes longer periods for maturity.

Due to its indeterminate habit of growth, oil palm requires evenly distributed rainfall of 150 mm $month^{-1}$ to meet evapotranspiration demand. Under Indian conditions, several months in a year have little or no rainfall, while a few months have abundant rainfall. Therefore, oil palm is promoted as an irrigated crop in India as the rainfall is much less than 2000 mm in majority of locations. The crop requires at least 5 h of bright sunlight day^{-1}, and it comes up well between 29 and 33 °C of mean maximum and 22–24 °C mean minimum temperature. High level of relative humidity, viz. >80%, is favourable for better growth and development of oil palm. Soil available nutrients and soil organic carbon contents were more in the plantations where proper crop residue management practices were followed besides resorting to improved water and nutrient management technologies (Behera et al. 2022; Manorama et al. 2021a).

4 Indicators of Soil Health and Soil Health Assessment

Indicators of soil quality can be categorized into four general groups: visual, physical, chemical, and biological (Yachi and Loreau 1999). Some examples of visual indicators which clearly indicate that the soil quality is threatened or changing are exposure of subsoil, change in soil colour, ephemeral gullies, ponding, runoff, plant response, weed species, blowing soil, and deposition. Physical indicators generally reflect limitations to root growth, seedling emergence, infiltration, or movement of water within the soil profile. Examples are topsoil depth, bulk density, porosity, aggregate stability, texture, crusting, and compaction. Chemical indicators include measurements of pH, salinity, organic matter, phosphorus concentrations, cation-exchange capacity, nutrient cycling, and concentrations of elements that may be potential contaminants (heavy metals, radioactive compounds, etc.) or those that are needed for plant growth and development. The soil's chemical condition affects soil-plant relations, water quality, buffering capacities, availability of nutrients and water to plants and other organisms, mobility of contaminants, and some physical conditions, such as the tendency for crust to form. Biological indicators include measurements of micro- and macro-organisms, their activity, or by-products. Earthworm, nematode, or termite populations have been suggested for use in some parts of the country. Respiration rate can be used to detect microbial activity, specifically microbial decomposition of organic matter in the soil. Ergosterol, a fungal by-product, has been used to measure the activity of organisms that play an important role in the formation and stability of soil aggregates. Measurement of decomposition rates of plant residue in bags or measurements of weed seed numbers or pathogen populations can also serve as biological indicators of soil quality. The main indicators of soil health are soil organic matter and the soil organisms along with their diversity. It has been well established that soil biodiversity promotes simultaneous multiple ecosystem functions as it is a habitat for various organisms that contribute to different functions in the soil (Bender et al. 2016; Kibblewhite et al. 2008; Singh and Ryan 2015). Soil microorganisms are also known to respond very quickly to the changes occurring in the soil environment which signifies their role in bioremediation and excellent candidates of bioindicators (Bhardwaj et al. 2014).

No single property can be used as an index of soil quality. Monitoring soil quality should be directed primarily towards the detection of trend changes that are measurable over a 1–10-year period. Establishing acceptable ranges, examining trends and rates of change over time, and including estimates of the variance associated with the measurements are important in interpreting indicators. Evaluations before and after, or with and without intervention, are also needed to develop appropriate and meaningful relationships for various kinds of soils and the functions that are expected of them.

5 Soil Fertility Constraints; Issues of Soil Contamination

In tropical regions, soil acidity and nutrient depletion are two major constraints in oil palm cultivation (Auler et al. 2017; Fageria and Baligar 2008). The cultivation of oil palm on sloppy lands/eroded lands in hilly tracts with high and intensive rainfall pose problem for its performance. Similarly, soils with salinity grade of >4 (EC in $dS\ m^{-1}$) are not suitable for its cultivation as it is highly sensitive to soil salinity.

Soils with high clay content as well as soils which are too sandy are considered as less suitable for oil palm cultivation. Deep soils with friable structure are more suitable for oil palm, and a minimum of 75 cm soil depth is more preferred. Plain land is preferable if available. In case of scarcity of plain land, sloppy lands also can be utilized for oil palm cultivation. The high rainfall in hill slopes causes soil erosion and runoff, loss of nutrients, and fertile top soil. In Malaysia many of the plantations are located in hill slopes also. The management practices need to be altered to suit to the lands on hill slopes. Depending upon the degree of slope, the method of planting needs modification. If slope is between 12% and 23% (6° and 12°), circular platforms or half-moon terracing is required to grow the crop, and at the slope of 23–38% (12–20°), crop can be grown only on bench terraces. If the slope exceeds 38% (20°), it is not recommended to grow oil palm (Paramananthan 2000). The land at lower elevations may be considered for planting first and then venture to higher elevations, if necessary.

Soil is a non-renewable resource that is vulnerable to degradation, depending on the factors and complexity of interactions between the processes throughout the years (Lal 2015). High rainfall leads to soil erosion which subsequently results in leaching of soil nutrients, organic matter and soil biota, loss of soil structure, and changes in infiltration rates (Lord and Clay 1999). Studies also showed that the soil degradation of oil palm in the inter-space was higher than in the row as there was a reduction of carbon (C) content which was more prominent in older oil palm plantations (Guillaume et al. 2016). Removal of nutrients by FFB and immobilization of nutrients in trunk and roots are several contributing factors in the soil degradation as oil palm continuously consumes nutrients from the soil for its growth and production. Continuous compaction on the soil reduces porosity and aeration, limiting water and nutrient availability to plants, and also increases the risk of resistance to root penetration (Yahya et al. 2010b). This may lead to poor crop yields and, over time, increases the pollution on land. General soil types of oil palm cultivation in Southeast Asia and their characteristics along with management to be followed for overcoming the constraints are presented in Tables 1 and 2.

In Malaysia, oil palm is largely cultivated in tropical Oxisols and Ultisols (Kurniawan et al. 2018; Li et al. 2019). Oil palm production in these soils get affected as they are highly acidic with high Al saturation, high P fixation capacity due to heavy rainfall which causes excessive loss of basic cations (Soumare et al. 2020). Liming is recommended to neutralize this soil acidity and improve the nutrient availability through changes in biogeochemical processes. After 48 months of application, surface liming raised the soil pH to a depth of 20 cm, and the effect has spread through the entire soil profile after 60 months of repeated lime application (Da Costa and Crusciol 2016). Given below (Table 3) are the types of soils in different oil palm cultivated locations in India with details of pH, EC, SOC, mean rainfall, and average FFB yield.

In most of the locations, soil condition is suitable for oil palm cultivation. However, in locations with low SOC content and higher pH and EC, proper recycling of plantation waste which is generated in the plantations itself could ameliorate the soils. In a field study in farmers' plantations of West Godavari district of

Table 1 General soil types of oil palm-cultivated areas in Southeast Asia, their characteristics, and management options

Pedon	Type of soil	Parent material	Topography	Drainage	Limitations	Advantage	Treatment
1	Terric Troposaprist	Marine clay	Flat	Poor	Poor anchorage; subsidence; K deficiency	Constant water availability	Hole in hole planting; water management
2	Typic Sulfaquept	Estuarine clay	Flat	Imperfect	Acidity; Stagnant water	Constant water availability	Keep water >20 cm below surface
3	Typic Hapludox	Granite	Steep (30%)	Well drained	P fixation; moisture stress; erosion; P and K deficiency	No stagnant water	P placement; EFB mulch, terraces
4	Xanthic Kandiudox	Reworked materials	Gentle rolling	Well drained	High clay content; low fertility	Good moisture retention	Mulch (EFB); fertilization
5	Typic Paleudult	Subrecent terrace	Undulating	Imperfect	Moisture stress; low fertility; K and Mg deficiency	Easy to work	Mulch (EFB); fertilization
6	Typic Hapludult	Colluvium	Undulating	Moderately well	Low fertility; P, K, and Mg deficiency	Sufficient moisture	Mulch (EFB); fertilization
7	Typic Kandiudult	Granite	Rolling	Well drained	Low fertility; N, P, K, and Mg deficiency	Easy to work	FFB mulching; fertilization
8	Typic Melanudand	Volcanic ash	Gentle rolling	Well drained	K deficiency	Excellent structure	Fertilization of K

Adopted from Mutert (1999). Better Crops International 13(1): 36–38

Andhra Pradesh, India, mulching with the plantation waste and adoption of improved water and nutrient management practices could improve the SOC contents in top 0–30 cm soil of sole as well as intercropped (with cocoa) stands of oil palm to the tune of 40% and 56%, within a period of 6 years (Manorama et al. 2021a).

In Southeastern Nigeria, soils of oil palm cultivation are categorized as sedimentary sandstone soils (imperfectly drained) and basement complex soils (well drained). The soils derived from sedimentary sandstone had higher effective cation exchange capacity (ECEC) (5.66–17.13 cmol kg^{-1}) than the basement complex soils (4.73–11.88 cmol kg^{-1}). The exchangeable acidity values of all the soils fall within the range of 0.40–3.36 cmol kg^{-1}. In general, the soils have low nutrient status, and this decreased with soil depth. For sustained productivity of the oil palm, proper soil management aimed at increasing the nutrient status of these soils is advocated (Iren and Ibanga 2008). In Benin, about 60% of population live in the southern part of the country (INSAE 2013), which lies in the oil-palm agroforestry belt of humid tropical zone of West Africa. The major soil types of Klouekanme and

Table 2 Characteristics of topsoil (0–30 cm) in common soil types of oil palm cultivated areas in Southeast Asia

				N	P Bray	Exchangeable						
			Org C	total	II	Ca	Mg	K	Al	Clay	Silt	Sand
Pedon	Type of soil	pH	(%)	(%)	(g kg^{-1})	(cmol(+) kg^{-1})				(%)		
1	Terric Troposaprist	3.8	24.5	1.1	35	0.85	1.56	0.24	9.50	55	32	13
2	Typic Sulfaquept	4.1	2.5	0.2	18	0.18	0.20	0.32	12.50	72	21	7
3	Typic Hapludox	4.4	1.1	0.1	6	0.28	0.25	0.16	0.60	37	9	54
4	Xanthic Kandiudox	4.3	1.8	0.2	15	0.86	0.48	0.24	3.20	63	5	32
5	Typic Paleudult	4.4	1.2	0.1	12	0.16	0.03	0.09	1.40	18	8	76
6	Typic Hapludult	4.1	1.4	0.1	8	0.76	0.18	0.15	1.80	20	19	61
7	Typic Kandiudult	4.9	0.8	0.1	5	0.19	0.10	0.05	0.80	33	7	60
8	Typic Melanudand[a]	4.8	6.4	0.5	8	1.86	0.25	0.07	0.80	18	53	29

Adopted from Mutert (1999). Better Crops International 13(1): 36–38
[a] 12% allophane was present in the less than 2 mm soil mineral fraction

Table 3 Characteristics of oil palm cultivated soils in different states of India

Location	Soil order	Soil pH	Soil EC (dS m^{-1})	Soil OC content (g kg^{-1})	Average annual precipitation (mm)	Age of plantations (years)	FFB yield (t ha^{-1} $year^{-1}$)
West Godavari (Andhra Pradesh)	Alfisols	5.44–8.26	0.05–2.15	0.80–31.2	1094	12–18	7–32
North Goa, South Goa (Goa)	Alfisols	4.25–6.77	0.05–1.06	5.07–48.4	3005	5–21	1–22
Mysore, Hassan (Karnataka)	Alfisols Inceptisols	4.91–8.74	0.10–2.54	1.17–28.9	1126	4–18	3–53
Aizwal, Kolasib, Mamit (Mizoram)	Inceptisols Alfisols	4.41–5.41	0.03–0.48	3.12–35.9	1881	5–7	1–16
Surat (Gujarat)	Vertisols	6.38–8.28	0.26–1.16	6.24–16.8	1107	4–5	1–19
Thanjavur (Tamil Nadu)	Inceptisols Vertisols Entisols	4.76–7.97	0.01–0.31	0.10–7.80	940	4–19	4–26

Adopted from Behera et al. (2022). Nutritional imbalances and nutrient management in oil palm. In: Roy SS, Kashyap P, Adak T (eds) Natural Resource Management in Horticultural Crops. pp. 161–185

Toviklin districts in south-western Benin are Ferralsols (FAO 1998) or Eustrustox (USDA soil taxonomy), with good permeability and fairly fertile top layers (0–20 cm). On an average, they contain 5.3 cmol kg^{-1} CEC, 39 g kg^{-1} total C, 670 mg kg^{-1} N, and 7.3 mg kg^{-1} P with 100% base saturation (Igue et al. 2000).

In Latin America, in recent times, oil palm area has been expanded mostly on non-forested areas only, following a distinct trajectory of converting pastures. In eastern Amazon, Brazil, oil palm growing in agroforestry systems in combination with conservation practices could maintain adequate soil fertility for the crops cultivated in these systems (de Jesus Costa et al. 2023).

Nutrient disorders or deficiencies play a major role in achieving higher productivity in oil palm-growing countries, including India. However, by proper nutrient management, it is possible to overcome these problems. If one can correctly identify the deficiency/disorder, it can be rectified by the application of that particular nutrient either by foliar spray or soil application. Nutrient management through fertilizer application was reported as the most important contributor of FFB yield increment in oil palm, accounting for 26% (Prasad et al. 2012). From a large number of trials, it is deduced that large quantities of potassium (K), nitrogen (N), phosphorus (P), magnesium (Mg), and boron (B) are required for oil palm. In peat soils, fertilization with zinc (Zn) and copper (Cu) is essential, and specific deficiencies will lead to visible foliar symptoms (Broeshart et al. 1957). Each nutrient has a specific function and plays a vital role in plant metabolism. Its deficiency affects that particular process, which in turn is expressed as symptoms of deficiency. The symptoms manifested under deficiency of each nutrient are explained in Table 4 and Fig. 2.

Table 4 Symptoms of nutrient deficiencies/disorders in oil palm and their management

Nutrient element	Deficiency symptoms in nursery	Deficiency symptoms in main field	Management options	Remarks
Nitrogen	In nursery, the seedlings first become dull and turn to pale green. Under acute conditions, the seedling colour changes to lemon yellow, with reduced growth	In main field, chlorosis appears first on older fronds, later spreading to younger tissues in older palms. N deficiency symptoms are observed if the N content of frond 17 is below optimal level which depends on the location. Severe deficiency occurs if it contains N in deficient level. The tissues of the midrib become bright yellow or orange, which contrasts with the paler chlorosis of the laminar tissues. The swelling at the pinna base (apocone) also becomes yellow or orange in colour as N deficiency is pronounced. Unlike Mg deficiency, both ranks of pinnae become equally chlorotic, and there is no shading effect of the upper rank pinnae on the lower one. In the later stages of acute deficiency, dark-coloured discolouration begins to appear normally towards the pinnae and dieback develops from the pinnae tip and margin	In nursery, spraying of 2% urea solution at weekly or fortnightly intervals is recommended till deficiency symptoms disappear. Maintaining weed-free environment around the seedlings and improving drainage by making drains helps in preventing occurrence of N deficiency. In main field, application of N fertilizer @ 1200 g N adult^{-1} palm year^{-1} (equivalent to 3 kg of urea or 6 kg of ammonium sulphate or 2 kg of urea plus 1.5 kg of DAP adult^{-1} palm year^{-1}) in 3–4 splits is recommended as soil application. Under fertigation, 600 g of N fertilizer (equivalent to 652 g DAP +1065 g urea palm^{-1} year^{-1}) is recommended in 12 equal splits. Utilizing the information on optimal nutrient concentration ranges in fronds, through soil and leaf analysis, location specific nutrient requirements can be arrived and would help in precise application of nitrogen. Addition of organic manures like farm yard manure and neem cake also helps in overcoming the problem	Seen in lateritic soils which are poor in fertility and acidic in reaction. Conditions leading to denitrification, such as water logging, prolonged inundation, poor aeration etc., lead to N deficiency. Weed competition, insufficient application of N fertilizers, low pH inhibiting N mineralization, high salt content, and unfavourable soil temperature and water content for soil microbial activity are the other causes for N deficiency

Phosphorus	In small seedlings, the oldest leaves become dull and assume the olive green colour. Then the chlorotic condition increases in severity	Under field conditions, reduced frond length is the symptom of P deficiency. Stunted growth, with short dark fronds, appears in palms, and trunks of affected palms become narrow and tapered	Application of 600 g of P_2O_5 adult^{-1} palm year^{-1} (equivalent to 4 kg of SSP or 1.5 kg of DAP adult^{-1} palm year^{-1}) in 3–4 splits is recommended to meet the requirement of oil palm. In fertigation, 300 g of P (equivalent to 652 g of DAP is recommended in 12 equal splits). Precise requirement of P can be arrived through soil and leaf analysis utilizing leaf nutrient norms of particular location	Phosphorus deficiency occurs in acidic and P-deficient soils

(continued)

Table 4 (continued)

Nutrient element	Deficiency symptoms in nursery	Deficiency symptoms in main field	Management options	Remarks
Potassium		The most common symptoms include confluent orange spotting, mid-crown yellowing, orange blotch, and premature desiccation. The development of K deficiency symptoms occurs in the older fronds Symptoms first appear as pale green rectangular spots bounded by the venison, the pinnae of the older leaves. As the age increases, the outline of the spots becomes irregular, and colour changes through olive green to bright orange with coalescence between adjacent spots until a rectangular mass develops in the centre of older spots. Ultimately, all pinnae become chlorotic, which contain a mass of minute spots. The terminal results of K deficiency symptoms mentioned above, viz. confluent orange spotting, mid-crown yellowing, and orange blotch end in marginal or terminal dieback of the pinnae. The tissue becomes grey in colour, brittle, and desiccated. Finally, fragmentation occurs	To control K deficiency, application of 1200 g of K adult^{-1} palm year^{-1} (equivalent to 2 kg of MOP adult^{-1} palm year^{-1}) in 3–4 splits is recommended. For fertigation, 600 g of K palm^{-1} year^{-1} (equivalent to 1 kg MOP in 12 equal splits) followed by leaf and soil analysis for precise nutrient management is recommended	Seen in soils with low K status, peat soils, and soils with poor moisture retention capacity

Magnesium		Symptoms are seen on older leaves when the leaf Mg falls below critical level in frond 17 and also when the ratios of K:Mg and N:Mg exceed about 5 and 10, respectively. Olive green- and ochre-coloured areas appear on the pinnae of older leaves and merge gradually with the healthy green tissue. The yellow colour spreads down towards the frond midrib until the whole pinnae are affected, becoming a deep orange colour in later stages. "Shading effect" is the most apt diagnostic features of Mg deficiency, in which there is an absence of chlorosis in parts of affected pinnae, protected from direct sunlight. As the deficiency becomes more intense, the purplish brown lesions expand rapidly until the entire pinna is affected with distal and marginal tissues being the most affected	Application of 500 g magnesium sulphate adult^{-1} palm year^{-1} or spraying of 2% solution of Epsom salt 3–4 times at an interval of 2–3 weeks is recommended to overcome Mg deficiency. Under fertigation, the dose becomes 250 g of $MgSO_4$ in 12 equal splits	Mg deficiency appears when palms are grown on soils of granite origin, peat soils, and acid–sulphate soils. Heavy application of potash fertilizers also causes Mg deficiency
Zinc		Small, narrow white streaks on lower and mid-crown fronds occur. It may also produce blotchy leaf symptoms	Two to three applications of zinc sulphate (1 g Zn L^{-1}) as a foliar spray and balanced application of P fertilizer are recommended	Occurs in highly weathered acid and calcareous soils. Heavy application of phosphatic fertilizer also induces Zn deficiency

(continued)

Table 4 (continued)

Nutrient element	Deficiency symptoms in nursery	Deficiency symptoms in main field	Management options	Remarks
Boron		Hook leaf, fishbone leaf, rounded frond tip, blind leaf, leaflet shattering, and bristle tip are attributed to B deficiency	Boron @ 20 g $palm^{-1}$ $year^{-1}$ is recommended	Occurs due to high rainfall conditions, sandy and peat soils, low and high pH soils, and heavy applications of N and K fertilizers. Also, in general, when the level of boron falls below 10 ppm of dry matter, symptoms of incipient deficiency are observed
Copper	Copper-deficient plants exhibit severe stunting in the nursery	Initially appears as whitish yellow mottling of younger fronds. As the deficiency intensifies, yellow, mottled, inter-veinal stripes appear, and rusty, brown spots develop on the distal end of leaflets. Affected fronds and leaflets are stunted and leaflets dry up	Basal application of 50 g copper sulphate to affected palms or application of potash fertilizer eliminates the deficiency	Appears in peat soils (where Cu is complexed by organic compounds) and coarse-textured ferralitic and ferruginous soils poor in copper and calcareous soils. Heavy application of N and P fertilizers also induces Cu deficiency
Iron		Interveinal chlorosis appears on the youngest fronds, but leaf veins remain green. The youngest fronds later turn white, but older fronds remain yellow. Growth ceases, and death may occur after 1 year in severely Fe-deficient palms	It is recommended to provide two to three foliar applications of ferrous sulphate (0.5%) at weekly intervals to rectify the problem	Occurs in soils of very high pH, on or near termite mounds, overlying coral and on poorly drained soils. Heavy application of P fertilizers results in Fe deficiency

Manganese		Initially, discontinuous longitudinal interveinal chlorotic streaks appear on younger fronds. These l streaks eventually become chlorotic with a striped appearance. Canopy appears retarded as newly emerging fronds become progressively smaller and chlorotic. In severe cases, chlorosis and necrosis affect the newly emerging spear before frond pinnae have expanded. Here, the symptoms are found on young fronds. The symptoms are equally pronounced on upper (sun exposed) and lower (shaded) rank pinnae	Foliar spray of 50 g Mn L^{-1} is recommended to overcome Mn deficiency	Appears in highly leached tropical soils, deep peat soils, or where large amounts of limestone have been applied to sandy soils. Soil with high exchangeable Mg status also induces Mn deficiency
N/K imbalance (white stripe)		Narrow bands of chlorotic tissue appear in the pinnae. The white stripes are apparent in young fronds, as the spear opens, extending the entire length of pinnae. In the early stages, the chlorosis produces a stripe delineated from the adjacent green tissue. The stripe may be at any portion between the midrib and the edges of lamina. In severe cases, there is a change in the morphology of the crown, i.e. upright and ob-conical in shape	Application of 2–3 kg of potassium chloride $palm^{-1}$ $year^{-1}$ and temporary cessation of N fertilizer application are recommended	When N:K ratio in the leaves is more than 2.5, white stripes appear. B deficiency and excessive N application also induce development of white stripe

Fig. 2 Nutrient deficiencies/disorders in oil palm. (**a**) N deficiency. (**b**) P deficiency. (**c**) K deficiency. (**d**) Mg deficiency. (**e**) Mn deficiency. (**f**) Cu deficiency. (**g**) Fe deficiency. (**h**) B deficiency. (**i**) N:K imbalance. (Adopted from Manorama et al. 2021, ICAR-IIOPR technical bulletin on Best Management Practices for growing irrigated oil palm in India. ISBN No: 81-87561-61-0, pp. 1–28)

6 Impact of Climate Change and Plantation Soils

At present, 95% of oil palm plantations are located around 10^0 of N and S of equator. A study by Paterson et al. (2017) on climate change prediction impacts revealed that the areas now categorized as suitable for oil palm cultivation would turn unsuitable in most parts of the world. Using Climex model, projections were made for its suitability in 2050 and 2100. In this model, a parameter "Ecoclimatic index" is estimated which is the product of annual growth index and stress indices. These estimates indicate that in 2050, the area under unsuitable climate category increases by 6% at global level and that under highly suitable climate category decreases by 22%. By 2100 most of the highly suitable area becomes unsuitable in Americas. Continent-wise data indicates that in Latin America, Brazil and Columbia are projected to be the badly affected nations. In Africa, by 2100, Nigeria is expected to get badly affected but not Madagascar. Southeast Asia where the present giants of oil palm cultivation, i.e. Indonesia and Malaysia, are located may become totally unsuitable.

This is mainly because of stress factors, viz. heat, dry, and cold stresses. By 2050, the stress factors are very minimal in all the oil palm-growing areas, and India has none. But by 2100, much of India is likely to become heat stressed. But a few studies indicate that climate change has already occurred (Paterson et al. 2017). A recent study to evaluate the impact of climate change (on ten global crops, viz. barley, cassava, maize, oil palm, rapeseed, rice, sorghum, soybean, sugarcane, and wheat) for the period of 1974–2013 reported a change in yield levels ranging between −13.4% in oil palm and +3.5% in soybean. At global level, the change is quantified as −2551 (oil palm) to +982 (sugarcane) kg ha^{-1} $year^{-1}$. Among the top three global cereals, recent yields have decreased for rice (−0.3%) and wheat (−0.9%) and increased negligibly for maize (Ray et al. 2019). On positive side, a study on impact of CO_2 concentration on seedling growth indicated that the nursery period got reduced by 27 weeks instead of 52 weeks (Jaafar and Hafiz 2012).

The effect of climate change on oil palm plantations could be either beneficial or negative in different agro-climatic regions. Rising atmospheric CO_2 concentration, changing patterns of precipitation, higher temperature, and altered frequencies of extreme events significantly affect oil palm production with associated consequences for water resources and pest/disease distributions. As temperature increases up to 20 °C, photosynthetic activity also increases. Then it plateaus until 35 °C, then it begins to decline, and at 40 °C, photosynthetic activity completely ceases. Higher temperatures can cause abortion of bunches and bunch failure by affecting photosynthesis, pollination, and anthesis and production of female inflorescences. High temperature can also cause dehydration in oil palm. The leaflets get folded to minimize exposure to the sun and thereby resulting in reduced photosynthesis. The stomata on abaxial side of leaflets are closed to lessen the transpiration load; hence CO_2 uptake is reduced and eventually the photosynthesis. Under irrigated conditions, diurnal variations explain the sensitivity of stomata leading to decreased photosynthetic rates after 10.00 AM. Increased ambient and leaf temperatures may also cause this. In summer season, early (1–2 weeks) ripening of bunches in the same whorl (pseudo-ripening) may occur due to higher temperature (>38 °C), thus drastically reducing oil extraction rates (<8%). In oil palm, the consequences of any kind of stress (temperature/water deficits) can be visualized only after 1.5–2 years in terms of reduced bunch yield. In oil palm being perennial and indeterminate in growth, there is limited scope for scheduling crop calendar (Suresh et al. 2017). Oil palm plantations, like any vegetation, assimilate CO_2 from the atmosphere, acting as a carbon sink. Moreover, oil palm plantations assimilate more CO_2 and produce more biomass ha^{-1} each year than forests due to very high fruit production (Kotowska et al. 2015; Lamade and Boillet 2005). On mineral soils, Fowler et al. (2011) and Ishizuka et al. (2005) found only small fluxes of methane both in forest and oil palm plantations.

6.1 Climate Smart Strategies

The results of climate change prediction models indicate a shift in areas suitable for oil palm cultivation from the current existing locations towards poles. It is expected that tropical areas would become hotter with severe heat stress making them unsuitable to grow oil palm. Suitable areas are likely to move to temperate regions and in hills to upper elevations. Therefore, it is wiser to find out the most suitable locations for growing the crop from time to time through multicriteria overlay techniques and shift the crop accordingly with a well-planned strategy. Further, mitigation strategies need to be strengthened to continue the crop in present suitable areas which would become less suitable with climate change. Efforts need to be focused to reduce the ill effects of climate change with concerted co-operation from all the corners, with which it is possible to minimize climate change impacts for a considerable time lag.

To reduce the impacts of flooding, erosion, drought, heavy rain, and winds, management practices like minimum tillage, maintenance of permanent soil cover, etc. are to be promoted. Conservation agriculture, organic agriculture, and risk-coping production systems are to be explored. Intensive soil tillage reduces soil organic matter content (through aerobic mineralization), while low tillage and permanent soil cover increases it. Conservation agriculture and organic agriculture have the ability to increase soil organic carbon, reduce the use of mineral fertilizers, and reduce on-farm energy costs and hence are promising adaptation options in oil palm. For adaptation to climate change, higher resilience is required in oil palm cropping systems against both excess of water (due to high intensity rainfall) and lack of water (due to extended drought periods). Soil organic matter is the key element to respond to both problems. It improves and stabilizes the soil structure, because of which soils can absorb higher amounts of water without causing surface run off, which otherwise leads to soil erosion and, further downstream, to flooding. Soil organic matter also improves the water holding capacity of the soil during extended drought.

Low tillage in combination with AMF potentially ameliorate the effect of climate change on oil palm growth through nutrient uptake, water relations, and protecting against pathogens and toxic stress. Also, low tillage helps to decrease the GHG emissions from oil palm plantations (Paterson and Lima 2018). Among different wastes produced in oil palm plantations and mills, cut fronds contribute major portion (58.8% of solid wastes) (Dalton et al. 2017). From the results of field studies at MPOB (2014), systematic piling of fronds in between palms could recover 7.5 kg of N into the soil, with 106 kg of P, 9.81 kg of K, and 2.79 kg of Mg which also benefit soil enrichment. It reduces dependence on inorganic fertilizer, minimizes fertilizer costs, and improves environmental conservation. Composting of oil palm fronds is a better option for recovering the nutrients and to reduce the environmental impacts of other recycling methods. For composting of oil palm fronds, white rot fungus (WRF) is a good option, as it produces various types of enzymes such as lignin peroxidase, manganese peroxidase, and laccase for the breakdown of lignin, cellulose, and hemicellulose. Apart from WRF, sewage sludge is considered to have the potential inoculum for organic waste composting. Co-composting of lignocellulosic biomass with sewage sludge produces high-quality compost, which can be used as a natural fertilizer for crops. Also, the use of thickened palm oil mill effluent (POME) anaerobic sludge as microbial seed may reduce composting time while increasing compost production (Sabiani and Awang 2022).

For each t of CPO extracted, about 1 t of empty fruit bunches (EFB) is generated (Caliman et al. 2001). It is estimated that each t of EFB is comparable to 6–7 kg of urea, 1.7 kg of triple super phosphate or 2.8 kg of rock phosphate,

16.3 kg of KCl or 19.3 kg of MOP, and 3–4.4 kg of kieserite (Caliman et al. 2001; Singh et al. 1999). Feeding activity of soil fauna is enhanced by EFB applications and associated with increased concentrations of base cations and soil moisture. This improved biological activity assists in ecosystem functions such as litter decomposition, nutrient cycling, organic carbon stabilization, and ultimately oil palm productivity. The use and presence of earthworms may increase the effectiveness of growing oil palm, as they can contribute to soil turnover and structure formation and serve as a fertility enhancer and, again, increase biodiversity (Paterson and Lima 2018). It is suggested that use of AMF could ameliorate the effects of climate change on oil palm. AMF addition will increase biodiversity within plantations (Paterson and Lima 2018). From a 10-year study, it was found that EFB application also improved soil organic carbon, total N, CEC, and available P (Comte et al. 2013). Through EFB application, it is possible to have better soil permeability (Caliman et al. 2001; Carron et al. 2015b), higher water retention at field capacity, better soil aggregate stability, and hence lower erodibility (Zaharah and Lim 2000).

In order to address the impact of climate change, the possible regional impacts of climate change on oil palm needs to be investigated. The impact of increasing temperature and precipitation patterns is projected to be different over the various agro-ecological regions. Hence the effect of climate change will be different in different oil palm-growing regions. Crop-based and agro-ecological region-based adaptation strategies are to be developed based on vulnerability of individual crop and the region by integrating all the measures available for sustainable productivity.

7 Fertility Management for Healthy Soils: Fertilizer Recommendations

Understanding fertility of the soil is a complex process as several factors influence the system. One among them is the nutrient component of the soil. Nutrients influence soil health and biotic component of the soil. Recycling of N and P from and within the soil system affects the nutrient dynamics. According to Kibblewhite et al. (2008), soil health is described to be influenced by fertilizer application. The limited supply of nutrients may cause severe damage to crop health. Rapid degradation of soil in the form of nutrient and organic matter depletion can be avoided by judicious fertilizer application (Kibblewhite et al. 2008; Singh and Ryan 2015).

Plants become resistant to drought, cold, and insect, pest, and disease attack through balanced use of fertilizers. At the same time, efficient water management also contributes greatly for fertility management of soil as the applied nutrients become available only when there is sufficient moisture in the root zone. Conversely, excess water in the root zone depletes applied nutrients to deeper layers of the soil and makes them unavailable to the plant. So, a balance needs to be maintained with optimal moisture levels in the soil for proper management of nutrients.

7.1 Nutrient Management in the Nursery

Oil palm is normally propagated by seedlings which are raised in poly bags for 12 months in a double-stage nursery. The growth and productivity of oil palm are mainly dependent on the quality of seedlings. The growth and vigour of oil palm seedlings are mainly decided by nutrient management as it is a heavy feeder and requires huge amounts of nutrients. The best potting mixture for growing the seedlings should be a medium containing soil, sand, and organic manure in 1:1:1 ratio. Diammonium phosphate (DAP) and neem cake powder @ 100 g each should be mixed with the potting mixture before filling the poly bags meant for main nursery. Oil palm factory wastes like POME sludge, decanter cake/muck, fruit fibre, nut shells, and empty fruit bunches are cost-effective and easily available in huge quantity as compared to regular organic manures. POME sludge @ 10% in a potting mixture can be an optimum dose for raising good quality oil palm seedlings. During the first few weeks of growth, the seedling is entirely dependent on the supply of stored food material in endosperm. The amount of fertilizers required for the growth of seedlings depends on the composition of potting mixture and variety. When the mixture contains soil, sand, and manure in equal proportion, fertilizer dose of 10 g N, 5 g P, and 10 g K seedling^{-1} applied in equal splits at an interval of 3 months is ideal under rainfed conditions. Application of N, P, and K @ 30, 38, and 25 g, respectively, at monthly intervals has been found optimum for better growth of seedlings. Manual application of fertilizer is laborious and expensive. Chemical fertilizers like DAP and NPK (17:17:17) complex can be applied @ 50% of the recommended dose of fertilizers/seedling through drip system of irrigation as they showed promising results with respect to important characters like seedling height, leaf and root production, stem girth, and total dry biomass production (Ramachandrudu et al. 2018).

7.2 Nutrient Management in the Plantations

Supplying adequate nutrients in balanced proportions for ensuring efficient nutrient uptake, healthy growth, and profitable FFB yields while minimizing the impacts of over-fertilization on soil and environment are the prime objectives of nutrient management in oil palm. This also includes correct timing, placement, and method of fertilizer application and right source of fertilizers. The fertilizer recommendations require a good understanding of general principles governing mineral nutrition of oil palm and methods to maximize fertilizer use efficiency. To achieve reasonable productivity levels and for maintaining good soil condition by supplementing the natural supply of soil nutrients, building up soil fertility, and compensating nutrients removed by harvested products or lost to the environment, oil palm cultivation requires significant amount of fertilizers. The nutrients removed by oil palms producing 25 t FFB ha^{-1} year^{-1} are estimated as 192, 11, 209, 36, and 71 kg ha^{-1} year^{-1} of N, P, K, Mg, and Ca, respectively (Goh and Hardter 2003). For achieving targeted yields with healthy growth of palms and to meet the growth and yield demands, each essential nutrient needs to be provided in adequate quantities (Corley and Tinker 2016).

For satisfying the growing food demands with minimal environmental impact, augmenting nutrient use efficiency (NUE) (Lobell 2007) is a necessity. However, right source and right rate of fertilizers and application at right time and right place make it possible.

Straight fertilizers are suitable to apply exact quantities of required nutrients, while mixtures are suitable where labour availability is a problem. For acidic soils, rock phosphate is better source of P, and for light-textured soils with high rainfall, polymer-coated fertilizers are preferred as they minimize the nutrient loss through runoff, leaching, and denitrification. Selection of fertilizer depends on many factors, viz. market availability which again depends on local policies, soil condition, preference of the growers, i.e. straight or blended fertilizers, etc.

For applying exact quantities of different fertilizers, nutrient budgeting needs to be done. In oil palm, classic investigations were carried out during the 1960s by earlier workers, viz. Ng et al. (1968), Ng and Thamboo (1967) in Southeast Asia, and Tinker and Smilde (1963) in Africa. These findings need to be confirmed with field experiments before generating advisories. The field trials conducted in Malaysia on diverse soil types, environments (Foster et al. 1985; Goh 2011), and unfertilized conditions (Tarmizi et al. 1992) were used for this purpose. Different methods are available for estimating right quantities of fertilizers in oil palm, viz. Palm Oil Research Institute Malaysia-Oil Palm Efficient Nutrient System (PORIM-OPENS), French system, Foster system, Integrated Site-specific Fertilizer Recommendation System (INFERS), and Diagnosis and Recommendation Integrated System (DRIS). These systems are based on different approaches (soil analysis, leaf analysis, and both combined), and out of them, the complimentary approach based on leaf analysis has been reported to give a more reliable indication of nutrient responses. The French Agricultural Research Centre for International Development (CIRAD) through yield response curves and critical leaf nutrient levels determined economic rates of fertilizers utilizing huge data sets of 100 field experiments from Indonesia, Latin America, and West Africa initiated in the 1940s (Caliman et al. 1994). Diagnostic recommendation integrated system (DRIS) norms which are estimated based on nutrient ratios in leaves (Beaufils 1973) can be used to determine the ranges of nutrients. Location-specific DRIS norms and critical leaf nutrient concentrations are effective in identifying nutrient imbalances. But these norms are available only for certain locations of Colombia, India, and Brazil. INFERS used a complex nutrient balance approach considering the nutrient supply and demand to achieve predicted site yield potential. However, Corley and Tinker (2016) opined that though these estimates involve complexity, good agreement with actual rates of nutrient application was found especially for K followed by N, for achieving predicted yields. In addition to these quantitative approaches, balance sheet approach based on nutrient balance concept is a simple one and useful for the locations where no other information is available. Given below are the general rates of nutrient applications in different countries for oil palm plantations (Table 5).

Recommended nutrient schedules for oil palm grown under rain fed conditions of India are given in Table 6.

From the information generated by conducting field trials on oil palm nutrition in diverse agro-climatic locations of India, viz. Vijayarai (Andhra Pradesh), Gangavathi (Karnataka), Mulde (Maharashtra), and Aduthurai (Tamil Nadu), the optimum dose of fertilizer differs from location to location. In Tungabhadra

Table 5 General fertilizer application rates for oil palm in different countries

		N		P		K		Mg		B	
Country	Source	Amount (kg year^{-1})	Type	Amount (kg year^{-1})	Type	Amount (kg year^{-1})	Type	Amount (kg year^{-1})	Type	Amount (kg year^{-1})	Type
Brazil, Vertisol	Lauzeral (1980), Pacheco et al. (1985)	0.7–1.5	AS			1.2	KCl				
Brazil, Latosol	Lauzeral (1980)	0.5	Urea	1.5	TSP	1.5	KCl	0.75	Kieserite	0.075	Borax
Cameroon	Kouame et al. (2017)					1–2.5	KCl	0.5	Kieserite		
Colombia	Lauzeral (1980), Ollagnier et al. (1970)	0.25	Urea			1.25	KCl	0.65	Kieserite	0.075	Borax
Congo	Dubos et al. (1999)	0.5	AS	0.75–1	TSP	0.7–1	KCl				
Equador	Lauzeral (1980)	1.5	Urea					0.75	Kieserite	0.075	Borax
Ghana	Danyo (2013)	2	AS	1 kg/3 y	Granular RP	1.5	KCl	0.5	Kieserite		
India	Rao et al. (2014)	2.6	Urea	3.75	SSP	2	KCl	0.5	Kieserite	0.10	Borax
Indonesia	Ng (1986), Tampubolon et al. (1990)			2	RP	1.5–2.5	KCl				
Ivory coast	Kouame et al. (2017)					1–2	KCl				
Malaysia	Ng (1986)			1.5	RP	3	KCl	0.5	Kieserite		
Nigeria	Imogie et al. (2012), Ollagnier et al. (1987)					2–2.3	KCl				
Peru	Daniel and Ochs (1975)					0.5	KCl				

AS ammonium sulphate, *TSP* triple super phosphate, *SSP* single super phosphate, *RP* rock phosphate

Table 6 Recommended doses of nutrients for oil palm (g palm^{-1} year^{-1}) for soil application in India

Age of the palm (years)	N	P_2O_5	K_2O	$MgSO_4$
1	400	200	400	125
2	800	400	800	250
3 and above	1200	600	1200	500

Adopted from Rao et al. (2014). Nutrient management in oil palm. Technical Bulletin, Indian Institute of Oil Palm Research, Pedavegi. 24 p

command area of Karnataka, 1200–600–1200 g of N–P_2O_5–K_2O palm^{-1} year^{-1} was found to produce FFB yield of 12–15 t ha^{-1} (Reddy et al. 2009), whereas, under sandy loam soils of Andhra Pradesh, application of 800–400–1800–500 g N–P_2O_5–K_2O–$MgSO_4$ palm^{-1} year^{-1} recorded 18–19 t of FFB ha^{-1} (Rao 2009). This warrants for location-specific recommendations to achieve site yield potential and to maintain soil health. In B-deficient soils or whenever the deficiency symptoms of B is noticed, borax is recommended @ 100 g palm^{-1} year^{-1} (14.5% B). Also, organic nutrient sources, viz. farm yard manure (FYM) (50–100 kg palm^{-1} year^{-1}) or green manure (100 kg palm^{-1} year^{-1}), and neem cake (5 kg palm^{-1} year^{-1}) can also be applied. Application of organic manures compensates a portion of nutrient requirement to be supplied through chemical fertilizers. Under rainfed conditions of Kerala, India, the potential of nutrient supply through composting of oil palm wastes into organic manure (by vermicomposting) was found to be 175 kg N, 73 kg P_2O_5, 129 kg K_2O, 70 kg CaO, and 71 kg MgO ha^{-1} year^{-1} (Varghese et al. 2007). Recently, fertigation technique has been standardized at ICAR-IIOPR for oil palm, and the nutrient requirement in this technique has come down substantially to 50% of the above recommendation. This is mainly because of application of nutrients at monthly intervals in the form of readily available nutrient solution which enhances nutrient uptake and use efficiency. Here, the recommended nutrient doses are 600:300:600 g of N, P_2O_5, K_2O palm^{-1} year^{-1} in 12 equal splits at monthly interval. This technology is validated on farmers' fields in West Godavari district of Andhra Pradesh, India, and the results indicated that the nutrient use efficiency is almost doubled with a 15% improvement in net profit. In oil palm, fertigation through venturi system is highly effective and economical and helps in labour saving too. Further, application of nutrients based on soil and leaf nutrient analysis was more effective in economizing nutrient requirement and also improving soil bio-geochemical environment (Manorama et al. 2021a).

7.3 Fertilizer Sources

In oil palm plantations, generally, fertilizers like urea, ammonium sulphate, and DAP are used as a source of N. Urea is the most economical source of N. As a source of phosphorus nutrition, fertilizers like single super phosphate, rock phosphate, and DAP are used. Under acidic soil conditions, rock phosphate is a good source of P. However, under high pH conditions, application of rock phosphate is not recommended. Potassium may be applied as chloride or sulphate. Potassium chloride (Muriate of potash) is found to be cheaper and available in the local market. Magnesium sulphate can be applied as a source of Mg to oil palm. There are synergistic and antagonistic interactions between nutrients. The applied N fertilizer will not be efficiently utilized unless adequate P fertilizer is applied. Therefore, to ensure proper uptake of nutrients, balanced fertilization is to be followed.

In case of fertigation, NPK nutrients can be supplied through DAP, urea, and MOP fertilizers, as single super phosphate cannot be easily dissolved in water and so it clogs the irrigation system. In a Malaysian study by Vijiandran et al. (2020), application of compound fertilizers over straight fertilizer showed no significant differences in leaching losses of N, P, K, and Mg for matured oil palms planted on sedentary or marine alluvial soils. Therefore, application of compound fertilizers has minimal advantage with respect to leaching losses of N, P, K, and Mg nutrients in adult oil palms in comparison with straight fertilizers.

7.4 Time and Method of Fertilizer Application

Correct timing of fertilizer is crucial in both economic and environmental point of view because it influences loss of nutrients through leaching, runoff, and volatilization. Application of urea one day before moderate rainfall (10–20 mm day^{-1}) can reduce volatilization losses (Kee et al. 2005). Application of N and K fertilizers avoid nutrient imbalances. Fertilizer application timing should be based on weather conditions and plant requirements. Teoh and Chew (1984) reported that in mature oil palm, N and K fertilizers are usually applied once or twice a year on coastal marine, nutrient-rich clay soils, and at least twice year^{-1} on alluvial, nutrient-poor inland soils. Chan et al. (1993) opined that enhanced frequency of K application from 3 to 6 times could not record any significant yield improvement. But soil texture plays a key role in frequency of nutrient applications. In light-textured soils (with low cation exchange capacity), N and K fertilizers should be split into several applications at correspondingly lower rates to minimize leaching losses, and on the other hand, on heavy clays, lower application frequencies were found to be sufficient (Pushparajah et al. 1973). In peat soils with high and frequent rainfalls, timing of fertilizer application is crucial to keep leaching losses at bay, especially when applying B and K fertilizers (Melling et al. 2011). In general, uniform application of nutrients helps in minimizing losses due to leaching and evaporation. Goh and Hardter (2003) recommend split doses only when annual requirements are more than 0.5 kg of N palm^{-1}, 0.30 kg of Mg palm^{-1}, and 0.03 kg of B palm^{-1}.

The fertilizer application schedule depends on weather, soil conditions, and plant requirements. On alluvial, nutrient-poor inland soils of Malaysia, Teoh and Chew (1984) reported that in mature oil palm, N and K fertilizers are usually applied once or twice a year. On the other hand, on coastal marine, nutrient-rich clay soils, nutrient application should be at least twice year^{-1}. Conversely, Chan et al. (1993) opined that enhanced frequency of K application from 3 to 6 times @ 9 kg palm^{-1} year^{-1} could not record any significant yield improvement in Malaysia. Soil texture plays a key role in the frequency of nutrient applications. In case of light-textured soils (with low cation exchange capacity), N and K fertilizers should be applied at frequent intervals at correspondingly lower rates to minimize leaching losses. On the other hand, on heavy clays, lower application frequencies were found to be sufficient (Pushparajah et al. 1973). In peat soils, with high and frequent rainfalls, the timing of fertilizer application is very crucial. Particularly when applying B and K fertilizers, care needs to be exercised to avoid leaching losses (Melling et al. 2011). In general, uniform application of nutrients helps in minimizing losses due to leaching and evaporation. Goh and Hardter (2003) recommend split doses only when annual requirements are more than 0.5 kg of N palm^{-1}, 0.30 kg of Mg palm^{-1}, and 0.03 kg of B palm^{-1}.

Under irrigated conditions, four equal splits of recommended doses of nutrients at 3 months interval are recommended as soil application. Fertilizers are to be broadcast around palm base in weed free circles, leaving 1 m space from the trunk. After broadcasting, with the help of forks, fertilizers need to be incorporated into the soil. Irrigation should be provided immediately after fertilizer application. Under rainfed conditions, nutrients are to be applied in two equal splits, shortly before the onset of the rainy season, but not before or during the dry season or in the middle of the rains. Under normal circumstances, N and K fertilizers should be applied in close sequence followed by Mg, P, and B fertilizers. In the case of fertigation, fertilizer application at monthly intervals is recommended. Each fertilizer needs to be thoroughly mixed in water, and it has to be connected to an irrigation system for uniform distribution to the palms.

7.5 Precise Nutrient Management Approaches

Nutrient requirements can be estimated through factorial experiments in each location if the crop is a shorter duration one and requires minimal space for taking up the field trial. In oil palm or for that matter most of the plantation crops, it is quite difficult to conduct field trials in different agro-climatic conditions. Therefore, based on sound principles of soil fertility and mineral nutrition of plants, some alternate approaches need to be resorted to. In oil palm, soil and leaf analysis-based estimation of nutrient status gives a good direction to decide about the optimal doses of fertilizers. Through soil analysis, it is possible to assess the nutrient supplying capacity of the soil which of course depends on factors like parental material, climatic conditions, rainfall, anthropogenic interventions, etc. Soils are categorized into three classes based on their nutrient status (Table 7), and the fertilizer dose is dependent on this fertility class (Pushparajah and Chew 1979). Through leaf analysis, nutrient uptake capacity of the crop can be assessed. This is because of a significant relationship between leaf nutrient concentration and FFB yield at a site. It is further observed that the highest yield appears to be critically dependent on exact leaf nutrient composition. Each nutrient has a maximum concentration, and when all nutrients reach their highest values, the maximum yield is attained.

From leaf nutrient status, critical concentrations of leaves are estimated in Malaysia by von Uexkull and Fairhurst (1991) (Table 8).

Table 7 Fertilizer recommendation for oil palm based on soil nutrient status

Parameter/nutrient element	Soil fertility status		
	Low	Medium	High
Soil organic carbon (%) (SOC)	<0.50	0.50–0.75	>0.75
Available N (kg ha^{-1})	<280	280–560	>560
Available P_2O_5 (kg ha^{-1})	<20	20–50	>50
Available K_2O (kg ha^{-1})	<150	150–300	>300
Exchangeable Ca (meq 100 g^{-1})	<1.5	1.5	>1.5
Exchangeable Mg (meq 100 g^{-1})	<1.05	1.05	>1.05
Available B (mg kg^{-1})	<0.5 deficient		
	Fertilizer recommendation		
	Nutrient deficiency symptoms may occur. Fertilizer response is likely. Increase fertilizer dose by 25%	Hidden hunger is likely. May respond to fertilizer. Maintain fertilizer dose	No response to fertilizer. Reduce fertilizer dose by 25%
Soil reaction (pH): <6.5 is acidic, 6.5–7.5 is neutral, and >7.5 is alkaline			
Electrical conductivity (EC): <2 dS m^{-1} is considered as normal			

Adopted from Pushparajah and Chew (1979). The Proceedings of the Malaysian Seminar on the Fertility and Management of Deforested Land, Henry J. T. L. Liau et al. (eds.) pp. 77–79

Table 8 Critical leaf nutrient concentrations in young and adult oil palm plantations

Palm age	Nutrient	Deficient	Optimum	Excess
Young palms (<6 years)	N (%)	<2.50	2.60–2.90	>3.10
	P (%)	<0.15	0.16–0.19	>0.25
	K (%)	<1.00	1.10–1.30	>1.80
	Mg (%)	<0.20	0.30–0.45	>0.70
	Ca (%)	<0.30	0.50–0.70	>0.6
	S (%)	<0.20	0.25–0.40	>1.00
	Cl (%)	<0.25	0.50–0.70	>1.00
	B (ppm)	<8	15–25	>40
	Cu (ppm)	<3	5–8	>15
	Zn (ppm)	<10	12–18	>80
Adult palms (>6 years)	N (%)	<2.30	2.40–2.80	>3.00
	P (%)	<0.14	0.15–0.18	>0.25
	K (%)	<0.75	0.90–1.20	>1.60
	Mg (%)	<0.20	0.25–0.40	>0.70
	Ca (%)	<0.25	0.50–0.75	>1.00
	S (%)	<0.20	0.25–0.35	>0.60
	Cl (%)	<0.25	0.50–0.70	>1.00
	B (ppm)	<8	15–25	>40
	Cu (ppm)	<3	5–8	>15
	Zn (ppm)	<10	12–18	>80

Adopted from von Uexkull et al. (1991). Fertilizing for High Yield and Quality. The Oil Palm. IPI, Bern, 79 p

8 Fertilizers and Plant Protection Chemicals on Soil Health and Soil Microbiology

Soil health is an integrative property that reflects the capacity of soil to respond to agricultural intervention, so that it continues to support both the agricultural production and the provision of other ecosystem services (Kibblewhite et al. 2008). As in the case of any other crop, excessive application of fertilizers and other chemicals causes damage to soil health and microbial population of oil palm plantations also. But at the same time, if the crop removal of nutrients exceeds the fertilizer input, it leads to the nutrient depletion which is an unsustainable practice. Soil properties, status of organic matter, supply, and availability of nutrients are influenced by cultivation of plantation crops (Arevalo-Gardini et al. 2015; Buresh and Tian 1998; Jose et al. 2009). Crops extract nutrients from different soil layers to use in various metabolic activities. Through decomposition of fallen litter, added crop residues, and dead root tissues, the nutrients are returned to soil. Regular and ample fertilizer usage results in small but consistent increases in soil organic matter (SOM), mainly because of increased root biomass (Singh and Ryan 2015). There are also indications that excess usage of ammonium sulphate as a source of N increases the risk of acidification due to reduced base cation availability. Therefore, the rate of application of fertilizers plays the decisive role on maintenance of soil health. Soil acts both as source and sink of carbon, and it depends on management practices, biomass addition, and changes in biology as well as climatic conditions (Zomer et al. 2017; Gong et al. 2015; Santiago et al. 2017). Plantation crops like oil palm, rubber, and tea improve carbon sink potential and SOC stock of soil (Nath et al. 2018; Pulhina et al. 2014). They help to mitigate climate change effects through reduced GHG emissions and by sequestering considerable quantities of carbon. There are evidences of appreciable levels of carbon sequestration by oil palm in aboveground biomass (Lamade and Boillet 2005) and in soil as well (Suresh and Kiran Kumar 2011). Behera et al. (2020) reported that in comparison with fallow land, the soils of oil palm plantation recorded increased SOC content, soil pH, electrical conductivity, exchangeable calcium, and magnesium and available sulphur in 0–20, 20–40, and 40–60 cm depths of soil. Addition of crop residue resulted in increased soil pH, which is attributed to decarboxylation of organic acid anions causing proton consumption and release of OH-s, specific adsorption of organic molecules released during decomposition on to Al and Fe hydroxides with release of OH-ions and higher concentration of base cations in crop residue (Mokolobate and Haynes 2002). The higher concentration of available P has also been reported in some cultivated soils of India (Sanyal et al. 2015), European Union (Tóth et al. 2014), and the USA (Sharma et al. 2017) due to continued application of crop residues. The SOC stock in 6-, 12-, and 18-year-old oil palm plantations has enhanced by 20%, 18%, and 45%, respectively, in comparison with SOC stock in fallow land, indicating appreciable C sequestration due to oil palm cultivation (Behera et al. 2020).

Physical, chemical, and biological properties of the soil get changed due to excessive use of chemical fertilizers. This includes leaching, soil pH, soil acidification/alkalization, shift in the microbial biodiversity, reduction of beneficial insects, increase in disease susceptibility, exchangeable Ca, and loss of soil fertility (Kibblewhite et al. 2008; Mishra and Dash 2014). Kee (1995) reported that after 7 years of oil palm cultivation, soil pH decreased from 4.2 to 3.8. Excessive usage of the reactive N and P causes loss of nitrate and P to water bodies, eventually releasing high levels of ammonia and N oxides, which are GHG of 300 times more heat-trapping power compared to CO_2 (Mole 2014; Velthof et al. 2014). Nitrogenous fertilizers contribute to increase in soil acidity through nitrification process resulting from the release of hydrogen ions, thus leading to reduced microbial N immobilization (Singh and Ryan 2015). Nowadays, plantations rely on machinery for fertilizer application, and a study showed that 258 GJ of input energy is required to produce 20 t ha^{-1} year^{-1} of FFB. Out of this, 66% was contributed by application of chemical fertilizers (Azwan et al. 2016). Fuel and machinery were identified as the factors impacting the environmental issue related to GHG emission such as CO_2, based on life cycle analysis (LCA) conducted by the Malaysian Palm Oil Board (MPOB) (Zulkifli et al. 2010). It is thus evident from the studies that high N fertilization rates result in large N_2O emissions (Hassler et al. 2017) and excess fertilization causes deleterious environmental effects, such as GHG emissions and nutrient leaching losses (Kurniawan et al. 2018).

9 Soil Biodiversity and Biological Functions

Soil is a complex system with many living organisms interacting within physico-chemical conditions (Zaidi et al. 2017). Soil biota is very diverse, and they are the key drivers of majority functions (carbon transformation, nutrient cycling, structure maintenance, pest regulation) in soil. In India, oil palm is promoted on agricultural soils and other suitable waste lands. Unlike in other Southeast Asian countries where oil palm replaced tropical forests, it is grown either on cultivable lands or fallows without any damage to the existing forest. Replacing of forest ecosystems with oil palm plantations as done in certain Southeast Asian countries has resulted in severe loss of biodiversity (Foster et al. 2011) coupled with substantial GHG emissions (Bessou et al. 2014; Zulkifli et al. 2010). In addition to this, negative impacts were also noticed on certain ecosystem services, viz. erosion prevention, soil fertility, mitigation of extreme events, etc. Therefore, from an ecological point of view, conversion of natural forests and peatlands to oil palm plantations should be avoided. Conversely, the changes in soil physical, chemical, biological functions, and even other ecosystem functions of oil palm cultivation in India are quite different from that of most of the other SE Asian countries as oil palm is not replacing any forest. In a study to evaluate the activities of soil microorganisms as influenced by oil palm cultivation on coastal plain sand area of Akwa, Ibom State, Nigeria, it was found that microbial activities increased up to 39 years of age and then decreased (Chukwu et al. 2020).

For proper functioning of ecosystems, biodiversity needs to be augmented and maintained (Bender et al. 2016). Manures and bio-fertilizers help to achieve this. Ecological intensification can be accomplished through bio-fertilizer and organic fertilizers act as a source of nutrients for the inhabitation of the microorganisms. Many workers reported change in soil microbial count with the application of EFB. Carron et al. (2015b) reported that with the application of EFB, physical, chemical, and biological properties are most influenced in the A_0 humic soil layer. They observed three successive periods, viz. a disturbance period (0–6 months), a resilience period (6–18 months), and a final period (18–24 months), where most of the soil quality parameters were continued to improve. Soil K, pH, and base saturation improved immediately after EFB application, whereas others (soil bulk density, N, CEC, Ca, Mg, OC) improved gradually over the study period of 2 years. A sharp improvement in macrofauna abundance and taxonomic diversity was also observed. Application EFB led to an increased Shannon index compared with the conventional treatment, indicating a more balanced distribution between taxa. A sharp increase in nemato fauna abundance was noted during the second year. From an Indonesian study by Carron et al. (2015a), it is reported that EFB application improved soil quality, rise in macrofauna, and SOC content. Ants and coleopteran densities were reported to reach 349 ind. m^{-3} (individuals m^{-3}) and 95 ind. m^{-3}, respectively, along with bacterial feeder nematodes. On the contrary, the population of earthworms, millipedes, and nematodes followed a different trend, showing a sharp decrease. In the subsequent 12-month period, the density of earthworms, millipedes, nematodes, and bacterial feeder nematodes progressively increased. During the final 6 months of EFB decomposition, the increase in densities of macrofauna (333 ind. m^{-3}) and nemato fauna (1068 ind. 100 g^{-1} in the A_0 and 788 ind. 100 g^{-1} in the A_1 layer) was observed. Also, a reduction in soil bulk density with improved SOC, N, and Ca contents were observed, thus highlighting the correlations between physico-chemical and biological properties of the soil.

10 Microbiome and Metagenomic Approach

The soil is a complex and important ecosystem, with rich microbial diversity acting as abode for biogeochemical and nutrient cycles (Robe et al. 2013). Microbiome is the combined genetic material of the microorganisms in a particular environment. The differences in productivity of plantations are also attributed to diversity and abundance of soil microbial community (Situmorang et al. 2014). Soil physico-chemical changes influence bacterial soil communities (Tripathi et al. 2012). Understanding the existence and function of soil bacteria helps in better plantation management (Rolf 2005). Under sustainable agriculture practices, bio-control agents (BCA) and plant growth promoters (PGP) have gained popularity over the past three decades. Soil quality of oil palm plantation as micro-habitat and nutrient source can be estimated by evaluating and observing microbial community structure as soil bioindicator.

Metagenomics is the direct genetic analysis of genomes contained in an environmental sample. A metagenomic approach using molecular technique can reveal both culturable and unculturable microorganisms. Some of such widely used techniques are Amplified Ribosomal DNA Restriction Analysis (ARDRA), Terminal Restriction Fragment Length Polymorphism (T-RFLP), and Denaturing Gradient Gel Electrophoresis (DGGE) (Kirk et al. 2004; Malik et al. 2008). Among these techniques, T-RFLP has been widely used as it has higher reproducibility, resolution, and sensitivity (Fredricksson et al. 2014). Soil bacterial diversity among oil palm plantations has been investigated by Situmorang et al. (2016) using the metagenomic approach, T-RFLP, from plantation sites with heterogeneous soil texture and characteristics. All the sampling areas were grouped into two clusters. The diversity specific to soils was found both in bacterial diversity and in their abundance. The two restriction enzymes used, *Hha*I and *Hinf*I, influenced the number of culturable bacteria dominancy. In conclusion, bacterial community was influenced by soil characteristics. Specific bacteria, such as *Halobacillus* and *Propionibacillus*, which was present in each soil type including peat soil, was influenced by micro soil environment, where it lived and adapted to habitat conditions, such as anaerobic condition and high oxide-reduction activities.

In the larger context of sustainable management of natural resources, soil biodiversity and linkages between above- and belowground biota are important (Calugar and Ivan 2016). Lo and Chong (2020) estimated the microbial diversity of oil palm soil under disease-free, high and low basal stem rot incidences (in Malaysia) using Illumina MiSeq using V3–V4 region of *16S rRNA* gene. Bacteria (97.4%) and Archaea (0.2%) were in majority at kingdom taxonomy level. The most abundant phyla were *Proteobacteria*, *Acidobacteria*, *Actinobacteria*, and *Verrucomicrobia*. Higher alpha diversity of all species was observed among all tested soil samples from each estates. Beta analysis was carried out using non-phylogenetic UnifRac matrix and visualized using principal coordinates analysis (PCoA). This information is useful in designing biological control for basal stem rot disease.

Similarly, EFBs are also used as a mulching material, or they are composted to use as an organic source of nutrients. By identifying suitable microbes to degrade these EFBs, empty fruit bunches can be better utilized as a source of nutrients. Devit Purwoko et al. (2022), using Illumina MiSeq sequencing of the ITS2 for fungi, the V3 region of the *16S rRNA* gene, and the V4 region of the 18S rRNA gene for bacteria, developed the data on fungal and bacterial diversity in naturally decaying oil palm EFB in plantations. Bacterial diversity in decaying oil palm EFB was dominated by the phylum Planctomycetes (40–60%), whereas most of the fungal sequences belonged to Ascomycota (60–90%). Planctomycetes are reported to play a role in N and carbon biogeochemical cycles (Storesund et al. 2020). Ascomycota is reported to play a role in the lignin degradation process (Tahir et al. 2019).

On Earth, Isoprene (2-methyl-1,3-butadiene), with atmospheric emissions of 400–600 Tg year^{-1}, is the most abundantly produced biogenic volatile compound (BVOC). It plays a complex role in atmospheric chemistry and, hence, climate. The largest source of isoprene is terrestrial plants, and oil palms are among the highest isoprene-producing trees. DNA stable isotope probing (DNA-SIP) to study the microbial isoprene-degrading community associated with oil palm trees revealed novel genera of isoprene-utilizing bacteria including *Novosphingobium*, *Pelomonas*, *Rhodoblastus*, *Sphingomonas*, and *Zoogloea* in both oil palm soils and on leaves. The *isoA* (which encode the α-subunit of the isoprene monooxygenase (IsoMO), a key enzyme in isoprene metabolism)-containing bacteria are more abundant in soils than in the oil palm phyllosphere (Carrión et al. 2020). Soils and their inhabitants make fundamental contributions to the terrestrial ecosystem's functionality (Hättenschwiler et al. 2018; Hedde 2018). They provide different ecosystem services more particularly related to food provisioning (Conti 2015). In addition, soil biological processes determine the potential of soils for carbon sequestration. The predominant interactions within soil fauna are trophic. Microbes with beneficial effect on the crop have an additional advantage as they also increase the diversity of soil microorganisms through their interaction and colonization in soil. The results generated with respect to microbial diversity in relation to basal stem rot disease, EFB decomposition, isoprene degradation, and other processes through metagenomic research are highly constructive in designing suitable ecosystem services, to contribute in maintaining healthy soils and environment.

11 Technological Options to Promote Soil Health

Resorting to best management practices (BMPs) is an important approach to improve both economic and environmental sustainability. Among the BMPs for oil palm, the efficient and appropriate supply of nutrients is a crucial component.

Experiments on six commercial sites in Indonesia suggested potential yield increases of 1.5% to up to 23% if BMPs were followed (Donough et al. 2009). Both bunch numbers and bunch weight increased with such practices over a 5-year period and that the increase was progressive for the first 4 years. Under irrigated conditions of India, ICAR-IIOPR standardized BMPs for resource utilization especially water, nutrients, and space through scientific management of these resources. Under irrigation management, water requirement on weekly basis for oil palm has been estimated based on historical evapotranspiration data, for four states of India (Andhra Pradesh, Tamil Nadu, Karnataka, and Telangana), and drip method was found to be the most efficient one. For nutrient management, fertigation technique has been standardized based on soil and leaf analysis data of a particular location. In situ mulching with fronds and other plantation wastes was found to be effective in minimizing evaporation losses, reducing weed growth, adding organic matter to soil, etc. All these practices were tested in farmers' fields of two villages, viz. Challachintalapudi and Makkinavarigudem in West Godavari district of Andhra Pradesh, India, in comparison with conventional farmer adopted practices. Mulching in combination with improved water and nutrient management could improve the SOC, water holding capacity, available nutrients, FFB yield, and CEC after a period of 6 years (Manorama et al. 2021a). Availability of huge quantities of waste (more than 15 t ha^{-1} $year^{-1}$) in oil palm plantations is a positive feature in soil health management.

11.1 System Approach

In peninsular India, oil palm is grown as an irrigated crop under tropical conditions. Due to very less mean annual rainfall, it is grown as irrigated crop and supplied with abundant quantities of water and nutrients. This makes the system unsustainable with heavy leaching of nutrients to deeper layers combined with faster rate of degradation of organic carbon under high temperature. Crop intensification with suitable intercrops could be an appropriate option in such situations, as all the component crops add biomass to the soil, thereby improving SOC, available soil nutrients, and system productivity. A lot of work has been carried out in this direction at ICAR-IIOPR, and compatible intercrops have been identified for different situations. Suitable intercrops vary based on the stage of growth of the palm as canopy cover bring about differences in light interception in juvenile and adult gardens leading to changes in selection of intercrops. In juvenile plantations, crops like maize, banana, sunflower, tobacco, chilli, turmeric, cut flowers, etc. can be grown depending upon suitability and marketing facilities. But in adult plantations, shade-loving and proven models only should be adopted. Intercropping models recommended for Andhra Pradesh and plains are (1) oil palm + cocoa; (2) oil palm + banana; (3) oil palm + long pepper + bush pepper; (4) oil palm + red ginger + heliconia; and (5) oil palm + long pepper + cut foliage. Annuals like elephant foot yam, colocasia, turmeric, black turmeric, white turmeric, wild turmeric, mango ginger, ginger, red ginger, black ginger, galanga, bitter ginger, and brahmi and fodder crops like guinea grass and hybrid napier have been found to be suitable intercrops for adult oil palm plantations in Andhra Pradesh and Telangana states of India. Intercropping enhances the profitability nearly to the tune of 191–293% over sole oil palm depending upon the type of system adopted. Other enterprises like dairying, poultry, sheep, and goats can also be integrated to achieve better economic benefits from the oil palm which encourages area expansion under oil palm. Considering the nature of oil palm growth and light infiltration pattern through its canopy, intercropping in adult oil palm gardens appears to be feasible from seventh year onwards. However, the success of intercropping in oil palm gardens mainly depends on the selection of suitable crops and efficiency of management.

The prime objective of intercropping is production of more output by utilizing freely available inputs like land and light between two rows of the main crop. Cereals; pulse; oil seeds; vegetables; flowers; fruit crops like banana, papaya, and pineapple; tobacco; cotton; tapioca; fodder crops; etc. during the juvenile phase and cocoa, long pepper, and banana during mature phase are commonly grown inter-/mixed crops in oil palm grown in India. The points to remember while selecting suitable crops for intercropping are component crops should be complementary, having market value, no commonality of pests, differential rooting depths and growth habits, and profitability. It has been estimated that roots of fully grown oil palm extend beyond 4.5 m though most of the roots get concentrated around the bole to a radius of 2–3 m (Suresh and Reddy 2004). Vertically, roots get concentrated to a depth of 40–50 cm though some of roots are extended up to 5 m. In Andhra Pradesh, India, conditions, estimates show that light infiltration in 9–19-year-old oil palm gardens ranged from 14% to 38% (Suresh and Reddy 2009). Availability of inter-space and light are the major deciding factors for growing intercrops in adult oil palm gardens. Location-specific annuals, biennials like banana and sugarcane, and perennials like papaya, pineapple, and fodder crops can be cultivated without any problem during juvenile phase (1–3 years) of oil palm. Rice can also be an option in sloppy lands. In adult plantations, combination of cocoa and cinnamon in alternate inter-space, black pepper trailed on palm trunk, and *Anthurium* (flower crop) planted in intra-row spaces were found ideal for getting maximum net returns under Kerala, India, conditions (Varghese and Sunitha 2005). Among the five medicinal plants evaluated (*Justicia adhatoda* L., *Alpinia galanga* (L). Willd., *Kaempferia galanga* L., *Nilirianthes haenianus*, and *Asparagus racemosus* Willd.), *Kaempferia galanga* L. emerged as the most profitable crop under Kerala conditions with the highest cost benefit ratio (Jessy Kutty et al. 2005). Under Andhra Pradesh conditions, perennial crops, i.e. heliconia (*Heliconia psittacorum* L.f.), red ginger (*Alpinia purpurata* K.Schum), cocoa (*Cacao theobroma* L.), bush pepper (*Piper nigrum* L.), Java long pepper (*Piper chaba* Trel. & Yunck), banana (*Musa paradisiaca* L.), and ornamentals, viz. fish tail fern (*Nephrolepis falcata* (Cav.) C.Chr.), asparagus (*Asparagus densiflorus* (Kunth) Jessop), ti plant (*Cordyline fruticosa* (L.) A.Chev), dracaena (*Dracaena* sp.), philodendron (*Thaumatophyllum xanadu* (Croat, J.Boos & Mayo), dumb cane (*Dieffenbachia* sp.), and aglaonema (*Aglaonema* sp.), were found as successful intercrops in mature oil palm plantations. Similarly, annual crops like colocasia (*Colocasia esculenta* (L). Schott), turmeric (*Curcuma longa* L.), black turmeric (*Curcuma caesia* Roxb.), elephant foot yam (*Amorphophallus paeoniifolius* (Dennst.) Nicolson), ginger (*Zingiber officinale* Roscoe.), and mango ginger (*Curcuma amada* Roxbugh.) have been found as suitable intercrops (Ramachandrudu 2019). From these results, it can be deduced that only shade-loving and shade-tolerant crops can be grown along with oil palm since the availability of sun light in gardens is less than 30%.

Successful cropping models standardized for Andhra Pradesh conditions are oil palm + cocoa, oil palm + banana, oil palm + long pepper + bush pepper, oil palm + red ginger + heliconia, and oil palm + long pepper + cut foliage plants. Oil palm + cocoa/*Cinnamomum* + black pepper on *Gliricidia* + black pepper trailed on oil palm + *Anthurium*/*Kaempferia galanga* L. and oil palm + *Kaempferia galanga* L. are recommended for oil palm-based cropping systems under Kerala conditions (Ramachandrudu et al. 2018). Some successful intercrop combinations are presented in Fig. 3.

Some of the oil palm farmers adopt certain wrong practices like tying of oil palm leaves and cutting of basal leaves to generate more space for intercrops. All these practices are to be avoided as they reduce the photosynthetic area thereby FFB production (Ramachandrudu 2019). The number of arthropod orders was significantly greater in polyculture than monoculture in smallholdings in the state of Selangor, on the west coast of Peninsular Malaysia (Ghazali et al. 2016). To support biodiversity in oil palm plantations, polyculture farming is a useful strategy.

Integrated farming system (IFS) involves integration of animal husbandry enterprises along with cultivation of fodder or pasture in oil palm gardens. In this system, wastes of one component become resource for other component and ensure the overall productivity improvement of the system (CARDI 2010). In Indonesia, integration of cattle in IFS could save 25–50% of weeding costs, and the production of FFB increased by 16.7% (Henuk et al. 2018). Similarly, oil palm + cattle integrated farming system resulted in higher population and biomass of earthworms as well as abundance and diversity of soil mesofauna in comparison with non-integrated system. Several physical and chemical properties of soils are positively correlated with the presence of these soil fauna (Niswati et al. 2021). The "Sistem Integrasi Sapi-Kelapa Sawit" (SISKA) or "System Integration Cattle–Oil Palm Plantations" is a combination of oil palm plantations with cattle, which has been introduced in Indonesia in 2011. In this, cattle and/or buffalo can be used as labour for transporting FFB, as a source of organic fertilizer and as a weed eater. The waste from plantations and oil palm industries can be utilized as animal feed to produce meat (Setiadi et al. 2011).

In Uganda, it is reported that between planting and the first oil palm harvest, annual crops like maize and beans and perennials like banana and pineapple can be grown to improve food security and family nutrition. Intercropping with legumes (beans, peas, soybeans, etc.) also increases soil fertility through nitrogen fixation, and cover crops such as mucuna help to prevent soil erosion. For permanent intercropping, double-row avenue system is recommended with high-value crops like cacao, coffee, vanilla, and fruit or timber trees, where palms are planted closer together in double rows at 6 m spacing within rows while leaving 15-m-wide "avenues" (Namanji et al. 2021).

Fig. 3 Inter-/mixed cropping options in oil palm plantation. (**a**) *Theobroma cacao* L. (**b**) *Piper nigrum* L. c. *Piper chaba* Trel. & Yunck. (**d**) *Musa paradisiaca.* (**e**) *Heliconia psittacorum* L.f. f. *Alpinia purpurata* K.Schum. (**g**) *Cordyline fruticosa* (L.) A.Chev. (**h**) *Nephrolepis falcata* (Cav.) C.Chr. (**i**) *Dracaena* sp. (**j**) *Megathyrsus maximus* (Jacq.). (Adopted from Ramachandrudu et al. 2018, ICAR-IIOPR technical bulletin on Systems approach in oil palm for higher productivity and profitability, ISBN No.: 81-87561-53-X, pp 1–51)

11.2 Conservation Agriculture

Conservation agriculture practices would help in minimizing the use of natural resources and thereby enhancing system sustainability. Thus, conservation agriculture and related management practices are necessary for maximizing the production, maintaining soil fertility, and reducing the environmental impact (Conti 2015). Minimum tillage in combination with cover cropping has the potential to facilitate stability and increase production substantially and offer better soil conservation in cropping systems of tropical mountain areas (Hobbs 2007). The practice of EFB application contributes to sustainable production in oil palm and found beneficial in enhancing soil ecosystem functioning (Tao et al. 2016, 2017). In Indonesia, a study by Saijo et al. (2022) proved that the experimental plots treated with *Indigofera zollingeriana* + EFB retained more water (28.78% at a depth of 20 cm in July); accrued the highest carbon stocks of oil palm and soil (81.6 and 36.60 t ha^{-1}, respectively), the highest respiration (72.00 mg CO_2-C 100^{-1} g of soil day^{-1}), and the highest total microbial count (105 × 10^5 CFU g^{-1}); and recorded higher activity as well as population of soil microorganisms than in other treatments. In Malaysia, the application of 300 kg of EFBs palm^{-1} year^{-1} over a 10-year period led to an increase in soil organic matter from 1.5% to 2.7% and a 9% increase in yield compared with fertilizer treatments alone (Rosenani et al. 2011). In Indonesia, on a yellow-brown podzolic, applying 80 t ha^{-1} EFBs every 2 years was found to increase leaf nutrient levels to the optimum, while mineral fertilizer application alone resulted in deficient leaf K levels, and EFB alone caused low N, P, and Mg levels (BLRS 1997). Research on four management zones including (1) EFB application along the sides of harvesting paths, (2) chemical fertilization within palm circles, (3) understory vegetation with pruned fronds in inter-row areas, and (4) no input in the cleared part of harvesting paths showed significantly higher soil fauna feeding activity under the EFB application than other management practices (Tao et al. 2016).

Satriawan et al. (2017) studied soil and water conservation measures in Indonesia on different land capability classes and found that application of soil conservation techniques to each group of land capability could control surface runoff and erosion. On land capability class III, cover plants + manure was able to control runoff and erosion and reduce leaching of N, where soil conservation produced the lowest erosion (3.73 t ha^{-1}) and N loss by leaching (0.25%). On land capability class IV, sediment trap + cover plants+ manure was able to control runoff and erosion and reduce organic C loss and P leaching that soil conservation produced the lowest runoff (127.77 m^3 ha^{-1}), erosion (12.38 t ha^{-1}), C organic (1.14%), and P (1.28 ppm). On land capability class VI, soil and water conservation was able to control runoff, erosion, and soil nutrient leaching.

Mulching is application of materials on the surface of palm basin to reduce soil temperature, conserve soil moisture, improve physico-chemical and biological properties of soil, and suppress weed growth. It also supplies a good amount of organic matter and nutrients particularly N, P, K, and Mg after decomposition. The use of residues is the most important criterion in ensuring sustainable palm oil production (Stichnothe and Bessou 2017). Biomass of oil palm gardens such as leaves, male inflorescences, dry grass, oil palm mill wastes like EFB, pressed mesocarp fibre, nut shells and POME sludge, decanter cake, etc. can be used for mulching. EFB and leaf shreds can be used for mulching basins @ 250 and 150 kg palm^{-1} year^{-1}, respectively.

Mulching materials from oil palm residues such as pruned palm fronds (OPF), EFB, and Eco-mat (ECO; a compressed EFB mat) are often the recommended soil and water conservation practices (CP) for oil palm plantations on hill slopes. Another recommended CP is the construction of silt pits or trenches (SIL) across the hill slope to capture runoff and then return the water and nutrients into the surrounding soil. A 3-year field experiment was conducted by Moradi et al. (2015) in a non-terraced oil palm plantation in Malaysia, and soil samples from 0 to 0.15, 0.15 to 0.30, and 0.30 to 0.45 m depths were collected every 3 months and analysed for their soil physical properties. Soil water content up to 0.75 m depth was also measured daily. EFB released the highest amount of organic matter and nutrients into the soil compared to OPF, ECO, and SIL. Hence, EFB mulching was the most effective to increase soil aggregation, aggregate stability, soil water retention at field capacity, available soil water content, and the relative proportion of soil mesopores. Due to these improved soil physical properties, the highest soil water content was recorded in EFB mulching. Unlike ECO that concentrated more water in the upper soil layers, EFB distributed the soil water more uniformly throughout the whole soil profile, but SIL concentrated more soil water in the lower soil layers (>0.30 m) because the water levels in the pits were often below 0.30 m from the soil surface. The large opening area of the silt pits could have also caused large evaporative water losses from the pits. EFB mulching is recommended as the best CP, particularly for oil palm plantations on hill slopes.

In a field trial under irrigation by Manorama et al. (2021a) in the West Godavari district of Andhra Pradesh in India, mulching with plantation wastes (fronds, spent male inflorescences, and failed bunches) could enhance SOC content by 52% in a monocrop stand of 13-year-old oil palm plantation. Whereas, in a mixed crop stand (with cocoa), the corresponding improvement of SOC was 77% above its conventional cultivation, i.e. sole crop of oil palm without mulching. Mulching in combination with improved water and nutrient management in sole and intercrop stands could enhance the SOC content by 40 and 56%, respectively, after 6 years, over their corresponding levels at 13 years, in the top soil layer of 0–30 cm. Conventional cultivation of oil palm recorded a reduction in SOC by 4% during the same 6-year period. A similar trend was also observed in soil available nutrients (N,P,K, Ca, Mg, S, and B). The partial least squares (PLS)-path model confirmed that sustainable production of oil palm is possible through mulching and improved management.

In Indonesia, Tarigan et al. (2019) reported that the application of EFB gave a significant effect on chemical soil characteristics such as soil pH mainly in 0–30 cm layer, organic carbon, exchangeable K, exchangeable Al, and base saturation. Base saturation increased up to 75 cm soil depth. The distribution of fronds systematically in inter-rows brought down the rainwater runoff by >30%, even on 5% gentle slopes. Thus, soil loss through erosion got reduced up to 65% depending up on the slope intensity of the terrain. Consequently, nutrient loss was minimized, ensuing higher fertilizer efficiency. The presence of EFB could maintain soil moisture content and reduce soil compaction mainly in the upper layer. The ill effects of slope on water runoff, soil erosion, and nutrient losses could be minimized by application of EFB or fronds on the surface of the soil. The addition of organic materials such as EFB and fronds increased the value of soil fauna feeding activity, the earthworms, and soil insect abundance. In hilly terrains, mulching, formation of half-moon terraces and bench terraces, contour bunding and planting on the contour line, use of drip irrigation system, cover cropping, and construction of water storage tank for irrigation could be followed for soil and water management. As majority locations of oil palm cultivation in North Eastern region of India are on hill slopes, a special attention needs to be paid for conserving soil and water. Depending upon the slope of the area, planting of oil palm is to be taken up either on circular platforms or half-moon terraces or bench terraces.

11.3 Crop Residue Recycling

Oil palm produces plantation waste, in the form of fronds (leaves), male inflorescences, and bunches that fail to develop, to the tune of 15–17 t ha^{-1} $year^{-1}$ in mature oil palm plantations (Sunitha and Krishnakumar 2006). Recycling of this crop residue through mulching is recommended in oil palm. In a time span of 19 years, in situ mulching could improve the SOC levels by 52% and 77% in sole and intercrop stands, over conventional cultivation. In intercropping, large quantities of leaf litter, both from oil palm and cocoa, have contributed to enhanced SOC. A marginal improvement in clay content of the plots which received mulching under improved management practices also had contributed towards SOC improvement (Manorama et al. 2021a). The mean resident time (MRT) of SOC depends upon the degree of shield that is available on the soil surface. Soil organic carbon under oil palm production system is more than that under arable crops, as oil palm covers larger surface of the soil and adds lots of bio-waste to the soil (Senjobi 2007). High soil quality represents high productivity for a long period of time without causing any significant soil or environmental degradation (Karlen et al. 2006). Improved SOC content also contributed to higher CEC, in mulched plots with improved management practices, which could hold nutrients on exchangeable sites due to the formation of organo-mineral complexes. Nutrient availability (N, P, K, S, Mg, and B) and soil moisture retention have enhanced in plots covered with in situ mulching under improved management, largely because of higher levels of SOC under tropical conditions of southern India. As a result, FFB yield levels in sole and intercropped stands of oil palm improved to the extent of 27% and 26% (on an average) during 6 years against a meagre +3% change in yield of conventional cultivation (Manorama et al. 2021).

In a field study by Moradi et al. (2014), residue quality and rate of mass loss of the oil palm residues were determined to be in the decreasing order of leaflets > oil palm fronds > EFB > Eco-mat > rachis. One exception was that EFB's rate of mass loss was higher than that for fronds due to the low quality of rachis in the fronds.

Nutrient release followed an exponential pattern as they released faster and at larger amounts at the beginning than at the later stage of residue decomposition. Leaflets released the highest amount of nutrients (except for K) than other oil palm residues, after 8 months of decomposition, whereas the rachis released the lowest. Fronds with easily decomposable leaflets, which are rich in nutrients, also released total nutrient concentrations that were either significantly higher (for N and Ca) or not significantly different (for P and Mg) than EFB. Due to the high heat and high pressure process to convert EFB into Eco-mat, a loss of some nutrients with a decrease in residue quality of Eco-mat took place. EFB released significantly higher total amounts of N, K, and Mg nutrients than Eco-mat. This information would help to determine the accurate fertilizer requirement when these materials are used as a mulch.

The EFB is dominated by lingo-cellulosic material with 25% lignin, 25% hemicelluloses, and 50% cellulose in cell wall. As a mulch EFB is a good weed suppressor (Alam et al. 2009; Misson et al. 2009), AdeOluwa and Adeoye (2008) suggested that EFBs could be used as a cheaper organic manure in oil palm farms. In Malaysia, after 3 months of EFB application to palm basins @ 100 kg palm^{-1}, Kheong et al. (2010) noticed a good proliferation of roots at a depth of 30–45 cm. A significant improvement in the contents of total and exchangeable K and total Ca at 15–45 cm soil depth were associated with this proliferation. Additionally, considerable improvements in soil pH, soil moisture, and P at 0–15 cm soil depth might also have contributed for this proliferation of roots. This increased root mass entails better nutrient uptake leading to increased yields and improved nutrient status.

In Malaysia, to determine the quality parameters which describe the N mineralization, Zaharah and Lim (2000) developed a multiple regression model to predict soil N dynamics with EFB application as:

$$\min \mathrm{N}\, 20.45 - 25.02\,\mathrm{EFBN} + 0.38\,\mathrm{PP} + 0.47\,\mathrm{LIG} - 0.40\,\mathrm{LIG/N} \quad \left(r^2 = 0.80\right)$$

where
Min N is the soil mineral N at any time
EFBN is the nitrogen content of EFB
PP is the polyphenol content of EFB
LIG is the lignin content of EFB
LIG/N is the lignin to N ratio of EFB

Residue management, such as frond stack application practices, may also have contributed to increased crop yield (Tohiruddin and Foster 2013). Organic residues act as a food source for soil organisms (Lavelle and Spain 2001), thus helping to preserve biodiversity in terrestrial ecosystems. In Indonesia, application of fronds and enriched mulch (EMU) in BMP plots contributed to the recycling of nutrients which contributes to improve soil quality and yields (Rahman et al. 2021). The SOC stocks of plantations under BMP (68 t ha^{-1}) were higher by 31% and 18% than plantations under current management in which recommended fertilizers are applied but no EFB or compost is added (CMP) (57 t ha^{1}) and smallholder management practices (SHMP) (46 t ha^{-1}), respectively. The FFB yield improvement in BMP system was 33% higher than SHMP system. In conclusion, residue incorporation or retention as a part of BMP is an effective strategy for increasing SOC stocks as well as FFB yields of oil palm plantations.

The potential of oil palm EFB as a promising feedstock for bioconversion into value-added products has been evaluated by Rosli et al. (2017). The structural constituents included carbohydrates, viz. glucan, xylan, and arabinan, accounted for 31.2%, 18.7%, and 2.7%, respectively, and lignin accounted for 27.7% of its composition. Dermiyati et al. (2020) isolated and characterized microorganisms from oil palm EFB waste under aerobic, anaerobic, and facultative anaerobic conditions. These microbial communities directly or indirectly help plant growth by preventing diseases or stress (Babalola 2010; Singh et al. 2019).

11.4 Cover Cropping and Green Manuring

Growing of leguminous creepers in the inter-space of oil palm gardens is called cover cropping. It is mainly recommended for high rainfall areas. The main advantages of cover cropping are protection of soil from erosion, smothering of weeds, soil enrichment with N, and conservation of moisture. When incorporated into the soil, they improve the fertility of soil by adding considerable amount of organic matter and plant nutrients. The common cover crops which can establish well in oil palm gardens in India are *Pueraria phaseoloides*, *Calopogonium mucunoides*, *Centrosema pubescens*, *Mimosa invisa*, and *Mucuna bracteata*. Pueraria has been found to establish very well and spread fast in oil palm gardens of Kerala and Andhra Pradesh. In fact, the biodiversity increases as the cover crops and organic residues on the soil surface enhance the stability of soils and regulate temperatures (Conti 2015).

Leguminous cover crops are well-known for contribution of biologically fixed nitrogen through symbiosis with root nodule bacteria belonging to *Rhizobium/Bradyrhizobium* species in oil palm plantations. Additionally, they also curtail the soil and nutrient loss during immature phase of oil palm growth and improve top soil fertility by enhancing microbial activity, aeration, and nutrient retention capacities (Teo et al. 2010). Foster (1975) calculated that, by not considering the contribution from roots, legume crop contributes to 2 kg of total N palm^{-1} annually over 3 years, while during the decline of the cover, 124 kg N ha^{-1} year^{-1} is returned to the soil in phosphate fertilizer applied plots, against 12 kg N ha^{-1} in unfertilized plots. On the other hand, 150 kg of N is derived from the atmosphere (fixed) year^{-1}, plus the additional uptake from the soil of about the same amount in the early years of establishment (Agamuthu and Broughton 1985). Additionally, the cycling of organic materials from the cover crop can return about 30 kg P ha^{-1} and 80 kg K ha^{-1} (Foster 1975).

Green manuring includes growing of crops, mostly leguminous crops, and incorporating them when they attain maximum vegetative growth. Such crops are usually planted on unoccupied land between the main crops. In oil palm, growing of green manure crops is recommended around palm basins during early stage (up to 3 years), in two rows within a radius of 90 cm from palm basin. While growing, they act as ground cover, preserving the soil structure with their root system, preventing erosion and the washout of nutrients, suppressing the growth of weeds, and also enriching the soil with nitrogen. When oil palm seedlings are planted in the main field, weed growth would be more during initial stages (first 3 years) as the palm plants are at their initial stages of growth. To avoid competition from weeds, basins should be kept clean all the time. Moreover, lots of inter-space lies vacant between palms which facilitates growing green manure crop for weed control and to add organics to the soil when incorporated. Growing of green manure crops also helps to protect the young seedlings from hot weather conditions and damaging winds. Suitable green manure crops, viz. *Sesbania aculeata* (Willd.) (dhaincha), *Sesbania rostrata* Bremek. & Oberm (new dhaincha), *Crotalaria juncea* L. (sunhemp), *Vigna trilobata* (L.) Verdc. (pillipesara/jeeluga), and *Cyamopsis tetragonoloba* (L.) Taub. (cluster bean) for local conditions, can be grown and incorporated into the basin before flowering.

11.5 Bio-fertilizers, Bio-control agents, and Bio-stimulants

The microorganisms responsible for biological N fixation, phosphate solubilization, uptake of immobile elements, and production of plant growth-promoting substances are commonly used as bio-fertilizers. Bio-fertilizers are formulations containing living cells of beneficial microorganisms, multiplied in a suitable carrier. They are cost-effective and eco-friendly inputs in crop production. The rhizobacteria are also known as bio-enhancers as they are capable of increasing root growth and water and nutrient absorption by the host plant. Bio-fertilizers improve soil fertility and crop productivity as an alternative to conventional fertilizers, in sustainable agriculture (Zaidi et al. 2017).

Bio-control agents have been reported to suppress certain plant diseases, soil-borne diseases, and parasites due to their biological control ability (Bender et al. 2016; Mishra and Dash 2014; Nelson and Craft 1992). Unlike chemical fertilizers, bio-fertilizers and bio-control agents have little or no GHG emissions, thus mitigating global warming and climate change.

Plant growth-promoting rhizobacteria (PGPR) belonging to *Azospirillum* and *Bacillus* species, when used as bioinoculants in oil palm seedlings, increase growth and nutrient accumulation (Amir et al. 2005). With PGPR, there was a positive response in oil palm seedlings, leading to better accumulation of N, P, and K in the plant tissues, enhanced root dry weight, and shoot growth. The effectiveness of arbuscular mycorrhizal fungi (AMF) in combination with endophytic bacteria (EB) in the bio-control of the devastating basal stem rot (BSR) disease, caused by *Ganoderma orbiforme* (Fr.) Ryvarden (most commonly known as *Ganoderma boninense*), has been reported in nursery as well as in field conditions by Sundram et al. (2015). The application of two AMF with a combination of *Glomus intraradices* UT126 and *Glomus clarum* BR152B and endophytic bacteria *Pseudomonas aeruginosa* UPMP3 significantly reduced disease development, measured as the area under disease progression curve (AUDPC) and the epidemic rate of disease in the nursery. Though the application of *Glomus intraradices* UT126 singly reduced disease development of BSR, a combination of the two AMF with *P. aeruginosa* UPMP3 reduced the BSR disease incidence to 57% and 80%, respectively, in both nursery and main field. To develop an early associative interaction between the plant cells and bacteria, the introduction of diazotrophic rhizobacteria to oil palm tissues during the in vitro micropropagation process has been found to be an effective strategy (Om et al. 2009). In the association, the diazotrophs provide the host plants with phytohormones and fixed nitrogen. Plants inoculated with *Acetobacter diazotrophicus* (R12) had more root dry weight and volume than plants inoculated with *Azospirillum brasilense* (Sp7). The enhanced root hair formation and root development could be due to the release of plant growth regulators (PGR) such as polyamines, ethylene, indole-acetic acid, and amino acids in both combined-N and N-free cultures, by *Azospirillum* sp., as reported by Thuler et al. (2003).

The important role of phosphate-solubilizing bacteria (PSB) as bio-fertilizers in oil palm cultivation for promoting plant growth has been reported by many researchers, viz. Acevedo et al. (2014) and Situmorang et al. (2015). Phosphobacterin, mainly constituting of bacteria (*Bacillus* spp., *Pseudomonas* spp. and *Rhizobium* spp.) and fungi (*Aspergillus* spp.), inoculations enable insoluble P to become available to plants through a solubilization effect (Bhardwaj et al. 2014; Mishra and Dash 2014). Symbiotic interaction of endophytic bacteria isolated from the roots of oil palm (*Pseudomonas aeruginosa* UPMP3 and *Burkholderia cepacia* UMPB3), with AMF *Glomus intraradices* UT126 and *Glomus clarum* BR152B, and their antagonistic effect on *Ganoderma boninense* have been reported by Sundram et al. (2011). The effect of endophytic bacteria resembled that of mycorrhiza helper bacteria (MHB) association because the association significantly promoted AMF spore germination and hyphal length. However, the endophytic bacteria were extremely damaging to *G. boninense* PER 71. *Azospirillum brasilense* and *A. lipoferum* were found to be effective diazotrophic bioinoculants which enhanced root and shoot growth of the oil palm seedlings by 30% and 60%, respectively, compared to that in un-inoculated plants, and the ^{15}N analysis indicated that biological nitrogen fixation (BNF) could contribute up to 40% of the total N accumulated by the plants (Amir et al. 2001).

11.6 Organic Amendments and Integrated Approaches

Inadvertent use of chemical fertilizers leads to soil and environmental pollution thereby impacting microbial populations in soil and aquatic life. Organic manures enhance physico-chemical and biological properties of the soil, act as soil conditioners, and do not pollute the environment (Mishra and Dash 2014), but rather improve SOC content and contribute to the carbon sequestration (Yan and Gong 2010). However, to meet the entire nutrient requirement of crops through organic sources, huge quantities of organic manures are required. In Malaysia, application of 300 kg of EFB palm^{-1} year^{-1} over a 10-year period led to an increase in soil organic matter from 1.5% to 2.7% and a 9% increase in yield compared to fertilizer treatments alone (Rosenani et al. 2011). Under irrigated conditions in India, recycling of plantation wastes, viz. cut fronds, spent male inflorescences, and failed bunches through in situ mulching, improved soil physical and chemical properties. However, integrated management of nutrients proved better over chemical fertilizer application alone in most of the crops. The integrated application of different sources of nutrients, viz. chemical, organic, and bio-fertilizers, triggers different dynamics of nutrient availability compared to those of chemical fertilizer application alone. The integrated application of nutrients enhances microbial biomass, biological activity, and biodiversity in soil leading to improved soil health and sustainable crop productivity (Bokhtiar and Sakurai 2005; Chen 2006; Dutta et al. 2003). External application of chemical fertilizers accounts for emissions, and they also impact both upstream supply chain and in the field. In integrated system, the recycling of oil palm plantation wastes and palm oil by-products generated in the processing mills makes it possible to reduce the amounts of mineral fertilizers without affecting the productivity of palms. This also maintains nutrient levels in the soil. Studies on EFB application (Sect. 11.3) on soil quality showed improved soil nutrient resources, soil permeability, and soil biological activity. At the same time, the recycling of co-products reduced the polluting emissions and environmental impacts related to waste treatments. The co-composting of EFB with POME drastically reduced the methane emissions, during conventional anaerobic effluent digestion. So, application of chemical, organic sources along with bio fertilizers is the right approach for better soil health and sustaining the production in oil palm. Organic cultivation of oil palm also supported enhanced C and N sequestration (Mardegan et al. 2022).

In India, oil palm is cultivated under irrigated conditions in majority of the oil palm area, and nutrient management is recommended based on soil and leaf nutrient status. Complete organic farming is not in vogue, but integrated system of chemical and natural methods is being practiced by many growers. In some farmers' plantations of West Godavari district (in Makkinavarigudem village), Andhra Pradesh, Jeevamrutham (a cow-based organic nutrient slurry) has been used as organic nutrient source and also as an organic pesticide. But, the experience of farmers revealed that integrating organic system with minimal chemical-based farming is the most productive and economically viable system while safeguarding the soil health. The reason is that oil palm being a heavy feeder of nutrients is not able to meet out its nutrient requirements through organic sources alone. Therefore, the best management practices recommended by ICAR-IIOPR for oil palm, viz. right selection of suitable site, appropriate panting method in plain and sloppy areas, recycling of bio-waste for nutrient supply, soil and leaf analysis-based nutrient management through fertigation, mulching with plantation waste, composting of oil palm plantation and mill waste to supply nutrients, precise water management, weed control, intercropping combinations for different agro-ecological situations, and integrated pest management, contribute to sustainable production of palm oil. Improvement in soil physical properties, water holding capacity, SOC, CEC, available nutrient status, and improved productivity levels were recorded in farmers' plantations by following these practices (Manorama et al. 2021a).

11.7 Integrated Nutrient Management

It is well acknowledged that combined application of mineral and organic manures contributes to better soil health. However, application of mineral fertilizers is the most common practice for improving soil fertility. Continuous application of mineral forms of N fertilizers (urea and ammonium sulphate) lead to acidification, decreased base saturation, and a drop in soil pH (Phicot et al. 1981) resulting in reduced yields. In the soils of Okomu oil palm plantation in Nigeria, rigorous application of mineral fertilizers is reported to cause a cationic imbalance (Ogendengbe 1991).

Contrarily, organic manures contribute to improve biological and physical properties of the soil, thus helping them display the potential to correct the negative impacts of mineral fertilizers on soil. Application of compost promotes the availability of beneficial microorganisms in the soil. It improves the physical properties like structure and porosity, aids in better nutrient conversion, improves organic matter content in soil, and helps in nutrients and water retention. In SE Asia, composting of oil palm mill by-products (EFB, decanter cake, fibre, POME, ashes) is a novel development to sustain oil palm production. Conventional open system of composting is in vogue in majority of palm oil plants (Bukhari et al. 2014). For faster composting, addition of mature compost at the initial stage, followed by urea application 2 weeks later, was found effective in Indonesia (Salètes et al. 2004). By composting of EFB with simultaneous evaporation of oil mill waste water (POME), Schuchardt et al. (2002) obtained a C/N ratio of 1:15 within a period of 12 weeks, with a decrease in carbon by more than 60%. Decanter cake slurry addition can also significantly speed up the decomposing process of the POME and EFB mixture (Yahya et al. 2010a). The high initial C/N ratio in EFB can be lowered by the addition of nitrogen-rich products like urea or mature compost, which improves the activity of microorganisms for a faster processing. Addition of in situ compost minimizes the cost on inorganic sources of fertilizers. In Indonesia, 15 t ha^{-1} of compost applied could reduce the mineral fertilizers by up to 55% (Tohiruddin and Foster 2013). The co-composting of empty bunches together with POME drastically reduced methane emissions during the conventional anaerobic effluent digestion.

Bio-fertilizers help in enhanced availability of soil nutrients due to their solubilizing capacity. Under glass house conditions, about 20–50% of nitrogen requirement of oil palm seedlings was estimated (^{15}N isotope dilution approach) to be contributed through N_2 fixation by diazotrophic bacteria. Under field nursery conditions, inoculating these diazotrophic bacteria into the rhizosphere also improves nutrient accumulation of oil palm seedlings (Amir et al. 2005). In Malaysia, *Bacillus sphaericus* UPMB-10, a rhizobacterium isolated from oil palm roots, has been found to have the ability to fix atmospheric N, contributing 28% of the total N requirement of the plant (Amir et al. 2003). In field-planted young immature oil palm, soil inoculated with *Bacillus sphaericus* UPMB-10 contributed 63% of the total N uptake. Hence, *B. sphaericus* UPMB-10 has the potential to be formulated for use as a biofertilizer (Zakry et al. 2012). In nursery, de Carvalho et al. (2008) (through ^{15}N natural abundance technique) estimated that N derived from atmosphere (NDFA) could meet

50% of the plant's nitrogen requirement. This shows that, to meet the total nutrient needs of the crop, addition of chemical fertilizers is essential. Thus, combined application of bio-compost, bio-fertilizer, and inorganic fertilizers is the best approach, and it is expected to sustain the production as well as soil health.

However, for precise nutrient management, actual requirements need to be estimated. Leaf nutrient concentrations are influenced by several factors including leaf age, leaflet rank, leaf number, fruiting cycle, planting material, palm density, fertilizer treatment, rainfall, and soil properties. Therefore, it is recommended to develop fertilizer recommendations based on both soil and leaf analysis results of a particular location. The optimum nutrient ranges recorded for oil palm plantations of different locations of India (Table 9) can be utilized for interpretation of leaf analysis values and development of fertilizer recommendations for oil palm plantations in such areas.

Further, in addition to optimal nutrient concentrations in leaves, an attempt has also been made to derive soil sufficiency ranges for Krishna–Godavari basin area of Andhra Pradesh, India, where oil palm is cultivated in a very large area (Manorama et al. 2021b). The established nutrient norms are presented in Table 10. The spatial variability existing in nutrient contents of soil as well as oil palm leaf samples clearly indicates the need for site-specific nutrient management through soil and leaf analysis-based assessment. In this direction, an android-based mobile application (https://play.google.com/store/search?q=oil%20palm%20nutrient%20advisor&c=apps) has been developed by ICAR-IIOPR by incorporating the nutrient norms for seven Indian states, viz. Andhra Pradesh, Karnataka, Gujarat, Tamil Nadu, Mizoram, Telangana, and Goa, for generating nutrient advisories based on soil and leaf analysis-based information. This generates advisories for both soil application and fertigation.

11.8 Water Management

Water shortage upsets all facets of metabolism in oil palm, leading to stomatal closure and a considerable decline in the transpiration and photosynthetic rates, thus restricting the production of photo assimilates and growth regulators (Jazayeri et al. 2015; Rivera Mendez et al. 2012). But waterlogged conditions also hamper the absorption and transport of nutrients, particularly nitrogen, phosphorus, potassium,

Table 9 Optimum leaf nutrient ranges of oil palm plantations at different locations in India

State	N (%)	P (%)	K (%)	Mg (%)	B (mg kg^{-1})
Andhra Pradesh	1.57–2.63	0.08–0.16	0.48–0.88	0.25–0.71	22.6–60.2
Goa	1.64–2.79	0.31–0.52	0.37–0.75	0.35–0.63	3.10–13.9
Gujarat	2.63–2.85	0.16–0.18	0.56–0.88	0.34–0.84	9.10–32.5
Karnataka	2.34–2.97	0.08–0.40	0.78–0.91	0.25–0.98	5.71–31.0
Mizoram	1.91–2.95	0.46–0.65	0.63–1.00	0.48–0.88	9.41–31.0
Tamil Nadu	1.61–2.11	0.11–0.12	0.33–0.89	0.23–0.73	30.6–54.8

Adopted from Behera et al. (2022). Nutritional imbalances and nutrient management in oil palm. In: Subhra Saikat Roy, Poonam Kashyap, Tarun Adak (ed) Natural Resource Management in Horticultural Crops, pp 161–185

Table 10 Soil sufficiency ranges of nutrients and optimal concentrations of leaf nutrients

Soil			Leaf		
Nutrient/ parameter	Sufficiency range through BLA approach	Optimal range through DRIS approach	Nutrient	Sufficiency range through BLA approach	Optimal range through DRIS approach
N (kg ha^{-1})	250–653	134–403	N (%)	1.89–3.39	1.72–3.09
P (kg ha^{-1})	62–226	36–206	P (%)	0.10–0.23	0.09–0.22
K (kg ha^{-1})	53–723	56–728	K (%)	0.60–0.96	0.51–0.96
Ca (meq g^{-100})	2.2–6.2	1.94–6.49	Ca (%)	0.95–1.66	0.93–1.62
Mg (meq g^{-100})	0.94–4.3	0.81–4.14	Mg (%)	0.28–0.76	0.34–0.74
S (mg kg^{-1})	21.9–115.2	15.8–85.1	S (%)	0.10–0.23	0.12–0.23
B (mg kg^{-1})	1.3–15.5	1.1–12.24	B (mg kg^{-1})	38–101	27–100
pH	5.8–7.4	6.52–7.83			
EC (dS m^{-1})	0.11–0.39	0.13–0.38			

BLA boundary line approach
Adopted from Manorama et al. (2021b). Land Degrad Dev 1–13

calcium, magnesium, and copper. Rivera et al. (2016) detected that oil palm seedlings grown under waterlogged conditions developed root aerenchyma and pneumatophores, as adaptation mechanisms. Thereby, their gas exchange or micronutrient reduction and assimilation processes (except copper) did not get affected (Fig. 4).

On the other hand, under permanent waterlogging, the seedlings had poor growth because of higher leaf respiration rates and perhaps due to pneumatophore maintenance and limitations in the absorption and transport of macronutrients, while the plants under partial waterlogging exhibited a growth pattern similar to that observed under optimal soil moisture conditions (Rivera et al. 2016). Therefore, optimal water management pays a crucial role in nutrient uptake and thereby influences assimilate production and final FFB yield.

In majority countries, oil palm is cultivated as rainfed crop under ample rainfall conditions. But in India, it is largely grown as irrigated crop. Water requirement depends on age of the plantation, soil type, season, and agro-climatic condition of the location. If the palm does not get enough water, the number of unopened spear leaves increases and depending upon the stage at which the water stress occurred results in either increased maleness or aborted inflorescences. However, application of water without considering evapotranspiration demand, weather conditions, and soil type causes more harm than good, because excess application of water dissolves nutrients present in the root zone and leaches them out into deeper layers of the soil. This causes deprivation of nutrients to plants and consequently affects the yield levels. Therefore, application of water as per crop water demands is very much important in water management.

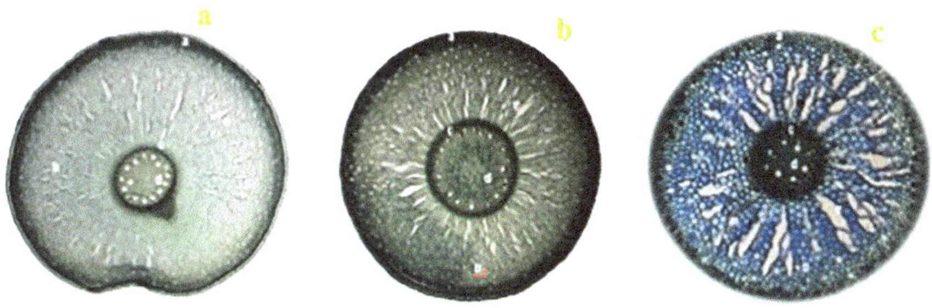

(Adopted from Rivera et al.2016, Agronomía Colombiana 34(2):163-171)

Fig. 4 Methylene blue-stained cross sections (120 µm thick and 3.5 ± 0.5 mm diameter) of primary roots of oil palm grown under waterlogged conditions—optical microscope image (10×). (**A**) Partial waterlogging; (**B**) continuous waterlogging; (**C**) pneumatophore: (a) rhizodermis (in the case of pneumatophores, it broke and the parenchyma became suberized on the edges); (b) parenchyma; (c) endodermis; (d) vascular bundle; (e) aerenchyma. A secondary root or pneumatophore appears from the endodermis of the root under partial waterlogging. (Adopted from Rivera et al. 2016, Agronomía Colombiana 34(2):163–171)

11.9 Plant Protection

Utilizing natural sources to develop antagonistic activity against harmful plant pathogens is recommended in sustainable agriculture to reduce the usage of chemical pesticides that are harmful to the environment (Sadoma et al. 2011). In oil palm, under insect pests, rhinoceros beetle (*Oryctes rhinoceros* L.) is a common pest, and it attacks majority plantations. For managing this problem, pheromone traps are widely used. The natural enemies include *Santalus parallelus*, *Harpalus* sp., *Scarites* sp., *Pheropsophus occipitalis*, and *Agrypnus bifoveatus* sp. nr. known to attack the grub stages of rhinoceros beetle as predators. Rugose spiraling whitefly (*Aleurodicus rugioperculatus* Martin) has emerged as a serious problem of both nursery and main plantations in India in recent times. An entomo-pathogenic fungi, *Isaria fumosorosea* pfu 5 identified by ICAR-NBAIR, Bengaluru, was found to be highly effective against the pest which can be applied at 5 mL L^{-1} of spray fluid along with 5% of any detergent powder. Installation of yellow sticky traps to trap the flying adults during dawn and dusk times can be followed. During early stages of infestation, application of neem oil @ 2.5 mL L^{-1} along with detergent was found effective.

Species of *Trichoderma* are widely renowned as bio-control agents (BCA) as they have the ability to produce antibiotics, capable of mycoparasitism, create competition for nutrients, and produce cell wall-degrading enzymes in defence against different kinds of plant pathogens (Angel et al. 2016; Shakeri and Foster 2007; Sundram 2013). On the other hand, a few species of *Trichoderma* also display other favourable effects like growth promotion, nutrient uptake, improved fertilizer use efficiency, and tolerance to abiotic stresses (Lorito et al. 2010).

Trichoderma has been extensively studied for its role to protect oil palm against basal stem rot disease caused by *Ganoderma. boninense* (Sundram 2013; Sundram et al. 2008). The antifungal metabolites secreted by *Trichoderma* caused physical deformation on the *G. boninense* hyphae during interaction (Angel et al. 2016; Siddiquee et al. 2009). Furthermore, the nursery and field assessment conducted by Sundram et al. (2015) showed the potential of isolated endophytic *Trichoderma* strains as suitable bio-fertilizers for the management of disease in the oil palm cultivation. Taking into consideration their multiple beneficial effects on the oil palm, they can be effectively used for the management of soil health in a sustainable way. Host tolerance to pathogen attack was found increased after the inoculation of AM (arbuscular mycorrhiza), by lowering the disease development in plants (Sundram et al. 2015). The ability of AM as a bio-control agent against *G. boninense*, a white rot fungal pathogen that causes serious disease to the oil palm, was reported (Sundram et al. 2011, 2015).

12 Carbon Sequestration

Plantation crops which are perennial trees represent sizeable carbon pools, as they hold appreciably more carbon per unit of land area than other vegetations and have great potential as carbon sinks. Under irrigated conditions, carbon contents in different fronds of a mature palm were observed to range from 0.413 to 1.314 kg (Suresh et al. 2008). Younger leaves contained higher carbon but no pattern was noticed among the middle and lower whorls. The carbon sequestration in adult oil palm hybrids ranged from 17.98 to 35.44 t C ha^{-1} under irrigated conditions. The amount of C sequestered in a 10-year-old oil palm plantation was 21.18 and 15.78 t ha^{-1} (Table 11), under irrigated and rainfed conditions, respectively. The amount of C sequestered was more in trunk followed by leaves and roots under both the situations (Suresh and Kiran Kumar 2011).

13 Organic Farming Systems

Organic farming relies on management practices that promote and enhance biodiversity, biological cycles, and soil biological activity (USDA 2019). It is considered as holistic production management system that promotes and enhances agroecosystem health, including biodiversity, biological cycles, and soil biological activity. It is one of the alternatives that can be used by farmers for

Table 11 Standing biomass (t ha^{-1}) and carbon sequestered (t C ha^{-1}) by different plant parts in a 10-year oil palm plantation grown under irrigated and rainfed conditions (143 palms ha^{-1})

System	Leaves	C content	Trunk	C content	Roots	C content	Total	C content
Irrigated	20.56	6.78	30.84	12.34	8.22	2.06	59.62	21.18
Rainfed	11.42	3.77	2262	9.04	11.86	2.97	45.89	15.78

Adopted from Suresh and Kiran Kumar (2011). Indian J Dry land Agric Res Dev 26(2): 55–57 sustainable soil management. For sustainable production of oil palm, RSPO (Roundtable on Sustainable Palm Oil), a voluntary global level organization, established in 2004 developed a set of standards for certified palm oil production in 2007, and they are being revised at periodical intervals. Under this, guidelines have been provided for producing certified palm oil by both large-sized estates and individual small holders. Standards are available for taking up new planting, replanting of existing plantations on steep slopes, ecosystem management through recycling (waste management), nutrient management with minimal use of chemical fertilizers, ground cover management for avoiding soil erosion, efficient water management, IPM practices for pest control, etc. (https://rspo.org). In continuation to it, Natural Habitats Group has established Palm Done Right, an educational platform dedicated to ethically producing palm oil, which has formulated the following points for organic production of oil palm:

- Cultivating diverse, non-commercial crops to attract beneficial insects.
- Planting leguminous (i.e. "nitrogen fixing") cover crops to nurture the soil and provide "green compost".
- Multi-cropping for both biodiversity and economic stability.
- Reusing palm leaves and cuttings for mulch and compost to naturally increase the soil's carbon content.
- Collecting grey water through drainage systems to reuse elsewhere on the farm.
- Opting for non-fossil fuel run equipment and transportation—like mules and oxen—as much as possible.
- Using empty palm bunches as fuel for the processing of the palm oil or as a nutrient-rich mulch.

Organic farming in oil palm has positively impacted the organic matter dynamics and soil properties in the South-Eastern Amazon region. Thus, it is considered as a viable alternative for oil palm producers in this region (Brandão et al. 2021). The oil palm growers of Nigeria trust that organic farming is more adaptive to their farming practices as it is in agreement with the social and cultural practices of the people. They also believe that it helps in preservation of the natural environment and suggested that considering the existing indigenous knowledge, organic farming technologies be developed, especially for weed control (76.2% farmers) and for replenishing the soil nutrients (72.4% farmers) (Solomon and Okolo 2008).

14 Future Strategies

Soil health management for oil palm cultivation has to be seen separately under two scenarios, viz. cultivation after replacing primary forests and cultivation on suitable normal land or waste lands. At present, most of the oil palm plantations are located around equator within 15^{0} N and15^{0} S latitudes. This is the hot spot of biodiversity, and majority of tropical forests are present in this range. Further, it also hosts the peatlands which are the largest carbon sinks. Therefore, comparing oil palm cultivation with existing primary forests (when replaced by oil palm) will definitely indicate top soil compaction and erosion (Foster et al. 2011), loss of biodiversity, decreased C and N amounts (Drescher et al. 2016; Goodrick et al. 2015; Hergoualc'h et al. 2017), poor C sequestration, altered nutrient cycling (Hamilton et al. 2016), more GHG emissions, etc., as tropical forests always outweigh oil palm in all these ecosystem functions.

Therefore, it is suggested to avoid expanding oil palm cultivation to the areas where forest cover is already there. However, under Indian conditions, it is being cultivated either as irrigated or rainfed crop on open lands without replacing any forest cover. It is assumed that a perennial plant cover protects the soil better than an annual crop (Jacks and Whyte 1939) and contributes to lesser land degradation than arable farming (Ruthenberg 1972). Oil palm cultivation is eco-friendly as it contributes huge quantities of biomass from plantations, organic wastes from mills, good C sequestration potential, improved biodiversity (Sabrina et al. 2009), less disturbance to soil, and low land degradation by providing protective cover throughout the year, because even in Africa, oil palm cultivation led to improved soil health over a long period of time and it performed the ecological functions similar to forest (Tao et al. 2016, 2017). Therefore, oil palm cultivation as such is not causing any damage to ecosystem, but it is the wrong selection of site and following unsustainable management practices which makes the difference. Moreover, with climate change predictions, it is expected that the suitable areas for oil palm cultivation are more likely to move towards poles in the near future.

Expansion of oil palm cultivation to non-forest areas is the right approach, which can be done through periodical assessment of climate change impacts and studying different scenarios. Though a compromise between economic interests and ecological sustainability is difficult to achieve, conservation agriculture practices are found to be a via media in recent times. Therefore, the best management practices including conservation agriculture practices, viz. mulching, waste recycling, precision application of nutrients, and water, need much emphasis to safeguard the soil health while protecting the economic interests of the growers. With oil palm plantation wastes and mill wastes being lignocellulosic in nature, efficient techniques need to be developed for their faster composting. Diversifying the sources of nutrients through chemicals, organic wastes, and bio-fertilizers (nutrient fixing and solubilizing) improves the nutrient availability, soil biodiversity, and thereby soil health and sustainability. Integrated management of nutrients, pests, and diseases is the priority by developing suitable location-specific packages. Research on molecular techniques for identifying microbial populations and their functions in soil (metagenomics) needs to be strengthened. Designing and development of efficient water and nutrient management systems by incorporating latest developments in the fields of artificial intelligence and Internet of things (IOT) would sufficiently contribute to soil health and sustainability by minimizing the use of chemicals and water.

References

Acevedo E, Galindo-Castaneda T, Prada F, Monica N, Hernan MR (2014) Phosphate-solubilizing microorganisms associated with the rhizosphere of oil palm (*Elaeis guineensis* Jacq.) in Colombia. Appl Soil Ecol 80:26–33

AdeOluwa OO, Adeoye GO (2008) Potential of oil palm empty fruit bunch (EFB) as fertilizer in oil palm (*Elaeis guineensis* L. Jacq.) nurseries. In: Neuhoff D, Halberg N, Alfldi T, Lockeretz W, Thommen A, Rasmussen IA, Hermansen J, Vaarst M, Lck L, Carporali F, Jensen HH, Migliorini P, Willer H (eds) Cultivating the future based on science: 2nd conference of the International Society of Organic Agriculture Research ISOFAR, Modena, Italy. http://orgprints.org/12808/

Agamuthu P, Broughton WJ (1985) Nutrient cycling within the developing oil palm—legume ecosystem. Agric Ecosyst Environ 13:111–123

Alam MZ, Mamun AA, Qudsieh IY, Muyibi SA, Salleh HM, Omar NM (2009) Solid state bioconversion of oil palm empty fruit bunches for cellulase enzyme production using a rotary drum bioreactor. Biochem Eng J 46:61–64. https://doi.org/10.1016/j.bej.2009.03.010

Amir HG, Shamsuddin ZH, Halimi MS, Ramlan MF, Marziah M (2001) Effects of Azospirillum inoculation on N fixation and growth of oil palm plantlets at nursery stage. J Oil Palm Res 13:42–49

Amir HG, Shamsuddin ZH, Halimi MS, Ramlan MF, Marziah M (2003) N_2 fixation, nutrient accumulation and plant growth promotion by rhizobacteria in association with oil palm seedlings. Pak J Biol Sci 6:1269–1272

Amir HG, Shamsuddin ZH, Halimi MS, Marziah M, Ramlan MF (2005) Enhancement in nutrient accumulation and growth of oil palm seedlings caused by PGPR under field nursery conditions. Commun Soil Sci Plant Anal 36:2059–2066

Angel LPL, Yusof MT, Ismail IS, Ping BTY, Azni INAM, Kamarudin NH, Sundram S (2016) An *in vitro* study of the antifungal activity of *Trichoderma virens* 7b and a profile of its non-polar antifungal components released against *Ganoderma boninense*. J Microbiol 54:732–744

Arevalo-Gardini E, Canto M, Alegre J, Loli O, Julca A, Baligar V (2015) Changes in soil physical and chemical properties in long term improved natural and traditional agroforestry management systems of cacao genotypes in Peruvian Amazon. PLoS One 10(7):132–147. https://doi.org/10.1371/journal.pone.0132147

Auler AC, Pires LF, Dos Santos JAB, Caires E, Borges JAR, Giarola NFB (2017) Effects of surface-applied and soil-incorporated lime on some physical attributes of a Dystrudept soil. Soil Use Manag 33:129–140. https://doi.org/10.1111/sum.12330

Auxtero EA, Shamshuddin J (1991) Growth of oil palm (*Elaeis guineensis*) seedlings on acid sulfate soils as affected by water regime and aluminum. Plant Soil 137:243–257

Azwan MB, Norasikin AL, Abd Rahim S, Norman K, Salmah J (2016) Analysis of energy utilisation in Malaysian oil palm mechanisation operation. J Oil Palm Res 28(3):485–495

Babalola OO (2010) Beneficial bacteria of agricultural importance. Biotechnol Lett 32(11):1559–1570

Beaufils ER (1973) Diagnosis and recommendation integrated system (DRIS). Soil Sci. Bull., vol 1. University of Natal

Behera SK, Shukla AK, Suresh K, Manorama K, Mathur RK, Kumar A, Harinarayana P, Prakash C, Tripathi A (2020) Oil palm cultivation enhances soil pH, electrical conductivity, concentrations of exchangeable calcium, magnesium, and available sulfur and soil organic carbon content. Land Degrad Dev 31:2789–2803. https://doi.org/10.1002/ldr.3657

Behera SK, Shukla AK, Suresh K, Mathur RK (2022) Nutritional imbalances and nutrient management in oil palm. In: Roy SS, Kashyap P, Adak T (eds) Natural resource management in horticultural crops. Today and Tomorrow's Printers and Publishers, New Delhi, pp 161–185

Bender SF, Wagg C, Van Der Heijden MGA (2016) An underground revolution: Biodiversity and soil ecological engineering for agricultural sustainability. Trends Ecol Evol 31(6):440–452. https://doi.org/10.1016/j.tree.2016.02.016

Bessou C, Chase LDC, Henson IE, Abdul-Manan AFN, Canals LM, Agus F, Sharma M, Chin M (2014) Pilot application of palm GHG, the round table on sustainable palm oil green house gas calculator for oil palm products. J Clean Prod 73:136–145

Bhardwaj D, Ansari MW, Sahoo RK, Tuteja N (2014) Bio-fertilizers function as key player in sustainable agriculture by improving soil fertility, plant tolerance and crop productivity. Microb Cell Factories 13:66

BLRS Annual Report (1997) Agronomy, crop protection, vol 1. P.T.P.P. London Sumatra Indonesia Tbk Pusat Penelitian Bah Lias—Bah Lias Research Station, Sumatra

Bokhtiar SM, Sakurai K (2005) Effects of organic manure and chemical fertilizers on soil fertility and productivity of plant and ratoon crops of sugarcane. Arch Agron Soil Sci 51:325–334

Brandão F, Schoneveld G, Pacheco P, Vieira I, Piraux M, Mota D (2021) The challenge of reconciling conservation and development in the tropics: lessons from Brazil's oil palm governance model. World Dev 139:105268. https://doi.org/10.1016/j.worlddev.2020.105268

Broeshart H, Ferwerda JJD, Kovachich WG (1957) Mineral deficiency symptoms of the oil palm. Plant Soil 8:289–300

Bukhari NA, Loh SK, Nasrin AB, Sukiran MA, Ngatiman M, Daryl JT, Nordiana MS, Soon LW, Choo YM (2014) Composting of oil palm biomass: current status in Malaysia. Bioresour Technol 100:2867–2873

Buresh RJ, Tian G (1998) Soil improvement by trees in sub-Saharan Africa. In: Nair PKR, Latt CR (eds) Directions in tropical agroforestry research. Springer, Dordrecht, pp 51–76

Caliman JP, Daniel C, Tailliez B (1994) Oil palm mineral nutrition. Plant Réch Dév 1:36–54

Caliman JP, Budi M, Salètes S (2001) Dynamics of nutrient release from empty fruit bunches in field conditions and soil characteristics changes. In: Proceedings of the 2001 PIPOC international palm oil congress. MPOB, Bangi, pp 550–556

Calugar A, Ivan O (2016) Soil microarthropods and their bioindicator value regarding the bioedaphic conditions in forest ecosystems of Danube Delta. Studia Universitatis Vasile Goldis: Seria Stiintele Vietii 26:215–219

CARDI (2010) A manual on integrated farming system. Caribbean Agricultural Research and Development Institute, Ministry of Economic Development, Belize, pp 1–58

Carrión O, Gibson L, Elias DMO, McNamara N, Alen TAV, Op den Camp HJM, Supramaniam CV, McGeity TJ, Murrell JC (2020) Diversity of isoprene-degrading bacteria in phyllosphere and soil communities from a high isoprene-emitting environment: a Malaysian oil palm plantation. Microbiome 8:81. https://doi.org/10.1186/s40168-020-00860-7

Carron MP, Pierrat M, Snoeck D et al (2015a) Temporal variability in soil quality after organic residue application in mature oil palm plantations. Soil Res 53(2):205–215. https://doi.org/10.1071/SR14249

Carron MP, Auriac Q, Snoeck D, Villenave C, Ribetre F, Suhardi, Marichal R, Caliman JP (2015b) Spatial heterogeneity of soil quality around mature oil palms receiving mineral fertilization. Eur J Soil Biol 66:24–31

Chan KW, Lim KC, Ahmad A (1993) Fertilizer efficiency studies in oil palm. In: Basiron Y, Jalani BS, Chang KC, Cheah SC, Henson IE, Kamaruddin N, Paranjothy K, Rajanaidu N, Tayeb D (eds) Presented at the 1991 PORIM international palm oil conference—agriculture. PORIM, Kuala Lumpur, pp 302–311

Chen JH (2006) The combined use of chemical and organic fertilizer and/or bio-fertilizer for crop growth and soil fertility. In: Proceedings of the international workshop on sustained management of the soil—rhizosphere system for efficient crop production and fertilizer use, Bangkok, Thailand, pp 1–11

Chukwu ED, Udoh BT, Okoli NH, Nnabuihe E (2020) Soil microbial activities influenced by oil palm cultivation in a coastal plain sands area of Akwa Ibom State, Nigeria. Int J Sci Res Publ 10(5):373–379. https://doi.org/10.29322/IJSRP.10.05.2020.p10143

Comte I, Colin F, Grünberger O, Follainc S, Whalena JK, Caliman JP (2013) Landscape-scale assessment of soil response to long-term organic and mineral fertilizer application in an industrial oil palm plantation, Indonesia. Agric Ecosyst Environ 169:58–68

Conti FD (2015) Conservation agriculture and soil fauna: only benefits or also potential threats? A review. ECronicon Agric 2:473–482

Corley RHV, Tinker PB (eds) (2016) The oil palm, 5th edn. John Wiley & Sons, Hoboken, NJ

Da Costa CHM, Crusciol CAC (2016) Long-term effects of lime and phosphogypsum application on tropical no-till soybean–oat–sorghum rotation and soil chemical properties. Eur J Agron 74:119–132

Dalton OS, Mohamed AF, Chikere AO (2017) Status evaluation of palm oil waste management sustainability in Malaysia. OIDA Int J Sustain Dev 10:41–47. http://www.ssrn.com/link/OIDA-Intl-Journal-Sustainable-Dev.html

Daniel C, Ochs R (1975) Amélioration de la production des jeunes palmiers_a huile du Pérou par l'emploi d'engrais chloré. Oléagineux 30:295–298

Danso I, Okuere SA, Larbi E, Danso F, Nuertey BN (2019) Assessment of nutrients status of areas supporting optimum oil palm (*Elaeis guineensis* Jacq. L.) cultivation in Ghana. Ghana J Agric Sci 54(2):1–14. https://doi.org/10.4314/gjas.v54i2.1

Danyo G (2013) Commercial oil palm cultivation in Ghana: an outline. ProJ Agric Sci Res 1:22–41

de Carvalho ALV, Alves BJR, Baldani VLD, Reis VM (2008) Application of 15N natural abundance technique for evaluating biological nitrogen fixation in oil palm ecotypes at nursery stage in pot experiments and at mature plantation sites. Plant Soil 302:71–78

de Jesus Costa LR, de Matos GSB, Gomes MF, Kato OR, Castellani DC, Guedes RS, Vasconcelos SS (2023) Soil fertility in oil palm agroforestry systems in the Eastern Amazon, Brazil. Agrofor Syst 97:865–881. https://doi.org/10.1007/s10457-023-00832-4

Dermiyati D, Suharjo R, Telaumbanua M, Yosita R, Sari AW, Andayani AP (2020) Abundance and characterization of microorganisms isolated from oil palm empty fruit bunches waste under aerobic, anaerobic, and facultative anaerobic conditions. Biodiversitas J Biol Divers 21(9):4213–4220. https://doi.org/10.13057/biodiv/d210936

Donough CR, Witt C, Fairhurst TH (2009) Yield intensification in oil palm plantations through best management practice. Better Crops Plant Food 93:12–14

Doran JW, Parkin TB (1994) Defining and assessing soil quality. In: Doran JW, Coleman DC, Bezdieck DF (eds) Defining soil quality for a sustainable environment, SSSA special publication no. 35. Soil Science Society of America, Madison, WI, pp 3–21

Drescher J, Rembold K, Allen K, Beckschäfer P, Buchori D, Clough Y, Faust H, Fauzi AM, Gunawan D, Hertel D, Irawan B, Jaya INS, Klarner B, Kleinn C, Knohl A, Kotowska MM, Krashevska V, Krishna V, Leuschner C, Lorenz W, Meijide A, Melati D, Nomura M, Pérez-Cruzado C, Qaim M, Siregar IZ, Steinebach S, Tjoa A, Tscharntke T, Wick B, Wiegand K, Kreft H, Scheu S (2016) Ecological and socio-economic functions across tropical land use systems after rainforest conversion. Philos Trans R Soc Lond B Biol Sci 371:20150275. https://doi.org/10.1098/rstb.2015.0275

Dubos B, Caliman JP, Corrado F, Quencez P, Suyanto S, Tailliez B (1999) Importance of magnesium nutrition in oil palm e results of several years' experiments. Plant Rech Dév 6:313–325

Dutta S, Pal R, Chakraborty A, Chakrabarti K (2003) Influence of integrated plant nutrient supply system on soil quality restoration in a red and laterite soil. Arch Agron Soil Sci 49:631–637

Fageria N, Baligar V (2008) Chapter 7: Ameliorating soil acidity of tropical oxisols by liming for sustainable crop production. Adv Agron 99:345–399

FAO (1998) World reference base for soil resources, by ISSS–ISRIC–FAO. World soil resources report no. 84, Rome, p 88

Foster HL (1975) The effect of ground management on oil palm nutrition—a review (no. 4), MARDI report. Institiut Penyelidikan dan Kemajuan Pertanian Malaysia (MARDI), Malaysia

Foster HL, Chang KC, Mohd Tayeb D, Tarmizi AM, Zakaria ZZ (1985) Oil palm yield responses to N and K fertilizers in different environments in Peninsular Malaysia. In: PORIM occasional paper. Palm Oil Research Institute of Malaysia, p 23

Foster WA, Snaddon JL, Turner EC, Fayle TM, Cockerill TD, Ellwood MDF, Broad GR, Chung AYC, Eggleton P, Khen CV, Yusah KM (2011) Establishing the evidence base for maintaining biodiversity and ecosystem function in the oil palm landscapes of South East Asia. Philos Trans R Soc Lond B Biol Sci 366:3277–3291. https://doi.org/10.1098/rstb.2011.0041

Fowler D, Nemitz E, Misztal P, Marco CD, Skiba U, Ryder J, Helfter C, Neil Cape J, Owen S, Dorsey J, Gallagher MW, Coyle M, Philips G, Davison B, Langford B, Mackenzie R, Muller J, Siong J, Dari-Salisburgo C, Carl PD, Aruffo E, Giammaria F, Pyle JA, Hewitt CN (2011) Effects of land use on surface-atmosphere exchanges of trace gases and energy in Borneo: comparing fluxes over oil palm plantations and a rainforest. Philos Trans R Soc B Biol Sci 366:3196–3209

Fredricksson NJ, Hermansson M, Wilén BM (2014) Impact of T-RFLP data analysis choices on assessments of microbial community structure and dynamics. BMC Bioinformatics 15:360. https://doi.org/10.1186/s12859-014-0360-8

Ghazali A, Asmah S, Syafiq M, Yahya MS, Aziz N, Tan LP, Norhisham AR, Puan CL, Turner EC, Azhar B (2016) Effects of monoculture and polyculture farming in oil palm smallholdings on terrestrial arthropod diversity. J Asia Pacific Entomol 19:415–421. https://doi.org/10.1016/j.aspen.2016.04.016

Goh KJ (2011) Fertilizer recommendation systems for oil palm: estimating the fertilizer rates. In: Goh KJ, Chiu SB, Paramananthan S (eds) Agronomic principles and practices of oil palm cultivation. Agricultural Crop Trust (ACT), Selangor, pp 338–388

Goh KJ, Hardter R (2003) General oil palm nutrition. In: Fairhurst TH, Hardter R (eds) Managing oil palms for large and sustainable yields. Potash & Phosphate Institute (PPI), Potash & Phosphate Institute of Canada (PPIC) and International Potash Institute (IPI), Singapore, pp 191–230

Gong L, Ran Q, He G, Tiyip T (2015) A soil quality assessment under different land use types in Keriya river basin, Southern Xinjiang, China. Soil Tillage Res 146:223–229. https://doi.org/10.1016/j.still.2014.11.001

Goodrick I, Nelson PN, Banabas M, Wurster CM, Bird MI (2015) Soil carbon balance following conversion of grassland to oil palm. Glob Change Biol Bioenergy 7:263–272. https://doi.org/10.1111/gcbb.12138

Guillaume T, Holtkamp AM, Damris M, Brummer B, Kuzyakov Y (2016) Soil degradation in oil palm and rubber plantations under land resource scarcity. Agric Ecosyst Environ 232:110–118

Hamilton RL, Trimmer M, Bradley C, Pinay G (2016) Deforestation for oil palm alters the fundamental balance of the soil N cycle. Soil Biol Biochem 95:223–232. https://doi.org/10.1016/j.soilbio.2016.01.001

Hartemink AE (2005) Soil fertility decline in the tropics—with case studies on plantations. https://doi.org/10.1079/9780851996707.0000. Accessed on 10 Oct 2022

Hassler E, Corre MD, Kurniawan S, Veldkamp E (2017) Soil nitrogen oxide fluxes from lowland forests converted to smallholder rubber and oil palm plantations in Sumatra, Indonesia. Biogeosciences 14:2781–2798. https://doi.org/10.5194/bg-14-2781-2017

Hättenschwiler S, Barantal S, Ganault P, Gillespie LM, Coq S (2018) Quelsenjeuxsontassociés à la biodiversité des sols? Innov Agron 69:1–14

Hedde M (2018) Indicateursbasés sur la faune des sols: des outils pour l'agricultureinnovante? Innov Agron 69:15–26

Henuk YL, Hasnudi, Yunilas, Ginting N, Mirwandhono E, Hasanuddin, Ginting J, Bakti D, Rosmayati, Purba E, Hafid H, Kapa MMJ (2018) The integrated farming systems between cattle and oil palm plantation in Indonesia. In: Conference paper, 17th ADRI international conference, 23–24 Apr 2018. Grand Daffam Bela Ternate Hotel, Ternate-Maluku Utara

Hergoualc'h K, Hendry DT, Murdiyarso D, Verchot LB (2017) Total and heterotrophic soil respiration in a swamp forest and oil palm plantations on peat in Central Kalimantan, Indonesia. Biogeochemistry 135:203–220. https://doi.org/10.1007/s10533-017-0363-4

Hobbs PR (2007) Conservation agriculture: what is it and why is it important for future sustainable food production? J Agric Sci Cambridge 145(2):127. https://doi.org/10.1017/S0021859607006892

Hoffmann MP, Castaneda Vera A, van Wijk MT, Giller KE, Oberthur T, Doough C, Whitbread AM (2014) Simulating potential growth and yield of oil palm (*Elaeis guineensis*) with PALMSIM: model description, evaluation and application. Agric Syst 131:1–10

Igue AM, Floquet A, Stahr K (2000) Land use and farming systems in Benin. Adapted farming in West Africa: issues, potentials and perspectives, pp 227–238
Imogie AE, Oviasogie PO, Ejedegba BO, Udosen CV (2012) Effect of potassium (K) source on oil palm yield at Okomu oil palm Plc, Ovia North East LGA of Edo state. Int J Plant Res 2:35–38
INSAE (2013) Quatrième Recensement Général de la Population et de l'Habitat. INSAE, Cotonou
IPNI (2015) Oil palm best management practices in Ghana. International Plant Nutrition Institute mid term report. Available at: http://ssa.ipni.net/ipniweb/region/africa.nsf/0/438DA94552CC5B6085257F15004A8C5D/$FILE/Oil%20palm%20project%20Mid%20term%20report.pdf. Accessed on 20 Oct 2022
IPOA (2022) Palm oil industry's performance in 2022. Available at: https://gapki.id/en/news/22601/palm-oil-industrys-performance-in-2022. Accessed on 21 May 2023
Iren OB, Ibanga IJ (2008) The morphological, physical and chemical properties of soils in selected oil palm plantations in Southeastern Nigeria. J Sustain Agric Environ 10(2):160–171
Ishizuka S, Iswandi A, Nakajima Y, Yonemura S, Tsuruta H, Murdiyarso D (2005) The variation of greenhouse gas emissions from soils of various land-use/cover types in Jambi province, Indonesia. Nutr Cycling Agroecosyst 71:17–32
Jaafar HZE, Hafiz M (2012) Photosynthesis and quantum yield of oil palm seedlings to elevated carbon dioxide. In: Advances in photosynthesis—fundamental aspects. InTech. https://doi.org/10.5772/26167
Jacks GV, Whyte RO (1939) The rape of the Earth—a world survey of soil erosion. Faber and Faber, London
Jazayeri SM, Mendez YDR, Camperos JE, Romero HM (2015) Physiological effects of water deficit on two oil palm (Elaeis guineensis Jacq.) genotypes. Agron Colomb 33:164–173. https://doi.org/10.15446/agron.colomb.v33n2.49846
Jessy Kutty PC, Jayachandran BK, Pandurangan AG (2005) Oil palm based cropping system: an approach to medicinal plants conservation and eco-restoration. In: XVII kerala science congress. KFRI, Peechi, pp 25–27
Jose CT, Bhat R, Ismail B, Jayasekhar S (2009) Spatial smoothing technique in field experiments. J Ind Soc Agric Stat 63(2):151–157
Karlen DL, Hurley EG, Andrews SS, Cambardella CA et al (2006) Crop rotation effects on soil quality at three northern corn/soybean belt locations. Agron J 98(3):484–495. https://doi.org/10.2134/agronj2005.0098
Kee KK (1995) Regional rainfall pattern and climatic limitations for plantation crops in peninsular Malaysia. Planter 71(827):67–78
Kee KK, Goh KJ, Chow C, Teo L (2005) Improvement of efficiency of fertilizer applications. In: Chew PS, Tan YP (eds) Proceedings of agronomy and crop management workshops. Presented at the 2004 MOSTA best practices workshop. Malaysian Oil Scientists' and Technologists' Association (MOSTA), Petaling Jaya, pp 269–291
Kheong LV, Zaharah AR, Hanafi MM, Hussein A (2010) Empty fruit bunch application and oil palm root proliferation. J Oil Palm Res 22:750–757
Kibblewhite M, Ritz K, Swift M (2008) Soil health in agricultural systems. Philos Trans R Soc Lond B Biol Sci 363:685–701. https://doi.org/10.1098/rstb.2007.2178
Kindohounde NS, Nodichao L, Adoloukpe NSH, Saidou A (2021) Mapping of soil nutrient deficiency in oil palm plantations of southern Benin. African Crop Sci J 29(1):141–156
Kirk JL, Lee AB, Hart M, Moutoglis P, Klironomos JN, Lee H, Trevors JT (2004) Methods of studying soil microbial diversity. J Microbiol Methods 58(2):169–188
Kotowska MM, Leuschner C, Triadiati T (2015) Quantifying above- and belowground biomass carbon loss with forest conversion in tropical lowlands of Sumatra (Indonesia). Glob Chang Biol 21:3620–3634
Kouame K, Camara B, Kone B, Ballo K, Sekou D, Ake S (2017) Bibliographical review about mineral nutrition and fertilization of palm tree (*Elaeis guineensis* Jacq) at production stage. Int J Agric Environ Res 3:2040–2064

Kurniawan S, Corre MD, Utami SR, Veldkamp E (2018) Soil biochemical properties and nutrient leaching from smallholder palm oil plantations, Sumatra-Indonesia. Agrivita J Agric Sci 40(2):257–266. https://doi.org/10.17503/agrivita.v40i2.1723

Lal R (2015) Restoring soil quality to mitigate soil degradation. Sustainability 7:5875–5895

Lamade E, Boillet JP (2005) Carbon storage and global change: the role of oil palm. Oleagineux Corps Gras Lipides 12:154–160

Lauzeral A (1980) Les sols d'Amérique Latine et la culture du palmier_a huile. Oléagineux 35:477–490

Lavelle P, Spain A (2001) Soil ecology. Kluwer Academic Publisher, Dordrecht

Li Y, Cui S, Chang SX, Zhang Q (2019) Liming effects on soil pH and crop yield depend on lime material type, application method and rate, and crop species: a global meta-analysis. J Soils Sediments 19:1393–1406

Lo RKS, Chong KP (2020) Metagenomic data of soil microbial community in relation to basal stem rot disease. Data Brief 31:106030. https://doi.org/10.1016/j.dib.2020.106030

Lobell DB (2007) The cost of uncertainty for nitrogen fertilizer management: a sensitivity analysis. Field Crop Res 100:210–217

Lord S, Clay L (1999) Environmental impacts of oil palm—practical considerations in defining sustainability for impacts on the air, land and water. New Britain Palm Oil Ltd. Centre, Dami Oil Palm Research Station, 38 p

Lorito M, Woo SL, Harman GE, Monte E (2010) Translational research on Trichoderma: from omics to the field. Annu Rev Phytopathol 48:395–417

Malik S, Beer M, Megharaj M, Naidu R (2008) The use of molecular techniques to characterize the microbial communities in contaminated soil and water. Environ Int 34(2):265–276. https://doi.org/10.1016/j.envint.2007.09.001

Manorama K, Behera SK, Suresh K, Prasad MV, Mathur RK, Harinarayana P (2021a) Mulching and technological interventions avoid land degradation in an intensive oil palm (*Elaeis guineensis* Jacq.) production system. Land Degrad Dev 1:1–13. https://doi.org/10.1002/ldr.3886

Manorama K, Behera SK, Suresh K (2021b) Establishing optimal nutrient norms in leaf and soil for oil palm in India. Ind Crop Prod 174:114223. https://doi.org/10.1016/j.indcrop.2021.114223

Mardegan SF, Castro AFD, Chaves SSNF, Freitas RSDS, Avelar MS, Filho TAOT (2022) Organic farming enhances soil carbon and nitrogen dynamics in oil palm crops from Southeast Amazon. Soil Sci Plant Nutr 68(1):104–113. https://doi.org/10.1080/00380768.2022.2031285

Melling L, Chua KH, Lim KH (2011) Agro-management of peat soils under oil palm in Sarawak. In: Goh KJ, Chiu SB, Paramananthan S (eds) Agronomic principles and practices of oil palm cultivation. Agricultural Crop Trust (ACT), Selangor, pp 695–728

Mishra P, Dash D (2014) Rejuvenation of bio-fertilizer for sustainable agriculture and economic development. Consilience J Sustain Dev 11(1):41–61

Misson M, Haron R, Kamaroddin MFA, Amin NAS (2009) Pre-treatment of empty palm fruit bunch for production of chemicals via catalytic pyrolysis. Bioresour Technol 100(11):2867–2873. https://doi.org/10.1016/j.biortech.2008.12.060

Mokolobate MS, Haynes RJ (2002) Increases in pH and soluble salts influence the effect that additions of organic residues have on concentrations of exchangeable and soil solution aluminium. Eur J Soil Sci 53:481–489. https://doi.org/10.1046/j.1365-2389.2002.00465.x

Mole B (2014) Fertilizer produces far more greenhouse gas than expected. Science News: Magazine Soc Sci Public. https://www.sciencenews.org/article/fertilizerproduces-far-more-greenhouse-gas-expected

Moradi A, Teh CBS, Goh KJ, Husni MHA, Ishak CF (2014) Decomposition and nutrient release temporal pattern of oil palm residues. Ann Appl Biol 164:208–219

Moradi A, Teh CBS, Goh K, Hanif AHM, Ishak CF (2015) Effect of four soil and water conservation practices on soil physical processes in a non-terraced oil palm plantation. Soil Tillage Res 145:62–71. https://doi.org/10.1016/j.still.2014.08.005

MPOB (2014) Oil palm & the environment. Malaysian Palm Oil Board. Available at: http://www.mpob.gov.my/en/palm info/environment/520

MPOB (2022) Overview of the Malaysian palm oil industry. Malaysian Palm Oil Board. Available at: https://bepi.mpob.gov.my/images/overview/Overview2022.pdf

Murphy DJ, Goggin K, Paterson RRM (2021) Oil palm in the 2020s and beyond: challenges and solutions. CABI Agric Biosci 2:39. https://doi.org/10.1186/s43170-021-00058-3

Namanji S, Ssekyewa C, Slingerland M (2021) Intercropping food crops into oil palm plantations—experiences in Uganda and why it makes sense. Policy brief. Ecological Trends Alliance, Tropenbos International, Kampala, Ede, 4 p

Nath AJ, Brahma B, Sileshi GW, Das AK (2018) Impact of land use changes on the storage of soil organic carbon in active and recalcitrant pools in a humid tropical region of India. Sci Total Environ 624:908–917. https://doi.org/10.1016/j.scitotenv.2017.12.199

Nelson EB, Craft CM (1992) Suppression of dollar spot on creeping bentgrass and annual bluegrass turf with compost-amended top dressings. Plant Dis 76:954–995

Ng SK (1986) La nutrition en phosphore et la fertilisation des palmiers_a huile. Oléagineux 41:307–313

Ng SK, Thamboo S (1967) Nutrient contents of oil palms in Malaya. 1. Nutrients required for reproduction: fruit bunches and male inflorescence. Malays Agric J 46:1–45

Ng SK, Thamboo S, De Souza P (1968) Nutrient contents of oil palms in Malaya. 2. Nutrients in vegetative tissues. Malaysian Agric J 46:61–113

Niswati A, Romelah S, Dermiyati, Tugiyono (2021) Integrated farming system of cattle and oil palm plantation increasing population and diversity of soil fauna in Ultisols soils. IOP Conf Ser Earth Environ Sci 648:012172. https://iopscience.iop.org/article/10.1088/1755-1315/648/1/012172/meta. Accessed on 25 Oct 2022

Ogendengbe HD (1991) The nutrient status of Okomu oil palm plantation. MSc thesis. University of Ibadan, Nigeria

Ollagnier M, Ochs R, Martin G (1970) La fumure du palmier_a huile dans le monde. Fertilité 36:64

Ollagnier M, Daniel C, Fallavier P, Ochs R (1987) The influence of climate and soil on potassium critical level in oil palm leaf analysis. Oléagineux 42:446–450

Om AZ, Ghazali AHA, Keng CL, Ishak Z (2009) Microbial inoculation improves growth of oil palm plants (Elaeis guineensis Jacq.). Trop Life Sci Res 20:71–77

Pacheco AR, Tailliez BJ, Rocha de Souza EJ (1985) Les déficiences minérales du palmier_a huile dans la région du Bel_e m, Par_a (Brésil). Oléagineux 40:295–309

Paramananthan S (2000) Soil requirements of oil palm for high yields. In: Goh KJ (ed) Managing oil palm for high yields: agronomic principles. Malaysian Society of Soil Science and Param Agricultural Surveys, Kuala Lumpur, pp 18–38

Paterson RRM, Lima N (2018) Climate change affecting oil palm agronomy, and oil palm cultivation increasing climate change, require amelioration. Ecol Evol 8(1):452–461

Paterson RRM, Kumar L, Shabani F, Lina N (2017) World climate suitability projections to 2050 and 2100 for growing oil palm. J Agric Sci 155(5):689–702

Phicot JMP, Sedego MP, Arrivers HF (1981) Evaluation de la fertilite dun sol ferrugineux tropical sous l influence de femures milnerales et oganiques. L' Agron Trop 36(2):122–133

Prasad MV, Sairam CV, Arulraj S, Jameema J (2012) Estimation of cost of production of oil palm in Andhra Pradesh. In: Abstracts of paper, plantation crops symposium XX, Coimbatore, p 136

Pulhina FB, Rodel DL, Urquiola JP (2014) Carbon sequestration potential of oil palm in Bohol, Philippines. Ecosyst Dev J 4(2):14–19

Pupathy UT, Paramananthan S (2014) Agro-management for oil palms planted on acid sulphate soils. Sel Papers Soil Sci 1:85–101

Pupathy UT, Sundian N (2020) Key agronomic management factors for maximising oil palm (*Elaeis guineensis* Jacq.) yields on acid sulphate soils in Malaysia and Indonesia. IOP Conf Series Earth Environ Sci 454:012171. https://doi.org/10.1088/1755-1315/454/1/012171

Purwoko D, Safarrida A, Tajuddin T, Rupaedah B, Suyono A, Wahid A, Sugianto M, Suja I (2022) Metagenomic data of microbial in natural empty fruit bunches degradation. Data Brief 41:107967. https://doi.org/10.1016/j.dib.2022.107967

Pushparajah E, Chew PS (1979) Utilization of soil and plant analyses for plantation agriculture. In: The proceedings of the malaysian seminar on the fertility and management of deforested land. Society of Agricultural Scientists, Sabah, Kota Kinabalu, pp 177–199

Pushparajah E, Ng SK, Ratnasingam K (1973) Leaching losses of nitrogen, potassium and magnesium on Peninsular Malaysian soils. In: Preprint of Conference of fertilizer and chemistry in tropical soils. Presented at the Conference of fertilizer and chemistry in tropical soils. Malaysian Society of Soil Science, Kuala Lumpur, pp 121–129

Rahman N, Giller KE, de Neergaard A, Magid J, Van de van G, Bruun TB (2021) The effects of management practices on soil organic carbon stocks of oil palm plantations in Sumatra, Indonesia. J Environ Manag 278:111446. https://doi.org/10.1016/j.jenvman.2020.111446

Ramachandrudu K (2019) Oil palm based systems approach for higher net income. In: Manorama K, Suresh K (eds) Recent advances in oil palm production and special emphasis on emergence of new pest and its management, pp 120–126

Ramachandrudu K, Suneetha V, Suresh K, Rao BN, Manorama K (2018) Systems approach in oil palm for higher productivity and profitability. Technical bulletin, Indian Institute of Oil Palm Research, Pedavegi, Andhra Pradesh, pp 1–52

Rao KK (2009) Effect of different methods of irrigation and nutrient requirement on yield of oil palm. Int J Oil Palm 6(1):31–34

Rao BN, Suresh K, Behera SK, Ramachandrudu K, Manorama K (2014) Nutrient management in oil palm. Technical bulletin. Directorate of Oil Palm Research, Pedavegi, pp 1–24

Ray DK, West PC, Clark M, Gerber JS, Prishchepov AV, Chatterjee S (2019) Climate change has likely already affected global food production. PLoS One 14(5):e0217148. https://doi.org/10.1371/journal.pone.0217148

Reddy BGM, Patil DR, Pattara AS, Shetty H (2009) Irrigation and NPK requirement of oil palm for Tungabhadra command area. Int J Oil Palm 6(1):35–38

Rivera Mendez YD, Chacon LM, Bayona CJ, Romero HM (2012) Physiological response of oil palm interspecific hybrids (*Elaeis oleifera* H.B.K. Cortes versus *Elaeis guineensis* Jacq.) to water deficit. Braz J Plant Physiol 24:273–280. https://doi.org/10.1590/S1677-04202012000400006

Rivera YD, Cuenca JC, Romero HM (2016) Physiological responses of oil palm (*Elaeis guineensis* Jacq.) seedlings under different water soil conditions. Agron Colombiana 34(2):163–171. https://doi.org/10.15446/agron.colomb.v34n2.55568

Robe P, Nalin R, Carpellano C (2013) Extraction of DNA from soil. Eur J Soil Biol 39:183–190

Rolf D (2005) The metagenomics of soil. Nat Rev Microbiol 3:470–478

Rosenani AB, Zauyah S, Kulaseharan S, Jamaluddin N (2011) Effects of ten year application of empty fruit bunches in an oil palm plantation on soil chemical properties. Nutr Cycl Agroecosyst 89:341–349

Rosli NS, Harun S, Jahim JM, Othaman R (2017) Chemical and physical characterization of oil palm empty fruit bunch. Malaysian J Anal Sci 21(1):188–196

Ruthenberg H (1972) Farming systems in the tropics. Clarendon Press, Oxford, 344 p

Sabiani NHM, Awang N (2022) Aerobic composting of oil palm fronds—a review. Appl Ecol Environ Res 20(3):2509–2533. https://doi.org/10.15666/aeer/2003_25092533

Sabrina DT, Hanafi MM, Aziz NAA, Mohamed MTM (2009) Earthworm populations and cast properties in the oil palm plantations. Malaysian J Soil Sci 13:29–42

Sadoma MT, El-Sayed ABB, El-Moghazy SM (2011) Biological control of downy mildew disease of maize caused by *Peronosclerospora sorghi* using certain biocontrol agents alone or in combination. J Agric Res 37:1–11

Saijo S, Yahya S, Hidayat Y, Rosawanti P (2022) Application of empty fruit bunches of oil palm and *Indigofera zollingeriana* for conservation of oil palm plantation. Planta Trop J Agrosains 10(2):160–168. https://doi.org/10.18196/pt.v10i2.15467

Salètes S, Siregar FA, Caliman JP, Liwang T (2004) Ligno-cellulose composting: case study on monitoring oil palm residuals. Compost Sci Util 12(4):372–382

Santiago BB, Magdalena LU, Vanwalleghem T, Machuka MAH (2017) Effects of land use change on soil quality indicators in forest landscapes of the western Amazon. Soil Sci 182(4):128–136

Sanyal SK, Dwivedi BS, Singh VK, Majumdar K, Datta SC, Pattanayak SK, Annapurna K (2015) Phosphorus in relation to dominant cropping sequences in India: chemistry, fertility relations and management options. Curr Sci 108(7):1262–1270

Satriawan H, Fuady Z, Agusni (2017) Soil conservation techniques in oil palm cultivation for sustainable agriculture. Jurnal Pengelolaan Sumberdaya Alam dan Lingkungan 7(2):178–183

Schuchardt F, Darnoko D, Guritno P (2002) Composting of empty oil palm fruit bunch (EFB) with simultaneous evaporation of oil mill waste water (POME). In: Proceedings of international oil palm conference, Nusa Dua, Bali, Indonesia, 8–12 Jul 2002. Bali, Nusa Dua

Senjobi BA (2007) Comparative assessment of the effect of land use and land type on soil degradation and productivity in Ogun State, Nigeria. Published PhD thesis submitted to the Department of Agronomy, University of Ibadan, Ibadan, 161 p

Setiadi BK, Diwyanto W, Puastuti IGAP (2011) Peta Potensi dan Sebaran Areal Perkebunan Kelapa Sawit di Indonesia: Sistem Integrasi Sapi-Kelapa Sawit (SISKA). In: Pusat Penelitian dan Pengembangan Peternakan, Badan Penelitian dan Pengembangan Pertanian, Kementerian Pertanian, Jakarta

Shakeri J, Foster HA (2007) Proteolytic activity and antibiotic production by Trichoderma harzianum in relation to pathogenicity to insects. Enzyme Microb Technol 40(4):961–968

Sharma P, Laor Y, Raviv M, Medina S, Saadi I, Krasnovsky A, Vager M, Levy GJ, Bar-Tal A, Borisover M (2017) Compositional characteristics of organic matter and its water-extractable components across a profile of organically managed soil. Geoderma 286:73–82. https://doi.org/10.1016/j.geoderma.2016.10.014

Siddiquee S, Yusuf UK, Hossain K, Sarwar J (2009) In vitro studies on the potential Trichoderma harzianum for antagonistic properties against *Ganoderma boninense*. J Food Agric Environ 7:970–976

Singh B, Ryan J (2015) Managing fertilizers to enhance soil health. Paris, International Fertilizer Industry Association (IFA), p 24

Singh G, Kow DL, Lee KC, Loong SG (1999) Empty fruit bunches as mulch. In: Singh G (ed) Oil palm and the environment—a Malaysian perspective. Malaysian Palm Oil Growers' Council, Kuala Lumpur, pp 171–183

Singh M, Singh D, Gupta A, Pandey KD, Singh PK, Kumar A (2019) Plant growth promoting rhizobacteria: application in bio-fertilizers and biocontrol of phytopathogens. In: PGPR amelioration in sustainable agriculture. Elsevier, pp 41–66. https://doi.org/10.1016/B978-0-12-815879-1.00003-3

Situmorang EC, Nugroho YA, Wicaksono WA (2014) Impact of empty fruit bunches application on soil bacterial biodiversity in oil palm plantation. In: ICOPE 2014: oil palm cultivation: becoming a model for tomorrow's sustainable agriculture, Cirad, PT SMART and WWF, Bali, Indonesia, 12–14 Feb 2014

Situmorang EC, Prameswara A, Sinthya HC, Mathius TN, Liwang T (2015) Indigenous phosphate solubilizing bacteria from peat soil for an eco-friendly bio-fertilizer in oil palm plantation. KnE Energy 1:65–72

Situmorang EK, Nugroho YA, Prameswara A, Andarini E, Hartono, Setyobudi RH, Toruan-Mathius N, Liwang T (2016) The bacterial diversity investigation in oil palm plantation using terminal restriction length polymorphism. AIP Conf Proc 1744:020017. https://doi.org/10.1063/1.4953491

Solomon O, Okolo C (2008) Small scale oil palm farmers perception of organic agriculture in Imo state, Nigeria. J Environ Extension 7:67–71

Soumare A, Boubekri K, Lyamlouli K, Hafidi M, Ouhdouch Y, Kouisni L (2020) From isolation of phosphate solubilizing microbes to their formulation and use as biofertilizers: status and needs. Front Bioeng Biotechnol 7:425

Stichnothe H, Bessou C (2017) Challenges for life cycle assessment of palm oil production system. Indonesian J Life Cycle Assess Sustain 1:1–9

Storesund JE, Lanzèn A, Nordmann E-L, Armo HR, Lage OM, Ovreas L (2020) Planctomycetes as a vital constituent of the microbial communities inhabiting different layers of the

Meromictic Lake Sælenvannet (Norway). Microorganisms 8(8):1150. https://doi.org/10.3390/microorganisms8081150

Sundram S (2013) The effect of Trichoderma in surface mulches supplemented with conidial drenches for the suppression of Ganoderma basal stem rot in oil palm. J Oil Palm Res 25:314–325

Sundram S, Abdullah F, Zainal AM, Umi KY (2008) Efficacy of single and mixed treatments of Trichoderma harzianum as biocontrol agents of Ganoderma basal stem rot in oil palm. J Oil Palm Res 20:470–483

Sundram S, Sariah M, Idris AS, Radziah O (2011) Symbiotic interaction of endophytic bacteria with arbuscular mycorrhizal fungi and its antagonistic effect on Ganoderma boninense. J Microbiol 49:551–557

Sundram S, Meon S, Seman IA, Othman R (2015) Application of arbuscular mycorrhizal fungi with Pseudomonas aeruginosa UPMP3 reduces the development of Ganoderma basal stem rot disease in oil palm seedlings. Mycorrhiza 25:387–397

Sunitha S, Krishnakumar T (2006) Biomass production and potential nutrient contribution from oil palm at felling. J Plantn Crops 34(3):309–311

Suresh K, Kiran Kumar M (2011) Carbon sequestration potential of oil palm under irrigated and rainfed condition. Indian J Dry Land Agric Res Dev 26(2):55–57

Suresh K, Reddy VM (2004) Studies on nutrient and water management of oil palm. NRCOP annual report, pp 43–45

Suresh K, Reddy VM (2009) Oil palm based cropping systems under irrigated conditions-Indian scenario. In: Proceedings of national conference on oil palm, Vijayawada, 2–4 Feb 2008, pp 155–161

Suresh K, Reddy VM, Sarma KN (2003) Quantification of root biomass in oil palm grown under basin irrigation. Int J Oil Palm 2(3–4):39–41

Suresh K, Reddy VM, Kochu Babu M (2008) Biomass, carbon and nitrogen distribution with leaf age in oil palm. In: Poster presented in the third Indian horticultural congress, Bhubaneswar, 6–9 Nov 2008

Suresh K, Behera SK, Manorama K, Rao BN (2017) Oil palm. In: Hebbar KB, Naresh Kumar S, Chowdappa P (eds) Impact of climate change on plantation crops. Astral International Pvt. Ltd., New Delhi, pp 101–122

Tahir AA, Mohd Barnoh NF, Yusof N, Said NNM, Utsumi M, Yen AM, Hashim H, Noor MJMM, Akhir FNM, Mohamad SEM, Sugiura N, Othman NA, Zakaria Z, Hara H (2019) Microbial diversity in decaying oil palm empty fruit bunches (OPEFB) and isolation of lignin-degrading bacteria from a tropical environment. Microbes Environ 34(2):161–168. https://doi.org/10.1264/jsme2.ME18117

Tampubolon FH, Daniel C, Ochs R (1990) Réponses du palmier_a huile aux fumures azotées et phosphorées_a Sumatra. Oléagineux 45:475–486

Tao HH, Slade EM, Willis KJ, Caliman JP, Snaddon JLS (2016) Effects of soil management practices on soil fauna feeding activity in an Indonesian oil palm plantation. Agric Ecosyst Environ 218:133–140. https://doi.org/10.1016/j.agee.2015.11.012

Tao HH, Snaddon JL, Slade EM (2017) Long-term crop residue application maintains oil palm yield and temporal stability of production. Agron Sustain Dev 37:33. https://doi.org/10.1007/s13593-017-0439-5

Tarigan S, Suhardi, Wahyuningsih R, Pujianto PA, Wahyuningsih RM, Caliman JP (2019) The role of oil palm biomass recycling on soil health in oil palm plantations. IOP Conf Series Earth Environ Sci 336:012019. https://doi.org/10.1088/1755-1315/336/1/012019

Tarmizi AM, Dolmat MTH, Zakaria ZZ (1992) Maximum yield of oil palm in peninsular Malaysia: yield response and efficiency of nutrient recovery. In: Presented at the 1990 ISOPB international workshop on yield potential in the oil palm, Oct 29–30. Palm Oil Research Inst of Malaysia, Phuket, pp 145–153

Teo L, Muhammad Shukri R, Ong KP, Zainurah A (2010) Alternative oil palm fertilizer sources and management. Oil palm bulletin no. 61, pp 11–32

Teoh KC, Chew PS (1984) Investigations on areas and frequencies of fertilizer application in mature oil palms. In: Bachik AT, Pushparajah E (eds) Proceedings of the international conference on soils and nutrition of perennial crops. Presented at the international conference on soils and nutrition of perennial crops, 13–15 Aug 1984. Malaysian Society of Soil Science (MSSS), Kuala Lumpur, pp 375–387

Thuler DS, Floh EIS, Handro W, Barbosa HR (2003) Plant growth regulators and amino acids released by *Azospirillum* sp. in chemically defined media. Lett Appl Microbiol 37:174–178

Tiemann TT, Donough CR, Lim YL, Hardter R, Norton R, Tao HH, Jaramillo R, Satyanarayana T, Zingore S, Oberthur T (2018) Chapter 4: Feeding the palm: a review of oil palm nutrition. Adv Agron 152:149–243

Tinker PB (1976) Soil requirements of the oil palm. In: Corley RHV, Hardon JJ, Wood BJ (eds) Developments in crop science. I: Oil palm research. Elsevier, Amsterdam, pp 165–174

Tinker PBH, Smilde KW (1963) Dry-matter production and nutrient content of plantation oil palms in Nigeria: II. Nutrient content. Plant Soil 19:350–363

Tohiruddin L, Foster HL (2013) Superior effect of compost derived from palm oil mill by-products as a replacement for inorganic fertilizers applied to oil palm. J Oil Palm Res 25(1):123–137

Tóth G, Guicharnaud R-A, Tóth B et al (2014) Phosphorus levels in croplands of the European Union with implications for P fertilizer use. Eur J Agron 55:42–52. https://doi.org/10.1016/j.eja.2013.12.008

Tripathi B, Kim M, Singh D, Lee-Cruz L, Lai-Hoe A, Ainuddin AN, Rusea G, Rahim RA, Husni MHA, Chun J, Adams JM (2012) Tropical soil bacterial communities in Malaysia: pH dominates in the equatorial tropics too. Microbial Ecol 64:474–484. https://doi.org/10.1007/s00248-012-0028-8

USDA (2019) Agricultural Marketing Service—National Organic Program. National Organic Standards Board (NOSB) policy and procedures manual. Available at: https://www.ams.usda.gov/about-ams/programs-offices/national-organicprogram

USDA (2022) Oilseeds and products update. Available at: https://apps.fas.usda.gov/newgainapi/api/Report/DownloadReportByFileName?fileName=Oilseeds%20and%20Products%20Update_Jakarta_Indonesia_ID2022-0021. Accessed on 21 May 2023

Vargas LEP, Laurance WF, Clements GR, Edwards W (2015) The impacts of oil palm agriculture on Colombia's biodiversity: what we know and still need to know. Trop Conserv Sci 8:828–845. https://doi.org/10.1177/194008291500800317

Varghese PT, Sunitha S (2005) Natural resource management-soil and water conservation techniques and land use systems. NRCOP annual report, pp 44–45

Varghese PT, Sunitha S, Nampoorthiri KUK (2007) Integrated plant nutrient management in oil palm plantation through recycling of palm wastes. In: Recent advances in plantation crops research, pp 130–133

Velthof GL, Lesschen JP, Webb J, Pietzak S, Miatkowski Z, Pinto M, Kros J, Oenema O (2014) The impact of the nitrates directive on nitrogen emissions from agriculture in the EU-27 during 2000-2008. Sci Total Environ 468–469:1225–1233

Verheye W (2010) Growth and production of oil palm. In: Land use, land cover and soil sciences. UNESCO-EOLSS Publishers

Vijiandran JR, Husni MHA, Teh CBS, Zaharah AR, Xaviar A, Ho SH (2020) Leaching losses of nutrients on matured oil palms fertilized with straight and compound fertilizers. In: Malaysian soil science conference 2020 (Soils 2020), Johor Bahru, Malaysia

von Uexkull HR, Fairhurst TH (1991) Fertilizing for high yield and quality. The oil palm. IPI, Bern, 79 p

Wigena IDP, Sudrajat Sitorus SRP, Siregar H (2009) Soil and climate characterization and its suitability for nucleus smallholder oil palm at Sei Pagar, Kampar District, Riau Province. J Soil Clim 30:1–16

Yachi S, Loreau M (1999) Biodiversity and ecosystem productivity in a fluctuating environment: the insurance hypothesis. Proc Natl Acad Sci U S A 96:1463–1468. https://doi.org/10.1073/pnas.96.4.1463

Yahya A, Sye CP, Ishola TA et al (2010a) Effect of adding palm oil mill decanter cake slurry with regular turning operation on the composting process and quality of compost from oil palm empty fruit bunches. Bioresour Technol 101:8736–8741

Yahya Z, Husin A, Talib J et al (2010b) Soil compaction and oil palm (*Elaeis guineensis*) yield in a clay textured oil. Am J Agric Biol Sci 5:15–19

Yan X, Gong W (2010) The role of chemical and organic fertilizers on yield, yield variability and carbon sequestration-results of a 19-year experiment. Plant Soil 331:471–480

Zaharah AR, Lim KC (2000) Oil palm empty fruit bunch as a source of nutrients and soil ameliorant in oil palm plantations. Malaysian J Soil Sci 4:51–56

Zaidi A, Khan MS, Saif S, Asfa R, Ahmed B, Shahid M (2017) Role of nitrogen-fixing plant growth-promoting rhizobacteria in sustainable production of vegetables: current perspective. In: Microbial strategies for vegetable production. Springer, Berlin/Heidelberg, pp 49–79

Zakry FAA, Zulkifli HS, Khairuddin A, Zakaria ZZ, Rahim AA (2012) Inoculation of Bacillus sphaericus UPMB-10 to young oil palm and measurement of its uptake of fixed nitrogen using the N isotope dilution technique. Microbes Environ 27(3):257–262. https://doi.org/10.1264/jsme2.ME11309. http://wwwsoc.nii.ac.jp/jsme2/

Zomer RJ, Bossio DA, Sommer R, Verchot LV (2017) Global sequestration potential of increased organic carbon in cropland soils. Sci Rep 7:15554. https://doi.org/10.1038/s41598-017-15794-8

Zulkifli H, Halimah M, Chan KW, Choo YM, Wahid MB (2010) Life cycle assessment for oil palm fresh fruit bunch production from continued land use for oil palm planted on mineral soil (part 2). J Oil Palm Res 22:887–894

K. Manorama is a Principal Scientist of Agronomy at ICAR-Indian Institute of Oil Palm Research with 10 years of experience in nutrient and water management and resource use efficiency in oil palm.

S.K. Behera is a Soil Scientist with 5 years of experience in DRIS-based nutrient management in oil palm plantations and presently works at the ICAR-Indian Institute of Soil Science, Bhopal, India.

K. Suresh is the Director of ICAR-Indian Institute of Oil Palm Research (ICAR-IIOPR) at Pedavegi, Andhra Pradesh, India. He is a Plant Physiologist, since the last 26 years, and his specialization includes physiological and biochemical basis of growth and yield of oil palm.

Arecanut (*Areca catechu* L.)

Ravi Bhat, S. Sujatha, Bhavishya, U. K. Priya, Alka Gupta, and Anok Uchoi

Abstract

The physical, chemical and biological health of the soil is important for sustained crop yield and plant health in perennial crops. Acidic laterite soils support a vast majority of arecanut grown in India. High rainfall, undulating topography, leaching of potassium (K+) and calcium (Ca_2+), phosphorus (P) fixation and high zinc (Zn) fixing capacity are some of the important characteristics of laterite soil. Nutrient concentrations in soils, soil pH, cation exchange capacity (CEC), soil organic carbon (SOC), varieties and age of the plants affect nutrient uptake by plants. Arecanut, being a perennial crop, needs better management throughout the growing period. Thus, the soil health plays a major role in growth and productivity of arecanut. The soil health should be assessed regularly, and the crop need is met as per the requirement for sustained production. Crop waste recycling through vermicomposting, use of microbial consortia, soil test-based nutrient application through organic and inorganic sources, practices of conservation agriculture and growing more than two crops per unit area for diversity in plant population are some of the technologies useful for maintenance of soil health and higher productivity. Studies conducted in arecanut-growing regions indicated improvement in soil health due to adoption of cropping/farming system approach, recycling of wastes through vermicomposting, and other practices like mulching,

R. Bhat (✉) · A. Gupta
ICAR-Central Plantation Crops Research Institute, Kasaragod, Kerala, India

S. Sujatha
ICAR-Indian Institute of Horticulture Research, Bengaluru, Karnataka, India

Bhavishya · U. K. Priya
ICAR-Central Plantation Crops Research Institute, Regional Station, Vittal, Karnataka, India

A. Uchoi
ICAR-Central Plantation Crops Research Institute, Research Center, Kahikuchi, Assam, India

G. V. Thomas, V. Krishnakumar (eds.), *Soil Health Management for Plantation Crops*, https://doi.org/10.1007/978-981-97-0092-9_4

green manure application, etc. which were measured in terms of microbial activity in soil, increased soil fertility parameters and yield of crops. These practices also helped in overcoming negative impact of climate change.

Keywords

Arecanut · Soil health · Soil fertility · Cropping system · Waste recycling · Climate

Abbreviations

µg	Microgram
ABCS	Arecanut-based cropping system
ABIFS	Arecanut-based integrated farming system
ACZ	Agro-climatic zone
Al_2O_3	Aluminium oxide
As	Arsenic
B	Boron
C	Carbon
Ca	Calcium
CA	Conservation agriculture
CaO	Calcium oxide
Cd	Cadmium
CEC	Cation exchange capacity
CO_2	Carbon dioxide
Cr	Chromium
Cu	Copper
EC	Electrical conductivity
Fe	Iron
Fe_2O_3	Ferric oxide
g	Gram
ha	Hectare
HDMSCS	High-density multispecies cropping system
K	Potassium
K_2O	Potassium oxide
kg	Kilogram
MgO	Magnesium oxide
Mn	Manganese
MOP	Muriate of potash
N	Nitrogen
OMR	Organic matter recycling
P	Phosphorus
P_2O_5	Phosphorus pentoxide
Pb	Lead

S	Sulphur
SOC	Soil organic carbon
t	Ton
YLD	Yellow leaf disease
YLD-AHR	Yellow leaf disease-apparently healthy rhizosphere
YLD-DIR	Yellow leaf disease—disease-intensive rhizosphere
Zn	Zinc

1 Introduction

In general, crop productivity and income generation are given more importance for food and nutritional security of burgeoning population and self-sustenance of small and marginal farmers. In this process, soil health management is given least importance resulting in proliferation of biotic and abiotic stresses especially in perennial crops. However, the importance of soil health is being realized not only for sustaining the crop productivity but also for maintaining plant health and in turn the health of human beings and animals. The soil health mainly depends on the soil type, crops grown and input supply. Soils must be managed in a sustainable way to maintain the functions of all global ecosystems (Bockstaller et al. 2009; Hillel 2009; Rao and Rogers 2006) and to improve the productivity of agricultural crops (Lal 2009). Low fertility of many tropical agroecosystems has resulted in meagre crop yields. Lack of adequate soil fertility is the single most important factor that limits crop yields (Fageria 2007). Intensive land use and high rainfall pose greater challenges to maintain soil health and quality in perennial crops like arecanut. The nutrient removal by traditional local varieties in arecanut is estimated at 79 kg N, 28 kg P_2O_5 and 79 kg K_2O ha^{-1} (Rethinam 1990). For high-yielding arecanut varieties with more than 2 kg dry kernel yield $palm^{-1}$, the nutrient requirement is still higher (Sujatha et al. 1999) which is validated through nutrient removal studies (Bhat and Sujatha 2012). Balanced nutrient application based on crop nutrient demand and soil fertility status would be ideal in the present scenario of climate change and environmental pollution. Supplementation of crop nutrient demand through organic and/or inorganic sources is necessary in order to arrest decline in soil fertility and crop productivity.

There is widespread decline in fertility of tropical soils (Hartemink 2006). Though the plantation belts show less decline in soil fertility (Hartemink 2003), different plantation belts show considerable differences in the dynamics of soil chemical properties (Hartemink 2005). Soil health and quality are mainly affected by excessive use of machinery, nutrient imbalance, soil pollution due to excess nutrition, soil loss due to erosion and cultivation of exhaustive crops. The sustenance of soil health in perennial plantation crops needs regular management interventions. This can be achieved through regular application of organic materials to soil along with inorganic sources of nutrients. The following pages highlight different aspects of soil health in arecanut garden.

2 Importance of Arecanut

Arecanut (*Areca catechu* L.) belonging to the family *Palmae*, is a palm growing tall and erect with typically unbranched stem. It is an important plantation crop in India due to high economic value of dry kernel obtained from the fruit. The palms attain a height of 12–30 m and a diameter of 30–45 cm (Raghavan and Baruah 1958). It is the source of arecanut/betel nut/*supari* which is mainly used for the masticatory purpose. Traditionally arecanut is consumed either alone or in combination with betel leaves (*Piper betle* L.) and slaked lime. It also has economic, sociocultural, religious and medicinal importance in India and other Southeast Asian countries.

2.1 Area and Production

Globally, arecanut is cultivated in an area of 1.57 million ha with a production of 2.44 million tons as per 2021–2022 statistics (www.faostat.com). Major arecanut-growing countries, viz. India, Bangladesh, Indonesia, and Myanmar, contribute to 94.5% of the total area and 90% of the total production in the world. India is the leading producer of arecanut with a production of 1.56 million tons from cultivated area of 0.79 million ha during 2021–2022. India accounts for 50.6% of the total area and 64.2% of the global arecanut production (www.faostat.com). India is a major market for arecanut as it is the leading producer and consumer of arecanut products. During 2019–2020, India exported 2639.92 tons of arecanut to 25 countries, mainly Sri Lanka, the Maldives, the UAE, the USA and Nepal (www.dasd.gov.in). Interestingly, India also imported 16,761 tons of arecanut worth Rs. 34,256 lakhs mainly from Sri Lanka and Indonesia. In India, 96.7% of arecanut area is concentrated in the states of Karnataka, Kerala, Assam, Meghalaya, Mizoram and West Bengal. Among these six important states, Karnataka state alone contributes significantly to arecanut cultivation to the tune of 68.4% of area as per 2019–2020 statistics. In recent years, area under arecanut is increasing in nontraditional regions of India such as the northern part of Karnataka, Tamil Nadu and Andhra Pradesh.

2.2 History and Uses

The use of arecanut has long and diverse history and has been mentioned in various ancient Sanskrit texts (650–1300 BC). It has been quoted in 'Anjana Chaitra' by Sisu Mayana as early as 1300 BC (Bhat and Rao 1962). Arecanut cultivation and chewing have been mentioned in the works of Magha (650 BC) in 'Shishupala Vadha' and 'Raghuvamsa' by Kalidasa (fourth Century AD). The popularity of the arecanut in sixth century AD has been described by Amar Simha in his 'Amara Kosha'. Pictorial references to arecanut dating back to the second and fourth century AD can be seen in Ajanta caves (Mohan Rao 1982). Use of arecanut against leukoderma, leprosy, cough, epilepsy, worms, etc. has been described by Vagbhatta in 500 AD. Arecanut has also been used as stimulant and appetizer by Bhavamista in thirteenth century AD. Arecanut in powder form was used as antihelminthic for dogs (Watt 1889). Increase in production of saliva, gastric juices and strengthening of gums is noted with chewing of arecanut. Similarly, the burned arecanut husk ash has been used as a dentifrice in Malaya.

Traditionally, arecanut along with betel leaves is offered to God as material of worship in religious ceremonies. The tradition of 'Tambula' (betel leaf with arecanut) offering was probably prevalent as early as Gupta period. Arecanut is chewed in tender, ripe or processed form either alone or in combination with betel leaves or 'pan' (*Piper betle* L.), lime, tobacco and camphor. The habit of chewing arecanut is also seen in other countries like Nepal, Sri Lanka, Myanmar, Thailand, the Philippines, Africa, Arabia, South China, Pakistan and Bangladesh. Arecanut is common throughout Asia and is spreading increasingly in the Western society where Asian communities have settled. In areas such as Eastern India and southern Sri Lanka, fermented arecanut is also popular. In Taiwan, arecanut is often used in the unripe stage when it is green.

Some of the uses of arecanut stem include inexpensive posts, rafters, trenails, beams, pillars, joints, scaffolding, water channel, spear handles, bows, walking sticks and furniture (Raghavan and Baruah 1958). Eco-friendly products like disposable plates and cups are made out of arecanut leaf sheath (Jagannathan et al. 2011), so also caps and hats for farm workers. Disposable plates and cups are popular in India and are also exported to other countries. It is also used for making sole tops in home footwear and cheap summer wear sandals as they absorb moisture (Annamalai et al. 2004). Arecanut leaf sheath can also be used as cattle feed as they have less silica and lignin (Gowda 2016) which reduces the cost of cattle feed mix by Rs. 1.5–2.0 kg^{-1} without reducing the milk yield. The arecanut leaf sheath can also be used as substrate on commercial scale for the production of oyster mushroom (*Pleurotus sajor-caju*) (Moorthy and Chandramohanan 1991).

The tannins in arecanut (*Chogaru*) obtained as a byproduct during boiling in water are used for dyeing clothes, rope, etc. and for tanning leather in Southeast Asian and Pacific Ocean countries (Baens 1941; Burkill 1935; Padre 2023; Watt 1889). The process of hydrochloric acid hydrolysis followed by mechanical fibrillation of arecanut husk fibre leads to highly crystalline and thermally stable cellulose nanofibrils with very high aspect ratio (Julie Chandra et al. 2016) which is used for many industrial applications. The husk can be used after composting as a potassium-rich manure. Fallen dried leaves can be easily converted into nutrient-rich manure using earthworms (Chowdappa et al.1999). The recycling potential of different plant parts of arecanut is described in detail in a recent review (Sujatha et al. 2015).

3 General Soil Characteristics and Requirements

Traditional arecanut-growing area is seen in red/lateritic soils of Western Ghats region, West Bengal and Assam and clay loams in plains of Karnataka (Mohapatra 1977; Mohapatra and Bhat 1982). Aiyer (1966) listed non-suitable soils for arecanut cultivation as sticky clay, sandy, alluvial, brackish and calcareous soils. Hot, moist, rich alluvial areas of the coastal belt are seen to support arecanut in Malaysia and Fiji. The soils of arecanut-growing regions of India have been characterized by Bhat and Mohapatra (1971) during 1970s–1980s (Table 1).

Table 1 Important soil types in arecanut-growing areas of India (Bhat and Mohapatra 1971)

Location	Soil group	pH
Vittal (South Canara)	Lateritic	5.2
Peechi (Kerala)	Alluvial, lateritic	5.6–6.8
Palode (Kerala)	Lateritic	4.2–5.0
Mohitnagar (West Bengal)	Alluvial	4.5–6.0
Kahikuchi (Assam)	Lateritic	4.4–4.8
Hirehalli (Karnataka)	Clay loam	6.2

Texturally the arecanut soils vary from clay loam to sandy clay loam to sandy loam (Jyothi et al. 2009; Mohapatra 1977). Granite, gneisses and schists are the parent material for the soils in the Western Ghats region of India (Babu 1981). Abundant sesquioxides and bases combined with poor native soil fertility are some of the characteristics of arecanut-growing belt. Highly mobile nutrients like N and K are prone to leaching losses due to heavy rainfall and immobile nutrient like P will be fixed. Acidity of the soil is one more character which is caused by leaching of bases due to high rainfall. These soils have low CEC because of the dominant kaolinite clay leading to low K fixation sites. The arecanut-growing soils of Kolaba and Ratnagiri districts have been characterized as lateritic, mildly acidic, rich in total N and micronutrients and low in bases, P and K (Khadilkar et al. 1964). However, the coastal region has alluvial soils which are neutral, base-saturated and rich in organic matter. The major arecanut-growing area in Karnataka was characterized as laterite soil which is acidic in nature with high Cu and Fe, while other micronutrients, viz. Zn, Mn and B, ranged between deficient to adequate (Badrinath et al. 1993).

The data on fertility status of arecanut-growing regions have been collected from Arecanut Research Station located in different states in India (Mohapatra 1977; Mohapatra and Bhat 1982). Medium available P was noticed at Peechi (Kerala), Mohitnagar (West Bengal) and Kahikuchi (Assam), while, it was low in soils of Vittal, Hirehalli (both in Karnataka) and Palode (Kerala). Medium to high available K was noticed in all the regions except Mohitnagar. The soils of Hirehalli had neutral to alkaline pH, whereas all other centres reported acidic to neutral pH. The soils from Peechi, Mohitnagar, Kahikuchi and Hirehalli showed higher total CaO and MgO contents. All the soils had higher Al_2O_3 content compared to Fe_2O_3 (Table 2).

Sirsi Taluk in Karnataka is having the highest area under the traditional arecanut-growing belt. The soil has been characterized for fertility status (Teena et al. 2023). The soil reaction was slightly acidic to neutral, nonsaline with high content of organic carbon. All the soils studied had low available N, available P was low in 24% of soils too high in 74% of the soils and available K was of medium in 99% of the soils. The exchangeable Ca and Mg were sufficient in all the soils, whereas available S was medium to high in the soils studied. The Zn deficiency was seen in 28% of the soils, and soil was having sufficient availability of Fe, Cu and Mn.

The arecanut-growing soils from five different agro-climatic zones (ACZs), viz. eastern dry zone, southern dry zone, southern transition zone, hilly zone and coastal zone of Karnataka, India, were characterized (Vasundhara et al. 2020, 2021). In all the agro-climatic zones, the soils were deep to very deep, non-gravelly and

Table 2 Nutrient status of soils (0–25 cm depth) of experimental stations under arecanut (Mohapatra 1977)

Stations	pH	Organic carbon (%)	Total N (%)	Available (ppm)		Cao (%)	MgO (%)	Fe_2O_3 (%)	Al_2O_3 (%)
				P_2O_5	K_2O				
Vittal (Karnataka)	5.3–5.6	0.7–1.1	0.05–0.09	3.8–7.1	34–85	0.07–0.30	0.6–1.7	8.8–12.0	12.0–21.4
Hirehalli (Karnataka)	6.5–8.2	0.3–1.4	0.04–0.19	Trace-5.5	30–108	0.30–0.38	0.6–1.1	7.2–19.2	11.3–39.0
Peechi (Kerala)	5.1–5.6	1.0–2.0	0.06–0.16	50–81	115–130	0.53	1.7–2.3	8.0–11.2	17.1–24.6
Palode (Kerala)	4.9–5.0	0.7–1.4	0.06–0.13	Trace-3.0	81–91	0.15–0.23	0.6–1.1	4.8–7.2	18.0–30.9
Mohitnagar (West Bengal)	5.7–6.2	0.1–2.2	0.14–0.22	9.2–69.1	8–55	0.23–0.30	1.7–2.9	4.0–7.2	4.8–14.5
Kahikuchi (Assam)	5.1–5.3	0.5–1.8	0.03–0.12	9.6–29.0	70–190	0.15–0.30	1.1–2.3	2.4–8.0	6.5–13.6

Table 3 Classification of the arecanut soils in India (Vasundhara et al. 2020)

Agro-climatic zone	Serial no.	Location	Soil classification
Eastern dry zone	Pedon 1	Kunigal	Fine-loamy, mixed, isohyperthermic, Rhodic Kanhaplustalfs
	Pedon 2	Hesaraghatta	Fine, mixed, semi-active, isohyperthermic, Typic Haplustalfs
Southern dry zone	Pedon 3	T. Narasipura	Fine, mixed, semi-active isohyperthermic, Rhodic Paleustalfs
	Pedon 4	Maddur	Clayey, skeletal, mixed, active isohyperthermic, Typic Haplustalfs
Southern transitional zone	Pedon 5	H D Kote	Fine, smectitic, superactive isohyperthermic, Vertic Haplustepts
	Pedon 6	Hassan	Fine, mixed, semi-active, isohyperthermic, Typic Haplustepts
Hilly zone	Pedon 7	Mudigere	Fine, kaolinitic, sub-active, isohyperthermic, Typic Rhodustalfs
	Pedon 8	Sirsi	Clayey-skeletal, kaolinitic, isohyperthermic, Ustic Kanhaplohumults
Coastal zone	Pedon 9	Belthangady	Fine, kaolinitic, sub-active, isohyperthermic, Kanhaplic Haplustults
	Pedon 10	Kumta	Fine, mixed, sub-active, isohyperthermic, Typic Paleustults

well-drained. The texture was sandy clay loam in surface to clay in sub-surface. The soils of coastal and hilly zones were strongly acidic with high organic carbon. However, near neutral to moderately alkaline soils with medium to high organic carbon was found in southern transition zone to southern and eastern dry zone. Among the different zones, the cation exchange capacity (CEC) ranged from 2.5 to 32.6 cmol (p+) kg^{-1} with a base saturation ranging from 21.9 to 99.6%. The soil taxonomy was identified at sub-group level (Table 3).

Arecanut needs well-drained and deep soil for better root proliferation. It cannot tolerate water stagnation. However, with special care and microsite management, arecanut can also be grown in fertile clay loam soils.

4 Indicators of Soil Health and Soil Health Assessment

Specific physical, chemical and biological properties of soil are used to evaluate soil health. Soil should be viewed as integrated system for measuring soil health through physical, chemical and biological properties. These three properties are interrelated in a way that physical and chemical properties are shaped by biological activity, and biological activity is enhanced or limited by chemical and physical conditions.

4.1 Biological Properties

Soil organic matter, respiration, microbial biomass and mineralizable N are some of the parameters for biological indicators of soil quality. Soil enzyme activities are considered as integrative indicators of soil health. The soil health under different systems can be compared by measuring the soil indicators. Higher CO_2-C flush has been recorded in arecanut-based high-density multispecies cropping system (ABHDMSCS) compared to arecanut monocropping measured through microbial biomass. ABHDMCS recorded 806.7 kg C ha^{-1} compared to 708.9 kg C ha^{-1} in arecanut monocropping. The ABHDMSCS also recorded higher nitrogen and phosphorus in the biomass flush (134.4 and 16.1 kg N and P ha^{-1}, respectively) as compared to arecanut monocropping system (118.2 and 14.2 kg N and P ha^{-1}, respectively) (Bopaiah 1991). SOC enrichment occurs in plantation belt due to in situ decay of roots that are rich in recalcitrant materials like cellulose and lignin in combination with reduced soil temperatures.

4.2 Chemical Properties

Soil nutrient balance or optimum soil fertility is important to achieve high crop yields. The important chemical soil health indicators include soil pH, organic carbon, EC of soil and nutrient concentration in soil. Apart from playing a key role in soil quality, water-holding capacity and susceptibility of soil to degradation (Feller et al. 2001), the soil organic matter also serves as a source or sink to atmospheric CO_2 (Lal 1997), and a higher microbial biomass and elevated respiration indicate an increase in the soil C content (Sparling et al. 2003). The soil organic matter is also the principal reserve of nutrients such as N in the soil. Top 2 m depth contains large quantities of mineral N in some of the tropical soils (Havlin et al. 2005). The deficit or excess of nutrient concentrations in soil may be identified through optimum nutrient norms developed for different soils which can help in adopting precision agricultural practices that might save inputs by way of skipping nutrient application (Bhat et al. 2012).

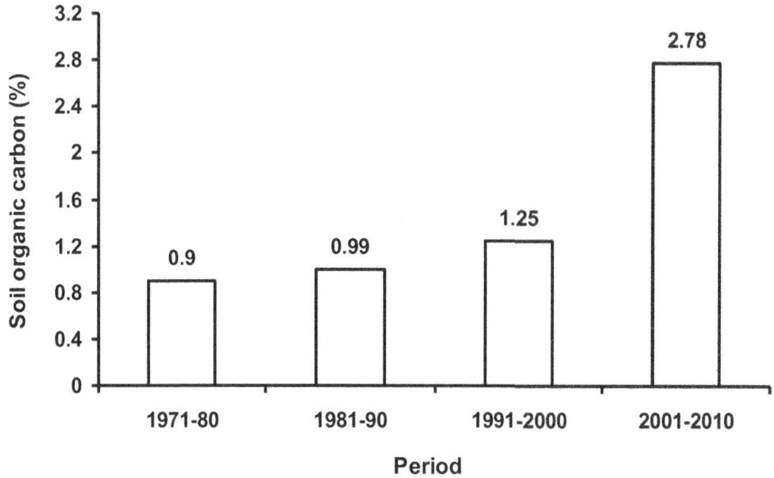

Fig. 1 Soil organic carbon status in arecanut tract

The data on SOC status in arecanut production systems during 1970–2010 is depicted in Fig. 1. There was a steep increase in the SOC in arecanut belt from 0.9% during 1970–1980 to 2.78% during 2001–2010. The increase in SOC may be attributed to green manuring and recycling of wastes (Bhat and Sujatha 2007; Sannamarappa 1987). Enhancement of SOC in crop land soils serves dual benefits, viz. an option for offsetting the increasing atmospheric CO_2 concentration and improving soil fertility for crop production (Smith 2004).

4.3 Physical Properties

The sensitivity of arecanut to both deficit and excess water makes soil physical properties such as texture, bulk density, porosity and water-holding capacity as important parameters for measuring soil health. Bulk density reduced with application of organic manure in arecanut, while application of organic manures increased the water-holding capacity (Acharya et al. 2015). Digging the soil of arecanut garden has improved the productivity of arecanut by 73% compared to no digging indicating the need to loosen the soil for better aeration (Sannamarappa et al. 1976). The poor performance of arecanut in clay soil belt is attributed to soil compaction, poor aeration and water stagnation leading to poor root production. The increasing area under arecanut cultivation in clay soil regions of nontraditional areas in Karnataka has also shown increased problems like disorders and poor root growth due to water stagnation (Bhat and Sujatha 2014). Reduced tillage, crop diversification, intercropping of cover crops, mulching and application of organic amendments have been recommended for improving soil health in these soils.

5 Soil Fertility Constraints and Issues of Soil Contamination

High rainfall, undulating topography, presence of kaolinite clay mineral, leaching of K^+ and Ca_2^+, P fixation and high Zn fixing capacity are some of the important constraints of gravelly laterite soil of red clay which encompasses major arecanut-growing areas. Heavy rainfall during monsoon season causes nutrient loss through run-off. Low nutrient retention and moisture holding capacities of soil are due to low cation exchange capacity (3–15 c mole kg^{-1}), faster infiltration rate, high hydraulic conductivity and porous nature. Laterite soils of Karnataka, India, have shown deficiency of N and K (Badrinath et al. 1993). Water stagnation, poor soil aeration and drainage in clay soils pose problems to monocot palms.

In recent years arecanut is being cultivated on sloppy lands, paddy fallows and eroded lands. There is a need to adopt special management strategies under these situations for better crop production. Soil and water conservation measures like terracing and planting of vetiver on slopes are essential practices before planting arecanut in sloppy lands. Similarly special management strategies like improving soil aeration and breaking hard pans are needed in paddy fallow soils. The problems of fertile clay loam soils like water stagnation, poor soil aeration and slow infiltration can be overcome by microsite improvement. In paddy fallows, the plant growth and nutrient uptake of arecanut is affected by seepage of water and continuous water stagnation above field capacity in the root zone (Bhat and Sujatha 2014; Bhat et al. 2016). Low N use efficiency, and susceptibility to drought, pests and diseases are reported due to K deficiency.

Rock phosphate is one of the important fertilizers used in acidic soil as a source of P in arecanut. Phosphatic fertilizers especially rock phosphate are known to contain Cd, As, Cr, Pb and Zn. Earlier reports have shown that rock phosphate contains the highest levels of heavy metals (Azzi et al. 2017; Sabiha-Javied et al. 2009). Thus, there is a likelihood of addition of heavy metals to the soil through application of rock phosphate which will lead to accumulation of these in plants, and in turn they will enter the human food chain.

Application of blanket doses of fertilizers, leaching of nutrients, excess soil fertility and indiscriminate use of pesticides may cause soil pollution in plantation belt. Excess fertility status in soil may hinder the uptake of important nutrients due to antagonistic nutrient interactions in soil.

6 Impact of Climate Change on Arecanut Soils

The major climatic constraints in arecanut-growing regions of humid tropics are variations in heavy rainfall, high relative humidity and temperature. In other regions, low rainfall and high temperatures are the major problems. High relative humidity leads to proliferation of pests and diseases, whereas heavy rainfall causes leaching of nutrients. Softness of kernel and low nut recovery are problems at high altitude with low temperature. The traditional arecanut-growing regions experience untimely and heavy rains. Heavy rainfall events in laterite soil belt of humid regions with undulating topography leading to water stagnation, run-off, soil erosion and leaching of nutrients causes reduced resource use efficiency. Yield loss of about 15–75% has been reported in arecanut due to drought (Bhat and Sujatha 2004). Heavy rainfall events inflict a yield loss up to 90% due to fruit rot (Jose et al. 2008). The climate change pattern in humid tropics is analysed from long-term weather data, and the effects of weather variations on phenology and yields of arecanut are reported (Sujatha et al. 2017). These findings indicate the need to revise the present recommendations of fertilizer applications and the importance to be given for soil fertility management during flowering and fruit developing periods. This emphasizes the need for thorough understanding of the impact of climate change to streamline the adaptation strategies for sustainability of perennial plantations.

7 Fertilizer Recommendations for Healthy Soils

The presence of component crops, management practices and soil type decide the fertilizer recommendation in arecanut plantation. Fertility management through drip fertigation in arecanut has the benefit of improved yield, profits, soil health and savings in input cost (Bhat and Sujatha 2006). The advantage of drip fertigation like higher root production, increased nutrient use efficiency and reduced nutrient loss due to frequent addition in small quantity is highlighted by several reports (Bhat et al. 2007; Bhat and Sujatha 2009; Sujatha and Bhat 2013a, 2015a). Long-term studies on drip fertigation have shown increase in soil pH and organic carbon content in soils under arecanut and arecanut-cocoa system (Bhat and Sujatha 2009; Sujatha and Bhat 2015a). Recycling of waste biomass through vermicompost is one of the management strategies in arecanut for sustaining soil health. This practice has improved soil fertility status in terms of pH, SOC and the availability of soil test Ca, Mg, Mn and Cu. However, a decrease in the availability of P, K and soil test Fe and Zn is reported (Sujatha and Bhat 2012, 2016). Thus, the scientific approach of need-based nutrient application and organic manures can help in improving soil health and plant productivity.

The integrated application of organic and inorganic sources of nutrients have improved the soil fertility parameters, viz. pH, SOC, P, K, Ca, Mg and micronutrients (Bhat et al. 2007). Arecanut yield has shown significant positive correlation with soil K ($r = 0.31$) indicating the need for continuous application of K to soil. The optimum nutrient limits for laterite soils (Bhat et al. 2012) give a base for adopting need-based nutrient application for better management of soil fertility which can reduce cost of production. Several studies have indicated positive as well as negative effect of crops grown as intercrop in arecanut on soil quality parameters like soil pH and SOC content and soil fertility (Abdul Khader et al. 1992; Bhat and Sujatha 2007, 2011a,b; Manikandan et al. 1987; Muralidharan 1980, 1990; Sujatha et al. 2011b). Increase in soil pH was reported with intercropping in arecanut, whereas the SOC content in soil improved with dicot plants, and it reduced when the intercrops grown were grasses and rhizomatous crops (Sujatha et al. 2011a).

The scarcity and ill effects of chemical fertilizers have to be overcome by locally available organic nutrient sources. One of them is recycling of waste biomass available from the crops. Residues of annual crops are generally used as cattle feed making them unavailable for recycling. Earlier report says that exploring the agricultural potential of soils is possible with recycling of organic wastes (Tiesson et al. 1994). Recycling of waste biomass and addition of dead and decayed roots improve soil health and fertility. Intercropping with leguminous green manure crops as cover crops and incorporating them in the garden led to nutrient recycling and prevented soil erosion thereby helped in maintenance of soil health and fertility (Bopaiah 1982; Mohapatra and Bhat 1982). Intercropping *Pueraria javanica* and *Mimosa invisa* in arecanut gardens and recycling added about 10 kg green manure $palm^{-1}$ which could meet 69–89% of N, 28–43% of P and 29–38% of K requirement of arecanut. The results of several studies showed the importance of cropping system approach for soil fertility management in arecanut-growing regions through nutrient and carbon cycling. Cocoa produces considerable biomass when intercropped with arecanut as litter fall and pruned biomass which can be recycled effectively (Abdul Haris et al. 1999; Balasimha 2007).

The long-term field trials conducted in arecanut-growing region of laterite soil in India during 1970s and 1980s indicated that application of 100 g N, 40 g P_2O_5 and 140 g K_2O along with 14 kg of green leaf $palm^{-1}$ $year^{-1}$ for arecanut with chali or kernel yield of around 2 kg $palm^{-1}$ is optimum. Higher quantity (200:80:240 g of N, P_20_5 and K_20/palm/year) of nutrients are recommended for high-yielding varieties with yield level of above 3 kg kernel $palm^{-1}$ $year^{-1}$ (Sujatha et al. 1999). The recommendation for clay soil belt is 50 g N, 40 g P_2O_5 and 140 g K_2O $palm^{-1}$ $year^{-1}$.

Arecanut is heavy feeder of N and K, and the total uptake of macronutrients in arecanut is in the order of N > K > Ca > P > Mg (Bhat and Sujatha 2012). Since good amount of nutrients are found in leaf and other wastes of arecanut, recycling them in the garden is necessary which avoids nutrient removal from the system. Continuous application of only organics has resulted in K deficiency in arecanut-growing soils (Sujatha and Bhat 2013b, 2016) which in the long run may lead to serious problems to soil and plant health. Antagonistic reactions of nutrients in soil are also reported due to imbalanced nutrient application to arecanut soils (Bhat and Sujatha 2014). Arecanut shows nutrient disorders, viz. crown choking, crown bending and oblique nodes, due to imbalanced nutrient application. Very high soil test P in clay soils resulted in reduced Zn uptake leading to widespread crown-choking disorder. Crown-bending disorder is due to the combined deficiency of Zn and Ca. Deficiency of K has resulted in nut splitting in arecanut. The above facts indicate the importance of balanced nutrient application for better soil health.

Drip fertigation maintains soil fertility status at or near optimum level in the rhizosphere zone of arecanut. Some of the beneficial points from adopting drip fertigation are reduced P fixation, improvement in mobility of P and K beyond 30 cm soil depth and distance from dripping point, improvement in agronomic nutrient use efficiency, saving in nutrient input and water use, better root growth and enrichment of SOC and reduced cost of production (Bhat and Sujatha 2006, 2008; Bhat et al. 2007; Sujatha and Abdul Haris 2000).

Sustaining the yield and soil fertility in perennial system needs proper nutrient balance in soil-plant system. Soil fertility, leaf nutrient status and yield level are the important parameters to be considered for proper planning of nutrient management strategies in arecanut. Plant and soil analysis should be considered for identifying deficiency/toxicity level of nutrients before visual symptoms are noticed. Since excess soil fertility status is more dangerous than deficiency, soil testing plays major role in assessing the soil fertility. Plant nutrient uptake is affected by the antagonistic nutrient interactions in soil. Soil test values of P and K below 30 kg P_2O_5 and 300 kg K_2O ha^{-1}, respectively, in arecanut basins indicate deficit level of particular nutrient in soil leading to the reduction in yield and other problems if nutrients are not applied sufficiently at this stage. When the plants show symptoms of deficiency or toxicity despite good soil fertility, the need for leaf sampling arises for understanding the problem in arecanut. Leaf samples should be collected from middle of the fourth leaf on the crown for nutrient analysis.

8 Organic Farming Approach and Soil Health

Organic methods of production are recognized as environment-friendly and sustainable in terms of crop and soil health (Palm et al. 2001; Roberts et al. 2007; Te Pas and Rees 2014). In north eastern region of India, where arecanut is grown in almost every household, the importance of organics is enormous as application of chemical fertilizer is almost nil (Acharya et al. 2015). Several long-term trials in arecanut indicated enrichment of soil with SOC both in on-station and on-farm situations due to the adoption of practices like green manuring, mulching, cropping systems, biomass recycling, etc. (Bhat 2002; Bhat and Sujatha 2004, 2007, 2009, 2011b, 2012; Bhat et al. 2007; Sujatha and Bhat 2010, 2012; Sujatha et al. 2011a).

9 Cropping System Approach and Soil Health Management

Continuous cultivation of perennial crops like arecanut, banana and cocoa as cropping system model on the same land exhausts the soil in the long run which makes the soil fertility management most important and challenging. Several reports have highlighted the impact of arecanut-based cropping system on soil quality indicators (Bhat and Sujatha 2011a; Sujatha and Bhat 2010, 2013a, 2015b; Sujatha et al. 2011a). The data presented in Table 4 clearly highlight the conspicuous variations in soil fertility parameters in different component crops of various cropping systems. The chemical properties of soil under different plantation crops show differences among them (Hartemink 2006). In general, the inter–/mixed cropping in arecanut has shown increase in soil pH which can be adopted as a management strategy against low pH of laterite soils in arecanut belt. The increase in SOC content in soil under arecanut can be attributed to higher fine root biomass production and in situ root decay in arecanut (Bhat and Sujatha 2008). The intercrops like

Table 4 Soil fertility status as influenced by different component crops in arecanut-based cropping systems

Cropping system	Crop	pH	SOC	P	K	Reference
Arecanut + cocoa	Arecanut	5.7	1.1	–	–	Manikandan et al. (1987)
	Cocoa	5.3	0.8	–	–	
Arecanut + cocoa	Arecanut	5.65	2.63	6.6	144	Sujatha and Bhat (2015a)
	Cocoa	5.37	1.84	11.6	168	
Arecanut + cocoa + banana	Arecanut	5.14	1.87	64.5	62	Bhat et al. (2007)
	Cocoa	5.00	1.62	26.2	100	
	Banana	5.38	–	–	–	
Arecanut + vanilla	Vanilla	6.00	2.45	23.2	202	Sujatha and Bhat (2010)
Arecanut + medicinal and aromatic plants	Medicinal and aromatic plants	5.9–6.5	1.88–2.10	–	–	Sujatha et al. (2011a)
Arecanut + fodder	Arecanut	–	2.40			Sujatha and Bhat (2015b)
	Fodder (napier hybrid bajra	–	2.57–3.09			

vanilla and fodder crop maintain higher SOC status due to adoption of organic farming approach. The arecanut + cocoa system increases available P in arecanut basins compared to sole arecanut. However, the reduction in Zn under the system indicates the need for inclusion of Zn in the fertilizer schedule of arecanut + cocoa system (Manikandan et al. 1987). In all cropping systems, the soil test K levels are less compared to the optimum K levels for laterite soils (Bhat et al. 2012).

10 Influence of Fertilizers and Plant Protection Chemicals on Soil Health and Soil Microbiology

Inorganic fertilizers applied without organic fertilizers resulted in slight acidification of soils in arecanut tract (Mohapatra and Bhat 1990). However, regular application of organic manures resulted in congenial soil conditions for microbial growth as indicated by proliferation of bacteria, fungi and actinomycetes in laterite soils (Bopaiah and Bhat 1981). Above optimum levels of soil organic carbon and soil test nutrients except K were found in arecanut-growing regions of Karnataka, India (Bhat and Sujatha 2014; Bhat et al. 2012). These reports give indications of possible savings in inputs by avoiding blanket applications and using soil test values as guidelines for nutrient management.

Adoption of ABHDMSCS has shown improvement in soil fertility status in terms of soil pH, SOC and available P over years (Bhat and Sujatha 2007). However, there was depletion of soil available K in all the crop rhizospheres during the same period. The arecanut-based cropping system (ABCS) can be made self-sustainable by organic matter recycling over a long-term period in terms of only N and P availability (Bhat and Sujatha 2007). Acidic laterite soils are generally low in available P. Thus, recycling of organic residues is important which improves P availability because organic phosphates are less readily fixed than inorganic P. In ABCS, application of chemical fertilizer in combination with organic matter recycling (OMR) results in higher availability of P in all the crop basins (Bhat and Sujatha 2007) due to higher phosphatase activity (Bhat et al. 2008). Even the arecanut-cocoa system also showed negative balance in available K (Manikandan et al. 1987). The results indicated that OMR alone cannot meet the K requirement of the crops in the system as all the crops included are heavy feeders of K. In another study application of organic manures improved the N content in soil, whereas P and K content in soil reduced drastically over chemical fertilizer applications (Acharya et al. 2015). The population of earthworms and microbials, viz. bacteria, fungi and actinomycetes, was higher with organic manure application.

Bordeaux mixture is a chemical regularly sprayed in arecanut gardens as a preventive for fruit rot caused by *Phytophthora*. The main constituent of Bordeaux mixture is copper sulphate and lime. One school of thought is that the application of Bordeaux mixture over the years has resulted in accumulation of Cu in soil. However, another school of thought is that the quantity used is very low, and it will not have any impact on Cu accumulation in soil. The study in this line is lacking in arecanut soils. However, a meta-analysis of published articles on Cu ecotoxicity indicated that application of more than 400 kg Cu ha^{-1} $year^{-1}$ reduced the microbial activity by 30%. Application of copper up to 3200 kg ha^{-1} $year^{-1}$ did not negatively affect nematodes population, whereas application of 200 kg Cu reduced the earthworm biomass by 15%. The microbial respiration reduced by 50% when soil Cu level was higher than 200 kg Cu ha^{-1}. The final conclusion from the study was that the soil biological quality and functions were not affected with application of 4 kg Cu ha^{-1} $year^{-1}$ (Karimi et al. 2021). Copper content in soil up to 217 mg kg^{-1} did not cause their accumulation in food to cause dietary risk. Further, the total richness and diversity of microorganisms were not reduced with Cu content up to 1700 mg kg^{-1}. The beta diversity and correlation analysis between taxa and Cu content have shown that the Cu contamination exerted significant selection pressure on the soil microbiota (Fagnano et al. 2020). The Cu content in soils of arecanut-growing regions ranges from 2 to 35 mg kg^{-1} (Bhat et al. 2012, 2016; Jayaprakash et al. 2022; Niranjana et al. 2016).

11 Soil Biodiversity, Microbial Synergy and Function-Specific Microbial Communities

The maintenance of soil fertility depends upon the soil biochemical processes which is measured by studies on soil microflora and enzyme activities. The crops grown and the management practices adopted influence the nature and activity of microflora in a given environment. The arecanut rhizosphere has rich diversity of microorganisms (Bopaiah 1982). They are *Aspergillus*, *Chaetomium*, *Fusarium*,

Penicillium, *Rhizoctonia*, *Rhizopus*, *Trichoderma* and *Alternaria* among fungus, *Bacillus* and *Pseudomonas* among bacteria and *Micromonospora*, *Nocardia* and *Streptomyces* among actinomycetes. The asymbiotic N-fixers, viz. *Azotobacter*, *Derxia* and *Beijerinckia*, and P solubilizers, viz. *Aspergillus*, *Penicillium* and *Pseudomonas*, were also enumerated from arecanut rhizosphere. Many of the microbes have cellulose, hemicelluloses, lignin and pectin-degrading properties which are important in decomposing organic matter. Distribution of microflora in arecanut soil rhizosphere has been reported in detail by Bopaiah and Bhat (1981) and Bopaiah and Koti Reddy (1982). Chowdappa et al. (1999) have reported that the vermicompost prepared from arecanut wastes is rich in nutrient content and microbial load. Application of chemical fertilizer along with OMR resulted in higher bacterial abundance in crop basins. Application of inorganic NPK and OMR in the form of vermicompost in ABCS has shown overall increase in microbial biomass C, soil microflora and enzymatic activities (Bhat et al. 2008). Changes in microbial performance caused by environmental conditions is measured by the ratio of soil microbial biomass C to organic C (Anderson and Domsch 1989). The microbial quotient (C_{mic}/C_{org}) is improved over a 20-year period in all component crops of ABCS.

12 Application of Microbiome and Metagenomic Approaches

The rhizosphere bacterial diversity of a plant is considered to play an essential role in mediating plant as well as soil health. Metagenomic approach indicated the presence of various beneficial microorganisms, viz. *Proteobacteria*, *Actinobacteria*, *Acidobacteria*, *Firmicutes* and *Bacteroidetes* in arecanut rhizosphere of Wayanad region of Kerala, India (Mahesh Mohan et al. 2019). Metagenomic approach was used to understand the variation in bacterial phyla for differentiating the arecanut rhizosphere in yellow leaf disease (YLD) affected area. The bacterial community composition was distinct between the YLD endemic apparently healthy rhizosphere soil (YLD-AHR) and YLD endemic disease-intensive rhizosphere soil (YLD-DIR). This study has revealed great genetic richness and abundance of bacterial communities in the rhizosphere niche environment, and *Proteobacteria*, *Bacteroidetes*, *Firmicutes*, *Acidobacteria*, *Planctomycetes*, *Patescibacteria* and *Actinobacteria* were the dominant phyla in the rhizosphere of the arecanut palms in YLD endemic region of India. Significant abundance of chemoautotrophic potential taxa, viz. *Rhizobiales*, *Xanthobacteraceae*, *Myxococcales*, *Enterobacteriaceae*, *Pseudomonadales*, *Nitrosomonadaceae*, *Burkholderiaceae*, *Firmicutes*, *Acidobacteria* and *Patescibacteria*, and their predominant genera, viz. *Anaeromyxobacter*, *Rhodopseudomonas*, *Pseudolabrys*, Xanthobacteria, *Burkholderia*, *Bradyrhizobium*, *Bacillus*, *Acinetobacter*, *Haliangium*, *Geobacter*, *Acidibacter*, actinomycetes, *Parcubacteria*, *Saccharimonadia* and *Candidatus Udaeobacter*, signify their role in N, S and Fe cycling and carbon cycle in rhizosphere (Paulraj et al. 2021, 2022).

13 Soil Health and Management in Yellow Leaf Disease Affected Arecanut Gardens

Yellow leaf disease (YLD) is yellowing of leaves in arecanut to varying intensity in sporadic patches of arecanut plantations in laterite tract. Nayar (1976) reported that the symptoms of YLD are pronounced immediately after the cessation of southwest monsoon, that is, August–September. Previous studies in farmer's gardens indicated that application of nutrients sustains the palms (Chandramohanan 2005). Comparison of earlier reports (Mohapatra et al. 1975, 1976; Yadava et al. 1973) and results published by Bhat et al. (2016) and the established nutrient limits for arecanut plant and soil (Bhat and Sujatha 2013; Bhat et al. 2012) indicated imbalanced soil and plant nutrient status. The YLD-affected areas showed higher SOC, low availability of P and K in soil, above optimum leaf P (0.3%) and critical leaf K levels. Bhat et al. (2016) reported that YLD-affected palms showed deficit of major nutrients and accumulation of micronutrients. It was suggested to apply nitrogen and potassium in four to five splits in post-monsoon season to alleviate the problem of N and K deficiency.

14 Technologies to Promote Soil Health

Achieving ecologically and economically sustainable farming needs a management system which promotes and enhances agroecosystem health. Sustainable soil management with a goal of improving the productivity along with soil health is important. Intensive land use in small and marginal holdings poses great challenge to maintain soil quality and health in perennial crops like arecanut. However, research conducted over six decades has resulted in technologies for sustainable soil health along with improved crop productivity. Resource recycling and better input use efficient techniques can be adopted as per the crop demand by the farming community for sustainable production of arecanut while maintaining the soil health. Nutrient imbalances, excessive fertilization, soil pollution and soil loss processes lead to deterioration in soil health and quality. Shahane and Shivay (2021) have listed the potential options for improving the soil health, viz. the diversification of nutrient sources with emphasis on organic sources, adoption of principles of conservation agriculture, enhancement of soil microbial diversity, efficient resource recycling through the integrated farming system and amendment addition for correcting soil reactions.

14.1 Cropping and Farming System Approach

The benefits of cropping/farming system could be listed as providing balanced food, regular employment, sustaining soil health, mitigating aberrant weather situation and increasing the farm productivity and income. Efficient utilization of the inputs, improving soil fertility parameters and safeguarding the environment are the production processes optimized through cropping system that help in achieving sustainability.

Arecanut is a tall growing palm, and there is an excellent opportunity for a temporal and spatial distribution of other crop species in arecanut because of the rooting pattern and light interception by the palm and microclimate available in the plantation. Resource use efficiency can be improved by inter–/mixed cropping in arecanut which gives ample scope to overcome the soil, weather and crop constraints in arecanut production (Thomas et al. 2011). Congenial microclimate, enrichment of SOC, increased productivity and income are the benefits accrued due to cropping system approach (Sujatha et al. 2020).

Several long-term trials have shown that nutrient cycling, soil fertility and carbon cycling are influenced by inter–/mixed cropping. Cocoa biomass in terms of litter fall and pruned biomass has great potential for recycling as reported by Abdul Haris et al. (1999) and Balasimha (2007). Improvement in terms of soil pH, SOC and available P and reduction in available K have been reported with intercropping of cocoa (Manikandan et al. 1987; Shama Bhat 1983) and vanilla (Sujatha and Bhat 2010). However, intercropping with grasses and rhizomatic medicinal and aromatic plants depleted SOC (Sujatha et al. 2011a).

The intensive cropping system involving arecanut, cocoa, black pepper, clove and banana is a successful crop combination that generates considerable quantity of organic wastes. The ABHDMSCS produced 14.3 t ha^{-1} recyclable biomass (Bavappa et al. 1986). However, different cropping models produce differing quantity of recyclable biomass which varies from 8.72 to 10.35 t ha^{-1} $year^{-1}$. The reasonable quantity of biomass available from arecanut gardens can be converted to vermicompost with a recovery of 82–87% and nutrient content of 1.71% N, 0.21% P and 0.43% K (Biddappa et al. 1996; Bhat and Sujatha 2007; Sujatha et al. 2015). The ABHDMSCS adopted in a clay loam soil at Kahikuchi, Assam, India, showed increase in SOC, available N, P and K and decline in soil pH (Paul et al. 2020).

Arecanut-based integrated farming system (ABIFS) model comprising of arecanut, fodder grass (cv. C03) as intercrop and sole crop, dairy of four to six milking cows and fishery component (264 m^2) was economically sustainable especially for the small and marginal farmers, and recycling of wastes could meet the nutrient requirement of the crop, except K (Sujatha and Bhat 2015b). Dairy unit alone contributed 87% of the organic waste production from arecanut + dairy system which produced 13.7 t ha^{-1} of organic wastes. The waste recycled from ABIFS supplied 218 kg N, 51.8 kg P and 33 kg K which could meet N demand of 1.7 ha and P demand of 2.2 ha of arecanut. The resource use efficiency of the farm can be increased by using the hard laterite soil available on the farm for dairy, fishery and fodder cultivation which otherwise would have been kept fallow.

14.2 Conservation Agriculture Practices

Conservation agriculture (CA) is the name given to the practice adopted to regenerate degraded lands and prevent losses of arable lands. The main components of conservation agriculture are maintenance of permanent soil cover, minimum soil disturbance and diversification of plant species. Adopting CA practices leads to improvement of soil productivity by improving the SOC, in situ soil water conservation and soil structure apart from savings on input cost.

Inter-cultivation was a regular practice in traditional arecanut-growing region as the soil becomes hard after monsoon. It was a process involving significant cost on labours. However, the practice of mulching during monsoon has reduced the requirement of inter-cultivation in arecanut. Some reports also showed negative effect of inter-cultivation on the production of arecanut (Sadanandan et al. 1985).

Arecanut is sensitive to moisture stress, which causes reduction in growth and yield to a large extent. Even though rainfall is high, its distribution is poor in traditional arecanut-growing areas. Heavy rainfall coupled with run-off of soil requires adoption of suitable conservation measures. In sloppy arecanut gardens in north eastern region of India, catch pit with one row of pineapple in the downstream side improved growth and yield of palms effectively and conserved soil moisture (Borah et al. 2006). Other methods like trench with one layer of coconut husk, bunding with pineapple, staggered bunding with soil and coconut husk and half-moon bunding were also effective in conserving the soil moisture and improving the yield compared to control (Borah et al. 2006). On hill slopes, planting arecanut on contour is adopted which will reduce the soil erosion and water run-off. Intercropping with leguminous green manure crops or cover crops is recommended to maintain soil health through recycling of nutrients in the soil profile, preventing soil erosion and improving soil fertility (Bopaiah 1982; Mohapatra and Bhat 1982).

Mulching is generally being followed in the arecanut gardens of Malnad and submontane region of Karnataka, India. Scythed grasses, green leaves with twigs and leaf litter from the nearby forest, fallen arecanut dried leaves and arecanut husk are usually used for mulching the garden or palm basin (Bhat and Sujatha 2004). This checks evaporation and conserves soil moisture during summer and erosion during rainy season, keeps the weeds under check and adds humus and manure to the soil. Addition of organic manure and green leaves, as well as mulching with dried leaves/green leaves with twigs/arecanut husk is followed by the farmers in traditional arecanut-growing region as an effective method to conserve soil and soil moisture.

The potential of HDMSCS in reducing leaching losses as against arecanut monocrop was studied at Vittal (Mohapatra and Bhat 1990). Leaching losses of water-soluble nutrients in HDMSCS are estimated at 0.1 kg P, 116 kg K, 81 kg Ca and 37 kg Mg ha^{-1} as against 6.78 kg P, 190 kg K, 185 kg Ca and 34 kg Mg ha^{-1} in arecanut monocrop. This indicates that the adoption of HDMSCS can reduce the leaching losses of precious nutrients from the soil.

14.3 Cover Cropping and Green Manuring

The green leaves are usually collected from the nearby forest for mulching the palm basin and/or interspaces. In recent years the availability and collection of green leaf manures in required quantity have become a challenge. Hence, green manure or cover crops can be grown in the arecanut gardens wherever it is feasible as it will improve the soil fertility, reduce soil erosion, enhance biological activity, suppress the weed growth and improve the soil physical properties (Mohapatra and Bhat 1982). Studies conducted at Vittal, Hirehalli and Mohitnagar concluded that that *Pueraria javanica* and *Mimosa invisa* are suitable green manure crops in arecanut gardens form the point of view of their green manure yield and nutrient addition capacity (Mohapatra et al. 1970). *P javanica* and *M. invisa* could add on an average 10 kg green manure per palm when grown as intercrops in arecanut gardens which could meet 69–89% of N, 28–43% P and 29–38% of K requirement of arecanut.

Sesbania speciosa is also a green manure crop that can withstand water logging and drought. It is suitable for the areas receiving high rainfall, viz. valleys of Assam, Karnataka and Kerala. Prevention of soil erosion is important in hilly slopes, and *Calopogonium* and *Pueraria* may be used as cover crops for this purpose. *Tephrosia* is used mainly in some parts of Tamil Nadu, India, for its hardiness and high N content. It comes up well in loamy soils also. A comparative study conducted in Maidan tract of Karnataka, India, concluded that *M. invisa* and *Centrosema pubescens* are better green manure crops for arecanut garden over *Pueraria javanica* (*P. phaseoloides*), *Calopogonium mucunoides*, *Crotalaria anagyroides* and *Sesbania speciosa* in terms of production of green manure (Sannamarappa 1987). However, best improvement in SOC status (0.56%, which is 22% higher compared to control) was reported with *M. invisa* and *C. mucunoides*. Cowpea (*Vigna unguiculata*), sunn hemp (*Crotalaria juncea*), dhaincha (*Sesbania aculeata*) and pillipesara (*Phaseolus trilobus*) could be grown for green manuring purpose (Bopaiah 1982). *Gliricidia* can be planted on the border which can be used as a green leaf manure.

14.4 Organic Matter Recycling

The quantity of organic waste generation from arecanut ecosystem is immense, and most of the biomass is not properly utilized by the farming community. Recycling of organic wastes is crucial for preserving natural resources and preventing environmental hazards. They can be used in arecanut ecosystems to improve the soil fertility and productivity. On-farm waste recycling efficiently would be helpful to the arecanut farmers for reducing the dependence on fertilizers and organic manure from external sources, and it can be a key factor for sustainable farming.

Arecanut produces recyclable biomass in the form of leaf, leaf sheath, rachis (bunch wastes) and husk which amounts to 10–12 t ha^{-1} every year. The total recyclable biomass produced by arecanut is 7.9 to 9.4 million t of organic wastes annually in India. The recyclable leaf biomass produced by the palm is higher than husk, leaf sheath and rachis (Bhat and Sujatha 2012). The husk biomass which accounts for 16.5–17% of the total weight of the dry nuts of arecanut (Bhat and Mohapatra 1989; Bhat and Sujatha 2012) is not properly utilized.

Weed infestation in arecanut-growing areas especially with heavy rainfall events is very high. Weed biomass can be effectively used for conserving the soil and has potential to recycle the nutrients. Bhat and Mohapatra (1989) quantified the fresh weed biomass in arecanut plantation as 14 kg per palm which contributed 32 g N, 7 g P_2O_5 and 52 g K_2O. Annual and biennial crops during initial years and perennials such as cocoa, banana, clove, black pepper and acid lime during bearing stage are intercropped in arecanut gardens (Bhat and Sujatha 2011a; Sujatha et al. 2011b). The system generates considerable quantity of organic wastes which amounts as high as 14.3 t ha^{-1} $year^{-1}$ in arecanut-based HDMSCS (Bavappa et al. 1986). There is a wide range in the nutrient composition in these wastes depending upon the type of material, and the reported ranges of nutrients are 0.62–1.59% N, 0.07–0.16% P and 0.75–1.25% K (Bhat and Mohapatra 1989; Bhat and Sujatha 2012). The nutrients supplied by the recyclable biomass in arecanut were estimated at 95 g N, 10 g P_2O_5 and 110 g K_2O $palm^{-1}$ $year^{-1}$ (Bhat and Mohapatra 1989). Arecanut husk being rich in K content can be utilized as a potential source of K in organic farming of arecanut (Bhat and Sujatha 2012). These wastes can supply substantial quantity of nutrients if recycled efficiently and meet major crop nutrient demands.

The available recyclable biomass in arecanut garden is with a high C/N ratio of 37:62 and has high lignin and polyphenol content. This waste when applied directly to soil will lead to N immobilization and fail to meet immediate nutrient demand of the crop (Biddappa et al. 1996; Thomas et al. 1998) . Thus, there is necessity to convert these wastes into composts which can be achieved through earthworms, *Eudrilus eugeniae* and *Eisenia fetida* (Chowdappa et al. 1999). Fine granular and odourless vermicompost from these wastes can be produced within 60 days with a recovery of 80–87%. During this period the earthworm population also doubles. The report also indicated that recycling organic wastes is beneficial through vermicomposting than normal composting. Farmers can make profit from making vermicompost as commercial venture from the wastes available from arecanut gardens.

The cost of production can be reduced through on-farm recycling of waste biomass of arecanut-based cropping system for nutrient management through vermicomposting which produces good-quality compost (1.71% N, 0.21% P and 0.43% K) (Bhat and Sujatha 2007). The recycled arecanut wastes through vermicomposting could meet 50% N, 32.5% P and 26% K requirement of arecanut indicating low P and K supplementation potential of organic wastes. However, the vermicompost contained higher concentrations of secondary and micronutrients than in normal compost and substrate. The leachates derived from vermicomposting popularly known as vermiwash have high concentration of plant nutrients and can be used as liquid fertilizer (Sujatha and Bhat 2013a). The vermiwash obtained from mixing vermicompost and water at 1:10 ratio contains 250 mg kg^{-1} NO_3-N, 132 mg kg^{-1} P and 88–133 mg kg^{-1} K, and this liquid can be successfully used for drip fertigation. Organic matter recycling will reduce the need for chemical fertilizer in arecanut-based cropping system, and only 2/3 of the recommended dose is sufficient for all the crops involved in the cropping system (Bhat and Sujatha 2007). Sujatha and Bhat (2012) have reported that application of vermicompost sustained yield levels of arecanut at 2700 kg ha^{-1} in long run.

14.5 Integrated Nutrient Management Approach

The recent trends in agricultural research are minimization of external inputs and maximization of resource use efficiency. Organics have supplementary and complementary role in preserving the soil quality and environment. Potassium depletion in lateritic and shallow black soils of India (Naidu et al. 2011) and in plantation ecosystems have been reported (Sujatha and Bhat 2012, 2013b, 2016). Thus, crop needs have to be considered to sustain crop productivity along with maintenance of soil health.

Soils of traditional arecanut-growing regions have higher organic carbon. In general, 12 kg each of compost and green leaves are recommended if the SOC is below 2%. Addition of green leaves significantly improves the SOC especially at 0–50 cm soil depth but did not significantly influence the leaf N. Waste biomass recycled in the form of vermicompost to arecanut improved soil fertility status in terms of pH, SOC and the availability of soil test Ca, Mg, Mn and Cu. However, there was a decrease in the availability of major nutrients like P and K and soil test Fe and Zn (Sujatha and Bhat 2012, 2016). N and P nutrient demand of arecanut can be met with application of 4 kg vermicompost palm^{-1} year^{-1} (Sujatha and Bhat 2013b). Bhat and Sujatha (2007) reported optimum mineral N in crop basins of all component crops in ABCS indicating self-sustainable nature of the system with respect to availability of mineral N. Further they reported that INM approach with OMR in ABCS results in higher availability of P and similar nutrient status in all crop basins. This implies the need for integrating organic and inorganic source of nutrients to sustain productivity and maintain soil health.

Application of organics alone improved the soil fertility status in terms of pH, SOC, P, Ca, Mg and micronutrients and depleted soil test K, whereas, inorganic nutrition alone maintained soil fertility status in terms of SOC and optimum nutrient availability except for slight decrease in soil pH (Table 5) (Sujatha and Bhat 2012).

Table 5 Impact of chemical fertilizer, vermicompost and integrated application of both on soil fertility parameters in arecanut garden (Sujatha and Bhat 2012)

	pH	SOC (%)		P (ppm)		K (ppm)	
Treatment	2009	2003	2009	2003	2009	2003	2009
Vermicompost (VC) 100% N equivalent	6.3^{a}	2.36^{ab}	2.85^{ab}	4.9^{cd}	23.3^{b}	102^{c}	82^{c}
VC 200% N equivalent	6.3^{a}	2.30^{ab}	3.00^{a}	5.8^{cd}	35.3^{a}	97^{c}	92^{c}
CF 100%NPK	5.8^{bc}	1.97^{bc}	1.72^{e}	9.1^{ab}	8.5^{d}	162^{abc}	285^{a}
CF 200% NPK	5.7^{c}	1.61^{cd}	1.89^{de}	5.9^{cd}	15.4^{bcd}	187^{ab}	260^{a}
CF 50% NPK + VC 50%	5.8^{bc}	2.11^{ab}	2.36^{bcd}	9.5^{a}	13^{cd}	226^{a}	173^{b}
CF 100% NPK + VC 100%	5.8^{bc}	2.27^{ab}	1.89^{de}	4.8^{cd}	14.6^{cd}	170^{ab}	133^{bc}
CF 50% NPK + VC 150%	6.1^{ab}	2.46^{a}	2.55^{abc}	6.4^{bc}	20.2^{bc}	127^{bc}	141^{bc}
CF 150% NPK + VC 50%	5.7^{c}	2.00^{abc}	2.13^{cde}	5.8^{cd}	17.9^{bc}	222^{a}	189^{b}
Control (no fertilizer or vermicompost)	6.0^{abc}	1.50^{d}	1.96^{de}	2.9^{d}	8.7^{d}	101^{c}	127^{bc}

Note: The numbers with same alphabet are not significant
NPK was at 100 g N, 40 g; P_2O_5, 140 g; K_2O, palm^{-1} year–^{1}

Inorganic nutrition improved root growth and the SOC status due to in situ root decay in perennial crops (Bhat et al. 2007; Bhat and Sujatha 2008). A significant positive correlation between yield and soil K in arecanut ($r = 0.31$) suggests the need for continuous application of K. The sources of potassium may include muriate of potash (MOP), composted arecanut husk, composted coconut coir pith or gliricidia. Naidu et al. (2011) have reported deficiency of K and Zn in Indian soils. These constraints should be considered for formulating the management strategies, and practices like organic manure application, mulching and drip fertigation should be adopted to improve resource use efficiency. Higher K use efficiency can be achieved by adopting either frequent application of K fertilizers or drip fertigation.

The integrated application of inorganic fertilizer and recycling organic wastes also showed significant improvement in microbial population, microbial biomass C, microbial quotient, phosphatase and dehydrogenase activities in basins of various component crops in arecanut-based cropping system compared to interspaces (Bhat et al. 2008). The microbial C was 11 times higher in crop basins (440.1 μg C g^{-1} dry soil) as compared to interspaces (40.8 μg C g^{-1} dry soil).

15 Carbon Sequestration

Being perennial in nature, arecanut palms remain in field for more than three decades and accumulate carbon in both above- and belowground parts. Arecanut can store considerable carbon due to higher growth and biomass increment (Sujatha et al. 2017). Arecanut plantations of different ages recorded carbon stocks varying from 144 to 153 t ha^{-1} (Sujatha and Bhat 2015b). Further, soil carbon stocks (119–137 t ha^{-1}) at 0–30 cm soil depth are higher compared to aboveground carbon stocks (10–21 t ha^{-1}).

Balasimha and Naresh Kumar (2013) reported that 5- to 15-year-old plantations of arecanut—cocoa system recorded a standing biomass of 23.15–87.10 t ha^{-1} indicating high carbon sequestration potential of arecanut-cocoa system. In another study a 12-year-old arecanut plantation recorded 41 t ha^{-1} standing aboveground biomass (Bhat and Sujatha 2012), whereas it was 50 t ha^{-1} in 15-year-old arecanut plantation (Sujatha and Bhat 2016). Annual increment in carbon stock was 1.4–3.0 t ha^{-1} $year^{-1}$ in arecanut (Balasimha and Naresh Kumar 2013).

Significant number of arecanut plantations in traditional arecanut belt in Karnataka and northeast states of India have been established by converting natural forests. Though this is a matter of great concern, but arecanut plantations can sequester terrestrial carbon by accumulating carbon in woody biomass (Brahma et al. 2018) which is comparable to forest trees (Balasimha and Naresh Kumar 2013). However, the total stored biomass carbon is lesser in arecanut monoculture than traditional agroforestry systems and natural forest in Northeast India (Brahma et al. 2018). The enrichment of SOC status is also lesser in arecanut plantations than in forest (Das et al. 2021). Hence, it is necessary to adopt inter–/mixed cropping system to get the benefit of carbon sequestration.

Adoption of cropping/farming system will not only increase the total biomass but also increase the SOC content (Abdul Khader et al. 1992; Bhat and Sujatha 2007; Manikandan et al. 1987; Muralidharan 1980; Sujatha and Bhat 2015b). Growing green manure/cover crop in arecanut plantation will also increase the SOC. Arecanut-based mixed farming system involving fodder component for dairy enriches SOC and increases carbon sequestration (Sujatha and Bhat 2015b).

16 Future Strategies

Increasing population and dwindling natural resources warrant judicious and efficient use of natural resources for the sustained production of the crop. Soil health plays a major role in achieving sustained production in the long run. The management of soil health should form the focus in the crop production function to achieve sustainability in the context of the challenges faced by the crop. A change is required from the reliance on inorganic fertilizers to integrated or organic approach with focus on increasing biodiversity and biological functions. Promotion of soil microbiome requires management of microbial abundance, community composition and function. Climate change mitigation requires strategies for sequestration of carbon in soil and reduction of greenhouse gas emissions. The application of organics should be made compulsory, and the application of inorganic nutrients should be need-based. Improving soil health and reducing soil and water erosion in the plantations need adoption of recycling of organic wastes and inter- and mixed cropping. The requirement of soil should also be considered while planning nutrition of crops. Better management strategies, viz. cropping systems or mixed farming approach, need to be adopted for maintenance of soil health in arecanut ecosystem in view of climate change and ecological imbalance.

References

Abdul Haris A, Balasimha D, Sujatha S, Bhat R, Khader KBA (1999) The influence of drip irrigation and fertilizer on yield and photosynthetic characteristics of cocoa in mixed cropping system with arecanut. J Plantn Crops 27:131–135

Abdul Khader KB, Balasimha D, Bhat NT (1992) Resource use in arecanut based high density multispecies cropping system. J Plantn Crops 20:19–24

Acharya GC, Paul SC, Chakrabarty R, Ray AK (2015) Effect of organic and inorganic sources of nutrients on soil fertility status of arecanut (*Areca catechu*) in north-east India. Indian J Agric Sci 85(10):1335–1341

Aiyer AKYN (1966) Field crops of India, 6th edn. Bangalore Printing and Publishing Co. Ltd., Bangalore, p 564

Anderson TH, Domsch KH (1989) Ratios of microbial biomass carbon to total organic carbon in arable soils. Soil Biol Biochem 21:471–479

Annamalai SJK, Azeez S, Nayar NM (2004) Alternative uses of arecanut and utilisation of by-products. In: Balasimha D, Rajagopal V (eds) Arecanut. CPCRI, Kasaragod, pp 224–258

Azzi V, Kazpard V, Lartiges B, Kobeissi A, Kanso A, El Samrani AG (2017) Trace metals in phosphate fertilizers use in eastern Mediterranean countries. CLEAN – Soil Air Water 45(1):1500988. https://doi.org/10.1002/clen.201500988. 29 pages

Babu PVLP (1981) Laterite as an unconformity plane in the evolution of the Indian peninsula - a synthesis. In: Anon (ed) Laterisation processes. Oxford & IBH Publishing Co., New Delhi. 450 p

Badrinath MR, Gajendragad, Balakrishna Rao K (1993) Distribution of micronutrients in laterite soils of Puttur in relation to some soil properties. In: Sehgal J, Sarma VAK, Batta RK, Gajbhiye KS, Nagabhushna SR, Venugopal KR (eds) Red and lateritic soils of India – resource appraisal and management. NBSSLUP, Bangalore, pp 271–274

Baens L (1941) Possible utilization of tannins from the kernels of green betel nuts. Rev Filipinia Med Farm 32:143–145

Balasimha D (2007) Efficacy of pruning in enhancing bean yield of cocoa. J Plantn Crops 35:201–202

Balasimha D, Naresh Kumar S (2013) Net primary productivity, carbon sequestration and carbon stocks in areca-cocoa mixed crop system. J Plantn Crops 41(1):8–13

Bavappa KVA, Kailasam C, Abdul Khader KB, Biddappa CC, Khan HH, Kasturi Bai KV, Ramadasan A, Sundararaju P, Bopaiah BM, Thomas GV, Misra LP, Balasimha D, Bhat NT, Shama Bhat K (1986) Coconut and arecanut based high density multispecies cropping systems. J Plantn Crops 14:74–87

Bhat R (2002) Climate, soil and agronomy. In: Balasimha D (ed) Cocoa. Central Plantation Crops Research Institute, Kasaragod, pp 48–69

Bhat NT, Mohapatra AR (1971) Evaluation of interrelationship between organic carbon and available nitrogen in some soils where arecanut grown in India. Indian J Agric Sci 41:663–665

Bhat NT, Mohapatra AR (1989) Effect of supplying nutrients through organic manures, inorganic fertilisers and their combination on arecanut crop. J Plantn Crops 16(Suppl):443–447

Bhat PSI, Rao KSN (1962) On the antiquity of arecanut. Arecanut J 13:13–21

Bhat R, Sujatha S (2004) Crop management. In: Balasimha D, Rajagopal V (eds) Arecanut. Central Plantation Crops Research Institute, Kasaragod, pp 76–102

Bhat R, Sujatha S (2006) Cost-benefit analysis of ferti-drip irrigation in arecanut (*Areca catechu* L.). J Plantn Crops 34:263–267

Bhat R, Sujatha S (2007) Soil fertility status as influenced by arecanut based cropping system and nutrient management. J Plantn Crops 35:158–165

Bhat R, Sujatha S (2008) Root distribution of arecanut (*Areca catechu* L.) as influenced by drip fertigation in a laterite soil. Indian J Agric Sci 78(9):804–807

Bhat R, Sujatha S (2009) Soil fertility and nutrient uptake by arecanut (*Areca catechu* L.) as affected by level and frequency of fertigation in a laterite soil. Agric Water Manag 96:445–456

Bhat R, Sujatha S (2011a) Arecanut based high density multispecies cropping/farming systems. In: Thomas GV, Krishnakumar V, Maheshwarappa HP, Bhat R, Balasimha D (eds) Arecanut based cropping/farming systems. Central Plantation Crops Research Institute, Kasargod, pp 27–44

Bhat R, Sujatha S (2011b) Organic matter recycling in arecanut based high density multispecies cropping system. In: Thomas GV, Krishnakumar V, Maheshwarappa HP, Bhat R, Balasimha D (eds) Arecanut based cropping/farming systems. Central Plantation Crops Research Institute, Kasargod, pp 6–26

Bhat R, Sujatha S (2012) Influence of biomass partitioning and nutrient uptake on yield of arecanut (*Areca catechu* L.) grown on a laterite soil. Commun Soil Sci Plant Anal 43:1757–1767

Bhat R, Sujatha S (2013) Establishing leaf nutrient norms for arecanut (*Areca catechu* L.) by boundary line approach. J Plant Nutr 36(6):849–862

Bhat R, Sujatha S (2014) Soil fertility status and disorders in arecanut (*Areca catechu* L.) grown on clay and laterite soils of India. Commun Soil Sci Plant Anal 45:1622–1635

Bhat R, Sujatha S, Balasimha D (2007) Impact of drip fertigation on productivity of arecanut (*Areca catechu* L.). Agric Water Manage 90(1–2):101–111

Bhat R, Sujatha S, Thomas George V (2008) Soil microbiological and enzymological response to organic matter recycling and chemical fertilization in arecanut based cropping system. J Plantn Crops 36(3):281–289

Bhat R, Sujatha S, Jose CT (2012) Assessing soil fertility of a laterite soil in relation to yield of arecanut (*Areca catechu* L.) in humid tropics of India. Geoderma 189-190:91–97

Bhat R, Sujatha S, Jose CT (2016) Role of nutrient imbalance on yellow leaf disease in smallholder arecanut systems on a laterite soil in India. Commun Soil Sci Plant Anal 47:1738–1750

Biddappa CC, Upadhyay AK, Hegde MR, Palaniswami C (1996) Organic matter recycling in plantation crops. J Plantn Crops 24:71–85

Bockstaller C, Guichard L, Keichinger O, Girardin P, Galan M, Gaillard G (2009) Comparison of methods to assess the sustainability of agricultural systems- a review. Agron Sustain Dev 29:223–235

Bopaiah BM (1982) Beneficial micro-organisms in the root region of arecanut palm. In: Proceedings, Silver Jubilee Symposium on Arecanut Research and Development. CPCRI (RS), Vittal, 13–14, Dec. 1982. pp 74–77

Bopaiah BM (1991) Soil microflora and VA-mycorrhizae in areca based high density multispecies cropping and areca monocropping systems. J Plantn Crops 18(Suppl):224–228

Bopaiah BM, Bhat NT (1981) Effect of continuous application of manures and fertilizers on rhizosphere microflora in arecanut palm. Plant Soil 63:497–499

Bopaiah BM, Koti Reddy M (1982) Distribution of microflora population in the rhizosphere of arecanut. J Plantn Crops 10:127–128

Borah AS, Ray AK, Maheshwarappa H, Subramanian P, Krishnakumar V, Barman I (2006) Effect of soil and water conservation measures on growth and yield of arecanut under sloppy terrain of Assam. J Plantn Crops 34(2):135–137

Brahma B, Nath AJ, Sileshi GW, Das AK (2018) Estimating biomass stocks and potential loss of biomass carbon through clear-felling of rubber plantations. Biomass Bioenergy 115:88–96

Burkill IH (1935) A dictionary of the economic products of the Malay Peninsula, vol 1, London. 600 p

Chandramohanan R (2005) Management of yellow leaf disease in areca gardens and production of elite planting materials. Final report of Ad-Hoc scheme from AP Cess fund, Indian Council of Agricultural Research, CPCRI, Kasaragod, 39 p

Chowdappa P, Biddappa CC, Sujatha S (1999) Efficient recycling of organic wastes in arecanut (*Areca catechu*) and cocoa (*Theobroma cacao*) plantation through vermicomposting. Indian J Agric Sci 69:563–566

Das M, Nath PC, Sileshi GW, Pandey R, Nath AJ, Das AK (2021) Biomass models for estimating carbon storage in areca palm plantations. Environ Sustain Indic 10. https://doi.org/10.1016/j.indic.2021.100115. 8 pages

Fageria NK (2007) Green manuring in crop production. J Plant Nutr 30(5):691–719

Fagnano M, Agrelli D, Pascale A, Adamo P, Fiorentino N, Rocco C, Pepe O, Ventorino V (2020) Copper accumulation in agricultural soils: risks for the food chain and soil microbial populations. Sci Total Environ 734:139434. https://doi.org/10.1016/j.scitotenv.2020.139434. 11 pages

Feller C, Albrecht A, Blanchart E, Cabidoche YM, Chevallier T, Hartmann C, Eschenbrenner V, Larré-Larrouy MC, Ndandou JF (2001) Soil organic carbon sequestration in tropical areas. General considerations and analysis of some edaphic determinants for Lesser Antilles soils. In: Managing organic matter in tropical soils: scope and limitations: Proceedings of a workshop organized by the Center for Development Research at the University of Bonn (ZEF Bonn)—Germany, 7–10 June, 1999, pp 19–31

Gowda NKS (2016) Areca sheath as an alternate dry fodder for livestock. In: Broadening Horizons N°29, May 2016, Feedipedia. pp 1–2

Hartemink AE (2003) Soil fertility decline in the tropics – with case studies on plantations. ISRIC-CABI Publishing, Wallingford. 360 p

Hartemink AE (2005) Plantation agriculture in the tropics: environmental issues. Outlook Agric 34(1):11–21

Hartemink AE (2006) Assessing soil fertility decline in the tropics using soil chemical data. Adv Agron 89:179–225

Havlin JL, Beaton JD, Tisdale SL, Nelson WL (2005) Potassium. In: Soil fertility and fertilizers: an introduction to nutrient management, 7th edn. Pearson Prentice Hall, Upper Saddle River, NJ, pp 199–218

Hillel D (2009) The mission of soil science in a changing world. J Plant Nutr Soil Sci 172(1):5–9

Jagannathan D, Yadukumar N, Sarath Babu S (2011) Arecanut plate making: a profitable business. Indian J Arecanut Spices Med Plants 13:3–6

Jayaprakash SM, Faijalahmadmulla, Thippeshappa GN, Naveen NE, Chaitanya HS, Pruthvi Raj HM, Lakshman (2022) Characterization of soils of micronutrients under arecanut based cropping systems in the coastal region of Udupi District. Pharma Innovation J 11(12):4234–4239

Jose CT, Balasimha D, Kannan C (2008) Yield loss due to fruit rot (Mahali) disease of arecanut in Karnataka. Indian J Arecanut Spices Med Plants 10(2):45–51

Julie Chandra CS, Georgea N, Narayanankutty SK (2016) Isolation and characterization of cellulose nanofibrils from arecanut husk fibre. Carbohydr Polym 142:158–166

Jyothi TV, Vishwanatha Shetty Y, Dinesh Kumar M (2009) Characterization and DTPA- Zn status in traditional arecanut garden soils of South Karnataka. Karnataka J Agric Sci 22(5):1013–1015

Karimi B, Masson V, Guilland C, Leroy E, Pellegrinelli S, Giboulot E, Pierre-Alain M, Ranjard L (2021) Ecotoxicity of copper input and accumulation for soil biodiversity in vineyards. Environ Chem Lett 19:2013–2030

Khadilkar VB, Badhe NN, Pandya HA (1964) Physico-chemical studies in the coastal and hilly soils under arecanut palm of Bombay-Deccan. Arecanut J 15:7–13

Lal R (1997) Residue management, conservation tillage and soil restoration for mitigating greenhouse effect by CO_2-enrichment. Soil Till Res 43(1–2):81–107

Lal R (2009) Challenges and opportunities in soil organic matter research. Eur J Soil Sci 60:158–169

Manikandan P, Joshi OP, Khan HH, Mohapatra AR, Biddappa CC (1987) Nutrient profile in an arecanut-cacao system on a laterite soil. Tropical Agric 64:13–16

Mohan Rao M (1982) Introduction. In: Bavappa KVA, Nair MK, Prem Kumar T (eds) The Arecanut Palm. CPCRI, Kasaragod, pp 1–9

Mohan M, Girija D, Surendra Gopal K, Sureshkumar P (2019) Microbial insight into rhizosphere of arecanut palms of Wayanad using metagenomics. J Plantn Crops 47(3):189–196

Mohapatra AR (1977) Soil fertility research in arecanut: a review. J Plantn Crops 5:114–120

Mohapatra AR, Bhat NT (1982) Crop management. A. soils and manures. In: Bavappa KVA, Nair MK, Prem Kumar T (eds) The Arecanut Palm. CPCRI, Kasaragod, pp 97–104

Mohapatra AR, Bhat NT (1990) Soil research on arecanut in the last decade. J Plantn Crops 17(suppl):54–63

Mohapatra AR, Bhat NT, Aboobaker VO (1970) Some aspects of green manuring for arecanut. Arecanut Spices Bull 2:6–10

Mohapatra AR, Bhat NT, Devaraju C (1975) Evaluation of micronutrient status of soils from healthy and yellow leaf disease affected arecanut gardens. J Indian Soc Soil Sci 23(1):71–75

Mohapatra AR, Bhat NT, Harish Kumar P (1976) Yellow leaf disease of arecanut and soil fertility studies. Arecanut Spices Bull 8:27–31

Moorthy VK, Chandramohanan R (1991) Evaluation of various organic substrates for the cultivation of *Pleurotus sajor-caju*. J Plantn Crops 19:65–69

Muralidharan A (1980) Biomass production, plant interaction and economics of intercropping in arecanut. Ph.D. thesis, UAS, Bangalore. 271 p

Muralidharan A (1990) Intercropping in arecanut. J Plantn Crops 17:25–38

Naidu LGK, Ramamurthy V, Sidhu GS, Sarkar D (2011) Emerging deficiency of potassium in soils and crops of India. Karnataka J Agric Sci 24:12–19

Nayar R (1976) Yellow leaf disease of arecanut: virus pathological studies. Arecanut Spices Bull 8:25–26

Niranjana KS, Yogendra K, Mahadevan KM (2016) Chemical characterisation and nutrient rating of soils of arecanut *(Areca catechu L.)* gardens of hilly zone of Karnataka. Int J Agric Sci Res 6(3):25–30

Padre S (2023) Adike chogaru- hosa nirikshegala chiguru. Krishi Madhyama Kendra and Farmer First Trust, Puttur. 140 p

Palm CA, Giller KE, Mafongoya PL, Swift MJ (2001) Management of organic matter in the tropics: translating theory into practice. Nutr Cycl Agroecosyst 61:63–75

Paul SC, Acharya GC, Chakraborty R, Maheswarappa HP, Hussain M, Ray AK (2020) Sustainability of soil health and system productivity through arecanut based cropping system in the NE region of India. J Plantn Crops 48(2):111–119

Paulraj S, Bhat R, Rajesh MK, Ramesh SV, Priya UK, Pandian RTP, Vinayaka H, Chowdappa P (2021) Data of 16S rRNA gene amplicon based metagenomic signatures of arecanut rhizosphere soils in yellow leaf disease (YLD) endemic region of India. Data Brief 38:107443. 7 pages

Paulraj S, Bhat R, Rajesh MK, Ramesh SV, Priya UK, Pandian RTP, Vinayaka H, Chowdappa P (2022) Metagenomic analysis reveals the predominance of Candidatus *Patescibacteria* in the rhizosphere of arecanut palms in yellow leaf disease (YLD) endemic areas of India. J Plannt Crops 50(1):26–34

Raghavan V, Baruah HK (1958) Arecanut: India's popular masticatory —history, chemistry and utilization. Econ Bot 12:315–345

Rao NH, Rogers PP (2006) Assessment of agricultural sustainability. Curr Sci 91(4):439–448

Rethinam P (1990) Potassium in plantation crops. In: Proceedings of Seminar on Potassium for Plantation Crops. Potash Research Institute of India. Nov. 6–8, pp 7–15

Roberts P, Edwards-Jones G, Jones DL (2007) Yield responses of wheat (*Triticum aestivum*) to vermicompost applications. Compost Sci Util 15(1):6–15

Sabiha-Javied MT, Choudhry MM, Tufail M, Irfan N (2009) Heavy metal pollution from phosphate rock used for the production of fertilizer in Pakistan. Microchem J 91(1):94–99

Sadanandan AK, George CK, Muralidharan A, Abraham KJ (1985) Effect of inter cultivation on production of arecanut, *Areca catechu* L. In: Proceedings, Silver Jubilee Symposium on Arecanut Research and Development, CPCRI (RS), Vittal, 13–14 Dec. 1982. pp 68–69

Sannamarappa M (1987) Effect of green manuring on the production of arecanut. Indian Cocoa Arecanut Spices J 10:67–70

Sannamarappa M, Kumar SNS, Nagaraja B (1976) Effect of tillage operations in arecanut plantations of *maidan* parts of Karnataka on the productivity of palms. J Plantn Crops 4(2):58–59

Shahane AA, Shivay YS (2021) Soil health and its improvement through novel agronomic and innovative approaches. Front Agron 3:680456. https://doi.org/10.3389/fagro.2021.680456. 31 pages

Shama Bhat K (1983) Plant interactions in a mixed crop community of arecanut (*Areca catechu* L.) and cacao (*Theobroma cacao* L.), Ph.D. thesis, University of Mysore, Mysore. 158 p

Smith P (2004) Carbon sequestration in crop lands: the potential in Europe and the global context. Eur J Agron 20:229–236

Sparling G, Parfitt RL, Hewitt AE, Schipper LA (2003) Three approaches to define desired soil organic matter contents. J Environ Qual 32:760–766

Sujatha S, Abdul Haris A (2000) Root distribution as influenced by different methods of irrigation in young arecanut (*Areca catechu* L.) palms. J Plantn Crops 28:117–122

Sujatha S, Bhat R (2010) Response of vanilla (*Vanilla planifolia* A.) intercropped in arecanut to irrigation and nutrition in humid tropics of India. Agric Water Manag 97(7):988–994

Sujatha S, Bhat R (2012) Impacts of vermicompost and nitrogen, phosphorus, and potassium application on soil fertility status in arecanut grown on a laterite soil. Commun Soil Sci Plant Anal 43(8):2400–2412. https://doi.org/10.1080/00103624.2012.708074

Sujatha S, Bhat R (2013a) Impact of drip fertigation on arecanut-cocoa system in humid tropics of India. Agrofor Syst 87:643–656

Sujatha S, Bhat R (2013b) Impact of vermicompost and nitrogen-phosphorus-potassium application on biomass partitioning, nutrient uptake and productivity of arecanut (*Areca catechu* L.). J Plant Nutr 36(6):975–988

Sujatha S, Bhat R (2015a) Soil fertility changes due to drip fertigation in arecanut cocoa system. Commun Soil Sci Plant Anal 46:1981–1997

Sujatha S, Bhat R (2015b) Resource use and benefits of mixed farming approach in arecanut ecosystem in India. Agric Syst 141:126–137

Sujatha S, Bhat R (2016) Impact of organic and inorganic nutrition on soil-plant nutrient balance in arecanut (*Areca catechu* L.) on a laterite soil. J Plantn Nutr 39(5):714–726

Sujatha S, Bhat R, Reddy VM, Abdul Haris A (1999) Response of high yielding varieties of arecanut to fertilizer levels in coastal Karnataka. J Plantn Crops 27:187–192

Sujatha S, Bhat R, Kannan C, Balasimha D (2011a) Impact of intercropping of medicinal and aromatic plants with organic farming approach on resource use efficiency in arecanut (*Areca catechu* L.) plantation in India. Ind Crop Prod 33(1):78–83

Sujatha S, Bhat R, Balasimha D, Apshara ES (2011b) Arecanut based inter/mixed cropping systems. In: Thomas GV, Krishnakumar V, Maheshwarappa HP, Bhat R, Balasimha D (eds) Arecanut based cropping/farming systems. CPCRI, Kasaragod, pp 6–26

Sujatha S, Bhat R, Chowdappa P (2015) Recycling potential of organic wastes of arecanut and cocoa in India: a short review. Environ Technol Rev 4(1):97–102

Sujatha S, Bhat R, Chowdappa P (2017) Soil health management in arecanut and cocoa. In: Maheshwarappa HP, Chowdappa P (eds) Soil health Management in Plantation Crops. Today & Tomorrow's Printers and Publishers, New Delhi, pp 1–25

Sujatha S, Bhat R, Bhavishya (2020) Arecanut based cropping/farming systems. In: Chowdappa P (ed) Arecanut. Today and Tomorrow's Printers and Publishers, pp 145–175

Te Pas CM, Rees RM (2014) Analysis of differences in productivity, profitability and soil fertility between organic and conventional cropping systems in the tropics and sub-tropics. J Integr Agric 13(10):2299–2310

Teena R, Shankar M, Nagaraj MS, Yogeeshapa H, Hadimani HP, Shreepad V (2023) Soil fertility mapping of arecanut growing areas in Sirsi Hobli of Uttara Kannada district. Karnataka Pharma Innovation J 12(3):1234–1238

Thomas GV, Prabhu SR, Reeny MZ, Bopaiah BM (1998) Evaluation of lignocellulosic biomass from coconut palm as substrate for cultivation of *Pleurotus sajor-caju* (Fr) singer. World J Microbiol Biotechnol 14:879–882

Thomas GV, Krishanakumar V, Maheshwarappa HP, Bhat R, Balasimha D (2011) Arecanut based cropping/farming systems. Central Plantation Crops Research Institute, Kasaragod, p 138

Tiesson HE, Cuevas E, Chacon P (1994) The role organic matter in sustaining soil fertility. Nature (London) 371:783–785

Vasundhara R, Prakash NB, Anil Kumar KS, Rajendra H (2020) Characterisation and classification of arecanut-growing soils of Karnataka, India. J Plantn Crops 48(2):91–102

Vasundhara R, Prakash NB, Anil Kumar KS, Rajendra H, Dharumarajan S (2021) Soil fertility status of coconut and arecanut growing soils. J Plantn Crops 49(2):83–93

Watt G (1889) A dictionary of the economic products of India. Periodicals Experts, Delhi. 559 p

Yadava RBR, Mathai CK, Vellaicham YK (1973) Note on the investigation mineral accumulation of arecanut palm affected by yellow leaf disease. Indian J Agric Sci 43(9):892–894

Dr. Ravi Bhat is currently the Principal Scientist (Agronomy) at ICAR-CPCRI, Kasaragod. He has more than 30 years of experience in agronomy of arecanut, coconut and cocoa with specialization in cropping/farming systems, nutrient management, soil and water management and organic farming. He standardized the technology of fertigation for arecanut, nutrient requirement for different yield levels, soil fertility management and developed various arecanut and coconut based cropping/farming system models. He is recipient of Best Arecanut Scientist Award. He has published more than 100 research papers in national/international journals.

Dr. S. Sujatha has a doctorate in Agronomy. She is currently Principal Scientist at ICAR-Indian Institute of Horticultural Research, Bengaluru, Karnataka, India. She started her carrier at ICAR-Central Plantation Crops Research Institute, Regional Station, Vittal, Karnataka, and contributed on crop production aspects of arecanut and cocoa for 20 years. Her areas of research are crop production aspects of plantation crops, flower crops, medicinal crops and fruit crops with specialization in cropping/farming systems, nutrient management, soil fertility, plant nutrition, water management, drip fertigation and organic farming. She has published 50 research papers in international/national journals.

Mr. Bhavishya has a doctorate in Horticulture and currently serving as scientist at ICAR-Central Plantation Crops Research Institute, Regional Station, Vittal, Karnataka. His areas of research are nutrient management in cropping/farming systems of arecanut and cocoa, enhancing input use efficiency, biotic stress-nutrient interaction and carbon sequestration.

Dr. U. K. Priya is a soil scientist specialized on soils and nutrition of arecanut and cocoa. She is currently serving at ICAR-Central Plantation Crops Research Institute, Regional Station, Vittal, Karnataka. Her areas of interest are recycling organic wastes, carbon sequestration and soil fertility management.

Dr. Alka Gupta is currently the Principal Scientist at ICAR-CPCRI, Kasaragod, India. She has 29 years of experience on biofertilizer technology, biomass recycling, plant growth-promoting rhizobacteria and arbuscular mycorrhizal fungi associated with coconut, arecanut and cocoa. She has 200 publications in national/international journals, reviews and book chapters. She is a recipient of ICAR Award for Outstanding Interdisciplinary Team Research in Agriculture and Allied Sciences as a Team Member.

Dr. Anok Uchoi is serving as a scientist at ICAR-Central Plantation Crops Research Institute, Research Centre, Kahikuchi, Assam, India. His areas of research include cropping systems, nutrient management and carbon sequestration in arecanut.

Spices

V. Srinivasan, M. Murugan, and R. Dinesh

Abstract

Black pepper (*Piper nigrum* L.) and small cardamom (*Elettaria cardamomum* Maton) are the two important spice crops grown under tropical humid climatic conditions, and they form essential components in the western and eastern cuisines as food additives and preservatives. Soil health management assumes great significance in spice crops as the current productivity levels of these crops show a disturbing downward trend. Besides biotic stresses, important factor contributing to reduced yields in spices is poor soil quality; a reflection of low carbon sequestration coupled with alarmingly higher nutrient mining from these acidic soils. Poor nutrient management strategies resulted in depletion of C reserves, imbalanced nutrient levels, and decreased nutrient cycling and processes due to decreased biological activity. Low pH and lower levels of organic carbon, K, Ca, Mg, B, and Zn are the major constraints in spice-growing soils. This, coupled with high soil erosion and associated nutrient losses in high-rainfall spice-growing areas, has further aggravated the problem. Long-term integrated nutrient management (INM) trials indicated that fertilizer input efficiency can be increased through rational use of inputs and scientific management. Adoption of INM techniques will not only enhance crop productivity but will also increase the total productivity and the efficiency of fertilizer use in these soils. While doing so there is an urgent need to augment supplies of soil amendments, organic manures, and fortified and customized fertilizers supplying secondary and micronutrients and bio-agents to sufficiently support nutrient supply in spice production systems. The chapter discusses the soil and climate suitability for these two

V. Srinivasan (✉) · R. Dinesh
ICAR-Indian Institute of Spices Research, Kozhikode, Kerala, India

M. Murugan
Cardamom Research Station, Kerala Agricultural University, Pampadumpara, Kerala, India

G. V. Thomas, V. Krishnakumar (eds.), *Soil Health Management for Plantation Crops*, https://doi.org/10.1007/978-981-97-0092-9_5

major spices and soil fertility status of the major growing tracts, crop uptake potential, and the measures to meet the gap in nutrient supply through various fertilizer management strategies. Targeted yield equations for predicting nutrient requirements to get fixed yields in soils with varying fertility levels were standardized for site-specific nutrient management, and pepper-specific micronutrient application enhanced the yield and quality. The chapter also covers major cropping systems that are prevalent for these two crops.

Keywords

Black pepper · Cardamom · Soil fertility · Fertilizer requirement · Integrated nutrient management · Organic manures · Biofertilizers · Yield response · Cropping and farming systems

Abbreviations

AEC	Anion exchange capacity
AICRPS	All India Coordinated Research Project on Spices
AM	Arbuscular mycorrhizal
BS	Base saturation
CC	Coir pith compost
CE	Cation exchange capacity
cfu	Colony forming unit
CHR	Cardamom hill reserve
DAP	Di-ammonium phosphate
DHA	Dehydrogenase activity
DRIS	Diagnosis and recommendation integrated system
FYM	Farm yard manure
GAP	Good agricultural practice
ICAR-IISR	ICAR-Indian Institute of Spices Research
ICH	Indian Cardamom Hills
INM	Integrated nutrient management
MOP	Muriate of potash
MSE	Mean biased error
MSL	Mean sea level
NAA	Naphthalene acetic acid
NRA	Nitrate reductase activity
PGPRs	Plant growth-promoting rhizobacteria
RMSD	Root mean square deviation
SOM	Soil organic matter
SSP	Single super phosphate
STV	Soil test value
TOC	Total organic carbon
VAMF	Vesicular arbuscular mycorrhizal fungi

1 Introduction

Spices are natural products used to improve taste, aroma, and color in food preparations, as preservatives in food, and are also used in the traditional medicine, pharmaceutical, cosmetic, and perfumery products. India the "Land of Spices," is the world's largest producer, consumer, and exporter of spices. Black pepper (*Piper nigrum* L.), small cardamom (*Elettaria cardamomum*), clove (*Syzygium aromaticum*), nutmeg (*Myristica fragrans*), cinnamon (*Cinnamomum verum*), and vanilla (*Vanilla planifolia*) are the perennial spice crops grown under tropical humid climatic conditions. Black pepper, the "king of spices" belongs to the family *Piperaceae*, is a perennial vine grown for its berries that is extensively used as spice and medicine. Small cardamom is a perennial, herbaceous rhizomatous monocot belonging to the family *Zingiberaceae*. Small cardamom is known as "queen of spices," and the dried fruit (capsule) of cardamom plant is the product of commerce. Soil health management aspects of black pepper and small cardamom are dealt in this chapter.

1.1 Black Pepper

Black pepper has originated in the tropical evergreen forests of Western Ghats of South India and is cultivated in 26 countries located in the humid tropical belt of the globe with adequate rainfall and humidity. Major cropped area is in India (2.83 lakh ha), Indonesia (1.98 lakh ha), Brazil (20,000 ha), Vietnam (56,500 ha), and Malaysia (15,000 ha) with a production of 65, 27, 34, 106, and 16.5 thousand tonnes in 2014, respectively (www.ipcnet.org). The productivity of black pepper in Thailand is the highest (2870 kg ha^{-1}) followed by Vietnam (1870 kg ha^{-1}) because of its high-density planting and intensive cultivation. Low productivity in other growing countries is mainly related to poor inherent soil fertility, use of low level of agronomic inputs like fertilizers and crop protection chemicals, and crop loss due to diseases.

Although each state in India produces one or the other spices, the states that are the major producers of black pepper and cardamom mostly encompass acid soils. The cultivation of these two spices is mainly confined to the states of Kerala, Karnataka, and Tamil Nadu including the northeastern states like Assam, Arunachal Pradesh, and Sikkim. Among the major spice crops, the area under black pepper during 2021–2022 was 283 thousand ha with a production of 70,000 t, while cardamom is cultivated in an area of 69.19 thousand ha, with a production of 23,340 t. Kerala state has 59% of total cardamom area and contributes 70% of production; Karnataka state has 34% of area and 23% of production, and Tamil Nadu state has 7% of area as well as production (Tables 1 and 2).

1.1.1 Black Pepper-Growing Soils

Black pepper, though grows well on a wide variety of soils, prefers well-drained virgin red, lateritic, and alluvial soils rich in humus with friable nature and adequate water-holding capacity. In Malaysia, most of the plantations are in soils of alluvial

Table 1 Area, production, and productivity of black pepper in India (2021–2022)

State	Area (ha)	Production (t)	Productivity (kg ha^{-1})
Karnataka	1,90,000	39,000	205
Kerala	76,351	21,000	275
Tamil Nadu	6973	2000	286
Others	10,638	8000	752
Total	2,83,962	70,000	246

Source: Spices Board, Kochi (2022)

Table 2 Area, production, and productivity of cardamom in India (2021–2022)

State	Area (ha)	Production (t)	Productivity (kg ha^{-1})
Kerala	39,143	21,270	543
Karnataka	25,135	697	28
Tamil Nadu	4912	1373	279
Total	69,190	23,340	337

Source: Spices Board, Kochi (2022)

origin having kaolinitic type of clay with poor buffering capacity and nutrient status (DeWaard 1969). In Indonesia, black pepper plantations are raised in all types of soils, ranging from rich volcanic to clay loam. In Sri Lanka, red clay loam and sandy loam are preferred by planters (Srinivasan et al. 2012b). In India, it is grown mainly in west coast tract, where the soils are acidic in nature with low levels of P and K, secondary nutrients, and micronutrients especially zinc. The major black pepper-growing soils of India can be classified into four orders – red loam (Alfisols 70%), forest loam (Mollisol 10%) and laterite (Oxisol 6%), and alluvium (Entisol 4%) (Sadanandan 1993). Increase in productivity of black pepper was noticed in soils with near neutral pH, high soil organic matter (SOM), and base saturation (Mathew et al. 1995). However, Sangakkara (1989) observed the highest growth of the variety Panniyur-1 at pH 7.8 and 8.1. The soils that are not suitable for cultivation of black pepper were waterlogged and diseased (Nambiar et al. 1965). Black pepper plantations are established on a wide variety of soils with varying texture from sandy loam to clayey loam that are acidic (pH 5.0–6.2) and in coastal sandy soils (Khan et al. 2002).

Hamza et al. (2004) reported that well-managed black pepper gardens are sandy- to loam-textured with near neutral pH, high in exchangeable bases, organic carbon, and micronutrients especially zinc. Hydroponic experiment to examine the effects of increasing acidity (pH 7.0, 5.5, 4.0, and 3.5) on seedling growth revealed that the low pH directly inhibited the root development and function; limited K, Ca, and Mg absorption; and reduced the seedling growth (Zu et al. 2014). At pH 5.5, black pepper attained maximum growth, while the minimum growth occurred at pH 3.5. A soil fertility survey carried out in the major black pepper-growing tracts in the states of Kerala and Karnataka indicated that 10% of the gardens were low ($<$ 120 kg ha^{-1}), 31% medium (120–280 kg ha^{-1}), and 59% high ($>$ 280 kg ha^{-1}) in available K status (Hamza et al. 2007).

Studies on the soil status of main spice-growing areas of Kerala indicated a disturbing trend of nutrient imbalance. Soils were generally acidic with >70% of samples falling below 6.0 pH. The mean available P level was 114.35 ± 83 kg ha^{-1} and 9% of the samples had high (25–35 kg ha^{-1}), 30% had very high (36–100 kg ha^{-1}), and 25% of the soils recorded extremely high (> 100 kg ha^{-1}) available P. The exchangeable K levels were low-medium (69%), and 38% of samples were low in exchangeable Ca. A major share of the samples was low in exch. Mg (44% low and 13% very low), available S (29%), B (32%), and Zn (29%) (IISR 2011). Soil acidification is alarming, and 90% of the soils showed slightly to higher acidity pressing immediate intervention through amendments to restore the productivity. Unscrupulous application of P fertilizers has resulted in very high buildup of soil P interfering with the uptake of other essential nutrients. Extensive deficiency of secondary nutrients Ca and Mg to the extent of 40 and 75%, correspondingly in soils of Kerala, warrants immediate correction. The deficiency of B is extensive (60%) requiring adoption of suitable remedial measures (Dinesh et al. 2014; KSPB 2013).

1.1.2 Environmental Factors

Black pepper adapts successfully between 20° north and 20° south of the equator and up to 1500 m above MSL under humid tropical conditions throughout the year for its successful growth and yield (Fig. 1) and cannot tolerate excessive heat and dry spell. In India, black pepper-growing tracts receive 1500 to 4000 mm rainfall in two spells during the southwest monsoon (June–September) and northeast monsoon

Fig. 1 A high-yielding black pepper plantation in India (Photo: V Srinivasan)

(October–December) with a peak in July–August. Rainfall after a period of stress induces profuse flowering and the growth of fruit-bearing lateral shoots (plagiotrops) are formed during the peak rainfall period of June–July (Mathai 1983). Black pepper plant prefers a relative humidity of 60–95% and tolerates temperature between 10 and 40 °C. However, the ideal temperature range is 23–32 °C with an average of 28 °C. Soil temperature of 26–28 °C is found to be optimum for root growth of black pepper. Among weather variables, mean maximum temperature in March first fortnight and June second fortnight, total rainfall during March second fortnight, and September second fortnight (RAIN > TMIN > TMAX) were found to be significantly correlated with the black pepper yield (Kandiannan et al. 2011).

Even though black pepper is considered to be a day-neutral plant, vines exposed to direct sunlight developed physiological disorders, even under optimum soil moisture conditions (Vijayakumar et al. 1984a). The vines maintained under shade (7% incident light) remained green and healthy than those exposed to sunlight, which developed necrotic patches during summer (Vijayakumar and Mammen 1990). Fifty-percent shade boosted the growth of black pepper cuttings in the nursery. In a few varieties of black pepper, Mathai (1983) observed decrease in carbon fixation with light intensity above 50,000 lux (900 mol s^{-1} m^{-2}). The top portion of the black pepper vine receives more sunlight, which decreases gradually downward of the vine due to its own and mutual shading (Mathai and Chandy 1988; Ramadasan 1987).

In order to allow sufficient light for crop growth, shade regulation of live standards is essential during rainy/cloudy weather; otherwise the yield may be reduced to 50% or even more (Ramadasan 1987). Berry set percentage significantly reduced due to 75% leaf area reduction of subtending leaves (Krishnamurthy et al. 2000). Use of lime or China clay can be practiced to reduce stress during summer months. Spraying of lime over the leaf surface has enhanced the chlorophyll content (Vijayakumar et al. 1984b); but it reduced the yield compared to China clay (Sadanandan et al. 2000). Leaf photosynthesis, nitrate reductase activity (NRA), and stomatal conductance showed a significant positive correlation, while leaf temperature had significant negative correlation with productivity. High yielders maintained considerably higher NRA, stem and leaf carbohydrate content, and photosynthetic rate compared to low yielders during juvenile stage (Krishnamurthy and Chempakam 2009).

1.1.3 Issues in Soil Fertility

An early study by Sadanandan (1993) showed that 57.0% of the soils in spice-growing areas of Kerala, India, were low in available P, while 52.0% from Karnataka state and 9.0% sample from Kerala state were deficient in available K. Among micronutrients, deficiency of available Zn was more predominant in acid soils of India with highest deficiency rate of 57.0% in acid soils of Meghalaya followed by Jharkhand, Odisha, and West Bengal (23.0–54.0%). The studies at Calicut and Wayanad districts, the main spice-growing areas of Kerala, showed an alarming trend of nutrient imbalance. Soils were generally acidic with >70% of samples falling below 6.0 pH. The mean available P level was 114.35 ± 83 kg ha^{-1}, and 9% of the samples had high (25–35 kg ha^{-1}), 30% had very high (36–100 kg ha^{-1}), and

25% of the soils recorded extremely high (> 100 kg ha^{-1}) available P. The exchangeable K levels were low-medium (69%), and 38% of samples were low in exchangeable Ca. A major share of the samples were low in exchangeable Mg (44% low and 13% very low), available S (29%), B (32%), and Zn (29%) (KSPB 2013). In Malaysia, many black pepper farms are owned by smallholders, who often apply inadequate quantity of fertilizers (Rosli et al. 2013). Similar trend was observed across the cropping systems involving coconut, arecanut, banana, and vegetables in other districts of Kerala.

Soil acidification was alarming, and 90% of the soils were found to be slightly to highly acidic warranting immediate intervention through amendments to restore the productivity. Indiscriminate P fertilizers application has resulted in very high buildup of soil P interfering with the uptake of other essential nutrients (Dinesh et al. 2014). Extensive deficiency of secondary nutrients like Ca and Mg to the extent of 40 and 75%, respectively, in soils of Kerala necessitate immediate correction. The deficiency of B is extensive (60%) requiring urgent recommendations (KSPB 2013). As per the International Crops Research Institute for the Semi-Arid Tropics (ICRISAT) survey conducted in 2011, the soils of major black pepper-growing tracts of Karnataka, India (Uttara Kannada, Hassan, Chikkamagaluru, Shimoga, Kodagu districts), were acidic, high in organic carbon (OC) content, medium to sufficient in available P, low to medium in K, and just above critical limits in soil available Zn and B (Fig. 2a–d). The spice-growing districts of north-eastern hill states, Tamil Nadu and Andhra Pradesh, also showed extensive deficiency of major and micronutrients that can become yield-limiting.

Indiscriminate and long-term use of N and P straight fertilizers, which are generally free from micronutrients, ignoring potential soil amelioration with liming materials, has raised concern about preferential accumulation of P and imbalance of other nutrients and created widespread deficiencies of secondary and micronutrients especially Mg, B, and Zn in major spice-growing soils. The alleviation of such deficiency by use of corrective soil amendments and micronutrient fertilizers is also limited due to nonavailability of standard fertilizer materials and its awareness by farmers. Efforts to correct this imbalance have to be made through promotion of site-specific nutrient management taking into consideration the initial soil fertility status.

1.1.4 Nutrient Management

Nutrient Demand

In India agriculture production leaves an estimated negative nutrient balance of 10.0 M t $year^{-1}$ of NPK which is likely to increase. At an estimated nutrient removal of 37.5 M t of NPK (11.9 N + 5.3 P_2O_5 + 20.3 K_2O M t) to meet food production by 2025, the nutrient balance identifies excess use of N and P_2O_5 and deficit use of nearly 18 M t of K which accounts for 55% of NPK removal (Parama and Munawery 2012). Under this scenario, the total estimated fertilizer consumption for spices production is only 348.7 thousand tons (hardly 1.2%), of which the major spices that are being grown under acid soil environments are applied with 89.54 thousand tons

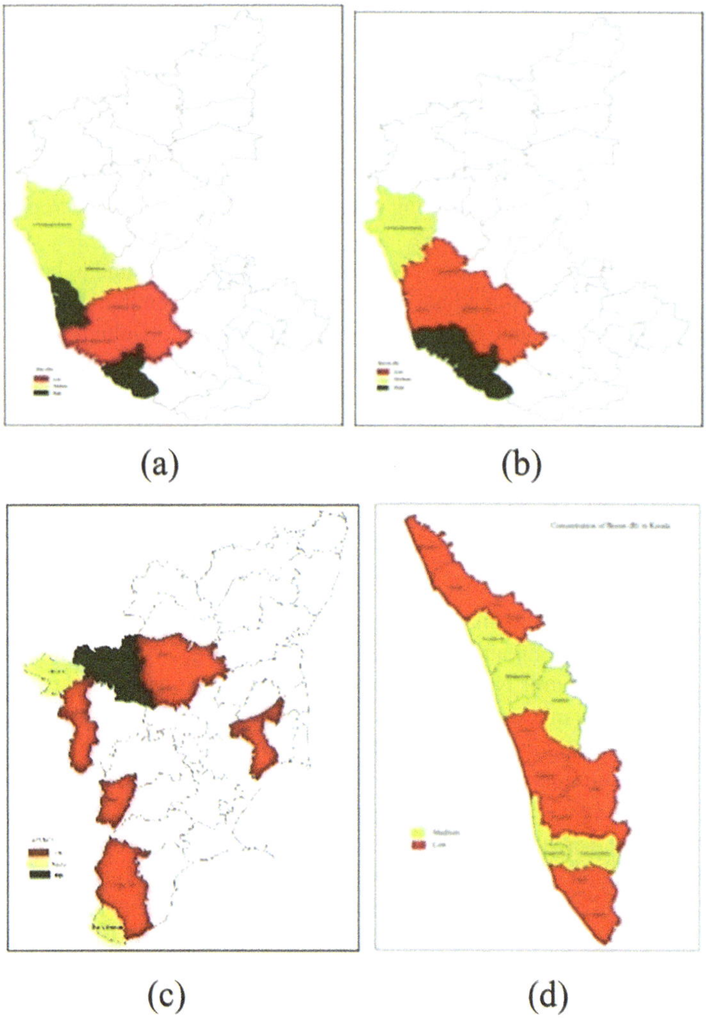

Fig. 2 Deficiency status of black pepper-growing soils (**a**) Zn, (**b**) B status in Karnataka districts, (**c**) Zn status in Tamil Nadu districts, (**d**) B status in Kerala districts (red, deficient; yellow, low; and green, adequate) (Source: IISR (2016))

of NPK. But their estimated nutrient removal by the way of harvesting is 134.36 thousand tons, identifying an excess use of P_2O_5 and deficit of N and K_2O to the tune of −14 and − 41 thousand tons, respectively (Table 3). The nutrient demand for production of major spices to achieve the yield goals of 2025 will be 183.15 thousand tons (Srinivasan et al. 2012a).

Paduit et al. (2018) have estimated the N, P, and K uptake by black pepper with the present yield scenario of 1.0 t ha^{-1} (world average as per FAO statistics 2016) and for projected yield levels of 2.1 t ha^{-1} (world average) with improved good

Table 3 Nutrient-wise fertilizer consumption and balance in spice crops

	Fertilizer consumption ('000 t)			
	N	P_2O_5	K_2O	Total
2010–2011				
All spice crops	193.48	97.51	57.68	348.67
Major spices[a]	43.35	23.32	22.87	89.54
Crop uptake[a]	57.52	12.96	63.88	134.36
Balance	−14.17	10.36	−41.00	−44.82
2025				
Crop uptake[a]	80.29	17.31	85.55	183.15

Source: Srinivasan et al. (2012a)
[a]Spices like black pepper, cardamom, ginger, and turmeric are considered

agricultural practices. With the projected yield levels under the adoption of good agricultural practices (GAP) (i.e., 2.1 t ha^{-1}), the nutrient removal of about 21,900 t of N, 2000 t of P and 15,200 t of K is expected from the top five major producing countries, viz., Vietnam, Indonesia, India, Brazil, and China of which Vietnam alone contributes about 50% of the total nutrient removal.

An unfavorable ratio of consumption can restrict the crop yields and can lower the productivity, thus hampering the fertilizer use efficiency. Additionally, N and P losses in the nutrient cycle resulting from excessive erosion and leaching under heavy rainfall in spice-growing areas also threaten the environment. Even though the fertilizer consumption pattern in India is grossly imbalanced with a consumption ratio of 1:0.4:0.2 in 2009–2010 as against favorable ratio of 1:0.5:0.25 (or 4:2:1), the consumption ratios in spice-growing agroeco-regions are near to optimum with 1:0.5:0.33 in NE hills, 1:0.5:0.43 in southern plateau, and 1:0.6:0.6 in coastal zones (NAAS 2009). But this balanced nutrient supply ratio also seems to be insufficient for major spices as their nutrient removal pattern is 1:0.23:1.12 of NPK, making a deficit of K supply by about 47% of the total crop removal. A large removal of K_2O by rhizomatous spices with little replenishment in soil through inputs is a major concern. The efficacy of balanced fertilization also remains low due to the widespread deficiencies of secondary (Ca, S, Mg) and micronutrients (Zn, B, Mn, Fe). In addition to balanced chemical fertilization, amendments (especially lime in acid soils) also need to be applied to these soils. The crop residue management and loss of plant nutrients through erosion and leaching are key factors for monitoring plant nutrient balance in major spice cropping and farming systems.

Nutrient Uptake

Nutrient composition and uptake by black pepper vines vary with age, variety, soil type, and management. An adult black pepper vine (of 5–8 years) removes per year 233.44 g N, 16.8 g P, 171.9 g K, 18.3 g Mg, 75.0 g CA, 365 mg Fe, 281 mg Mn, 104 mg Zn, 89 mg Cu, and 60 mg B (Sadanandan 2000). So, the major nutrients should be applied in amounts sufficient to meet the uptake in addition to the allowance made for nutrient losses through leaching. For proper growth of black pepper, leaf nutrient concentration of 2.7, 0.1, 3.0, 1.0, and 0.2% of N, P, K, Ca, and Mg, respectively, is essential, below which the soil is deficient (Mathew et al. 1995). The optimum time of leaf sampling for foliar diagnosis is just before flushing, and the youngest fully matured leaf is the ideal leaf to analyze. The nutrients that are present in the leaves of Panniyur-1 and Karimunda ranged from 1.75 to 3.12% N, 0.12 to 0.16% P, 2.61 to 3.55% K, 1.40 to 2.37% Ca, 0.15 to 0.27% Mg, 78 to 139 ppm Fe, 161 to 542 ppm Mn, 19 to 30 ppm Zn, 248 to 800 ppm Cu, 53 to 92 ppm Al, and 1.24 to 4.73 ppm Mo (Sadanandan 1994). The nutrient removal by the var. Kuching was 252 kg N, 32 kg P_2O_5, and 224 kg K_2O ha^{-1} (De Waard and Sutton 1960). De Waard (1969) reported a critical level of 2% K in the plant as a limit for K deficiency. Total biomass increases linearly and reaches maximum at 22 months after planting, after which a decrease in dry matter was observed due to fruit export and fallen leaves toward harvest. The biomass of black pepper vine increases steadily indicating continuous vegetative growth for the next flowering season. At 30 months, the black pepper had removed 293.08 kg of nitrogen, 46.41 kg of phosphorus, 264.95 kg of potassium, 35.4 kg of magnesium, and 74.82 kg of calcium (Ann 2012a). The nutrient uptake rates were lower than nutrient applied, which suggested that fertilizer had been overused for black pepper production.

Nutrients like P and K have a major role in improving the yield (Nybe et al. 1989). The order of nutrient concentration in leaves, branches, stems, and fruits/spikes were N > K > Ca > Mg > P (Nagarajan and Pillay 1975), whereas in white pepper it was in the order K > N > Mg > Ca > P and in flowers, N > K > Ca > P > Mg (Adzemi et al. 1993). The magnitude of nutrient removal by way of harvested produce followed the order of N > K > Ca > Mg > P > S > Fe > Mn > Zn (Mathew et al. 1995). Sankar et al. (1988) recorded that 1 kg of dry black pepper removed 26.6 g N, 1.5 g P, 21 g K, 8.8 g Ca, 5.2 g Mg, 0.9 g S, 0.1 g Fe, 0.017 g Zn, 0.11 g Mn, and 0.028 g Cu. In the early years of yield (3–5 years), the removal by berries (27–30%) followed by leaves (24–30%), stem (18–22%), and branches (10–18%) constitute the major part of N, P, K, Ca, and Mg uptake by the plant. But in aged vines (of 8 or more years) major sink for NPK was observed to be leaves (50–62%) followed by stem (16–24%), root (11–16%), and berries (8–10%) (Srinivasan et al. 2007). Ann (2012b) recorded a nutrient uptake of 37% N, 40% P, 36% K, 48% Mg and 35% Ca to the total plant uptake by the fruit and spike at the stage of first harvest (after 3 years of planting). Bush pepper can also be used to investigate the fertility status and nutrient supplying capacity of soil (Geetha et al. 1996).

Nutrient Management

Being a surface feeder, most of the feeding roots are distributed on the top 50–60 cm layer of the soil. Growth and development as well as yield of the crop were the best in humus-rich soils. In India, black pepper is grown predominantly as intercrop in lateritic soils, which are poor in fertility and nutrient retention. In Indonesia and Sri Lanka also, black pepper is intercropped with short-duration crops and plantation crops like coffee and coconut. The soils of the western coastal tract receiving high rainfall are also low to medium in P and K status. The hills of Western Ghats are very much suitable for growing black pepper as monocrop or as an intercrop in coffee, cardamom, or tea plantations (Sadanandan 2000).

Table 4 Fertilizer recommendations developed under All India Coordinated Research Project for black pepper in India*

Region/state	NPK (g $vine^{-1}$)
North Kerala districts	50:50:200
Medium fertile soil	100:40:140
Arecanut and black pepper mixed cropping system for heavy rainfall Sirsi region of Karnataka	200:80:280
Karnataka (under irrigation)	150:60:210
Andhra Pradesh (under rainfed conditions)	50:50:150
Chintapalli area of Andhra Pradesh	100:60:160
Laterite soils of Kerala where soil N and K are low	140:55:270

Note: *Application of 1/3 of the above dose is recommended for the first year and 2/3 during the second year and full dose from third year onward. Fertilizers are best applied in two splits, first half in May–June with the receipt of rains and second half in August–September after cessation of heavy rainfall. Application of micronutrients – zinc, boron, and molybdenum – @ 5, 2, and 1 kg ha^{-1} in soil in September increased the yield and quality of black pepper

**The P dose may be either increased to 100 g $vine^{-1}$ $year^{-1}$ or decreased to 25 g vine $^{-1}$ $year^{-1}$ depending on P status of soil (Source: Ravindran and Johny 2000)

Judicious use of organic manures and inorganic fertilizers is the most efficient and economical way to obtain higher production in black pepper (Raj 1978). Fertilizer recommendations for black pepper developed for different regions under All India Coordinated Research Project on Spices (AICRPS) are given in Table 4.

As a perennial crop, black pepper exploits a fixed volume of soil having limited nutrient supply for longer period. Hence, the fertility assessment should be based on the buffering capacity of the soil for long-term supply and intensity factor for immediate availability of any nutrient (Nair 2004). However, the quantity of fertilizers required to get high yield varied largely among locations. Targeted yield equation for predicting nutrient requirements to get fixed yields in soils with varying fertility levels was standardized in black pepper based on long-term field trials at ICAR-IISR, Kozhikode. The nutrient contribution from soil was up to 26% for N, 15% for P_2O_5, and 10% for K_2O for the removed berry yield, and that from the fertilizer sources was worked out to be 50% for N, 20% for P_2O_5, and 35% for K_2O. The nutrient uptake ratio of NPK in black pepper has followed 1.10:0.17:1.0. The soil test-based fertilizer recommendation model for NPK for different yield targets of black pepper was developed as.

$$N\ (kg\ ha^{-1}) = (Target\ yield \times 100 \times 0.016) - (STV\ for\ N \times 0.211)$$
$$P_2O_5\ (kg\ ha^{-1}) = (Target\ yield \times 100 \times 0.014) - (STV\ for\ P_2O_5 \times 0.697)$$
$$K_2O\ (kg\ ha^{-1}) = (Target\ yield \times 100 \times 0.052) - (STV\ for\ K_2O \times 0.227).$$

The fertilizer recommendations based on these established equations when validated for varying yield targets at different soil fertility levels resulted in realized yield levels with a minimum deviation of −2.9 to +40% and low root mean square deviation (RMSD) and mean biased error (MSE) values, and the yield increase was

Table 5 Soil test-based fertilizer recommendations for black pepper yield target levels of 3 and 6 t ha^{-1}

Soil test value for available nutrients (kg ha^{-1})	Fertilizer nutrient recommended (kg ha^{-1}) for yield targets 3.0 t ha^{-1}	6.0 t ha^{-1}
Nitrogen		
< 150	50	100
150–250	25	80
250–400	10	55
>400	–	20
Phosphorus (P_2O_5)		
< 10	40	80
10–30	30	70
30–50	10	55
>50	–	30
Potassium (K_2O)		
< 110	150	310
110–300	125	275
300–500	80	250
>500	35	110

Source: Devasahayam et al. (2015)

significantly higher to the blanket fertilizer recommendation (Srinivasan et al. 2011a, 2013a). Fearnside (1980) developed a regression equation to predict the yield from soil fertility (Y = 12.119+ 0.292A+ 0.382B- 0.0552 where Y = black pepper yield, A = pH of soil, B = carbon (% dry weight), and C = phosphate (ppm) with high correlation ($P < 0.0001$).

The recommended doses of nutrients for varying soil test values of N, P, and K and for varying yield targets are given in Table 5.

Nutrient Interaction

Interaction of nutrients is important in formulating efficient nutrient recommendations for high and sustainable production. Sadanandan and Rajagopal (1989) found significant correlation between leaf P index (P/N + P + K + Ca) of the youngest matured leaf with the yield of black pepper (r = 0.81**). Interaction studies by Sadanandan and Hamza (1992) showed negative correlation between plant P vs Zn (r = −0.56**), P vs Cu (r = −0.72*), and P vs Mo (r = −0.76**). Significant negative correlation was reported in soil K and leaf Mg (r = −0.30**). A negative correlation exists between leaf K and Mg (r = −0.63**) showing the antagonistic effect of K on Mg. Significant positive correlation was obtained between iron and manganese (r = −0.77**), negative correlation between Cu and Mo (r = −0.77**), and positive correlation between Zn and Mo (r = 0.43**). HNO_3 extractable K has shown a significant positive correlation (r = 0.688**) with K uptake by bush pepper (Srinivasan et al. 2002b). The pepper yield was positively correlated with DTPA extractable Fe (r = 0.55*) and Cu (r = 0.41**) in soil and with leaf iron (r = 0.62**), leaf manganese (r = 0.27**), and leaf Cu (r = 0.37**) indicating the importance of trace elements in black pepper nutrition. Studies by Srinivasan et al. (2013b) indicated that yield had highly significant positive correlation with soil available K (0.416**) followed by soil Zn and S. Soil available P, Ca, Mg, and Mn also exhibited positive correlations with yield. The quality parameters like piperine and oleoresin were mainly influenced by the availability of micronutrients like Fe, Mn, S, and Zn, with Mn and Fe showing significantly the highest positive influence on oleoresin (r = 0.854** and r = 0.738**) and piperine (r = 0.652** and r = 0.537**).

Nutrient Deficiencies and Toxicity

Nutrient deficiencies often result in predisposing the black pepper plants to diseases. Deficiencies of nutrients like P and K have been indicated as the reason for diseases (De Waard 1969; Harper 1974). Slow decline of black pepper is attributed to nematode and fungal infection, soil moisture stress, and deficiency of K and P (Nambiar and Sarma 1977). Wahid and Kamalam (1982) found that K levels of the healthy vines were considerably higher than those of diseased ones and concluded that K deficiency as a cause for slow decline of black pepper vines.

Nambiar et al. (1965) worked out tentative ratios of K_2O (total)/N, K_2O (available)/N and $CaO + K_2O + MgO/N$ in soil and found that slow decline of black pepper occurred when these ratios were below 14.1, 0.05 and 3.8, respectively. Similarly, Wahid and Kamalam (1982) reported that foliar yellowing and necrosis of distal ends of lamina in slow wilt-affected gardens were due to N and K deficiencies, respectively. In Brazil, Duarte and Albuquerque (1991) hypothesized that deficiency of Zn, Fe, and B and the lack of equilibrium between P and K in relation to Ca and Mg were the predisposing factors to *Fusarium* wilt infection.

Mineral Fertilizers

While applying fertilizers agroecological condition, soil nutrient status, soil physical properties, cropping and farming system, and disease incidence should be considered. The critical stages of nutrient requirement are initiation of flower primordia, flower emergence, berry formation and development (Raj 1978), and at the time of fertilizer application soil moisture availability is to be ensured. Fertilizer is to be applied in ¾ circles around the vine or in alternating parallel bands on either side of the vine underneath the edge of the canopy. Over 90% of the feeder root activity was found within a 30 cm radius around the vine with the uniform vertical distribution down to a soil depth of 40 cm. Radiotracer studies on the field-planted bush pepper showed that majority of the roots are concentrated in the surface layer of the soil up to 20 cm depth and 30 cm lateral distance. In the deeper layers of 40 and 60 cm, the root spread was within a lateral distance of 10 cm. About 60% of the roots were distributed in 20 cm depth and about 30% in 20–40 cm depth constituting about 70–87% roots within 40 cm depth and the remaining down below. Similarly, 79–93% roots were distributed within 30 cm lateral distance (Thankamani et al. 2003). Fertilizers are applied in two splits in India. ICAR-IISR recommends one-third at first year, two-third at second year, and full dose from third year of field planting. Fertilizer application is to be restricted to a lateral distance of 30 cm in full circle area around the vine (Geetha et al. 1993). Raj (1978) also found that fertilizer application advocated should contain 11–13% N, 5–7% P_2O_5, 6–18% K_2O, 4–5% MgO and trace elements. In India, the amount of fertilizer used is very low, and only 4–10% of the farmers use fertilizers for black pepper (Anandaraj et al. 1989), as more than 95% are cultivated in homesteads, where multispecies cropping system is prevalent that helps in improvement of soil fertility (Bavappa et al. 1986). Pillai et al.(1987) reported 200 kg K_2O as optimum for black pepper var. Panniyur-1 in a K-poor laterite soil, whereas application of 270 g K_2O vine^{-1} year^{-1} with recommended N and P resulted in significant increase in soil K availability and optimized the yield and quality (Sadanandan and Hamza 2001). Among the sources, potassium sulfate was found to have a positive influence on quality (oleoresin 13.5% and piperine 6.1%) of black pepper. Application of potash @ 150 g vine^{-1} year^{-1} reduced spike shedding. Half of recommended dose of K as SOP for soils of high K status and 100% recommended K as SOP + 2% foliar spray for soil of low K status was also recommended (Srinivasan et al. 2008).

Fertilizer use efficiency enhanced with slow-release "nimin" (nitrification inhibitor)-coated urea (Sadanandan and Hamza 1993) and Mussoorie rock phosphate (Sadanandan 1986) to supply N and P, respectively. Application of NPK @ 150, 60, and 270 together with Zn, B, and Mo @5, 2, and 1 kg ha^{-1} produced the highest yield (IISR 1996). The additional nutrients used for an increase of 1.0 kg black pepper berry are 6.35 g N, 6.33 g K, 1.11 g Ca, 0.47 g Mg, 0.44 g P, 0.29 g S, 42.82 mg Fe, 34.45 mg Mn, and 4.2 mg Zn. This requirement was irrespective of yield potential indicating that plants with higher production potential require higher doses of nutrients. Long-term studies (18 years) on NPK fertilization of black pepper in a laterite soil of Kerala, India, revealed that the economical optimum was at 140-55-270 kg ha^{-1} N, P_2O_5 and K_2O, respectively, with 15 t green leaf ha^{-1} year^{-1} (Sadanandan and Hamza 2002). The application of 400–600 g plant^{-1} NPK and Mg (12:12:17:2) three times a year, with 500 g plant^{-1} dolomite applied two times a year was reported to be the best fertilizer dose for productive black pepper plantations in Bangka (Wahid et al. 1990). Veloso et al. (2000) reported increased yield of black pepper due to fertilizer application @ 72–98 kg ha^{-1} N, 60 kg ha^{-1} P, and 13 to 42 kg ha^{-1} K.

Studies in China showed that the nutrient uptake rates were lower than the nutrient doses applied suggesting over use of fertilizer for black pepper production. Based on the uptake studies, optimum fertilizer dosages of 62-10-62-6-18 kg ha^{-1}, 237-22-246-22-65 kg ha^{-1}, and 390-62-352-47-100 kg ha^{-1} of N-P-K-Mg-Ca for the year 1, year 2, and year 3 of cropping year were suggested by Ann (2012b). In China, to obtain higher yields, growers use higher rates of N and P fertilizers, whereas organic and K fertilizers have been ignored for a long time. Applying excessive amounts of N and P fertilizers not only caused the fertilizers ineffective but also caused the soil pH to decrease significantly. In addition, the rhizosphere soil available P content increased and was approximately ten times higher than that in the non-rhizosphere soil. However, the decline in SOM content in both the rhizosphere and non-rhizosphere soils warranted more organic fertilizer application in long-term monoculture of black pepper (Li et al. 2016).

Hamza and Sadanandan (2002) reported that under Mo-deficient conditions, black pepper in field responds up to 0.9 kg ha^{-1} of Mo, and application of Mo at 0.25 mg kg^{-1} soil or foliar spray of 0.1% sodium molybdate is economical in increasing the yield and quality in potted bush pepper (Hamza and Sadanandan 2005a). Hamza and Sadanandan (2005b) suggested that application of Zn at 5 mg kg^{-1} for potted bush pepper and Zn at 6.2 kg ha^{-1} for field condition was optimum for increasing yield and quality. Application of 2.5 kg ha^{-1} Zn for field-planted pepper as Zn-EDTA was found to increase soil Zn availability, berry yield, oleoresin, and piperine. Among sources of Zn and methods of application, soil application of Zn as $ZnSO_4$ gave the highest benefit-cost ratio of 5.27 followed by foliar spray of $ZnSO_4$.

Muthumanickam (2003) observed an increased yield (9.2 kg $vine^{-1}$) and oleoresin content (12.34%) of black pepper in acid soils of Yercaud, India, due to soil application of 40 g P_2O_5 as rock phosphate and spraying of 0.5% $ZnSO_4$. Among growth parameters, canopy compactness (56.67 laterals/m^2), leaf area (585.50 cm^2), and leaf area index (2.62) were higher in the variety Panniyur-1 with Zn (0.5%) foliar nutrition, and among yield attributes, number of spikes $vine^{-1}$ (627.00) and dry weight of berries (2.61 kg $vine^{-1}$) were higher for the variety Panniyur-1 with Zn (0.5%) foliar spray (Divya et al. 2018).

Based on diagnosis and recommendation integrated system (DRIS) norms for getting an optimum yield (average around 2350 g $vine^{-1}$), Sadanandan et al. (2000) proposed that the black pepper-growing soils should have an optimum pH of 5.45 and organic carbon of 4.65%; Bray extractable P of 42.5 ppm; exchangeable K, Ca, and Mg of 190, 730, and 96 ppm, correspondingly; and extractable Fe, Mn, Zn, and Cu of 32, 15, 3.5, and 12 ppm, respectively, (Sadanandan et al. 1996).

1.1.5 Management Strategies to Enhance Soil Health

Organic Manures

In the early stages of the black pepper growth, prawn dust, fish and bone meal, farm yard manure, liquid cattle manure, compost, cakes made out of castor, soybean, cotton seed, coconut, and groundnut were used, besides green leaves to replenish the soil nutrient status. Liberal manuring with a mixture of cow dung, compost, and oil cakes increased the yield under Assam, India conditions (Choudhury 1947). Organic manures from goat, poultry, and pig (190 to 202 g pot^{-1}) gave similar results as NPK fertilizer (215 g pot^{-1}) in bush pepper. Farm yard manure (FYM) was the next in order (169 g), and least was 95 g pot^{-1} obtained with no manure application. The availability of NPK from soil was high with goat and poultry manures. The availability of Ca was the highest in the pig manure (627 ppm), which was shown to be superior to check (575 ppm) but on par with the rest of the treatments. All the organic manures provided higher Mg (71 to 82 ppm) to the plants (Sadanandan and Hamza 1999).

Biofertilizers

Biofertilizers and vermicompost application also enhanced the growth, biomass, and nutrient uptake of black pepper (Bopaiah and Khader 1989; Kandiannan et al. 1998). Govindan and Chandy (1985) reported increased growth of black pepper by *Azospirillum* inoculation. Inoculation with a combination of two/three biofertilizers, viz., *Azospirillum* (20 g), phosphobacteria (*Bacillus megaterium* var. *phosphaticum*, 20 g), and VAM (*Glomus* sp., 100 cc) mixed with 5 kg potting mixture consisting of soil/sand/FYM (1:1:1), enhanced the plant height, leaf area, biomass and dry matter production, and nutrient content of black pepper cuttings (Kandiannan et al. 2000a). Different sources of *Azospirillum* (10 g in 5 kg of soil) registered enhanced growth parameters (plant height, leaf area, shoot) and dry matter production of black pepper 90 days after treatment (Kandiannan et al. 2000b).

Several plant growth-promoting rhizobacteria (PGPRs) isolated from the rhizosphere of black pepper were found to promote growth, suppress diseases, and solubilize soil P (Ramachandran et al. 2007). Leaf compost and vermicompost supplementation supported free-living N-fixing bacteria and P-solubilizing bacterial populations (Parthasarathy et al. 2006). Application of *Pseudomonas fluorescens* and inclusion of *Trichoderma harzianum* in potting mixture for rooting black pepper cuttings were found to increase the biomass production by influencing nutrient uptake through P solubilization (Thankamani et al. 2005), increasing feeder root production, and thereby enhancing the absorptive surface area and phytohormone and siderophores production (Diby et al. 2005a). Diby et al. (2005b) reported noticeable uptake of N and P and enhanced vigor of black pepper when inoculated with *Pseudomonas fluorescens* strains, due to higher root proliferation and nutrient mobilization especially that of P in the rhizosphere. *P. fluorescens* strains enhanced P uptake by 122% over control (non-bacterized plants) and N uptake by 65% resulting in increased root growth and biomass production. Application of *Azospirillum* sp. in combination with half the recommended N, recommended superphosphate, and potash resulted in higher yield over application of 100% of the fertilizer doses (Thankamani et al. 2011).

Tiing Lau et al. (2020) reported several rhizobacteria that are potent to be developed not only as biocontrol agents for black pepper disease management but also as biofertilizers to improve growth in Sarawak, Malaysia. The isolated bacteria were identified as *Bacillus subtilis*, *Bacillus siamensis*, *Brevibacillus gelatini*, *Pseudomonas geniculata*, *Pseudomonas beteli*, *Burkholderia ubonensis*, and *Burkholderia territorii* and were effective in production of antifungal siderophore (53.4%–73.5% per 0.5 mL of cell-free supernatants), chitinase (1.19 to 1.76), and high phosphate-solubilizing index (1.61 to 2.01) and efficient in ammonia (60.3 mM–75.3 mM) production, thereby stimulating black pepper plant growth and root development. Arbuscular mycorrhizal (AM) fungi-inoculated black pepper plants produced significantly higher mycorrhizal root colonization (95%) and spore numbers (312/50 g of sample). The effect of AM fungi was excellent in improving root biomass than aboveground biomass. Nutrient accumulations were higher in AM-inoculated plants, and activities of acid phosphatase and dehydrogenase enzyme were significantly higher in AM-inoculated soils. AM inoculation at the earlier stage of plant development could improve symbiosis and increase plant growth in the nursery which may improve the plants performance after planting in the field (Sarathambal et al. 2023a), and the AM colonization enhances the resistance to *Phytophthora capsici* in black pepper plants by alterations in defense enzymes and gene expression and root exudates composition (Sarathambal et al. 2023b).

Integrated Nutrient Management

Hamza and Sadanandan (2005a) observed increased quality of pepper in terms of its piperine content in potted bush pepper with application of lime at 20 g pot^{-1} and significant yield increase under field condition with high benefit-cost ratio (4.58) at an application rate of 0.5 t ha^{-1}. They also reported an increased soil Mo availability with the application of 0.5 t ha^{-1} lime alone or 0.5 kg ha^{-1} Mo + 0.5 t ha^{-1} lime. Sadanandan and Hamza (2002) recorded an increase in almost all the nutrients and yield due to the adoption of integrated nutrient management (INM) involving recycling of farm yard manure. Mathew and Nybe (2006) observed that combined application of 50% N as FYM + *Azospirillum* + 50% P as inorganic + phosphobacteria +100% K as MOP resulted in the highest berry yield comparable with 50% N and P as inorganic + biofertilizers +100% K as inorganic and recommended dose of chemical fertilizers. The foliar and soil NPK contents also showed higher status with 50% N and P as inorganic + biofertilizers +100% K as inorganic. In addition, cattle manure or chicken manure @ 1–2 kg $vine^{-1}$ for the first and the second year and 3–4 kg $vine^{-1}$ in subsequent years is also applied. A study by Stephen and Nybe (2003) also indicated that treatments involving complete organic (FYM, wood ash) and biofertilizer (*Azospirillum*, AMF, phosphobacteria) + organic + inorganic combinations exhibited higher values of soil nutrients (N, P, K, Ca, Mg). Application of organics like castor, cotton, and carap nut cakes @ 1–2 kg per plant or poultry waste @ 1–2 kg or cattle manure @ 3–5 kg, 200 to 300 g NPK (12:12:17) mixture, 500 g of lime, and 300 g thermo-phosphate $plant^{-1}$ $year^{-1}$ in addition to foliar sprays of micronutrients is the general practice followed by the growers in Brazil (Duarte and Albuquerque 1991). For the Johore area in Malaysia, the recommendation is 120 kg N, 80 kg P_2O_5, and 50 kg K_2O ha^{-1} for the first year; 200 kg N, 100 kg P_2O_5, and 100 kg K_2O ha^{-1} for the second year; and 250 kg N, 100 kg P_2O_5, and 250 kg K_2O ha^{-1} for mature plants to give sustainable high yield (Azmil and Yau 1993). In Sarawak, Malaysia, the elite growers use 110 g $plant^{-1}$ of either Nitrophoska or Nitrophoska special in the first year and 230 g Nitrophoska $plant^{-1}$ in the second year at bimonthly intervals. From third year onward the fertilizer dose is increased to 1.8–2.3 kg $plant^{-1}$ in four applications mainly during July to January (Sadanandan 2000). In Sarawak, the recommendation is 2.04 kg fertilizer mixture of 16 N/6 P_2O_5/18 K_2O/4 Mg per yielding plant plus 454 g dolomite and 28 g trace elements, and farmers are reported to apply 4 to 6 kg fertilizer per plant in seven to eight splits (George 1982).

The highest yield was obtained for the application of coir pith compost (2.5 t ha^{-1}) with FYM (10 t ha^{-1}) + ½ NPK (50:20:140 kgha-1 in field) or 1:05:2 g in pot^{-1}) + *Azospirillum* (20 kg ha^{-1}), which was on par with coir pith compost

(CC) + FYM + BF (*Azospirillum*) and NPK alone. Significant increase in quality parameters, viz., piperine and oleoresin content, was registered with the application of CC + ½ NPK + BF, and the increase in piperine content was significant with the increase in the quantity of CC applied. Contents of piperine and oleoresin were found to be higher in potted bush pepper than field, and higher oleoresin content was in combined application of FYM and CC along with ½ NPK and BF (Srinivasan et al. 2000). Srinivasan et al. (2005) recommended application of coir pith compost at 1.25 t ha^{-1} integrated with half the recommended fertilizer dose and *Azospirillum* sp. @ 20 g $vine^{-1}$ for higher yield and fertility buildup in black pepper gardens. Economic analysis of experimental results using appropriate tools identified application of ½ prevailing package of practices + Zn is a risk-free rational technology for better returns from black pepper (Madan et al. 2007). Depletion of non-exchangeable K was minimum with the addition of FYM and vermicompost or leaf composts as their addition significantly increased the available K content by 3.8 times over no organic application (Srinivasan et al. 2002b). Kumar and Sathiamoorthy (2002) reported that the number of developed berries per spike and the total number of berries per spike were the highest with the application of 50 ppm naphthalene acetic acid (NAA). Berry set was the highest with the application of 150 ppm NAA, whereas green berry yield was the highest with the application of 50 ppm NAA.

ICAR-Indian Institute of Spices Research (IISR), Kozhikode, Kerala, has developed crop-specific/designer micronutrient foliar formulations for black pepper to correct the micronutrient deficiencies in the crop and to meet its physiological and metabolic requirement by maintaining optimal nutrient ratio of secondary (Mg) and micronutrients (Zn, B) in the leaves. These formulations supply essential nutrients like Zn, B, Mg, and K and is recommended as foliar spray (@ 5 g l^{-1}) at critical stages of crop growth like flowering and berry development. The trials conducted at farmer's field in Kerala and Karnataka have given a clear yield advantage of 15–30% in sprayed fields as compared to control (non-sprayed) fields and significant improvement in the bulk density of the berries (Srinivasan et al. 2017a).

Field trials conducted in the farmer's field at various locations in Karnataka on mild and moderately virus affected vines showed that integrated application of FYM, soil test-based NPK, foliar micronutrient (IISR Black pepper special) @ 5 g l^{-1} twice (June and September), and PGPRs twice (June and September) either fortified with FYM (and applied 10 kg $vine^{-1}$) or as drenching (2–3 l $vine^{-1}$) boosted the health of both mild and moderately virus-infected vines resulting in more number of leaves, well-developed canopy, spike intensity per unit area, and increased fresh yield (30–50% higher) in virus-infected vines as compared to no application. The health status of the vines also scored high, 2.9–3.2 (on the scale of 1–5 from severely infected to very healthy based on visual scoring) under both categories showing improved health of the vines, thus masking the virus symptoms. The fresh yield was higher for integrated application of FYM + NPK + PGPR + micronutrients (3.6–7.1 kg $vine^{-1}$) significantly higher than check (2.4–4.4 kg $vine^{-1}$) (Srinivasan et al. 2017b).

Organic Management

For nutritional management under organic farming, a judicious application of FYM (10 kg), burnt earth (10 kg) or wood ash (2 kg) and *Azospirillum* (50 g) or *Phosphobacteria* (50 g) vine^{-1} increased the yield by 37–173% over control on Panniyur-1 at Kerala (AICRPS, 2005). The studies at ICAR-IISR showed that soil pH increased to neutrality, and P and Zn availability increased significantly under organic system of black pepper cultivation with the highest average fresh yield of 2.9 kg vine^{-1} as compared to INM and chemical systems. The available K levels were the lowest in the organic sites compared to chemical and INM sites. The microbial population also was high in organic plots with a bacterial population of 24–80 × 10^5 cfu as compared to 2x10^5 cfu in chemical system. The activity of various enzymes like acid phosphatase, cellulase, dehydrogenase, and phosphodiesterase were relatively high in organic and integrated system of management (Srinivasan et al. 2011b).

Field experiment on black pepper conducted at IISR during 2007–2012 under different management systems, viz., fully organic, fully inorganic, and integrated systems, revealed that organic system of management significantly increased the soil pH, organic carbon, N, P, Zn, and Cu content on par with that of integrated system. Maximum yield was recorded under organic management system that was significantly higher than inorganic and integrated systems of management. Studies in Malaysia showed that black pepper yield from organic farming was lower compared with that of the conventional practice of using chemical-based farming inputs (Paulus and Wan 2011). Butit provided higher benefit-cost ratio of 1.67 compared with that of conventional practice on the production cost per yield basis. Organic management system also recorded significantly higher acid phosphatase, alkaline phosphatase, and dehydrogenase activity in soil, whereas integrated management system had significantly higher phosphodiesterase activity.

Even though the influence of different management systems on oil and oleoresin contents in black pepper was not significant, significantly higher piperine content was recorded under organic system of management on par with that of integrated system. For organic production, application of 10 kg FYM + 500 g neem cake +500 g wood ash+2 kg vermicompost with biofertilizers *Azospirillum* and P-solubilizing bacteria (20 g) has been standardized for meeting its nutrient requirement (Srinivasan et al. 2017c, 2017d). Application of PGPR consortia formulation mixed with farmyard manure (or well-decomposed cow dung) and applied @ 1–2 kg vine^{-1} in the basin or applying soil drenching 1–5 vine^{-1} of the dissolved biocapsule formulation (one capsule dissolved in 200 l water) can also be used while establishing the garden to enhance the soil and plant health. The soil organic amendments consisted of fermented juices, biochar, and compost showed positive influence on soil bulk density, soil porosity, pH, CEC, total organic carbon (TOC), C/N ratio, available P, exchangeable Ca, soil respiration, and soil microorganism count (bacteria, actinomycetes, and fungi). The effect of soil organic amendments on fresh berry yield was comparable to that of NPK fertilizer, and the economic viability also was comparable to the conventional NPK fertilization in Sarawak, Malaysia (Sulok et al. 2021).

Intercropping

John (2000) described about the prospects of black pepper as an intercrop in coffee plantations in Karnataka, India (Fig. 3). The feasibility of intercropping medicinal plants in black pepper garden was evaluated by Thankamani et al. (2012). According to Sewwandi et al. (2019), intercropping black pepper in tea plantations has been a common practice of smallholders in the low- and mid-country regions of Sri Lanka. The possibility of intercropping black pepper in coconut gardens of West Bengal, India (Ghosh 2009), and lower Brahmaputra valley of Assam, India (Nath et al. 2021), is reported.

1.1.6 Soil Biological Properties

Studies on the influence of monocropping of black pepper for 10, 21, and 55 years and on soil physiochemical properties, enzyme activities, bacterial abundance, and bacterial community structures in tropical China showed that long-term continuous cropping led to a significant decline in soil pH, SOM contents, and enzymatic activities and resulted in a decreased soil bacterial abundance and altered soil microbial community membership and structure, which in turn resulted in poor growth of black pepper. Pyrosequencing analysis of 16S rRNA genes revealed that the *Acidobacteria* and *Proteobacteria* were the main phyla in the replanted black pepper orchard soils, comprising up to 73.82% of the total sequences. The relative abundances of *Bacteroidetes* and *Firmicutes* phyla decreased with long-term continuous cropping; and at genus level, the *Pseudomonas* abundance significantly

Fig. 3 Black pepper + coffee cropping system (photo: V Srinivasan)

depleted after 21 years continuous cropping. In addition, bacterial diversity significantly decreased under 55 years black pepper continuous cropping, suggesting duration of monocropping was the major determinant for bacterial community structure (Xiong et al. 2015).

The studies by Li et al. (2016) revealed that the relative abundance of the dominant bacterial phyla in rhizosphere soil increased considerably with long-term consecutive monoculture but decreased in non-rhizosphere soil with a significant decline in *Firmicutes*. For fungi, an increasing trend over time was observed in both rhizosphere and non-rhizosphere soils, with the abundance of the pathogenic fungi *Fusarium* increasing significantly accompanied by a decrease in the bacteria *Pseudomonas* and *Bacillus* that are beneficial for black pepper. Increasing pathogen and decreasing antagonistic bacteria might cause black pepper to be more vulnerable to diseases on long-term basis. Continuous monocropping, especially for 38 years, considerably decreased soil microbial diversity. Additionally, the rhizosphere soil pH, SOM, and available K contents decreased with increasing planting duration, though available N and P increased. All soil nutrient contents and microbial diversity indices were higher in rhizosphere soil compared to non-rhizosphere soil.

The role of using beneficial microorganisms in the sustainable production of black pepper is increased because many people are being aware of the hazards associated with consuming products with chemical substances. The average total density microorganism was 10^6 $cfug^{-1}$ soil of which the beneficial microbes contributed to the density of 14–20 × 10^4 $cfug^{-1}$ soil in major black pepper-growing soils of Vietnam (Nguyen et al. 2020). The use of organic fertilizers is common to the farms where high microbial densities were recorded. There existed a negative correlation between the density of beneficial microorganisms (nitrogen-fixing bacteria, phosphorus-soluble bacteria, cellulose degradation bacteria) and the density of fungal pathogens and nematodes with $R = -0.84$ and $R = -0.81$, respectively. The density of beneficial microorganisms in the soil and the incidence of diseases in orchards were also negatively correlated ($R = -0.69$).

1.2 Small Cardamom

1.2.1 Altitude and Climatic Requirements

The most favorable altitude range for growing Indian cardamom (normally called as cardamom) is from 600 to 1500 m above mean sea level (MSL). Its cultivation in India is restricted to the South Western Ghats regions, which constitute an extensive chain of hills that lie parallel to the west coast of peninsular India. Based on adaptability, nature of panicle, and shape and size of capsules, the cardamom cultivars are categorized into Malabar, Mysore, and Vazhukka. The cultivar Malabar with a prostrate panicle (panicles spreading on ground) is widely grown in Karnataka state, while the cultivar Mysore, characterized with erect panicles is extensively cultivated in Kerala and parts of Tamil Nadu. Whereas, the cultivar Vazhukka, a natural hybrid between Malabar and Mysore types with a distinct semi-erect (pendent) panicle, is the most popular cultivar in Kerala state. Type Malabar, the traditional cardamom of

Karnataka, possesses the capacity to be productive at lower elevations of 500–700 m MSL. The other types, Mysore and Vazhukka, are not productive below 700 m elevation (Abraham and Tulasidas 1958). At lower elevation, while vegetative growth is satisfactory, capsule and seed production are poor. In Guatemala, cardamom is grown at varying altitudes, ranging from 90 to 1200 m above MSL. Cardamom is highly sensitive to elevation, and the wrong choice of cultivar or wrong location can severely affect the growth and productivity.

Cardamom is very sensitive to temperature fluctuations, and in India, optimum growth and development are observed in the warm and humid conditions at a temperature range of 10–35 °C (Anon 1976; Jacob et al. 1995). The upper temperature limit will normally be around 31–35 °C for a brief period. In the eastern side of the Western Ghats, a combination of desiccating winds passing from the hinter lands of east and low humidity leads to desiccation and drying of plants. In such areas, protective irrigation would be essential for retention of humid conditions for setting of capsules (Korikanthimath 1991). It is noticed that rate of spread of "katte" disease (a viral disease) will be more during summer than in rainy season. Cold conditions result in low setting of capsules. Thus, extremes of temperature and wide diurnal variations are not conducive for healthy growth of cardamom plants. The Indian average yield of cardamom is only 149 kg ha^{-1} as against 300 kg ha^{-1} in Guatemala and Papua New Guinea. One of the main reasons for the increased yield in Guatemala is the well-distributed rainfall (Krishna 1997; Mohanchandran 1984). In India, 75–80% of total area under cardamom is rainfed (Charles 1986).

Cardamom is grown in South India under rainfall conditions ranging from 1500 to 5750 mm. Nearly, 75–78% of total rainfall in received from the southwest monsoon which lasts for 3 to 4 months from June–October. Northeast monsoon and pre-monsoon showers (summer rains) are not well-distributed and dependable. Therefore, cardamom suffers from both excessive moisture and moisture stress. Dry spells in these tracts usually occur during December–May. The Indian Cardamom Hills (ICH) enjoys typical monsoon rainy climate except the easternmost slopes where the annual mean maximum and minimum temperatures attain correspondingly 38 °C and 16 °C. The mean annual rainfall in the easternmost slopes was reported to be 1.500 mm, while the west facing slopes in Elappara range (Kerala, India) experienced a maximum of 7000 mm. The changes in climatic elements mainly temperature and rainfall for the last century in the ICH have been studied (Murugan et al. 2000; Murugan et al. 2007) in relation to cardamom productivity.

As the extensification of crop agriculture in the perennial crop agroecosystems like cardamom and tea systems is unlikely in the ICH, intensification is the way that has been practiced in the ICH agroecosystems for the past 30 years (since 1995). Cardamom was thought to be very sensitive to climatic variables and used to be grown as a shade-loving crop under multistoried (three to four layers) thick evergreen forest canopy (Figs. 4 and 5).

Recently under changing climate situation, cardamom behaves like summer annuals, tolerating heat to a significant extent. The tree cover in the tropical mountains has been used to modify the heat budget of a land surface by its influence on incoming and outgoing radiation, as well as by its direct modification of humidity

Fig. 4 Cardamom under natural shade trees (photos: M. Murugan)

Fig. 5 High-yielding cardamom plant

and temperature below canopy. Mean and maximum temperatures were lower under the forest canopy than under the sparse cover, but the minimum temperatures were higher in the forest 10 cm above the forest vegetation. The forest canopy functioned effectively as a screen for short and long wave radiation. Temperature fluctuations were thus dampened, and the daily climatic cycle was less contrasted in forested tropical mountains (Murugan et al. 2012a).

There has been a drastic reduction in tree cover due to deforestation, felling, and excessive opening of canopy during the last decade, and this could be the reason for the increase in minimum temperature (Korikanthimath et al. 1998a). In hills especially, removal of canopy cover may have an adverse effect on soil temperature since high-altitude sunlight has high-energy light particles which can increase the soil temperature quickly than in the plains (Garbell 1947). Under shade-less conditions or less shaded areas, elevated soil temperature (36 °C) and the strong wind movement increases evapotranspiration, and the soil dries up quickly. All these interrelated changes alter microclimate that would affect growth and development of cardamom. An annual rainfall of 2000 mm with uniform distribution was considered ideal for cardamom (Murugan et al. 2000, 2012a).

1.2.2 Soil Requirements

Cardamom is cultivated under shade trees, and the soils in general have high fertility status due to addition of leaf litter (Zachariah 1978). In India, soils of major cardamom-growing areas come under the order Alfisols, formed under alternate wet and dry conditions and the suborder Ustalfs derived from schists, granite, and gneiss and lateritic in nature (Sadanandan et al. 1990). Soils most favorable for growth and development of cardamom are red lateritic loam with layers of organic debris present in the evergreen forests, although it grows on a variety of soils with only a shallow zone of humus accumulation. Cardamom generally grows well in forest loamy soils with good drainage. The cardamom-growing soils of Karnataka, India are mostly clay loam (Kulkarni et al. 1971). These soils are generally acidic in nature with pH from 4.2 to 6.8, high in organic matter and N, low to medium in available P, and medium to high in available K. Nair et al. (1980) observed that shade trees bring to the soil surface the nutrients they take up from lower layers by way of leaf shed and thus help to maintain a high fertility status and soil pH in surface soils. Even though on an average about 5.5 t ha^{-1} of organic material as leaf litter, weeds, and pruned plant parts are recycled in a cardamom plantation in a year, the nutrients are mainly in the organic form and are available to the crop only by the process of mineralization. As the rate of mineralization is always low, the nutrients that become available to the crop will be able to sustain only average growth and production (Srinivasan et al. 1993a).

As a consequence of tillage operations in steep slopes and large-scale deforestation, the decline in soil fertility as well as the conducive microclimatic condition necessary for the normal growth and development of cardamom had undergone near irreversible alteration (Miniraj and Murugan 2000). Several studies have shown that the fertility status of cardamom hill reserve (CHR) soils was generally high due to closed nutrient cycling which occurred in almost in all forest ecosystems (Korikanthimath et al. 2001; Murugan et al. 2008). The CHR soils are acidic due to inherent soil-forming factors and the continuous application of fertilizers and chemicals. Consequently, the concerns on the sustainability of the cardamom agroforestry system have been increasing because of the never-ending agricultural intensification and degradation of tropical evergreen forest cover (Murugan et al. 2017).

Korikanthimath et al. (2000a) collected soils from various cardamom plantations in Kodagu district (Karnataka, India) and found that they were acidic (5.0 to 6.6 pH) in nature with high organic carbon content, which decreased with depth of the soil. The soils were medium (56.82%) to high (43.18%) in available N, low (79.55%) in available P, and low (38.63%) to medium (52.27%) in available K. The soil profile collected from cardamom fields indicated relatively higher level of available N and organic matter than from uncultivated area due to recycling of organic materials in the former. The organic carbon content correlated positively and significantly with available nitrogen content.

A study by Srinivasan et al. (2002a) showed that the S status of cardamom-growing soils of South India, ranged from traces to 36 ppm in Kerala, traces to 27.5 ppm in Karnataka, and 15–36 ppm in Tamil Nadu, categorizing 43.3 and 50% of Kerala and Karnataka soil samples deficiency with respect to S, respectively.

Srinivasan et al. (1993b), investigated the micronutrient status of cardamom-growing soils of South India, and the study indicated that the status of available Fe ranged from 14.6 to 65.8 ppm, Mn from 1.3 to 44.8 ppm, Cu from 0.66 to 32.2 ppm, and Zn from 0.01 to 11.1 ppm. Based on the critical limits prescribed for various micronutrients, it was observed that 68% of area sampled was deficient in available Zn, 49% in available boron, 28% in available molybdenum, and 9% in available manganese. None of the soils recorded deficiency of available iron and copper. Micronutrient survey conducted by the Indian Cardamom Research Institute (ICRI), Myladumpara, Idukki, (Kerala), showed that zinc deficiency is widespread in cardamom soils and B deficiency is observed in certain areas. So, it is recommended that Zn may be applied as a foliar spray as zinc sulfate @ 250 g100 l^{-1} of water during April–May and September–October. Besides Zn, application of B seems to be imperative in soils deficient in B. In such cases, soil application of B in the commercial grade borax @ 7.5 kg ha^{-1} is recommended in B-deficient areas which needs to be applied in two doses along with NPK fertilizers.

Korikanthimath et al. (2000b) reported the uptake pattern of nutrients in different parts of 1-year-old cardamom. The N, P, and K uptake increased with increase in dry matter production. Of the various nutrients, the total K uptake was more followed by N and P, and it suggests that crop requires more of K than N and P. Among the different plant parts, the highest contribution to the total N uptake was by green leaves (44.69%) and tillers (29.6%) in 1-year-old and prepotents, respectively. The highest uptake of P and K in 1-year-old plants was by roots and grown-up tillers, respectively.

Chemical Properties of Soils

Century-old continuous cultivation of cardamom in the CHR has been reported to reduce the soil pH because of constant crop removal and uptake of alkali elements like Ca and Mg as well as heavy leaching and runoff due to incessant intense rainfall and topography. Since the CHR massif was dominated by charnockite, input from acidic parent material was the bare minimum (Murugan et al. 2012b).

Cardamom soils have small buffering capacity due to little clay fraction and are highly prone to acidification and cation depletion (Noble et al. 2000). Currently, the nutrient-poor cardamom soils depend totally on external application of fertilizers to maintain its fertility status. Disappointingly, acidification of cardamom soils will be difficult to correct as long as excessive levels of N fertilization continue. Goulding and Annis (1998) found that each ~50 kg ha^{-1} of added ammonium-N generates 4 kmole H^+ ha^{-1} $yr.^{-1}$ and requires 500 kg $CaCO_3$ ha^{-1} $year^{-1}$ to neutralize in field conditions. Decreased soil pH coupled with intense rainfall during monsoon months increased the incidence and intensity of rot diseases in cardamom system. Low shade levels in combination with higher levels of frequent application of fertilizers increased the prevalence of sucking and chewing types of insect pests in cardamom resulting in additional requirement of insecticides for their management. The greater variability observed among the soil biophysicochemical properties between forest soils and cardamom soils could be attributed to application of larger quantities of fertilizers to the cardamom soils (Murugan et al. 2012b; Noble et al. 2000). Continuous application of higher dose of manures and fertilizers can result in soils becoming excessively enriched with some nutrients, particularly P, K, Ca, and Mg in top soil layers and with nitrate-N, Ca, and Mg in subsoils. The accumulation of excessive levels of Ca, Mg, and K in soils is unlikely to pose any environmental risk. However, levels of P and N above those required for optimal crop production will increase the potential for nutrient runoff and leaching (Ledgard et al. 2000; Murugan et al. 2017; Sharpley et al. 1994).

Ongoing soil and atmospheric warming along with medium shade level on cardamom might increase the uptake of N and P from soil due to higher growth rate and aboveground biomass in response to climate change-induced nutrient uptake (Farrar and Jones 2000; Skillman et al. 2005). Increased nutrient uptake with enhanced photosynthetic activity resulted in the increase of biomass and yield of cardamom variety *Njallani*. This particular variety is highly responsive to fertilizers and other chemicals but highly susceptible for insect pests (thrips and capsule borers) and diseases (*Rhizoctonia*, *Phytophthora*, and *Fusarium* spp). Higher nutrient uptake reported for this variety could be due to the higher root cation exchange capacity (CEC) (62.2 meq 100 g^{-1}) as well as higher N content (leaves and capsule) of the variety (Murugan et al. 2012b).

Cardamom prefers ammoniacal form of N than nitrate form, and planters use more ammoniacal form of N to cardamom. Application of farm yard manures, cow dung, cow dung slurry, neem cake, and compost from municipalities and cities has been very common. Decomposition of these organic materials by organisms has been accelerated by elevated soil temperatures and moisture leading to higher mineralization of organic materials, which brings more N and P into soil solution besides contributing to soil acidity. Liming is done both for tropical and temperate acid soils to alleviate nutrient deficiencies through routine fertilizer amendments. The continuous liming has had positive effects on soils. It increased base saturation (BS), cation exchange capacity (CEC), and nitrogen mineralization and decreased anion exchange capacity (AEC) and acid weathering of primary and secondary minerals in cardamom soils. Except a few plantations, liming had helped increasing cardamom yields and soil quality. In a few occasions because of over liming and dispersing properties of Ca and Mg, soil permeability was affected (Murugan et al. 2012b).

The soil pH of cardamom hill soils ranged from 3.6 to 6.7, with nearly 60% of samples (99 samples out of 167) having pH > 4.6 and the remaining having a pH value of <4.6. Approximately 20% of soil samples had pH values ranging from 5.5 to 6.7. Undisturbed forest soils showed narrow range of pH (4.1–4.9). The yield improvement for the application of dolomite (2 kg plant^{-1}) was 68.37% over the no lime application. The highest dry (500 g plant^{-1}) capsule yield was observed in dolomite-applied plots. The enhanced growth and yield of cardamom were likely due to the positive impact of liming on soil chemical and biological properties, particularly the rise of soil pH (Murugan et al. 2022).

The organic matter content of cardamom soils and forest soils ranged 1.1–5.7% and 3.0–7.9%, with an average value of 3.5% and 4.9%, respectively (Murugan et al. 2012b). The available N varied greatly for cardamom soils (1.4 to 83.4 ppm) as well as forest soils (8.2 to 64.4 ppm) with a corresponding mean value of 16.6 and 31.8 ppm. Despite the greater variability of available N in cardamom soils, 66% of samples had registered above mean level. The available P (Bray 1 P) in soils of cardamom plantations, and undisturbed forest changed considerably. It ranged from 18.3 to 95.6 ppm in forest soils; however, in cardamom soils up to 200 ppm had been reported in the survey. Furthermore, the mean available P was very high for cardamom soils (108.6 ppm) than that for forest soils (67.6 ppm). Noticeably, the variability of available P was much higher for cardamom soils (CV 72.0%) than those of forest soils (CV 39.9%). Liming increased the availability of secondary plant nutrient (Ca, Mg and S) contents of the soil. Drastic variation in soil available K was noticed; and it ranged from 47 to 785 ppm for cardamom soils and 28 to 490 ppm for forest soils. The available K (average) for forest soils was higher (377.0 ppm) than those for cardamom soils (331.8 ppm). The available Ca concentration was found to vary strongly between the two soil categories. Ca saturation percentage ranged from 45.2 to 79.6 in cardamom soils with a mean value of 64.2. In comparison, cardamom soils (223–1614 ppm) had wider range than that of forest soils (485–1145 ppm). The mean available Ca for forest soils was 810 ppm, while it was 883.7 ppm for cardamom soils (Murugan et al. 2012b).

The mean available Mg content was higher for forest soils (217.8 ppm) than that for cardamom soils (158.1 ppm). Similar trend was noticed for available S, wherein the mean available S for forest soils ranged from 5.2 to 32 ppm, while for cardamom soils it varied from undetectable level to 89.9 ppm. Upper limit of available Na was much higher for cardamom soils compared to forest soils. The trend of percentage saturation of all these four nutrient elements in both forest and cardamom soils was in the order of Ca > Mg > K > Na (Murugan et al. 2012b).

Available Micronutrients

Soil micronutrients have greater importance in the health of soil, crop, human, and animal ecosystem. Most of the elements in the soils are biologically active, and some are necessary for life processes and are therefore called essential nutrients. Essential available micronutrient concentrations for cardamom soils (mean) were in the following order: Fe (207.6 ppm) > Mn (47.1 ppm) > Cu (17.4 ppm) > Zn (3.8 ppm) > B (0.3 ppm). The forest soils also had a similar trend except the third place Cu, which was replaced by Zn. However, there has been two- and fourfold increase, respectively, for Fe and Cu in cardamom soils. Both N and K had the highest range values in cardamom leaf samples which were in that order of 2.7–3.7% and 2.9–3.5%, respectively. P values (0.09–0.1%) were small and the lowest. The highest concentration of K was reported in capsule rind (43,487.9 ppm) followed by whole capsule (17,778.4 ppm) and seed (10,538.8 ppm). This shows that cardamom is a high K-loving crop (Murugan et al. 2012b).

Presence of higher concentrations of heavy metals and rare metals in both cardamom soils and capsules can be attributed to excessive application of manures (prepared from city and town wastes) that contain almost all of these heavy and rare metals. Majority of the fungicides and micronutrient sprays are given directly on the capsules, and, therefore, the concentrations of heavy and rare metals in capsules are always higher than the leaves of cardamom. Comparable results of some of these metals on cardamom have been reported by Ozkultu et al. (2006). Of all the metal elements studied, the most important to consider in terms of food chain contaminations are arsenic (As), cadmium (Cd), mercury (Hg), lead (Pb), and selenium (Se) (McLaughlin et al. 1999).

Cardamom has been cultivated in the ICH for many centuries, and to increase its yields, planters applied large quantities of fertilizers, manures, and liming materials continuously for decades. Application of higher levels of fertilizers and manures to already nutrient-poor soils resulted in further decrease in soil pH aggravating the soil fertility and environmental problems like leaching of base cations from soil (Murugan et al. 2012b). The land use and management factors were considered to be playing important role in regulating microbial communities in tropical soils than in many temperate systems, where favorable climatic factors, combined with fertilizers, helped to buffer changes in soil environment (Giller et al. 1997).

Biological Properties

The soil habitat is considered to be highly heterogeneous environment for microbiota inhabiting it. Therefore, microorganism resident in soils was exposed to varied abiotic and nutritional conditions. These organisms collectively are responsible for underlying catalysis of the biochemical process in soils including those resulting in disease suppressiveness at the scale of microhabitats and organismal biosphere. And these processes are susceptible to major changes in the surroundings. Microorganism in soils plays a critical role in the maintenance of soil function in both natural and managed agricultural soils because of their involvement in key processes such as soil structure formation, decomposition of organic matter, toxin removal, and the recycling of carbon, N, P, and S, besides suppressing of soil borne plant diseases and promoting plant growth and development (Doran et al. 1996).

The population of bacterial and fungal feeding nematodes was minimum in cardamom soils which is the indication of very poor soil biodiversity. Soil CO_2 evolution and dehydrogenase activities were higher for cardamom soils than forest soils causing enhanced decomposition and loss of SOM in cardamom soil (Murugan et al. 2017).

Higher population of bacteria (48.2×10^7 colony forming units (cfu) g^{-1} soil) and fungi (22.9×10^4 CFU g^{-1} soil) and higher activities of dehydrogenase (DHA) (0.04 g triphenyl formazan g^{-1} h^{-1}) and CO_2 evolution were recorded for cardamom rhizosphere soils than the forest soils. Average root CEC of cardamom (*Njallani* variety) was 61.6 meq100 g^{-1}, and the pH of the soil samples ranged from 5.5 to 6.0. Higher population of antagonistic bacteria (*Pseudomonas* spp.) and *Trichoderma* (4.2–4.3×10^5cfu g^{-1}) was observed in intensively managed cardamom soils with lower population of pathogenic bacteria and fungus at this pH range (Murugan et al. 2017).

1.2.3 Nutrient Management

Response to Fertilizers

Cardamom is generally grown in the rich fertile soils of the forest ecosystem. The nutrient analysis data obtained from cardamom-growing tracts in South India revealed that the limiting leaf nutrients were in the order Zn > K > P > Ca > Mg > Cu > Mo > Fe > Mn > N (Hamza et al. 2009). Being a perennial crop, sucker production is throughout the year, and initiation of panicles and development of capsules are spread over a period of 8 to 9 months in a year. A steady absorption and utilization of plant nutrients take place throughout the life cycle of cardamom which warrants regular application of nutrients for higher yields.

Nutrient uptake studies in cardamom have indicated the uptake of N, P, K, Ca, and Mg to be in the ratio of 6:12:3:0:8, respectively, and for the production of 1 kg of cardamom capsules, 0.122 kg N, 0.014 kg P, and 0.2 kg K are removed by the plant (Kulkarni et al. 1971). Korikanthimath et al. (2000b) noticed that N, P, and K uptake by cardamom increased with increase in dry matter production. Of the various nutrients, the total K uptake was more followed by N and P in both the stages of crop suggesting that the crop requires more of K than N and P. Among the different plant parts, the highest contribution to the total N uptake was by green leaves (44.7%) and tillers (29.6%) in 1-year-old and prepotent plants, respectively. The highest P and K uptake was by roots and grown-up tillers in 1-year-old plants and green leaves and tillers in prepotent plants, respectively. The broad nutrient N, P, and K uptake ratio per plant was 9:1:17. Among the secondary nutrients, the highest uptake of Ca was observed in shoots, whereas, for Mg it was the leaves which showed the highest values. Uptake of Ca by shoots at harvest stage was more than that of pre-bearing stage. The reverse was the trend in case of Mg (Krishnakumar and Potty 2002).

The application of organic manures such as neem cake @ 1–2 kg $plant^{-1}$ or poultry manure/FYM/compost/vermicompost @ 5 kg $plant^{-1}$ may be done once in a year during May–June. If necessary, application of Mussoorie rock phosphate or bone meal may be done based on soil analysis. Application of 3–4 t of well-rotten cattle or sheep manure apart from 75 kg each of N, P_2O_5, and 100 kg K_2O ha^{-1} was found to greatly enhance the yield of cardamom. As 70% of cardamom roots were confined to shallow depth of 5–40 cm and 30–50 cm radius, for maximizing the efficiency of applied manures, it would be necessary to apply at a radius within 50 cm. Field studies in cardamom have shown that application of recommended NPK nutrients as organic fertilizer (50% each as FYM and neem cake +50% P each as bone meal and rock phosphate +50% K as wood ash) was effective in increasing the yield and quality of cardamom (Sadanandan and Hamza 2006).

In cardamom-growing soils with high status of N and K, a maintenance dose of 30:60:30 kg N, P_2O_5 and K_2O ha^{-1} was recommended for healthy and vigorous growth of plants (Zachariah 1978). Based on further studies and the factors affecting the availability of nutrients in the soil, a fertilizer dose of 75:75:150 kg N, P_2O_5, and K_2O $ha^{-1}$was recommended under rainfed situation for normal yield of 100 kg dry capsules ha^{-1} (Murugan et al. 2007). For realizing higher yield, the fertilizer doses are to be increased proportionately. Additional fertilizer dose of 0.65 kg N, P_2O_5, and 1.3 kg K_2O is to be applied for increase in yield of every 2.5 kg dry capsules over normal yield (Kologi 1977). Srinivasan et al. (1998) indicated that significant yield increases and savings in fertilizer requirements would be possible by way of soil-cum-foliar application method. In Karnataka, India, N, P_2O_5, and K_2O @37.5:37.5:75 kg ha^{-1} through soil and 2.5% urea +0.75% single super phosphate (SSP) + 1.0% muriate of potash (MOP) through foliar gave 64 kg yield increase ha^{-1} over no fertilizer application. The yield increase recorded was 43 kg ha^{-1} under Tamil Nadu condition, where the dosages required were 20:40:20 kg N, P_20_5, and K_2O ha^{-1} through soil and 3% urea+1% SSP +2% MOP through foliage. In irrigated plantations, the general fertilizer recommendation is to apply N, P_2O_5, and K_2O ha^{-1} @ 125:125:250 kg ha^{-1} to soil in three splits (Anon 1997; Murugan et al. 2007). A fertilizer dose of 37.5:37.5:75 kg of NPK ha^{-1} for 1-year-old plants during September and 75:75:150 kg of NPK ha^{-1} from second year onward in two split doses (during May and September) was found to be optimum (Murugan et al. 2007).

But when plant population was increased to 5000 ha^{-1} under trench method of planting, a fertilizer dose of 125:125:250 kg N, P_2O_5, and K_2O ha^{-1} in three splits was recommended (Korikanthimath et al. 1998b). Application of FYM at 5 kg $plant^{-1}$ + 50% recommended N + *Azospirillum* yielded 141.20 kg ha^{-1} similar to that of FYM at 5 kg $plant^{-1}$ + 75% recommended N (146.34 kg ha^{-1}), thereby providing 25% saving in inorganic N (Kumar et al. 2009). Kumar et al. (2000) reported that the highest application of 150:150:225 kg NPK ha^{-1} recorded the highest green capsule yield. Contrarily, Murugan et al. (2007) recorded that increasing the levels of N, P, and K increased the yields of cardamom up to 125:125:200 kg ha^{-1}. Under low rainfall situations prevailing in the lower Palani tracts of Tamil Nadu, India, a fertilizer dose of 40:80:40 kg N, P_2O_5, and K_2O ha^{-1} was found to be optimum (Natarajan and Srinivasan 1989). Ranganathan and Natesan (1985) observed beneficial effect of application of $ZnSO_4$ to nursery plants by incorporating it along with fertilizer mixture. Sivadasan et al. (1991) found the beneficial effect of liming of cardamom soils in correcting acidity and enhancing the rate of nitrification resulting in better growth. Liming acid soils dissolves calcium and magnesium phosphates and sulfates and allows nutrients to be more available for plant uptake, leading to higher yield (Murugan et al. 2012b). However, subsequent observations (Anon 1997; Nair et al. 1988) indicated that liming is not required as a routine practice in cardamom plantations. The fertilizer schedule for cardamom-growing states of India is given in Tables 6 and 7.

It is necessary to give one-third of the recommended dose (NPK 75:75:150 kg ha^{-1}) during the first year of plant growth both under rainfed and irrigated conditions. During the second year of plant growth, the dose may be increased to one-half of the recommended dose of NPK 75:75:150 kg ha^{-1} and 125:125:250 kg NPK ha^{-1} for rainfed and irrigated conditions, respectively, and fertilizers at full dose may be applied from third year onward.

Among the N forms, ammoniacal forms of N were found to be the best for acidic soils to enhance nutrient use efficiency. Thus, complex fertilizers such as 20:20:15 or 10:26:26 or 18:18:18 or DAP are much useful. In the case of P source, acid-soluble form of P as either rock phosphate or bone meal is better for acid soils. A mixture containing N/P/S association in the form of ammonium phosphate sulfate enhances the efficiency of applied nutrients. Spray of water-soluble form of P, viz., polyfeed (13:40:13 NPK), monoammonium phosphate (0:52:53 NPK) is helpful for profuse flowering and yield. Similarly, water-soluble form of K, viz., potassium nitrate (13:0:45 NPK), sulfate of potash (0:0:50:20 NPKS) polyfeed (13:40:13 NPK), and monopotassium phosphate (0:52:34 NPK) would help in getting bold capsules with high oil content (Subbiah et al. 2008).

Table 6 Fertilizer schedule for cardamom-growing states

Region	Soil application (NPK kg ha^{-1})	Soil-cum-foliar application[a]	Time of application	
			Soil	Foliar
Karnataka	75:75:150	NPK 37.5:37.5:75 kg ha^{-1} Urea (2.5%) SSP (0.75%) MOP (1.0%)	May–June August–September	September November January
Kerala	75:75:150	NPK 37.5:37.5:75 kg ha^{-1} Urea (2.5%) SSP (0.75%) MOP (1.0%)	May–June August–September	September November January
Tamil Nadu	40:80:40	NPK 20:40:20 kg ha^{-1} Urea (3.0%) SSP (1.0%) MOP (2.0%)	May–June August–September	September November January

[a]kg ha^{-1} (NPK) for soil application and % for foliar application
Note: Instead of straight fertilizers, 2% each of diammonium phosphate (DAP) and MOP can also be used. Source: Krishnakumar and Potty (2002)

Table 7 Stage of cardamom growth for fertilizer application

Time of application	Water-soluble fertilizers/micronutrients/growth hormones (dosage for 200 l water)
Pre-monsoon spray during panicle initiation and development stage	Monoammonium phosphate (12:61:0 NPK) @ 1000 g + zinc sulfate @ 500 g or EDTA chelated zinc @ 200 g + naphthalene acetic acid @ 8 g (40 ppm) + borax @ 300 g or boric acid @ 200 g
Mid-monsoon spray during capsule setting, development and maturation stage	Potassium nitrate @ 1000 g + zinc sulfate @ 500 g or EDTA chelated zinc @ 200 g + borax @ 300 g or boric acid @ 200 g
Capsule development and maturation stage	Sulfate of potash (SOP) @ 1000 g (0:0:50:20 – NPKS) + zinc sulfate @ 500 g or chelated EDTA zinc @ 200 g + magnesium sulfate @ 500 g + borax @ 300 g or boric acid @ 200 g

Source: Subbiah et al. (2008)

Targeted yield equations for recommending nutrient requirements for cardamom in soils with varying fertility levels were standardized based on field trials. The nutrient contribution from soil (up to 53% for N, 7–10% for P_2O_5 and 20% for K_2O) for the capsule yield of cardamom and that from the fertilizer sources (60% for N, 10% for P_2O_5 and 40% for K_2O) were established. The soil test value (STV) and crop response-based fertilizer recommendation model for NPK for different yield targets (q ha^{-1}) of cardamom was developed as.

N (kg/ha) = (Target yield (q ha^{-1}) × 100 × 0.528) – (STV for N × 0.87)
P_2O_5 (kg/ha) = (Target yield×100 × 0.41) – (STV for P2O5 × 0.73)
K_2O (kg/ha) = (Target yield×100 × 0.68) – (STV for N × 0.49)

The fertilizer recommendations based on these established equations when validated for varying yield targets in var. *Njallani* and Appangala-1 at different soil fertility levels result in yield levels 0.62, 0.83, and 0.82 kg $plant^{-1}$ with a minimum deviation from the targets 0.4, 0.6, and 0.8 kg $plant^{-1}$ with the positive mean deviation of 55, 38, and 2%, respectively, in both varieties. The yield gain was 36–80% higher as compared to farmers' practice in both varieties studied. Based on the studies, the economic response in cardamom can be expected up to the fertilizer dosage of 340 kg of N, 230 kg of P_2O_5, and 530 kg of K_2O ha^{-1}, with a plant population of 1200 plants, at varying soil fertility levels (ICAR- IISR 2020).

As the soil fertility will vary with the agroecological conditions or management systems, site-specific nutrient management for yielding gardens based on their soil test results for major nutrient is advocated (Ankegowda et al. 2022). The recommended dose of nutrients for varying soil test values of N, P, and K is given as follows (Table 8).

The ICAR- IISR, Kozhikode has successfully field-tested and commercialized a crop-specific micronutrient formulation for cardamom (Patent No. 413017). This micronutrient mixture for cardamom was developed to fulfil the requirement of the crop by maintaining optimal leaf nutrient ratio of secondary and micronutrients (especially Mg, Zn and B). The crop-specific micronutrient formulation is

Table 8 Soil test-based fertilizer recommendations for cardamom capsule yield target levels of 650 kg and 1.0 t ha^{-1}

	Fertilizer nutrient recommended (kg ha^{-1}) for yield targets	
Soil test value for available nutrients (kg ha^{-1})	650 kg ha^{-1}	1.1 t ha^{-1}
Nitrogen		
< 150	255	400
150–250	200	370
250–400	100	320
>400	50	150
Phosphorus (P_2O_5)		
< 25	225	275
25–50	175	225
>50	100	175
Potassium (K_2O)		
< 250	350	550
25–500	280	500
500–800	180	450
>800	50	250

Source: Ankegowda et al. (2022)

recommended for cardamom as a foliar spray (two to three foliar sprays @ 5.0 g l^{-1} water) during flowering, capsule formation, and maturation stages of the crop at monthly intervals. The application of this micronutrient mixture enhances the yield by 10–15% and increases boldness of the capsules while contributing to the health of the plants through balanced nutrition (Srinivasan et al. 2017a).

Method and Time of Fertilizer Application

Nearly 70% of cardamom roots are confined to shallow depth of 5–40 cm and 30–50 cm radius (Khader and Sayed 1977; Nair 1988). Therefore, for the maximum efficiency of applied fertilizers, it would be necessary to apply them at a radius within 50 cm, and being a surface feeder, deep placement of fertilizers is not advisable. Before application of fertilizers, remove the mulch from plant base, coil the panicle, and apply the fertilizer followed by farm yard manure or mix with soil by forking. After mulching, panicles may then be allowed to spread on the ground for Malabar types to facilitate honeybee movement for better pollination and capsules setting.

Sudharshan et al. (1988) observed that cardamom growth is influenced by seasonal conditions, especially rainfall pattern, and vegetative buds emerge from the bases of tillers almost throughout the year. However, majority of vegetative buds are produced after the rainy period. It was shown that linear growth rate of tillers increases with the onset of southwest monsoon and growth rate slows down with the cessation of rain. Peak flowering and fruit set period coincide in cardamom and nearly 70–90% of flower production was recorded between May and August. For this reason, efficient utilization of fertilizers and proper time of application are very important. Application of fertilizers in May and later in September is found to be the best (Pattanshetty and Nursath 1973). However, under irrigated conditions, tiller and panicle initiation are continuous processes, and therefore more split applications are beneficial. Srinivasan et al. (1998) stated that cardamom is grown under shade trees, and therefore complete utilization of applied fertilizers by the crop may not take place because of root interference. For yielding cardamom plants, soil-cum-foliar application will be an effective method. Soil application of fertilizers is two rounds during May–June and August–September and subsequent applications through foliage during September, November, and January.

In Guatemala, almost all growers fertilize cardamom. There are different fertilizer mixtures available there. One such fertilizer mixture is 15:15:15, which is marketed under brand name Barco Vikingo, and the other mixture which contains calcium and magnesium in addition to NPK is 18:6:12:4:0.2 with the brand name of Agrovet SA. Fertilizer application starts from the second year of planting with 2 oz. $plant^{-1}$. This is gradually increased to 3 oz. of urea over and above this normal fertilizer application. The recommended dose is applied two to four times a year (George 1990).

1.2.4 Management Strategies to Enhance Soil Health

Soil and Water Conservation

Though cardamom tracts receive heavy rainfall (1500–4500 mm), the availability of soil moisture during summer months is a limiting factor due to undulating topography of the plantations, and 75% of rainfall falls in 3 months (June, July, and August) (Hegde and Korikanthimath 1996). In low to medium rainfall areas (1200 to 2000 mm), trench system of planting, i.e., planting cardamom in 0.45 × 0.45 m dimension trenches by filling with farm yard manure, forest soil, and top soil, was found to conserve more soil moisture and supplement the plants even during dry spells (Korikanthimath 1987). In area having a slope of more than 7%, of all along the contour, bench terracing of about 2 m width has to be made for taking up cardamom planting. The timely shade regulation and keeping the cut biomass between the rows across the slope act as a barrier to check soil erosion and to prevent surface runoff losses of rainwater for effective soil and moisture conservation (Korikanthimath 1989). Agronomical practices like shade regulation, trench system of planting, provision of shelter belts, cultural measures like mulching, earthing up, contour staggered trenches, vegetative barriers, intercropping, half-moon, and continuous half-moon terracing help in better growth and development of cardamom (Ankegowda et al. 2002).

Biofertilizers

Roots of cardamom plants growing in experimental plots with three fertilizer levels were examined before and after the southwest monsoon. Three species of vesicular arbuscular mycorrhizal fungi (VAMF) such as *Glomus macrocarpum*, *G. fasciculatum*, and *Gigaspora coralloidea* were recorded. The colonization in roots varied from 40 to 100% in pre-monsoon samples and 63–94% in post-monsoon samples.

Degree of infection varied from 4.5 to 76.0 in pre-monsoon and 30.0 to 70.0 in post-monsoon samples. Increasing doses of fertilizer tended to decrease the intensity and incidence of the VAMF population in soil and roots (Iyer et al. 1989).

Sreeramulu and Bagyaraj (1998) tested the growth response of cardamom seedlings (cv. Malabar) to 13 different vesicular arbuscular mycorrhizal fungi under mat house conditions at Bangalore, India. In general, seedlings inoculated with the fungi grew taller and had more leaves and tillers and greater seedling biomass and uptake of nutrients than control seedlings. Among the various mycorrhizal fungi tested, seedlings inoculated with *Gigaspora margarita* and *Glomus monosporum* exhibited significantly greater growth with increased uptake of nutrients. Inoculation with VAMF significantly improved the growth of cardamom plants, even when co-inoculated (either prior or simultaneously) with root knot nematodes. The VAMF inoculation could effectively colonize the roots mass of cardamom when inoculated prior or simultaneously with nematodes and reduced the nematode population significantly. *G. margarita* and *Glomus fasciculatum* were found to be more effective in improving the growth of cardamom seedlings even in the presence of root knot nematodes (Thomas et al. 1989).

Cardamom-Based Cropping System

In intercropping systems, plants compete for light, water, soil, and space. Competition will occur when the immediate supply of a single necessary factor falls below the combined demands of the plants (Donald 1963). In some perennial crops, the leaf canopies of the components of crop combinations occupy a differential vertical layer with tallest components having foliage requiring or tolerating shade and/or high humidity. Better utilization of environmental resources is the prime objective of these planting systems. In mixed cropping, even if all other conditions are favorable, light is the limiting factor through interception and absorption by the taller canopy (Balasimha and Yadukumar 1993; Nair 2014). In this situation, shade-loving species which require low light intensities are likely to perform better under high-density multispecies cropping system (Kasturi Bai et al. 1991; Rethinam and Venugopal 1994). Srinivasan et al. (1999) conducted surveys and economic analysis of various cardamom plantations adopting mixed cropping systems in Kerala and Karnataka and indicated that the inclusion of black pepper (*Piper nigrum*) or coffee (*Arabica coffee*) and arecanut (*Areca catechu*) was highly remunerative giving higher benefit-cost ratios.

In arecanut plantations situated at higher elevations, especially in Uttara Kannada district of Karnataka (India), arecanut occupies the top most canopy, and shade-loving crop like cardamom is grown under its shade. Arecanut has a fairly deep root system which can tap water and nutrients from deeper layers, whereas cardamom is a shallow-rooted crop, which can absorb water and nutrients from top layers. Korikanthimath et al. (2000a) studied the microclimate and photosynthetic characteristic in arecanut and cardamom-mixed cropping systems and reported that arecanut received higher amount of light (59.0%) compared to cardamom (40.5%) due to arecanut crown forming the top most canopy and cardamom constituting the lower canopy in the cropping system.

Normally, cardamom is not cultivated as a mixed crop beneath the coconut canopy in the higher elevation Western Ghats, where coconut is cultivated comparatively in smaller units. Cultivation of cardamom under coconut resulted in higher net income and employment generation throughout the year (Korikanthimath et al. 1998c, 1998d). In coffee plantations, cropping strategy has to be on the line of multi-storied cropping system, which aims at raising compatible combination of crops having varying morphological frame and rooting habits, their canopies tap sunlight at varying heights, and roots absorb moisture and nutrients from different horizons (Rethinam and Venugopal 1994).

Korikanthimath et al. (1999) reported that mixed cropping generated higher and continued income for the farmers over a period of 10 months (July–April) and gainful employment for family members all year round. Robusta coffee mix cropping with cardamom provided higher economic returns and employment (Korikanthimath et al. 1997). Cardamom and arabica coffee were found to compete each other for environmental factors, and therefore, they may not be suitable for mixed cropping system in the long run because of height factor and growth habit even though net returns are higher in mixed cropping system. Tree spices like nutmeg, clove, and *Garcinia* can be grown as shade trees in cardamom plantations provided; they are spaced distantly.

Integrated Nutrient Management

One of the factors for low productivity in cardamom is inadequate application of manures either in the organic or inorganic form. To meet the nutrient requirements of cardamom, it is necessary to integrate both organic and inorganic fertilizers to obtain higher yields and also to maintain soil health and fertility. Organic manures are considered essential for improving physical characteristics of soil, apart from their nutritional values, and they are indispensable for cardamom irrespective of whether fertilizers are applied or not. Application of organic manures such as compost or farm yard manure @ 5 kg plant^{-1} may be made once in a year in May–June along with Mussoorie phosphate and MOP (Anon 1997). Use of 3–4 t of well-rotten cattle or sheep manure apart from 30 kg each of N, P_20_5, and 40 kg K_2O acre^{-1} greatly enhanced the yield in cardamom (Shanmugavelu and Rao 1977). Thimmarayappa et al. (2000) reported that integrated nutrient management with 25% organic manure +75% inorganic fertilizers and 50% organic manure +50% inorganic fertilizers also recorded yields at par with 75% organic manure +25% inorganic fertilizers.

2 Conclusions

Black pepper and cardamom require a friable soil rich in humus and essential plant nutrients, with good drainage. In addition to major (N, P, and K) and secondary (Ca and Mg) nutrients, micronutrients especially B and Zn are most important for better growth, development, and yield of these crops. In soils with low pH, liming could enhance the yield. It is necessary to develop sound soil fertility management program that encompasses nutrient recommendations based on robust soil tests. Crop management programs that are location-specific should form the basis for manure and fertilizer management that optimizes economic return while protecting the environment. Overall, it is apparent that in majority of the soils, the reasons for low productivity of these crops are low pH, low levels of organic carbon, and low CEC, K, Ca, Mg, Zn, and B. Inorganic fertilizer schedules for different agroecological locations are available for these crops. Similarly organic nutrition schedules have also been devised for these crops. However, among the different nutrient management strategies, INM through judicious use of inorganic fertilizers, biofertilizers, biosolids, biological nitrogen fixation, and other nutrient cycling mechanisms that enhance nutrient use efficiency appears to be the best bet. Investigations on the effect of rhizosphere organisms on mobilizing the insoluble sources of plant nutrients leading to development of efficient nutrient utilization systems is the need of the hour. Equally, clear insight into the effect of chemical inputs on the nature and diversity of the rhizosphere microbiome will be of immense help in formulating effective management strategies for sustainable soil health and crop production.

References

Abraham P, Tulasidas G (1958) South Indian cardamom and their agricultural value. ICAR Tech Bull. 79, Indian Council of Agricultural Research, New Delhi, pp 1–27

Adzemi MA, Hamdan J, Sjahril JS (1993) Macro-nutrient removal studies on black pepper (*Piper nigrum* L.) in Sarawak. In: Ibrahim MY, Bong CFJ, Ipor B (eds) The pepper industry: problems and prospects. UnivPertanian Malaysia, Malaysia, pp 104–109

AICRPS (2005) Annual report, All India Coordinated Research Project on Spices, IISR Calicut, India, pp 6–8

Anandaraj M, Abraham J, Sarma YR, Balakrishnan R (1989) Incidence of foot-rot disease of black pepper (*Piper nigrum* L.) in Kerala in relation to cultivation practices. Indian J Agric Res 59:751–753

Ankegowda SJ, Chethan CV, Korikanthimath, VS (2002) Soil and water conservation measures in cardamom cultivation, In: National symposium on soil and water conservation measures and sustainable land use systems with special reference to the Western Ghats region, invited and contributory papers, 16–17 November, 2002 ICAR Research Complex for Goa, Old Goa, India, pp 47–49

Ankegowda SJ, Bhat AI, Jayasree E, Leela NK, Senthil Kumar CM, Alagupalamuthirsolai M, Akshitha HJ, Faisal MP (2022) Cardamom. In: Thomas L, Rajeev P, Dinesh R (eds) . ICAR-Indian Institute of Spices Research, Kozhikode, India, p 29

Ann YC (2012a) Determination of nutrient uptake characteristics of black pepper (*Piper nigrum* L.). J Agric Sci Tech 6:86–89

Ann YC (2012b) Impact of different fertilization methods on the soil, yield and growth performance of black pepper. Malaysian J Soil Sci 16:71–87

Anonymous (1976) Cardamom in Karnataka. UAS Tech. Series No.14. University of Agricultural Sciences, Regional Research Station, Mudikgeri, p 5

Anonymous (1997) Cardamom-package of practices. Kochi, Spices Board, p 19

Azmil IAR, Yau PY (1993) Improvements in agronomic practices for pepper cultivation in Johore. In: Ibrahim MY, Bong CFJ, Ipor IB (eds) The pepper industry: problems and prospects. University Pertanian Malaysia, Sarawak, Malaysia, pp 15–23

Balasimha D, Yadukumar N (1993) Effect of plant density on photosynthesis in cashew. J Plant Physiol 36:5–7

Bavappa KVA, Kailasam C, Khader KBA, Biddappa CC, Khan HH, Bai K, Ramadasan A, Sundaraju P, Bopaiah BM, Thomas GV, Mishra LP, Balasimha D, Bhat NT, Sham Bhat K (1986) Coconut and arecanut based high density multispecies cropping systems. J Plant Crops 14(2):74–87

Bopaiah BM, Khader KB (1989) Effect of biofertilizer on growth of black pepper (*Piper nigrum* L.). Indian J Agric Sci 59:682–683

Charles JK (1986) Productivity in cardamom-an insight. Indian Cardamom J 19(12):13–20

Choudhury S (1947) A sclerotial disease of black pepper. Indian Farming 8:557–559

De Waard PWF, Sutton CD (1960) Toxicity of aluminium to black pepper in Sarawak. Nature 195:1129

Devasahayam S, John Zacharaiah T, Jayashree E, Kandiannan K, Prasath D, Eapen SJ, Sasikumar B, Srinivasan V, Suseela Bhai R (2015) Black pepper- extension pamphlet. In: Rajeev P, Thomas L (eds) . ICAR-Indian Institute of Spices Research, Kozhikode, Kerala, India, p 24

DeWaard PWF (1969) Foliar diagnosis, nutrition and yield stability of black pepper (*Piper nigrum* L.) in Sarawak. Comm. No. 58. Royal Tropical Institute, Amsterdam, The Netherlands, p 150

Diby P, Anandaraj M, Kumar A, Sarma YR (2005a) Antagonistic mechanisms of fluorescent pseudomonads against *Phytophthora capsici* in black pepper (*Piper nigrum* L.). J Spices Aromat Crops 14:94–101

Diby P, Sarma YR, Srinivasan V, Anandaraj M (2005b) *Pseudomonas fluorescens* mediated vigour in black pepper (*Piper nigrum* L.) under green-house cultivation. Annals Microbiol 55(3):171–174

Dinesh R, Srinivasan V, Hamza S, Anandaraj M (2014) Massive phosphorus accumulation in soils: Kerala's continuing conundrum. Curr Sci 106(3):343–344

Divya SB, Hegde NK, Hegde L, Manju MJ, Shivakumar KM (2018) Effect of foliar nutrition with zinc and boron on the performance of black pepper (*Piper nigrum* L.) under hill zone (Zone-9) of Karnataka, India. Intl J Curr Microbiol Appl Sci 7(7):517–522

Donald DM (1963) Competition among crop and pasture plants. Adv Agron 15:1–118

Doran JW, Sarrantonio M, Liebig MA (1996) Soil health and sustainability. Adv Agron 56:2–54

Duarte MLR, Albuquerque FC (1991) Fusarium disease of black pepper in Brazil. In: Sarma YR, Premkumar T (eds) Diseases of black pepper. National Research Center for Spices, Calicut, India, pp 39–54

Farrar JF, Jones DL (2000) The control of carbon acquisition by roots. New Phytol 147:43–53

Fearnside PM (1980) Black pepper yield prediction for the transamazon highway of Brazil. Turrialba 30:35–42

Garbell MA (1947) In: tropical and equatorial meteorology. Pitman Publishing Corp, New York

Geetha CK, Aravindakshan M, Wahid PA (1993) Influence of method of fertilizer application on nutrient absorption by black pepper vines. South Indian Hort 41:95–100

Geetha CK, Aravindakshan M, Wahid PA (1996) Comparison of nutrient response curves in bush pepper and vine pepper (*Piper nigrum* L.). J Plant Crops 24:135–138

George CK (1982) Pepper cultivation in Malaysia. Indian Cocoa Arecanut Spices J 5:75–76

George CK (1990 Report of the visit of the delegation for mutual understanding in cardamom trade with Guatemala, p 12

Ghosh DK (2009) Performance of black pepper (Piper nigrum L) as intercrop with coconut in the alluvial plains of West Bengal. Indian Cocon J 51(11):4–7

Giller KE, Cadisch G, Ehaliotis C, Adams E, Sakala WD, Mafongoya PL (1997) Building soil phosphorous capital in Africa, Buresh RJ et al. (ed.) replenishing soil fertility in Africa. SSSA Spec. Publ. 51, SSSA, Madison, WI, pp 151–192

Goulding, KWT, Annis B (1998). Lime and liming in UK agriculture. Proceedings No 410. The Fertilizer Society. York, UK

Govindan M, Chandy KC (1985) Utilization of the diazotroph Azospirillum for inducing rooting in pepper. Curr Sci 54:1186–1188

Hamza S, Sadanandan AK (2002) Response of black pepper (*Piper nigrum* L) to zinc. Indian J Agric Sci 72(12):745–746

Hamza S, Sadanandan AK (2005a) Response of black pepper (*Piper nigrum L.*) to molybdenum nutrition. Indian J Agric Sci 75(5):54
Hamza S, Sadanandan AK (2005b) Comparison of source and method of zinc application on yield and quality of black pepper (*Piper nigrum* L). J Spices Aromat Crops 14:117–121
Hamza S, Sadanandan AK, Srinivasan V (2004) Influence of soil physico-chemical properties on black pepper yield. J Spices Aromat Crops 13:6–9
Hamza S, Srinivasan V, Dinesh R (2007) Nutrient diagnosis of black pepper (*Piper nigrum* L.) gardens in Kerala and Karnataka. J Spices Aromat Crops 16:77–81
Hamza S, Srinivasan V, Dinesh R (2009) Nutrient diagnosis of cardamom (*Elettaria cardamomum*) gardens in South India. Indian J Agric Sci 79:429–432
Harper RS (1974) Pepper in Indonesia. World Crops 26:130–133
Hegde R, Korikanthimath VS (1996, 7–8 February) Agronomic approaches for the drought management in cardamom. In: Proceeding of the National Seminar on drought management in plantation crops, Satheesan KV (eds), Kottayam, Kerala, pp 75–79
Indian Institute of Spices Research (IISR) (2016) Annual report (2015–16). ICAR-Indian Institute of Spices Research, Kozhikode, Kerala, India, p 100
Indian Institute of Spices Research (IISR) (1996) Ann Rep (1995–96). Indian Institute of Spices Research, Calicut
Indian Institute of Spices Research (IISR) (2011) Annual Report 2010–11. IISR, Kozhikode, p 91
Indian Institute of Spices Research (IISR) (2020) Annual report. In: Praveena et al (eds), ICAR-IISR, Kozhikode, Kerala, pp 33–34
Iyer R, Sastry RK, Korikanthimath VS (1989) Effect of fertilizer levels on vesicular arbuscular mycorrhiza colonization in cardamom. J Plant Crops 16(Suppl):429–432
Jacob JA, Joseph A, Murugan M (1995) Cardamom and climate. Spice India 8(6):12–14
John STD (2000) Prospects of pepper as an intercrop in coffee plantations in Karnataka. Indian Coffee 64(2):7–9
Kandiannan K, Thankamani CK, Sivaraman K, Anandaraj M, Krishnamurthy KS (1998) Effect of biofertilizer and vermicompost on growth of black pepper (*Piper nigrum* L.) In: First International Agronomy Congress on Agronomy, Environment and Food Security-Extended Summaries Indian Society of Agronomy, New Delhi, p 206
Kandiannan K, Sivaraman K, Anandaraj M, Krishnamurthy KS (2000a) Growth and nutrient content of black pepper cuttings as influenced by inoculation with biofertilizers. J Spices Aromat Crops 9(2):145–147
Kandiannan K, Sivaraman K, Thankamani CK, Dake GN, Anandaraj M, Krishnamurthy KS (2000b) Effect of *Azospirillum* on growth of black pepper. In: Muraleedharan N, Raj Kumar R (eds) Recent advances in plantation crops research. Allied Publishers Ltd, New Delhi, pp 404–407
Kandiannan K, Utpala P, Krishnamurthy KS, Thankamani CK, Srinivasan V, Aipe KC (2011) Modeling the association of weather and black pepper yield. Indian J Hort 68(1):96–102
Kasturi Bai KV, Voleti SR, Ramadasan A, Kailasam C (1991) Growth and dry matter production in the component crops under monocropping and high density multispecies cropping systems. J Plant Crops 18(Suppl):151–155
Kerala State Planning Board (KSPB) (2013) Soil fertility assessment and information management for enhancing crop productivity in Kerala. Rajashekaran P, Nair KM, Rajasree G, Suresh P, Narayankutty MC (eds), Kerala State Planning Board, Trivandrum, p 514
Khader KBA, Sayed AAM (1977) Fertilizing cardamom – its importance. Cardamom 9(12):13–14
Khan HH, Upadhyay AK, Palaniswami P (2002) In: Proceedings of PLACROSYM XIV. National Research Centre for Oil Palm, Pedavegi, Andhra Pradesh, pp 9–22
Kologi SD (1977) In towards higher yield in cardamom. Central Plantation Crops Research Institute, Kasargod, India, pp 28–29
Korikanthimath VS (1987) Systems of planting-cum-fertilizer levels for cardamom under rainfed conditions. NRC for Spices, Annual Report 1987, Calicut, Kerala, India, pp 30–31
Korikanthimath VS (1989) Systems of planting-cum-fertilizer levels in cardamom under rainfed conditions. NRC for spices, Annual report, 1989, Calicut, Kerala, India, pp 24–26

Korikanthimath VS (1991) Shade management for high productivity in cardamom (*Elettaria cardamomum* Maton.). Spice India 4(2):15–21

Korikanthimath VS, Hiremath GM, Hegde R, Mulge R, Hosmani MM (1997) Input use pattern in mixed cropping of Robusta coffee with cardamom. Karnataka J Agric Sci 10(3):653–658

Korikanthimath VS, Gowda SJA, Yadukumar N, Hegde R, Hosmani MM (1998a) Microclimatic and physiological analysis in coffee-cardamom mixed cropping system. J Spices Aromat Crops 7(2):133–135

Korikanthimath VS, Hegde R, Mulge R, Hosmani MM (1998b) Growth and yield parameters of cardamom (*Elettaria cardamomum* Maton) as influenced by nutrition and planting density. J Spices Aromat Crops 7:39–42

Korikanthimath VS, Hiremath GM, Hegde R, Mulge R, Hosmani MM (1998c) Relative requirement of inputs in the mono v/s mixed cropping of coconut with cardamom. Madras Agric J 85(3/4):186–188

Korikanthimath VS, Kiresur V, Hiremath GM, Hegde R, Mulge R, Hosmani MM (1998d) Economics of mixed cropping of coconut with cardamom. Crop Res-Hisar 15(2/3):188–195

Korikanthimath VS, Ankegowda SJ, Yadukumar N, Hegde R, Hosmani MM (1999) Micro climatic and physiological characteristics of Robusta coffee-cardamom, Coorg mandarin and pepper in mixed cropping system. Indian J Agron 44(2):289–293

Korikanthimath VS, Ankegowda SJ, Yadukumar N, Hegde R, Hosmani MM (2000a) Microclimatic and photosynthetic characteristics in arecanut and cardamom mixed cropping systems. J Spices Aromat Crops 9(1):61–63

Korikanthimath VS, Gaddi AV, Ankegowda SJ (2000b) Status of major nutrients in soils of cardamom (*Elettaria cardamomum* Maton) plantations in Kodagu District Karnataka, India. J Spices Aromat Crops 9(2):117–122

Korikanthimath VS, Hegde R, Gaddi AV, Chandravanshi P (2001) Nutrient management in cardamom. Fert News 46:37–48

Krishna KVS (1997) Cardamom plantations in Papua New Guinea. Spice India 10(7):23–24

Krishnakumar V, Potty SN (2002) Nutrition of cardamom. In: Ravindran PN, Madhusoodanan KJ (eds) Cardamom-the genus Elettaria. Taylor and Francis, NY, pp 129–142

Krishnamurthy KS, Chempakam B (2009) Investigation on the influence of seedling's physiological attributes on productivity in black pepper. Indian J Hort 66:95–100

Krishnamurthy KS, Kandiannan K, Ankegowda SJ (2000) Is the subtending leaf sole source of nutrients for the developing spike in black pepper? In: Muraleedharan N, Raj Kumar R (eds) Recent advances in plantation crops research. Allied Publishers, New Delhi, pp 238–240

Kulkarni DS, Kulkarni SV, Suryanarayana Reddy BG, Pattanashetty HV (1971) Nutrient uptake by cardamom (*Elettaria cardamomum* Maton.). In: Proceedings of International Symposium on Soil Fertility Evaluation, New Delhi, Vol.1, pp 293–296

Kumar AR, Sathiamoorthy S (2002) Effect of plant growth regulators on berry set and yield in black pepper (*Piper nigrum* L.) var. Panniyur 1. South Indian Hort 50:691–694

Kumar MD, Shanthaveerabhadraiah SM, Ravishankar CR (2000) Effect of fertilizer levels on the yield of small cardamom under natural shade. Spices and aromatic plants: challenges and opportunities in the new century, Centennial conference on spices and aromatic plants, Calicut, Kerala, India, pp 179–180

Kumar MD, Devaraju KM, Madaiah D, Shivakumar KV (2009) Effect of integrated nutrient supply on yield and uptake of cardamom (*Elettaria cardamomum* L. Maton.). Karnataka J Agric Sci 22:1016–1019

Ledgard SF, Klein D, Crush JR, Thorrold BS (2000) Dairy farming, nitrogen losses and nitrate-sensitive areas. Proc N Z Soc Anim Prod 60:256–260

Li Z, Zu C, Wang C, Yang J, Yu H, Wu H (2016) Different responses of rhizosphere and non-rhizosphere soil microbial communities to consecutive *Piper nigrum* L. monoculture. Sci Rep 6:35825

Madan MS, Srinivasan V, Thankamani CK, Hamza S (2007) Economic analysis of low input technology for increased black pepper (*Piper nigrum* L.) production. Indian J Agric Sci 77:445–447

Mathai CK (1983) Growth and yield analysis in black pepper varieties (*Piper nigrum* L.) under different light conditions. Ph.D Thesis, University of Agricultural Sciences, Bangalore, India

Mathai CK, Chandy KC (1988) Yield response of black pepper varieties to varying growth light regimes. Indian Cocoa Arecanut Spices J 11:85–88
Mathew J, Nybe EV (2006) Integration of nutrient sources on nutrient status, yield and quality of black pepper. J Plant Crops 34:340–343
Mathew PG, Wahid PA, Sreekandan Nair G (1995) Soil fertility and nutrient requirement in relation to productivity in black pepper. J Plant Crops 23:109–115
Mclaughlin MJ, Parker DR, Clarke JM (1999) Metals and micronutrients-food safety issued. Field Crop Res 60:143–163
Miniraj N, Murugan M (2000) Ecological decline of cardamom hills – an analysis. In: Ramana KV et al (eds) Spices and aromatic plants-challenges and opportunities in the new century. Indian Society for Spices, Calicut, pp 172–176
Mohanchandran K (1984) Cardamom exports – today and tomorrow. Cardamom 17(11):5–8
Murugan M, Miniraj N, Joseph C (2000) Changes in climatic elements and their impact on production of cardamom (*Elettaria cardamomum* M.) in the cardamom hills of Kerala, India. J Spices Aromat Crops 9(2):157–160
Murugan M, Backiyarani S, Josephrajkumar A, Hiremath MB, Shetty PK (2007) Yield of small cardamom (*Elettaria cardamomum* M) variety PV1 as influenced by levels of nutrients and neem cake under rain fed condition in southern Western Ghats, India. Caspian J Environ Sci 5(2):19–25
Murugan M, Mukund V, Ramesh R, Hiremath MB, Josephrajkumar A, Shetty PK (2008) Centennial rainfall variation in semi-arid and tropical humid environments in the cardamom hill slopes, southern Western Ghats, India. Caspian J Environ Sci 6(2):31–39
Murugan M, Shetty PK, Raju R, Anandhi A, Rajkumar AJ (2012a) Climate change and crop yields in the Indian Cardamom Hills, 1978–2007 CE. Clim Change (Springer Nature) 110:737–753
Murugan M, Panigrahy BK, Shetty PK, Subbiah A, Raju R (2012b) Effect of heavy metal and nutrient uptake by soils in Indian Cardamom Hills. J Soil Sci Environ Manage 3(8):196–206
Murugan M, Ravi R, Anandhi A, Kurien S, Dhanya MK (2017) Pesticide use in Indian cardamom needs change in cultivation practices. Curr Sci 113(6):1058–1063
Murugan M, Dhanya MK, Aswathy TS, Ashokkumar K, Mathews N, Sathyan T, Surya R (2022) Liming tropical high upland acid soils improves soil fertility and cardamom productivity. J Curr Opin Crop Sci 3(1):1–5
Muthumanickam D (2003) Influence of different phosphorus sources and zinc spray on yield and quality of black pepper under acid soils. J Spices Aromat Crops 12(1):15–18
NAAS (2009) Crop response and nutrient ratio, policy paper no. 42. National Academy of Agricultural Sciences, New Delhi, p 17
Nagarajan M, Pillay NG (1975) A note on the nutrient composition of fruiting branches of black pepper (*Piper nigrum* L). J Madras Agric 62:87–89
Nair CK (1988) Phosphatic fertilizers of small cardamom In: Proceedings of the Seminar on the use of Rock Phosphate in West Coast Soils UAS Bangalore and PPCL, p 79
Nair KPP (2004) The agronomy and economy of black pepper (*Piper nigrum* L) – the king of spices. Adv Agron 82:271–389
Nair PKR (2014) Agroforestry: practices and systems. Encycl Agric Food Syst:270–282
Nair CK, Srinivasan K, Zachariah PK (1980) Proc. seminar on diseases of plantation crops and manuring of plantation crops. Tamil Nadu Agricultural University, Madurai, India, pp 50–53
Nair CK, Srinivasan K, Sivadasan CR (1988) Soil acidity in cardamom plantations. Spice India 1(6):40–41
Nambiar KKN, Sarma YR (1977) Wilt diseases of black pepper. J Plant Crops 5:92–98
Nambiar EB, Nair TJ, Money NS (1965) Preliminary studies on the incidence of wilt disease of pepper and its relationships to nitrogen and base status of soils. Indian J Agric Sci 37:276–281
Natarajan P, Srinivasan K (1989) Effect of varying levels of N, P and K on yield attributes and yield of cardamom (*Elettaria cardamomum* Maton). South Indian Hort 37:97–100
Nath JC, Phukon RM, Sumitha S, Maheswarappa HP, Patil B (2021) Performance of black pepper varieties as intercrop in coconut gardens in the. J Plant Crops 49(3):176–181. https://doi.org/10.25081/jpc.2021.v49.i3.7451

Nguyen TH, Nguyen AD, Vinh NQ (2020) Biodiversity of soil microorganisms and their effects on disease management at black pepper farms in Gia Lai Province. Asian J Biol 9(4):1–11
Noble AD, Gillman GP, Ruaysoongnern S (2000) A cation exchange index for assessing degradation of acid soil by further acidification under permanent agriculture in the tropics. European J Soil Sci 51(2):233–243
Nybe EV, Nair PCS, Wahid PA (1989) Relationships of foliar nutrient levels with yield in black pepper (*Piper nigrum* L). Trop Agric Trinidad 66:345–349
Ozkutlu F, Sekeroglu N, Kara M (2006) Monitoring of cadmium and micronutrients in spices commonly used in Turkey research. Journal of Agriculture and Biological Sciences 2(5):223–226
Paduit N, Pampolino M, Aye TM, Oberthur T (2018) Nutrient uptake and distribution in black pepper. Better Crops 102(4):24–27
Parama RVR, Munawery A (2012) Sustainable soil nutrient management. J Indian Inst Sci 92(1):1–16
Parthasarathy VA, Srinivasan V, Dinesh R (2006) Organic spices- present status and scope. National Seminar on appropriate Technologies for Sustainable Horticulture Souvenir, Annamalai Univ. Tamil Nadu, India, pp 34–42
Pattanshetty HV, Nursath R (1973) Early August is the most suitable time for planting cardamom in Mudigere area. Curr Res 2(7):47–48
Paulus AD, Wan AL (2011) Pepper production Technology in Malaysia (eds: Fong LK, Liang SS), Malaysian Pepper Board, Sarawak, Malaysia
Pillai VS, Sasikumaran S, Nambiar PKV (1987) N P K requirements of black pepper. Agric Res J Kerala 25(1):74–80
Raj HG (1978) A comparison of the system of cultivation of black pepper – Piper nigrum L. in Malysia and Indonesia. In: Silver Jubilee Souvenir – Pepper Research Station, Panniyur. The Directorate of Extension Education. Kerala Agricultural University, Trichur, India, pp 65–74
Ramachandran K, Srinivasan V, Hamza S, Anandaraj M (2007) Phosphate solubilizing bacteria isolated from the rhizosphere soil and its growth promotion on black pepper (*Piper nigrum* L.) cuttings. First International Meeting on Microbial Phosphate Solubilization, Developments in Plant and Soil Sciences, Springer Netherlands. 102: 325–331
Ramadasan A (1987) Canopy development and yield of adult pepper vines in relation to light interception. J Indian Cocoa Arecanut Spices 11:43–44
Ranganathan V, Natesan S (1985) Cardamom soils and manuring. Planters. Chronicle 80:233–236
Ravindran PN, Johny AK (2000) Black pepper. In: Peter KV (ed) Handbook of herbs and spices. Wood head Publishing Ltd., Cambridge UK, pp 62–110
Rethinam P, Venugopal K (1994) Cropping systems in plantation crops. In: Adv. Hort. Vol.9: plantation and spice crops, part 1. Chadha KL, Rethinam P (eds). Malhotra Publishing House, New Delhi, pp 603–620
Rosli A, Radam A, Rahim KA (2013) Technical efficiency of pepper farms in Sarawak, Malaysia: an application of data envelopment analysis. Int J Bus Soc Sci 4(7):227–234
Sadanandan AK (1986) Management of wilt affected pepper gardens. In: Nair MK, Nambiar KKN (eds) PLACROSYM VI. Indian Society for Plantation Crops, Kasaragod, Kerala, pp 287–297
Sadanandan AK (1993) In: Singh M, Mishra MK (eds) Potassium for plantation crops. Potassium Research Institute of India, Haryana, India, pp 75–88
Sadanandan AK (1994) Nutrition of black pepper. In: Chadha KL, Rethinam P (eds) Advances in horticulture volume 9 – part 1. Malhotra Publishing House, New Delhi, pp 423–456
Sadanandan AK (2000) Agronomy and nutrition of black pepper. In: Ravindran PN (ed) Black pepper. Harwood Academic Publishers, New Delhi India, pp 163–223
Sadanandan AK, Hamza S (1992) Micronutrients interacting with yield of black pepper (*Piper nigrum* L.) and inter-relationship of nutrients (Abs) National Seminar on Development in Soil Science, CRIDA, Hyderabad, pp 109–110
Sadanandan AK, Hamza S (1993) Comparative efficiency of slow–release nitrogenous fertilizers on transformation on nitrogen and yield response of black pepper in an Oxisol. J Plant Crops 21(Suppl):58–66

Sadanandan AK, Hamza S (1999) Effect of organic farming on soil quality, crop productivity and quality of spices. In: Proc. 3rd IFOAM-Asia Scientific Conference and General Assembly on Food Security in Harmony with Mature, pp 90–94

Sadanandan AK, Hamza S (2001) Studies on heavy metal residues due to continuous application of rock phosphate for sustainable black pepper production. In: Gupta AK (ed) Proceedings of the national symposium on combating pollution accumulation in ecosystem for sustainable agriculture. Allahabad Agricultural Institute, Allahabad, India, pp 75–83

Sadanandan AK, Hamza S (2002) In: Rethinam et al (eds) Plantation crops research and development in the new millennium. Coconut Development Board, Kerala, India, pp 336–341

Sadanandan AK, Hamza S (2006) Effect of organic farming on soil quality, nutrient uptake, yield and quality of Indian spice. In: Abstracts of 18th World Congress of Soil Science, July 9–15, 2006, Philadelphia, USA

Sadanandan AK, Rajagopal CK (1989) Inter-relation between P content of leaf and yield of pepper in an Alfisol. J Indian Soc Soil Sci 35:197–199

Sadanandan AK, Korikanthimath VS, Hamza S (1990) Potassium in soils of cardamom plantations. In: Seminar on Potassium in Plantations Crops, University of Agricultural Sciences, Bangalore, India

Sadanandan AK, Bhargava BS, Hamza S (1996) Leaf nutrient norms for black pepper (*Piper nigrum* L) using DRIS. J Plant Crops 24(suppl):53–59

Sadanandan AK, Hamza S, Bhargava BS, Raghupathi MB (2000) Diagnosis and recommendation integrated system (DRIS) norms for black pepper growing soils of South India. In: Recent advances in plantation crops research (eds) Muraleedharan N, Raj Kumar R (PLACROSYM-XIII) Allied Publishers Limited, New Delhi, India, pp 203–205

Sangakkara VR (1989) Effect of pH of propagation medium on root and shoot growth of black pepper. J Plant Crops 17:54–57

Sankar SJ, Wahid PA, Kamalam NV (1988) Absorption of soil applied radio phosphorus by black pepper vine and support tree in relation to their root activities. J Plant Crops 16:73–87

Sarathambal C, Sivaranjani R, Srinivasan V, Alagupalamuthirsolai M, Subila KP, Anamika B (2023a) Effect of arbuscular mycorrhizal inoculation on growth, mineral nutrient uptake, photosynthesis and antioxidant activities of black pepper cuttings. J Plant Nutr 46(10):2508–2524

Sarathambal C, Jeevalatha A, Sivaranjani R, Biju CN, Charles S, Srinivasan V, George P, Peter B, Radhika R (2023b) Arbuscular mycorrhizal colonization alters biochemical, molecular defense responses and root exudate composition against *Phytophthora capsici* infection in black pepper. Rhizosphere 25:100651

Sewwandi I, Mahindapala J, Thrikawala S (2019) Intercropping pepper with tea: farmers' perception, technology adoption and issues faced by the tea smallholders of Imbulpe divisional secretariat in Balangoda, Sri Lanka. In: Open University Research Session 2019, held at Nawala, Nugegoda, Sri Lanka, pp 1–6

Shanmugavelu KG, Madhava Rao VN (1977) Spices and plantation crops. Popular book Depot, Chennai, India

Sharpley AN, Chapra SC, Wedepohl R, Sims JT, Daniel TC, Reddy KR (1994) Managing agricultural phosphorus for protection of surface waters: issues and options. J Environ Qual 32:437–451

Sivadasan CR, Nair CK, Srinivasan K, Mathews K (1991) Zinc requirement of cardamom (*Elettaria cardamomum* Maton). J Plant Crops 18:171–173

Skillman JB, Garcia M, Virgo A, Winter K (2005) Growth irradiance effects on photosynthesis and growth in two co-occurring shade-tolerant neotropical perennials of contrasting photosynthetic pathways. Am J Bot 92:1811–1819

Sreeramulu KR, Bagyaraj DJ (1998) Response of cardamom seedlings to vesicular arbuscular mycorrhizal fungi. J Spices Aromat Crops 7(2):89–94

Srinivasan K, Rama Rao KVV, Naidu R (1993a) Organic matter addition and its nutrient contribution in cardamom plantations. J Tropic Agric 31:119–121

Srinivasan K, Vadiraj BA, Krishnakumar V, Naidu R (1993b) Micronutrient status of cardamom growing soils of South India. J Plant Crops 21(Suppl):67–74

Srinivasan K, Krishnakumar V, Potty SN (1998) Evaluation of fertilizer application methods on growth and yield of small cardamom (*Elettaria cardamomum* Maton). In: Muralitharan N, Rajkumar R (eds) Recent advances in plantation crops research. Allied Publishers, New Delhi, India, pp 199–202

Srinivasan K, Siddagangaiah BAK, Krishnakumar V (1999) Mixed cropping systems in cardamom. J Spices Aromat Crops 8(1):63–66

Srinivasan V, Sadanandan AK, Hamza S (2000) An IPNM approach in spices with special emphasis on coir compost. International Conference on Managing Natural Resources for Sustainable Agricultural Production in the 21st Century, IARI New Delhi, pp 1363–1365

Srinivasan K, Vadiraj BA, Krishnakumar V, Bindumol GB, Potty SN (2002a) Available Sulphur status of cardamom growing soils of South India. In: Proceedings of PLACROSYM XIV. Indian Society for Plantation Crops, Kasaragod, India, pp 342–344

Srinivasan V, Rubina MR, Hamza S, Ramachandran K (2002b) Forms of soil potassium as influenced by organic fertilizers in bush black pepper (*Piper nigrum* L). J Potassium Res 18:99–104

Srinivasan V, Hamza S, Sadanandan AK (2005) Evaluation of composted coir pith with chemical and biofertilizers on nutrient availability, yield and quality of black pepper (*Piper nigrum* L.). J Spices Aromat Crops 14:15–20

Srinivasan V, Dinesh R, Hamza S, Parthasarathy VA (2007) Nutrient management in black pepper (*Piper nigrum* L.) CAB reviews: perspectives in agriculture, veterinary science, nutrition and natural. Resources 2(62):1–14

Srinivasan V, Kandiannan K, Hamza S (2008) Sulphate of potash (SOP) as potassium source for increasing nutrient availability, yield, quality and profitability of black pepper. In: Krishnamurthy KS, Prasath D, Kandiannan K, Suseela Bhai R, Saji KV, Parthasarathy VA (eds) National Seminar on Piperaceae – Harnessing Agro-technologies for accelerated production of economically important piper species. Indian Institute of Spices Research, Calicut, India, p 346

Srinivasan V, Dinesh R, Krishnamurthy KS, Hamza S (2011a) Nutrition and physiology. In: Singh HP, Parthasarathy VA, Srinivasan V, Saji KV (eds) Piperaceae crops: production and utilization. Westville Publishing House, New Delhi, pp 101–121

Srinivasan V, Thankamani CK, Kandiannan K, Dinesh R, Ankegowda SJ, Krishnamurthy KS (2011b) Production system and health management in black pepper. In: Krishnamurthy KS et al (eds) Souvenir and abstracts, National Symposium on spices and aromatic crops (SYMSAC VI): exploiting spices production potential of the Deccan region. Indian Society for Spices, Kozhikode, Kerala, India, pp 139–144

Srinivasan V, Dinesh R, Hamza S (2012a) Management of acid soils for sustainable production of spices in India. In: Sudhir et al (eds) 8th Intenational symposium on soil – plant interactions at low pH. UAS, GKVK, Bangalore, pp 153–168

Srinivasan V, Dinesh R, Krishnamurthy KS, Hamza S (2012b) Black pepper health management: water and nutrients. Focus on Pepper 4(2):17–32

Srinivasan V, Dinesh R, Ankegowda SJ, Hamza S, Thankamani CK, Krishnamurthy KS (2013a) Soil test-based site-specific fertilizer recommendation and its economic optimum for targeted yield of black pepper. In: Sasikumar et al (eds) Souvenir & Abstracts – SYMSAC VII: post-harvest processing of spices and fruit crops. ISS ICAR-IISR, Kozhikode Kerala, p 202

Srinivasan V, Kandiannan K, Hamza S (2013b) Efficiency of Sulphate of potash (SOP) as an alternate source of potassium for black pepper (*Piper nigrum* L). J Spices Aromat Crops 22(2):120–126

Srinivasan V, Dinesh R, Hamza S, Sheeja TE. (2017a) Designer foliar micronutrient formulations for enhanced yield and quality in major spice crops, Invention Intelligence Sep-Oct, pp 47–49

Srinivasan V, Bhat AI, Dinesh R, Ankegowda SJ, Hamza S, Biju CN, Krishnamurthy KS (2017b) Rejuvenation of virus affected black pepper plantations through soil and plant health management. Indian J Arecanut Spice Med Plants 19:24–27

Srinivasan V, Dinesh R, Hamza S (2017c) Soil health management in spice crops. In: Maheswarappa HP, Chowdappa P (eds) Soil health management in plantation crops. Today & Tomorrow's Printers and Publishers, New Delhi, pp 177–211

Srinivasan V, Thankamani CK, Hamza S, John Zacharia T, Dinesh R, Praveena R (2017d) Organic spices production – black pepper, ginger and turmeric. ICAR-Indian Institute of Spices Research, Kozhikode, India, pp 21–32

Stephen F, Nybe EV (2003) Organic manures and biofertilizers on nutrient availability and yield in black pepper. J Tropic Agric 41:52–55

Subbiah A, Murugan M, Ramesh Kumar A, Jagadeesan R, Prabhu M (2008) Nutrient management in small cardamom (*Elettaria cardamomum* M). Asian J Hort 3:449–455

Sudharshan MR, Kuruvilla KM, Madhusoodanan KJ (1988) Effective pollination for better fruit set in cardamom. Spice India 1(6):14–21

Sulok KMT, Ahmed OH, Khew CY, Lai PS, Zehnder JAM, Wasli ME, Aziz ZF (2021) Use of organic soil amendments to improve soil health and yield of immature pepper (*Piper nigrum* L.). Org Agr 11:145–161

Thankamani CK, Ashokan PK, Kamalam NV (2003) Root distribution pattern of bush – black pepper (*Piper nigrum* L) employing radiotracer technique. J Nucl Agric Biol 32:23–28

Thankamani CK, Sreekala K, Anandaraj M (2005) Effect of *Pseudomonas fluorescens* (IISR-6) and *Trichoderma harzianum* (P-26) on growth of black pepper (*Piper nigrum* L.) in the nursery. J Spices Aromat Crops 14:112–116

Thankamani CK, Srinivasan V, Krishnamurthy KS, Kandiannan K (2011) Effect of *Azospirillum sp.* and nutrients on yield of black pepper (*Piper nigrum* L.). J Spices Aromat Crops 20(1):9–13

Thankamani CK, Kandiannan K, Madan MS, Raju VK, Hamza S, Krishnamurthy KS (2012) Feasibility of intercropping medicinal plants in black pepper garden. J Spices Aromat Crops 21(2):113–117

Thimmarayappa M, Shivashankar KT, Shanthaveerabhadraiah SM (2000) Effect of organic manure and inorganic fertilizers on growth, yield attributes and yield of cardamom (*Elettaria cardamomum* Maton.). J Spices Aromat Crops 9:57–59

Thomas GV, Sundararaju P, Ali SS, Ghai SK (1989) Individual and interactive effects of VA-mycorrhizal fungi and root-knot nematode, *Meloidogyne incognita* on cardamom. Trop Agric (Trinidad) 66(1):21–24

Tiing Laua E, Tanib A, Khewa CY, Chuac YQ, Hwang SS (2020) Plant growth-promoting bacteria as potential bio-inoculants and bio-control agents to promote black pepper plant cultivation. Microbiol Res 240:126549

Veloso CAC, Carvalho JG, Malavolta E, Muraoka T (2000) Response of black pepper cultivars to NPK on a yellow latosol of eastern Amazonia. Sci Agric 57:343–347

Vijayakumar KR, Mammen G (1990) Effect of contact shading on leaf chlorophyll content and yield performance in black pepper (*Piper nigrum* L.). In: Proceedings of the International Congress of Plant Physiology, vol 2 Society of Plant Physiology and Biochemistry, New Delhi, India, pp 935–938

Vijayakumar KR, Unni PN, Vasudevan VK (1984a) Physiological changes in black pepper (cv. Panniyur-1) associated with water logging. J Agric Res Kerala 22(1):96–99

Vijayakumar KR, Unni PN, Vamadevan VK (1984b) Prevention of photo-induced chlorophyll loss by the use of lime reflectant on the leaves of black pepper (*Piper nigrum* L). Agric Forest Meteorol 34:17–20

Wahid PA, Kamalam NV (1982) Mineral nutrition of slow wilt affected pepper (*Piper nigrum* L.). J Plant Crops 10:21–25

Wahid P, Zaubin R, Nuryani Y (1990) Effect of fertilizer on the yield of black pepper (*Piper nigrum* L.) in Bangka. Ind Crops Res J 16:43–49

Xiong W, Li Z, Liu H, Xue C, Zhang R, Wu H, Li R, Shen Q (2015) The effect of long-term continuous cropping of black pepper on soil bacterial communities as determined by 454 pyrosequencing. PLoS One 10(8):e0136946

Zachariah PK (1978) Fertilizer management for cardamom. In: Proceedings of the First Annual Symposium on Plantations Crops, Rubber Research of India, Kottayam, Kerala, India, pp 141–156

Zu C, Li Z, Yang J, Yu H, Sun Y, Tang H, Yost R, Wu H (2014) Acid soil is associated with reduced yield, root growth and nutrient uptake in black pepper (*Piper nigrum* L.). Agric Sci 5(5):466–473

Dr V Srinivasan is the Principal Scientist and Head, Division of Crop Production and Post-harvest technology, at ICAR-IISR, Kozhikode, India. He has 26 years of experience on soil science and plant nutrition aspects of spices and spice-based cropping systems. He was involved in the development of six micronutrient formulations for major spices, received 5 patents, and licensed all formulations. He is also the inventor of a unique lime-based microbial formulation and has developed soil test-based yield response functions for targeted yield in spices and decision support system for fertilizer recommendation. His major contributions include formulation of organic farming packages for major spices, its long-term effects on soil and crop quality, plant-microbe interactions, and their influence on nutrient use efficiency of spice crops. He has published 75 research papers and edited 3 books. e-mail: srisoilv@gmail.com.

Professor Muthusamy Murugan started his professional career at the Cardamom Research Station, Kerala Agricultural University, Pampadumpara during 1993. He received his M.Sc (Soil Science and Agrl Chemistry) degree from the Tamil Nadu Agricultural University, Coimbatore, and Doctoral degree in Agroecology and Biosecurity from the National Institute of Advanced Studies (NIAS, Tata Institute), Indian Institute of Science, Bangalore. During his postgraduate research programs, he enjoyed reputed fellowships from Govt. of India and TATA. He also served as an academic faculty member at the NIAS during 2011–2012. He is currently the Professor and Head at the Cardamom Research Station, Pampadumpara, and focuses research aspects on soils, pesticides, and agricultural sustainability, which are highly inter and transdisciplinary in nature, particularly in the Indian Western Ghats tropical mountain situation. He has published more than 100 articles in leading science research publishers like Springer Nature, Elsevier, Taylor and Francis, IGI, Frontiers, Current Science, Cambridge University and American Chemical Society Journals, etc. e-mail: muthupeyan@gmail.com.

Dr R Dinesh is the Director of ICAR-Indian Institute of Spices Research, Kozhikode, Kerala. He made significant contributions in soil fertility/soil biochemistry/soil microbiology with major focus on plant-microbe interactions and their influence on nutrient cycling in soils under perennial horticulture. He started his career as a scientist at ICAR-CIARI, Port Blair, in 1994, where he worked on arable soils and soils under tropical forests and mangrove ecosystems. He is one of the inventors of the smart and novel delivery system for plant beneficial microbes through encapsulation (biocapsules) which is the first of its kind in the world biofertilizer industry. He also co-developed designer micronutrient and PGPR formulations specific to spice crops which have been licensed to many private agencies. His products and processes have six patents and have been nonexclusively licensed to several national and international companies. He was also involved in the development of organic and integrated nutrient management plans and GAPs for various spices and soil-based land-use plans for Kerala state. He is a Fellow of NAAS, New Delhi, and has received many awards including Recognition Award; Young Scientist Award of NAAS, New Delhi; and Young Scientist Award of the Indian Society of Soil Science, New Delhi. He has also published many research articles in high impact journals. e-mail: rdinesh2005@gmail.com.

Cashew (*Anacardium occidentale* L.)

S. Mangalassery, J. D. Adiga, G. L. Veena, N. K. Binitha,
and K. S. Anil Kumar

Abstract

Soil is an important natural resource supporting the life on earth. Majority of the agricultural activities depend on soil. Maintaining soil health requires scientifically sound management of the soil in order to preserve its ability to function as a living ecosystem that supports bacteria, humans, animals, and plants. It became apparent that soil and land resource was negatively impacted by both intensive agricultural systems and careless and ineffective agricultural methods that did not take soil health into account. Cashew (*Anacardium occidentale* L.), a tree nut crop that grows in tropical regions of the world, is treasured for its very edible nut that is traded all over the world. The soil resources of cashew-growing regions across the globe are reported to suffer from the problems of soil acidity and deficiencies of major and micronutrients. The soils in cashew-growing regions exhibit low base status, water-holding capacity, and organic carbon status due to the region's distinctive climate and inadequate implementation of management practices. In order to stop the deterioration of soil resources under cashew farming, it is crucial to evaluate the soil conditions and adjust its acidity. It has been demonstrated that site-specific soil and water conservation techniques increase crop productivity and retain moisture in addition to keeping the topsoil from being washed away during periods of heavy rainfall. Application of

S. Mangalassery (✉) · J. D. Adiga · G. L. Veena
ICAR-Directorate of Cashew Research, Puttur, Karnataka, India
e-mail: shamsudheen.m@icar.gov.in; Jd.Adiga@icar.gov.in; Veena.G@icar.gov.in

N. K. Binitha
College of Agriculture, Kerala Agricultural University, Kasaragod, Kerala, India
e-mail: binitha.nk@kau.in

K. S. Anil Kumar
ICAR-National Bureau of Soil Survey and Land Use Planning, Regional Centre, Bangalore, Karnataka, India

G. V. Thomas, V. Krishnakumar (eds.), *Soil Health Management for Plantation Crops*, https://doi.org/10.1007/978-981-97-0092-9_6

manures, soil test-based inorganic fertilizer application, biofertilizers, green manuring, and recycling of the biomass available in the plantations helped to improve the soil health in cashew plantations and prevented further degradation. There is a need to spread awareness about the importance of managing soil health for sustainable crop production among all the stakeholders.

Keywords

Cashew · Biofertilizers · Biomass recycling · Integrated nutrient management · Intercropping · Organic farming · Soil conservation · Soil health management

Abbreviations

AMF	Arbuscular mycorrhizal fungus
BCR	Benefit/cost ratio
CAN	Calcium ammonium nitrate
FYM	Farm yard manure
HDP	High-density planting
ICTs	Information and communication technologies
MOP	Muriate of potash
RDF	Recommended dose of fertilizers
VAM	Vesicular arbuscular mycorrhiza

1 Introduction

Water, air, organic matter, and minerals make up soil. Because it is home to millions of species, soil is filled with life. Soil is habitat to a variety of biological activities that are vital to life (Nannipieri et al. 2017). Soil serves a variety of environmental purposes, including chemical, biological, and physical activities. These functions are useful to the plants and other living organisms through processes such as nutrient cycling, aeration, anchorage, and water retention (Al-Kaisi et al. 2017). Due to the rising demand for food, fodder, and fibers, consequent to the massive expansion in human population has put pressure on agricultural production processes (Scholes 2016). The health and quality of the soil have declined as a result of increasing pressure on land and soil resources.

An agricultural enterprise seeking to be both profitable and sustainable must have healthy soil. Reducing soil erosion, enhancing biological activity and nutrient cycling, optimizing water retention, and raising fertility are all part of maintaining and/or enhancing soil health (El-Ramady et al. 2014). Implementing agricultural techniques that preserve or improve the physical, chemical, and biological characteristics of soil as well as promote soil functions is another aspect of managing soil health (Idowu et al. 2019). Creating crop and soil management plans that improve soil health is crucial to achieving agricultural yield that is sustainable. "The continued capacity of soil to function as a vital living ecosystem that sustains plants, animals, and humans" is the definition of soil health given by Lehmann et al. (2020). To put it briefly, soil health refers to the soil's capacity to deliver ecosystem services.

Originally from Brazil, cashew (*Anacardium occidentale* L) is a tree nut crop that was brought to India in the sixteenth century by Portuguese explorers as a soil-binding crop to prevent soil erosion in coastal areas (Mangalassery et al. 2019a). Its commercial value and ability to withstand unfavorable soil and environmental conditions were identified, and cashew was commercially cultivated along India's east and west coasts. Currently, 1.19 million ha of cashew are grown in 17 states in India, producing 0.79 million tons of the crop annually (DCCD 2022). 34 tropical nations grow cashew worldwide (Mangalassery et al. 2019c). The total area planted with cashew is 6.57 million ha, and 3.71 million t of nuts are produced globally (FAOSTAT 2023). A broad variety of soil types, including red latosols, laterite soil, sandy to sandy loam, and loam, are used for cashew cultivation in India (Rupa 2017; Venugopal and Abdul Khader 1991). Owing to its resistance to drought, cashew is commonly grown on sloped and degraded areas where it would be impractical to produce other crops profitably. In terms of nitrogen (N), base status, cation exchange capacity, and micronutrients like zinc (Zn) and boron (B), the majority of soils used for cashew cultivation have low soil fertility (Rejani et al. 2013). The basic cations in the soil are washed away by the excessive precipitation in the coastal regions where cashew is produced, increasing the acidity of the soil. A portion of the nutrients are rendered inaccessible for cashew because of the high soil acidity, which also reduces the plant's ability to absorb nutrients (Mangalassery et al. 2021). Selecting cashew as a crop in such an area helps prevent further deterioration of soil resources, enhancing soil fertility and other ecosystem services including carbon sequestration and biological activity. To enhance the soil health, appropriate management techniques need to be adopted.

2 Properties of Soils in Cashew Orchards

Due to the continuous nutrient removal by the trees and lack of replenishment, soil nutrients in India's cashew-growing regions have been depleted. About 2.7% of samples in Puttur, Karnataka; 20.0% of samples in Vengurla, Maharashtra; 81.4% of samples in Bhubaneswar; and 92.9% of samples in Bapatla had low levels of organic carbon (Mangalassery et al. 2021). In 94.3% of the Puttur samples, 37.1% of the Vengurla samples, 94.3% of the Bhubaneswar samples, and 75.7% of the Bapatla samples, the available N was found to be low. In Pilicode, Kerala, cashew orchards were found to have significant levels of available N and organic carbon. About 65.7% of the samples in Puttur, 41.4% in Vengurla, 58.6% in Bhubaneswar, 22.9% in Bapatla, and 17.1% in Pilicode were determined as having low available potassium (K). The available phosphorus (P) was low in such soils. Additionally, these soils were deficient in magnesium (Mg) and calcium (Ca). The micronutrients like manganese (Mn) and iron (Fe) were adequate. But overall, there was a lack of zinc (22.9 to 57.14% samples in various regions) and copper (Cu) (8.57 to 32.9%

samples in various regions) in the soils. These findings point to the growing nutrient inadequacies in the soils used for cashew production on the nation's east and west coasts. In such a scenario, nutrient management based on soil tests can stop soil degradation in addition to enhancing cashew growth and production. In the cashew-growing soils of Brazil, the need for suitable management options for increasing the contents of Ca, Mg, K, and P was emphasized by Salim et al. (2018). The lack of adequate K in cashew-growing soils of Brazil was also reported by Silva et al. (2020). Barthélémy et al. (2021) reported low to very low levels of N, P, Ca, Mg, and Zn in cashew-growing soils of Ivory Coast, with adequate levels of soil P and high to very high contents of Fe, Cu, and Mn.

Falade (1977) conducted research on the fertility status of Nigerian soils used for cashew cultivation. In the Kogi state of Nigeria, low levels of clay and organic matter in cashew soils had a detrimental effect on the soil's fertility status, which was reflected in the minerals and nutrients found in the leaves and nuts. C, N, P, and Mg were reported to be the limiting nutrients in cashew soils of Nigeria (Aikpokpodion et al. 2009). It indicates that the soil fertility issues more or less follow similar patterns in the different cashew-growing tracts in the tropical regions. The soils in the tropics are in the extreme stages of weathering due to the high temperature, alternate wet and dry cycles, and nature of parent materials making them low in base status and cation exchange capacity (Mangalassery et al. 2019a).

Cashew is generally grown in poor fertile soils. However, deep, well-drained sandy loam soils without a hard pan are ideal for maximizing the crop's potential. The best soil types for cashew crop include red sandy loam, lateritic soils, and coastal sands with a pH of mildly acidic (Mangalassery et al. 2019b). Flooding and stagnant water is not conducive to the proper growth of cashews. Cashew cultivation is not suited on heavy clay soils that have poor drainage or on soils that have a pH of greater than 8.0 (Bhat et al. 2016). Its growth is further inhibited by very saline and alkaline soils (Salam and Peter 2009).

3 Soil Constraints and Management

3.1 Soil Salinity

Seed germination, growth, and seedling establishment are crucial for maintaining the species in harsh conditions. An excessive amount of salt in the soil can affect the germination and growth of many different types of plants. In cashew seedlings raised on a variety of substrates, salinity involving NaCl reduces the K+ levels in the plant while also promoting the buildup of Na + and Cl-1 to fatal levels. According to Kramer and Boyer (1995), a soil water shortage affects the architecture of the root system, reduces nutrient uptake from the soil, and inhibits the solubilization of nutrients in the root environment. According to Shahid et al. (2020), salinity can result in an excess of harmful ions and a deficit of vital nutrients, depending on the types of salts that are predominant in the agricultural environment.

During the establishing period, dwarf cashew seedlings are more susceptible to salinity. Long-term exposure to salinity was reported by Viegas et al. (2001) to significantly lower K^+ content in tissues, especially in roots. In this study they found that the concentrations of Na^+ and Cl^-ions in leaves reached lethal levels in response to time and increasing salinity. From a study, Viegas et al. (2004) discovered that salinity reduced transpiration rates, which in turn affected the activity of the enzymes that reduce nitrate. In contrast, salinity caused an increase in glutamine synthetase activity in leaves, which was subsequently accompanied by a rise in the concentration of soluble amino acids. They claimed that in plants exposed to salinity, the assimilatory NO_3 reduction has assumed a prominent function.

According to Alves et al. (2013), salinity slowed down seedling emergence, germination, and growth. Reduced mobilization of cotyledon reserves, increased buildup of Na^+ in the seedling parts, and decreased K^+ content of roots were all caused by the excess NaCl concentration of the growth medium. The osmotic and ionic components of salt stress appear to be responsible for preventing the depletion of cotyledon reserves and delaying the dwarf cashew seedlings' growth and development. Salinity was also found to postpone seedling emergence and establishment (Marques et al. 2013). With increasing salinity, the concentrations of Na^+ and Cl^- ions rose; this was particularly noticeable in established seedlings, where the content of K^+ decreased.

3.2 Soil Acidity

Soils used to cultivate cashews are typically acidic. However, the best suitable pH for growth of cashew is in the range of 6.0 to 7.3. Nutrients including N, P, Ca, Mg, B, and molybdenum (Mo) become less available under very acidic soil conditions, and nutrients like iron, manganese, and aluminum may increase to harmful levels and impact plant growth (Mangalassery 2018). Liming is the process of using lime, dolomite, or other liming materials to adjust the acidity of soil. Liming should be done while establishing a plantation and then again on a regular basis depending on the results of soil tests. Lime can be administered at any time of the year, but in order to maximize its effectiveness, it should be applied soon after intense rain, as moisture is necessary for the reaction between lime and soil.

4 Fertility Management for Healthy Soils

4.1 Nutrient Requirement and Management

While cashew thrives in soils that most other crops cannot, it does better when external nutrients are applied. Rainfed cashew is mostly grown on lateritic and red soils. The bulk of cashew farming regions hardly ever receive the adequate and required fertilizer nutrients. The amount of nutrients in the cotyledons provides the necessary nutrients for the cashew plant throughout its early stages of growth. N > K > Ca > Mg > P > S is the order in which nutrients are extracted from the cotyledons (Bezerra et al. 2007). Ximenes (1995) reported that for up to 75 days after planting, the cashew accession CCP 76 cotyledons could supply approximately 54% of N, 45% of P, 17% of K, 1% of Ca, 16% of Mg, and 36% of S needed for seedling development. When the seedlings are 1-month-old after emergence, these contributions will be even greater. According to Hammed et al. (2011), the nutritional content of cashew seedlings increases over a period of time, and nutrients get absorbed in the following order: N > Ca > P. The size of the initial nuts used for developing the seedlings decides how much of each nutrient it absorbed. The most uptake of N, P, K, Ca, and Mg was seen in seedlings raised from jumbo (> 16 g) nuts, followed by medium-sized (4 to 8 g) and small (< 2 g) nuts. This suggests that in order to maintain the increased growth and vigor of the seedlings in the nursery, a suitable nutrient supplementing scheme should be used while raising the cashew seedlings utilizing the jumbo nut size.

A mature cashew tree takes up 2.847 kg N, 0.75 kg P, and 1.265 kg K, according to earlier studies (Mohapatra et al. 1973). According to Babu et al. (2015), cashew is a crop that responds well to fertilizers. A suitable nutrient management program has been shown to enhance output by 50–100%. The cashew's established response to nitrogen is evident in all cashew-growing regions (Falade 1984; Latha et al. 1994; Reddy et al. 1982; Veeraraghavan et al. 1985). The substantial nutrient uptake revealed by leaf analysis further emphasizes the need for a steady and adequate supply of nutrients for cashew (George et al. 1984; Rupa et al. 2014). Plant age, leaf age, nutrient mobility within the plant, season, environmental factors, and genotype under investigation all affect the mineral concentrations in the leaves of adult cashew plants. Significant differences were noted by Kumar et al. (1982) in the N, P, and K concentrations in cashew leaves growing in a dry, non-fertilized environment. They also showed that these nutrients are depleted from the leaf during the fruit development process, possibly as a result of ions being transferred to the fruit. Throughout a production cycle, the leaf content of macronutrients in dwarf cashew plants grown in dry, unfertilized soil varied. According to Fragoso et al. (1999), the variances were linked to leaf age and the plant's annual phenological fluctuations. Up to the start of the reproductive phase (rainy season), they observed rising levels of N, P, K, Mg, and S. After that, they found falling levels (K remaining constant) during flowering and fruit development (dry season). Additionally, the amounts of Ca tended to rise as the leaf grew older. The leaves of a dwarf cashew plant that was given larger dosages of N and K fertilizer did not change in terms of these nutrients. According to Fragoso et al. (1999), the cashew nuts and the apples removed macronutrients in the following order: N > K > Mg > P > S > Ca and K > N > Mg > P > S > Ca, respectively. In their study, Crisostomo et al. (2004) examined the accumulation of N and K in mature cashew trees subjected to varying dosages of these two nutrients. They found no discernible variation in the leaf content between plants not receiving N fertilizer and those receiving up to 173 kg N ha^{-1}. Conversely, plants that received 73.4 kg K_2O ha^{-1} as fertilization showed higher leaf K concentrations than plants that did not receive any K. From the first to the seventh year of farming, nut output increased. The seventh year saw the highest nut output of 1753 kg ha^{-1} with 55.83 and 31.20 kg ha^{-1} $year^{-1}$ of N and K_2O, respectively. Nevertheless, the treatment of K and N had no effect on the quality of the nuts.

According to reports, adding P to soils lacking in P increases nut yield (Richards 1992). Numerous authors have shown a significant favorable impact of K on cashew development and yield (Ghosh and Bose 1986; Veeraraghavan et al. 1985). NPK fertilizer use has been shown to have a considerable impact on cashew nut characteristics and production (Harishu Kumar and Sreedharan 1988; Rao et al. 1984). In the first year, the highest yield of 3.53 kg $tree^{-1}$ and the lowest disease infection (90 inflorescence panicles infected) were obtained by applying urea and MOP @ 120 and 48 kg ha^{-1}, respectively. In the second year, the highest yield was obtained by applying NPK fertilizer in the form of urea and SSP @ 60 and 144 kg ha^{-1} or MOP @ 24 kg ha^{-1}, which also significantly reduced the disease infection (125 inflorescence panicles infected). Thus, in Ochaja, Nigeria's acid sand soils, proper administration of NPK fertilizers can lower the infection of inflorescence blight and raise cashew output (Adejumo 2010).

The general nutrient recommendation for inorganic fertilizers for the several cashew-growing regions in India is given in Table 1. Based on the findings of the soil test, the age of the plant, and the spacing used, these fertilizer doses need to be modified. The chart provides guidelines for cashew plants spaced at 7.5 m x 7.5 m, which is the standard spacing. The recommended fertilizer under a high-density cashew planting technique is reasonable up to 80–100% canopy coverage, which is often reached in the first 6–8 years after planting. Due to the nutrient buildup in the soil caused by the decomposing cashew biomass fallout, it may be essential to reduce the prescribed doses of fertilizers per plant after a particular stage of the crop.

After heavy rains have stopped, weeds must be removed, and the bases of individual trees must be cleared before applying fertilizer. The secret to increasing fertilizer use efficiency is timing of fertilizer applications to coincide with the crop's need for growth and periods of high root activity (Carr 2013). Increased root activity

Table 1 Recommended doses of fertilizers to cashew in different states in India

	Nutrient dose for mature cashew plantations (5th year of planting) (g $tree^{-1}$ $year^{-1}$)		
State	N	P_2O_5	K_2O
Kerala	500	125	125
	750	325	750
Karnataka	500	250	250
	750	125	125
Tamil Nadu	500	200	300
Andhra Pradesh	500	125	125
	1000	125	125
Maharashtra	1000	250	250
Odisha	500	250	250
West Bengal	1000	250	250

Source: Mangalassery et al. (2019b)

Fig. 1 Fertilizer application to a grown-up cashew tree

during the flushing and early flowering phase (September to December) promotes better soil nutrient absorption. As a result, an attempt should be made to time the administration of fertilizer with this stage (Rupa 2017). As the tree approaches the reproductive phase, there is a high internal demand for nutrients during the flushing phase. Therefore, during this growth phase, application of balanced fertilizer is crucial. It is best to apply fertilizer to cashew in two separate dosages, the first at the start of the monsoon and the second in the post-monsoon season. If a single application is only feasible due to labor shortages or other circumstances, it can be carried out in the post-monsoon season when there is enough soil moisture. In laterite soils in areas with high rainfall in the west coast region, circular trenches of 25 cm in depth and 15 cm in width should be opened at intervals of 0.5, 0.75, 1, and 1.5 m from the trunk in the first, second, third, and fourth years, respectively, following planting (Fig. 1). Fertilizers can be applied in 50 cm circular strips in loamy soils on east coast regions with minimal rainfall. After fertilizers are applied, the trench needs to be sealed off right away using green leaves as mulch. During 1st, 2nd, 3rd, 4th and 5th year of planting 1/5, 2/5, 3/5, 4/5, and full quantity of recommended dose need to be applied (Mangalassery et al. 2020b).

4.2 Micronutrients

For healthy growth and development, the amount of micronutrients in the soil should be at its ideal level. Correct dosage of micronutrient fertilizers, either by soil application or foliar spray, is required to address the widespread incidence of micronutrient deficiencies identified in cashew-growing soil (Mangalassery et al. 2023). Micronutrient deficits can be quickly and easily corrected with foliar nutrition treatment. Better fruit set and increased nut yield are ensured by foliar spraying of

Fig. 2 Foliar applications of micronutrients to cashew tree

fertilizers at the emergence of the flush, panicle initiation, and fruit-set stages (Fig. 2) (Ghosh 1988). Foliar sprays of 0.5–1.0% $FeSO_4$, 0.5–1.0% $MnSO_4$, 0.5% $ZnSO_4$, 0.1% $CuSO_4$, 0.1% boric acid, and 0.1% Mo salts can be given with the emergence of the flush, panicle initiation, and fruit-set stages to remedy the deficiencies of Fe, Mn, Zn, Cu, B, and Mo (Rupa 2017). According to Jena et al. (2018), using $ZnSO_4$ @ 0.5% as foliar application will increase the quantity of hermaphrodite flowers and sex ratio. Treating cashew plants with $ZnSO_4$ @ 0.5% + borax @ 0.1% proved to be the best treatment in terms of yield features (nut weight, nut length, kernel weight, and number of nuts kg^{-1}), out of different growth-promoting compounds they studied. The cashew apple with the maximum length and weight was found to be with $ZnSO_4$ @0.5% and NAA significantly impacted cashew apple quality measures such as TSS, starch, and sugar content.

4.3 Nutrient Management in High-Density Planting Systems

Cashew growth and production are significantly influenced by planting density and fertilizer amounts. A system known as high-density planting (HDP) involves planting in less space than traditional spacing in order to include more plants per unit area and boost productivity in any type of crop, ideally with some changes to agro-techniques, particularly canopy management, nutrient requirements, and irrigation management. The rate of fertilizer dosages under a HDP scheme must be modified in accordance with the number of trees per unit area. Fertilizer recommendations tailored to high-density planting are provided for some localities. Poduval and

Yadukumar (2011) found that applying fertilizer at a rate of 75 kg N + 25 kg P_2O_5 + 25 kg K_2O ha^{-1} year^{-1}, with a tree density of 400 plants ha^{-1}, produced the highest benefit/cost ratio (BCR) of 2.26. The maximum cumulative yield of 15 t ha^{-1} with a BCR of 3.04 was observed in grafted cashew plants of variety BPP 8 (H 2/16), closely followed by 14 t ha^{-1} with a BCR of 2.61, at HDP (500 plants ha^{-1}) and moderately to high fertilizer dose (150–50-50 to 225–75-75 kg NPK ha^{-1}) at eleventh harvest under Odisha conditions (Tripathy et al. 2015). Rajamanickam et al. (2020) found that the highest values of yield parameters, such as nut weight (59.11 g), yield tree^{-1} (11.14 kg), estimated yield (2.25 t ha^{-1}), and BCR (2.41) of cashew, were recorded when soil application of recommended dose of fertilizers (RDF)(NPK @ 280:160:240 g tree^{-1}) along with soil application of micronutrients ($FeSO_4$ @ 200 g tree^{-1}). In Karnataka, India, the best results were obtained with 100% RDF (500:250:250 g) + 10 kg FYM applied in Maidan tracts, combined with foliar spraying of 3% urea +0.5% MOP +1% K_2SO_4, 0.5% $ZnSO_4$ + 0.1% Solubor as boron +0.5% $MgSO_4$ for nut yield tree^{-1} of the Chintamani-1 cashew variety (Rajendra et al. 2023).

5 Technological Options to Promote Sustainable Soil Health

5.1 Cropping Based on Land-Use Capability

The first criterion in managing soil health is the correct crop selection based on soil capabilities (Guo 2021). To evaluate the land use and land capabilities, various indices and systems are available. Following a thorough evaluation of the soil's capability class, the suggested management strategies should be implemented in order to address the short comings of specific soil class. Soil erosion and a decline in soil health are caused by the unscientific usage of the land (de la Rosa and Sobral 2008). The assessment of land suitability is dependent on the crop requirements as well as the features of the land and soil. A suitability index in the form of ranks is produced by matching crop requirements with several characteristics, such as climate, terrain, and soil, in accordance with the FAO framework for land suitability assessment (FAO 1976). According to Sehgal (1966), the different classifications are N1 (not suitable, but potentially suitable; uneconomical for usage), S1 (very suitable), S2 (moderately suitable), and S3 (marginally suitable). Rejani et al. (2013) reported a criterion for choosing an appropriate location for cashew based on soil physical properties, geographical features, climatic, and environmental factors. Using GIS, a suitability map was also created for the cashew-growing regions of India based on the established criteria. Shamsudheen et al. (2004) evaluated the suitability of soil sites for cashew in the northern Karnataka plateau landscapes using the FAO framework for land evaluation. Similar suitability evaluation maps have been created for a few additional cashew-growing locations, including Malawi (Benson et al. 2016), Indonesia (Sutandi et al. 2014), and the Ivory Coast and Ghana (CIAT 2011). A growing number of software and mobile apps are being developed using the land use and crop requirements criteria to make them quick and simple for end users to utilize. An effort to evaluate the cashew site suitability in India is the cashew site suitability evaluator, an Android-based smartphone app (Mangalassery et al. 2022a). Though the cashew area is anticipated to grow globally, it is expected that some of the presently acceptable places may become unsuitable under the changing climate scenario and vice versa (Grüter et al. 2022).

5.2 Soil and Water Conservation Measures

High-intensity rainfall is a defining feature of the cashew-growing regions, causing runoff and soil erosion. According to Rejani and Rupa (2017), cashew undergoes moisture stress from December/January to May, which falls between the blooming and fruit-setting season of the plant. This causes flower drying and immature nut drop. Furthermore, there are no water sources available for irrigation in the traditional cashew production sites, which include hillocks and sloppy laterite terrains. Despite being a hardy crop, research shows that cashew reacts favorably to management of nutrients and water (Rupa 2017). Adopting region-specific soil and water conservation techniques can help control the crop's water deficit to some extent in locations without access to irrigation water. Adoption of such measures is part of cashew production technology in sloppy locations to reduce surface runoff and soil erosion (Rejani and Yadukumar 2006). Modified crescent bunds, staggered trenches with coconut husk burial, and reverse terraces are a few soil and water conservation techniques that work well for cashew orchards (Mangalassery et al. 2020a). Another common soil conservation strategy that is suggested for cashew is the use of catch pits in a terrace. It was discovered that by using these techniques, pre-monsoon rainwater could be harvested, and the cashew output could be increased by 32–35%. Reduced soil loss and runoff velocity, improved soil moisture retention, and groundwater recharge are other advantages. In order to boost agricultural income and land production, sloppy or barren areas can be converted to cashew farming by implementing such soil and water conservation techniques. In addition to the obvious advantages of conserving soil and water, research indicates that these actions enhance soil quality and soil carbon sequestration under cashew (Mahajan et al. 2021).

5.2.1 Trenches/Engineering Measures

Continuous bench terraces: Terraces have the advantage of offering a level platform for cashew planting, which further reduces the likelihood of erosion. They help stop the movement of soil and water down slope.

Continuous contour trench: These trenches are dug over the whole length of the field along the contour in sloppy areas with a 7–8% slope. It is advised to dig a trench that is 0.5 m by 0.6 m.

Modified crescent bund: According to Rejani and Yadukumar (2010), the modified crescent bund is to be placed upstream of the cashew terrace. It is crescent-shaped, measuring 6 m in length, 1 m in width, and 0.5 m in height at a radius of 2 m. This will aid to retain water and litter (Figs. 3 and 4).

Fig. 3 Modified crescent bund for soil and water conservation in cashew orchards

Fig. 4 Terrace with crescent bund in cashew orchards

Fig. 5 Reverse terrace for soil and water conservation in cashew orchards

Staggered trench: To improve water retention, staggered trenches measuring 5 m in length, 1 m in width, and 0.5 m in depth should be dug across the slope, between two rows of cashew plants, or in the center of four plants.

Reverse terraces: Designed to slope from the periphery to the center, reverse terraces should have the following recommended dimensions: 2 m length, 2 m breadth, and 0.7 m depth (Fig. 5).

Catch pits: To catch and hold runoff and improve percolation, catch pits should be built upstream of cashew-planted terraces with recommended dimensions of 3 m in length, 0.5 m in width, and 0.5 m in depth.

Tree base terrace: It was discovered that the creation of a tree base terrace, extending two meters surrounding the plant, was advantageous for preserving moisture after 3 years of planting. The process involves removing soil from the slope's upper side and filling up the lower part. The overhang must be positioned to provide a catch pit for soil deposit and moisture conservation.

5.2.2 Bioengineering Measures

Coconut husk burial: Using coconut husk burial in conjunction with soil and water conservation methods, such as staggered trenches and modified crescent bunds, helps the soil retain water for extended periods of time (Rejani and Yadukumar 2006). It is also possible to implement this coconut husk-burying technique next to cashew plants. Husks are to be buried in trenches that are opened across the slope between two rows of cashew and measure 3.5 m in length, 1 m in width, and 0.5 m in depth. Three to four layers of husks can be buried in these ditches, with the convex side of the first layer touching the earth. The convex side should be positioned higher on the last layer of husks. Between layers of husks, a small layer of soil and leaf materials can be inserted. After that, the trench can be filled with dirt, leaving a depth of roughly 10 cm.

Use of bigger pits and mulching: When establishing cashew plantations, this procedure must be adhered to. One-meter pits should be excavated at the prescribed intervals after terracing and other soil and water conservation techniques. The recommended filling rate for these pits is 1/3 depth of topsoil, organic manure, and rock phosphate. Place the graft in the middle of the pit and make sure you mulch it properly. In cashew plantations, coir-pith compost mulching has also been proven to be advantageous.

Trenches with vegetative barriers: Staggered trenches (in reversely sloppy terrain) and continuous contour trenches, when combined, can significantly reduce runoff and soil loss. Among the suggested vegetative barriers are *Stylosanthes hamata* and *Vetiveria zizanioides*. In addition to decreasing runoff and soil erosion, the vegetative barriers have the potential to yield extra revenue through harvesting.

Green manuring and mulching: *Gliricidia* is a type of green manure crop that can be grown in borders and empty spaces to give green biomass for mulching. Green mulch mulching the tree basin contributes to soil moisture conservation.

Circular trench with leaf litter and coconut husk: Coconut husks and leaf litter are buried in circular trenches that are 0.3 m wide and 0.5 m deep, with the opening occurring 2 m from the cashew trunk. This technique is often advised for east coast regions.

5.3 Intercropping in Cashew Plantation

In India, cashew has commonly been planted with 7.5 m between each other. In the early years of a cashew plantation, the pace of canopy expansion is modest. Wider planting spacing thereby wastes both vertical and horizontal space as well as nutrients, moisture, and solar radiation (belowground resource). Furthermore, a serious weed infestation will result from the unutilized interspacing. Under standard density planting conditions, it takes 8 to 10 years for the cashew canopy to fully cover itself (Saroj 2015). When cashew plantations are first established, they can be planted with appropriate crops in the interspaces during the first few years of normal density planting. By serving as a cover to protect the soil in otherwise exposed areas during the early stages of a plantation, this integration of different crops can help reduce soil erosion and runoff, conserve soil moisture, and inhibit the growth of weeds (Adiga and Kalaivanan 2017). The main intercrops planted in India include pepper, vanilla, yams, ginger, pine apple, and turmeric. During the early stages of plantation establishment, cashew can be successfully interplanted with other crops including sweet potatoes and vegetables like amaranthus, cluster beans, coleus, cucumber, bottle gourd, pumpkin, and brinjal (Nayak et al. 2020). In China, the legume cover crops such as *Calopogonium mucunoides* and *Macrophyllum atropurpureius* and intercrops such as peanut, sweet potato, and beans are grown (Kangde et al. 1998). In Nigeria, Aliyu and Hammed (2008) reported cultivation of intercrops such as maize, cassava, cowpea, and cocoa along with cashew. Abeysinghe (2009) observed that intercropping in cashew plantation was beneficial to the both the main crop and intercrops and soil health in Sri Lanka.

5.4 Use of Organic Manures

The traditional tropical soils, where cashew is grown are generally low in organic matter. To maintain the organic matter levels, additional organic matter, manures, compost, oil cakes, biofertilizers, etc. can be supplied (Table 2).

Research on applying organic sources of nutrients showed higher cashew yields and higher levels of residual fertility. Yadukumar et al. (2012) found an increase in cashew yield over a period of years by using chicken manure to replace the 50% N requirement. When 75% of the prescribed N dose for cashew is substituted with compost made of organically recyclable biomass, they saw a rise in soil N. It has been claimed that improving the nutrient utilization efficiency in laterite soil can be achieved by applying 10 to 15 kg of FYM plant^{-1} in addition to the required dosage of inorganic fertilizers (Salam et al. 2008). Green manuring is a substitute that can be used in place of FYM. Green manure crops like sunn hemp, sesbania, and glyricidia can be planted between cashew rows and along boundaries. The amount of organic matter in soil is increased by applying green manure. Additionally, it strengthens the structure of the soil and lessens soil erosion and runoff (Mangalassery et al. 2022b).

Table 2 The sources of organic manures and oil cakes and their nutrient contents

Source	Nitrogen (N) (%)	Phosphate (P_2O_5) (%)	Potash (K_2O) (%)
Farm yard manure (FYM)	0.5–1.5	0.4–0.8	0.5–1.0
Compost (urban)	1.0–2.0	1.0	1.5
Compost (rural)	0.4–0.9	0.3–0.6	0.7–1.0
Green manure	0.5–0.7	0.1–0.2	0.6–0.8
Castor cake	5.5–5.8	1.8–1.9	1.0–1.1
Cotton seedcake (undecorticated)	3.9–4.0	1.8–1.9	1.6–1.7
Mahua cake	2.5–2.6	0.8–0.9	1.8–1.9
Karanje cake	3.9–4.0	0.9–1.0	1.3–1.4
Neem cake	5.2–5.3	1.0	1.4–1.5
Safflower cake (undecorticated)	4.8–4.9	1.4–1.5	1.2–1.3
Poultry manure	2.2	1.8	1.1

Source: Mangalassery et al. (2019b)

5.5 Biomass Recycling

Cashew plantations have a large amount of biomass available that might be recycled. According to Guruprasad et al. (2009), a mature cashew plantation yields 1.38 to 5.2 t of cashew biomass fallout (leaves, twigs, blossoms, and apples) ha^{-1} annually in plantations that are 10–15 to 25–40 years old. Cashew biomass may be composted by adding 20% cow dung slurry, with 60–65% recovery, and after 6 months, the composted biomass can be utilized as mature compost. Usually to prepare the compost, three layers of recyclable cashew biomass are combined with 20% cow dung slurry (20% of the biomass's total weight). To keep the moisture content at 50% during composting, water is to be sprayed. According to Yadukumar and Nandan (2005), turning decomposing biomass once every 30 days is necessary to allow for aeration and the release of heat produced in the compost heaps.

Vermicompost preparation with earthworms (*Eudrilus* sp.) from cashew biomass is possible (Kalaivanan et al. 2017). About 3.5 t of compost, or vermicompost, has been reported to be recovered from 5.5 t of cashew biomass. Use of earthworms in composting contributes micronutrients, improves subsurface aeration, promotes microbial activity, and accelerates the breakdown of waste products.

When vermicompost is applied to soil, it provides a large population of beneficial microorganisms, including bacteria (39 x 10^6 cfu), fungi (39 x 10^5 cfu), actinomycetes (28 x 10^5 cfu), and *Azospirillum* (0.52 x 10^4 cfu). In addition, the application of vermicompost supplies 1.2% N, 0.89% P, 0.59% K, 2.75% Ca, 0.82% Mg, 29.6 ppm Zn, 24.3 ppm Mn, 12.2 ppm Cu, and 162 ppm Fe (Rupa 2017). It is predicted that recycled cashew biomass vermicompost/compost can meet approximately half of the cashew's nutrient requirements. When vermicompost is added to soil, it not only provides a significant amount of plant nutrients but also adds a large amount of organic matter.

5.6 Biofertilizers

The soils used for cashew cultivation were also shown to benefit from the application of biofertilizers that contained live strains of advantageous microorganisms like fungi, bacteria, and algae. The germination percentage of nuts and seedling growth has been found to rise with the application of biofertilizers including *Azospirillum*, *Azotobacter*, and vesicular arbuscular mycorrhizal (VAM) fungi in the potting mixture (Sinish et al. 2005). The introduction of nitrogen-fixing, P-solubilizing, and other beneficial microbes improve soil fertility in cashew-growing areas (Kalaivanan and Rupa 2017). For cashew nurseries, the application of biofertilizer is a particularly effective way to manage the nutrients. A constraint in the cultivation of cashew is the inadequate establishment of seedlings and grafted plants. To ensure their survival and subsequent establishment in primary fields, high-quality grafted plants must be produced and established in nurseries. The use of biofertilizers, such as arbuscular mycorrhizal fungus (AMF), phosphate-solubilizing bacteria, *Azospirillum*, and *Pseudomonas fluorescens*, can help overcome these problems. All of them have the capacity to raise soil fertility in low fertile soils and boost plant nutrition with minerals.

According to Ibiremo (2010), the application of organic fertilizer, phosphate fertilizer, and AMF inoculation significantly improved the growth of cashew seedlings in Nigeria. However, the influence of single super phosphate was not as strong as rock phosphate in all the parameters recorded. Four months after cashew nuts were sown, Weber et al. (2004) reported that symbiotic plant associations with exotic fungi of the commercial product (MycoGold) and native fungi of the first community (*Glomus etunicatum*, *G. glomerulatum*, *Scutellospora* sp., and *Acaulospora foveata*) produced better plant growth. The development of cashew and the chemical characteristics of the soil were favorably impacted by the application of organic fertilizer supplemented with phosphate fertilizer and AMF inoculation. For the production of cashews, Nigerian Sokoto rock phosphate was proven to be a feasible substitute for single super phosphate (Ibiremo et al. 2005; Ibiremo 2010; Ibiremo et al. 2012). According to Shankarappa et al. (2017), inoculating consortia of three biofertilizers (*Azospirillum*, *P. fluorescens*, and AMF) enhanced the cashew graft's nutrient uptake and plant growth metrics. AMF can be used to increase the drought tolerance of cashew seedlings and boost their growth by increasing the availability of nutrients in the soil. Along with plants' natural defenses, consortia biofertilizers helped cashew seedlings exhibit less signs of water stress. Additionally, the AMF increased the amount of mineral nutrients available in the soil, which enhanced the growth of inoculated plants in both water-stressed and regularly watered environments (Assih et al. 2022). According to research done at the Cocoa Research Institute of Nigeria, Nduka et al. (2019) found that applying compost manure and biochar to the soil, either separately or in combination, had a major positive agronomic impact. Among the effects of biochar were improved nutrient uptake, enhanced vegetative growth indices, and changes to the physicochemical characteristics of the soil.

5.7 Integrated Nutrient Management

For long-term soil health, it is crucial to apply organic manures, biofertilizers, and inorganic fertilizers in combination. Enhancing the physical, chemical, and biological qualities of soil can be facilitated by employing an integrated strategy. It has been observed that applying organic amendments either alone or in conjunction with inorganic fertilizers for 5 years to cashew in laterite soils in India improves the biological characteristics and soil aggregate stability (Mangalassery et al. 2019a). The results of the study showed that adding organic nutrition sources considerably raised the percentage of macro-aggregates in soil. About 2.2 to 12.7% in the silt + clay size fraction, 14.6 to 37.0% in micro-aggregates, 18.4 to 51.7% in small macro-aggregates, and 17.7 to 50.9% in big macro-aggregates were the increases in SOM that resulted from the application of organic amendments. There was improvement in terms of soil microbial biomass carbon, soil microbial biomass nitrogen, and soil enzymes like dehydrogenase, acid phosphatase, and ß-glucosidase, with application of the organic amendments in the soil.

Opoku-Ameyaw and Appiah (2000) investigated the effects of cow dung treatment and sulfate of ammonia on the growth of young cashew seedlings among mature sheanut (*Vitellaria paradoxa*) plants in the northern Ghanaian savanna ecological zone. Cow dung applied @ 7 kg plant^{-1} had no effect on plant height, girth, or number of leaves plant^{-1}; in contrast, MOP and triple superphosphate applied @ 22 and 21 g plant^{-1}, together with ammonium sulfate applied @70 g plant^{-1}, increased these parameters. Research by Adewale et al. (2013) also showed that different combinations of soil nutrients had different effects on the genotype-specific growth and development trends of cashews.

In a greenhouse experiment carried out at the Cocoa Research Institute of Nigeria, Ibadan, Akanbi et al. (2013) found that instead of relying solely on expensive and limited chemical fertilizers, farmers could achieve optimal and sustainable growth performance of cashew seedlings by using inexpensive, easily accessible organic fertilizers like chicken manure and cocoa pod husk ash. In the greenhouse at the Cocoa Research Institute of Nigeria, Ibadan, Ipimorotimi and Akanbi (2015) examined the effects of using NPK (15:15:15) and four organic nutrient sources, namely, cow dung, chicken droppings, cocoa pod husk, and kola pod husk, on the growth performance of cashew seedlings. They discovered that while organic fertilizers derived from plants were inferior to those derived from animals, organic fertilizers were better than NPK.

Among the different organic manures evaluated for cashew (Vengurla-4) in Konkan region of Maharashtra, India, use of 50% N as FYM + biofertilizers consortium (200 g tree^{-1}) + rock phosphate produced the highest number of panicles m^{-2}. This was followed by the RDF + 10 kg FYM and in situ green manuring/green leaf manuring to meet 100% N by keeping litter and planting cowpea (Gajbhiye et al. 2016). With 10 kg FYM + RDF applied, the highest pooled yield of 4.8 kg tree^{-1} was achieved. Plant growth indices including trunk girth and canopy spread decreased as the rate of N delivery increased, but plant height increased (Patil et al. 2020). With 100% RDF level, however, the cashew nut output was higher, and the flowering laterals were at their maximum. Higher yields were obtained with 25% FYM + 75% urea (8.53 kg plant^{-1}) and 100% RDF through calcium ammonium nitrate (CAN), among other sources of N. 125% RDF applied through CAN was preferable to other sources of combination when it came to level and source interaction, followed by 125% RDF applied through vermicompost. For cashew variety Balbhadra, in Ranchi, India, the mean number of flowering laterals as well as mean annual and cumulative nut yield improved with combined application of 25% N as FYM, recycling organic residue in the cashew garden, in situ green manuring/green leaf manuring, and applying biofertilizers @ 200 g plant^{-1}. On the other hand, the combination of recycling organic residue and adding 20% cow dung slurry ranked second in terms of mean nut weight and mean apple weight (Jha et al. 2021). Apple weight, nut weight, kernel weight, shelling percentage, and quality aspects of cashew cultivar BPP-8 improved in Odisha, India, when the nutrients were managed using organic sources (e.g., 25% N as FYM), recycling of organic residue, in situ green manuring/green leaf manuring, and biofertilizer consortia (200 g plant^{-1}) (Pattnaik et al. 2021). In terms of recovery and growth parameters of grafted plants in the field, Mere et al. (2022) reported that the organo-mineral formulations $N_{40}P_{80}K_{72}$ + 2 kg of chicken droppings and $N_{40}P_{80}$ + 2 kg of poultry droppings applied as bottom manure gave superior outcomes. Following field planting, these fertilizer formulations gave the grafted cashew plants a survival rate of 85 and 80%, respectively. Additionally, it enhanced vigor, growth, and leaf production. For cashew cultivation in a Nigerian southern guinea savanna, site-specific conservation farming, a combination of organic and inorganic fertilizer application, and the application of calcite and dolomite at the proper dosages are recommended (Ofem et al. 2020).

6 Organic Farming in Cashew Plantation

Organic cashew farming is frequently regarded as one of the sustainable agricultural methods; when adopted properly, it can yield substantial long-term benefits. With the adoption of organic farming, the environmental pollution problem can be reduced. The fertility of the soil and the availability of nutrients are key determinants of crop yield. According to reports, nutrients are barely applied to 20% of the cashew cultivation area. Despite reports that cashew farms generate 1.38 to 5.20 t ha^{-1} of biomass from cashew leaf litter, they are not sufficiently recycled on the plantations (Bhat et al. 2016). Kumar and Hegde (1999) reported on the advantages of recycling nutrients using cashew leaf litter. During harvest season, the leaf litter is removed to make nut picking easier. The leaf litter is often burned or composted during the other months. Sometimes, the decomposed material is used for other crops, such as coconuts and arecanuts. Nutrient depletion in the soil would result from these cultural practices. There is increased awareness among the consumers on the health benefits of organically grown cashew. Concerns about the environment and government legislation are contributing to the rising cost of fertilizers. In India, cashew is grown as a rainfed crop on poorly fertile soils with little care. Therefore, by default, the majority of cashews produced are organic. The fundamental issue with the adoption of organic farming in cashew is that they are unable to produce the potential output, primarily because there are no organic pest management methods available. So, in endemic areas the yield loss can occur.

Reusing biomass produced in an established plantation can supply around half of the nutrients needed by cashews. Furthermore, it is recommended to apply 5–8 kg of castor cake, 30 kg of FYM, 10 kg of poultry manure, or 15 kg of vermicompost (depending on local availability) in addition to 50 g of biofertilizers (Mangalassery et al. 2019b). Green manure crops, like *Gliricidia maculata*, which produce 60 kg of green biomass per tree and three cuttings year, can be planted in the areas between cashew trees by farmers as an alternative to external manure application (Yadukumar et al. 2008).

For organic cultivation, mid- and late-season (February–April) types are advised because they fend off attacks from the main pest, the tea mosquito bug (Saroj 2015).

a b

Fig. 6 Some mid-season cashew varieties in India suitable for organic farming (a = Bhaskara, b = Dhana)

The mid- and late varieties' late blooming and fruiting will result in a reduced insect population in the field. However, given the recent climatic changes, it is also occasionally expected that there would be a quick outbreak of the pest at times. The cashew varieties Bhaskara, Dhana, Dhanashree, Amrutha, Priyanka, V-7, and BPP-8 are examples of mid-season varieties in India. Ullal-1, Chinthamani-1, and Madakkathara-2 are the late-season cultivars (Fig. 6).

7 Awareness on Soil Nutrient Management

The main obstacle to effective nutrient management in cashew is the growers' little attention to the crop. It is due to the notion that cashew, being hardy crop, can come up in any kind of soil (Poduval and Yadukumar 2011). Extension efforts are to be further strengthened to create awareness among the growers about the importance of nutrient management in soil health.

8 Information and Communication Technologies for Nutrient Management

Growers' lack of focus on cashew nutrition management is one of the several obstacles to cashew's potential yield realization. Applying the appropriate amount of fertilizer at the appropriate time is essential for prudent resource management, as well as for attaining the highest possible yield and benefit. Farmers may need to rely on scientists and extension agents for information on the proper dosages because of the great variation in field conditions, fertilizer availability, and choice. Farmers are unable to accurately identify the appropriate amount of fertilizer to be administered.

Farmers' resistance to implementing nutrient management techniques may be caused by a number of things, including a lack of resources, difficulty adopting the methods, and access to technical expertise (Daxini et al. 2018; Mahajan and Gupta 2009). Nearly 60% of farmers are thought to be unable to obtain information on cutting-edge technologies (Raj and Bhattacharjee 2014). Information and communication technologies (ICTs) are becoming more and more common as a means of giving farmers timely, accurate, and customized information. Smartphone use and reasonably priced Internet access have made it possible for farmers to use mobile apps to find solutions to such problems. Government and business organizations are developing mobile apps and software to reach farmers in rural areas more quickly and effectively in a variety of industries.

9 Conclusions

The ability of the soil to deliver the desired ecosystem services is indicated by its health. The health of the soil determines the viability of agriculture. Soils used for cashew cultivation often have low levels of N, P, K, and micronutrients. There is also widespread deficiency in copper, boron, and zinc micronutrients. Other problems with this soil are its low carbon content and poor water-holding capacity. Research shows that cashews need a steady and sufficient amount of nutrients to grow and thrive. But nutrient management aspects have not been given enough attention, which has a negative impact on the health of the soil. Under cashew plantations, maintaining and enhancing soil health necessitate careful planning from field selection to adoption of appropriate soil and water conservation techniques, nutrient recycling, and integrated application of inorganic, organic, and biofertilizer fertilizers. In cashew cultivation, implementing excellent agricultural techniques in accordance with scientific recommendations guarantees the right balance of nutrients in the soil and adequate nutrient supply to the crop. Micronutrient needs and supplementation must be taken into consideration in the nutrient management programs. Above all, preserving and enhancing the soil health in the cashew ecosystem depend heavily on stakeholder participation and awareness.

References

Abeysinghe DC (2009) Effect of intercropping of young cashew (*Anacardium occidentale* L.) on land productivity. In: Cashew Research Workshop. Wayamba University, Sri Lanka, 20 November 2009. p 23

Adejumo TO (2010) Effects of NPK fertilization on yield and inflorescence blight of cashew. J Agric Biotechnol Sustain Dev 2(5):66–70

Adewale BD, Ibiremo OS, Odoh NC, Adeyemi EA (2013) Genetic estimates and trend analysis of some growth parameters of cashew (*Anacardium occidentale* L.) as influenced by nine nutrient combinations. J Agric Biotechnol Sus Dev 5(1):6–11. https://doi.org/10.5897/JABSD12.035

Adiga JD, Kalaivanan D (2017) Cashew based cropping systems. In: Saroj PL (ed) Cashew: improvement, production and processing. Astral International Pvt Ltd, New Delhi, pp 277–293

Aikpokpodion PE, Uloko B, Edibo G (2009) Nutrient dynamics in soil and cashew (*Anacardium occidentale* L.) leaf and kernel in Kogi State, Nigeria. J Appl Biosci 25:1573–1578

Akanbi OSO, Famaye AO, Ojeniyi SO, Nduka BA, Taiwo N, Olasupo FO, Azeez OM, Edibo GO, Gbemisola DO (2013) Growth response of cashew seedlings to applied organic fertilizers of animal and plant origin on a degraded soil in Ibadan, South-West, Nigeria. Int Res J Agric Sci Soil Sci 3(3):93–98

Aliyu OM, Hammed LA (2008) Nigerian cashew economy: a review of nut production sector. In: Proceedings of the International Academy of African Business and Development (IAABD) Conference. May 20–24, 2008. University of Florida, Gainesville, USA

Al-Kaisi MM, Lal R, Olson KR, Lowery B (2017) Fundamentals and functions of soil environment. In: Soil Health and Intensification of Agroecosytems. Elsevier, London, pp 1–23

Alves FAL, Ponte LFA, Silva SLF, Maia JM, da Silveira JAG (2013) Germinação e estabelecimento de plântulas de cajueiro-anãoprecoce (*Anacardium occidentale* L.) emfunção da salinidade. Agrária - Revista Brasileira de CiênciasAgrárias 8(2):197–204. https://doi.org/10.5039/agraria.v8i2a2062

Assih A, Nenonene AY, Bokobana A (2022) Mycorrhizal inoculation effect on water deficit tolerance of cashew seedlings (*Anacardium occidentale* L.) and soil nutrients availability. Asian J Agric Hort Res 9(4):112–122

Babu V, Reddy MNN, Rajanna KM, Vidya M (2015) Performance of cashew to NPK fertilizers. International Society for Horticultural Science (ISHS), Leuven, pp 291–294. https://doi.org/10.17660/ActaHortic2015.1080.37

Barthélémy DL, Sibirina S, Dofoungo K, Paul AKD (2021) Diagnosis of the chemical fertility of soils under cashew groves (*Anacardium occidentale* L.) in the department of Korhogo in the north of Côte D'ivoire. Int J Innov Sci Res 53(1):50–60

Benson T, Mabiso A, Nankhuni F (2016) Detailed crop suitability maps and an agricultural zonation scheme for Malawi: Spatial information for agricultural planning purposes. International Food Policy Research Institute (IFPRI), East Lansing, MI. (Feed the Future Innovation Lab for Food Security Policy Research Paper). Report No. 17

Bezerra MA, Lacerda CF, Gomes Filho E, de Abreu CE, Prisco JT (2007) Physiology of cashew plants grown under adverse conditions. Brazilian J Plant Physiol 19:449–461

Bhat MG, Nagaraja KV, Rupa TR (2016) Cashew research in India. J Hort Sci 5(1):1–16

Carr MKV (2013) The water relations and irrigation requirements of cashew (Anacardium occidentale L.): A review. Exp Agric 50(1):24–39. https://doi.org/10.1017/S0014479713000392

CIAT (2011) Predicting the Impact of Climate Change on Cashew Growing Regions in Ghana and Côte d'Ivoire. International Center for Tropical Agriculture, Managua. 2011 Sep. (Decision and Policy Analyses Program)

Crisostomo LA, Rosseti AG, Pimentel CRM, Barreto PD, Lima RN (2004) Produtividade, atributosindustriais e avaliaçãoeconômica de castanhaemcajueiro-anãoprecoceadubado com doses crescentes de nitrogênio e potássio, emcultivo sob sequeiro. Rev Ciên Agron 35:87–95

Daxini A, O'Donoghue C, Ryan M, Buckley C, Barnes A, Daly K (2018) Which factors influence farmers' intentions to adopt nutrient management planning? J Environ Manag 224:350–360. https://doi.org/10.1016/j.jenvman.2018.07.059

DCCD (2022) Area and production of cashew 2022–23. Directorate of Cashewnut and Cocoa Development, Kochi

de la Rosa D, Sobral R (2008) Soil Quality and Methods for its Assessment. In: Braimoh AK, Vlek PLG (eds) Land Use and Soil Resources. Springer, Dordrecht, pp 167–200. https://doi.org/10.1007/978-1-4020-6778-5_9

El-Ramady HR, Alshaal T, Amer M, Domokos-Szabolcsy É, Elhawat N, Prokisch J, Fári M (2014) Soil quality and plant nutrition. In: Ozier-Lafontaine H, Lesueur-Jannoyer M (eds) Sustainable Agriculture Reviews 14. Springer, Cham, pp 345–447

Falade JA (1977) Cashew growing soils in Nigeria. East African J Agric 43(2):100–105

Falade JA (1984) Effect of lime on the efficiency of nitrogen fertilizers applied to cashew in acid sandy soil. J Plant Crops 12(2):140

FAO (1976) A Framework for Land Evaluation. Soils Bulletin No. 32. FAO, Rome

FAOSTAT (2023). Available at https://www.fao.org/faostat/en/#data/QCL. Accessed 01 Aug 2023
Fragoso H, Bezerra F, Melo F, Hernadez F (1999) Macronutrient removal by both cashew nut and apple of two dwarf cashew clones. Rev Bras Ciên Solo 23:603–608
Gajbhiye RC, Salvi SP, Pawar SN (2016) Effect of organic manures on growth and yield of cashew cv. Vengurla-4 under Konkan region of Maharashtra. Asian J Soil Sci 11(1):159–165. https://doi.org/10.15740/HAS/AJSS/11.1/159-165
George T, Veeraraghavan P, Rao D (1984) Studies on the leaf nutrient content of cashew (*Anacardium occidentale* L.) in relation to methods of fertilizer application. Indian Cashew J 16:11–13
Ghosh SN (1988) Effect of nitrogen, phosphorus and potassium on flowering duration, yield and shelling percentage of cashew (*Anacardium occidentale* L.). Indian Cashew J 19:19–23
Ghosh SN, Bose TK (1986) Nutritional requirement of cashew in laterite tract of West Bengal. Indian Cashew J 18(1):11–16
Grüter R, Trachsel T, Laube P, Jaisli I (2022) Expected global suitability of coffee, cashew and avocado due to climate change. PLoS One 17(1):e0261976. https://doi.org/10.1371/journal.pone.0261976
Guo M (2021) Soil Health Assessment and Management: Recent Development in Science and Practices. Soil Syst 5(4):61
Guruprasad T, Hanumanthappa M, Sudhir K, Dhananjaya B (2009) Technologies for improvement of senile cashew gardens. Cashew Cocoa J 1(3):18–22
Hammed LA, Laval BA, Kolapo KA (2011) Growth and nutrient uptake of cashew seedlings as affected by nut-size in the nursery. African J Agric Res 6(17):3962–3971. https://doi.org/10.5897/AJAR10.836
Harishu Kumar P, Sreedharan C (1988) Nut characters as influenced by different levels of NPK in cashew (*Anacardium occidentale* L.). Indian Cashew J 18(2):15–17
Ibiremo OS (2010) Effect of organic fertilizer fortified with phosphate fertilizers and arbuscular mycorrhizal fungi inoculation on the growth of cashew in two ecologies in Nigeria. J Agri Sci 1(2):101–107
Ibiremo O, Fagbola O, Ogumlade M, Iloyanomon C (2005) Performance of cashew seedlings as influenced by AM fungi inoculations and phosphate fertilizers. Niger J Hort Sci 10(1):47–51
Ibiremo O, Ogunlade M, Oyetunji O, Adewale B (2012) Dry matter yield and nutrient uptake of cashew seedlings as influenced by arbuscular mycorrhizal inoculation, organic and inorganic fertilizers in two soils in Nigeria. ARPN: J Agric Biol Sci 7(3):196–205
Idowu OJ, Sultana S, Darapuneni M, Beck L, Steiner R (2019) Short-term conservation tillage effects on corn silage yield and soil quality in an irrigated, arid agroecosystem. Agron 9(8):455. https://doi.org/10.3390/agronomy9080455. 17 pages
Ipimorotimi RR, Akanbi OSO (2015) Comparison of organic nutrients sources with NPK for cashew seedlings growth towards organic cashew farming in Nigeria. Int J Plant Soil Sci 4(2):185–195. https://doi.org/10.9734/IJPSS/2015/11581
Jena C, Panda PK, Karna AK (2018) Effect of growth promoting substances on flowering behaviour of cashew cv. BPP- 8 grown in the coastal region of Odisha. Int J Chem Stud 6(6):908–911
Jha PK, Sengupta S, Dash AP, Meher R (2021) Effect of organic manures, inorganic manures and biofertilizers on growth and yield of cashew cv. Balbhadra at Jharkhand region. Int J Chem Stud 9(2):508–511. https://doi.org/10.22271/chemi.2021.v9.i2h.11867
Kalaivanan D, Rupa TR (2017) Organic cultivation of cashew. In: Saroj PL (ed) Cashew: improvement, production and processing. Astral International Pvt Ltd, New Delhi, pp 295–322
Kalaivanan D, Adiga JD, Lakshmipathi (2017) Compost from cashew leaf litter and cashew apple waste. Kisan World A J Agric Rural Dev 44(11):48–50
Kangde L, Shibang L, Suisheng D (1998) Integrated Production Practices of Cashew in China. In: Papademetriou MK, Herath EM (eds) Integrated Production Practices of Cashew in Asia. Food and Agriculture Organization of The United Nations, Regional Office for Asia and the Pacific, Bangkok
Kramer PJ, Boyer JS (1995) Water Relations of Plants and Soils. Academic Press, San Diego, p 495

Kumar DP, Hegde M (1999) Recycling of plant nutrients through cashew leaf litter. The Cashew 13(April–June):2–8

Kumar PH, Nair BP, Rakhiappan P, Nagabhushanam S, Mohan E (1982) Variation in mineral composition of leaves of cashew (*Anacardium occidentale* L.) as affected by season, position and age. Indian Cashew J 14:7–10

Latha A, John PS, George M (1994) Productivity of cashew as influenced by cholorphyll and leaf nitrogen content. J Trop Agric 32(1):86–88

Lehmann J, Bossio DA, Kögel-Knabner I, Rillig MC (2020) The concept and future prospects of soil health. Nat Rev Earth Environ 1(10):544–553

Mahajan A, Gupta RD (2009) Constraints in the adoption of INM system. In: Mahajan A, Gupta RD (eds) Integrated nutrient management (INM) in a sustainable rice—wheat cropping system. Springer Netherlands, Dordrecht, pp 185–191. https://doi.org/10.1007/978-1-4020-9875-8_12

Mahajan GR, Das B, Manivannan S, Manjunath BL, Verma RR, Desai S, Kulkarni RM, Latare AM, Sale R, Murgaonkar D, Patel KP, Morajkar S, Desai A, Barnes N, Mulla H (2021) Soil and water conservation measures improve soil carbon sequestration and soil quality under cashews. Int J Sediment Res 36(2):190–206. https://doi.org/10.1016/j.ijsrc.2020.07.009

Mangalassery S (2018) Soil and water management strategies for increasing income from cashew orchards and conservation of natural resources. Paper presented at the International Conference on emerging issues in agricultural, environmental and applied sciences for sustainable development (EIAEASSD-218), Sam Higginbottom University of Agriculture, Technology and Sciences (SHUATS), Allahaad, UP, India, 27–29 November 2018

Mangalassery S, Kalaivanan D, Philip PS (2019a) Effect of inorganic fertilisers and organic amendments on soil aggregation and biochemical characteristics in a weathered tropical soil. Soil Tillage Res 187:144–151. https://doi.org/10.1016/j.still.2018.12.008

Mangalassery S, Nayak MG, Adiga JD, Preethi P, Muralidhara BM (2019b) Nutrient management in cashew., Technical Bulletin No. 1/2019. ICAR-Directorate of Cashew Research, Puttur. https://doi.org/10.13140/RG.2.2.13756.28806

Mangalassery S, Rejani R, Singh V, Adiga JD, Kalaivanan D, Rupa TR, Philip PS (2019c) Impact of different irrigation regimes under varied planting density on growth, yield and economic return of cashew (*Anacardium occidentale* L.). Irrig Sci 37(4):483–494. https://doi.org/10.1007/s00271-019-00625-7

Mangalassery S, Nayak MG, Preethi P, Mualidhara BM (2020a) Soil and water conservation and irrigation management in cashew., Technical Bulletin No. 31. ICAR-Directorate of Cashew Research, Puttur, Karnataka, p 34

Mangalassery S, Preethi P, Mog B, Nayak MG (2020b) Integrated nutrient management in cashew. In: Eradasappa E, Mog B, Savadi S, Mangalassery S (eds) Training manual on cashew production and post harvest technologies. ICAR-Directorate of Cashew Research, Puttur, Karnataka, pp 39–50

Mangalassery S, Nayak MG, Prabha Susan P, Rupa TR, Behera SK, Srinivasan V (2021) Delineating the nutrient constraints and developing nutrient norms for cashew (*Anacardium occidentale* L.) in coastal India. J Plant Nutr 44(17):2627–2639. https://doi.org/10.1080/01904167.2021.1921198

Mangalassery S, Adiga JD, Mog B, Savadi S, Vanitha K, Raviprasad TN, Rajasekhara H (2022a) Software and Mobile Apps on cashew developed under RKVY project. ICAR-Directorate of Cashew Research, Puttur

Mangalassery S, Adiga JD, Mog B, Thondaiman V (2022b) Nutrient management in cashew. In: Eradasappa E, Rao V, Chandrakumar A (eds) Cashew production and post harvest technologies. National Institute of Agricultural Extension Management (MANAGE), Hyderabad. ICAR-Directorate of Cashew Research, Puttur, pp 37–47

Mangalassery S, Adiga JD, Veena GL, Mog B, Manjesh GN, Thondaiman V, Bhagya HP, Manjunatha K (2023) Soil, water and nutrient management in cashew for enhancing productivity. In: Vanitha K, Rajashekara H, Raviprasad TN, Mangalassery S, Adiga JD (eds) Training manual on improved crop production technologies in cashew with major emphasis on integrated pest management. ICAR-Directorate of Cashew Research, Puttur, Karnataka, pp 65–78

Marques E, Freitas P, Alencar N, Prisco JT, Gomes-Filho E (2013) Increased Na+ and Cl− accumulation induced by NaCl salinity inhibits cotyledonary reserve mobilization and alters the source-sink relationship in establishing dwarf cashew seedlings. Acta Physiol Plant 35. https://doi.org/10.1007/s11738-013-1254-5

Mere SBJTO, Hermann MB, Djaha AJB, Amanoudo MJ, Tassiki A, Akounnou JDF (2022) Reactions of cashew grafted seedlings to different formulas of mineral and organic fertilizers as bottom dressing in plantation. Int J Biol Chem Sci 16(1):98–111. https://doi.org/10.4314/ijbcs.v16i1.9

Mohapatra AR, Vijayakumar K, Bhat NT (1973) A study on nutrient removal by the cashew tree. Indian Cashew J 9(2):19–20

Nannipieri P, Greco S, Ceccanti B (2017) Ecological significance of the biological activity in soil. Soil Biochem:293–356. https://doi.org/10.1201/9780203739389-7

Nayak MG, Mangalassery S, Preethi P, Muralidhara BM, Mog B (2020) Intercropping and cashew based cropping/farming system. ICAR-Directorate of Cashew Research, Puttur, Karnataka

Nduka BA, Ogunlade MO, Adeniyi DO, Oyewusi IK, Ugioro O, Mohammed I (2019) The influence of organic manure and biochar on cashew seedling performance, soil properties and status. Agric Sci 10:110–120. https://doi.org/10.4236/as.2019.101009

Ofem K, Abua SO, Umeugokwe CP, Ezeaku V, Akpan-Idiok AU (2020) Characterization and suitability evaluation of soils over sandstone for cashew (*Anacardium occidentale* L) production in a Nigerian southern Guinea savanna. Int J Sci Res Pub 10(7):353–368. https://doi.org/10.29322/IJSRP.10.07.2020.p10342

Opoku-Ameyaw K, Appiah MR (2000) Improving the growth of Cashew (*Anacardium occidentale*) seedlings interplanted into mature sheanut stands in Northern Ghana. Ghana J Agric Sci 33(2):159–164. https://doi.org/10.4314/gjas.v33i2.1865

Patil SV, Nagaraja G, Rajkumar GR, Lingaraju NN, Adiga JD (2020) Nitrogen management strategies to improve growth and yield of cashew. Int J Chem Stud 8(5):611–614. https://doi.org/10.22271/chemi.2020.v8.i5i.10360

Pattnaik RK, Panda PK, Jena C, Nayak S (2021) Influence of organic manures and bio-fertilizers on qualitative and qualitative traits of cashew cv. BPP-8 in Odisha condition. Chem Eng 2(4):37–41. https://doi.org/10.53709/CHE.2021.v02i04.00

Poduval M, Yadukumar N (2011) Effect of different doses of fertilizers on different densities of cashew plantation. J Plantn Crops 39(1):35–40

Raj S, Bhattacharjee S (2014) Mobile phone applications for agricultural extension in India. In: Raj S (ed) Mobile phone for agricultural extension: Worldwide mAgri innovations and promise for future. New India Publishing Agency, New Delhi, pp 1–74

Rajamanickam C, Rathinasamy A, Indirani R, Anbarasu M (2020) Effects of soil application of micronutrients in cashew (*Anacardium occidentale* L.). Int J Current Microbiol Appl Sci 9(3):2661–2666. https://doi.org/10.20546/ijcmas.2020.903.30

Rajendra BN, Ramachandra RK, Reddy AN, Honnabyraiah MK, Vishnuvardhana JSL, Reddy AB, Ramesh M, Subramanyam B (2023) Role of differential nutrient management on growth and yield attributes of cashew (*Anacardium occidentale* L.) in maidan tracts of Karnataka. Int J Bio-Res Stress Manage 14(2):237–242. https://doi.org/10.23910/1.2023.3312a

Rao MA, Satyanarayana RK, Rao RR (1984) Effect of NPK on nut yield of cashew nut. Cashew Causeric 6:3–4

Reddy AV, Narasimha Rao PV, Ankaiiah S, Subba Rao IV (1982) Cashew NPK nutrition in relation to growth under graded doses of nitrogen fertilization. Indian Cashew J 14(4):15–21

Rejani R, Rupa TR (2017) Soil and water conservation practices. In: Saroj PL (ed) Cashew: improvement, production and processing. Astral International Pvt Ltd, New Delhi, pp 171–185

Rejani R, Yadukumar N (2006) Effect of soil and water conservation techniques for cashew plantations grown on steep slopes. J Plantn Crops 34(3):304–308

Rejani R, Yadukumar N (2010) Soil and water conservation techniques in cashew grown along steep hill slopes. Scientia Hort 126(3):371–378. https://doi.org/10.1016/j.scienta.2010.07.032

Rejani R, Rupa TR, Nayak MG (2013) Suitability of cashew growing areas in India - an appraisal using GIS. J Agrometeorol 15:123–128

Richards NK (1992) Cashew tree nutrition related to biomass accumulation, nutrient composition and nutrient cycling in sandy red earths of Northern Territory, Australia. Scientia Hort 52(1–2):125–142. https://doi.org/10.1016/0304-4238(92)90015-5
Rupa TR (2017) Nutrient and water management. In: Saroj PL (ed) Cashew: improvement, production and processing. Astral International Pvt Ltd, New Delhi, pp 233–252
Rupa TR, Kalaivanan D, Reshma RP (2014) Nutrient content in the leaves of cashew (*Anacardium occidentale* L.) in relation to variety. J Plantn Crops 42(1):145–150
Salam MA, Peter KV (2009) Cashew-A monograph. Studium Press (India) Pvt. Ltd., New Delhi
Salam MA, John PJ, Joseph M, Poduwal M, Kumar P, Yadukumar N, Bhat MG (2008) Quantitative estimation of soil fertility and fertiliser recommendations (QUEFC) for cashew (*Anacardium occidentale* L.). J Plantn Crops 36(2):86–94
Salim M, Miller R, Ticona-Benavente C, Leeuwen J, Alfaia S (2018) Soil fertility management in indigenous homegardens of Central Amazonia, Brazil. Agrofor Syst 92. https://doi.org/10.1007/s10457-017-0105-6
Saroj PL (2015) Cashew based cropping systems. Advances in cashew production technology. ICAR-Directorate of Cashew Research, Puttur, Karnataka
Scholes RJ (2016) Climate change and ecosystem services. Wiley Interdiscip Rev Clim Chang 7(4):537–550
Sehgal J (1966) Pedology - concepts and applications. Kalyani Publishers, Ludhiana
Shahid MA, Sarkhosh A, Khan N, Balal RM, Ali S, Rossi L, Gómez C, Mattson N, Nasim W, Garcia-Sanchez F (2020) Insights into the Physiological and Biochemical Impacts of Salt Stress on Plant Growth and Development. Agronomy 10:938, 34 pages. https://doi.org/10.3390/agronomy10070938
Shamsudheen M, Dasog GS, Patil PL (2004) Evaluation of soil site suitability for cashew (*Anacardium occidentale*) in some selected plateau landscapes in the northern region of Karnataka. Paper presented at the National symposium on input use efficiency in agriculture – issues and strategies Vellanikkara, Thrissur, Kerala, India, 25–27 November 2004
Shankarappa TH, Mushrif SK, Subramanyam B, Sreenatha A, Prasad BNM, Reddy NA (2017) Effect of biofertilizers on growth and establishment of cashew grafts under nursery condition. Int J Curr Microbiol App Sci 6(8):1959–1965. https://doi.org/10.20546/ijcmas.2017.608.232
Silva JN, Oliveira JC, Martins ÊK, Nunes Júnior ES, Irineu TH, Chaves AP, Linhares AC (2020) Soil fertility evaluation of agricultural areas of cashew community of Catolé Rocha-PB. Bulg J Agric Sci 26(1):186–191
Sinish M, Mercy G, John P (2005) Organic methods for cashew root stock production. Cashew 19(1):8–15
Sutandi A, Iswandi A, Daras U, Hikmat M, Krisnohadi A (2014) Establishing land suitability criteria for cashew (*Anacardium occidentale* L.) in Indonesia. Appl Environ Soil Sci 2014:1
Tripathy P, Sethi K, Patnaik AK, Mukherjee SK (2015) Nutrient management in high density cashew plantation under coastal zones of Odisha. Int J Bio-resource and Stress Manage 6(1):93–97
Veeraraghavan PG, Celine VA, Balakrishnan S (1985) Study on the fertiliser requirements of cashew (*Anacardium occidentale* L.). Cashew Causeie 7(2):6–8
Venugopal K, Abdul Khader KB (1991) Effect of soil and climate on the productivity of cashew. Indian Cashew J 20(3):19–24
Viegas RA, Silveira JAG, Lima Junior AR, Queiroz JE, Fausto MJM (2001) Effects of NaCl-salinity on growth and inorganic solute accumulation of young cashew plants. Rev Bras Eng Agric Amb 5(2):14–20
Viegas RA, Silveira JAG, Silva LMM, Viégas PRA, Queiroz JE, Rocha IMA (2004) Reduçãoassimilatória de nitratoemcajueiroscultivadosemmeiosalinizado. Rev Bras Eng Agric Amb 8(2–3):189–195
Weber OB, de Souza CCM, Gondin DMF, Oliveira FNS, Crisostomo A, Caproni AL, Saggin OJ Jr (2004) Inoculation of arbuscular micorrhizal fungi and phosphate fertilization on dwarf cashew seedlings. Pesq Agropec Bras, Brasília 39(5):477–483

Ximenes CHM (1995) Adubação mineral de mudas de cajueiroanãoprecocecultivadasemdiferentessubstratos. Fortaleza, Universidade Federal do Ceará. M. Sc. dissertation p 82

Yadukumar N, Nandan SL (2005) Recycling organic wastes of cashew plantations by aerobic composting. J Plantn Crops 33:99–102

Yadukumar N, Rejani R, Nandan SL (2008) Studies on green manuring in high density cashew orchards. J Plantn Crops 36(3):265–269

Yadukumar N, Rejani R, Nandan SL (2012) Effect of organic and inorganic nutrition in cashew. J Plantn Crops 40(1):1–8

Dr. S. Mangalassery is currently working as Senior Scientist at ICAR-Directorate of Cashew Research, Puttur, India. His main research interests are on soil and ecosystem management, soil carbon sequestration, different tillage practices, soil biology, and abiotic stress tolerance. He published 44 research papers in national and international journals, edited 3 books, and contributed 28 book chapters. He was awarded International Research Excellence Scholarship from the University of Nottingham and the Lamming award of the University of Nottingham.

Dr. J. D. Adiga is the Director, ICAR-Directorate of Cashew Research, Puttur, Karnataka and a Doctorate in Horticulture. Also served as Professor of Fruit Science at College of Horticulture, University of Horticultural Sciences, Bagalkot, Karnataka. He has published more than 100 research papers in journals of repute, authored 2 books and several book chapters. He is the recipient of two awards and serves as member of the editorial committee of a journal.

Dr. G. L. Veena has a Doctorate in Horticulture and presently a scientist at ICAR-DCR, Puttur, Karnataka. Her main research interests are identification of nutritionally rich genotypes, germplasm conservation and utilization, and value addition of cashew apple. She has published 25 research articles in national international journals, 2 books, 15 book chapters, 20 popular articles and 6 e-publications.

Dr. N. K. Binitha has a doctorate in Soil Science and currently Assistant Professor, Soil Science and Agricultural Chemistry at the College of Agriculture, Kerala Agricultural University, Padannakkad, Kerala. She has 12 years of teaching and research experience. She was instrumental in formulating a customized fertilizer-nutrient stick for vegetable cultivation, and micronutrient solution – micro sol for banana.

Dr. K. S. Anil Kumar is a former Principal Scientist (Soil Science) from ICAR-National Bureau of Soil Science and Land Use Planning, Hebbal, Bangalore with an illustrious career of 34 years. He also served as the Mission Director, State Horticultural Mission-Kerala. He has published 130 research papers in peer reviewed journals and 40 books/chapters.

Natural Rubber (*Hevea brasiliensis* Muell Arg.)

Mercykutty Joseph

Abstract

The major commercial source of natural rubber is the rubber tree (*Hevea brasiliensis*). Cultivation of NR tree is mainly confined to the regions experiencing warm humid equatorial monsoon climate. Thailand, Indonesia, Vietnam, Malaysia and India are the major natural rubber-producing countries. The soils under NR cultivation in general are acidic in pH and low in nutrients especially Ca and Mg. The soil site suitability criteria for NR cultivation have been developed. The top 0–15 cm layer of the soils under NR cultivation is rich in OC through the turnover of large quantity of organic matter from cover crops, crop residues from the intercrops/cropping system and litter addition from annual leaf fall. Cultivation of NR with leguminous cover crops or carefully designed intercrops suiting the local food habits in the young phase help in the addition of large quantity of crop residues favouring the buildup of soil organic matter, proliferation of microbial community and improvement in soil fertility and soil health. Liming to manage soil acidity and supplementation of Ca and Zn and B recommendations are recent inclusions in the fertility management of NR growing soils in India. Biofertilizers, especially P solubilizers and mobilizers and plant growth-promoting rhizobacteria like *Pseudomonas* got immense potential for the internal mobilization of nutrients in NR growing soils. Carbon sequestration potential of NR plantations is an added advantage in the global warming scenario. Repeated cycles of NR cultivation in the same soil in the traditional belt of cultivation and expansion of NR cultivation to more and more marginal areas having low soil fertility, in addition to the climatic constraints and uncertainties, warrant careful management of the soil health for sustained production of NR.

M. Joseph (✉)
Rubber Research Institute of India (Retd), Rubber Board, Kottayam, Kerala, India

G. V. Thomas, V. Krishnakumar (eds.), *Soil Health Management for Plantation Crops*, https://doi.org/10.1007/978-981-97-0092-9_7

Keywords

Crop residue · Natural rubber (*Hevea brasiliensis*) · Nutrient availability · Soil fertility management · Soil health · Soil microbial community · Soil biodiversity

List of Abbreviations

AHP	Analytical hierarchy process
AMF	Arbuscular mycorrhizal fungi
Al	Aluminium
B	Boron
BS	Base saturation
BI	Bioinoculants
C	Carbon
Ca	Calcium
CEC	Cation exchange capacity
CFU	Colony-forming units
FYM	Farm yard manure
GIS	Geographic Information System
H	Hydrogen
IAA	Indole acetic acid
LGP	Length of growing period
MAFS	Mature rubber-based agroforestry system
MCDM	Multi-criteria decision-making
Mg	Magnesium
N	Nitrogen
NBSS and LUP	National Bureau of Soil Survey and Land Use Planning
NR	Natural rubber
OC	Organic carbon
P	Phosphorus
PGPR	Plant growth-promoting rhizobacteria
PWAC	Percent water available capacity
RubSIS	Rubber Soil Information System
SMU	Soil management unit
SOC	Soil organic carbon
YAFS	Young rubber-based agroforestry system
Zn	Zinc

1 Introduction

Natural rubber (NR) tree (*Hevea brasiliensis*), a native of the rain forests of the Amazon basin of South America, is predominantly grown in the tropics, and the latex harvested from the tree is the major commercial source of natural rubber. In the world over, the major NR-producing regions lie between 15° north and 15° south (Thomas and Panikkar 2000). In a more recent review, Verheye (2010) reported that NR cultivation is presently expanded between 25° north (Yunnan highlands, China) and 21° south in Brazil. However, the main production zone worldwide is concentrated between 15° N and 15° S. Malaysia, Indonesia, Thailand, Sri Lanka, India, Cambodia, Vietnam, the Philippines, Papua New Guinea and Southern China are the NR-producing countries in Southeast Asia and the South Pacific region.

Large plantations are present in Central and West Africa (Congo, Cameroon, Cote d'Ivoire, Liberia). In tropical America NR plantations are concentrated over a small area between 10° N and 10° S. Natural rubber plantations have expanded faster than all other tree crops in Southeast Asia (Fox and Castella 2013), with a 1.8-fold increase in surface area over the last three decades (FAO 2012). Thailand, Indonesia, Malaysia, Vietnam and India account for >90% of the 11.5 million ha of rubber plantations worldwide (FAOSTAT 2016).

Rubber has a key role in numerous engineering and industrial applications. As an essential industrial raw material for manufacturing over 55,000 varieties of products, from balloons to giant-size tyres, NR has a direct impact on the quality of life of modern man. Various forms of rubbers have critical and strategic application in almost every aspect of human life including transportation, health and hygiene, construction, communication, aerospace, etc.

2 General Soil Characteristics and Requirements

Soils under NR cultivation, in general, were developed under warm humid equatorial monsoon or tropical wet and dry monsoon climates with variable duration of dry season extending from 3 to 5 months which favour high degree of laterization. In the humid tropical conditions with excessive rainfall and high temperature, ferrallitization is the dominant process responsible for soil formation involving the net loss of silica, the formation of kaolinite and relative accumulation of Fe and Al in soil (Bridges 1979; Eswaran and Bin 1978a, b; NBSS and LUP 1999).

In most of these soils, acidity is developed due to the high levels of exchangeable Al^{3+} and the loss of basic cations due to erosion, leaching and crop removal. The average value of exchangeable Al in the surface soil (0–25 cm depth) ranged from 0.36 to 0.9 cmol (+) kg^{-1}, and the highest concentration of 0.58–2.23 cmol (+) kg^{-1} was recorded at a depth of 50–75 cm (Suresh et al. 1994). Details on the acidity parameters of the major soil series of the traditional NR growing tract of India are reported (Syamala 2006). Joseph (2011) reported wide variation in the base saturation percent and exchangeable bases in relation to the variation in average annual rainfall. Soils of the high rainfall areas showed very low base saturation with extremely low status of K, Ca and Mg.

NR growing soils of South India are highly weathered acid soils and are subjected to repeated cycles of NR cultivation. These soils have medium to high organic matter status (Karthikakuttyamma et al. 2000; NBSS and LUP 1999). Indications of P buildup in older plantations were reported by NBSS and LUP (1999). Ulaganathan et al. (2010, 2012) reported imbalance in nutrient availability and declining soil fertility in the replanting fields from the traditional regions of NR cultivation in India. Mandal et al. (2010, 2012, 2013) reported the same results from the nontraditional regions of NR cultivation in India. Detailed description of the characteristics of the NR growing soils of Sri Lanka (Dissanayake 2001) and Malaysia (Webster and Baulkwil 1989) and India (Battacharya et al. 1996; Karthikakuttyamma et al. 2000; NBSS and LUP 1999) have been reported. An overview on the rubber-growing soils of India is available (Joseph 2016).

The NR tree can grow on the vast majority of the acid soils of the humid tropics. However, deep well-drained soils of pH below 6.5, free from underlying sheet rocks, are well suited for its performance. A minimum soil depth of 100 cm is essential for effective root growth and establishment of the plantation (Karthikakuttyamma et al. 2000; Webster and Baulkwil 1989). Soils of medium texture are best suited as per Pushpadas and Karthikakuttyamma (1980). Deficiencies of nutrients are not a major limitation for NR cultivation in these soils, and even if there is any deficiency, it can easily be corrected by application of appropriate fertilizers.

Sys (1975) reported the land capability classification system based on FAO principles for the NR growing areas of Malaysia. Principal criteria in grading of soil suitability classes for NR as desirable range and, to minor limitations, severe limitations, very severe limitations based on the degree of severity of limitations, were provided by Rubber Research Institute of Malaysia (1977). According to Saengruksawong et al. (2012), in Thailand only 102,000 ha of NR plantation has soil characteristics suitable for NR plantation. Chambon et al. (2016) reported that in Thailand, the majority of NR expansion has taken place on poor soils with low fertility. As per Blagodatsky et al. (2016), the rapid expansion of area under NR plantations in Thailand could partly be due to the ability of NR trees to grow on a wide range of soil types and pedoclimatic conditions, ranging from optimal tropical lowland to suboptimal environments such as low-fertility areas with distinct dry seasons or steep slope.

Soil site suitability conditions for NR cultivation was evaluated by selecting six major NR growing areas supporting excellent, moderate and marginal growth of NR in the main rubber-growing regions in South India and identified the major constraints for NR production, and the parameters influencing the yield of NR were found to be the period of moisture availability, (length of growing period) followed by soil depth, percent water available capacity (PWAC), slope, winter temperature and excess rain (Kharache et al. 1995). Based on the FAO guidelines for land evaluation (FAO 1976), the soil site suitability criteria for NR were assessed and reported by NBSS and LUP (1999). Land quality criteria, oxygen availability to roots (soil drainage), nutrient availability (texture, pH, CEC and BS), rooting conditions

Table 1 Soil site suitability criteria for natural rubber

Climate	
Rainfall	An evenly distributed rainfall of 2000–3000 mm in a year
Temperature	Warm tropical temperature of about 21–35 °C without wide variation
Land quality	
Effective soil depth	>100 cm
Oxygen availability to roots—soil drainage	Well-drained
Depth to water table	>3 m
Erosion hazard—slope	A slope of 10–30% is ideal
Soil	
Texture	Medium texture—sandy clay loam and loam
Coarse fragments (vol. %)	<35
pH	Range of 4.0–6.5
Cation exchange capacity (cmol(+) kg^{-1} soil)	>4
Base saturation (%)	<30

Adapted from NBSS and LUP (1999)

(effective soil depth, coarse fragments-loamy soils, clay content) and erosion hazard ratings for highly suitable, moderately suitable, marginally suitable and not suitable areas were provided (Table 1). Sandy clay loam or loam texture, CEC >4.0 cmol (P+) kg^{-1} and base saturation (%) <30 are the highly suitable nutrient availability conditions for NR. Depth of the soil is an important parameter influencing growth and yield of NR. Tap root development is restricted in shallow soils thus affecting proper anchorage of trees. Deep soils, having high clay content, can hold more moisture and are helpful under drought situations. The mapping units as per NBSS and LUP (1999) were regrouped according to soil depth, gravelliness and organic matter content and spatially merged into soil management units (SMUs) through categorical generalization in GIS. The soil management units ranged from SMU 1 with no limitation to SMU 7 with major limitations. The SMUs 1–4 are highly suitable for NR cultivation, with a few constraints. Soil management units (SMUs) 5 and 6 with root limiting layer within 100 cm of the surface with high proportion of gravel in the solum are only moderately suitable for NR cultivation. The SMU 7 is unsuitable for NR plantation as the soil depth is <50 cm (Naidu et al. 2008, 2014). Differences in productivity between locations influenced by edaphic factors were reported (Meenakumari et al. 2011; Meti et al. 2014).

3 Indicators of Soil Health and Soil Health Assessment

The vegetation cover and cultural practices with nearly zero tillage in the mature stage are conducive for accumulation of organic matter and buildup of SOC in NR plantations. Cultivation of leguminous cover crops in the initial years (Kothandaraman et al. 1989; Philip et al. 2003; Watson 1957, 1989) and litter addition through annual leaf fall (Varghese et al. 2001) favours high levels of organic residue accumulation in the NR growing soils. Cultural practices adopted in NR plantations with nearly zero tillage favour stabilization or even accumulation of organic matter (Krishnakumar and Potty 1992). Compared to jhumming/shifting cultivation, cultivation of NR for a period of 10 years increased the SOC content by 0.6% (Krishnakumar et al. 1991). Nutritional self-sufficiency of the well-maintained mature NR ecosystem in relation to latex yield and soil properties was reported (George and Joseph 2011; Sivanadyan et al. 1995).

4 Soil Fertility Constraints

Karthikakuttyamma (1997) found that continuous cultivation of NR for three cycles resulted in significant reduction in soil pH and the base reserves of the soil especially Ca, Mg and K and an increase in total Fe, Al and sesquioxides in the 0–60 cm layer of the soil. In one cultivation cycle, the highest depletion was noticed for Ca followed by Mg and continuous cultivation for over 70–80 years; about 2900 kg of Ca was lost from 0–60 cm layer of the soil. Continuous cultivation of NR resulted in reduction of total and available N and K and improvement in the P content of the soil (Karthikakuttyamma et al. 1998). Decline in carbon management indices in NR plantations in comparison to virgin forest was reported by Abraham (2015) in the traditional belt of NR cultivation in India. Based on the fertility capability classification system for the NR growing soils of South India, Sanchez et al. (1982) reported three categories, namely, type (top soil texture), substrata type (subsoil texture) and modifiers (fertility constraints), like, low CEC, pH <4.0 or >5.5, high P fixation, low K reserve, base saturation 35% or greater within 60 cm of the soil surface, low organic carbon and high gravel content (NBSS and LUP 1999).

Even though NR is reported to tolerate a pH range of 3.8–7.0, the extremes could affect its growth and productivity. In soils with pH <4.0, the toxicity or deficiency of mineral elements often limits crop production rather than the low pH per se. NR is grown in soils with CEC ranging from 3.55 to 18.02 cmol (+) kg^{-1} soil (Krishnakumar 1989; Krishnakumar and Potty 1992; NBSS and LUP 1999). Exchangeable bases especially in the upper horizons were low in the soils under NR cultivation because of leaching of appreciable quantities of cations, especially Ca, Mg and K down the profile due to high rainfall. Growth of young NR plants was reduced under extremely acidic soil pH, which may be due to the combined effect of extreme acidity and high Al^{3+} and H^+ ions and low availability of nutrients (Ambily and Joseph 2018). Available Ca and Mg status of the geo-referenced soil samples from the NR growing regions of South India were mapped geo-statistically using kriging interpolation technique and reported widespread deficiency of these secondary nutrients in the NR growing soils (Prasannakumari et al. 2020).

Widespread deficiency of micronutrients has not been reported from NR plantations in India or other major NR growing countries so far. But of late, Zn and B deficiency symptoms are noticed in replanting fields especially during summer months. The most common symptom of B deficiency noticed in young plants is

Fig. 1 (**a**) Boron deficiency symptoms. (**b**) Zinc deficiency symptoms

shortening of internodes and the bottle brush appearance of the leaf whorls, especially the middle whorls (Fig. 1a). In many instances, these symptoms are transient and disappear with the spread of root system and moisture availability. Zinc deficiency is characterized by the typical sickle leaf symptom wherein the leaf laminae are elongated with narrow width and sometimes curling of leaf laminae (Fig. 1b).

These symptoms are indicative of the low status/chances of hidden hunger. Low available Zn status of the soils in the traditional belt of NR cultivation in India was reported (Joseph et al. 1995; Joseph and Sudhakumari 2013; NBSS and LUP 1999). Very high spatial variability of plant available micronutrients such as Zn and B in the traditional NR growing regions and Konkan region of India was reported by Philip et al. (2020). Available Zn status ranged from 0.01 to 10.91 mg kg^{-1} and available B from 0.01 to 2.48 mg kg^{-1} (Philip et al. 2020).

5 Impact of Climate Change on Plantation Soils (Environmental Impact)

Equatorial monsoon type of climatic conditions characterized by mean monthly temperature of 25–28 °C and well-distributed rainfall with no marked dry period (Bradshaw 1977; Strahler 1969) are the climatic conditions at the natural habitat of NR. NR trees perform best in climates closely resembling that of its original habitat under commercial cultivation, For the optimum growth and production of NR trees, the conducive climatic conditions are rainfall of 2000 mm or more, evenly distributed without any marked dry season and with 125–150 rainy days per annum; maximum temperature of about 29–34 °C and minimum of about 20 °C or more with a monthly mean of 25–28 °C; high atmospheric humidity of the order of 80% with moderate wind; and bright sunshine amounting to about 2000 h $year^{-1}$@ 6 h day^{-1}

throughout all the months (Rao and Vijayakumar 1992; Webster and Baulkwill 1989). NR, though genetically adapted to tropical environment, has proved to thrive and commercially viable in subtropical environments of Northeast India and China where the mean minimum temperature from December to February is around 10 °C. A detailed review on agroclimatic requirement of NR is presented by Vijayakumar et al. (2000).

Global warming and extreme weather events such as unprecedented drought, flooding, landslides, cyclones, etc. will negatively affect the production of NR globally. Significant deviations in the distribution of annual rainfall in many parts of the world were reported due to the impact of global warming (Pendergrass et al. 2017). Occurrence of high-intensity rainfall has been on the rise leading to flash floods and landslides in the different Western Ghats states of India where NR is a major crop. Atmospheric temperature has been steadily going up in almost the entire NR belt along these states (Raj et al. 2011; Satheesh and Jacob 2011). Increase in the evaporative demand of the atmosphere and decrease in the water-holding capacity of the soil over the years were reported (Scheff and Frierson 2014).

NR cultivation may shift to more and more nontraditional regions due to global warming. Liu et al. (2015) using the maximum entropy model and climate scenario predicted that the land suitability for NR cultivation in China will increase by 2080. Future distribution of NR for Northeast India based on climate and socio-economic factors was predicted using Maxent models. The land areas suitable for NR cultivation will further increase in Northeast India under the projected climate scenario in 2050 (Ray et al. 2014, 2016). Based on the climate data, the land suitability for NR cultivation in Malaysia was analysed and projected for 2040, 2070 and 2100. Changing climates will become favourable for NR cultivation in Malaysia in the future (Hazir et al. 2018, 2020). Potential shift of NR cultivation as a consequence of climate change in the Greater Mekong Subregion areas, vulnerable to excessive heat levels, has been mapped and reported that more locations will be suitable for NR cultivation by 2030 (Golbon et al. 2018). Correlating the spatial and temporal increase of evapotranspiration with the expansion of NR plantations from 2000 to 2017 in Southwest China, Ling et al. (2022) revealed that evapotranspiration was increased due to the expansion of NR plantation and climate change. NR farmers in the upper Mekong region will be benefitted by adopting environmentally friendly operations and agricultural practices in NR plantations due to the increase in temperature in the region as per the survey conducted among 611 NR farmers (Min et al. 2020). Based on multi-criteria decision-making (MCDM) and analytical hierarchy process (AHP), a composite climate change vulnerability model has been developed which could be used as a geographic information system (GIS)-based scientific framework to assess the vulnerability of NR plantations to climate change impacts for NR-producing countries worldwide (Pradeep et al. 2022). There is no specific literature on the effect of climate change on the health of soils under NR cultivation. However, the shift in NR cultivation regions or expansion to new areas suggests the need for more in-depth studies on the soil suitability assessments and development of appropriate soil health management practices for achieving sustained high productivity from the NR plantations.

6 Fertilizers and Plant Protection Chemicals on Soil Health and Soil Microbiology

Pollution possibilities are limited in NR plantations. Even with annual application of Cu-based fungicides for the control of diseases, mainly spraying for abnormal leaf fall disease, accumulation of Cu was not recorded in soils under NR plantations (Prasannakumari et al. 2014). NR processing utilizes large quantity of water which when discharged as effluent is loaded with rubber particles, carbohydrates, proteins, lipids, organic and inorganic salts and processing chemicals are potential pollutants. The quantity of effluent discharged is approximately 20 l kg^{-1} of rubber but varies with the processing conditions. Treatment systems using anaerobic technology have evolved over the years, and details can be obtained from Mathew et al. (2000).

7 Soil Biodiversity, Microbial Synergy and Function Specific Microbial Communities

Establishment of NR monoculture plantations by converting primary or secondary forests reduced the soil macrofauna diversity (Gilot et al. 1995; Lavelle et al. 2014) and soil microbial activities (Abraham and Chudek 2008; Gilot et al. 1995) and modified the soil microbial community and structure (Krashevska et al. 2015; Schneider et al. 2015) and the soil prokaryotic population (Schneider et al. 2015). Occurrence of mycorrhizal spores in NR growing soils and colonization in roots of mature NR plants from different locations were studied and reported (Joseph 1997) that the spore population ranged from 214 to 428 in 50 g of soil. The common genera of AM fungi reported from NR growing soils include *Glomus* spp., *Gigaspora* spp., *Acaulospora* spp. and *Sclerocystis* spp. The frequency of occurrence was in the order *Glomus* spp. >*Acaulospora* spp. >*Sclerocystis* spp. Only a few spores of *Gigaspora* spp. were recorded (Joseph 1997; Joseph et al. 2002). The arbuscular mycorrhizal infection in roots ranged from 67 to 92% (Jessy 2004; Jessy et al. 2007; Joseph 1997; Joseph et al. 2002). On inoculation, AMF colonization in the roots ranged from 70 to 80% in the control as well as plants treated with AMF, indicating the abundance of native AMF in the soil (Joseph et al. 2015). Studies on the impact of conversion of fields under cassava cultivation to NR plantation on the AMF communities reported a modification in their composition which was attributed to a change in soil texture and nutrient contents in NR plantations (Herrmann et al. 2016b). The microbial composition or diversity was studied by a range of different methods with varying resolutions (Herrmann et al. 2016b; Krashevska et al. 2015; Schneider et al. 2015), and the macrofaunal community was studied using morphological techniques (Gilot et al. 1995; Lavelle et al. 2014). Soil microbiota as decomposers and nutrients transformers and soil macrofauna as ecosystem engineers contribute to key functions such as carbon transformation, soil structure maintenance and nutrient cycling (Lavelle et al. 2006) were directly affected by land-use conversion or age of NR plantation. The AMF colonization and the number of phosphobacteria harboured were more in the roots of NR in the intercropping system compared to NR grown as monocrop (Vimalakumari et al. 2001). The presence of diverse crops in the NR-based cropping system harboured more earthworms as indicated by higher number of earthworm casts (Jessy et al. 2015).

Peerawat et al. (2018) studied the soil biodiversity (bacterial, fungal and macrofaunal), microbial activities and pedoclimatic characteristics over a chronosequence of 1- to 25-year-old NR plantation and reported that conversion from cassava to young NR plantation (1–3 years) had a significant effect on microbial biomass and activities and fungal composition, but did not impact the bacterial and macrofaunal diversity. Most of the biotic parameters, composition, abundance and activities were significantly modified after 7 years of NR cultivation indicating that canopy closure is the major driver for the shift in soil biota composition. In older NR plantations, dominance of *Trichoderma* (fungi), *Firmicutes* (bacteria) and earthworms was recorded. NR plantations that are 23- to 25-year-old harboured the highest microbial and macrofaunal biomass. However, in old NR plantations, a significant decrease in bacterial population was noticed. Litter accumulation, increase in soil moisture content and buildup of OC content over a period of 25 years of NR plantation were found to be the major factors for soil biota proliferation rather than the change in land use.

8 Fertility Management for Healthy Soils: Fertilizer Recommendations

Soil fertility management through regular application of fertilizers is very essential to achieve optimum growth and productivity of NR plantations and to maintain the sustainability of the production system. General fertilizer recommendation of NR for India (Karthikakuttyamma et al. 2000), Malaysia (Webster and Baulkwil 1989) and Sri Lanka (Samarappuli 2001) is available.

In the immature phase of NR, proper manuring is essential for good growth and trees becoming ready for early tapping. Good management in the immature phase of NR improves the soil fertility status and reduces the fertilizer requirement in the mature phase especially in the first 2 or 3 years of tapping (Punnoose et al. 1976). However, the response to applied fertilizers in mature trees through yield increase is little or difficult to establish and differs widely with the nature of the soil and agro-management practices (Pushpadas et al. 1978), and inconsistent or no response was reported (Jessy et al. 2004; Joseph et al. 1998; Punnoose 1993). Lack of response to fertilizer P in NR, especially in the mature phase, was reported by many researchers (Jessy 2004; Jessy et al. 2004). In mature NR plantations, site-specific balanced fertilizer application on the basis of analysis of soil and leaf samples and assessment of soil fertility and plant nutrition status is scientific and economical (Karthikakuttyamma et al. 2000; Pushparajah 1977). Nutritional self-sustainability of the mature NR system and dependence of the response to fertilizers on the fertility of the soil were reported by George and Joseph (2011). According to them by adequate manuring during the immature phase as well as establishing leguminous cover crop, skipping of fertilizers for a short period is possible in well-maintained mature NR plantations grown under favourable terrain. The Rubber Soil Information

System (RubSIS) developed by Rubber Research Institute of India based on georeferenced data base of soil fertility parameters is a decision-making tool for making appropriate fertilizer schedule for NR plantations (Pradeep et al. 2019). Vrignon-Brenas et al. (2019) suggested nutrient budget method for improving the nutrient use efficiency and sustainability of NR plantations.

Increasing soil acidity and poor status of Ca and Mg in the NR growing soils as reported by NBSS and LUP (1999) lead to the conduct of field experiments on soil acidity management through liming. Liming significantly increased the dry matter production and nodulation of *Pueraria phaseoloides* and had a positive effect on the uptake of N, P, Ca and Mg (Philip et al. 1995). Significant improvement in pH and availability of Ca in the soil and reduction in exchangeable Al content in the soil were reported by Syamala et al. (2003). Experiments in NR nursery indicated that application of dolomite or quick lime in combination with fertilizer significantly improved the growth of seedlings and the availability of P and K status of the soil. However, improvement in pH achieved was temporary and reverted back to original after 1 year indicating the strong buffering capacity of the soil (Syamala and Joseph 2006). Field experiment conducted in young NR plantation for 7 years from planting indicated that liming at 75 or 100% of the lime requirement along with fertilizer was significantly superior in improving the growth of plants over control during the sixth and seventh year of experimentation and improved the exchangeable Ca status of the soil and changed the pH of the soil from very strongly acidic to moderately acidic. Similarly, exchangeable Al content in the soil was significantly reduced by lime alone or lime at 75 or 100% lime requirement along with fertilizer treatments (Joseph et al. 2009).

In the tapping phase, application of Mg along with NPK significantly improved the yield over NPK alone application. Similarly, Ca supplied through shell lime significantly improved the growth of plants. Lime application either through shell lime or dolomite significantly increased the pH of the soil and improved the availability of P, K and Ca in the soil. For a crop like NR which is more adapted to the acidic soil environment, the liming program is for the supply of Ca and Mg to maintain the base status of the soil and for a temporary change in pH which will improve the availability of P and K and the microbial activity in the soil (Joseph et al. 2008). Application of Zn @5.0 kg ha^{-1} or above through either zinc sulphate or zinc oxide increased the availability of Zn in the soil and improved the growth of NR plants (Joseph et al. 2007).

Recently, recommendations on micronutrients, Zn and B, were included in the fertilizer schedule for NR in India. For young NR plants in the initial 4 years, for areas low in available Zn (<1.00 μg^{-g} soil of 0.1 MHCL extractable Zn), single application of zinc sulphate @50 g $plant^{-1}$ is recommended. Similarly, for areas low in available B (<0.5 μg^{-g} soil of hot water extractable boron), single application of borax @ 20 g $plant^{-1}$ is recommended. In general, one-time application during the immature phase is recommended. In sloping areas (slope >20%), a second dose of zinc sulphate @25 g $plant^{-1}$ is recommended. For mature phase, in Zn-deficient areas, one-time application of zinc sulphate @25 kg ha^{-1} is recommended. Similarly, one-time application of borax @10 kg ha^{-1} is recommended for deficient soils.

Recommendation of lime to correct soil pH and to supplement Ca is also included recently in the fertilizer schedule in India. For areas having soil pH <5.0 and low in available Ca, powdered shell lime @ 250 g plant^{-1} or dolomite @ 500 g plant^{-1} at the time of pit filling or 2–3 weeks before planting, and thoroughly mixing with soil is recommended. In mature NR plantations with soil pH <5.0 (very strongly or extremely acidic soil and low in available Ca status), application of lime @ 1.0–1.5 kg tree^{-1} at an interval of 3 years, 2–3 weeks before application of the first dose of fertilizer, is recommended. Lime may be applied by broadcasting in between the two rows of trees.

9 Technological Options to Promote Soil Health

9.1 Intercropping and Cropping System Research

Extensive conversion of traditional land-use systems to monoculture NR plantations as a consequence of smallholder NR expansion led to profound impacts on biodiversity, soil health and sustainability. Natural rubber-based agroforestry systems are considered the best way to improve soil properties and the environmental quality of NR monoculture. By altering the planting density, several intercrops can be included in the NR plantations as an alternative to the monoculture plantations (Fig. 2), and reports from the different NR growing countries are available (Jessy et al. 1998, 2005, 2013a; Pathiratna and Perera 2006; Rodrigo et al. 2004; Wibawa et al. 2006; Xianhai et al. 2012). The planting system generally adopted in NR cultivation in India (6.79 × 3.4 m) allows intercropping with a variety of crops during the initial 3 years, where radiation availability is around 97%, 76% and 43% of the open during first, second and third year, respectively, and it gradually decreases to less than 10% by the seventh year (Joseph 1999). Banana, pineapple, vegetables, cassava, tuber crops, ginger and turmeric are common intercrops in young NR plantations in India (Datta et al. 2019; Jessy et al. 1998) (Figs. 3 and 4).

Intercropping is not commonly practiced in the mature phase due to the limited light availability and also due to concerns of negative effect on the performance of

Fig. 2 (**a**) young rubber plantation with cover crop (**b**) mature rubber plantation

Fig. 3 Intercropping in young rubber with (**a**) banana and (**b**) vegetables and banana

Fig. 4 Intercropping in young rubber with (**a**) elephant foot yam, pineapple and banana, (**b**) *Colocasia* and banana

NR. NR is seen grown in mixed stands with timber trees, fruit trees and different annual crops in Indonesia (Wibawa et al. 2006).

In Central America, coffee-based agroforestry systems are more common (Muschler and Bonnemann 1997) and rather than maximizing coffee production, farmers are focusing on a wider spectrum of ecosystem services, including cultivation of shade trees (van Oijen et al. 2010). Intercropping coffee with rubber improved the microclimate by lowering the irradiance level, reducing the temperature and increasing the relative humidity. Coffee yield was not negatively influenced by the proximity to NR trees (Partelli et al. 2014).

According to Jessy et al. (2016), diverse crops can be integrated with NR without any adverse effect on growth and yield of NR. Crop diversification generated additional income, mitigated drought and sustained soil fertility (Jessy et al. 2016). The growth of NR was significantly higher under mixed planting with coffee, vanilla on *Gliricidia* standards, *Garcinia* and nutmeg along with NR without altering the planting design and density of NR trees. Yield of NR was not influenced by the mixed planting. There was higher soil moisture status and soil microbial population in the mixed planting system compared to monoculture NR. Yield of all the intercrops was good during initial years and with the canopy closure and increased shade from NR plants, *Garcinia* perished and vanilla and coffee continued to yield. In a separate experiment on evaluation of the performance of nine shade-tolerant medicinal plants in mature NR plantation, all the plants established well and produced reasonable biomass, but among the nine, *Strobilanthes cuspida* and *Alpinia calcarata* performed better (Jessy et al. 2016). In another study, the feasibility of cultivation of short-duration vegetables like *Amaranthus* and salad cucumber during the annual leaf shedding period in mature NR plantations to meet part of the household requirement was reported by Jessy et al. (2016).

Maintenance of soil fertility status and buildup of soil available P through intercropping was reported by George and Meti (2018) and Jessy et al. (1998). Improvement in the major and micronutrient status of the soil through intercropping with perennial crops was recorded (Jessy et al. 2016), and the probable reason attributed by the authors was the addition of farm yard manure to intercrops. Intercropping with long-duration crops in NR plantations did not adversely affect the growth or yield of NR and in most cases improved the growth of NR plants. In addition to that, intercropping reduced the weed growth, conserved soil moisture and sustained or improved the soil fertility status (Jessy and Jacob 2020). Improvement in the total C, OC and SOC stock in young NR plantations intercropped with banana compared to the monoculture planting was reported by Joseph et al. (2022).

In the NR-based agroforestry system, significant increase in total SOC and N contents and macroaggregates compared to the NR monoculture treatment was recorded except in *H. brasiliensis-C. arabica* system. SOC and N contents in all water-stable aggregate fractions were significantly higher in NR-based agroforestry systems except *Coffea arabica* agroforestry system compared to NR monoculture. Soil surface cover with constant leaf litter addition and extensive root systems in the NR-based agroforestry systems might have contributed for the improvement in SOC and N, improved soil aggregation, reduction in soil erosion, decreased C and N loss and ultimately improvement in the C and N accumulation rates (Chen et al. 2017).

9.2 Conservation Agriculture Practices

Retaining natural flora/weeds in NR plantations increased soil pH, which might be due to the high cation content in the litter (Jessy et al. 2013b, 2020). Significant improvement in SOC status and availability of Ca and K were recorded when natural flora was retained in NR plantations (Abraham and Joseph 2015).

9.3 Crop Residue Recycling

NR is a deciduous tree, leaf fall occurs during the January-February months and approximately 6.0 t ha^{-1} of dry matter is added every year through annual litter fall facilitating recycling of nutrients releasing 94–130 kg of N, 5–6 kg of P and 22–25 kg of K to the system and is comparable to 8.3 t ha^{-1} under forests (Sivanadyan and Moris 1992). Leaf litter addition, decomposition and nutrient release recording from a 15-year-old NR plantation of clone RRII 105 revealed that 4824 kg ha^{-1} of litter was added and 92% of which decomposed in a period of 6 months releasing 88 kg N, 2.4 kg P, 45 kg K, 60 kg Ca, 16 kg Mg, 1.5 kg Mn and 0.25 kg Zn (Varghese et al. 2001).

9.4 Cover Crops and Green Manuring

Establishment of leguminous cover crops in young NR in the initial 4 years was the conventional practice till recently, before the introduction of intercropping for fetching additional income and other added advantages mentioned in the previous pages. The most commonly grown cover crop *Pueraria phaseoloides* has been reported to add 3.0 t ha^{-1} of organic matter, whereas *Mucuna bracteata* added 5–6 t ha^{-1} of organic matter to the soil. The nutrient addition to the system through these cover crops over a period ranged from 226 to 253 kg N, 18 to 27 kg P, 85 to 131 kg K and 15 to 27 kg Mg ha^{-1} (Kothandaraman et al. 1989). A comparative evaluation of the performance of *M. bracteata* and *P. phaseoloides* at different age groups recorded highest dry matter production in 2-year-old stage in both species, and it is higher in *Mucuna* (7.6 t ha^{-1}) compared to *Pueraria* (5.5 t ha^{-1}). Nutrient accumulated in a hectare over 2 years were 174, 13, 104, 65 and 18 kg of N, P, K, Ca and Mg, respectively, for *Pueraria* and 236, 15, 79, 56 and 15 kg of N, P, K, Ca and Mg for *Mucuna* (Philip et al. 2003). Liu et al. (2021) quantified the cumulative effects of perennial legume cover crop-*Flemingia macrophylla* over 7 years on rhizosphere P cycling and availability during the dry and wet seasons in young NR-based agroforestry system (YAFS) and mature NR-based agroforestry system (MAFS) in tropical China by analysing cover crop biomass and P content, as well as net rhizosphere changes in P-mobilizing capacity (pH, organic acids and phosphatase activity) and P fractions. Compared to monoculture NR, YAFS enhanced plant biomass and P stock in both seasons, whereas the MAFS exhibited an opposite trend, except for increased biomass and P stock of floor litter in the dry season. Young agroforestry system recorded significant effect on net rhizosphere properties by reducing pH and increasing organic acids concentrations and acid phosphatase activity, compared to MAFS wherein the influence was not reflected on the rhizosphere properties. Results from the study indicated that rhizosphere P cycling was more improved by cover crop in YAFS and cover crop residue contributed to regulating rhizosphere P cycling and availability more than rhizosphere P-mobilizing capacity. The study highlighted the importance of residue management of permanent cover crops on rhizosphere P cycling and availability in AFS with different ages.

Polthanee et al. (2016) reported that in Thailand, *Calopogonium caeruleum* was established in the third year of a NR plantation intercropped with banana. Philip and Jessy (2022) studied the feasibility of establishing *C. caeruleum* under partial shade, after the removal of the intercrop (pineapple), and reported that even though the growth of *C. caeruleum* was slow initially, it established well in the latter phase. Dry matter production of 3.5-year-old *C. caeruleum* under partial shade was 4 t ha^{-1}, and the nutrient accumulation was 108, 10.3, 79.5, 44.3 and 5.8 kg ha^{-1} of N, P, K, Ca and Mg, respectively.

9.5 Biofertilizer and Biostimulant Technologies

Establishment and maintenance of leguminous cover crops in the initial years of NR cultivation is an accepted agronomic practice which helps in the enhancement of N availability through symbiotic N fixation. Studies conducted at Rubber Research Institute of India, Kottayam, India, indicated that the population of the N-fixing bacteria (*Rhizobium*) was very low and the nodules were less effective (Kothandaraman et al. 1993). Similarly, NR growing soils in general were poor in *Azotobacter* population. Soil pH at the extremely acidic range negatively affects the growth and functioning of the bacteria. pH correction through liming was found to improve the growth and nodulation in leguminous plants (Philip et al. 1995). *Azotobacter* sp. isolated from NR growing soils on inoculation recorded improved growth of NR seedlings and improved population in the rhizosphere, and 50% reduction in N fertilizer could be achieved. Population of various microorganisms in the rhizosphere of NR seedlings was also increased by *Azotobacter* inoculation (Joseph et al. 1997).

Inoculation of phosphate-solubilizing bacteria improved the diameter and height of NR seedlings at reduced levels of phosphatic fertilizer (Joseph et al. 1996). Inoculation with efficient isolates of fluorescent rhizobacteria (*Pseudomonas*) improved the shoot weight, root weight, nodulation and nitrogenase activity of *Pueraria phaseoloides* (Joseph et al. 2003). The beneficial effect of combined inoculation of two AMF isolates (isolated from NR growing soils), viz. *Glomus* sp. and *Acaulospora* sp. at low levels of P fertilizer(rock phosphate) application during the early stages of growth of NR seedlings, was reported by Joseph et al. (2002).

Seventeen fluorescent rhizobacteria (*Pseudomonas*) collected from NR growing soils were assessed and among them, seven isolates were selected based on phosphate solubilization and antagonistic activity. These fluorescent rhizobacteria were found to be producing indole acetic acid (IAA) and gibberellin (Joseph et al. 2003). The positive effect of these plant growth-promoting rhizobacteria (PGPR) on the growth of NR seedlings in the nursery was reported (Joseph et al. 2015; Syamala et al. 2010, 2015). Inoculation of an efficient phosphofungi (*Penicillium* spp.) isolated from NR growing soils improved the growth of NR seedlings at the 50% of the recommended level of fertilizers (Joseph et al. 2017).

The effect of integration of chemical fertilizers with bioinoculants on soil fertility and growth of NR plants in the traditional NR growing region of South India experiencing humid tropical climate indicated that after 2–3 rounds of treatment incorporation, buildup of phosphobacteria was significant (Table 2). *Azotobacter* and *Pseudomonas* buildup was noticed in a few instances. Root infection of AMF was more than 90% in both control and treated plants indicating natural infection in NR roots (Joseph et al. 2020) (Table 2). Buildup of phosphobacteria in the soil contributed for enhanced solubility of the P either from the organic form or from the fixed forms. AMF colonization in the NR roots in promoting the mobilization of P from the soil and uptake is enhanced and could possibly be one of the reasons for the maintenance of optimum P concentration in the plant even when the soil is tested low in available P (Joseph et al. 2018). Increased solubility coupled with enhanced mobilization through the AMF colonized in the roots might have

Table 2 Effect of nutrient management treatments on the population of phosphobacteria (CFU g^{-1} soil) and root infection of AMF (%)

Treatments	Phosphobacteria (CFU g^{-1} soil)		AMF (%)	
	Location 1	Location 2	Location 1	Location 2
Control	20.3 (4.5)	9.3 (3.0)	96.0 (9.79)	97.3 (9.86)
100% N and P	18.7 (4.3)	12.3 (3.5)	93.3 (9.65)	100 (10.00)
25% N and P + BI	38.0 (6.2)	37.0 (6.1)	89.3 (9.45)	91.3 (9.54)
50% N and P + BI	27.0 (5.2)	30.7 (5.5)	96.0 (9.80)	97.3 (9.86)
75% N and P + BI	25.0 (5.0)	33.0 (5.7)	92.0 (9.58)	96.0 (9.79)
100% N and P + BI	38.7 (6.2)	34.7 (5.9)	92.0 (9.58)	97.3 (9.86)
BI alone	39.0 (6.2)	39.0 (6.2)	96.0 (9.79)	94.7 (9.73)
SE	0.2	0.2	0.17	0.15
CD ($P = 0.05$)	0.7	0.7	NS	NS

Figures in parenthesis are the square root transformed values
Location 1, Central Experiment Station, Chethakkal, Kerala; Location 2, Puthukkad, Trissur, Kerala; BI, bioinoculant (consortium of *Azotobacter* sp. *Bacillus* spp. *Pseudomonas* spp. and AMF) (Joseph et al. 2020)

contributed for enhanced uptake of P by the plants as reported in many tree crops. In the NR growing soils of South India, 64–86% of the total P was in the organic form (Prasannakumari et al. 2008), and solubilization of these organic forms through bioagents contributes to increased availability for plant uptake.

9.6 Organic Amendments and Integrated Approaches

Jessy et al. (1996) reported that application of farm yard manure (FYM) along with chemical fertilizers improved the soil physical properties, microbial population and cation exchange capacity. Philip et al. (2012) from their studies in immature rubber conducted from planting onwards for 8 years reported that integrated application of chemical fertilizer in combination with FYM enhanced the growth of NR plants and improved the soil properties. A combination of 25% recommended dose of fertilizer and 75% FYM was superior to all other treatments in improving the growth and soil properties.

In the seedling nursery, growth of plants with 50% reduced level of N and P fertilizer applied in combination with bioinoculants was on par with the standard recommended dose of chemical fertilizers alone or in combination with bioinoculants (Joseph et al. 2015). Field experiments in young NR plants were conducted in the traditional belt of NR cultivation (South India) and nontraditional region (Agarthala, Tripura, Northeast India) with the same set of treatments integrating bioinoculants with graded levels of inorganic fertilizer recorded difference in response to treatments. From field experiment with young NR plants in Tripura, Northeast India, where the soil OC status was only 0.92%, positive response to fertilizer application was recorded on the availability of nutrients and growth of plants (Table 3), and the possibility of reducing the chemical fertilizer by integration with bioinoculants was reported (Mandal and Joseph 2018). At the same time, results from the field experiment were conducted in two locations in South India,

Table 3 Effect of nutrient management treatments on the growth (girth at 150 cm height) of rubber plants (cm)

Treatments	Traditional belt of rubber cultivation in south India[a]		Northeast India[b]
	Central Experiment Station, Chethakkal, Kerala	Puthukkad, Thrissur, Kerala	Agartala, Tripura
Control	33.2	35.5	21.9
100% N and P	33.5	36.2	27.5
25% N and P + BI	35.0	37.7	26.4
50% N and P + BI	34.6	37.0	27.9
75% N and P + BI	37.9	36.6	27.4
100% N and P + BI	34.8	38.4	27.2
BI alone	35.6	38.2	22.4
SE	1.2	1.0	1.2
CD ($P = 0.05$)	NS	NS	2.9

BI, bioinoculant (consortium of *Azotobacter* sp. *Bacillus* spp. *Pseudomonas* spp. and AMF)
[a]OC status of the soil is >1.0%; [b]OC status of the soil is <1.0%
(Joseph et al. 2020; Mandal and Joseph 2018)

and growth of NR plants did not show significant difference between the treatments (Table 3) indicating lack of response of rubber plants to applied nutrients where the soil OC status was high (Joseph et al. 2020).

Results from the study clearly indicated that the response to fertilizers depend on the fertility status of the soil. The OC-rich soil in the first two locations in the traditional belt of NR cultivation in South India contributed towards the maintenance of sufficiency level of N and P in the soil in the control plots, and external application of chemical fertilizers alone or in combination with bioinoculants could not bring in additional advantage on the growth of plants compared to Agartala, where OC status was <1.0%. Natural rubber being a long-duration crop, accounting for the response to the effect of even inorganic fertilizers itself, is very difficult, and as such the effect of these bioinoculants on soil nutrient availability, release of growth hormones, enzymes and its effect on growth of plants is very difficult to establish.

10 Carbon Sequestration

Natural rubber plantations are a large reservoir of C stored in the biomass. Natural rubber wood used for making diverse long-term products and rubber sheets used for tyre manufacture is additional fixed C sink. Carbon sequestration potential of world's NR plantations is estimated to be around 0.0782 PgC year^{-1} (Jacob 2003). Jacob (2006) reported that during the 21 years of growth of a NR plantation grown in the traditional NR growing region of Central Kerala, India, the total amount of C sequestered added up to 66.7 t C acre^{-1}, and the mean annualized C sequestration rate during the 21-year period comes to roughly 3.17 t C $acre^{-1}$ $year^{-1}$ and was comparable with that of the forest ecosystems, especially with the above ground pool. Mandal et al. (2012) reported that over a period of 20 years, soils under NR in Tripura, Agartala, Northeast India, are experiencing subtropical climate accumulated about 10.2 t C ha^{-1} in the 0–60 cm depth, and it was found to be less than that of the adjacent forest ecosystem. Establishment of a tree plantation, like any other form of land-use conversion, brings in changes in the ecosystem characteristics and fluxes (Njar et al. 2011). According to Ziegler et al. (2012), NR plantations are considered as one of the top four best land uses in Southeast Asia for C stock. However, the effect of NR plantations on SOC balance and nutrient cycle depends on the previous land use, and it is positive if arable land was converted to NR plantation (Herrmann et al. 2016a, b; Njar et al. 2011; Yasin et al. 2010) and negative if forest was converted to NR plantation (Blagodatsky et al. 2016; de Blécourt et al. 2013).

Carbon sequestration potential of modern clones ranged from 54 t ha^{-1} to 114 t ha^{-1} because of the difference in growth pattern and biomass accumulation (Ambily et al. 2012). Ambily and Ulaganathan (2015) reported that the total biomass ranged from 1140 to 2045 kg $tree^{-1}$, C stock from 479 to 860 kg $tree^{-1}$ and C storage potential from 148 to 258 t carbon ha^{-1} for three clones at 30 years of age. Biomass C accumulation including latex C and C distribution in soil under NR plantations of five age groups, viz. 7, 13, 19, 25 and 47 years old were examined by Liu et al. (2017). Total biomass C stock (TBC) and total net primary productivity (NPP total) recorded quadratic relationship with age of the plantation. Regardless of the age of the plantation, around 68% of the C was stored in aboveground biomass, and latex contributed to approximately 18% of C sequestration. The total ecosystem C stock (TEC) across stands averaged 159.6, 174.4, 229.6, 238.1 and 291.9 Mg ha^{-1}, respectively, for the 7-, 13-, 19-, 25- and 47-year-old plantations, of which >45% was stored in the soil. Regression analysis showed that TEC for NR plantation at 22 years is comparable to a baseline of 230.4 Mg ha^{-1} for tropical forest in China and would reach the maximum value at around 54 years.

Conversion of natural forests to NR plantations reduces the soil C level due to change in litter dynamics, microbial activity and environmental conditions. Results from a reciprocal soil and litter translocation mesocosm experiment for 15 months in a seasonal rainforest and adjacent NR plantation reported that rate of soil heterotrophic respiration (Rh) was higher at NR plantation site. Changes in physical environmental conditions, rather than changes in litter input or soil biochemical properties, attributed to the elevated soil heterotrophic respiration in NR plantation, resulting in soil C loss (Zhang et al. 2021). Abraham and Philip (2022) reported the high C sequestration potential of NR-based cropping system based on their study comparing the effect of five cropping systems on total soil C and related characters with that of nearby forest. Compared to other cropping systems, in NR or NR-based cropping systems, more than 70% of the total C was in the finer fraction (<53 μM) indicating higher C sequestration potential. Intercropping banana in young NR plantation improved the total C, OC and SOC stock compared to monoculture of NR (Joseph et al. 2022).

11 Organic Farming Systems

The concept of organic farming in NR compared to organically cultivated vegetables/fruits or other consumable crops is not a viable option because rubber is only an industrial raw material. However, many NR growers are aware of the benefits of chemical-free, organic farming practices on soil properties and environment, and some of them are practicing the same. Natural addition of organic matter through leaf litter and crop residues promotes SOC buildup in soil and the SOC content is high in the NR growing soils compared to adjacent fields where other crops are grown. The present organic manure recommendation in NR cultivation is application of 2.5 t of compost or FYM ha^{-1} (25 kg per 100 m^2) to the seedling nursery and 12 kg pit^{-1} in the planting pits.

12 Future Strategies

The early NR plantations are now in the second or third cycle of replanting and hence grown in repeated cycles in the same soil. With the fast expansion of area, new plantations are being established in marginal areas with less fertile soil and less favourable environmental conditions. Increasing soil acidity, depletion of SOC, deficiency of secondary nutrients such as Ca and Mg and low availability of Zn and B through repeated cycles of NR cultivation in the same soil indicate chances of soil fatigue and point towards the necessity of adequate soil health management for sustaining the productivity. Assessment of spatial variability of soil fertility parameters through georeferenced soil sampling and analysis and development of fertility maps for the NR growing soils are very useful for better management of soil fertility for achieving sustained productivity. Establishment of leguminous cover crops along with intercrops or after the 4 years of intercropping (shade-tolerant leguminous cover crops) will be beneficial for improving the soil quality and health and may be promoted among the growers. Calcium supplementation for Ca nutrition and for temporary pH change may be resorted in extremely acidic soils for maintaining optimum Ca to Al ratio in the exchange complex. Zinc and B deficiencies of the NR growing soils are to be corrected. Awareness among the farmers is to be created in maintaining the under-flora vegetation in mature plantations through careful weed management strategies for promoting biodiversity which will be helpful in conserving soil moisture and recycling of crop residues. NR-based agroforestry systems suited to the local socio-economic conditions may be promoted among the growers for sustaining the production system with the added advantage of improving the health of the soil. More areas in the marginal lands are being brought under NR cultivation, and adequate attention for fine-tuning the agro-management practice for the establishment and uniform growth of NR in these marginal areas are to be developed. Predictions on climate change-associated global warming on shift in NR cultivation to newer areas require detailed studies on soil suitability assessment and development of specific soil health management protocols. Conservation of soil moisture by permitting other crops and weeds in NR plantations and subsequent mitigation of the risks associated with climate uncertainties may be promoted in NR plantations. In-depth studies on climate change-associated changes on soil quality and health are needed for better understanding of the system and development of management strategies and options. Carbon sequestration potential and the potential ecosystem services of the rubber plantations may be accounted for recovering carbon credit and economic gains for the NR growers.

Site-specific nutrient management approach is promoted among the farmers as a step towards reducing the cost on fertilizer application and at the same time ensuring the maintenance of health and productivity of the soil. An integrated approach involving biological management of nutrients, especially of N and P, is to be envisaged in NR plantations because of crop residue/litter recycling potential and P buildup in the soil over the years. Adequate attention for maintenance of health and productivity of the soils under NR cultivation is highly essential for achieving sustained higher productivity from NR plantations.

References

Abraham J (2015) Soil health in different land use systems in a tropical region of Kerala. Rubber Sci 28(1):68–71

Abraham J, Chudek JA (2008) Studies on litter characterization using ^{13}C NMR and assessment of microbial activity in natural forest and plantation crops' (teak and rubber) soil ecosystems of Kerala, India. Plant Soil 303(1 & 2):265–273

Abraham J, Joseph P (2015) A new weed management approach to improve soil health in a tropical plantation crop, rubber (*Hevea brasiliensis*). Exp Agric 52(1):36–50. https://doi.org/10.1017/s0014479714000544

Abraham J, Philip A (2022) Climate change mitigation: potential role of soils under rubber-based cropping systems. Rubber Sci 35(1):27–36

Ambily KK, Ulaganathan A (2015) Biomass production, carbon storage capacity and nutrient export in natural rubber. Rubber Board Bull 33(4):4–10

Ambily KK, Joseph M (2018) Effect of soil pH and base status on the growth of young natural rubber plants. Rubber Sci 31(3):232–242

Ambily KK, Meenakumari T, Jessy MD, Ulaganathan A, Nair NU (2012) Carbon sequestration potential of RRII 400 series clones of *Hevea brasiliensis*. Rubber Sci 25(2):233–240

Bhattacharya T, Seghal J, Sarkar D (1996) Soils of Tripura: their kinds, distribution and suitability to major field crops and rubber, detailed bulletin and data base for optimizing land use. NBSS Publication 65, Soils of India Series 6, National Bureau of Soil Survey and Land Use Planning, Nagpur. 149 p

Blagodatsky S, Xu J, Cadisch G (2016) Carbon balance of rubber (*Hevea brasiliensis*) plantations: a review of uncertainties at plot, landscape and production level. Agric Ecosyst Environ 221:8–19

Bradshaw MJ (1977) Earth, the living planet. Hodder and Stoughton, London. 322 p

Bridges EM (1979) World soils, 2nd edn. Cambridge University Press, Cambridge. 128 p

Chambon B, Ruf F, Kongmanee C, Angthong S (2016) Can the cocoa cycle model explain the continuous growth of the rubber (*Hevea brasiliensis*) sector for more than a century in Thailand? J Rural Stud 44:187–197

Chen C, Liu W, Jiang X, Wu J (2017) Effects of rubber-based agroforestry systems on soil aggregation and associated soil organic carbon: implications for land use. Geoderma 299. https://doi.org/10.1016/j.geoderma.2017.03.021

Datta B, Jessy MD, Dey SK (2019) Rubber-based cropping system studies in North East India. Rubber Sci 32(3):272–282

de Blécourt M, Brumme R, Xu J, Corre MD, Veldkamp E (2013) Soil carbon stocks decrease following conversion of secondary forests to rubber (*Hevea brasiliensis*) plantations. PLoS One 8(7):e69357

Dissanayake DMAP (2001) Rubber growing soils and their characteristics. In: Agronomy Tillekeratne LMK, Nugawela A (eds) Hand book of rubber, vol 1. Rubber Research Institute of Sri Lanka, Dartonfield, Sri Lanka, pp 12–25

Eswaran H, Bin WC (1978a) A study of a deep weathering profile on granite in Peninsular Malaysia: I. Physico-chemical and micromorphological properties. Soil Sci Soc Am J 42(1):144–149

Eswaran H, Bin WC (1978b) A study of a deep weathering profile on granite in Peninsular Malaysia: II. Mineralogy of the clay, slit, and sand fractions. Soil Sci Soc Am J 42(1):149–153

Food and Agriculture Organization of the United Nations (1976) Framework for land evaluation. FAO Soils Bulletin No. 32. Food and Agriculture Organization, Rome, 87 p

Food and Agriculture Organization of the United Nations (2012) World agriculture watch. Methodological framework. Summary version 2.6. Food and Agriculture Organization, Rome. 48 p

Food and Agriculture Organization of the United Nations (2016) FAOSTAT statistics database. https://search.library.wisc.edu/catalog/999882363002121

Fox J, Castella JC (2013) Expansion of rubber (*Hevea brasiliensis*) in Mainland Southeast Asia: what are the prospects for smallholders? J Peasant Stud 40(1):155–170

George S, Joseph P (2011) Natural rubber plantation: a nutritionally self-sustaining ecosystem. Nat Rubber Res 24(2):197–2002

George S, Meti S (2018) Cocoa and coffee as intercrops in mature rubber plantation: effects on growth and yield of rubber and physico-chemical properties of soil. Rubber Sci 31(1):31–40

Gilot C, Lavelle P, Blanchart E, Keli J, Kouassi P, Guillaume G (1995) Biological activity of soil, under rubber plantations in Côte d'Ivoire. Acta Zool Fenn 196:186–189

Golbon R, Cotter M, Sauerborn J (2018) Climate change impact assessment on the potential rubber cultivating area in the Greater Mekong Subregion. Environ Res Lett 13(8):084002. https://doi.org/10.1088/1748-9326/aad1d1

Hazir MHM, Kadir RA, Karim YA (2018) Projections on future impact and vulnerability of climate change towards rubber areas in Peninsular Malaysia. IOP Conf Ser Earth Environ Sci 169(1):1–9. https://doi.org/10.1088/1755-1315/169/1/012053

Hazir MHM, Kadir RA, Gloor E, Galbraith D (2020) Effect of agro-climatic variability on land suitability for cultivating rubber (*Hevea brasiliensis*) and growth performance assessment in the tropical rainforest climate of Peninsular Malaysia. Clim Risk Manag 27:100203. https://doi.org/10.1016/j.crm.2019.100203

Herrmann L, Bräu L, Robin A, Robain H, Wiriyakitnateekul W, Lesueur D (2016a) High colonization by native arbuscular mycorrhizal fungi (AMF) of rubber trees in small-holder plantations on low fertility soils in North East Thailand. Archiv Agron Soil Sci 62(7):1041–1048

Herrmann L, Lesueur D, Bräu L, Davison L, Jairus T, Robain H, Robin A, Vasar M, Wiriyakitnateekul MW, Öpik M (2016b) Diversity of root-associated arbuscular mycorrhizal fungal communities in a rubber tree plantation chronosequence in Northeast Thailand. Mycorrhiza 26:863–877

Jacob J (2003) Carbon sequestration potential of natural rubber plantations. In: Proceedings of IRRDB Symposium on Challenges for Natural Rubber in Globalization, Chang Mai, Thailand, 15–17 Sept 2003

Jacob J (2006) Carbon sequestration potential of natural rubber plantations. In: Mathew NM (ed) Jacob J. Koyoto protocol and the rubber industry, Rubber Research Institute of India, pp 165–176

Jessy MD (2004) Phosphorus nutrioperiodism in rubber. PhD thesis, Kerala Agricultural University, Vellanikkara, Thrissur, Kerala, India, 166 p

Jessy MD, Jacob J (2020) Conserving/improving biodiversity in rubber plantations in India: a review of possibilities and ecological implications. Rubber Sci 33(2):112–126

Jessy MD, Sethuraj MR, Joseph K (1996) Physico-chemical properties of soils under rubber with different nutrient management practices: a case study. Indian J Nat Rubber Res 2:134–136

Jessy MD, Philip V, Punnoose KI, Sethuraj MR (1998) Evaluation of a multispecies cropping system during immaturity phase of rubber. Indian J Nat Rubber Res 11(1 & 2):80–87

Jessy MD, Nair ANS, Joseph P, Prathapan K, Krishnakumar V, Nair RB, Mathew M, Punnoose KI (2004) Response of the high yielding *Hevea* clone RRII 105 to fertilizers. Nat Rubber Res 19(1 & 2):38–45

Jessy MD, Punnose KI, Nayar TVR (2005) Crop diversification and its sustainability in young rubber plantation. J Plantn Crops 33(1):29–35

Jessy MD, Meera Bai M, Nair ANS, Meti S (2007) Adaptability to low soil phosphorus in rubber trees: role of roots and arbuscular mycorrhizal fungi. J Plantn Crops 35:133–138

Jessy MD, Meti S, Nair UN (2013a) A cropping system for reduction of gestation period and enhanced yield of rubber trees (*Hevea brasiliensis*). Rubber Sci 26:210–216

Jessy MD, Syamala VK, Ulaganathan A (2013b) Ground covers differ in their effect on soil pH in rubber (*Hevea brasiliensis*) plantations. National Seminar on Developments in Soil Science. Central Arid Zone Research Institute, Rajasthan, India, 23–26 Oct 2013

Jessy MD, Joseph P, George S (2015) Establishing perennial intercrops in rubber plantations after the removal of pineapple: effect on growth and yield of rubber, soil moisture and nutrient status. Rubber Sci 28(2):138–146

Jessy MD, Joseph P, George S (2016) Possibilities of diverse rubber-based agroforestry systems for small holdings in India. Agrofor Syst 91(3):515–526

Jessy MD, Syamala VK, Ulaganathan A (2020) Natural flora improves properties of acid soil in rubber plantations. J Indian Soc Soil Sci 68(3):315–332

Joseph K (1997) Studies on vascular arbuscular mycorrhizal fungi in the growth improvement of *Pueraria phaseoloides* Benth. PhD Thesis, Mahatma Gandhi University, Kottayam, Kerala, India, 174 p

Joseph M (1999) Studies on some competing factors in the intercropping young rubber (*Hevea brasiliensis*). PhD thesis, Kerala Agricultural University, Vellanikkara, Thrissur, 95 p

Joseph M (2011) Exchange properties of soils in the traditional rubber growing tract in South India. Nat Rubber Res 24(1):97–105

Joseph M (2016) Rubber growing soils of India: an overview. Rubber Sci 29(2):119–139

Joseph M, Sudhakumari B (2013) Assessment of zinc availability in the ultisols of south India under rubber cultivation. J Indian Soc Soil Sci 61(1):67–71

Joseph M, Prasad MP, Antony PA, Punnoose KI (1995) DTPA extractable soil micronutrients in the traditional rubber growing regions in India. Indian J Nat Rubber Res 8(2):135–139

Joseph K, Kothandaraman R, Mathew J (1996) Phosphate solubilization by *Bacillus circulans*, a leap towards phosphatic fertilizer management in rubber plantations. National Seminar on Organic Farming and Sustainable Agriculture, Bangalore, India, 9–11 Oct 1996, p 69

Joseph K, Vimalakumari TG, Mathew J, Kothandaraman R (1997) Effect of Azotobacter inoculation on rubber seedlings. Indian J Nat Rubber Res 10(1 & 2):34–38

Joseph M, Nair RB, Mathew M, Punnoose KI (1998) Potassium nutrition of mature rubber. Indian J Nat Rubber Res 11(1 & 2):58–66

Joseph K, Kothandaraman R, Vimalakumari TG, Mathew J (2002) Occurrence of Arbuscular mycorrhizal fungi in rubber growing soils and their effect on growth of *Heveabrasiliensis* seedlings. Proceedings of Placrosym XV:317–325

Joseph K, Sinju V, Jinu G, Mathew J, Kuruvilla CJ (2003) Plant growth promoting rhizobacteria in rubber (*Hevea brasiliensis*) plantations. 6th international workshop on plant growth promoting rhizobacteria. Indian Institute of Spices Research, Calicut, Kerala, India, 5–10 Oct 2003

Joseph M, Sudhakumari B, Punnoose KI, Karthikakuttyamma M (2007) Response of rubber seedlings in the nursery to application of zinc. Nat Rubber Res 20(1 & 2):61–65

Joseph M, Syamala VK, Joseph P (2008) Response of mature rubber to lime and magnesium application. J Plantn Crops 36(3):357–359

Joseph M, Syamala VK, Punnoose KI (2009) Effect of liming on the availability of nutrients and growth of young rubber. Nat Rubber Res 22(1&2):55–63

Joseph M, Joseph K, Mathew J, Hareeshbabu G, Elias RS (2015) Nutrient management in rubber seedling nursery: studies on an integrated approach through incorporation of bioinoculants. Rubber Sci 28(1):70–75

Joseph K, Philip S, Jose G (2017) Isolation and selection of efficient phosphofungi from rubber plantations. Rubber Sci 30(3):262–268

Joseph M, Chandy B, Aneesh P (2018) Leaf nutrient status of the rubber plantations of South India. Rubber Sci 31(3):217–226

Joseph M, Joseph K, Hareeshbabu G, Elias RS (2020) Availability of nitrogen and phosphorus in the soil and growth of natural rubber plants under integrated nutrient management system. Rubber Sci 33(2):164–176

Joseph P, Jessy MD, Mohan M (2022) Soil carbon pools under rubber (*Hevea brasiliensis)* based agroforestry systems in South India. Agrofor Syst 96:1121–1133

Karthikakuttyamma M (1997) Effect of continuous cultivation of rubber (*Heveabrasiliensis*) on soil properties. PhD Thesis, University of Kerala, Trivandrum, India, 176 p

Karthikakuttyamma M, Suresh PR, Prasannakumari P, George V, Iyer RS (1998) Effect of continuous cultivation of rubber (*Hevea brasiliensis*) on morphological features and organic carbon, total nitrogen, phosphorus and potassium contents of soil. Indian J Nat Rubber Res 11(1 & 2):73–79

Karthikakuttyamma M, Joseph M, Nair ANS (2000) Soils and nutrition. In: George PJ, Jacob KC (eds) Natural rubber: agromanagement and crop processing. Rubber Research Institute of India, Kottayam, pp 170–198

Kharche VK, Sehgal JL, Challa O (1995) Evaluation of soil-site conditions for suitability of rubber. Agroped 5:69–78

Kothandaraman R, Mathew J, Krishnakumar AK, Joseph K, Jayarathnam K, Sethuraj MR (1989) Comparative efficiency of *Mucuna bracteate* D.C. and *Pueraria phaseoloides* Benth. on soil nutrient enrichment, microbial population and growth of *Hevea*. Indian J Nat Rubber Res 2(2):147–150

Kothandaraman R, Mathew J, Joseph K, Jayaratnam K (1993) The impact of *Bradyrhizobium* inoculation on nodulation, biomass production and nitrogen fixation in *Pueraria phaseoloides*. Indian J Nat Rubber Res 6(1 & 2):50–54

Krashevska V, Klarner B, Widyastuti R, Maraun M, Scheu S (2015) Impact of tropical lowland rainforest conversion into rubber and oil palm plantations on soil microbial communities. Biol Fertil Soils 51:697–705

Krishnakumar AK (1989) Soils under *Hevea* in India: a physical, chemical and Mineralogical study with reference to soil moisture contents influence on yield of *Hevea brasiliensis*. PhD Thesis, Indian Institute of Technology, Kharagpur, India

Krishnakumar AK, Gupta C, Sinha RR, Sethuraj MR, Potty SN, Eappen T, Das K (1991) Ecological impact of rubber (*Hevea brasiliensis*) plantations in North East India. 2. Soil properties and biomass recycling. Indian J Nat Rubber Res 4(2):134–141

Krishnakumar AK, Potty SN (1992) Nutrition of *Hevea*. In: Sethuraj MR, Mathew NM (eds) Natural rubber: biology, cultivation and technology. Elsevier, Amsterdam, pp 239–262

Lavelle P, Decaëns T, Aubert M, Barot S, Blouin M, Bureau F, Margerie P, Mora P, Rossi JP (2006) Soil invertebrates and ecosystem services. Eur J Soil Biol 42:S3–S15

Lavelle P, Rodríguez N, Arguello O, Bernal J, Botero C, Chaparro P, Gómez Y, Gutiérrez A, del Pilar HM, Loaiza S, Pullido SX, Rodríguez E, Sanabria C, Velásquez E, Fonte SJ (2014) Soil ecosystem services and land use in the rapidly changing Orinoco River Basin of Colombia. AgrEcosyst Environ 185:106–117

Ling Z, Shi Z, Gu S, Wang T, Zhu W, Feng G (2022) Impact of climate change and rubber (*Hevea brasiliensis*) plantation expansion on reference evapotranspiration in Xishuangbanna, Southwest China. Front Plant Sci 13:830519. https://doi.org/10.3389/fpls.2022.830519

Liu S, Zhou G, Shibo F, Zhang J (2015) Effects of future climate change on climatic suitability of rubber plantation in China. Chin J Appl Ecol 26:2083–2090

Liu C, Pang J, Jepson MR, Lu X, Tang J (2017) Carbon stocks across a fifty year chronosequence of rubber plantation in tropical China. Forests 8(6). https://doi.org/10.3390/f8060209

Liu C, Wang QW, Jin Y, Tang JW, Lin F, Olatunji OA (2021) Perennial cover crop biomass contributes to regulating soil P availability more than rhizosphere P-mobilizing capacity in rubber-based agroforestry systems. Geoderma 401(1). https://doi.org/10.1016/j.geoderma.2021.115218

Mandal D, Sharma AC, Dey SK, Baruah TC (2010) Status of DTPA-extractable micronutrients in rubber growing soils of Tripura. Nat Rubber Res 23(1 & 2):98–104

Mandal D, Pal TK, Dey SK, Jacob J (2012) Changes in organic carbon and some soil properties under rubber (*Hevea brasiliensis*) plantation in sub-tropical Tripura. Nat Rubber Res 25(1):13–20

Mandal D, Pal TK, Joseph M, Dey SK (2013) Fertility evaluation of the soils under rubber plantations in Tripura. Rubber Board Bull 31(4):4–9

Mandal D, Joseph M (2018) Integrated nutrient management practices for young rubber in Tripura, North East India. Rubber Sci 31(2):121–129

Mathew J, Kumaran MG, Joseph K, George ES (2000) Waste management. In: George PJ, Jacob CK (eds) Natural rubber, agromanagement and crop processing. Rubber Research Institute of India, Kottayam, Kerala, pp 493–506

Meenakumari T, Meenattoor JR, Soman TA, Dey SK, Das G, Raj S, Sailajadevi T, Nair RB, Raman KA, Gireesh T, Mydin KK (2011) Yield of modern *Hevea* clones and their response to weather parameters across diverse environments. Nat Rubber Res 24(1):44–53

Meti S, Pradeep B, Jacob J, Meerabai M, Jessy MD (2014) Spatio-temporal analysis of rubber area and its association with soil and topography in Kanyakumari district. Rubber Sci 27(2):182–192

Min S, Wang X, Jin S, Waibel H, Huang J (2020) Climate change and farmers' perceptions: impact on rubber farming in the Upper Mekong region. Climatic Change 163(1):451–480. https://doi.org/10.1007/s10584-020-02876-2

Muschler RG, Bonnemann A (1997) Potentials and limitations of agro-forestry for changing land-use in the tropics: experiences from Central America. For Ecol Manag 91:61–73

Naidu LGK, Nair KM, Srinivas S, Vadivelu S (2008) Analysis of soil variability in rubber growing areas of Kerala and Tamil Nadu and generation of district-wise thematic maps on soil qualities. NBSS publication 1014. National Bureau of Soil Survey and Land Use Planning, Nagpur

Naidu LGK, Nair KM, Srinivas S (2014) Rubber-growing areas of Kerala and Tamil Nadu: soil related constraints to rubber productivity and delineation of soil management units. J Indian Soc Soil Sci 62(1):67–70

NBSS and LUP (1999) Resource soil survey and mapping of rubber growing soils of Kerala and Tamil Nadu. National Bureau of Soil Survey and Land Use Planning, Nagpur. 295 p

Njar G, Iwara A, Ekukinam U, Deekor T, Amiolemen S (2011) Organic carbon and total nitrogen status of soils under rubber plantation of various ages, south-southern Nigeria. J Environ Sci Resour Manag 3:1–13

Partelli FL, Araujo AV, Vieira HD, Dias JRM, Menezes LFT, Ramalho JC (2014) Microclimate and development of 'Conilon' coffee intercropped with rubber trees. Pesqagropec bras, Brasilia 49(11):872–881

Pathiratna LSS, Perera MKP (2006) Intercropping medicinal plants under rubber (*Hevea*). Bull Rubber Res Inst Sri Lanka 47:28–33

Peerawat M, Blaud A, Trap J, Chevallier T, Alonso P, Gay F, Thaler P, Spor A, Sebag D, Chutinan P, Suvannang N, Sajjaphan K, Brauman A (2018) Rubber plantation ageing controls soil biodiversity after land conversion from cassava. Agric Ecosyst Environ. https://doi.org/10.1016/j.agee.2018.01.034

Pendergrass AG, Knutti R, Lehner F, Deser C, Sanderson BM (2017) Precipitation variability increases in a warmer climate. Sci Rep 7(1):17966

Philip A, Joseph M, Punnoose KI, Antony PA (1995) Effect of liming on nutrient uptake, biomass production and nodulation in *Pueraria phaseoloides*. Indian J Nat Rubber Res 8(2):130–134

Philip A, Philip V, George ES, Punnoose K, Mathew M (2003) Leaf litter decomposition and nutrient release in a fifteen-year-old rubber plantation. Indian J Nat Rubber Res 16(1 & 2):81–84

Philip A, Varghese M, Syamala VK, Joseph K, Jessy MD, Nair NU (2012) Integrating organic manure to reduce chemical fertilizer input and enhance growth in young rubber plantations. J Plantn Crops 40(3):158–162

Philip A, Ulaganathan A, Pradeep B, Prasannakumari P, Jessy MD, Jacob J, Abraham J, Syamala VK, Ambily KK, George S, Joseph P, Eapen T, Cyriac J, Mathews PM, Anilkumar KS, Nair

KM (2020) Spatial variability of plant available zinc and boron in the traditional rubber growing regions and Konkan region of India. Rubber Sci 33(2):152–163
Philip A, Jessy MD (2022) *Calopogonium caeruleum* – an alternate cover crop in rubber plantations. Rubber Sci 35(1):72–77
Polthanee A, Promkhambut A, Khamla N (2016) Seeking security through rubber intercropping: a case study from North eastern Thailand. Asia Pacific J Sci Technol 21(3):1–11
Pradeep B, Jacob J, Jessy MD (2019) Rubber Soil Information System (RubSIS): a decision making tool for skipping fertilizer application in rubber plantations. Rubber Sci 32(1):63–67
Pradeep B, Sylas VP, Jessy MD (2022) A framework for assessing the vulnerability of rubber plantations to the impacts of climate change with special reference to Kerala, India. J Rubber Res. https://doi.org/10.1007/s42464-022-00186-z
Prasannakumari P, Joseph M, Nair NU (2008) Organic phosphorus status of the major soil series under rubber cultivation in South India. Nat Rubber Res 21(1 & 2):98–103
Prasannakumari P, Jessy MD, Antorny PA, Chacko J, Jacob J (2014) Nutrient and heavy metal status of soils under rubber-pineapple intercropping in comparison to rubber-cover crop system and natural forest. Rubber Sci 27(1):84–90
Prasannakumari P, Jessy MD, Pradeep B, Jacob J, Abraham J, Philip A, Ambili KK, Joseph P, Geroge S, Syamala VK, Ulaganathan A, Eappen T, Mathews PM, Anilkumar KS, Nair KM (2020) Spatial variability of available calcium and magnesium in soils in the rubber growing regions of South India. Rubber Sci 33(1):18–32
Punnoose KI (1993) Interrelationship of applied nutrients on growth, productivity and latex flow characteristics of *Hevea brasiliensis* Muell Arg. PhD Thesis, Kerala Agricultural University, Trissur, India, 170 p
Punnoose KI, Potty SN, Mathew M, George CM (1976) Response of *Hevea brasiliensis* to fertilizers in South India. Proceedings, International Natural Rubber Conference, 1975, Kuala Lumpur, Malaysia, pp 84–107
Pushpadas MV, Karthikakuttyamma M (1980) Agro ecological requirements. In: Radhakrishna Pillai PM (ed) Handbook of natural rubber production in India. Rubber Research Institute of India, Kottayam, pp 87–109
Pushpadas MV, Subbarayalu G, George CM (1978) Studies on correlations between nutrient levels in soil and leaf and yield of *Hevea brasiliensis*. Rubber Board Bull 15(1 & 2):11–23
Pushparajah E (1977) Nutrition and fertilizer us in Hevea and associated covers in Peninsular Malaysia: a review. J Rubber Res Inst Sri Lanka 54(1):270–283
Raj S, Satheesh PR, Jacob J (2011) Evidence for climate warming in some natural rubber growing regions of South India. Nat Rubber Res 24(1):10–17
Rao PS, Vijayakumar KR (1992) Climatic requirements. In: Sethuraj MR, Mathew NM (eds) Natural rubber, biology, cultivation and technology. Elsevier, Amsterdam, pp 200–219
Ray D, Behera MD, Jacob J (2016) Predicting the distribution of rubber trees (*Hevea brasiliensis*) through ecological niche modelling with climate, soil, topography and socio-economic factors. Ecol Res 31:75–91. https://doi.org/10.1007/s11284-015-1318-7
Ray D, Behera MD, Jacob J (2014) Indian Brahmaputra valley offers significant potential for cultivation of rubber trees under changed climate. Curr Sci 107(3):1–9
Rodrigo VHL, Silva TUK, Munasinghe ES (2004) Improving the spatial arrangement of planting rubber (*Hevea brasiliensis* Muell Arg.) for long-term intercropping. Field Crops Res 89(2–3):327–335
Rubber Research Institute of Malaysia (1977) A soil suitability technical grouping system for *Hevea*. Planter's Bull Rubber Res Inst Malaysia 152:135–146
Sanchez PA, Couto W, Buol SW (1982) The fertility capability soil classification system: interpretation, applicability and modification. Geoderma 27(4):283–309
Saengruksawong C, Khamyong S, Anongrak N, Pinthong J (2012) Growths and carbon stocks in rubber plantations on Chakkarat soil series, Northeastern Thailand. Suranaree J Sci Technol 19(4):271–278
Samarappuli L (2001) Nutrition. In: Tillekeratne LMK, Nugawela A (eds) Hand book of rubber, Agronomy, vol 1. Rubber Research Institute of Sri Lanka, Dartonfield, Agalawatta, Sri Lanka, pp 156–175

Satheesh PR, Jacob J (2011) Impact of climate warming on natural rubber productivity in different agro-climatic regions of India. Nat Rubber Res 24(1):1–9

Scheff J, Frierson DMW (2014) Scaling potential evapotranspiration with greenhouse warming. J Climate 27(4):1539–1558

Schneider D, Engelhaupt M, Allen K, Kurniawan S, Krashevska V, Heinemann M, Nacke H, Wijayanti M, Meryandini A, Corre MD, Scheu S, Daniel R (2015) Impact of lowland rainforest transformation on diversity and composition of soil prokaryotic communities in Sumatra (Indonesia). Front Microbiol 6:1–12

Sivanadyan K, Moris N (1992) Consequences of transforming tropical rain forests to *Hevea* plantations. Planter 68(800):547–567

Sivanadyan K, Gandimathi H, Haridas G (1995) Rubber, a unique crop: the mature *Hevea* stand as a nutritionally self–sustaining ecosystem in relation to latex yield. Rubber Research Institute of Malaysia, Kuala Lumpur. 54 p

Strahler AN (1969) Physical geography, 3rd edn. Wiley, New York. 733 p

Suresh PR, Karthikakuttyamma M, Mathew M, Pothen M, Augusthy A (1994) Contribution of exchangeable aluminium to subsoil acidity in rubber growing tracts. Proceedings of the Sixth Kerala Science Congress, Thiruvanathapuram, India, January 1994, pp 140–142

Syamala VK, Joseph M, Punnoose KI (2003) Effect of liming on soil properties and growth of rubber seedlings. Indian J Nat Rubber Res 16(1 & 2):66–74

Syamala VK (2006) Characterization of acidity and its management in rubber growing soils. PhD Thesis, Mahatma Gandhi University, Kottayam, Kerala, 145 p

Syamala VK, Joseph M (2006) Effect of different liming materials on availability of nutrients and growth of rubber seedlings in the nursery. J Plantn Crops 34(3):356–360

Syamala VK, George S, Joseph K, Nair ANS, Nair NU (2010) Effect of inorganic fertilizers and PGPR in the growth of rubber seedlings (*Hevea brasiliensis*) in nursery. First Asian PGPR congress for sustainable Agriculture, Acharya N. G. Ranga Agricultural University (ANGRAU), Hyderabad, India, 21–24 June 2009, Abstract p 54

Syamala VK, George S, Joseph K, Idicula SP, Nair ANS, Nair NU (2015) Growth of direct seeded and budded stump polybag rubber (*Hevea brasiliensis*) plants under different nutrient management systems. Rubber Sci 28(1):76–81

Sys C (1975) Report on the ad hoc expert consultation on land evaluation. FAO World Soil Resources Report No. 45. Rome, pp 59–79

Thomas KK, Panikkar AON (2000) Indian rubber plantation industry: genesis and development. In: George PJ, Jacob KC (eds) Natural rubber: agro-management and crop processing. Rubber Research Institute of India, Kottayam, pp 1–19

Ulaganathan A, Gilkes RJ, Nair NU, Jessy MD, Swingman N (2010) Soil fertility changes due to repeated rubber cultivation. Abstracts of the 19th biennial symposium on Plantation Crops, Kottayam, Kerala, India, 7–10 Dec 2010, pp 135–136

Ulaganathan A, Syamala VK, Jessy MD (2012) Macro and micronutrient status of the traditional rubber growing regions of South India. Proceedings of the International Rubber Conference, Kovalam, Kerala, 28–31 Oct 2012, p 154

van Oijen M, Dauzat J, Lawson G, Harmand JM, Vaast P (2010) Coffee agroforestry systems in Central America: II. Development of a simple process-based model and preliminary results. Agrofor Syst 80(3):361–378

Verheye W (2010) Growth and production of rubber. In: Verheye W (ed) Land use, land cover and soil sciences. Encyclopaedia of life support systems. UNESCO-EOLSS Publishers, Oxford

Varghese M, Sharma AC, Pothen J (2001) Addition of litter, its decomposition and nutrient release in rubber plantations in Tripura. Indian J Nat Rubber Res 14(1):116–124

Vijayakumar KR, Chandrashekhar TR, Varghese P (2000) Agroclimate. In: George PJ, Jacob KC (eds) Natural rubber, agro-management and crop processing. Rubber Research Institute of India, Kottayam, Kerala, pp 97–116

Vimalakumari TG, Joseph K, Jessy MD, Kothandaraman R, Mathew J, Punnoose KI (2001) Influence of intercropping on the rhizosphere microflora of *Hevea*. Indian J Nat Rubber Res 14(1):55–59

Virgnon-Brenas S, Gay F, Ricard S, Snoeck D, Perron T, Mareschal L, Laclau J, Gohet E, Malagoli P (2019) Nutrient management of immature rubber: a review. Agron Sust Dev 39:11. https://doi.org/10.1007/s13593-019-0554-6

Watson GA (1957) Cover plants in rubber cultivation. J Rubber Res Inst Malaya 15(1):2–18

Watson GA (1989) Nutrition. In: Webster CC, Baulkwill WJ (eds) Rubber. Longman Scientific and Technical, UK, pp 291–348

Webster CC, Baulkwill WJ (1989) Rubber. Longman Scientific and Technical, Essex. 614 p

Wibawa G, Joshi L, Noordwijk MV, Penot E (2006) Rubber based agro-forestry systems (RAS) alternatives for rubber monoculture system. IRRDB: Annual Conference, Ho-Chi Minh city, Vietnam, p 745

Xianhai Z, Mingdao C, Weifu L (2012) Improving planting pattern for intercropping in the whole production span of rubber tree. Afr J Biotechnol 11(34):8484–8490

Yasin S, Junaidi A, Wahyudi E, Herlena S, Darmawan KK (2010) Changes of soil properties on various ages of rubber trees in Dhamasraya, West Sumatra, Indonesia. J Trop Soils 5(3):221–227

Zhang M, Feng W, Chen J, Zou X (2021) Litter and microclimate controls on soil heterotrophic respiration after converting seasonal rainforests to rubber plantations in tropical China. Agri Forest Meteorol 310:108623. https://doi.org/10.1016/j.agrformet.2021.108623

Ziegler AD, Phelps J, Yuen JQ, Webb EL, Lawrence D, Fox JM, Bruun TB, Leisz SJ, Ryan CM, Dressler W, Mertz O, Pascual U, Padoch C, Koh LP (2012) Carbon outcomes of major land-cover transitions in SE Asia: great uncertainties and REDD+ policy implications. Glob Biol 18(10):3087–3099

Mercykutty Joseph completed her M.Sc. (Agri) and Ph.D. in Soil Science from the Tamil Nadu Agricultural University, Coimbatore, India. She had more than three decades of research experience, and her major areas of research in rubber were on chemistry and nutrient availability of tropical soils with special reference to rubber, characterization of soil acidity and its management, secondary and micronutrient management, development of integrated nutrient management systems, soil enzymes, rhizosphere chemistry and plant nutrition. She was actively involved in soil site suitability evaluations for rubber cultivation and provided advisory services on soil health and nutrient management of rubber plantations. She had undergone training on Diagnosis and Recommendation Integrated System (DRIS) at the University of Georgia, USA, and on rhizosphere chemistry and plant nutrition research at the Institute of Plant Nutrition, University of Hohenheim, Stuttgart, Germany.

Cocoa (*Theobroma cacao* L.)

Ravi Bhat, Bhavishya, and S. Sujatha

Abstract

Cocoa is a commercial beverage crop being grown in several countries in Africa, South America, and Asia. The crop is raised as an understory component in agroforestry systems all over the world, while it is considered as suitable component crop in coconut, arecanut, or oil palm-based cropping systems in parts of Asia. Being grown on a variety of soils which are basically poor in nutrition, soil test-based nutrient management is vital not only to obtain optimum growth and productivity but also to arrest soil degradation. A review of the various studies in different countries revealed that soil fertility status, carbon sequestration, and ecosystem functions were improved when cocoa was a component crop in agroforestry systems or palm-based cropping systems. Research work done by various institutions on technological interventions to enhance soil health including integrated nutrient management, conservation measures, cropping system approach, crop residue recycling, biofertilizers, and carbon sequestration are reviewed and future strategies formulated.

Keywords

Cocoa · Agroforestry systems · Soil health management · Cropping systems · Carbon sequestration · Integrated nutrient management · Soil fertility

---

R. Bhat (✉)
ICAR-Central Plantation Crops Research Institute, Kasaragod, Kerala, India

Bhavishya
ICAR-Central Plantation Crops Research Institute, Regional Station, Vittal, Karnataka, India

S. Sujatha
ICAR-Indian Institute of Horticulture Research, Bengaluru, Karnataka, India

G. V. Thomas, V. Krishnakumar (eds.), *Soil Health Management for Plantation Crops*, https://doi.org/10.1007/978-981-97-0092-9_8

Abbreviations

AGC	Aboveground carbon
AMF	Arbuscular mycorrhizal fungi
CEC	Cation exchange capacity
LER	Land equivalent ratio
MSL	Mean sea level
OMR	Organic matter recycling
SOC	Soil organic carbon

1 Introduction

Soil is the basic requirement for crop production and has a multifunctional role as a reservoir of nutrients, source of water, and media for biological activities as well as for providing mechanical support for the crops. Anthropogenic activities such as intensive cultivation practices, excessive use of synthetic chemicals, lack of nutrient substitution in crop production process, aberrations in climate, and environmental factors are adversely affecting soil fertility and crop productivity. In the recent past, soil degradation has emerged as a major impediment for sustainable crop production. Therefore, soil health management assumes great significance to achieve sustainable productivity of the crops in the current and future scenarios.

2 Importance of Cocoa

Cocoa (*Theobroma cacao* L.) is the most ancient beverage crop and is grown predominantly in tropical countries. The pods of cocoa are mainly used in the production of highly traded commodities such as chocolates, cocoa butter, and other value-added products having global importance. Although cocoa is native to the Amazon basin, its cultivation has spread rapidly to more than 58 countries in regions like Central and South America, West Africa, and Southeast Asia. At present, it is cultivated in more than 10 million ha area with a production of 5.24 million t of dry beans (ICCO 2020). The four countries, viz., Cote d'Ivoire, Ghana, Cameroon, and Nigeria, account for 74% of worldwide cocoa production. The global demand for chocolate is expected to double by 2050. The crop also has ecological relevance as it can be grown under different tree canopies providing habitat for higher diversity of organisms. Cocoa agroforestry intensification leads to biodiversity conservation and ecosystem functioning.

In India, cocoa is grown in southern states of Karnataka and Kerala earlier as mixed crop in arecanut and coconut plantations. The increasing importance for the crop is evident from the area expansion in nontraditional areas such as Tamil Nadu and Andhra Pradesh in India. Cocoa is grown in 97,563 ha with a production of 27,072 t of dry beans (DCCD 2020). The crop is predominantly taken up as a component crop in palm-based ecosystem by small farmers with land holding of 2–5 ha for additional income, and only 5% of the production comes from large plantations

(Malhotra et al. 2016). The demand for cocoa-based products has been growing at a rapid rate of 15–20% per annum. The present demand for dry bean per annum in India for chocolate industry and confectionaries is more than 50,000 t (Malhotra et al. 2016). The current domestic production of cocoa beans is not sufficient to meet the demand of the industry. Hence, India imported large quantities of dry beans worth Rs. 20,210 million during 2020–2021 compared to the imports worth Rs. 1350 million in 2005–2006 (www.dccd.in, accessed on 12 June 2023). The consumption of chocolate is increasing in various countries on account of increasing prosperity coupled with shift in food habits. Due to increasing demand for cocoa products, India plans to increase the production and reduce the dependency on costlier imports. During 2020–2021, cocoa products, worth Rs. 11,084, were exported to other countries (www.dccd.in, accessed on 12 June 2023).

Cocoa is rich in polyphenols and has several health benefits. Flavonoids including catechin, epicatechin, and procyanidins in cocoa beans exhibit antioxidant activity (Badrie et al. 2015). The dry beans of cocoa are widely used in the preparation of chocolates, confectioneries, beverages, and several value-added products. Cocoa butter is used in the cosmetic industry. Value-added products like fruit drink, wine, and vinegar are made from cocoa pulp juice/mucilage. Cocoa pod husk is used as animal feed ingredient (Aregheore 2000; Sampath et al. 1990; Smith and Adegbola 1982) and as cost-effective feed for fish (Fagbenro 1992, 1995). Fallen cocoa leaves and pruned biomass can be converted to nutrient-rich manure using earthworms (Chowdappa et al. 1999). Pod husk is a rich source of potassium and can be used as a manure (Sujatha et al. 2015).

3 General Soil Characteristics and Requirements

Globally, cocoa is grown on a variety of soils. The soil orders suitable for cocoa cultivation are listed by Smyth (1980). Entisols and inceptisols are suitable for cocoa, but shallow soils and soils with water stagnation problem are not ideal. Cocoa is ideally grown in Ultisols in humid tropics. Cocoa farming has been taken up on a wide range of soils, but most of the soils are poor in fertility status (Wood and Lass 2001).

Generally, the rooting pattern of a crop indicates the soil requirement for that crop. The root distribution pattern in 12-year-old cocoa trees on a red yellow podzolic-red yellow latosol intergrade indicated the presence of 83% of the roots within 60 cm soil layer and 17% of the roots between 60 and 120 cm soil depth. In podzolized soil, about 91% of the roots are distributed within 60 cm soil depth in 5-year-old cocoa trees. However, Zevallos and Coral (1972) noticed that the root development of cocoa in both soil types is confined to lower depth compared to that in soils of Bahia and Espirito. In cocoa, maximum roots are concentrated within 25 cm radius from the tree trunk in monocropped situation, while 42% of total roots are confined to 25 cm radius in mixed cropping situation. The ramification of all roots rapidly decreased beyond 100 cm depth, though they could be traced up to 480 cm below ground level (CPCRI 1991). Root activity in rainfed cocoa indicated that the maximum absorption of applied ^{32}P occurred from a soil depth of 30 cm in

Kerala, India (Wahid et al. 1989). The density of active roots at 0.0–20.0 cm, 20–40 cm, 40–60 cm, and 60–350 cm soil depth was 23.3%, 33.5%, 18.5%, and 24.7%, respectively. The spread of active roots laterally was observed within 1 m distance from the plant, which accounted for 75% of the total root activity. An area of 1.5 m radius around the plant had about 85% of the root activity.

The comparison of 13 cocoa plantations on Oxisols and 11 on Alfisols/Ultisols in southeastern Bahia indicated that low yields of cocoa on Oxisols are due to low soil fertility, excessive or deficit shade, and poor management practices (Santana and Zevallos 1981). Poor soil fertility and imbalanced fertilization are the major production constraints contributing to yield reduction in cocoa (Ahenkorah et al. 1987; Aikpokpodion 2010; Appiah et al. 1997; Baligar et al. 2006).

In the initial stages of introduction of cocoa to India, it was grown mainly in Ultisols that are less fertile containing sesquioxides and lower concentrations of base cations (Babu 1981). The important constraints in laterite soil are loss of nutrients through runoff and leaching due to high intensity rainfall (Badrinath et al. 1993), low cation exchange capacity (CEC) of 3–14 c mol kg^{-1}, and poor nutrient retention (Shivaprasad et al. 1998). Based on comparison of soil factors in high-yielding (990 kg ha^{-1}) and low-yielding (335 kg ha^{-1}) areas, it is concluded that soil resistance (shear strength) and clay content (reduced aeration) are the most important factors affecting the root development and limiting the yield in areas of lower productivity (Cadima and Alvim 1973).

In West Java, the growth of cocoa was found better on Regosols due to higher CEC and presence of more bases. The latosols were found not suitable due to high clay content. When the slope was steeper, the efficiency of N and K fertilization was lower, and the possibility of erosion was higher (Hardjono 1986). The soils with depth of 1.5 m, 3.5% organic matter, C/N ratio of above 9, base exchange capacity of more than 12 meq 100 g^{-1} soil, and base saturation above 35% are considered ideal for cocoa. Thus, the studies suggest that the soil depth should be above 1.0 m, and the texture should be sandy or alluvial with good water-holding capacity. Cocoa requires adequate rainfall for proper growth.

4 Indicators of Soil Health and Soil Health Assessment

Soil physical, chemical, and biological parameters are important indicators from soil health point of view. The assessment of these indicators gives a fair idea of the suitability of the soil to grow crops. Though measured separately, they are interlinked in soil and affect the crop growth in long run. Optimum physical parameters are essential for sustainable growth of plants.

4.1 Physical Indicators

Soil physical indicators, viz., bulk density, porosity, water-holding capacity, and infiltration rate, are important for crop growth (Shahane and Shivay 2021). These indicators directly influence the rooting and availability of nutrients and water to plants and in turn the crop growth and yield. Cocoa needs soil with better water-holding capacity and good aeration for better growth and yield. The bulk density of 1.35 g cm^{-3} for better growth of cocoa is considered optimum (Falade 1975).

4.2 Chemical Indicators

Soil pH, soil organic carbon (SOC), electrical conductivity, and soil nutrient availability are some of the major soil chemical indicators that indicate the status of nutrient availability to the crop also. In tropical soils, it is very important to improve and maintain the status of soil organic carbon for sustaining crop production (Greenland 1994; Woomer et al. 1994). Earlier workers reported the carbon level in soil under cocoa ecosystem (Beer et al. 1990; Ekanade et al. 1991). In a span of 9 years, the organic carbon level is increased by 4 g kg^{-1} from 28 g kg^{-1} implying that cocoa ecosystem maintains SOC status over the years. Though the organic carbon content was less under cocoa ecosystem compared to forest ecosystem, the same was higher than under annual crops (Duguma et al. 2001; Kotto-Same et al. 1997). The soil pH is lower in cocoa system (5.9) than in forest soil (6.8) (Ekanade 1988). In India, soil pH was lower in cocoa basin compared to that in arecanut basin and interspaces in arecanut + cocoa cropping system, but the available P content and micronutrients were comparable in arecanut and cocoa basins. However, the available K was less indicating higher K requirement for both crops (Manikandan et al. 1987).

4.3 Biological Indicators

Soil microbial communities and enzymatic activities measure the biological status of the soil and influence the mineralization and availability of nutrients to crops. Management of biological health of the soil is vital for sustainable crop production. Several researchers have reported very high content of microbial biomass, plant beneficial microbial communities (Bopaiah and Shetty 1991), function-specific nitrogen-fixing *Azospirillum* (Govindan and Purushothaman 1985), and plant beneficial actinomycetes (Barreto et al. 2008) in the rhizosphere of cocoa. The addition of high amount of biomass by cocoa enhanced the microbial proliferation in cocoa basin. An effective endophytic symbiotic system in cocoa with arbuscular mycorrhizal fungi (AMF), which acts as extension of fine roots for extraction of nutrients from greater volume of soil, was observed in the organically grown cocoa plantation. This highlights the important role played by AMF in the nutrient cycling and nutrient uptake in acidic soil under humid tropical conditions (Iglesias et al. 2011).

5 Soil Fertility Constraints: Issues of Soil Contamination

For better crop growth and yield, loose soil with optimum potassium, nitrogen, and trace elements is ideal (Badrie et al. 2015). In India, cocoa is grown as component crop in arecanut, coconut, and oil palm-based multi-storied cropping systems. The inherent constraints of laterite soils in arecanut growing regions are high rainfall, undulating topography, low CEC due to kaolinite clay, leaching of bases, and P fixation. Heavy rainfall during monsoon season increases run-off losses. The laterite soils have poor nutrient and moisture retention capacity resulting in deficiency of N and K due to heavy rainfall events (Badrinath et al. 1993). Arecanut is grown on clayey soils also with cocoa as component crop, where problems due to water stagnation are common. Similarly, soils under coconut-cocoa cropping system also experienced problems due to poor nutrient retention. Furthermore, poor soil fertility in cocoa-growing regions of Malaysia (Ling 1986), nutrient retention problems in sandy soils of Venezuela (Aranguren et al. 1982), drainage problems in Costa Rica soils (Alpizar et al. 1986), and high fertility levels in soils of Brazil (de Oliveira Leite and Valle 1990) are some of the production constraints for cocoa cultivation.

Rock phosphate contains considerable levels of heavy metals (Azzi et al. 2017; Sabiha-Javied et al. 2009). Continuous application of phosphate rock adds heavy metals such as cadmium (Cd) and zinc (Zn) as impurities to the soils in cocoa production system. This, in turn, results in accumulation of heavy metals in different plant parts of cocoa (Hanafi and Maria 1998) and in the human food chain.

6 Impact of Climate Change

The weather parameters like rainfall, temperature, and relative humidity affect the crop growth and yield. The variation in temperature, solar radiation, and day length was very little in tropical countries where cocoa is a major cash crop. Seasonal variations in both rainfall and temperature influence the pod setting. Distribution of rainfall is more important than total annual rainfall for cocoa in the West Coast region of India (Sujatha et al. 2017). Intensity and distribution of rainfall decide the pattern of cropping in mature cocoa. The crop needs to be irrigated during the rainless period when the rainfall is less than 1250 mm, whereas, rainfall above 2500 mm may increase humidity and result in related problems such as black pod disease. High humidity is common in cocoa-growing areas with 100% during night and 70–80% during day. Cocoa can be successfully grown up to 300 m MSL. However, it can be grown up to 1100–1200 m altitude. Optimum range of mean monthly maximum for cocoa is 30–32 °C. The mean maximum temperature in most of the cocoa-growing areas is 30–32 °C, whereas mean minimum is 18–21 °C. Cambium growth, which is linked to flowering, is inhibited at low temperatures. High temperature reduces the pod-growing period and in turn affects the yield and the bean size. Thus, any change in climate will have serious implications on cocoa cultivation. A positive high correlation between rainfall and yield has been reported from West Africa, Brazil, other Latin American countries, Malaysia, and Sri Lanka. Temperature, light intensity, and day length are normally not considered as limiting factors for cocoa yields except in Brazil which experiences nearly 4 months of low temperature during winter season.

The projections of climatic scenarios in 2030 and 2050 in Ghana and Ivory Coast, the countries having maximum area and production of cocoa (Läderach et al.

2011), indicated that the yearly and monthly minimum and maximum temperatures would increase by 2030 and 2050 (up to 2.0 °C) which will result in reduced suitability of the present cocoa-growing areas in these countries. The optimum altitude for ideal cocoa growth which is currently between 100 and 250 m above MSL will increase by 2050 to an altitude between 450 and 500 m above MSL. However, in West Africa, studies indicated that transforming cocoa-based agroforestry systems by changing composition of shade tree species would increase the suitable area by 6% by 2060 under future climate scenarios (Salamanca et al. 2023).

7 Fertility Management for Healthy Soils: Fertilizer Recommendations

Soil fertility management should encompass measures to maximize productivity and rebuild long-term soil fertility to ensure protection of environment and natural resources. Nutrient management is important to replenish the soil nutrients removed during crop production process, the nutrients lost by leaching, and various other environmental processes. Improving and maintaining soil fertility levels at desired level are essential to achieve optimum growth and productivity of crops. Application of fertilizers and organic manures or recycling of wastes available in the garden improves nutrient availability apart from keeping the soil healthy by increasing the biological activity. Soil microbial community plays an active role in making the nutrients available to plants and conditions need to be created for the growth and proliferation of microbial community. The recommendation of fertilizers should meet the demand of crops and soil. Excess application of fertilizers may cause undesirable impact on plant and soil. Thus, soil analysis and application of fertilizers based on the soil test values need to be given importance. Several studies have been conducted worldwide to recommend fertilizers to cocoa based on soil fertility levels and crop need.

Fertilizer recommendation for optimum yield varies in different cocoa-growing countries. Type of soil and the cultivation practices also decide the nutrient requirement of cocoa. The nutrient content in soil, plant, or both are considered for deciding the nutrient application to the crop. The fertilizer requirement of cocoa was 200 kg N, 25 kg P, 300 kg K, and 140 kg Ca ha^{-1} in Malaysia (Thong and Ng 1978). The results of the fertilizer experiments conducted in Nigeria revealed that the annual fertilizer rates (kg ha^{-1}) for cocoa are 120–204 N, 30–60 P, 67 K, and 6 g $tree^{-1}$ B (Ojeniyi 1982). In Southern Bahia, Brazil, the cocoa yield was the highest with 90 kg N, 90 kg P_2O_5, and 90 kg K_2O ha^{-1} (Cabala et al. 1982), whereas, on soils of Amazonia, (yellow latosols), annual application of 60 kg each of N, P_2O_5, and K_2O ha^{-1} was optimum (Morais 1988). The fertilizer dosages for cocoa crop based on trials conducted in different countries are summarized in Table 1. Fertilizer application to cocoa may also depend on the nutrient removal by the crop. Nutrient removal by cocoa in different countries is given in Table 2. The quantities of N, P, and K removed by 1.0 t of dry beans will work out to 43.8 kg, 8.0 kg, and 64.3 kg, respectively. The average crop removal by pods would be around 85, 37, and 154 g

Table 1 Fertilizer recommendation for mature cocoa trees in different countries

Country	Fertilizer				Remarks	References
	N	P_2O_5	K_2O	Mg O		
Sarawak	27.3	27.3	38.5	4.5	g tree^{-1}	Ebon et al. (1978)
Nigeria	120–204	30–60	67	–	kg ha^{-1}	Ojeniyi (1982)
Brazil	50–100	25	85	15	kg ha^{-1}	Hardy (1960)
Malaysia	200	25	300	70	kg ha^{-1}	Thong and Ng (1978)
India	100	40	140	–	g tree^{-1}	Shama Bhat (1988)
Colombia	150	90	200		kg ha^{-1}	Uribe et al. (2001)

Table 2 Nutrients (kg) removed in a crop of 1.0 t dry beans

Country	Type of cocoa	N	P	K	Ca	Mg
Nigeria	Amazon	39.8	6.3	85.6	–	–
Nigeria	Amelonado	38.3	5.7	76.9	–	–
Cameroon	Trinitario	34.2	6.3	72.6	8.2	6.8
W. Malaysia	Amazon	31.0	4.9	53.8	4.9	5.2
India	Forestero	43.8	8.0	64.3	–	–

each of N, P_2O_5, and K_2O when the cocoa yield was about 2 kg of dry beans plant^{-1} (about 60 pods) year^{-1}. In India, 100:40:140 g of N, P_2O_5, and K_2O plant^{-1} for a year was given as fertilizer recommendation for cocoa under average management which is based on the crop removal figures (Shama Bhat 1988).

8 Influence of Fertilizers and Plant Protection Chemicals on Soil Health and Soil Microbiology

The application of fertilizers either organic or inorganic and plant protection chemicals have shown to affect the health of the soil and microbial diversity in cocoa production systems. The systems may be agroforestry or monoculture or mixed cropping in coconut, arecanut, or oil palm. Microbial biomass in soil from plots subjected to organic management was not significantly different from the microbial biomass in soil from plots subjected to conventional management in Bolivia (Alfaro-Flores et al. 2015). Further, they reported that in dry season, the soils under conventional agroforestry showed significantly higher microbial quotient compared to the soils under organic monoculture indicating importance of diversity in plants for better microbial wealth compared to monoculture even with organic practices. In Indonesia, cocoa-coconut agroforestry was more ecofriendly by avoiding contribution to CO_2 emission, acidification, and eutrophication, whereas the system resulted into highest organic carbon and soil organic matter which supported growth and activity of beneficial soil microbes (*Pseudomonas* sp. and *Trichoderma* sp.) (Utomo et al. 2016).

Copper fungicides are used regularly on cocoa for control of pod rot disease caused by the pathogenic fungus, *Phytophthora* sp. In a 3-year study conducted in

Cameroon, spraying copper fungicides did show reduction in earthworm castings in cocoa orchard. The copper fungicide spray can safely be recommended to farmers for the control of pod rot disease in cocoa as copper fungicides had no negative effect on either litter decomposition or soil bulk density. However, the spray must be accompanied by growing of leguminous cover crops or mulch to hasten litter decomposition rates and soil coverage (Norgrove 2007).

9 Soil Biodiversity, Microbial Synergy, and Function-Specific Microbial Communities

In plantation crops, numerous opportunities exist to introduce inter-/mixed crops to enhance aboveground biodiversity leading to belowground biodiversity which influences soil health and crop growth. Enhancing system productivity with significant benefits to soil biology and fertility is possible with the adoption of technologies, viz., cover cropping and green manuring; recycling of lignin-rich crop residues; application of biofertilizers and bio-stimulants; cropping system approach—inter-, mixed, and high-density multispecies cropping; and mixed farming integrating animal husbandry (Thomas et al. 2022).

Microbial communities in the rhizosphere can exert beneficial, neutral, or harmful effects on plant growth (Whipps 2001). The mode of action of microbes for stimulating the plant growth is increasing nutrient absorption, inducing and producing plant growth regulators and protection from the pathogens. Induced resistance mechanisms in plants are also activated by microbes. Production of antibiotics such as lipopeptides (fengycin, iturin, surfactin) by *Bacillus subtilis* is related to indirect effect of microbes. Conversion of insoluble phosphate to soluble phosphate is one of the important activities of microbes in soil. Some rhizobacteria such as *B. subtilis* have multiple functions, viz., solubilization of insoluble phosphates, the formation of biofilm, the production of indole acetic acid (IAA), and protection against infections (Sucuia et al. 2015). The isolates of *B. subtilis* from cocoa rhizosphere have plant growth-promoting activities like phosphate solubilization and were able to form biofilm and produce IAA (Koua et al. 2020).

Soil biodiversity in terms of beneficial microbial communities and biological activity, as evidenced by soil enzymatic activities, is essential for sustained production and better soil health. Application of manures and recycling the agro-residues in garden and crops grown are major determinants of soil beneficial microbial communities. A lower Gram-negative to Gram-positive ratio in bacterial community was observed when native vegetation was removed prior to planting cocoa compared to the removal of selected uneconomical vegetation prior to planting cocoa. Increased Gram-negative to Gram-positive bacteria ratio was achieved by cover cropping with centro (*Centrosema macrocarpum*) and perennial peanut (*Arachis pintoi*), while the fungal biomass was reduced by planting centro (Buyer et al. 2017).

The traditional cocoa cultivation practices in Venezuela were favorable to promote symbiotic arbuscular-mycorrhizal fungi (AMF) infection in cocoa plants. *Acaulospora scrobiculata* was the major AMF species associated with cocoa plants.

In Venezuela, the diversity of the AMF community in cocoa agroforestry was like that in natural undisturbed ecosystems (Cuenca and Meneses 1996).

The elevation and chemical properties of soil under cocoa influence the distribution of AMF community (Edy et al. 2022). The cocoa plants grown in lower elevation (<500 m MSL) recorded higher number of AMF spores. Chemical and physical factors of the soil in the form of phosphorus, potassium, sodium, magnesium, total nitrogen, clay, and texture significantly influenced the AMF community structure.

Microbial biomass C (μg C g^{-1} dry soil) was significantly higher in cocoa basin at 0–30 cm depth (536.2) and 30–60 cm depth (241.4) compared to arecanut, clove, coffee, and banana basins when cocoa was one of the component crops in arecanut-based cropping system. Cocoa registered a higher microbial quotient (3.90) over other crops (Bhat et al. 2008). Changes in microbial performance caused by environmental conditions can be measured by the ratio of soil microbial biomass C to organic C (Anderson and Domsch 1989). Organic production system of cocoa indicated better fungal richness over bacterial richness and organic and conventional management as well as agroforestry and monoculture showed distinct fungal community composition (Lori et al. 2022).

10 Root-Associated Microbiome

Microbiomes associated with crop roots have been receiving greater attention in recent years due to their potential to enhance plant health and productivity. It is now greatly realized that optimizing beneficial interactions with rhizosphere microorganisms helps to reduce reliance on external inputs, increase pathogen resistance, and alleviate abiotic stress. However, currently there is lack of adequate studies and knowledge on cocoa root-associated microbiome. Schmidt et al. (2022) have described the enormous opportunities available for enhancing ecosystem services with cocoa root-associated microbiome. Disease suppression can be achieved by adopting the technology of breeding for microbiome-mediated disease resistance, and breeding for resource use efficiency can be used for improving nutrient cycling and yield enhancement.

11 Technologies to Promote Soil Health

Soil health is very important for crops that occupy the land for decades like plantation crops to achieve sustained growth and yield. Cocoa grown with minimal fertilizer input for a long time, loss of nutrients through erosion, and leaching and removal by the harvested crop deplete the soil (Ahenkorah et al. 1987; Appiah et al. 1997; Baligar et al. 2006; Hartemink 2005). Low soil fertility in cocoa-growing regions is one of the primary causes of poor yield (Appiah et al. 1997; Hartemink 2005; Snoeck et al. 2016), and deficiencies of nitrogen (N), potassium (K), zinc (Zn), iron (Fe), and boron (B) are the most common deficiencies noted in cocoa gardens (Ahenkorah et al. 1974; Bhavishya et al. 2021; Manikandan et al. 1987; Nelson et al. 2011).

In Malaysia, about 20.4 kg N, 3.6 kg P, and 10.5 kg K are removed from the soil for each 1000 kg of dry beans harvested, while the amount of K removed increases to about 53.8 kg if pod husks are also removed from the field (Thong and Ng 1978). In humid tropics of India, cocoa grown with arecanut removed 48.4 kg N, 6 kg P, and 95 K kg per 1000 kg beans (Bhat and Sujatha 2011a). Hence, the nutrient requirement of cocoa for sustainable production is high. Poor soil fertility and lack of appropriate fertilization are major contributors among the production constraints for reduction in cocoa yields (Ahenkorah et al. 1987; Aikpokpodion 2010; Appiah et al. 1997; Baligar et al. 2006).

It is essential to consider the crop's need to sustain crop productivity and to maintain soil health. The management strategies should address these constraints, and adoption of measures like biomass recycling, mulching, soil and water conservation, drip irrigation, and fertigation is essential to improve resource use efficiency and maintain soil health.

11.1 Cropping System Approach

Increasing the farm productivity and income, mitigating inherent soil constraints and aberrant weather situation, providing balanced food, regular employment, and sustaining soil health are some of the benefits of cropping/farming system in plantation crops (Sujatha et al. 2020). Limited land availability for agriculture warrants utilization of interspaces for growing other crops. Sustainability is achieved in cropping system by the optimization of production process in cropping system through efficient utilization of the inputs, and improving soil fertility parameters and safeguarding the environment lead to sustainability (Thomas et al. 2011). Cocoa is predominantly grown in agroforestry system in Africa and Latin America and in palm-based ecosystem in Asia. It is also grown as mixed crop with rubber in Southeast Bahia, Brazil (Alvim and Nair 1986). Osei-Bonsu et al. (2002) stated that cocoa can be grown as mixed crop with coconut instead of *Gliricidia sepium* in Ghana. In Côte d'Ivoire, cocoa can be planted with orange and avocado trees at a distance of 6.5 m from these fruit trees, but closer planting will reduce the yield of cocoa (Koko et al. 2013).

11.1.1 Arecanut-Cocoa Mixed Cropping System

The study conducted at ICAR-CPCRI, India, showed that evaporation, wind velocity, soil and air temperature, and relative humidity were influenced by arecanut + cocoa cropping system (Bavappa 1990; Shama Bhat 1983). The mean wind velocity was only 9% in arecanut+cocoa system of that in open area. Initial studies have shown that growing cocoa as mixed crop in arecanut plantation increases returns per unit area (Bhagavan and Shama Bhat 1989; Shama Bhat and Bavappa 1972). The land equivalent ratio (LER) which is the measure for advantage of cropping system was 1.74 in arecanut + cocoa mixed cropping system indicating 74% yield advantage over arecanut monocropping (Balasimha 2001; Shama Bhat 1988). The arecanut + cocoa cropping system registered higher LER (2.18) compared to other cropping systems, viz., arecanut + black pepper (1.5) and arecanut + banana (1.45).

Arranging arecanut at 2.7 m × 2.7 m and cocoa at 2.7 m × 5.4 m was recommended over a spacing of 2.7 m × 2.7 m for both the crops considering the operational advantages and similar yield with both spacings (Shama Bhat 1988).

11.1.2 Coconut-Cocoa Mixed Cropping System

The coconut + cocoa mixed cropping system is popular in Papua New Guinea, Malaysia, and India (Bhat 2002). Cocoa is planted at 4.5 m intervals between and within the rows of coconut spaced at 9.0 m × 9.0 m which gives 360 trees ha^{-1} in Papua New Guinea. In Malaysia, two rows of cocoa seedlings are planted between the rows of coconut palms spaced at 8.0 m × 8.0 m or 9.0 m × 9.0 m. In South India cocoa is suitable for mixed cropping in coconut (Nair et al. 1975). One row of cocoa at a spacing of 2.7 m in between two rows of coconut spaced at 7.5 m × 7.5 m is recommended. However, two rows of cocoa can be accommodated if the spacing of coconut is wider. A study on mixed cropping at ICAR-CPCRI, Kasaragod, India, indicated 68% and 115.9% increase in yield of coconut when cocoa was planted as single hedge and double hedge, respectively, over the pre-experimental yield (Nair et al. 1975). Other studies recorded lower mean maximum temperature, higher relative humidity, reduced evaporative demand, and monthly soil temperature under coconut + cocoa system compared to sole coconut (Nair and Balakrishnan 1977; Varghese et al. 1978).

11.1.3 Oil Palm-Cocoa Mixed Cropping System

Cocoa put in significantly better growth and yield when planted with a spacing of 2.4 m in oil palm garden spaced at 9.9 m or 10.5 m in triangular method compared to oil palm spaced at 8.7 m (Amoah et al. 1995) indicating the need for wider spacing in oil palm for growing cocoa as mixed crop.

11.2 Conservation Agriculture Practices

Since cocoa is very sensitive to low moisture stress, soil-water conservation practices have direct effect on its productivity (Amfo et al. 2021) especially in areas where it is grown as rainfed crop. Hence, adoption of practices to improve water-holding capacity of soil, infiltration and aeration, and reducing the extent to which runoff and leaching occurs is essential in cocoa cultivation (Franzluebbers et al. 1995; Keil et al. 2008; Rice and Greenberg 2000; Zuidema et al. 2005).

The spatial and temporal distribution of water can be improved with the construction of micro-catchment systems (pits, contour ridges and furrows, terraces, earth bunds, and micro-basins) and macro-catchments (open ponds, cisterns, micro-dams, sand dams, and spate irrigation systems) (Biazin et al. 2012; Rockström et al. 2010). Shade trees have greater role to play in cocoa farms to avoid physical impact of precipitation, which can reduce soil erosion (Beer et al. 1998). The effectiveness of soil conservation approach in cocoa plantations is enhanced by the combination of plant and ridges (Satriavan et al. 2015). Soil conservation with ridge + growing green beans, maize, and ground nut was effective in significantly reducing the soil erosion and nutrient losses.

Early growth and survival of cocoa can be improved during dry periods with mulching which can maintain higher soil moisture levels. Apart from providing mulch, the litter from shade trees and cocoa trees supply organic matter to the soil. While mulching can increase aeration, infiltration and drainage, the organic matter results in a slow and steady release of nutrients into the soil rhizosphere (Rice and Greenberg 2000). Adoption of the practice of using litter as mulch also maintains soil moisture, reduces erosion, and serves as habitats for beneficial organisms (Mamani-Pati et al. 2012). Acheampong et al. (2019) observed significantly higher early cocoa yields under the mulching (plastic mulch or coffee mulch) and irrigation treatments compared to the control.

11.3 Crop Residue Recycling

The biomass production in cocoa depends on age and growing conditions. The biomass produced by cocoa per hectare is estimated at 8450 kg under arecanut and 13,000 kg under coconut (Bhat and Sujatha 2011a, b). The differences in population density and the growth habits result in such variations. Pruning is a commonly followed practice in cocoa, and about 8–10 kg pruned biomass tree^{-1} can be obtained (Abdul Haris et al. 1999). The system also produces enormous litter (Aranguren et al. 1982; Asigbaase et al. 2021) which can be recycled. However, climate (dry spells), age, type of the plantings, and shading density decide the rate of litterfall (Abou Rajab et al. 2016; Asigbaase et al. 2021; Hartemink 2005; Saj et al. 2021).

Litterfall, throughfall, and stem flow are the means of adding nutrient to soil by cocoa. The cocoa planted with coconut in single-hedge and double-hedge system shed about 818 kg and 1785 kg of litter ha^{-1} year^{-1} (dry weight), respectively (Varghese et al. 1978). The litterfall resulted in improved organic carbon content and soil fertility which was reflected in the enhanced yield of coconut. The mean annual litterfall production in different cocoa agroforestry systems ranged from 4.6 to 8.5 Mg ha^{-1}, while the amount of accumulated litter ranged from 7.7 to 16.8 Mg ha^{-1} in Brazil (Fontes et al. 2014). Cocoa leaf fall under the double-hedge system of planting in coconut garden returned about 50 kg N, 11 kg P_2O_5, and 35 kg K_2O ha^{-1} year^{-1} (Varghese et al. 1978). In southern Bahia of Brazil, litterfall in different cocoa agroforestry could add 84.2–174.6 kg N, 5.8–16.9 kg P, 14–48 kg K, 68.2–199.8 kg Ca, and 27.8–59.7 kg Mg ha^{-1} year^{-1} (Fontes et al. 2014). Hence, nutrient addition from litterfall and recycling of cocoa pod husk could make the system nutrient positive for cocoa in agroforestry system. The quantities in throughfall and stem flow were 20.1 kg N, 2.8 kg P, 74.7 kg K, 38.9 kg Ca, 9.6 kg Mg, 2.3 kg Fe, 1.0 kg Cu, and 0.7 kg Zn ha^{-1} (Opakunle 1989).

Even though the biomass available in the form of leaf litter and pruned plant part is quite high (Sujatha et al. 2015), the slow rate of decomposition due to their high carbon, lignin, polyphenol, and cellulose concentrations make them unsuitable for direct usage as organic manure (Kaba et al. 2021a). Thus, composting the cocoa leaves using earthworms is suggested before recycling back to the garden (Chowdappa et al. 1999; Coulibaly et al. 2019). The composted cocoa leaves were rich in macro- and micronutrients higher than the normal compost. A recovery of

74% was obtained, and about 6.0–9.0 t of vermicompost can be obtained by recycling the biomass produced in 1 ha. Apart from leaf litter and pruned biomass, pod husk is also available for recycling. Cocoa pod husks are the good source of nutrients in cocoa farms as they contain 15.6 g kg^{-1} N, 3.24 g kg^{-1} P, 25.2 g kg^{-1} K, 1.84 g kg^{-1} Ca, 2.62 g kg^{-1} Mg, and 1.36 g kg^{-1} S (Fidelis and Rajashekhar Rao 2017). Hence, they can be composted and returned to soil which adds nutrient back to soil (particularly for P and K) in cocoa farms and helps to build and maintain soil organic carbon (Amponsah-Doku et al. 2022). The available recyclable biomass from cocoa (which includes pruned biomass, leaf litter, and pod husk) in India was estimated at 0.7–0.8 million tons (Sujatha et al. 2015). Thus, cocoa has potential to supply considerable nutrients through recyclable biomass.

11.4 Cover Cropping and Green Manuring

Cover crops play a major role in sustainable crop production system, and they can make important component in cocoa system (Baligar and Fageria 2007). The soil erosion loss was brought down significantly from 11 Mg ha^{-1} $year^{-1}$ under monocropping through planting of cover crops like *Indigofera spicata* in Malaysia (Hashim et al. 1995). The soil loss was up to 70 Mg soil ha^{-1} $year^{-1}$ when cocoa was intercropped with banana due to the adoption of clean weeding with herbicide (Hudson 1986).

Leguminous shade trees improve the soil nitrogen (N) by fixing the atmospheric N. The general shade tree component in cocoa agroforestry system includes fast-growing, nitrogen-fixing legumes viz., *Erythrina* spp., *Gliricidia sepium*, *Cassia*, and *Inga* spp. (Rice and Greenberg 2000). *G. sepium* (Jacq. Kunth ex Walp.) is grown with cocoa for shade in agroforestry systems, and the biomass can be used as green manure (Falvey 1982; Subramanian et al. 2005). The pruned biomass of gliricidia interplanted with cocoa can add 31.4–38.0 kg N ha^{-1} (Kaba et al. 2019). This could reduce the need for N fertilizers for cocoa trees. Green manure crop *Calopogonium* spp. was effective in improving the plant growth (Ndubuaku and Kassim 2003), and growing green bean significantly reduced the soil erosion and nutrient losses in cocoa plantation (Satriavan et al. 2015).

11.5 Biofertilizers

The plant-associated microbes can benefit plants in growth and development through enhanced nutrient availability, phytohormone modulation, biocontrol of phytopathogens, and amelioration of biotic and abiotic stresses (Kumar et al. 2021). Several researchers have reported the symbiotic association of arbuscular mycorrhizal fungi (AMF) with cocoa plants (Edy et al. 2019; Iglesias et al. 2011; Luis-Alaya et al. 2023). AMF inoculum produced 125% taller cocoa seedlings than the control seedlings at 16 weeks after inoculation (Iglesias et al. 2011). Moreover, the symbiosis of AMF in cocoa plantation has potential for the preparation of biofertilizer which can take part in biostabilization of soil heavy metals in the root rhizosphere through the production of glomalin-related soil proteins (Luis-Alaya et al. 2023). In Côte d'Ivoire, endophytic bacteria were isolated from cocoa seedlings which has very good potential as biofertilizer as it solubilizes tricalcium phosphate (Ouattara et al. 2019). Increased bean production by 39.7% was reported with the application of 1000 ppm humic acid and 1500 ppm biofertilizer (Rohman et al. 2019). Adoption of these innovations could lead to better establishment of cocoa seedlings in farmers' field and higher yield realization (Kaba et al. 2021b). Researchers have also identified isolates of many endophytic fungi, especially, *Trichoderma* species (Bae et al. 2009) and bacteria, *Bacillus* spp. (Rachel et al. 2008), from cocoa with potential of plant growth promotion and pathogen suppression capacities. These PGPRs also have vast scope to be used as biofertilizers (Vessey 2003). In India, *B. subtilis* isolated from cocoa rhizosphere of different cocoa-growing regions were showing good PGPR properties and are being used as biofertilizers in cocoa gardens (Gopal et al. 2011).

11.6 Organic Amendments

Cocoa produces substantial amount of recyclable biomass (Bhat and Sujatha 2011a, b) which can be added to soil after composting as a good organic source of nutrients. Since cocoa is grown in agroforestry system or mixed cropped with perennial crops, the recyclable biomass availability is enormous. Improvement in chemical and biological soil quality and development of distinct soil fungal and bacterial communities are possible with organic management in either monoculture or agroforestry systems (Lori et al. 2022). Earlier studies have highlighted the potential of cocoa biomass to supply considerable quantity of nutrients to soil (Chowdappa et al. 1999; Sujatha et al. 2015). The drip fertigation with vermicompost extract and application of vermicompost to sole and arecanut + cocoa system resulted in similar soil fertility status (Sujatha and Bhat 2015). Cocoa pod husk is a rich source of nutrients including potassium, and its application after composting can improve soil fertility and cocoa plant growth (Doungous et al. 2018; Fidelis and Rajashekhar Rao 2017; Mulia et al. 2017).

11.7 Integrated Nutrient Management

Even though the carbon storage in cocoa system was high (Gama-Rodrigues et al. 2010; Norgrove and Hauser 2013; Smiley and Kroschel 2008), skipping fertilizer application resulted in soil fertility decline in cocoa farms (Aikpokpodion 2010; Hartemink 2005). Manikandan et al. (1987) reported reduced available K in the arecanut-cocoa system indicating higher K requirement of both crops. Moreover, significant positive correlation between yield and soil K in cocoa ($r = 0.273$) indicated the need for continuous application of K to both the crops (Bhat et al. 2012). Inorganic nutrition helps in better utilization of organic nutrition through better root growth leading to higher SOC content in soil and improved yield (Bhat et al. 2007; Bhat and Sujatha 2008). Hence, biomass recycling has to be complemented with inorganic nutrition in cocoa systems for better returns.

Potassium depletion in lateritic and shallow black soils of India (Majumdar et al. 2017; Naidu et al. 2011), variations in soil nutrient status (Bhat and Sujatha 2014), and heavy K demand by plantation crops (Bhat and Sujatha 2007, 2012; Khan et al. 2001; Manikandan et al. 1987) warrant supplementing organic farming with inorganic nutrition as K supply through organic sources is meager. (Sujatha et al. 2015). Nutrient recommendation given for cocoa grown as mixed crop in arecanut is 100:40:140 g of N, P_2O_5, and K_2O tree^{-1} year^{-1} in humid tropics of India (Abdul Haris et al. 1999). Studies conducted earlier have indicated that drip fertigation is ideal for both arecanut and cocoa in the system which is evident from increased yield, profits, and input-saving (Sujatha and Bhat 2013). The long-term studies on drip fertigation also indicated benefit in terms of increase in soil pH and organic carbon in arecanut and arecanut-cocoa system (Bhat and Sujatha 2009; Sujatha and Bhat 2015). In a study on individual and integrated application of different doses of NPK and organic matter recycling (OMR) in arecanut-based cropping system, OMR and 2/3 of recommended NPK level provided the best soil health benefits in terms of microbial biomass C, microbial quotient, and phosphatase activity. Increased dehydrogenase activity was noticed up to the application of 2/3 of recommended chemical fertilizer, whereas the activity reduced with full recommended dose of chemical fertilizer in all crop basins except in banana (Bhat et al. 2008).

12 Carbon Sequestration

The cocoa plantation occupies the land for more than three decades and accumulates significant quantity of carbon in both above- and belowground parts. Cocoa ecosystem provides many environmental benefits, including carbon storage (Alpizar et al. 1986; Arthur et al. 2022; Balasimha and Kumar 2013; Beer et al. 1990; Kotto-Same et al. 1997; Norgrove and Hauser 2013), biodiversity (Mortimer et al. 2018; Perfecto and Vandermeer 2008; Rice and Greenberg 2000; Saj et al. 2017), and soil fertility (Hartemink 2005), apart from economic return. The carbon sequestration potential of cocoa is substantial as a mixed crop and in agroforestry system. Growing cocoa in agroforestry system would sequester higher carbon than in mixed cropping or monocropping (Abou Rajab et al. 2016; Balasimha and Kumar 2013; Jacobi et al. 2014). However, carbon stock got reduced after conversion of the forest into cocoa-based agroforestry (Nijmeijer et al. 2019), and this could make the system carbon-negative (Kongsager et al. 2012; Vervuurt et al. 2022). Plant characteristics (tree species, age, crops, biodiversity, and tree density), system characteristics (structure, function, stability) and management factors (tillage, fertilization, residues, holding size, and harvesting regime) influence the sequestration potential of agroforestry systems (Feliciano et al. 2018). Thus, removal of shade trees would significantly reduce carbon stocks (Norgrove and Hauser 2013). Even though the carbon stock per tree was higher in older cocoa trees than younger trees (Gonas et al. 2022;

Nijmeijer et al. 2019; Saj et al. 2013), the net annual increment was higher in the initial years, and it got decreased with age (Balasimha 2011).

The shade trees in the agroforestry system influenced the carbon stock significantly (Abou Rajab et al. 2016; Middendorp et al. 2018; Wade et al. 2010). In Peru, the total aboveground carbon (AGC) stock in cocoa trees in agroforestry system was 6.06–8.71 Mg ha^{-1} in differently aged cocoa systems, whereas the total AGC stock in the system was 13.64–24.86 Mg ha^{-1} (Gonas et al. 2022). It was lesser compared to the cocoa grown in shaded system in Cameroon (Duguma et al. 2001; Kotto-Same et al. 1997). The total biomass carbon stock was 14.4 Mg ha^{-1} in cocoa trees, while it increased to 121.1 Mg ha^{-1} in the agroforestry system (Norgrove and Hauser 2013). In Ghana, the AGC varied between 39 and 131 Mg C ha^{-1}depending on type of cocoa farm. The traditional cocoa plantations with more than 25% shade cover stored three times more aboveground biomass than intensified cocoa plantations with less than 25% shade (Wade et al. 2010). It indicates that the contribution of cocoa to the total carbon stock in the system was smaller in agroforestry system compared to cocoa grown in cropping system with palms (Abou Rajab et al. 2016; Balasimha and Kumar 2013; Somarriba et al. 2013). Most of the studies indicate that African cocoa plantations were superior in terms of aboveground biomass carbon than their American counterparts (Somarriba et al. 2013).

The AGC stocks in cocoa ranged from 1.66 t ha^{-1} in the fourth year after planting to 13.34 t ha^{-1} by tenth year in arecanut-cocoa system in humid tropics of India (Balasimha 2011). Estimates of annual biomass increments varied from 1.38 to 2.66 t ha^{-1}, while the annual increment in carbon stock was 0.6–1.0 t ha^{-1} year^{-1} (Balasimha and Kumar 2013). However, the annual rate of accumulation of carbon in aboveground biomass in Central America was higher which ranged between 1.3 and 2.6 Mg C ha^{-1} year^{-1} (Somarriba et al. 2013) which included timber and fruit trees. Hence, the primary productivity of cocoa mixed cropped in arecanut plantation is comparable to cocoa in agroforestry systems (Alpizar et al. 1986; Balasimha and Kumar 2013; Beer et al. 1990). However, in Bahi (Brazil) cocoa in commercial rubber-cocoa mixed cropping system produced less AGC. The aboveground carbon accumulated in this system was 82 Mg ha^{-1} which included 68 Mg ha^{-1} in rubber and 3.8 Mg ha^{-1} in cocoa (Cotta et al. 2008).

The organic carbon enrichment happens in the cropping system through addition of recyclable organic waste and dead and decayed roots. With age of the cocoa plantations, generally soil carbon stock increases (Arthur et al. 2022). Soil carbon stock in 1 m soil depth was 90 Mg C ha^{-1} in shaded cocoa plantation in Cameroon (Norgrove and Hauser 2013), whereas in Peru, it was 119.96 Mg ha^{-1} to 131.96 Mg ha^{-1} at 30 cm soil depth in different agroforestry system. But higher soil carbon stock of 150 Mg C ha^{-1} was reported from Indonesia in cocoa-gliricidia agro-forests (Smiley and Kroschel 2008), while 30-year-old cocoa agroforestry systems in Brazil on Oxisol stored 320 Mg C ha^{-1} soil carbon stock (Gama-Rodrigues et al. 2010). Interestingly, carbon stored per unit area of soil was higher with cocoa planted at low plant density under shade than cocoa planted at high density without shade (Ofori-Frimpong et al. 2010).

Studies in different cocoa-growing regions indicate that cocoa in agroforestry system or mixed cropping system can store considerable amount of carbon, which is of greater significance at the time of climate change. However, in some countries interests of the farmers have shifted toward monoculture of cocoa to achieve higher productivity, which would result in the loss of biodiversity and importantly reduced carbon storage in cocoa plantations (Abou Rajab et al. 2016; Bisseleua et al. 2009; Rice and Greenberg 2000; Siebert 2002).

13 Future Strategies

In the future scenario, management of soil health to maximize productivity by rebuilding long-term soil fertility and ensuring protection of environment and natural resources is a challenge in various cocoa-growing regions of the world due to the deterioration of production environment accelerated by climate change. As majority of the production comes from the cocoa agroforestry systems, the adoption of appropriate species mixtures and density, along with strategies to promote supporting and regulatory ecosystem services, can greatly enhance the sustainability of the system. Site-specific fertilizer recommendations integrated with organic matter recycling can be an effective strategy for development of healthy production systems. Root-associated microbiome studies are of great significance to reduce dependence on external inputs, increasing pathogen resistance, to overcome abiotic stress, and to optimize beneficial interactions. The palm-based cocoa cropping systems, with ideal microclimate supporting the complementary and synergistic biological interactions, are important not only for the economic stability of the smallholder farming community but also for its potential for carbon sequestration in the scenario of climate change.

References

Abdul Haris A, Balasimha D, Sujatha S, Bhat R, Khader KBA (1999) The influence of drip irrigation and fertilizer on yield and photosynthetic characteristics of cocoa in mixed cropping system with arecanut. J Plantn Crops 27:131–135

Abou Rajab Y, Leuschner C, Barus H (2016) Cacao cultivation under diverse shade tree cover allows high carbon storage and sequestration without yield losses. PLoS One 11(2):1–22. https://doi.org/10.1371/journal.pone.0149949

Acheampong K, Daymond AJ, Adu-Yeboah P (2019) Improving field establishment of cacao (*Theobroma cacao*) through mulching, irrigation and shading. Exp Agric 55:1–15. https://doi.org/10.1017/S0014479718000479

Ahenkorah Y, Akrofi G, Adri A (1974) The end of the first cocoa shade and manurial experiment at the Cocoa Research Institute of Ghana. J Hortic Sci 49:43–51

Ahenkorah Y, Halm BJ, Appiah MR, Akrofi GS, Yirenkyi JEK (1987) Twenty years' results from a shade and fertilizer trial on amazon cocoa (*Theobroma cacao*) in Ghana. Exp Agric 23:31–39

Aikpokpodion PE (2010) Nutrients dynamics in cocoa soils, leaf and beans in Ondo state. Niger J Agric Sci 1(1):1–9

Alfaro-Flores A, Morales-Belpaire I, Schneider M (2015) Microbial biomass and cellulase activity in soils under five different cocoa production systems in Alto Beni, Bolivia. Agrofor Syst 89:789–798. https://doi.org/10.1007/s10457-015-9812-z

Alpízar L, Fassbender HW, Heuveldop J, Fölster H, Enríquez G (1986) Modelling agroforestry systems of cacao (*Theobroma cacao*) with laurel (*Cordia alliodora*) and poro (*Erythrina poeppigiana*) in Costa Rica. Agrofor Syst 4:175–189. https://doi.org/10.1007/BF02028353

Alvim R, Nair PKR (1986) Combination of cacao with other plantation crops: an agro-forestry system in Southeast Bahia, Brazil. Agrofor Syst 4:3–15. https://doi.org/10.1007/BF01834698

Amfo B, Ali EB, Atinga D (2021) Climate change, soil water conservation, and productivity: evidence from cocoa farmers in Ghana. Agric Syst 191:103172. https://doi.org/10.1016/j.agsy.2021.103172

Amoah FM, Nuertey BN, Baidoo-Addo K, Oppong FK, Osei-Bonsu K, Asamoah TEO (1995) Underplanting oil palm with cocoa in Ghana. Agrofor Syst 30:289–299

Amponsah-Doku B, Daymond A, Robinson S, Atuah L, Sizmur T (2022) Improving soil health and closing the yield gap of cocoa production in Ghana – as review. Sci Afr 15:e01075. https://doi.org/10.1016/j.sciaf.2021.e01075

Anderson TH, Domsch KH (1989) Ratios of microbial biomass carbon to total organic carbon in arable soils. Soil Biol Biochem 21:471–479

Appiah MR, Sackey ST, Ofori-Frimpong K, Afrifa AA (1997) The consequences of cocoa production on soil fertility in Ghana–a review. Ghana J Agric Sci 30:183–190. https://doi.org/10.4314/gjas.v30i2.1970

Aranguren J, Escalante G, Herrera R (1982) Nitrogen cycle of tropical perennial crops under shade trees. II. Cacao. Plant Soil 67:259–269

Aregheore EM (2000) Crop residues and agro-industrial by-products in four Pacific Island countries: availability, utilization and potential value in ruminant nutrition. Asian- Australasian J Animal Sci 13(Suppl B):266–269

Arthur A, Acquaye S, Cheng W, Dogbatse JA, Konlan S, Domfeh O, Quaye AK (2022) Soil carbon stocks and main nutrients under cocoa plantations of different ages. J Soil Sci Plant Nutr 68(1):99–103. https://doi.org/10.1080/00380768.2022.2029219

Asigbaase M, Dawoe E, Lomax BH, Sjogersten S (2021) Temporal changes in litterfall and potential nutrient return in cocoa agroforestry systems under organic and conventional management, Ghana. Heliyon 7(10):e08051. https://doi.org/10.1016/j.heliyon.2021.e08051

Azzi V, Kazpard V, Lartiges B, Kobeissi A, Kanso A, El Samrani AG (2017) Trace metals in phosphate fertilizers use in eastern Mediterranean countries. CLEAN – Soil, Air, Water 45(1):1500988. https://doi.org/10.1002/clen.201500988. 29 p

Babu PVLP (1981) Laterite as an unconformity plane in the evolution of the Indian Peninsula – a synthesis. In: Anon (ed) Laterisation processes. Oxford & IBH Publishing, New Delhi. 450 p

Badrie N, Bekele F, Sikora E, Sikora M (2015) Cocoa agronomy, quality, nutritional, and health aspects. Crit Rev Food Sci Nutr 55(5):620–659. https://doi.org/10.1080/10408398.2012.669428

Badrinath MR, Gajendragad, Balakrishna Rao K (1993) Distribution of micronutrients in laterite soils of Puttur in relation to some soil properties. In: Sehgal J, Sarma VAK, Batta RK, Gajbhiye KS, Nagabhushna SR, Venugopal KR (eds) Red and Lateritic soils of India – resource appraisal and management. NBSSLUP, Bangalore, pp 271–274

Bae H, Richard CS, Moon SK, Soo-Hyung K, Mary DS, Rachel LM, Bailey BA (2009) The beneficial endophyte *Trichoderma hamatum* isolate DIS 219b promotes growth and delays the onset of the drought response in *Theobroma cacao*. J Exp Bot 60:3279–3295

Balasimha D (2001) Growth and yield of cocoa grown under areca in relation to plant density and canopy architecture. In: 13th International cocoa conference, Kota Kinabalu, Malaysia, pp 365–372

Balasimha D (2011) Carbon sequestration potential in cocoa plantations. In: Proceedings of seminar on strategies for enhancing productivity of cocoa, CPCRI RS Vittal, India, pp 86–88

Balasimha D, Kumar NS (2013) Net primary productivity, carbon sequestration and carbon stocks in areca-cocoa mixed crop system. J Plantn Crops 41(1):8–13

Baligar VC, Fageria NK (2007) Agronomy and physiology of tropical cover crops. J Plant Nutr 30(8):1287–1339

Baligar VC, Fageria NK, Machado RC, Meinhardt LW (2006) Concentration and uptake of P, Zn and Fe as influenced by soil acidity, and levels and forms of N, P and Fe in cacao. In: 15th international cocoa research conference, COPAL, Costa Rica, 10 Sept 2006

Barreto TR, da Silva ACM, Soares ACF, de Souza JT (2008) Population densities and genetic diversity of actinomycetes associated to the rhizosphere of *Theobroma cacao*. Brazilian J Microbiol 39:464–470

Bavappa KVA (1990) Arecanut and cocoa mixed cropping. J Plantn Crops 17:12–24

Beer J, Bonnemann A, Chavez W, Fassbender HW, Imbach AC, Martel I (1990) Modelling agroforestry systems of cacao (*Theobroma cacao*) with laurel (*Cordia alliodora*) or poro (*Erythrina poeppigiana*) in Costa Rica. Agrofor Syst 12:229–249

Beer J, Muschler R, Kass D, Somarriba E (1998) Shade management in coffee and cacao plantations. Agrofor Syst 38:139–164

Bhagawan S, Shama Bhat K (1989) An appraisal of mixed cropping trial on arecanut and cocoa. J Plantn Crops 17:10–17

Bhat R (2002) Climate, soil and agronomy. In: Balasimha D (ed) Cocoa. CPCRI, Kasaragod, pp 48–69

Bhat R, Sujatha S (2007) Soil fertility status as influenced by arecanut based cropping system and nutrient management. J Plantn Crops 35(3):158–165

Bhat R, Sujatha S (2008) Root distribution of arecanut (*Areca catechu* L.) as influenced by drip fertigation in a laterite soil. Indian J Agric Sci 78(9):804–807

Bhat R, Sujatha S (2009) Soil fertility and nutrient uptake by arecanut (*Areca catechu* L.) as affected by level and frequency of fertigation in a laterite soil. Agric Water Manag 96:445–456. https://doi.org/10.1016/j.agwat.2008.09.007

Bhat R, Sujatha S (2011a) Nutrient uptake pattern in cocoa. In: Proceedings of seminar on strategies for enhancing productivity of cocoa. ICAR-Central Plantation Crops Research Institute Regional Station, Vittal, India, pp 86–88

Bhat R, Sujatha S (2011b) Organic matter recycling in arecanut based high density multispecies cropping system. In: Thomas GV, Krishnakumar V, Maheshwarappa HP (eds) Areca-nut based cropping/farming systems. ICAR-Central Plantation Crops Research Institute, Kasaragod, India, pp 63–76

Bhat R, Sujatha S (2012) Influence of biomass partitioning and nutrient uptake on yield of arecanut grown on a laterite soil. Commun Soil Sci Plant Anal 43:1757–1767. https://doi.org/10.1080/00103624.2012.684823

Bhat R, Sujatha S (2014) Soil fertility status and disorders in arecanut (*Areca Catechu* L.) grown on clay and laterite soils of India. Commun Soil Sci Plant Anal 45:1622–1635. https://doi.org/10.1080/00103624.2014.907910

Bhat R, Sujatha S, Balasimha D (2007) Impact of drip fertigation on productivity of arecanut (*Areca catechu* L.). Agric Water Manag 90(1–2):101–111

Bhat R, Sujatha S, Jose CT (2012) Assessing soil fertility of a laterite soil in relation to yield of arecanut (*Areca catechu* L.) in humid tropics of India. Geoderma 189–190:91–97. https://doi.org/10.1016/j.geoderma.2012.05.010

Bhat R, Sujatha S, Thomas George V (2008) Soil microbiological and enzymological response to organic matter recycling and chemical fertilization in areca-nut based cropping system. J Plantn Crops 36(3):281–289

Bhavishya NH, Priya UK, Sujatha S, Bhat R (2021) Micronutrient management in arecanut-cocoa ecosystem. Indian J Arecanut Spices Medicinal Plants 23(4):11–15

Biazin B, Sterk G, Temesgen M (2012) Rainwater harvesting and management in rainfed agricultural systems in sub-Saharan Africa – a review. Phys Chem Earth 47–48:139–151. https://doi.org/10.1016/j.pce.2011.08.015

Bisseleua D, Missoup A, Vidal S (2009) Biodiversity conservation, ecosystem functioning, and economic incentives under cocoa agroforestry intensification. Conserv Biol 23(5):1176–1184. https://doi.org/10.1111/j.1523-1739.2009.01220.x

Bopaiah BM, Shetty SH (1991) Soil microflora and biological activities in the rhizosphere and root regions of coconut based multi-storied cropping and coconut monocroping systems. Soil Biol Biochem 23:89–94

Buyer JS, Baligar VC, He Z, Arévalo-Gardini E (2017) Soil microbial communities under cacao agroforestry and cover crop systems in Peru. Appl Soil Ecol 120:273–280

Cabala-Rosand P, Santana CJL, Miranda ER (1982) Response of Catongo cocoa to fertilizer levels in Southern Bahia, Brazil. Revista Theobroma 12:203–216

Cadima ZA, Alvim PDT (1973) Some soil factors associated with cacao yield in Bahia, Brazil. Revista Theobroma 3:13–26

Chowdappa P, Biddappa CC, Sujatha S (1999) Efficient recycling of organic wastes in arecanut (*Areca catechu*) and cocoa (*Theobroma cacao*) plantation through vermicomposting. Indian J Agric Sci 69:563–566

Cotta MK, Jacovine LAG, Paiva HN, Soares CPB, Virgens Filho AC, Valverde SR (2008) Quantificacao de biomassa e geracao de certificados de emissoesreduzidas noconsorcio seringueira – cacau. Árvore 32(6):969–978

Coulibaly SS, Ndegwa PM, Ayiania M, Zoro IAB (2019) Growth, reproduction, and life cycle of *Eudrilus eugeniae* in cocoa and cashew residues. Appl Soil Ecol 143:153–160

CPCRI (1991) Annual report. Central Plantation Crops Research Institute, Kasaragod, Kerala, p 153

Cuenca G, Meneses E (1996) Diversity patterns of arbuscular mycorrhizal fungi associated with cacao in Venezuela. Plant Soil 183:315–322

DCCD (2020) Statistics. Directorate of Cashew nut and Cocoa Development, Kochi, India. https://www.dccd.gov.in/Content.aspx?mid=1072&tid=1. Accessed 20 June 2023

de Oliveira LJ, Valle RR (1990) Nutrient cycling in the cacao ecosystem: rain and throughfall as nutrient sources for the soil and the cacao tree. Agric Ecosyst Environ 32:143–154

Doungous O, Minyaka E, Longue EAM, Nkengafac NJ (2018) Potentials of cocoa pod husk-based compost on Phytophthora pod rot disease suppression, soil fertility, and *Theobroma cacao* L. growth. Environ Sci Pollut Res 25:25327–25335. https://doi.org/10.1007/s11356-018-2591-0

Duguma B, Gockowski J, Bakala J (2001) Smallholder cacao (*Theobroma cacao* Linn.) cultivation in agroforestry systems of West and Central Africa: challenges and opportunities. Agrofor Syst 51:177–188

Ebon AH, Lu ATT, Tiong RHC (1978) Early growth results of cocoa on some Sarawak soils. Proceedings of International Conference on Cocoa and Coconuts, Kuala Lumpur, In, pp 224–242

Edy N, Zakaria EK, Lakani I (2019) Forest conversion into cacao agroforestry and cacao plantation change the diversity of arbuscular mycorrhizal fungi. In: The 1st biennial conference on tropical biodiversity. IOP Conference Series: Earth and Environmental Science (vol 270). IOP Publishing, Makassar, Indonesia. https://doi.org/10.1088/1755-1315/270/1/012015

Edy N, Zakaria E, Anshary A, Saleh S, Lapanjang I, Barus HN, Lakani I (2022) Arbuscular mycorrhizal fungi in cocoa plantation affected by different elevations and soil physical-chemical properties. IOP Conf Ser Earth Environ Sci 1075:012005. https://doi.org/10.1088/1755-1315/1075/1/012005

Ekanade O (1988) The nutrient status of soils under peasant cocoa farms of varying ages in southwestern Nigeria. Biol Agricult Horticult 5:155–167

Ekanade O, Adesina FA, Egbe NE (1991) Sustaining tree crop production under intensive land use: an investigation into soil quality differentiation under varying cropping patterns in western Nigeria. J Environ Manag 32:105–113

Fagbenro OA (1992) Utilization of cocoa-pod husk in low-cost diets by the clariid catfish, *Clarias isheriensis* (Sydenham). Aquac Res 23(2):175–182

Fagbenro OA (1995) Evaluation of heat-processed cocoa pod husk meal as an energy feedstuff in production diets for the clariid catfish, *Clarias isheriensis* (Sydenham). Aquacult Nutr 1(4):221–225

Falade JA (1975) Soil bulk density-moisture supply interaction in Amazon cocoa. West Afr J Biol Applied Chem (Nigeria) 18(3):15–22

Falvey JL (1982) *Gliricidia maculata*—a review. Int Tree Crops J 2(1):1–14. https://doi.org/10.1080/01435698.1982.9752735

Feliciano D, Ledo A, Hillier J, Nayak DR (2018) Which agroforestry options give the greatest soil and above ground carbon benefits in different world regions? Agric Ecosyst Environ 254:117–129. https://doi.org/10.1016/j.agee.2017.11.032

Fidelis C, Rajashekhar Rao BK (2017) Enriched cocoa pod composts and their effects on hybrid cocoa seedlings. Int J Recycl Org Waste Agric 6:99–106

Fontes AG, Gama-Rodrigues AC, Gama-Rodrigues EF, Costa MG, Machado RCR (2014) Nutrient stocks in litterfall and litter in cocoa agroforests in Brazil. Plant Soil 383:313–335. https://doi.org/10.1007/s11104-014-2175-9

Franzluebbers AJ, Hons FM, Zuberer DA (1995) Tillage and crop effects on seasonal dynamics of soil CO_2 evolution, water content, temperature, and bulk density. Appl Soil Ecol 2:95–109

Gama-Rodrigues EF, Ramachandran Nair PK, Nair VD, Gama-Rodrigues AC, Baligar VC, Machado RCR (2010) Carbon storage in soil size fractions under two cacao agroforestry systems in Bahia, Brazil. Environ Manag 45:274–283. https://doi.org/10.1007/s00267-009-9420-7

Goñas M, Rubio KB, Rojas Briceño NB, Pariente-Mondragón E, Oliva-Cruz M (2022) Tree diversity in agroforestry systems of native fine-aroma cacao, Amazonas, Peru. PLoS ONE 17(10):e0275994. https://doi.org/10.1371/journal.pone.0275994

Gopal M, Thomas L, Gupta A, George P, Thomas George V (2011) Plant growth promoting rhizobacteria (PGPR) for growth improvement in cocoa. In: Elain Apshara S, Jaganathan D, Balasimha D (eds) Proceedings of seminar on 'Strategies for enhancing productivity of cocoa' CPCRI. Regional Station, Vittal, Karnataka, 28–29 Jan 2011, 200 p

Govindan M, Purushothaman D (1985) Association of nitrogen fixing bacteria with certain plantation crops. Natl Acad Sci Lett (India) 8:163–165

Greenland DJ (1994) Soil science and sustainable land management. In: Syers JK, Rimmer DL (eds) Soil science and sustainable land management in the tropics. CAB International, Wallingford, pp 1–15

Hanafi MM, Maria GJ (1998) Cadmium and zinc in acid tropical soils: III. Response of cocoa seedlings in a green house experiment. Commun Soil Sci Plant Anal 29(11–14):1949–1960

Hardjono A (1986) Properties of soils affecting cocoa growth on the Bunisari estate. Menara Perkebunan 54:45–54

Hardy F (1960) Cacao Manual. Inter-American Institute of Agricultural Sciences, Turrialba, Costa Rica

Hartemink AE (2005) Nutrient stocks, nutrient cycling, and soil changes in cocoa ecosystems—a review. Adv Agron 86:227–253

Hashim GM, Ciesiolka CAA, Yusoff WA (1995) Soil erosion processes in sloping land in the east coast of Peninsular Malaysia. Soil Technol 8:215–233

Hudson N (1986) Soil conservation. B T Batsford Limited, London

ICCO (2020) Cocoa stat. International Cocoa Organization, London. https://www.icco.org/wp-content/uploads/Production_QBCS-XLIX-No.-1.pdf. Accessed 20 June 2023

Iglesias L, Salas E, Leblanc HA, Nygren P (2011) Response of *Theobroma cacao* and *Inga edulis* seedlings to cross-inoculated populations of arbuscular mycorrhizal fungi. Agric Syst 83(1):63–73. https://doi.org/10.1007/s10457-011-9400-9

Jacobi J, Andres C, Schneider M, Pillco M, Calizaya P, Rist S (2014) Carbon stocks, tree diversity, and the role of organic certification in different cocoa production systems in Alto Beni, Bolivia. Agrofor Syst 88:1117–1132. https://doi.org/10.1007/s10457-013-9643-8

Kaba JS, Zerbe S, Agnolucci M, Scandellari F, Abunyewa AA, Giovannetti M, Tagliavini M (2019) Atmospheric nitrogen fixation by gliricidia trees (*Gliricidia sepium* (Jacq.) Kunth ex Walp.) intercropped with cocoa (*Theobroma cacao* L.). Plant Soil 435:323–336. https://doi.org/10.1007/s11104-018-3897-x

Kaba JS, Yamoah FA, Acquaye A (2021a) Towards sustainable agroforestry management: harnessing the nutritional soil value through cocoa mix waste. Waste Manag 124:264–272

Kaba JS, Abunyewa AA, Kugbe J, Kwashie GKS, Owusu Ansah E, Andoh H (2021b) Arbuscular mycorrhizal fungi and potassium fertilizer as plant biostimulants and alternative research for

enhancing plants adaptation to drought stress: opportunities for enhancing drought tolerance in cocoa (*Theobroma cacao* L.). Sustain Environ 7(1):1–12. https://doi.org/10.1080/27658511.2021.1963927
Keil A, Zeller M, Wida A, Sanim B, Birner R (2008) What determines farmers' resilience towards ENSO-related drought? An empirical assessment in Central Sulawesi, Indonesia. Climat Change 86:291–307. https://doi.org/10.1007/s10584-007-9326-4
Khan HH, Upadhyay AK, Palaniswami C (2001) Potassium in plantation crops. In: Proceedings of the international symposium on K in nutrient management for sustainable crop production in India. Potash Research Institute of India, New Delhi, pp 467–497
Koko LK, Snoeck D, Lekadou TT (2013) Cacao-fruit tree intercropping effects on cocoa yield, plant vigour and light interception in Cote d'Ivoire. Agrofor Syst 87:1043–1052. https://doi.org/10.1007/s10457-013-9619-8
Kongsager R, Napier J, Mertz O (2012) The carbon sequestration potential of tree crop plantations. Mitig Adapt Strat Glob Change 18(8):1197–1213. https://doi.org/10.1007/s11027-012-9417-z
Kotto-Same J, Woomer PL, Appolinaire M (1997) Carbon dynamics in slash-and-burn agriculture and land use alternatives of the humid forest zone in Cameroon. Agric Ecosyst Environ 65:245–256
Koua SH, N'golo DC, Alloue-Boraud WM, Konan F, Dje KM (2020) *Bacillus subtilis* strains isolated from cocoa trees (*Theobroma cacao L.*) rhizosphere for their use as potential plant growth promoting rhizobacteria in Côte d'Ivoire. Curr Microbiol 77:2258–2264. https://doi.org/10.1007/s00284-020-02027-x
Kumar S, Sindhu SS, Kumar R (2021) Biofertilizers: an ecofriendly technology for nutrient recycling and environmental sustainability. Curr Res Microbiol Sci 3:100094. https://doi.org/10.1016/j.crmicr.2021.100094
Läderach P, Eitzinger A, Martínez A, Castro N (2011) Predicting the impact of climate change on the cocoa-growing regions in Ghana and Cote d'Ivoire. Final report. International Center for Tropical Agriculture (CIAT), Palmira, Colombia. 29 p
Ling AH (1986) Litter production and nutrient cycling in a mature cocoa plantation on inland soils of peninsular Malaysia. In: Pushparajah E, Soon CP (eds) Proceedings of the International Conference on Cocoa and Coconuts, Kuala Lumpur, pp 451–465
Lori M, Armengot L, Schneider M, Schneidewind U, Bodenhausen N, Mäder P, Krause HM (2022) Organic management enhances soil quality and drives microbial community diversity in cocoa production systems. Sci Total Environ 834:155223. https://doi.org/10.1016/j.scitotenv.2022.155223
Luis-Alaya B, Toro M, Calsina R, Ogata-Gutiérrez K, Gil-Polo A, Ormeño-Orrillo E, Zúñiga-Dávila D (2023) Evaluation of the presence of Arbuscular Mycorrhizae and cadmium content in the plants and soils of cocoa plantations in San Martin, Peru. Diversity 15(2):246. https://doi.org/10.3390/d15020246
Majumdar K, Sanyal SK, Singh VK, Dutta SK, Satyanarayana T, Dwivedi BS (2017) Potassium fertiliser management in Indian agriculture: current trends and future needs. Indian J Fertilisers 13(5):20–30
Malhotra SK, Hubballi VN, Chowdappa P (2016) Augmenting cocoa production in India. Cashew Cocoa J 4:7–14
Mamani-Pati F, Clay DE, Clay SA, Smeltekop H, Yujra-Callata MA (2012) The influence of strata on the nutrient recycling within a tropical certified organic coffee production system. ISRN Agron 389290:1–8. https://doi.org/10.5402/2012/389290
Manikandan P, Joshi OP, Khan HH, Mohapatra AR, Biddappa CC (1987) Nutrient profile in an arecanut-cacao system on laterite soil. Trop Agric 64:13–16
Middendorp RS, Vanacker V, Lambin EF (2018) Impacts of shaded agroforestry management on carbon sequestration, biodiversity and farmers income in cocoa production landscapes. Landscape Ecol 33:1953–1974. https://doi.org/10.1007/s10980-018-0714-0
Morais FIO (1988) Effect of fertilizers and amendments on cocoa growth and production in soils of Amazonia, Brazil. Proceedings of the 10th International Cocoa Research Conference, Santo Domingo, Dominican Republic, pp 247–251

Mortimer R, Saj S, David C (2018) Supporting and regulating ecosystem services in cacao agroforestry systems. Agrofor Syst 92:1639–1657. https://doi.org/10.1007/s10457-017-0113-6(0123456789

Mulia S, Mcmahon PJ, Purwantara A, Bin Purung H, Djufry F, Lambert S, Keane PJ, Guest DI (2017) Effect of organic and inorganic amendments on productivity of cocoa on a marginal soil in Sulawesi, Indonesia. Exp Agric 55(1):1–20. https://doi.org/10.1017/s0014479717000527

Naidu LGK, Ramamurthy V, Sidhu GS, Sarkar D (2011) Emerging deficiency of potassium in soils and crops of India. Karnataka J Agric Sci 24:12–19

Nair PKR, Balakrishnan TK (1977) Ecoclimate of a coconut + cocoa crop combination in the West Coast of India. Agric Meteorol 18:455–462

Nair PKR, Varma R, Nelliat EV, Bavappa KVA (1975) Beneficial effects of crop combination of coconut and cocoa. Indian J Agric Sci 45(3):165–171

Ndubuaku UM, Kassim A (2003) Use of organic and inorganic fertilizers to improve the rate of cocoa seedling establishment in the field in Nigeria. For Trees Livelihoods 13(1):101–106. https://doi.org/10.1080/14728028.2003.9752446

Nelson PN, Webb MJ, Berthelsen S, Curry G, Yinil D, Fidelis C (2011) Nutritional status of cocoa in Papua New Guinea. ACIAR Technical Reports No. 76. Australian Centre for International Agricultural Research, Canberra, 67 p

Nijmeijer A, Lauri PÉ, Harmand JM, Saj S (2019) Carbon dynamics in cocoa agroforestry systems in Central Cameroon: afforestation of savannah as a sequestration opportunity. Agrofor Syst 93:851–868. https://doi.org/10.1007/s10457-017-0182-6

Norgrove L (2007) Effects of different copper fungicide application rates upon earthworm activity and impacts on cocoa yield over four years. Eur J Soil Biol 43:S303–S310. https://doi.org/10.1016/j.ejsobi.2007.08.031

Norgrove L, Hauser S (2013) Carbon stocks in shaded *Theobroma cacao* farms and adjacent secondary forests of similar age in Cameroon. Trop Ecol 54(1):15–22

Ofori-Frimpong K, Afrifa AA, Acquaye S (2010) Impact of shade and cocoa plant densities on soil organic carbon sequestration rates in a cocoa growing soil of Ghana. Afr J Environ Sci Technol 4(9):621–624

Ojeniyi SO (1982) Review of results of fertilizer trials on cocoa (*Theobroma cacao*) in Nigeria. Proceedings of the 8th International Cocoa Research Conference, pp 171–174

Opakunle JS (1989) Throughfall, stemflow and rainfall interception in a cacao plantation in South Western Nigeria. Trop Ecol 30:244–252

Osei-Bonsu K, Opoku-Ameyaw K, Amoah FM, Oppong FK (2002) Cacao-coconut intercropping in Ghana: agronomic and economic perspectives. Agrofor Syst 55:1–8. https://doi.org/10.1023/A:1020271608483

Ouattara A, Coulibaly K, Konate I, Ismaë B, Tidou AS, Filali-Maltouf A (2019) Selection of cocoa tree (*Theobroma cacao* Linn) endophytic bacteria solubilizing tri-calcium phosphate, isolated from seedlings grown on soils of six producing regions of Côte d'Ivoire. Adv Microbiol 9:842–852. https://doi.org/10.4236/aim.2019.99051

Perfecto I, Vandermeer J (2008) Biodiversity conservation in tropical agro-ecosystems: a new conservation paradigm. Ann N Y Acad Sci 1134:173–200

Rachel LM, Nina KZ, Bryan AB, Siela NM, Mark G, Paul AB (2008) Bacterial endophytes: *Bacillus* spp. from annual crops as potential biological control agents of black pod rot of cacao. Biol Control 46:46–56

Rice RA, Greenberg R (2000) Cacao cultivation and the conservation of biological diversity. Ambio A J Hum Environ 29(3):167–173

Rockström J, Karlberg L, Wani SP (2010) Managing water in rainfed agriculture—The need for a paradigm shift. Agr Water Manag 97(4):543–550. https://doi.org/10.1016/j.agwat.2009.09.009

Rohman F, Wachjar A, Santosa E, Abdoellah S (2019) Humic acid and biofertilizer applications enhanced pod and cocoa bean production during the dry season at Kaliwining plantation, Jember, East Java, Indonesia. J Trop Crop Sci 6(3):153–163. https://doi.org/10.29244/jtcs.6.03.153-163

Sabiha-Javied MT, Choudhry MM, Tufail M, Irfan N (2009) Heavy metal pollution from phosphate rock used for the production of fertilizer in Pakistan. Microchem J 91(1):94–99

Saj S, Jagoret P, Todem Ngogue H (2013) Carbon storage and density dynamics of associated trees in three contrasting *Theobroma cacao* agroforests of Central Cameroon. Agrofor Syst 87:1309–1320. https://doi.org/10.1007/s10457-013-9639-4

Saj S, Durot C, Mvondo Sakouma K (2017) Contribution of associated trees to long-term species conservation, carbon storage and sustainability: a functional analysis of tree communities in cacao plantations of Central Cameroon. Int J Agric Sustain 15:282–302

Saj S, Nijmeijer A, Nieboukaho JDE, Lauri PE, Harmand JM (2021) Litterfall seasonal dynamics and leaf-litter turnover in cocoa agroforests established on past forest lands or savannah. Agrofor Syst 95:583–597. https://doi.org/10.1007/s10457-021-00602-0

Salamanca AAJ, NavarroCerrillo RM, Quero Pérez JL, Gallardo Armas B, Crozier J, Stirling C, de Sousa K, GonzálezMoreno P (2023) Vulnerability of cocoa based agroforestry systems to climate change in West Africa. Sci Rep 13:10033. https://doi.org/10.1038/s41598-023-37180-3

Sampath SR, Kumar MNA, Sundareshan K (1990) Evaluation of cocoa bean by-products for ruminant feeding. Indian J Anim Nutr 7:45–48

Santana CJL, Zevallos AC (1981) Productivity of cacao in oxisols in southern Bahia, Brazil. Boletim Tecnico CEPLAC, No. 82, p 23

Satriawan H, Harahap EM, Rahmawaty Karim A (2015) Effectiveness of soil conservation to erosion control on several land use types. Agric (Poľnohospodárstvo) 61(2):61–68

Schmidt JE, Du Val A, Isaac ME, Hohmann P (2022) At the roots of chocolate: understanding and optimizing the cacao root-associated microbiome for ecosystem services. A review. Agron Sustain Dev 42:14. https://doi.org/10.1007/s13593-021-00748-2

Shahane AA, Shivay YS (2021) Soil health and its improvement through novel agronomic and innovative approaches. Front Agron 3:680456. https://doi.org/10.3389/fagro.2021.680456

Shama Bhat K (1983) Plant interactions in a mixed crop community of arecanut (*Areca catechu* L.) and cacao (*Theobroma cacao* L.). Ph.D. Thesis, University of Mysore, Mysore, p 158

Shama Bhat K (1988) Growth and performance of cacao (*Theobroma cacao* L.) and arecanut (*Areca catechu*) under mixed cropping system. In: Proceedings of the 10th International Cocoa Research Conference, Santo Domingo, Dominican Republic, pp 15–19

Shama Bhat K, Bavappa KVA (1972) Cocoa under palms. In: Proceedings of the conference on cocoa and coconuts, Kuala Lumpur, Malaysia, pp 116–121

Shivaprasad CR, Reddy RS, Sehgal J, Velayutham M (1998) Soils of Karnataka for optimising land use. NBSS Publication – 47b. National Bureau of Soil Survey and Land Use Planning, Nagpur

Siebert SF (2002) From shade-to sun-grown perennial crops in Sulawesi, Indonesia: implications for biodiversity conservation and soil fertility. Biodivers Conserv 11(11):1889–1902. https://doi.org/10.1023/A:1020804611740

Smiley GL, Kroschel J (2008) Temporal change in carbon stocks of cocoa-gliricidia agroforests in Central Sulawesi, Indonesia. Agrofor Syst 73:219–231

Smith OB, Adegbola AA (1982) Studies on the feeding value of agro-industrial by-products and the feeding value of cocoa pods for cattle. Trop Anim Prod 7:290–295

Smyth AJ (1980) Soil classification and the cocoa grower. Cocoa Growers' Bull 30:5–10

Snoeck D, Koko L, Joffre J, Bastide P, Jagoret P (2016) Cacao nutrition and fertilization. In: Lichtfouse E (ed) Sustainable agriculture reviews, vol 19. Springer, Cham. https://doi.org/10.1007/978-3-319-26777-7_4

Somarriba E, Cerda R, Orozco L, Cifuentes M, Dávila H, Espin T, Mavisoy H, Ávila G, Alvarado E, Poveda V, Astorga C, Say E, Deheuvels O (2013) Carbon stocks and cocoa yields in agroforestry systems of Central America. Agric Ecosyst Environ 173:46–57

Subramanian P, Dhanapal R, Sanil P (2005) Gliricidia (*Gliricidia sepium*) as green manure in improving soil fertility and productivity of coconut under coastal littoral sandy soil. J Plantn Crops 33:179–183

Sucuia OA, Florica C, Cornea CP (2015) Biodiversity of *Bacillus subtilis* group and beneficial traits of *Bacillus* species useful in plant protection. Rom Biotechnol Lett 20(5):10737–10750

Sujatha S, Bhat R (2013) Impact of drip fertigation on arecanut-cocoa system in humid tropics of India. Agrofor Syst 87:643–656

Sujatha S, Bhat R (2015) Soil fertility changes due to drip fertigation in arecanut-cocoa system. Commun Soil Sci Plant Anal 46(16):1981–1997

Sujatha S, Bhat R, Chowdappa P (2015) Recycling potential of organic wastes of arecanut and cocoa in India: a short review. Environ Technol Rev 4(1):91–102. https://doi.org/10.1080/09593330.2015.1077897

Sujatha S, Bhat R, Chowdappa P (2017) Soil health management in arecanut and cocoa. In: Maheswarappa HP, Chowdappa P (eds) Soil health management in plantation crops. Today & Tomorrow's Printers and Publishers, New Delhi, pp 1–25

Sujatha S, Bhat R, Bhavishya (2020) Arecanut based cropping/farming systems. In: Chowdappa P (ed) Arecanut. Today and Tomorrow's Printers and Publishers, New Delhi, pp 145–175

Thomas GV, Krishanakumar V, Maheswarappa HP, Bhat R, Balasimha D (eds) (2011) Arecanut based cropping/farming systems. Central Plantation Crops Research Institute, Kasaragod. 138 p

Thomas GV, Subramanian P, Gopal M, Gupta A, Prabhu SR (2022) Unraveling the potential of belowground and aboveground biodiversity for sustainable management of the health of plantation crop soils in coastal agro-ecosystem. In: Lama T, Burman D, Mandal UK, Sarangi SK, Sen H (eds) Transforming coastal zone for sustainable food and income security. Springer, Cham. https://doi.org/10.1007/978-3-030-95618-9_48

Thong KC, Ng WL (1978) Growth and nutrient composition of monocrop cocoa plants on inland Malaysian soils. In: Proceedings of the international conference on cocoa and coconuts, Kuala Lumpur, pp 262–287

Uribe A, Méndez H, Mantilla J (2001) Effect of balanced fertilization on cocoa yield. Better Crops International 15(2):3–5

Utomo B, Prawoto AA, Bonnet S, Bangviwat A, Gheewala SH (2016) Environmental performance of cocoa production from monoculture and agroforestry systems in Indonesia. J Cleaner Prod 134(Part B):583–591. https://doi.org/10.1016/j.jclepro.2015.08.102

Varghese PT, Nelliat EV, Balakrishnan TK (1978) Beneficial interactions of coconut-cocoa crop combination. In: Nelliat EV, Ranganathan V, Vishveswara (eds) Proceedings of PLACROSYM-I, Kasaragod, India, pp 383–392

Vervuurt W, Slingerland MA, Pronk AA (2022) Modelling greenhouse gas emissions of cacao production in the Republic of Côte d'Ivoire. Agrofor Syst 96:417–434. https://doi.org/10.1007/s10457-022-00729-8

Vessey JK (2003) Plant growth promoting rhizobacteria as biofertilizers. Plant Soil 255:571–586

Wade ASI, Asase A, Hadley P (2010) Management strategies for maximizing carbon storage and tree species diversity in cocoa-growing landscapes. Agric Ecosyst Environ 138(3–4):324–334

Wahid PA, Kamalam NV, Ashokan PK, Nair RV (1989) Root activity pattern of cocoa (*Theobroma cacao* L.). J Nuclear Agric Biol 18:153–156

Whipps JM (2001) Microbial interactions and biocontrol in the rhizosphere. J Exp Bot 52:487–5119

Wood GAR, Lass RA (2001) Cocoa, 4th edn. Blackwell Science, Oxford

Woomer PL, Martin A, Albrecht A, Resck DVS, Scharpenseel HW (1994) The importance and management of soil organic matter in the tropics. In: Woomer PL, Swift MJ (eds) The biological management of tropical soil fertility. Wiley, Chichester, pp 47–80

Zevallos AC, Coral FJ (1972) Root system of cacao trees in two soil types of Sao Paulo. Revista Theobroma 4:16–22

Zuidema PA, Leffelaar PA, Gerritsma W, Mommer L, Anten NPR (2005) A physiological production model for cocoa (*Theobroma cacao*): model presentation, validation and application. Agric Syst 84(2):195–225. https://doi.org/10.1016/j.agsy.2004.06.015

Ravi Bhat is currently Principal Scientist (Agronomy), ICAR-CPCRI, Kasaragod. He has more than 30 years of experience in agronomy of arecanut, coconut, and cocoa with specialization in cropping/farming systems, nutrient management, soil and water management, and organic farming. He standardized the technology of fertigation for arecanut, nutrient requirement for different yield levels, soil fertility management, and developed various arecanut and coconut based cropping/farming system models. He is recipient of Best Arecanut Scientist Award. He has published more than 100 research papers in national/international journals.

Bhavishya has a doctorate in Horticulture and currently serving as scientist at ICAR-Central Plantation Crops Research Institute, Regional Station, Vittal, Karnataka. His areas of research are nutrient management in cropping/farming systems of arecanut and cocoa, enhancing input use efficiency, biotic stress-nutrient interaction and carbon sequestration.

S. Sujatha has a doctorate in Agronomy. She is currently Principal Scientist at ICAR-Indian Institute of Horticultural Research, Bengaluru, Karnataka, India. She started her carrier at ICAR-Central Plantation Crops Research Institute, Regional Station, Vittal, Karnataka, and contributed on crop production aspects of arecanut and cocoa for 20 years. Her areas of research are crop production aspects of plantation crops, flower crops, medicinal crops, and fruit crops with specialization in cropping/farming systems, nutrient management, soil fertility, plant nutrition, water management, drip fertigation, and organic farming. She has published 50 research papers in international/national journals.

Coffee (*Coffea* spp.)

S. A. Nadaf, P. Shivaprasad, C. Babou, N. Hariyappa, N. Chandrashekar, Prafulla Kumari, P. R. Sowmya, S. B. Hareesh, N. Rajib Pati, J. S. Nagaraja, K. Chandrappa, and M. Senthil Kumar

Abstract

Coffee farming regions on the globe along the equatorial zone between the Tropic of Cancer and the Tropic of Capricorn are generally called as the "Bean Belt." Coffee plant prospers well in volcanic, red-lateritic, alluvial and colluvial soils. Acidic soil reaction, high organic matter content, and better potassium status are the general characteristics of coffee soils. The productivity potential of coffee plantations depends on the timely cultural operations and soil health. Soil erosion, soil acidity, and low nutrient availability are some of the major constraints. Soil conservation measures such as contour planting, opening of cradle pits, scuffling, mulching, and terracing in steep slopes to avoid erosion are important cultural practices. Timely application of liming materials and fertilizers in two to three splits based on the soil test is required for nutrient management under sustainable coffee production. Nitrogen is essential for coffee plants in all seasons, and it has to be provided three to four times a year. The phosphorus requirement is higher for the coffee plants at the time of blossom season. Similarly, the potassium requirement is higher at the time of post-monsoon season. Coffee in India has been cultivated under the canopy of two-tier different shade trees. Leaf litter from different shade trees, periodical pruning practices for coffee plants add considerable amount of biomass to the coffee soil. This type of coffee cultivation practice will help to improve organic carbon status in the coffee soil. Hence, technological options to improve soil health including farm biomass recycling, application of biofertilizers, cover cropping, integrated nutrient management, agro-forestry systems for higher carbon sequestration, and opportunities for organic farming are attractive options for sustainable coffee farming.

S. A. Nadaf (✉) · P. Shivaprasad · C. Babou · N. Hariyappa · N. Chandrashekar · P. Kumari · P. R. Sowmya · S. B. Hareesh · N. R. Pati · J. S. Nagaraja · K. Chandrappa · M. S. Kumar
Central Coffee Research Institute, Coffee Board of India, Chikkamagaluru, Karnataka, India

G. V. Thomas, V. Krishnakumar (eds.), *Soil Health Management for Plantation Crops*, https://doi.org/10.1007/978-981-97-0092-9_9

Keywords

Coffee crop · Soil health · Liming · Sustainable nutrient management · Soil biodiversity · Bio-fertilizers · Organic farming

Abbreviations

$	Dollar
%	Percent
@	At the rate
<	Lesser than
>	Greater than
$^{\circ}$N	Degree North
$^{\circ}$S	Degree South
A.M.	Ante Meridiem
AFS	Agroforestry system
Al^{3+}	Exchangeable aluminum
Ara	Arabica coffee
AWC	Available water content
B	Boron
C	Carbon
Ca	Calcium
$CaCO_3$	Calcium carbonate
CaO	Calcium oxide
CaOH	Calcium hydroxide
CEC	Cat-ion exchange capacity
cm	Centi meter
cmol(+)	Centi mole
CO_2	Carbon dioxide
Cu	Copper
DRIS	Diagnosis and recommendation integrated system
et al.	And others
Fe	Iron
FUE	Fertilizer use efficiency
g	Gram
K	Potassium
K_2O	Available potassium
kg / acre / yr.	Kilogram per acre per year
kg / ha	Kilogram per hectare
kg / plant	Kilogram per plant
kg	Kilogram
m	Meter
m^3/ha	Cubic meter per hectare
Mg m^{-3}	Mega gram per meter cube

Mg	Magnesium
$MgCO_3$	Magnesium Carbonate
mm m^{-1}	Millimeter per meter
mm	Millimeter
mmhos cm^{-1}	Millimohos per centimeter
Mn	Manganese
Mo	Molybdenum
MOP	Muriate of potash
N	Nitrogen
N	North
NAR	Net assimilation rate
NBSSLUP	National Bureau of Soil Survey and Land Use Planning
NER	North Eastern Coffee Growing Region of India
NTA	Nontraditional Coffee Growing Area of India
°C	Degree celsius
Oct – Dec	October to December
P	Phosphorus
P_2O_5	Available phosphorus
pH	Soil reaction/soil acidity
ppm	Parts per million
RDF	Recommended dose of fertilizer
Rob	Robusta coffee
S	South
S	Sulfur
SOM	Soil organic matter
SOP	Sulphate of potash
SSP	Single super phosphate
tonne / ha	Tonnes per hectare
tons / ha	Tonnes per hectare
US	United States
VAM	Vesicular arbuscular mycorhiza
Zn	Zinc
$ZnSO_4$	Zinc sulphate

1 Introduction

The seeds of berries from the three coffee varieties, Arabica *(Coffea arabica* L.*)*, Robusta *(Coffea canephora),* and Liberica (*Coffea liberica* L), are used to prepare the beverage coffee. In the equatorial regions (Bean Belt), coffee crop is cultivated in over 80 countries. Green coffee beans (hulled) are of great importance as one of the most traded agricultural commodities at the international level. Among the commonly grown coffee types, Arabica has high flavor and acidity, while Robusta has high caffeine and is stronger. Once coffee fruits are ripe, the coffee beans are picked, processed, and dried. Harvested coffee beans are roasted in different degrees of temperature, depending on the preferred flavour before being ground. Roasted beans after grinding is brewed and served.

Coffee plants are originated first in Ethiopia. As per the credible evidence, drinking of coffee was seen during the middle of the fifteenth century in the Sufi shrines of Yemen. Coffee reached Europe in the 1600s where it was considered as an exotic new drink. In 1699, the Dutch managed to start a coffee plantation in their colony, Java now part of Indonesia. Baba Budan, a Muslim saint, introduced coffee in India during the sixteenth century and is said to have brought seven coffee beans from Arabia and grown near the Dattatreya Peeta/Baba Budengiri of Chikkamagaluru district of Karnataka in India. Commercial coffee cultivation started during the late 1820s by the British enterprise in South India. Area under coffee cultivation increased rapidly during the early 1860s, and at the same time, many coffee farms experienced outbreak of pests and diseases. During this period, leaf rust epidemic brought the flourishing coffee industry to a close in Sri Lanka. Arabica gave way to Robusta in Java-Indonesia due to pest and disease problems. Frantic efforts were made to introduce disease-tolerant selections, the first in this series being "Coorgs" developed during the 1870s, which soon became susceptible to leaf rust. Some enterprising planters introduced seeds from countries such as Brazil, Venezuela, Costa Rica, and Jamaica and tried to select and evolve disease-resistant Arabica with commercial qualities for large-scale plantations. Robusta was introduced to India from Indo-China region and its cultivation was started at lower elevations during the end of the nineteenth century. Today, commercial coffee cultivation is practiced in more than 55 countries in South America, Asia, and Africa. However, the top coffee-producing countries in the world are Brazil, Vietnam, Colombia, Indonesia, India, and Ethiopia. Furthermore, more than 3 billion cups of coffee are consumed in a single day in the world.

On an average, 2.68 million metric tons of coffee is produced every year by Brazil, which stands at the top in the world in production of coffee followed by Vietnam. In case of Robusta coffee production system in the world, Vietnam produces the largest Robusta coffee with an average of 1.5 million metric tons year^{-1}. Vietnam contributes around 40% Robusta beans in the world, and its coffee is famous for bitterness and low acidity, which are perfect for coffee blends. Colombia stands in third place in coffee production and it is nearly about 760 thousand metric tons in a year. Indonesia is also a major coffee producing country and it stands fourth place with a production of more than 668 thousand metric tons of coffee year^{-1}.

Ethiopia, considered the birthplace of *Coffea arabica*, is endowed with a rich variety of coffee and diverse origins. Ethiopia has effectively advertised premium Ethiopian coffee beans in the world market and coffee started being traded out of the country as early as the seventeenth century. Now the Ethiopian coffee is known for its unique taste, flavor, and aroma universally. Coffee of Colombia is known in the world for its aromatic, slightly fruity flavor, medium-bodied with a nutty aroma that gives a hint of citrus acidity, and the plants are grown in the highlands about 1500–2000 m above mean sea level (MSL). Colombia has the perfect terrain and favorable weather situation to produce the best Arabica beans. Indonesian coffee, also known as Java coffee, is well-known for its woody and earthy flavor as well as rich in cup quality. Such type of coffee is used widely in making different blends of coffee due to its earthy and complex spicy flavor properties.

2 Coffee Cultivation

The coffee-growing regions lie between latitude of 22° N and 26° S within the Bean Belt. Environmental factors such as temperature, number of rainy days, annual rainfall, elevation of terrain, shade pattern, total sunshine hours, wind velocity and direction, soil, and topography of the land are the important factors influencing coffee cultivation. South Indian states are the major coffee producer in India with Karnataka producing around 70% of coffee followed by Kerala and Tamil Nadu (CCRI 2014). Alluri Sitharamaraju district in Andhra Pradesh and Koraput district of Odisha are the major Arabica growing regions of nontraditional area of India. India occupies the eighth position among the coffee-producing countries and produces about 348 thousand metric tons of coffee every year. The global market contribution of Indian coffee is around 2.5–3.0%.

Growing healthy coffee plants involves balancing of sunlight, water, air, and essential nutrients from the healthy soil. A total of 16 elements are required for successful development of coffee plants. Deficiency of any of them can adversely affect coffee growth, development, and yield. Coffee trees absorb carbon, hydrogen, and oxygen from the surrounding air and water; nitrogen (N), potassium (K), phosphorus (P), calcium (Ca), magnesium (Mg), and sulfur (S) from the soil. Certain micronutrients such as iron (Fe), zinc (Zn), copper (Cu), manganese (Mn), boron (B), chlorine (Cl), and molybdenum (Mo) are required only in much smaller amounts.

The ideal soil texture for successful coffee production is a combination of sand, silt, and clay, creating a soil condition ideal for growing coffee plants. This can result in larger coffee cherries that can have higher cupping values and potentially better market prices. Soil reactions, chemical environment, and bioavailability of soil nutrients, especially micronutrients, have a significant impact on coffee production and quality. Coffee soils are red, lateritic, deep, heavy, and generally poor in drainage and also are acidic in soil reaction. Poor status of P, Ca, and Mg is the characteristic of coffee-growing soils of India. The pH of the soil should be optimal (6.1–6.5) so that the coffee plant can absorb more nutrients and perform better.

For optimum growth and yield of coffee, maximum plant biodiversity should be maintained within the coffee plantation, and sufficient species of shade trees should be made available. More importantly, intercrops such as black pepper, mandarin, cardamom, and avocado, besides cover crops may also be planted in coffee plantation. All of these different shade trees, and intercrops provide optimal shade for coffee plants and help for plant development. They also help to improve the soil carbon content by addition of leaf litter of the shade trees, which ultimately improves the soil health and soil productivity.

3 Climatic and Soil Requirements for Coffee Cultivation

3.1 Temperature

The fluctuations in temperature have a significant impact on coffee production. Brazil has the perfect climatic conditions for coffee production with adequate levels of regular rainfall, sunlight, ideal altitude, and stable temperatures throughout the year. The optimal temperature for Arabica coffee in Indian environment is 15 °C at night and 25 °C during the day, while that for Robusta coffee is between 20 and 30 °C (CCRI 2014). Temperatures above 30 °C can cause reduction in photosynthesis, and prolonged exposure to high temperatures can cause leaf chlorosis. At low temperatures below 10 °C, the leaves may turn to white or yellow. These symptoms appear more severe when daytime temperatures are high followed by low nighttime temperatures (Descroix and Snoeck 2004). In some parts of traditional regions of India, high summer temperatures and poor soil moisture may be limiting factors for coffee growth and development. In certain locations of nontraditional regions, day–night temperature fluctuations are common and they also have more impact on coffee cultivation. Such damages can be minimized to some extent by using appropriate cultural practices like regular shade management and providing soil mulching.

3.2 Elevation

Among the coffee species, Arabica prefers highlands with a cool climate, while Robusta prefers lowlands with a hot and humid climate throughout the year in its life cycle. The optimum altitude for Arabica is 1000–1500 m above MSL, while for Robusta it is 500–1000 m above MSL. Arabica varieties grown in lowlands are more likely to be affected by disease like coffee leaf rust and insect like coffee stem borer.

3.3 Rainfall

The optimum annual rainfall for coffee cultivation is 1600 to 2500 mm for Arabica and 1000 to 2000 mm for Robusta coffee. Most of the coffee grown in Karnataka and Kerala states of India is influenced by the southwest monsoon rainfall, where about 80% of such rain being received during June to September. Though Tamil Nadu state of India receives only a small portion of southwest monsoon showers, coffee cultivation in this region depends on northeast monsoon rains (October to December). The cultural operations such as fertilizer application, and pest and disease management are to be modified accordingly in this region. Very high rainfall areas (>3000 mm) are considered as less suitable for coffee cultivation as there will be more fruit drop due to soil saturation and development of diseases during monsoon period. However, by adopting good cultural practices like providing surface and vertical drainage and appropriate disease management, the crop loss can be minimized.

Drought after harvest of coffee beans promotes good flowering in coffee farming. This drying period of about 2–3 months coinciding with the harvest season is ideal in coffee. Arabica coffee can survive dry periods of up to 4 months if the soil is deep, fertile and has good water-holding capacity. As Robusta is grown at lower altitudes with less shade canopy, it results in soil moisture loss. Therefore, the dry season should not exceed 3 months in case of Robusta. Rainfall is must for flowering and fruit set in coffee and rains received during this period are known as blossom and backing showers. Blossom showers of 25 to 40 mm during March–April for Arabica and February–March for Robusta followed by backing showers of 50 to 75 mm after 20 days of blossom showers is very critical for coffee cultivation. In case of insufficient blossom or backing showers, artificial irrigation has to be given through sprinkler method to induce blossom. A 12-year study about the water requirements for Arabica in Kenya found that a total of 951 mm of rainfall is ideal for coffee when dry seasons are normal and soil water-holding capacity is high (Blore 1966; Wallis 1963). However, in areas with an annual rainfall of approximately 1100 mm, Arabica can be grown satisfactorily without irrigation. Awatarmani and Satynarayana (1973) reported that the morning hour rain on the day of blossom in Arabica coffee adversely affected the blossom-to-berry ratio, whereas no such adverse effect was noticed with the afternoon showers.

Radhakrishnan et al. (1993) studied the distribution, stability, and trends of precipitation in the coffee-producing regions in south India from 1950 to 1989 and found that all coffee-producing regions of Karnataka, Kerala, and Tamil Nadu had similar characteristics in monthly distribution and seasonal influence of rainfall. The annual precipitations in all coffee-growing regions ranged from 1400 to 3371 mm, with 80 to 127 rainy days. However, July was the wettest month in all parts of coffee-growing regions of India except Nilgiris, Purnees, and Sheroy (Tamil Nadu). Most of the regions receive rainfall during the southwest monsoon season (80% of the annual precipitation) during June and September, while the summer showers account for 9 to 17% of the annual rainfall in coffee-growing regions in south India. Rainfall from April to May is found to be more stable, with coefficients of variation ranging from 39 to 68%, indicating a higher probability of blossom showers. The probability of blossom showers in February is >50% in Purunes and Nilgiris regions of Tamil Nadu. It was found that 25 mm of precipitation was almost guaranteed in all coffee-growing regions in April, indicating good precipitation stability from June to November.

3.4 Shade Pattern and Sunlight

In many countries, coffee plantations are established in open environments without shade trees, and such plants are directly exposed to sunlight to increase coffee production. Indian coffee plantations are ecologically natural with shade-loving (Fig. 1 a, b, c), and coffee flowering is promoted by rain, sunlight, and appropriate agriculture lime and manure application. In Indian conditions, where there is a dry period

Fig. 1 (**a**) Robusta coffee grown under shade trees in India. (**b**) Arabica coffee grown under shade trees in India. (**c**) Profuse bearing of Arabica coffee under ideal management

of about 4 to 5 months, invariably coffee needs to be cultivated under shade to sustain the soil moisture and minimize evapotranspiration.

White stem borer is a major pest of Arabica in India, and its attack is severe under less shade or open condition. Maintaining adequate shade trees helps to reduce the adverse effects of high temperature and provide ideal microclimate. Shade also lowers the risk of soil erosion, restricts weed growth and generates mulch which protects and enriches soil organic matter (SOM) by leaf litter of the shade trees. Another advantage of shade is that the cupping quality of coffee will be better from plants grown under mixed shaded condition. Robusta coffee is resistant to coffee leaf rust and white stem borer and requires less shade than Arabica coffee.

Coffee is shade-adapted species, and at maximum light intensity, the stomata of coffee plants are closed, and so coffee leaves may not be able to utilize all the absorbed energy (Muller 1966) during the photosynthesis process. On sunny days in the tropics, the net photosynthetic rate of exposed coffee leaves is usually low because leaf temperature reaches 35–40 °C (Castnao and Robledo 1978). Mulching can provide nutrients, control soil temperature, improve soil available water content (AWC) and soil microbiota, and manage the erosion of surface soil (Robinson and Hosegood 1963).

Venkataramanan and Govindappa (1987) studied the shade requirement and productivity of coffee seedlings for 2 years by providing 25, 75 and 100% daylight. Coffee seedlings kept in the shade had more leaves, plant height, primary branches, and total dry weight than those exposed to 100% sunlight. The increase in total dry matter assimilation was due to an increase in net assimilation rate (NAR) and improvement of other growth parameters in coffee. Exposure to 100% daylight appears to decrease the net photosynthetic capacity of coffee leaves. The results support the practice of coffee cultivation under shade in South Indian conditions. Gopal (1974) based on the result of exhaustive field and greenhouse studies concluded that the malady of dieback (death of coffee plants) is induced by prolonged exposure of coffee plants to high sunlight intensities, high atmospheric temperature, low relative humidity, and soil moisture stress.

In recent years, some coffee planters have experimented with growing coffee in open conditions without shade by adopting fertigation and drip irrigation with an intention of maximizing yield per unit area. Majority of coffee plantations in India are situated in the biological hot spots of Western Ghats. A study estimated that coffee agroforestry system in Karnataka has an average water recharging capacity of 1739.86 m^3 ha^{-1} and protects water ways by acting as buffer zones for streams originating from Western Ghats. Cultivation of coffee under shade is a self-sustainable agricultural practice in India as it favors growing intercrops like black pepper and many other fruit and spice crops. The income from black pepper and timber from temporary shade trees will compensate for the lower yield from coffee (Basanna 2015).

3.5 Wind

Strong wind-prone areas are not suitable for coffee plantations as they cause damage to branches, leaf defoliation, and lodging of plants. Establishing coffee plantations in such areas needs basic precautions such as planting wind barriers, topping plants to a lower height, and providing support in the initial years of establishment. This situation usually occurs at high altitudes. Although not very strong, winds that last for a long time can cause physiological disadvantages such as excessive water evaporation from the leaf leading to death of the plant. The use of shade trees and windbreaks is necessary under such conditions.

4 Soil Requirements

The optimal soil for coffee farming should have more deep profile, high permeability, slightly acidic in reaction and with porous nature (well-structured soil), and good texture. Coffee roots require a lot of oxygen, so heavy, poorly drained soils are not suitable for growing coffee. The soil depth should be at least 1.5 m to allow the development of tap roots and ensure necessary water supply for the plant during dry periods. The presence of hard rocks in the bottom can affect plant establishment. Because most of the feeder roots are concentrated in the top 30 cm of the soil, the physical properties of the surface soil are important for coffee. In general, soil suitable for growing coffee should contain not more than 20–30% coarse sand and 70% of clay in the top layer of 30–50 cm (CCRI 2014).

4.1 Topography

Flat land and slightly sloppy hills are ideal for growing coffee, as they generally provide deep soils with high water-holding capacity. Coffee can be grown even on steep slopes; however, steep slopes make farming more expensive and difficult for mechanization. Gowda et al. (1965) reported that coffee in India grown in the foothills of the Western Ghats has a slope of 10–50%. Kharche (1996) suggested that 5 to 8% slope is ideal for Arabica coffee and obtained a significant negative correlation between slope and coffee yield. Wellman (1961) reported that in Brazil, flat and slightly rolling lands are dedicated to coffee while in Kenya, flat land is chosen for coffee cultivation. Gentle to moderate slopes are ideal for Arabica, while mild to fairly flat soils are suitable for Robusta. Both species prefer north and northeast exposure, but strong west exposure should be avoided (CCRI 2014).

5 General Properties of Soils Cropped to Coffee

5.1 Physical Properties

A study of soils at 46 sites in India spread across the coffee-growing regions of Karnataka, Kerala, and Tamil Nadu enabled synthesis of the general properties of coffee-growing soils in traditional area (NBSS LUP 2017). Most of India's coffee plantations are located in the Western Ghats and its foothills, and due to the humid climate and narrow range of day and night temperature, fluctuations in soil of coffee-growing region are not large. Soils range from deep to very deep and have a loamy or clayey texture. The surface soils are often dark colored due to organic matter and sub-soils have shades of brown and red. Coffee is seldom raised in low lands with limited drainage, and hence, coffee-growing soils are mostly well drained. The laterite soils of high lands do not have any significant amount of iron stone gravel or plinthite as in the case with laterite soils of Western Coastal Plain. The coarse fragments in coffee-growing soils, wherever found, are mostly rock fragments or resistant quartz fragments. Bulk density of surface soils is low with mean values for surface and subsoils at 1.22 and 1.30 Mg m^{-3}, respectively. The coffee plantations with two tiers of shade trees and almost complete coverage of land with the spread of coffee canopy generate a large volume of litter. Combined with no-tillage, litter recycling ensures that the organic matter content of coffee-growing soils is relatively high. The organic carbon content of coffee soils ranged from 1.03 to 4.76% at the surface and from 0.12 to 1.58% in the subsoil (NBSS LUP 2017). Organic carbon not only contributes to the supply of mineral nutrients to plants but also increases soil cation exchange capacity (CEC) by improving exchange sites and lowering the zero-point charge of soil colloids. Higher CEC realized for organic matter-rich surface soils helps to retain nutrients against leaching losses in high rainfall regions.

Soil pH in coffee soils ranges from slightly to moderately acidic on the surface and in the sub soils soil pH is moderately to strongly acidic. The soils of Western Ghats are naturally acidic due to their development under high rainfall conditions and dense forest vegetation. The subsoils in humid tropics are more acidic due to the depletion of bases from lower soil layers by deep-rooted vegetation. Regular application of external agriculture inputs in coffee plantation is also responsible for reduction of acidity in surface coffee-growing soils. However, $CaCO_3$ and $MgCO_3$, used as liming materials, are difficult to dissolve in water, and hence, only a small amount of calcium and magnesium reaches the subsoil layer.

Regular lime application in coffee plantations is also good practice to neutralize acidity caused by the application of high-analysis fertilizers and high rainfall. CEC of coffee soils ranged from 10 to 15 cmol(+) kg^{-1} in surface soil and 7 to 12 cmol(+) kg^{-1} in subsoil (NBSS LUP 2017). Low-activity clay soils that develop under humid tropical climates contain primarily kaolinite-type clay minerals. The CEC of kaolinite clay mineral is very low. The relatively high CEC of surface soil is due to the contribution of organic colloids. Base saturation of coffee-growing soils ranged from 55 to 75% in surface soil and 42 to 60% in subsoils, reflecting generally adequate reserves of basic cations such as calcium, magnesium, and potassium. Only soils from Kerala and Tamil Nadu had fair amount of Al^{3+} in surface and subsoils.

Coffee plants grow not only in volcanic formations but also under alluvial and colluvial soils in favorable soil texture. The climatic regions and soil classifications of Karnataka, Kerala, and Tamil Nadu of India belonged to four soil orders as per the standards set by Soil Classification – Soil Survey Staff – 1999. Soils of the Ultisol order are the most widely distributed soil type in coffee-growing areas of South India. Soils of Mollisol order are also found to occur significantly in this region. Other soil orders represented in coffee-growing areas are Alfisols and Inceptisols and these two orders are found in a limited extent (Nair et al. 2017). Coffee is being grown worldwide in different types of soil (Descroix and Snoeck 2004).

While characterizing soils under coffee cultivation in Karnataka, Kharche et al. (2000) noted that the soils had experienced a long period of weathering and development. Very high rainfall favored leaching of bases irrespective of the parent material. Patil and Dasog (1999) studied the genesis of six ferruginous pedons from the Western Ghats region and reported that laterization is the dominant pedogenic process along with illuviation. A study conducted by the Coffee Board in collaboration with ICAR National Bureau of Soil Survey and Land Use Planning during 2017 has enabled the identification of the main soil types in coffee plantations in India. The soils used for growing coffee in India belong to the red and lateritic hilly soils. The texture differed from sandy loam to clay loam, and the color varied from light gray to deep red. Soils are usually rich in organic matter and exhibit an acidic to neutral pH.

5.1.1 Soil Depth

Anil Kumar et al. (2010) reported that a deep profile preferably over 180 cm is desirable for coffee cultivation in India. Sys et al. (1991) suggested that soils that have >150 cm depth are ideal for Robusta coffee, while soils having a depth of >200 cm are ideal for Arabica coffee (Anil Kumar et al. 2010; Kharche et al. 1999). Zuviria and Valenzuela (1994) found that effective soil depth < 30 cm with low available water holding capacity is unsuitable for coffee growing. Reports are available that coffee roots can reach depths of >100 cm in the absence of physical limitations and the presence of appropriate chemical conditions such as sufficient calcium, phosphorus, and boron levels in soil (Carducci et al. 2014a; Silva et al. 2014; Favarin et al. 2013). Fiber roots are mainly responsible for nutrient absorption in coffee and typically reach a depth of 40 cm (Motta et al. 2006).

5.1.2 Soil Texture

Heavy clays are undesirable for coffee cultivation due to poor drainage (Wilson 1985). In years with too much rain, crop yields are significantly reduced. Coffee cannot grow in heavy clay or excessively light soils (Muller 1966; Nair et al. 2017). Coffee cultivation soils in the Indian state of Kerala consist of very deep clay to gravelly clay loams with low available water content (Krishnan et al. 1996). Similarly, Natarajan et al. (1997) found that the soils in which coffee is grown in Tamil Nadu are deep to very deep to moderately shallow. Ananthanarayana and Ravi (1997) studied the geological formations of coffee soils in Kodagu, Chikkamagaluru, and Hassan districts of Karnataka and found that the clay content ranged from 15 to 48%. Based on the morphological characterization of coffee-growing soils in these parts, Kharche et al. (2000) reported that these soils are very deep, reddish-brown to dark red in color, and loamy to clayey texture with an angular block structure, most of which was revealed to have a clay crust.

5.1.3 Soil Structure

Non-clumped soils with good aggregates are suitable for coffee cultivation (Landon 1984; Nair et al. 2017). In India, Sys et al. (1993) and Anil Kumar et al. (2010) noticed that soils with higher organic carbon content had better soil aeration, cohesion, and structure and soil fertility. Soil structure is generally angular and clumpy below ground and granular or weakly angular and clumpy in the organic-rich topsoil. The good structure of the soil comes from the abundance of SOM and iron and aluminum oxides, and the low bulk density and good structure ensure rapid water infiltration and permeability.

5.1.4 Coarse Fragments

Coarse debris reduces the water-holding capacity of the soil. The upper limit is 55%, above which the soil becomes unsuitable for coffee cultivation (Nair et al. 2017; Sys et al. 1993). Kharche (1996) stated that soils containing 0 to 5% coarse debris are the most suitable for Arabica, while soils containing >75% coarse debris are unsuitable for coffee cultivation.

The biomass of the calcium-loving coffee root system itself is also a source of SOM and contributes to increasing CEC. Silva (2017) found that in a layer of 20–40 cm of soil, even after 5 years of coffee cultivation, soil management has resulted in different classes of pores such as 9.0 to 2.9, 2.9 to 0.6, and 0.6 to 0.2 μm (mesopores). In Brazil, the preparation of the site for growing coffee plants includes deep trenches, which more effectively relieve the physical and chemical limitations of the soil, and, therefore, there will be development of stronger and deeper roots. However, such an approach leads to severe destruction of soil aggregates and affects the physical quality of the soil. Therefore, after these civil works, some agricultural management practices should be adopted to improve soil aggregation and increase the sustainability of coffee production. Da Silva et al. (2020) presented evidence that the application of gypsum and grass biomass residues promoted soil aggregation in planting furrows, resulting in better porous soil spatial arrangement for coffee roots.

5.1.5 Soil Moisture

Coffee has been reported to have low drought tolerance (Landon 1984; Young 1975), and moist soils are essential for good growth and yield (Pillai 1984). Coffee varieties differ in drought tolerance, and Arabica is reported to be more drought tolerant than Robusta (Wellman 1961). Extreme rainfall conditions have significantly increased the complexity of coffee on above- and below-ground. If drainage is poor during the rainy season, the bottom of the coffee plants may get wet and the fruit may fall. Landslides and erosion caused by heavy rains change the physical, chemical, and biological properties of coffee soil. Therefore, considering the changing climate situation, the use of climate-resilient technologies in the coffee sector is of great importance.

5.1.6 Available Water Holding Capacity

Kharche (1996) found a significant positive relationship between coffee Arabica yield and available water holding capacity (WHC). Kharche et al. (1999) reported that the available WHC in coffee-growing soils in Kodagu and Chikkamagaluru districts of Karnataka ranged from 65 to 160 mm m^{-1}, while successful Arabica coffee requires >150 mm m^{-1} as optimum (Nair et al. 2017).

5.2 Chemical Properties

There are several factors that affect the yield and quality of coffee berries. For example, nutrition, plant water status, soil properties, etc. Important soil chemical properties that influence coffee growth and yield are soil reaction, aluminum saturation, soil salinity and alkalinity, CEC and base saturation, and organic carbon.

5.2.1 Soil Reaction

Coffee is an acid-loving plant. Soil response in coffee cultivation is, therefore, of great importance in assessing production potential. In highly acidic soils, the toxicity of manganese and aluminum and the lack of available phosphorus led to abnormal coffee growth and yield. Soil response studies in different coffee-producing countries showed that plants are being grown in different pH ranges from 4.0 to 8.0. Neither of these extremes is good for economic performance. Robinson and Hosegood (1963) reported that a pH of 5.2 to 6.2 is optimal for Arabica coffee, while a soil pH above 4.5 is ideal for Robusta coffee (Forestier 1968). Krishnappa et al. (1988) recorded high coffee yields in areas with normal soil pH. Ananthanarayana and Ravindra (1998) found that coffee cultivation soils are generally acidic (4.0–6.0).

Similar results have been reported in coffee cultivation soils in India (Shivaprasad et al. 2018; Shivaprasad et al. 2020). Nagaraja (2003), as part of a new approach study of Diagnostic and Recommendation Integrated System (DRIS) for coffee cultivation in Karnataka (India) found that the optimum pH value for Arabica is 5.5 to 6.1. As per CCRI (2014), the optimal pH range for coffee cultivation in India is slightly acidic reaction (5.5–6.5).

5.2.2 Aluminum Saturation

Aluminum contributes to the exchangeable acidity of the soil, and high concentrations in the soil lead to root damage and prevent the absorption of other elements. Ananthanarayana and Ravi (1997) reported that Al saturation in soils where coffee is grown in Kodagu, Chikkamagaluru, and Hassan (Karnataka, India) ranged from 10.5 to 85.0 of the effective CEC of the soil.

5.2.3 Soil Salinity and Alkalinity

The coffee grown in Indian soils is mainly located in areas with high rainfall, and hence the possibility of salt accumulation in these soils is low. Sys et al. (1993) reported that electrical conductivity (EC) above 2 mmhos cm^{-1} is inappropriate for both Arabica and Robusta. A study conducted by Ananthanarayana and Ravi (1997) revealed that soil EC values in coffee-growing areas of Karnataka ranged from 0.02 to 0.56 mmhos cm^{-1}.

5.2.4 Cation Exchange Capacity and Base Saturation

The soil in which coffee is grown in India is considered to be a low-activity clay soil, and the contribution of organic matter is a very important factor for determining the CEC of the soil. A soil CEC of 16–24 cmol (p+) kg^{-1} and base saturation of approximately 50–80% is considered as optimal for cultivation (Sys et al. 1991, 1993). A study conducted by Anil Kumar (2002) found that CEC of topsoil in coffee growing areas of Karnataka ranged from 3.6 to 30.7 cmol (p+) kg^{-1}, with an average of 15.7 cmol (p+) kg^{-1}. Similar results were also reported by Nair et al. (2017). Soil base saturation in coffee soils ranged from 56 to 93%, with an average of 78.1%. This study also showed that most of the soils analyzed had optimal CEC and sufficient base saturation for coffee cultivation.

5.2.5 Organic Carbon

In India, coffee is grown under shade of the trees, so addition of fallen leaves from shade trees accumulates a large amount of organic matter in the soil. According to a study conducted at the Central Coffee Research Institute in Balehonnur, India, an average of about 10 tons of fallen leaves are collected per ha each year in coffee plantation, which builds the SOM. Additionally, when the fruit is harvested, a high percentage of coffee leaves are separated from the coffee plant. Natural falling of leaves during the dry and rainy seasons, and regular pruning of coffee plants contribute organic matter into the soil. Typically, the organic matter content of the soil in which coffee is grown is between 1.5 and 4.0%, and Kharche (1996) reported that such a SOM content is ideal for Arabica coffee. Ananthanarayana and Ravindra (1998) found that most of the acidic soils in Chikkamagaluru and Coorg districts, where coffee is grown, had high organic carbon content, and similar results were reported by Shivaprasad et al. (2018), and Shivaprasad et al. (2020) in Indian coffee soils. Organic matter affects the physical, chemical, and biological properties of the soil and, therefore, it plays an important role in the productivity of coffee. SOM also provides aeration, increases water infiltration, reduces soil erosion and also helps to control acidity caused by certain nitrogen-containing fertilizers. This is important because a high level of acidity in the soil reduces microbial activity, and further leads to the development of toxicity caused by the presence of Al and Mn.

6 Major Nutrients

6.1 Available Nitrogen

If the available nitrogen in the soil is <280 kg ha^{-1}, it is low; from 280 to 560 kg ha^{-1} is medium; and > 560 kg ha^{-1} N in soil is considered as high (CCRI 2014). Nitrogen availability status in coffee-growing areas is usually classified as moderate to high. Nitrogen fertilization provides excellent results as long as the cation balance is right and the coffee plantation is properly maintained. Nitrogen deficiency is more common during the dry season, especially in exposed plantations after high yields.

6.2 Available Phosphorus

The soil available phosphorus content of <9 kg ha^{-1} is classified as low; 9–22 kg ha^{-1} as moderate and > 22 kg ha^{-1} as high (CCRI 2014). Soil analysis in the coffee-growing region of Karnataka, India over the past 10 years has shown that about 50% of the soil falls into the high category and < 10% falls into the low category. This indicates that soil phosphorus has become enriched as a result of continuous application of phosphatic fertilizers over a period of time. Reducing fertilizer dosage when soil test results were high had no negative impact on yield. Direct reactions to the application of P to aged coffee are very rare. However, phosphorus is often used in small amounts (Shivaprasad et al. 2017, 2018, 2020). Kimew (1974) found a positive response to P application in Kenya when soil available P was <20 ppm. The correlations between soil available P and coffee yield and between fertilizer P and coffee yield were found to be nonsignificant.

6.3 Available Potassium

Potassium availability in soils below 125 kg ha^{-1} is considered to be low; 125–250 kg ha^{-1} as medium and > 250 kg ha^{-1} as high (CCRI 2014). Most of the coffee growing area in the Indian state of Karnataka falls under the intermediate to high category. It is known that high K content in the soil hinders the uptake of Ca and Mg, and the K content in coffee is closely related to the activity of Ca and Mg. Carvajal (1985) found that the Ca + Mg/K ratio is optimal when its value is 10 or slightly lower, and that above 24, the probability of reaction increases dramatically when K is added to coffee soils. Nagaraja (2003) found that the Ca + Mg/K ratio in the soil of Chikkamagaluru district was between 9.3 and 11.1, and similar results have been reported from other Indian coffee-growing soils (Shivaprasad et al. 2018, 2020).

7 Secondary Nutrients

Secondary nutrients such as calcium, magnesium, and sulfur are essential elements that plants require in amounts greater than 1 mg g^{-1} dry matter. They influence the growth and sustainable development of coffee plants.

7.1 Available Calcium, Magnesium, and Sulfur

Calcium is involved in several physiological processes that influence plant growth, development, and productivity. It is also important for the development of terminal buds and flowers. Magnesium is one of the components of chlorophyll. Sulfur is known to improve the quality of coffee. According to the DRIS approach, Nagaraja (2003) reported that soil exchangeable Ca levels in coffee soils ranged from 700 to 1400 ppm, and soil Mg levels ranged from 185 to 224 ppm and are ideal for coffee cultivation. The study found that soil exchangeable magnesium is very low in most coffee plantations in Karnataka. Ananthanarayana and Ravindra (1998) found that exchangeable Ca and Mg contents of 1000 ppm and 375 ppm, respectively, are considered as critical limits, below which they are likely to be responsive to Ca and Mg applications. Krishnamurthy and Alwar (1984) reported that 15–20 ppm of thermo-soluble sulfur is considered important for coffee cultivation. This study found that S well above the critical value was detected in approximately 70% of the soils investigated. This may be due to the high organic carbon content in coffee soil and the concomitant use of copper sulfate and zinc sulfate. The S requirement for producing 1 ton of clean coffee varies between 2.3 and 3.5 kg (Alwar and Krishnamurthy 1992).

8 Micronutrients

In general, the problem of micronutrients is less severe under acidic soil conditions due to the high mobility of micronutrients, especially Fe, Mn, and Cu. In acidic soils, toxicity of not only aluminum but also iron and manganese can be expected if poorly managed. Dhruvakumar and Rao (1982) observed very high iron and manganese content in the soil, which can cause toxicity, in the Koppa belt of Chikkamagaluru district, Karnataka, India. Analysis of coffee-growing soils shows that these elements are insufficient in the soils. DTPA extractable Zn, Cu, Fe, and Mn values in soil of Chikkamagaluru district under Arabica coffee are between 0.1 and 4.7, 1.1 and 74, 14 and 152, and 12 and 122 ppm, respectively, based on soil nutrient status. Optimal nutrient levels for high-yielding crops were calculated. The results show that about 53% of the samples had very low Zn content, while there were no samples with very low Fe, Mn, and Cu contents (Nagaraja 2003). A study by Raju and Deshpande (1987) found that widespread zinc deficiency in Arabica plantations was associated with high levels of phosphorus, potassium, calcium and high soil pH in coffee soils. There is limited literature on the B and Mo status of Indian coffee cultivation soils. Symptoms of B deficiency are very rarely observed in coffee plantations. Experiments conducted have shown that foliar spraying of boric acid corrects deficiencies. The micronutrients are released when organic matter decomposes to cover the requirements of the coffee plant. However, coffee grown in open conditions or in areas with less shade may respond better to micronutrient applications.

9 Nutrient Deficiency Symptoms and Corrective Measures

Failure to apply balanced fertilizers or overexploitation of the soil without providing adequate nutrients commonly results in nutrient deficiencies in plants. This can be recognized by discoloration or malformation of the plant's leaves. Nutrient deficiencies affect not only plant growth but also yield (Table 1).

Table 1 Deficiency symptoms of major, secondary, and micronutrient elements and corrective measures in coffee

Nutrient	Symptoms	Corrective measures
N	Yellowing of older leave, defoliation, and die-back in severe cases. Poor and stunted growth	Foliar spray of 0.5–2.0% urea
P	Irregular patches of yellow color and mottled appearance in older leaves. Leaf turns red or violet and drops under severe condition	Foliar spray of 0.5% SSP or 1.0 DAP
K	Necrosis or scorching of leaf tip and margins of older leaves	Foliar spray of 0.1% MOP
Ca	Death of growing tip. Brittle leaves with yellowish appearance.	Liming of soil with agricultural lime
Mg	Yellowing of older leaves between the lateral veins on either side of midrib. Yellow area surrounded usually by green lines	Foliar spray of 0.1% $MgSO_4$ or soil application of dolomite
S	Yellowing of young leaves, stunted shoot	Soil application of S-containing fertilizer
Zn	Short internodes, stunted and chlorotic small elongated leaves. Rosette-shaped leaf bunch	Foliar spray of 0.25% Zn SO_4, properly neutralized
B	Death of terminal shoot, leathery leaves with malformation of small leaves	Foliar spray of 0.05% boric acid
Mo	Downward bending of the lamina along the mid-rib; large brown, yellow spots	Foliar spray of 0.025% ammonium molybdate
Fe	Chlorosis of leaves	Provide good drainage

Source: (CCRI 2014)

10 Soil Constraints and Management Strategies

The productivity potential of a coffee farm depends on timely implementation of cultural measures and adoption of good soil management practices. Soil management practices adopted to improve the soil characteristics to create ideal condition/medium for plant growth. Regular cultural practices are performed to modify the soil properties and to create favorable growing conditions for the plants. Soil erosion, soil acidity, low nutrient, and WHC are some of the limitations in coffee cultivation. Adoption of soil conservation measures and timely application of organic matter, agriculture lime and fertilizers are the important practices to obtain good yields. On steep slopes, soil conservation measures such as contour planting and terrace planting are usually recommended to prevent erosion. Drilling trenches and holes, tidying up the soil, and mulching soil with suitable materials have been suggested as moisture management measures (CCRI 2014).

10.1 Soil Acidity Management

Coffee is grown in areas of southern India that receive a lot of rainfall, and the soil reaction varies from slightly acidic to strongly acidic. The optimum pH value for growing coffee is between 5.5 and 6.1. Based on the soil test report, it is mandatory to apply lime every 2 to 3 years. Agricultural lime ($CaCO_3$) and dolomite (Ca and Mg CO_3) are the two most commonly used liming materials by farmers to correct soil pH. The application rate is 1.25 to 5.0 tons ha^{-1}. Quicklime (CaO) and slaked lime (CaOH) have been shown to more effectively neutralize soil acidity, but they also have corrosive effects and can have negative effects on soil microflora and fauna, as well as soil structure, and hence, it is not recommended due to rapid changes in the coffee soil ecosystem. Because coffee is grown in the shade, the large amount of leaf litter that accumulates each year gives the soil a high buffering capacity, and large changes in soil pH usually go unnoticed when lime materials are applied. However, applying lime will speed up the decomposition of fallen leaves and provide Ca and Mg to plants and soil microorganisms. It is recommended to broadcast lime over the entire site. December to March is considered as best time to apply, and the neutralization value of lime raw materials is important in evaluating quality of liming materials. Lime materials with a neutralization value of 80% or higher are considered ideal. If the neutralization value is below 80, it will be less effective in neutralization of soil acidity. Regular application of lime every 2 to 3 years may reduce the incidence of root diseases in coffee plantations and avoid soil nutrient imbalances. Correcting soil acidity is not the only requirement for growing coffee but also to make necessary nutrients present in the proper proportions on the surface of the soil. Application of organic fertilizer @ 2 to 4 kg plant every 2 years during the pre- and post-monsoon periods improves soil health and fertilizer efficiency (FUE) in coffee plantations.

The coffee is one of the most nutrient-intensive crops in tropical agriculture, and most Brazilian soils are characterized by high acidity and very low amounts of phosphorus and basic cations in all horizons and layers. Therefore, in addition to regular liming and fertilization in the surface layer, it is often necessary to apply some chemical remediation to the deeper layers (>20 cm deep). A commonly used option is phosphogypsum. Gypsum is formed as a by-product of phosphate fertilizer production and is applied to the soil surface as a physical amendment for most crops. Phosphogypsum is highly mobile throughout the soil profile, moving downward as ion pairs and subsequently releasing Ca_2+ and sulfate (SO_{4^-2}). This also reduces Al_3+ activity through binding to SO_{4^-2} (Caíres and Guimaraes 2018). Improving subsurface soil fertility promotes root development (Carducci et al. 2014b; Silva et al. 2016a) and promotes the uptake and circulation of nutrients, especially N, Mg_2+, and K+ (Ramos et al. 2013). Phosphogypsum can be applied to the soil surface and covered with a mixture of soil resulting in deeper roots forming and increased organic matter input (Ramos et al. 2013; Serafim et al. 2011), which further improves soil aggregation.

10.2 Land Management

Because coffee production is primarily rain-fed, droughts and unexpectedly long dry seasons can have a significant impact on coffee productivity and bean quality. Hence, innovations to promote efficient use of soil water are needed in coffee farming. The most common techniques are soil amendments to allow deeper root development and mulching to reduce evaporative losses and soil surface temperatures. At the same time, soil preparation work is also carried out to increase and maintain soil fertility (Serafim et al. 2011, 2013; Silva et al. 2019a). Management techniques also include the preparation of deep trenches (0.60 m) and fertilization and the cultivation of grasses (*Brachiaria* sp.) between the rows to increase organic matter content and protect the soil from erosion (Carducci et al. 2015).

11 Impact of Climate Change on Coffee Soils

Climate change is characterized by higher concentrations of CO_2 in the atmosphere (>400 ppm), higher temperatures (>2–4 °C), drastic changes in daily, seasonal and annual temperature, increased humidity, changes in dry environmental conditions, heavy rainfall, severe storms, prolonged droughts, and heat waves are expected to have significant impacts on terrestrial coffee ecosystems, affecting coffee, soil properties, surface moisture, and river flows (Patterson and Sim 2013), which also affects soil water quality and terrestrial water cycle (Pangle et al. 2014). Coffee is a very climate-sensitive crop, growing best in a narrow temperature range and the right amount of sunlight and rainfall distribution pattern. Soils are closely connected to the atmospheric climate system, which affects coffee through C, N, and water cycles.

Climate change affects SOM dynamics and soil microbial populations. The reduction in shade cover/coffee cultivation with only shady silver oak coupled with recent changes in climatic conditions will affect coffee production and quality in

India (Chengappa and Devika 2016; Chengappa et al. 2017; Vaast and Raghuramulu 2012). Impact of climate change on coffee production has also been investigated in several other coffee-producing countries. Colombia faced a 30% decrease in coffee production between 2008 and 2012 due to excessive rainfall and coffee leaf rust (Cristancho et al. 2012). Excessive rainfall in the Indian context has many effects on topsoil erosion, which can lead to leaching of essential nutrients and create saturated soil conditions over long periods of time, impacting nutrients available to coffee plants. Using a global multi-model ensemble, Bunn et al. (2015) predicted that rising temperatures are likely to reduce yields of both Arabica and Robusta coffees, especially due to greater temperature fluctuations during intra-seasonal temperatures.

Gomes et al. (2020) used a bioclimatic model to identify areas suitable for coffee production in 2050, where coffee is grown in unshaded plantations and agroforestry systems. The researchers found that the model predicts that average annual temperatures in the mountainous southeastern region of Brazil will increase by 1.7 °C ± 0.3 °C, resulting in an area suitable for growing coffee in unshaded plantations by 60% by nearly 2050. However, the introduction of agroforestry systems with 50% shade coverage can reduce the average temperature and keep 75% of the area suitable for coffee cultivation, especially at 600–800 m altitude. Their findings suggest that major changes in areas suitable for coffee cultivation will occur within 30 years, potentially leading to land conflicts for coffee cultivation and conservation. Incentives that donate to the development of coffee agroforestry conditions at appropriate locations may be essential to safeguard coffee production in the southeast of Brazil.

Less shade can affect soils by exposing them to direct sunlight, increasing soil temperature, reducing soil moisture, and reducing microbial populations. High soil temperature and low moisture content affect coffee yield. During the past decade, Central American countries producing Arabica suffered from crop losses mainly due to microclimate changes caused by reduced shade (Avelino et al. 2015; Lin et al. 2007). Compared to other coffee-producing countries, climate change does not have a significant impact on coffee production, mainly due to the presence of shade trees in India and Ethiopia. High temperatures, high solar radiation, and low precipitation are not considered optimal environment for coffee production, so shade or shelter trees are advantageous (DaMatta and Ramalho 2006; Vaast et al. 2005).

Trees buffer the effects of bad weather and create a microclimate more suitable for coffee production. In addition, leaf litter of the shade trees in coffee plantation acts as mulch and reduces the impact of raindrops hits on the surface soil, conserving soil moisture and reducing soil erosion. Recent research has shown that growing coffee under shade trees can delay ripening, which results in larger coffee berries and improved biochemical composition compared to coffee grown in direct sunlight (Vaast et al. 2006) and improve the quality of coffee. Coffee agroforestry provides many ecosystem services compared to unshaded coffee, including increasing soil fertility, reducing soil erosion, and improving water quality. Coffee production with shade trees is important to improve growth and quality of coffee, sustain the natural vegetation, and restore agro-ecology and nature-based agroforestry practices.

There is evidence that climate change is having a significant impact on areas suitable for growing Arabica coffee in major growing regions, including the East African Plateau region (Craparo et al. 2015; Ovalle Rivera et al. 2015). Intercropping coffee with shade trees is one of the promising adaptation methods that can compensate for the increase in temperature within the canopy by 2–3 °C (Ovalle Rivera et al. 2015). Guillemot et al. (2018) stated that the introduction of the alien tree species (*Grevillea robusta*) into agroforestry systems (AFS) in the Western Ghats of India could pose a threat to both climate change mitigation and tree diversity conservation. Therefore growing coffee under the mixed shade trees (local forest tree species) is one of the sustainable agriculture practices in coffee ecosystem.

12 Fertility Management for Healthy Soils – Fertilizer Recommendations

Economically grown coffee plants have the dual function of ripening of current-year fruits and producing a fresh frame of cropping wood for the next year. The application of essential nutrients needed to maintain optimal plant growth and yield is called balanced fertilization. The most economical amount of N, P, and K to add in a particular block depends on the estimated yield, crop yield performance over the past five seasons, and soil test data. Nutrient dosages applied without considering soil test values are inaccurate, uneconomical and can also cause many environmental problems in the coffee ecosystem. Therefore soil test based fertilizer application is the best agricultural practice in coffee to manage the soil fertility.

12.1 Nutrient Requirement

It is estimated that 1 ton of clean Arabica coffee removes approximately 40, 7, and 45 kg of nitrogen, P_2O_5, and K_2O, respectively, while for Robusta, it is 45, 9 kg, and 58 kg, respectively (CCRI 2014). The nutrient requirements for biomass to produce 1 ton of clean coffee (about 6.0 ton of Arabica fruit or 5.0 ton of Robusta fruit) are approximately 13, 2.3, and 15 kg of N, P_2O_5, and K_2O, respectively (CCRI 2014). Nutrient dosages required to achieve target yields were determined by considering crop nutrient losses due to nutrient removal, leaching, fixation, and FUE. For mature Arabica coffee where the coffee yield does not exceed 2.5 tons ha^{-1}, a mandatory/maintenance dose of 50:50:50 kg $N:P_2O_5:K_2O$ ha^{-1} every year should be applied in areas where harvesting is carried out. If the production is from 2.5 to 3.75 tons, the dose is to be adjusted as 75:75:75 kg. These general recommendations are subject to change based on the soil analysis report (CCRI 2014). Krishnamurthy (1991) estimated the requirement to be 120–175 kg N, 80–120 kg P_2O_5, and 120–170 kg K_2O for both Arabica and Robusta coffees per unit area for producing 1.0–1.25 ton of clean coffee. Jayarama et al. (1993) reported a positive correlation between coffee yield and fertilizer recommendations based on soil testing. A significant positive correlation was observed between soils available K and yield.

Amaral et al. (2011) found that low application rates of NPK fertilizer affected coffee root production, while regular and high application rates were found to be beneficial to different types of coffee trees in Brazil. A variety grown at low fertilization levels was found to have the highest root production efficiency per unit of N, P, K, Ca, Mg, and S uptake. Under the same conditions, another variety (Rubi Mg-1192) recorded the lowest efficiency per unit of fertilization. It was concluded that the efficiency of nutrient utilization for root and shoot production differs among coffee cultivars.

12.2 Nutrient Management

There are 16 essential plant nutrients that are very important for coffee plant growth. Carbon, hydrogen, and oxygen make up 94% of plant tissue come from air and water, whereas the remaining 13 elements are taken up from soil and are broadly classified into three categories: macro, secondary, and micro; however, these terms do not refer to the meaning of the elements. For normal plant growth, macronutrients are required in greater amounts than secondary and micronutrients. These essential minerals are directly involved in the metabolism of plant systems. A plant cannot complete its life cycle without an element, and the action of an element must be specific and no other element can completely replace it. The role of essential minerals in coffee plants is given in Table 2.

Soil and plant tissue analyses are important to understand the content of these essential elements in plant and soil. While soil analysis provides the information of essential soil nutrients, leaf analysis gives exactly what the plant has successfully taken up from the soil.

Table 2 Essential minerals and their role in the coffee plant

Mineral/Element	Main functions in coffee plant
Nitrogen (N)	Plant growth; proteins; enzymes; hormones; photosynthesis
Phosphorus (P)	Energy compounds; root development; ripening; flowering
Potassium (K)	Fruit quality; water balance; disease resistance
Calcium (Ca)	Cell walls; root and leaf development; fruit ripening and quality
Magnesium (Mg)	Chlorophyll (green color); seed germination
Sulfur (S)	Amino acids and proteins; chlorophyll; disease resistance; seed production
Copper (Cu)	Chlorophyll; protein formation
Zinc (Zn)	Hormones/enzymes; plant height
Manganese (Mn)	Photosynthesis; enzymes
Iron (Fe)	Photosynthesis
Boron (B)	Development/growth of new shoots and roots; flowering, fruit set and development
Molybdenum (Mo)	Nitrogen metabolism
Chloride (Cl)	Photosynthesis; gas exchange; water balance

Source: Winston et al. (2005)

12.3 Soil Analysis

The aim of soil sampling is to obtain representative soil samples from two ha blocks.

To reduce costs, samples in blocks of 2 to 4 ha are sufficient if the plants are of the same age and appearance. Detailed information on the upper and lower limits of soil organic carbon and the availability of nitrogen, P_2O_5, and K_2O, as well as the optimum amount of secondary and micronutrients in coffee-growing soil, are listed in Tables 3 and 4.

12.4 Leaf Analysis

Leaf analysis is another method to determine the soil composition and actual nutrients of the plant itself at the time of sampling. The use of analysis technology is based on the idea that the higher the content of certain nutrients in plants, the more effective they are in the soil. Leaf analysis is often recommended on coffee plants to confirm nutrient deficiencies and determine the sufficiency/deficiency/toxicity of

Table 3 Upper and lower limits of organic carbon and available N, P_2O_5, and K_2O in coffee soils in India

Rating	OC (%)	Available N	Available P_2O_5	Available K_2O
		(kg ha^{-1})		
Low	< 1.0	< 280	< 23	< 141
Medium	1.0–2.5	280–560	23–56	141–336
High	> 2.5	> 560	>56	>336

Source: CCRI (2014)

Table 4 Optimum range of secondary and micronutrients in soil for coffee cultivation in India

Nutrient	Range (ppm)
Available calcium (Ca)	700–1400
Available magnesium (Mg)	180–375
Available sulfur (S)	16–32
Available copper (Cu)	4–40
DTPA-iron (DTPA-Fe)	16–72
DTPA-manganese (DTPA-Mn)	15–80
DTPA-zinc (DTPA-Zn)	1.5–4.0
Available boron (B)	0.5–2.0

Source: NBSS LUP (2017)

nutrients in the plant. Therefore, analysis results can be used to guide treatment before nutrient deficiency/toxicity occurs in the leaves. The leaf analysis for routine fertilizer recommendation in coffee plantations is not generally necessary unless the plants exhibit deficiency/toxicity symptoms.

Pre-flowering is the preferred sampling time if only one sample is taken each year. More frequent sampling (every 4 months) is highly desirable for large plantations, especially if nutritional problems occur. The nutrients present in coffee plants and its threshold values in coffee leaves are presented in Table 5.

Generally, in Indian conditions, 2–3 applications of fertilizers are recommended depending on the nutrient content. If the annual dose is less than 150:125:150 kg ha^{-1}, two applications per year are possible (CCRI 2014). If the amount of fertilizer exceeds this dosage, it is necessary to fertilize three times. Split fertilizer application reduces the loss of nutrients, especially nitrogen, through leaching. Nitrogen is important for coffee plants all year round and is to be supplied at least three times a year, such as pre-monsoon, mid-monsoon, and post-monsoon periods. Coffee requires more phosphorus during flowering, hence, soluble phosphorus needs to be supplemented with nitrogen and potassium after flowering, preferably with diammonium phosphate (DAP) or single super phosphate (SSP) during flowering. Applying 115 kg of urea or 250 kg of ammonium sulfate and potassium chloride (MOP) @ 65 kg ha^{-1} to coffee plants during the monsoon season is beneficial as it promotes growth. Application of potassium fertilizer during post-monsoon is very important as coffee trees require more potassium during this period. It accelerates fruit ripening, bean filling, and curing in coffee.

Urea is the most commonly used nitrogen fertilizer for coffee plants. It can be used on all types of soil and easily mixed with phosphate and potash fertilizers. If the soil pH is above 6.0, ammonium sulfate is recommended as a nitrogen source. DAP provides nitrogen and phosphorus in water-soluble form and can be used in coffee soils with pH 5.5–6.0. Water-soluble phosphates are especially beneficial for

Table 5 Threshold values of nutrients in coffee leaves

Nutrient	Range
Nitrogen (N)	2.5–3.5%
Phosphorus (P_2O_5)	0.1–0.20%
Potassium (K_2O)	1.8–3.0%
Calcium (Ca)	0.9–2.0%
Magnesium (Mg)	0.2–0.4%
Sulfur (S)	0.1–0.2%
Manganese (Mn)	101–260 ppm
Iron (Fe)	110–240 ppm
Boron (B)	40–50 ppm
Copper (Cu)	21–32 ppm
Zinc (Zn)	15–25 ppm
Molybdenum (Mo)	0.1–0.5 ppm

Source: CCRI 2014

young coffee plants. However, these water-soluble phosphates become rare forms when applied to very acidic soils. Therefore, rock phosphate is beneficial for acidic soils with pH below 5.0 Potassium sulfate (SOP) is also suitable for coffee.

Salamanca et al. (2017) reported that the highest percentage of plant nitrogen taken up from applied fertilizer occurred 2 months after application in areas with high rainfall and high SOM. Fertilizer N provided approximately 20–29% of the total nitrogen content of the plants, but after 4 months, the N obtained from fertilizers was only an average of 5% of the total application. Though there were differences in fertilizer nitrogen content, no effect on the application method, especially on the dry weight and macronutrient content of the various plant components (roots and shoots) was noticed. Fertilization with drip irrigation improves contents of sugar, fat, protein, and cup qualities like caffeine and chlorogenic acid content. It is recommended to apply N, P, and K in three splits during flowering period (350 kg ha^{-1} of N), berry expansion period (250 kg ha^{-1} of P_2O_5), and at berry ripening period (150 kg ha^{-1} of K_2O) in Arabica coffee in China (Li et al. 2023).

12.5 Nutrients and Fertilizer Recommendation

Fertilizer application to coffee plantations is recommended in drip circles. Using a fork, remove the mulch under the coffee plant to about 50 cm from the stem and spread fertilizers uniformly all around. After application and spreading, refill the soil, and cover with mulch again. This method is very efficient as it avoids volatilization and leaching of nutrients. When growing coffee on steep or uneven ground, fertilizer can be spread in a semicircular arc (horseshoe shape) or crescent shape from the top of the field down the slope.

According to Silva et al. (2016a) and Silva et al. (2019a), the use of soil amendments; adjustment for proper management of phosphorus consumption by the plant; soil and leaf analysis to monitor coffee nutrition; balanced nutrient supply; and availability of sufficient soil moisture are some of the main challenges for modern competitive coffee cultivation. Therefore, sustainable coffee production requires higher nutrient use efficiency, higher fertilizer recovery, lower 2-year storage, and building soil fertility to achieve higher yields. In Brazil, Guimaraes et al. (2018) estimated the dry matter nutrient content of each coffee brushing operation as 75 kg N; 20.6 kg P_2O_5; K_2O 193 kg; 24.4 kg CaO; 20.8 kg MgO; 3.5 kg SO_4; 90 g B; 55 g copper; 1 kg iron; 475 g manganese; and 400 g zinc. Dry matter calcification is required to make these nutrients available in the crop cycle. This depends on sufficient soil moisture, temperature, and the presence of microorganisms in the soil, as some nutrients such as nitrogen and phosphorus remain partially present over long periods of time (Ribeiro et al. 1999).

The fertilizer schedule for young Arabica and Robusta coffee recommended by the Central Coffee Research Institute of India is given in Table 6.

The dosage of nutrients required for targeted yield of clean coffee considering the absorbed nutrients by coffee plants from soil, nutrient loss through leaching, the

Table 6 Fertilizer schedule for young Arabica and Robusta coffee plants

	Arabica	Robusta
Age of coffee plants (years)	$N:P_2O_5:K_2O$ (kg ha^{-1})	
1	50:50:50	75:75:75
2	50:50:50	75:75:75
3	63:63:63	100:75:100
4	63:63:63	100:75:100

Source: CCRI (2014)

Table 7 Dosage of nutrients required to realize targeted yield in coffee plants

Yield clean coffee (kg ha^{-1})	Annual dose $N:P_2O_5:K_2O$ (kg ha^{-1})	Yield clean coffee (kg ha^{-1})	Annual dose $N:P_2O_5:K_2O$ (kg ha^{-1})
250	75:68:75	1500	200:155:200
500	100:85:100	1750	225:173:225
750	125:103:125	2000	250:190:250
1000	150:120:150	2250	275:208:275
1250	175:138:175	2500	300:225:300

Source: CCRI (2022) and CCRI (2014)

approximate fixation and FUE of fertilizers, as well as the soil test values, is given in Table 7.

A long-term experiment conducted at CCRI, Balehonnur, India to study the effect of inorganic manure application on coffee yield indicated that inorganic manure (NPK) application had practically no effect on N, organic carbon, humus, and CEC of the coffee soil (Raju 1978). But application of inorganic manures improved the soil fertility of coffee soils (availability of P and K). A similar study under Wayanad condition (Kerala, India) revealed better yields and a positive correlation between nutrient application and yield. A positive correlation was observed between soil-available potassium, applied K fertilizer, and coffee yield. Similarly, correlation was noticed in coffee between soil available P and yield by application of phosphatic fertilizer (Raju et al. 1982). Hence, balanced fertilizer application in coffee soil is very important for soil health and sustainability in coffee production.

However, the timing and frequency of fertilization have a significant impact on coffee growth and yield. Improving FUE requires timely application of fertilizers that coffee plants need most. Research has shown that the peak nutritional needs of coffee plants occur during flowering, fruit setting, fruiting, and crop maturity (CCRI 2014). Therefore, depending on the suitable climatic conditions, it is recommended to apply fertilizer at four intervals (CCRI 2014) such as the flowering period (February/March), pre-monsoon (May/June), 2 to 3 days break in monsoon (August), and post-monsoon season (September/October).

Li et al. (2023) from China reported that increasing the fertilization rate during the flowering and fruit expansion stages obviously improved the average leaf photosynthesis rate, stomatal conductance, carboxylation efficiency, and transpiration rate of coffee. They recommended to apply N, P_2O_5, and K_2O (20:20:20) @350 kg ha^{-1} during flowering, 250 kg ha^{-1} during fruit expansion, and 150 kg ha^{-1} during fruit ripening stages.

12.6 Foliar Nutrition

To manage drought, foliar application of nutrients is very important in coffee plantation. Urea (1 kg) + SSP (1 kg) + MOP (0.75 kg) + $ZnSO_4$ (1 kg) + 70 ml of Planofix with available wetting agents dissolved in 200 l of water is ideal for coffee. The nutrient mixture spray has to be done in two times; the first spray at 45 days after the last rainfall and the second at 45 days after the first spray. This will mitigate the effect of drought when soil moisture is deficient in coffee soils (CCRI 2014).

13 Fertilizers and Plant Protection Chemicals on Soil Health and Soil Microbiology

Chemical fertilizers play an important role in increasing crop productivity and soil fertility. They exist in different forms: nitrogen-containing fertilizers, phosphate-containing fertilizers, potassium-containing fertilizers, and micronutrients. Fertilizers not only increase crop productivity but also change the chemical and biological properties of the soil. However, continuous application of chemical fertilizers, coupled with deterioration of soil quality and microbial population, causes a decrease in SOM content. Excessive use of chemical fertilizers hardens the soil, reduces soil fertility, causes air, water, and soil pollution, reduces available nutrients in the soil, and poses a danger to the environment. Continuous use of chemical fertilizers can change soil pH (CCRI 2014). Chemicals remain in the soil for long periods of time, impacting soil health and soil biodiversity. Soil organisms such as earthworms that fertilize the soil are also destroyed by improper application of chemical fertilizers to the soil. In Karnataka, India, Kulandaivelu (2016) reported that soil electrical conductivity and bulk density increased by 34% and 21%, respectively, in conventional farming systems compared to long-term organic farming plots. Significant increases in organic carbon were observed in plots subjected to long-term organic farming. Both systems were found to have reduced WHC. The main nutrients (nitrogen and potassium) increased in both conventional and organic farming, but the trend was more pronounced in conventional farming systems. Although both systems increased phosphorus, availability was found to be higher in the conventional system. Regarding biological properties, soil respiration, and fluorescein diacetate activity (used to determine microbial activity) were found to be higher in organic farming compared to conventional farming. Higher soil urease activity was observed in conventional farming systems. The long-term effects of organic farming systems were clearly evident in microbial traits, with higher macrofauna (31.4%), microbial abundance (34%), and microbial diversity index compared to conventional systems.

Experiment on continuous application of inorganic fertilizers in Arabica growing soil revealed that long-term manuring had practically no effect on organic carbon, N, and CEC of coffee soils. Soil reaction (soil pH) and exchangeable Ca and Mg showed a significant decrease with simultaneous increase in exchangeable Al (Raju 1978) in coffee soils. Hence, application of fertilizers alters the chemical composition under long-term experiment. Lime application is one of the major amendment practices in coffee plantation. Varying amounts of $CaCO_3$ had practically no effect on the organic carbon content, CEC, and available potassium in the coffee-growing soils in India (CCRI 2014). In the soil of pH 4.2 an initial increase in available P (Bray 1) was observed; but further addition of $CaCO_3$ resulted in significant reduction in available P. In the soil of pH 5.6, liming caused reduction in P availability. However, this study further suggested that when the available P level in soil is in the high range and liming did not adversely affect P availability (Raju and Krishnamurthy 1978). The addition of $CaCO_3$ @ 2.5 to 20.0 t ha^{-1} significantly raised the soil pH with simultaneous decrease in DTPA extractable Zn. Increases in exchangeable Ca consequent to liming resulted in higher leaf Ca, in Arabica coffee seedlings grown in differentially limed soils.

Application of lime boosted the growth of seedlings in spite of decreasing extractable Zn, possibly due to the nutritional effect of Ca in promoting abundant root growth. Total Zn content was not altered appreciably by liming (Raju and Deshpande 1986) in coffee-growing soils. The results from CCRI, Balehonnur, India indicated that continuous NPK fertilizers supplement over a period increased Fe and Mn in soil without influencing the Cu and Zn levels in coffee soils (Raju 1985; Raju and Deshpande 1985). In other words, soil quality was affected due to more supply of urea, MOP, and SSP.

One more important practice in coffee plantation is the application of Bordeaux mixture. Copper accumulation in coffee soil has been observed due to continuous spraying of copper fungicide. Storage of a high amount of extractable Cu was observed on the surface of the soil and its accumulation decreased with depth (Venkataramaiah et al. 1974). Studies on microbial community revealed that the population of fungi, yeast, and bacteria was higher in the rhizosphere soils. Among the fungi isolated, *Aspergillus* spp. were the predominant species followed by *Fusarium* sp., and *Rhizotonia solani*, the incident of collar rot of coffee seedlings, both in rhizosphere and rhizoplane of seedlings from untreated seeds but the pathogenic fungi were absent when fungicide treated seeds were sown. Seed treated with carboxin had less fungal flora associated with the roots (Venkatasubaiah and Safeeulla 1984) which indicated that application of chemicals in coffee plantation altered the microflora and fauna of soil.

14 Soil Biodiversity, Microbial Synergy, and Function-Specific Microbial Communities

Indian coffee soil is rich in organic matter and acts as a reservoir for billions of different microorganisms, most of which are beneficial to plants. Growing coffee under different (mixed and mono) shade trees is a unique practice in India. This cultivation practice improves the health of the coffee soil by enriching organic matter, biodiversity, and nutrient availability. The impact of conversion of natural forests to cocoa and coffee forests and intensification of agroforestry to cocoa and coffee plantations on biodiversity in Africa, Latin America, and Asia were assessed. De Beenhouwer et al. (2013) found that forest species richness and total species richness were significantly lower in intensively managed systems than in natural cultivation. However, the decline in total species richness was greater when comparing woodlands with plantations (−46%) than when comparing forests with woodlands (−11%). Biodiversity responses to intensification varied between Asia and Latin America and between different species groups. In addition, enhanced management reduced the provision of ecosystem services by 37% when comparing forest with arable forest and 27% when comparing arable forest with plantations. A study conducted in the coffee growing regions of Chikkamagaluru, Kodagu, and Wayanad, India, showed vesicular-arbuscular mycorrhizal (VAM) colonization in both Arabica and Robusta coffee roots to be 21–64% and 10–68% (Rangeshewaran et al. 1990). Chikkamagaluru and Kodagu zones were dominated by *Glomus macrocarpum* and *Sclerocystis rubiformis*. The main species isolated from Wayanad coffee belt were *Gigaspora gigantea* and *Gigaspora heterogama*. Korikanthimath et al. (2000) reported a higher number of VAM spores (32 and 34 spores 10 g^{-1} soil) in rhizosphere soil of cardamom and coffee in a cardamom-robusta-coffee intercropping system in Kodagu compared to monoculture systems (21 and 32 spores). Similarly, population growth of P-solubilizing fungi was observed in a coffee-based cropping system in the Kodagu district of Karnataka, India (Korikanthimath et al. 2000). Mahdhi et al. (2017) reported that the colonization of the coffee root system by arbuscular mycorrhizal fungi (AMF) was high in Saudi Arabia. Microscopic observation showed a higher mycorrhizal intensity (24.3%) and spore density (256 spores 100 g^{-1} soil) at 1400 m altitude, and *Glomus* was the dominant genus in the coffee cultivation system. Liliana et al. (2021) reported from Mexico that AMF plays an important role in soil improvement as well as increasing the growth and yield of host plants. The diversity, abundance, and richness of AMFs were assessed in eight coffee plantations with different degrees of erosion. A total of 26 AMF morphotypes distributed in different genera were obtained. The highest diversity and richness were found in non-eroded agroecosystems, followed by sites with minimum erosion, and the lowest values within the highly eroded agroecosystems. These results suggest that erosion of soil affects AMF fungi, indicating the need to adopt cultural practices to preserve soil biodiversity in coffee plantations. Arias et al. (2012) studied the AMF community in different coffee production systems (0–15 cm depth) in Veracruz, Mexico, and reported that *Glomus clarum* and *Glomus* sp. 3 were the dominant morphospecies in all the sites. There was no significant difference in the number of AMF spores in different areas of coffee growing region. However, during the dry season, number of AMF spores differed between the forest type shaded with the traditional rustic system and the shaded simple system, with the latter having the most spores. With the exception of only one forest species, the coffee production systems shared the same AMF species as of the forest.

The presence of 13 spore morphotypes of AMF was reported by Hakim et al. (2013), with the genus *Glomus* dominating with 53.84%, followed by *Acaulospora* (30.76%) and *Gigaspora* (15.38%). In addition, it was found that the inoculation of a combined endomycorrhizal fungal culture on Arabica coffee seeds enabled the establishment of an efficient mycorrhizal symbiosis in coffee plants. Significant increases in root fresh weight, shoot weight, shoot height, stem diameter, and number of leaves were observed in inoculated plants compared to the control. Paudel and Kang (2018) reported that algae were useful biofertilizers for Arabica seedlings. The use of blue-green algae (BGA) as a biofertilizer helps to protect the environment by reducing the use of inorganic chemical fertilizers on coffee plantations. Yamel et al. (2019) demonstrated the effectiveness of inoculating coffee plants with a consortium of AMF and P-solubilizing fungal inoculations (*Aspergillus niger* and *Penicillium brevicompactum*) in increasing plant growth parameters and available soil phosphorus.

14.1 Soil Rhizosphere Management

SOM supports a large community of microorganisms capable of converting soil compounds into readily available mineral forms for root uptake. Leguminous shade trees in coffee plantations produce enormous amounts of leaf litter and provide nutrients to the plants after decomposing (CCRI 2014). Positive results were observed for application of bio-fertilizers to coffee seedlings in nursery. The coffee soils of India are generally rich in organic matter (CCRI 2014), and hence, the native microbial population would take care of the mineralization of nutrients under optimum moisture conditions.

15 Microbiome and Meta-Genomic Approaches

The microbiome of coffee plants has attracted interest in recent years because coffee plants are perennial tropical species of enormous importance, especially for developing countries. A better understanding of the microorganisms associated with coffee plants can help mitigate the adverse effects of climate change on crops, improve plant health, and make the system more sustainable (Leandro et al. 2022).

Metagenomics is a genomic technique used to study soil microbiomes by sequencing their genetic material. Metagenomic approaches were used to determine the microbial communities present in coffee plantations. The structure and function of the coffee rhizosphere microbiome were investigated by deep-shot metagenomic sequencing of the rhizospheres of five different species such as *C. arabica, C. canephora, C. stenophylla, C. racemosa,* and *C. liberica* (Leandro et al. 2022) in Brazil, The rhizosphere core microbiome includes genera such as *Streptomyces, Mycobacterium, Bradyrhizobium, Burkholderia, Sphingomonas, Penicillium, Trichoderma,* and *Rhizophagus*. Among these genera are several microbes that may be beneficial to plants. The microbial communities were dominated by *Streptomyces* and mycorrhizal fungi.

Fuentes-Ramírez et al. (2001) isolated two new N2-fixing species associated with Arabica coffee plants from the Acetobacteraceae family, *Gluconacetobacter johannae* sp. nova and *Gluconacetobacter azotocaptans* sp. nova. Caldwell et al. (2015) investigated differences in prokaryotic soil microbiology in three Brazilian coffee farms practicing intensive, organic agriculture and one in transition to intensive organic agriculture. They found two groups of prokaryotes associated with the nitrogen cycle, the archaea *Candidatus nitrososphaera* and the bacterial class rhizobiales were abundant. Profiling of the V4 regions of the 16S rRNA gene on all three farms revealed high prokaryotic diversity and thousands of species-level functional taxonomic units and many nitrogen-fixing genera (e.g., *Rhizobium, Agrobacter, Acetobacter, Rhodospirillum, Azospirillum)* known to favor plants growth were identified in each farm. Prates Jr. et al. (2019) compared three management systems and found that the agroecological management system was more similar to a natural forest with high biodiversity indices than traditional management.

16 Technological Options to Promote Soil Health

16.1 Soil Conservation Measures

Adoption of soil conservation methods in coffee farming are the most important practices for sustainable coffee production. Coffee is often grown at higher altitude in a high rain fall area which poses a significant challenge on occurrence of soil erosion, availability of soil moiture and status of organic matter in the soil. To address the soil erosion problem in coffee plantations contour planting and terracing planting practices are recommended. To conserve the soil moisture in coffee plantation cover digging, scuffling, leaf litter/ cherry husk mulching, opening of trenches or cradle pits are recommended and to conserve the soil organic matter in the coffee plantation, growing of coffee under the shade tree is very much essential along with the practicing of all the above said practices in the coffee ecosystem. Recent past increase in the extreme climatic conditions on the earth, diversified production systems and adoption of climate change mitigation measures to protect the crops and soil from such extremes, are most important. Growing of coffee under the two-tier shade (native forest species and exotic tree species) trees in coffee plantation helps in preventing soil erosion by avoiding direct splash of rain water on the surface of the soil pedon and improve the SOM by deposition of leaf litter on the surface soil and improve the soil moisture and available nutrient status and nutrient holding capacity of the soil (Nadaf and Bora 2021). Therefore, coffee cultivation, in mixed shade trees have a greater impact on soil fertility and soil productivity. Growing coffee in a monoshade system with exotic canopies has been characterized by lower soil carbon content, which in turn reduces soil microbial activity and soil porosity in coffee plantation (Nadaf and Bora 2021; Nadaf et al. 2021). The undulating terrain of the coffee regions and the heavy monsoon rains make it imperative to prevent soil erosion. However, the double-layer canopy of shade trees are the first part to prevent the soil erosion in coffee plantations. Growing of shade tree reduces the rate of raindrops, contour planting, embankment, and staggered planting/cover digging across the slope are the best soil management practices in coffee plantations. Although the coffee regions receive abundant annual rainfall of 2500–3000 mm, the regions experience a 3–6-month dry period from November to March/April. Consequently, the lack of sufficient soil moisture during dry periods becomes a limiting factor for coffee growth, therefore maintaining soil moisture on site is mandatory. In established farms, several measures to conserve soil moisture are generally recommended, such as post-monsoon harrowing, cover digging, opening lull holes across the slope (step digging) and opening of cradle pits. These cradle pits and trenches across a slope in a coffee farm act as mini-compost pits to break up the foliage of shade trees (CCRI 2014). The coffee soils of India, which are structurally classified as sandy and clay loams, are generally red and lateritic. These soils are prone to erosion in established plantation areas. Therefore, it is recommended to use erosion control measures such as embankment, contour planting, planting of shade trees, and mulching with organic wastes to prevent loss of topsoil. Circular planting is adapted for areas with a slope of 10–20%. In contour plantings, soil erosion is minimal and the coffee plants are located at the same level along a certain line of the slope (CCRI 2014). If the slope is >20%, a permanent terrace must be maintained to grow coffee as terrace. A terrace helps prevent excessive runoff of soil nutrients, prevents soil erosion, and helps water penetrate deep into the soil (CCRI 2014). Soil moisture conservation technologies for coffee cultivation include opening trenches/furrows, digging the soil, mulching, and mulching. In the case of new coffee plantations (new clearings), the field is fully dug at the end of the monsoon (October/November) to a depth of 30–45 cm.

Mulching of young coffee plants with organic crop residue will help to maintain optimal soil temperature, conserve soil moisture, control weed growth, and prevent soil erosion (CCRI 2014) in coffee ecosystem. In the post-monsoon period, in undulating lands, trenches/ pits, or cradle pits should be opened across the slope and these are gradually dug between the coffee rows along the entire contour. Cradle pits are short pits, about 1–1.5 m long 30 cm wide and 45 cm deep and length can be extend up to suitable length. Trench and cradle pits in the coffee estates are act as the compost pits, in addition to protecting against soil moisture loss and soil erosion. In these pits fallen leaf litter of shade trees, weed biomass, and Dadap (*Erythrina variegate*) and other crop residues can be decomposed. In addition, for better performance in sloppy areas, cradle holes are usually opened in October–November. These pits fill up during the northeast and southwest monsoons and therefore have to be renovated every 2–3 years. All deposits such as washed soil, leaf litter, weed biomass, and wood waste should be removed from these pits and spread near the surrounding plants. Cultivation of suitable inter crops such as black pepper, orange, and banana would also promote shade, soil fertility and improve the microclimate of coffee plantations. The two-tier system also acts as a kind of insurance against weather and price fluctuations in the international market (CCRI 2014).

16.2 Green Manuring

To meet the need for nutrients and prevent soil erosion, it is recommended to grow green manures in coffee plantations. Green manure plants such as *Crotolaria* sp. and legumes such as cowpea (*Vigna unguiculata*), horse grass (*Macrotyloma uniflorum*) can be grown initially for 2–3 years. Green manure is also cultivated with the aim of covering the land surface with vegetation and especially on hillsides during the rainy season to prevent soil erosion and runoff of soil (CCRI 2014). Growing of *Mimosa invisa* in coffee plantations was found to suppress weed growth considerably since *Mimosa* puts up bushy growth very fast. It prevents soil erosion by binding the soil and checking runoff water to a considerable extent. Mohan Kumar (1996) reported a higher value of N by incorporation of green manure crop in coffee soil. Further, it supports the production of a greater number of green flush and branches on coffee plants. Glyricidia is also used as shade tree and when it attains excessive growth, the plant could be recycled as manure.

16.3 Use of Organic Manures, Cover Cropping

If recycled, coffee husk and the bark can be a potential source of K for plants (Rosolem et al. 2003). To improve soil fertility, the coffee waste available in the plantation should be effectively recycled through aerobic/bag composting and such compost applied @ 2.5 t ha^{-1} $year^{-1}$ to maintain soil fertility. This will also reduce fertilization costs and prevent environmental pollution. Based on soil test results, soil pH should be adjusted with locally available agricultural lime or dolomite to reduce the soil acidity. The phosphorus requirement must be met with rock phosphate, bone meal, wood ash, and other approved products (CCRI 2014). The pongamia cake pellets are a very good organic fertilizer and are a rich source of NPK in the ratio 4:0.5:0.5, which improves soil fertility (CCRI 2014). Compressed cake pellets are also useful especially against nematodes. It is a natural fertilizer and can be mixed with neem cake granules for better coffee results.

Cover crops in coffee plantations have the potential to conserve soil moisture, prevent soil erosion, and improve soil health. Cover cropping is an effective method of weed control, and leguminous crops can reduce weeding costs and thereby increase crop profitability (Shivaprasad et al. 2005). Legumes such as *Crotalaria ochroleuca*, green leaf desmodium (*Desmodium intortum*), silver leaf desmodium (*Desmodium uncinatum*), Mucuna (white), Mucuna (mottled), and Lablab establish well in coffee system and produce large amounts of biomass. Weed control efficiency of horse gram (82%) and cowpea (68%) has been reported to be better compared to other cover crops (Keshavaiah et al. 1998; Shivaprasad et al. 2007). The approach of inter-row cover crops protects the soil from raindrops, increases soil biodiversity, produces more organic input, and increases root colonization with positive effects on aggregation and nutrient cycling (Nzeyimana et al. 2013). The interaction between coffee plants and flag grass (*Urochloa* or *Brachiaria* sp.) can increase SOC stocks and thus remove the greenhouse gas (CO_2) from the atmosphere (Silva et al. 2013). In such management systems, the decomposition of large organic material inputs promotes soil aggregation (Silva et al. 2014, 2016b, 2016c) and wildlife activity, promoting the formation of biopores that enhance root growth and reduce compaction problems (Neto et al. 2015).

16.4 Mulching

Field studies showed better efficiency of related deep furrowing, organic mineral fertilizer, and plastic mulching in soil aggregation (Souza et al. 2018). Plastic mulching covering the soil surface reduces evaporation, which leads to higher soil moisture levels as well as improves plant growth. A better development of vegetation leads to a higher concentration of soil organic matter, which is a very important factor influencing the stability and water retention of the aggregate. Due to the formation of stronger aggregates, soil macroporosity increases and water availability for plant roots improves (van der Meulen et al. 2006). Nzeyimana et al. (2020) found that different organic cover materials strongly improved soil nutrient levels, but the amount of nutrients released into the soil is controlled by the amount and type of biomass applied and the environmental conditions of the specific study site. Thus, organic mulching by smallholder coffee farmers should be considered as part of an improved land management technology aimed at improving soil fertility and yield. Nascimento et al. (2020) reported that polyethylene mulching in coffee plantations had a positive effect on productivity, ripening, grain shape, and beverage quality.

16.5 Intercropping

Intercropping is an agricultural practice where two or more crops are planted in the same field at the same time. Intercropping offers better crop stability and less yield loss during drought than monocropping; it also regulates the problem of water stress by conserving moisture within crops (Altieri and Koohafkan 2008). N_2-fixing trees (e.g., fruit trees: Inga) and annual crops (e.g., soybeans) combined with coffee can restore soil fertility and produce various agricultural products such as fruits, firewood, and timber, which can increase rural food security and community incomes (Souza et al. 2010). Cultivation in coffee plantations improves soil fertility because organic matter from trees or crops is integrated into the soil, which improves soil quality and allows farmers to save on the cost of expensive nitrogen fertilizers (Pumariño et al. 2015). Because intermediate plants cover a significant part of the soil, they produce more shade, conserve soil moisture, and create more buffering microclimatic conditions (Siles et al. 2010). It also improves both physical (porosity) and chemical properties (soil pH, enzymes) of the soil (Wang et al. 2015). It also increases microbial diversity, such as VAM populations, which provide nutrients to plants (Bainard et al. 2011). Intercropping is also a solution to maintain soil fertility by reducing soil erosion (splash effect, dissolution, and leaching) (Bucagu et al. 2013). In addition to income generation and food security, the intercropping system of coffee plantations is an approach to organic coffee production (Bacon et al. 2008).

Balota and Chaves (2010) intercropped coffee with different leguminous trees in Paraná State, Brazil, and *Leucaena leucocephala* L. significantly increased soil enzyme activity (urease, arylsulfatase, and phosphatase) and carbon and nitrogen mineralization rates both under the tree. Harelimana et al. (2018) reported that common bean (*Pheseolus vulgaris* L.) and soybean (*Glycine max* L.) were the most common intercrops with coffee in the Kamonyi region of Rwanda, increasing soil nitrogen through N_2 fixation. Van Asten et al. (2011) reported the agronomic and economic benefits of intercropping coffee and bananas in smallholder systems in Uganda. Jassogne et al. (2013) explained coffee/banana intercropping as an opportunity for smallholders in Uganda, Rwanda, and Burundi. In terms of hardiness, bananas provide valuable shade cover for very climate-sensitive coffee plants. In addition, bananas can retain water under drought stress, which reduces competition for water during drought compared to other shade trees. This increases farmers' incomes, improves resilience to climate impacts, and sequesters greater amounts of carbon compared to monoculture systems (van Asten et al. 2015). However, it is possible that the intercropping system removes more nutrients from the soil and, in the long run, coffee may eventually displace the banana. Therefore, the success of intercropping systems requires the identification of key production constraints – primarily soil fertility – and the development of site-specific recommendations to address them.

After a few years of planting coffee and adequate fertilization, mobile nutrients such as nitrogen (N), potassium (K), sulfur (S), and boron (B) can be leached from the absorbent roots. Therefore, almost year-round interaction with deep-rooted plants helped bring these nutrients back to the surface of the soil-plant system. Hence, inter-row management in collaboration with proper management of soil fertility will certainly give higher coffee production (Guimaraes and Reis 2010), because the nutritional efficiency of systematic production is higher. Similar results were reported by Silva et al. (2019a) in coffee farming in Brazil, where *Brachiaria* is a catch crop, plant residues are recycled and used as nutrients to feed the coffee. The amount and regularity of crop residue addition are more important than synchronizing coffee release and nutrient requirements, as organic matter content increases over the years. Brachiaria is more efficient than coffee plants in sequestering phosphorus from the soil, which gradually becomes available through the decomposition/mineralization of the canopy straw. Over the years, the grass also binds this nutrient deeply as its root system develops on a larger farm scale.

16.6 Biomass Recycling

In coffee plantations, various biodegradable agricultural wastes are available in the form of shade tree leaf litter, chopped coffee leaves and branchesm (prunings), weed biomass, and coffee processing waste (cherry or parchment skins) as coffee grounds. In addition, a high percentage of coffee leaves are separated from the tree during fruit harvesting, and defoliation also occurs in the dry and rainy seasons, adding to the organic matter status of the soil. It is estimated that under ideal shade conditions, shade trees produce approximately 10 ton of leaf litter (Fig. 2) ha^{-1} $year^{-1}$, contributing to the organic matter status of the soil. When completely decomposed, it is capable of releasing about 40 to 60 kg of N, 30 to 33 kg of P_2O_5, and 40 to 60 kg of K_2O into the soil (CCRI 2014).

Fig. 2 (**a**) Leaf litter in Coffee Estate at CRSS, Chettalli (**b**) Leaf litter in Coffee Estate at Coffee Research Sub Station, Chettalli

However in nature, not all available biomass is recycled effectively, and the decomposition process is slow. The layer of fallen leaves covers the soil and prevents the direct impact of rainwater, thereby preventing soil erosion. Wherever wet processing is applied in the field, the fruit husk/pulp obtained after pulping the coffee cherries can be used as organic fertilizer after composting. To produce each ton of clean coffee, about 5 to 6 tons of coffee beans must be ground. This will produce about 1–1.5 tons of fruit husk, which can be properly decomposed to obtain high-quality organic fertilizer (Salakinakop et al. 2004, 2005; Shivaprasad et al. 2003). Composting is a process in which larger particles are broken down into smaller particles by the action of microorganisms and macrofauna in the soil. The final end product is called compost. Recycling crop residue helps reduce the use of inorganic fertilizers. In addition, organic fertilizers also have a number of benefits such as improving soil structure, reducing soil erosion by binding soil particles, increasing the soil's ability to hold water and nutrients, and providing an environmentally friendly environment ideal for the growth of beneficial microorganisms in the soil. Finished compost can also be enriched with biological agents such as phosphorus-solubilizing bacteria (PSB), biopesticides, Trichoderma, and others by adding cultures of these organisms to compost at the end of the composting process (CCRI 2014).

In Andisol, which has a very low P availability of 1.17–2.54 ppm, P availability can be increased by using phosphate-solubilizing microbes and coffee husk compost. Mariani et al. (2020) reported that phosphate-soluble microbes and coffee skin compost increased plant height and plant dry weight by 35 and 32.17%, respectively, in Mount Sinabung under Robusta coffee plants. Jiang et al. (2023) reported that application of coffee compost @720 kg ha^{-1} produced the biggest effect on improving soil fertility (soil organic carbon, organic matter, and total nitrogen), microorganisms (bacteria, fungi, and actinomycetes), and enzyme (phosphatase and urease) activities in coffee plantation. Besides, there was improvement in the net photosynthetic rate, transpiration rate, and stomatal conductance of Arabica coffee plants.

16.7 Bio-Fertilizers

Bio-fertilizers are live microbial preparations that are applied to seeds or added to soil to improve soil fertility and productivity. These bio-fertilizers are products composed of selected beneficial microorganisms known to improve plant growth by providing plant nutrients and growth substances. Bio-fertilizers (also known as microbial inoculants) improve the fertility of the soil and promote the absorption of nutrients and water in deficient soil, which contributes to better rooting of plants. Bio-fertilizers are of various categories such as N_2 fixers, phosphate solubilizers, potassium mobilizing bacteria, plant growth promoting bacteria, arbuscular mycorrhizae, and organic matter degrading bacteria. Rhizobium, Azotobacter, Azospirillum, Sinilevä and Azolla are N_2-fixing biofertilizers. Bacillus and Pseudomonas are phosphate solubilizers. Bio-fertilizers also release growth-promoting substances and fungicides and improve seed germination and root growth. Phosphate-solubilizing microbes are seeded into the coffee pods along with rock phosphate while composting is being done to enrich the compost with available phosphorus.

From the field study with Biophos, P-solubilizing microbes such as *Pseudomonas straita, Aspargilus awamori, Candida* sp., and *Bacillus* sp., it was shown that the application of 10 kg ha^{-1} biofertilizer increased yield by 22% and available phosphorus by 32% (Shivaprasad et al. 2005). Chattopadhyay et al. (2006) reported that plants treated with Azospirillium + full N showed maximum growth, leaf NPK content, and endophyte formation. Muleta et al. (2013) studied phosphate-solubilizing rhizobacteria associated with Arabic coffee in the forests of southwestern Ethiopia. They reported that the production of organic acids by coffee-associated phosobacteria could be considered as the main mechanism involved in the solubilization of insoluble hydroxyapatite/tricalcium phosphate in coffee and concluded that an Arabica coffee-associated rhizobacterium with direct plant-promoting properties enabled its utilization using native microbes as microbial biofertilizer.

Arbuscular mycorrhizal fungi (AMF) are important components of agroecosystems because they form symbiotic associations with most crops. The mycorrhizal association is usually mutualistic and can provide many benefits to the host plant, including increased nutrient uptake, improved water conditions, and protection from pathogens (Bainard et al. 2011) and endomycorrhizal fungi improve the weight of the root, diameter of stem and number of leaf in coffee (Areqi et al. 2014) Swarupa et al. (1997) investigated the effect of a combined application of VAM and *Azospirillum brasilense* on coffee seeds and found that the best germination (>93%)

was observed after 60 days in seeds treated with fosobacterium either alone or in combination with VAM and/or *Azospirillum* (Salakinakop et al. 2003).

Similarly, increased uptake of nitrogen and trace elements (Fe, Cu, Zn, and Mn) was reported in a nursery experiment conducted by Premkumari and Balasubramanian (1993) at the Tamil Nadu Agricultural University, Coimbatore, India. Combined inoculation of *Azospirillum brasilense* (SP.7) with VAM fungi (*Glomus fasciculatum, Gigaspora margarita,* and *Acaulospora laevis*) significantly increased growth and also increased root VAM colonization to approximately 44% compared to the uninoculated control. Inoculation of S 795 in Arabica coffee with *Azotobacterium* was found to have a positive effect on shoot dry weight, root spread, and seedling zircon (Azizuddin 1984).

Six dominant original VAM isolates were collected from soils of Hassan, Chikkamagaluru, Mudigere, and Koppa coffee growing regions in India. *Glomus radiate* and *Acaulospora myriocarpa* increased plant growth compared to other isolates tested by Sudha and Balakrishna (2004). Identification of VAM fungi isolated from the rhizosphere of coffee plants in Yemen revealed the presence of AM species, namely, *Glomus proliferum, G. etunicatum, Acaulospora sporocarpia, Archeospora* sp. and *Scutellospora nigra* (Hakim et al. 2013). The diversity of AMF that occurs naturally in coffee fields can be selected and used for reforestation and restoration of degraded ecosystems, and even to improve the production of viable coffee plants.

In coffee soils (acidic), fixation of applied P is a major problem and drastically reduces the availability of plant nutrients. Certain P-solubilizing microorganisms contribute to the release of fixed P in acid soils and thus help to reduce the application rate of P fertilizers. De Beenhouwer et al. (2015) showed that high soil phosphorus availability decreases with AMF diversity, and that soil pH, nitrogen, and phosphorus availability strongly influence AMF community composition. Abawari et al. (2021) reported from greenhouse experiments that addition of inorganic phosphate to soil and inoculation with native PSB and PSF significantly increased plant growth by increasing uptake of nutrients (N, P, and K) in coffee. In addition, the results show that phosphate-solubilizing microbes can be better used as biofertilizer under greenhouse conditions (Abawari et al. 2021). Application of 10 kg ha^{-1} commercial biofertilizer (biofoss) containing PSB, i.e., armorial, *Candida* sp. and *Bacillus* sp. in addition to two-thirds of the recommended dose of phosphate fertilizer improved the availability of phosphorus nutrients to coffee plants as well as increased the crop yield (CCRI 2014). Vijay Singh et al. (2018) reported that PSB can solubilize various insoluble forms of P sources and thus act as a biofertilizer in sustainable organic farming systems. Studies involving 26 phosphate-solubilizing bacterial isolates from the rhizosphere of coffee plants showed varying degrees of phosphate-solubilizing ability of the isolates. The results of in vitro and greenhouse experiments showed a significant improvement in the growth of coffee seedlings treated with PSB isolates. Isolate PSB 20 had significant levels of phosphate solubilization and plant growth promotion and therefore has great potential for use as a biofertilizer for coffee seedlings. The use of rhizospheric phosphate-solubilizing bacteria can be a reliable alternative for sustainable P management in low-capital agriculture.

16.8 Integrated Nutrition Management

Strategies that promote optimal and efficient plant nutrition are widely needed to achieve increased production of existing crops. The goal should be to develop and implement more efficient and sustainable production systems that have the least impact on the environment. Organic sources and recycling alone are not sufficient to meet the increased food demand of a given land area. On the other hand, for possible environmental and economic reasons, the nutritional needs of plants often cannot be satisfied by mineral fertilizers alone. That is why we promote the rational combination of mineral fertilizers with organic and biological nutrients. Such integrated applications are not only complementary but also synergistic, as organic inputs have beneficial effects in addition to nutritional content. Use of newer high-yielding varieties, thinning of shade to achieve higher production, farming practices dependent on chemical fertilizers along with lack of farm waste recycling strategies have contributed to the decline in soil fertility. The shortage of certain chemical fertilizers and their rising prices have aggravated the problem. Integrated nutrient management (INM), which involves the judicious and combined use of different nutrient sources such as chemical fertilizers, organic fertilizers, and biofertilizers, is effective not only for high productivity but also for environmental benefits.

Abel Chemura (2014) reported the healthy growth of coffee in Zimbabwe using integrated nutrient sources in Arabica coffee. The use of integrated fertility management could be the most attractive option because it not only reduces the cost of inorganic fertilizers but also has an environmental impact by reducing the amount of compost needed for efficient coffee growth. The results of tests carried out at various research stations in the coffee growing regions of South India showed that a reduction of the recommended dose of fertilizer i.e., 120:90:120 kg NPK to 50% was possible by replacing either organic or bio-inoculants without significant reduction in coffee yield (CCRI 2015). The importance of biofertilizers as cheap and environmentally friendly production inputs in today's plant cultivation practice is increasing. The application of consortium of inorganic + farm yard manure (FYM) + bioinoculants achieved a significantly higher yield and maximum net coffee yield by adopting INM practices in coffee plantation (Nadaf et al. 2019).

16.9 Agro-Forestry Systems and Carbon Sequestration

Soil carbon, or SOC, is a part of SOM, which contains other important elements such as carbon, hydrogen, oxygen, and nitrogen. SOM consists of plant and animal materials in various stages of decomposition. Undecomposed materials on the Earth's surface, such as sand, are not organic matter until they begin to decompose.

Large variations in SOC have been reported in different coffee-growing regions. The reason may be changes in management methods and soil characteristics. SOC is considered as one of the most important sinks of the terrestrial carbon stock (Chandrappa 2005; Schroeder and Winjum 1995).

Carbon sequestration of Indian Arabica/Robusta agroforestry systems (AFS) was evaluated against adjacent reference forest areas in different connectivity zones with mixed shade and monoshade trees. Research reports have shown that coffee AFS with native species sequesters large amounts of C (190–200 t ha^{-1}, above and below ground) as well as forest. Robusta AFS shaded by the fast-growing exotic species *Grevillea robusta* emitted only 2/3 less carbon compared to native AFS or forest. Carbon sequestration decreases from dry to very wet conditions in all systems, including forests. Tree biomass and soil are the two most important carbon stocks, accounting for more than 90% of all carbon, while coffee plants and waste have a small share. These preliminary results confirm the strong potential of coffee AFS under the shade of native tree species to operate in the largest coffee-producing regions. The results of study on the carbon storage and shade tree diversity in native forests and coffee AFS under contrasting management (native vs. exotic shade trees, Robusta vs. Arabica systems) in 67 plots across a 3500 mm rainfall gradient in the Cauvery Basin, India are presented by Guillemot et al. (2018). It was reported that tree shade was significantly lower in the original AFS compared to forest (from 90% to 32% in the high-precipitation area). The C stocks in the natural AFS were lower (max 228 MgC ha^{-1}) compared to tree species (234 MgC ha^{-1}) in the forest system (max 44 and 45 species, respectively). The introduction of *G. robusta* had a negative effect on both variables, especially in Robusta coffee systems (max. 158 MgC ha^{-1}, 12 species). The current trend toward the spread of *G. robusta* (exotic agroforestry) in coffee AFS of the study area negatively affects C supply and tree diversity, especially in Robusta coffee systems. Agroforestry has been shown to be a promising carbon sequestration mechanism in India (Singh et al. 2000) and Sub-Saharan Africa (Unruh et al. 1993), and it is estimated that a coffee farm with large forest trees grown as shade of one ha can capture 70–80 tons of carbon ha^{-1}, which roughly corresponds to the carbon stored in the same forest area. One ha of full sun or open-ground coffee can only store less than 10 t of carbon. Leaf litter accumulated on the ground was measured in coffee plantations grown in the shade of native trees and exotic trees. Under evergreen vegetation, leaf litter accumulated 3.43 t ha^{-1}, while it was only 1.84 t ha^{-1} under exotic moist deciduous plants. A significant difference was observed between native and exotic trees in both evergreen conditions, but not in the moist deciduous vegetation. SOC concentrations showed no significant differences for any of the vegetation types studied. Coffee AFS is one of the most efficient land use systems that contribute enormously to carbon sequestration and thus helps curb global climate change. This is largely due to the various farming systems used in coffee cultivation in India. One of the most important intangible benefits of wood is its ability to sequester carbon. It is a well-established fact that trees in the earth's ecosystem are among the largest sinks of carbon. It was found that native trees have a significantly higher amount of litter added to the soil and carbon content than exotic tree species. This trend was seen in both evergreen and deciduous vegetation types. Therefore, when looking at increasing the carbon stock of a coffee plantation both in standing trees and in the soil, the proportion of native trees appears to be higher than that of exotic species.

Decomposition of waste not only increases carbon content but also increases other mineral nutrients necessary for plant growth. The decomposition rate of silver oak is generally low due to the higher salicylic acid content of the leaves. However, decomposition of waste depends on many factors; this is largely due to the microbial population in the soil. Prastowo and Mukaromah (2020) by investigating short-term microclimate variability and leaf litter SOC formation in different coffee and cocoa cultivation systems in Indonesia, reported that SOC formation in different cropping systems was related to short-term microclimate variability and leaf litter availability. Greater cumulative organic carbon in the coffee system was associated with greater litterfall.

17 Organic Farming and Its Impact

Organic agriculture is a production system that avoids or eliminates the use of synthetic fertilizers and pesticides and is based entirely on crop rotation, the use of animal and green manure, and biological pest control to maintain soil productivity and plant nutrient availability and control of insects, weeds, and other pests. The history of organic coffee production is relatively recent. Finca Ireland of Mexico is known as the first certified organic coffee producer in the world. In the mid-1980s, many small farmers organized the production and marketing of organic coffee in Mexico and Central America (Van Elzakker 2001). Organic coffee production has been reported in more than 40 countries worldwide, with Peru, Ethiopia, and Mexico accounting for most production. Coffee is mostly grown on deep, fertile jungle lands under a two-layer mixed canopy consisting of evergreen legumes and nonlegumes. Growing in the shade has several advantages. Shade trees provide a natural habitat for a large population of birds and natural enemies of pests/diseases, help reduce soil erosion, promote coffee soil fertility by recirculating nutrients found deep in the soil in the form of leaf litter, and ultimately protect coffee bushes from the vagaries of changing weather conditions. Traditional agricultural practices such as manure application, composting, and hand weeding are in vogue on most small farms.

In organic farming, it has been found necessary to increase skilled labor for labor-intensive activities such as manual weeding, shade regulation, and soil conservation measures. Horticultural practices in Indian coffee plantations are considered the best in the world, with a strong emphasis on microclimate manipulation and plant health to reduce over-reliance on agrochemicals. The production and consumption of organic coffee has become extremely important among coffee-importing countries, and organic coffee consumption is about 4% worldwide. Continuous application of organic fertilizer (cattle manure, Leucaena leaves, farm residues, and Sesbania) in tropical soils has been shown to increase SOC.

It has also been reported that continuous application of organic fertilizer increases the activity of soil enzymes and microbes. Cultivation of leguminous green manure crops such as Crotalaria, Tephrosia, and cowpea and horse grass in the first year of planting can be used to improve soil fertility. Green manure should be sown in May–June and applied to the soil before flowering. Bambang et al. (2020) reported that a single application of cattle slurry manure improved soil chemical properties such as organic C, N, P, CEC, Ca, and Mg, while soil chemical properties in pH, K, and Na decreased. Dual application of liquid fertilizer improved soil pH, N, Ca, Mg, and saturated basic chemical properties in coffee cultivation in Indonesia. This means that the use of liquid and solid animal manure as fertilizer, whether in single or double use, is beneficial. Application of liquid fertilizer significantly affected the height of coffee plants at the beginning of plant growth. Fertilizing the coffee plant with liquid and solid livestock manure had a good effect on both coffee and *Virga singularis planta*, as well as soil health and plant production. Organic farming methods can help to grow and improve soil fertility in terms of chemical, biological, and microbiological diversity in coffee farms compared to traditional coffee cultivation methods both on and off the soil. The working hypothesis stated that in organic coffee production, the use of biofertilizers would promote the growth of Acidobacteria, which is positively correlated with organic matter, while significantly contributing to carbon sequestration.

The soil-biofertilizer-biome scenario considered in the study (Dębska et al. 2016; Fatunbi and Ncube 2009) indicated an increase in stable fractions of organic matter, making it more resistant to decomposition, while increasing soil carbon sequestration and improving plant function, growth, stimulating root development and uptake of nutrients and water. Coffee waste from coffee plantation should be recycled and used as compost @ 2.5 t ha^{-1} $year^{-1}$ in two splits. Soil pH is adjusted with natural liming materials such as agricultural lime or dolomite based on soil test values. Similarly, nutrients should be distributed through rock phosphate, bone meal, wood ash, and microbial consortia to effectively manage soil fertility.

The organically grown coffee must be certified by recognized agencies to declare the product as organic coffee. Organic coffee is produced in about 40 countries worldwide (CCRI 2014). Mexico is the largest producer of organic coffee, with one-third of total production certified as organic coffee. The majority of organic coffee produced worldwide comes from small farms. The main consumers of organic coffee are the United States, Japan, and European countries (CCRI 2014). Coffee growing in India offers great opportunities for organic coffee production because the conditions are more favorable than in any other coffee-producing country (CCRI 2000).

Some of India's natural advantages for producing organic coffee are:

- Coffee is grown on fertile soil under the mixed shade of evergreen and non-bean legumes.
- Traditional agricultural practices such as using manure, composting, and manual weeding are common on the vast majority of small farms.

- The cultural methods adopted by growers are among the best in the world, with a major focus on microclimate regulation to reduce over-reliance on agrochemicals.
- The Indian coffee industry is dominated by small farms in Karnataka, Kerala, and Tamil Nadu.

Tribal farms in Andhra Pradesh and Northeastern regions are mostly organic by default as they do not use any chemical fertilizers or crop protection chemicals.

The Coffee Board of India has encouraged organic coffee production by providing guidelines (CCRI 2001) and providing subsidies to cover part of the cost of certification over the past 10 years. Most coffee farmers do not tend to convert their plantations to organic farming; this may be due to a number of limitations such as high certification costs, long conversion time, and also lack of premium price for the organic produce and lack of adequate marketing arrangements for organic coffee.

18 Future Strategies

Coffee plants do well in red, lateritic, alluvial, and colluvial soils with a favorable texture, such as volcanic formations. Each climate zone has specific flora, fauna, and soil. Coffee is grown in three major states of India viz. Karnataka, Kerala, and Tamil Nadu with dominant soil types. Soils of the Ultisol class were found to be the most common soil type in coffee-growing areas, as Mollisol species were also found to be significantly present. The soil is usually rich in organic matter and the reaction is acidic to neutral. Coffee soil is well supplied with phosphorus and potassium.

The productivity potential of coffee plantations depends on the timely adoption of good land management practices, timely application of lime, nutrients, and crop management practices. In addition, temperature, altitude, rainfall, shade pattern, sunshine hours, wind, soil, and soil topography are important environmental factors affecting coffee cultivation. Soil erosion, soil acidity, low soil nutrient levels, and water retention capacity are limitations to coffee cultivation. Protective measures and timely application of lime materials are important to obtain a good harvest. Soil conservation measures such as contour planting on steep slopes and embankments are generally recommended to prevent soil erosion. Opening trenches/holes, raking and mulching are recommended measures to conserve moisture in coffee plantations.

Coffee is a perennial crop that needs nutrients during different stages for optimum growth and yield. In general, it is recommended to give fertilizers in 2–3 splits depending on the amount of nutrients removed by the plants. Scatter application of fertilizers reduces leaching of nutrients, especially nitrogen, during heavy rains. Nitrogen is essential for coffee plants throughout the year and should be supplied at least 3–4 times a year, especially during flowering, pre-monsoon, monsoon, and post-monsoon breaks. Similarly, coffee plants require a higher amount of phosphorus during flowering. Application of potassic fertilizer during the post-monsoon period is very important because coffee plants require this nutrient during the ripening of berries, filling of beans, and healing.

Coffee plantations have a variety of biodegradable farm waste in the form of shade tree leaf litter, pruning (leaves and branches), weed biomass, and coffee processing waste (coffee pulp and cherry skins) that add organic matter to the soil. It is estimated that shade trees and coffee plants in the ideal shade of coffee plantations produce about 10 t of leaf litter ha^{-1} $year^{-1}$, which increases the state of soil organic matter. Cover cropping in coffee plantation prevents the direct impact of rainwater and therefore soil erosion. However, in nature, not all available biomass is efficiently recycled and decomposition occurs slowly. Therefore, the microbiome of coffee plants has attracted interest in recent years because coffee and shade trees are perennial plants and produce huge biomass and have a considerable importance in carbon sequestration and should be studied from a sustainability perspective. At this stage, sequencing of the coffee plant microbiome/soil microbiome genetic material must be considered from a soil health angle. Correct management of soil amendments and treatments, dose adjustment, and systemic phosphorus use, as well as nutrient consumption balance and leaf analysis to monitor coffee nutrition are the main challenges of modern and competitive coffee cultivation for a better use of available water for soil-plant system (Silva and Guimaraes 2016). Therefore, for sustainable coffee production, it is necessary to build soil fertility to achieve better nutrient use efficiency, higher fertilizer recycling rates, lower 2-year carrying capacity, and higher yields.

References

Abawari RA, Assefa F, Muleta D (2021) Effect of phosphate solubilizing bio-inoculants and vermicompost application on mineral uptake and growth of coffee (*Coffea arbica L.)* seedlings under greenhouse condition SINET. Ethiop J Sci 44(2):135–150. https://doi.10.4314/sinet.v44i2.1

Altieri MA, Koohafkan P (2008) Enduring farms: climate change, smallholders and traditional farming communities, vol 6. Third World Network (TWN), Penang, Malaysia, p 63

Alwar Anand RP, Krishnamurthy Rao W (1992) Sulphur requirement of Arabica cultivars. J Coffee Res 22(2):123–125

Amaral JFTD, Martinez HEP, Laviola BG, Filho EIF, Cruz CD (2011) Nutrients use efficiency by coffee cultivars. Cienc Rural, Santa Maria 41(4):621–629

Ananthanarayana R, Ravi MV (1997) Nature of soil acidity of coffee growing soils of Karnataka. J Ind Soc Soil Sci 45(2):384–385

Ananthanarayana R, Ravindra MR (1998) Technical bulletin on soil acidity and liming in Karnataka. Department of Soil Science and Agricultural Chemistry, Agriculture College, GKVK, Bengaluru, India, p 112

Anil Kumar KS (2002) Characterization, classification and suitability evaluation of coffee growing soils of Karnataka, Ph.D., thesis submitted to Department of Soil Science and Agricultural Chemistry, UAS, Bangalore, India

Anil Kumar KS, Krishnan P, Natarajan A, Nair KM (2010) Integration of physiological and phenological attributes of coffee species with land and climate to maximize coffee production in Karnataka. Bio Technol Indian J (BTAIJ) 4(2):62–70

Areqi MC, Amina OT, Benkirane R, Douira A (2014) Effect of a composite endo mycorrhizal inoculum on the growth of *Coffea Arabica* seedlings. Int J Plant Animal Environ Sci 4(1):185–194

Arias RM, Heredia-Abarca G, Sosa VJ, Fuentes-Ramıirez LE (2012) Diversity and abundance of arbuscular mycorrhizal fungi spores under different coffee production systems and in a tropical montane cloud forest patch in Veracruz, Mexico. Agrofor Syst 85:179–193. https://doi.10.1007/s10457-011-9414-3

Avelino J, Cristancho M, Georgiou S, Imbach P, Aguilar BG, Morales C (2015) The coffee rust crises in Colombia and Central America (2008–2013): impacts, plausible causes and proposed solutions. Food Secur 7(2):303–321. https://doi.org/10.1007/s12571-015-0446-9

Awatarmani NA, Satynarayana MS (1973) Effect of rain on coffee blossom. Indian Coffee 37:1–2

Azizuddin, Krishnamurthy Rao W (1984) Field and nursery experiments with Azotobacter in Coffee. Proceedings of 6th Symposium of Plantation Crops, Rubber Research Institute of India, Kottayam, 16–20 December 1984 pp 289–295

Bacon CM, Menendez VE, Gliessman SR, Goodman D, Fox JA (2008) Confronting the coffee crisis: fair trade, sustainable livelihoods and ecosystems in Mexico and Central America. Glob Environ Polit 8(3):159–161. https://doi.org/10.1162/glep.2008.8.3.159

Bainard LD, Klironomos JN, Gordon AM (2011) Arbuscular mycorrhizal fungi in tree-based intercropping systems: a review of their abundance and diversity. Pedobiologia 54(2):57–61. https://doi.org/10.1016/j.pedobi.2010.11.001

Balota EL, Chaves JCD (2010) Enzymatic activity and mineralization of carbon and nitrogen in soil cultivated with coffee and green manures. R Bras Ci Solo 34(5):1573–1583

Bambang RW, Novia L, Cahya S (2020) The effects of livestock's manure utilization as fertilizer on coffee plant's growth. Adv Food Sci, Sustain Agric Agroind Eng 3(2):53–67

Basanna HR (2015) Emergence of sun coffee in the Western Ghats. Indian Coffee LXXXIX(7):6–7

Blore TWD (1966) Further studies of water use by irrigated and unirrigated Arabica coffee in Kenya. J Agric Sci (Cambridge) 67:145–154

Bucagu C, Vanlauwe B, Giller KE (2013) Managing Tephrosia mulch and fertilizer to enhance coffee productivity on smallholder farms in the Eastern African highlands. European J Agron 48:19–29. https://doi.org/10.1016/j.eja.2013.02.005

Bunn C, Läderach P, Jimenez JGP, Montagnon C, Schilling T (2015) Multiclass classification of agroecological zones for arabica coffee: an improved understanding of the impacts of climate change. PLoS One 10:e0140490. https://doi.org/10.1371/journal.pone.0140490

Caires EF, Guimaraes AM (2018) A novel phosphogypsum application recommendation method under continuous no-till management in Brazil. Agron Jl 110:1–9. https://doi.org/10.2134/agronj2017.11.0642

Caldwell AC, Silva LCF, da Silva CC, Ouverney CC (2015) Prokaryotic diversity in the rhizosphere of organic, intensive, and transitional coffee farms in Brazil. PLoS One 10:e0106355. https://doi.org/10.1371/journal.pone.0106355

Carducci CE, Oliveira GC, Curi N, Rossoni DF, Costa AL, Heck RJ (2014a) Spatial variability of pores in oxidic latosol under a conservation management system with different gypsum doses. Ciência e Agrotecnologia 38:445–460. https://doi.org/10.1590/S1413-70542014000500004

Carducci CE, Oliveira GC, de Lima JM, Rossoni DF, Costa AL, Oliveira LM (2014b) Spatial distribution of coffee roots and pores of two latosols under conservationist management. Revista Brasileira de Engenharia Agrícola e Ambiental 18:270–278. https://doi.org/10.1590/S1415-43662014000300005

Carducci CE, Oliveira GC, Curi N, Heck RJ, Rossoni DF, de Carvalho TS, Costa AL (2015) Gypsum effects on the spatial distribution of coffee roots and the pores system in oxidic Brazilian Latosol. Soil Tillage Res 145:171–180. https://doi.org/10.1016/j.still.2014.09.015

Carvajal J (1985) Potassium nutrition of coffee. In: Robert DM (ed) Potassium in agriculture. American Society of Agronomy, Madison, Wisconsin, USA, pp 955–979

Castnao FJO, Robledo AJ (1978) Effecto del deficit de humedad en el suelosobre La temperature delsuelo y de lanhojas en plantas de Coffeacanephora y C. Arabica. Cenicafe 29:121–134

CCRI (2000) Package of practices for organic coffee. Published by Director of Research, Central Coffee Research Institute, Coffee Board, Karnataka, India, p 13

CCRI (2001) Guidelines for production of organic coffee in India. Published by Director of Research, Central Coffee Research Institute, Coffee Board, Karnataka, India, p 2

CCRI (2014) Coffee guide. Published by Director of Research, Central Coffee Research Institute, Coffee Board, Karnataka, India

CCRI (2015) Annual reports for the year 2014–15 Coffee Board Research Department published by Coffee Board, Karnataka, India pp 48–49

CCRI (2022) Soil testing report of Central Coffee Research Institute, Coffee Board, Karnataka, India pp 1–2

Chandrappa K (2005) Nutrient recycling and carbon sequestration in selected coffee growing soils. Ph.D. Thesis submitted to University of Agricultural Sciences, Bangalore, Karnataka, India

Chattopadhyay N, Swain S, Hore JK (2006) Response of coffee seedlings to nitrogen fixing biofertilizers. Agric Sci Digest 26:103–106

Chemura A (2014) The growth response of coffee (*Coffea arabica* L.) plants to organic manure, inorganic fertilizers and integrated soil fertility management under different irrigation water supply levels. Int J Recycl Org Waste Agric 3:59. https://doi.org/10.1007/s40093-014-0059

Chengappa PG, Devika CM (2016) Climate variability concerns for the future of coffee in India: an exploratory study. Intl J Environ Agric Biotechnol 1(4):819–826

Chengappa PG, Devika CM, Rudragouda CS (2017) Climate variability and mitigation: perceptions and strategies adopted by traditional coffee growers in India. Clim Dev 9:593–604. https://doi.org/10.1080/17565529.2017.1318740

Craparo ACW, van Asten P, Laderach P, Jassogne L, Grab S (2015) Coffea arabica yields decline in Tanzania due to climate change: global implications. Agric Forest Meteor 207:1–10. https://doi.org/10.1016/j.agrformet.2015.03.005

Cristancho MA, Rozo Y, Escobar C, Rivillas CA, Gaitan AL (2012) Outbreak of coffee leaf rust (*Hemileia vastatrix*) in Colombia. New Dis Rep 25(19) Retrieved 29 May 2015

da Silva EA, Benevenute PAN, de Oliveira GC, Zinn Y, Silva BM, de Melo LBB, Reis THP, de Oliveira CHC, Guimarães PTG, Carducci CE (2020) Soils under plastic and grass cover: effects on soil aggregation and nutrient cycling in Brazilian coffee growing. Agri Res Tech 24(2):74–82. https://doi.org/10.19080/ARTOAJ.2020.25.556265

DaMatta FM, Ramalho JDC (2006) Impacts of drought and temperature stress on coffee physiology and production: a review. Brazilian J Plant Physiol 18(1):55–81. https://doi.org/10.1590/S1677-04202006000100006

De Beenhouwer M, Aerts R, Honnay O (2013) A global meta-analysis of the biodiversity and ecosystem service benefifits of coffee and cacao agroforestry. Agric Ecosyst Environ 175:1–7. https://doi.org/10.1016/j.agee.2013.05.003

De Beenhouwer M, Van Geel M, Ceulemans T, Muleta D, Lievens B, Honnay O (2015) Changing soil characteristics alter the arbuscular mycorrhizal fungi communities of Arabica coffee (*Coffea arabica*) in Ethiopia across a management intensity gradient. Soil Biol Biochem 91:133–139. https://doi.org/10.1016/j.soilbio.2015.08.037

Dębska B, Długosz J, Długosz AP, Szott MB (2016) The impact of a bio-fertilizer on the soil organic matter status and carbon sequestration-results from a field-scale study. J Soils Sediments 16:2335–2343. https://doi.org/10.1007/s11368-016-1430-5

Descroix F, Snoeck J (2004) Environmental factors suitable for coffee cultivation, Book Coffee: Growing, processing sustainable production (ed:.Jean Nicolas Wintgens). GmbH &Co. KGaA, Weinheim, p 164–177

Dhruvakumar K, Rao W (1982) Distribution of manganese status in soils cropped to coffee of Koppa zone in Karnataka State. In: KVA B, Iyer RD, Das PK, Sethuraj MR, Ranganathan V, KKN N, Chacko MJ (eds) Proceedings Fifth Annual Symposium on Plantation Crops (PLACROSYM -V). Indian Society of Plantation Crops, Kasaragod, pp 375–378

Fatunbi AO, Ncube L (2009) Activities of effective microorganisms (EM) on the nutrient dynamics of different organic materials applied to soil. Am Eurasian J Agron 2(1):26–35

Favarin JL, Almeida REM, Salustio PEB, Pedrosa AW (2013) New concept in soil preparation for coffee plantations. Visão Agrícola 3(12):20–22

Forestier J (1968) Potassium and the Robusta coffee tree. Fertilite 30:3–63

Fuentes-Ramírez LE, Cristales BR, Hernández TA, Salgado JT, Wang ET, Romero ME (2001) Novel nitrogen-fifixing acetic acid bacteria, *Gluconacetobacter johannae* sp. nov. and *Gluconacetobacter azotocaptans* sp. nov., associated with coffffee plants. Int J Syst Evol Microbiol 51:1305–1314. https://10.1099/00207713-51-4-1305

Gomes LC, Bianchi FJJA, Cardoso IM, Fernandes RBA, Filho EIF, Schulte RPO (2020) Agroforestry systems can mitigate the impacts of climate change on coffffee production: a spatially explicit assessment in Brazil. Agric Ecosyst Environ 294:106858. https://doi.org/10.1016/j.agee.2020.106858

Gopal NH (1974) Some physiological factors to be considered for stabilization of Arabica coffee production in South India. Indian Coffee XXXVIII (7):168–175

Gowda BSP, Rangasetty HT, Chokkanna NG (1965) A note on soil conservation in coffee plantations. Indian Coffee XXIX(9):9–11

Guillemot J, Le Maire G, Munishamappa M, Charbonnier F, Vaast P (2018) Native coffee agroforestry in the Western Ghats of India maintains higher carbon storage and tree diversity compared to exotic agroforestry. Agric Ecosyst Environ 265:461–469. https://doi.org/10.1016/j.agee.2018.06.002

Guimaraes PTG, Reis THP (2010) Nutrição e Adubação do Cafeeiro. In: Reis PR, Cunha RL (eds) Café Arábica: do plantio à colheita. Epamig SM, Belo Horizonte, pp 343–414

Guimaraes PTG, Dias KG, Oliveira CHC (2018) Uso de Braquiária nas entrelinhas do cafeeiro. Circular Técnica Epamig 2018

Hakim A, AL-Areqi NA, Chliyeh M, Sghir F, Touhami AO, Benkirane RR, Douira A (2013) Diversity of arbuscular mycorrhizal fungi in the rhizosphere of *Coffea arabica* in the Republic of Yemen. J Appl Biosci 64:4888–4901

Harelimana A, Goff GL, Rukazambuga DTN, Hance T (2018) Coffee production systems: evaluation of intercropping system in coffee plantations in Rwanda. J Agric Sci 10(9):17–28. https://doi.org/10.5539/jas.v10n9p17

Jassogne L, Nibasumba A, Wairegi L, Baret PV, Deraeck J, Mukasa D, Wanyama I, Bongers IG, van Asten PJA (2013) Coffee/banana intercropping as an opportunity for smallholder coffee farmers in Uganda, Rwanda and Burundi. In: Blomme G, van Asten PJA, Vanlauwe B (eds) Banana Systems in the Humid Highlands of Sub-Saharan Africa. CABI, pp 144–149. https://doi.org/10.1080/14735903.2012.714576

Jayarama J, Krishnappa Naik CS, Shankar BN, Thimma Reddy H (1993) Correlation between soil fertility parameters and coffee yield in small coffee holdings of Chikamagalur zone. J Coffee Res 23(1):25–31

Jiang Z, Lou Y, Liu X, Sun W, Wang H, Liang J, Guo J, Li N, Yang Q (2023) Combined application of coffee husk compost and inorganic fertilizer to improve the soil ecological environment and photosynthetic characteristics of arabica coffee. Agronomy 13(1212):2–19. https://doi.org/10.3390/agronomy13051212

Keshavaiah KVY, Raghuramalu VB, Suresh Kumar, George Daniel (1998) Influence of cover cropping on weed smothering in young coffee plantations. Paper presented in thirteen plantation crop symposium, Coimbatore December 16–18

Kharche VK (1996) Developing soil site suitability criteria for some tropical plantation crops. Ph.D. Thesis, Panjabrao Deshmukh Krishividyapeeth, Akola, p 218

Kharche VK, Sehegal JL, Challa O (1999) Evaluation of land resources for coffee plantations- a case study from Karnataka. J Ind Soc Soil Sci 47(4):754–760

Kharche VK, Sehgal J, Challa O (2000) Characterization of coffee growing soils in karnataka. Agropedology 10:59–66

Kimew BS (1974) Response to soil applied NPK compound fertilizers in coffee small holdings. Kenya Coffee 39:222–228

Korikanthimath VS, Gayathri AG, Gowda SJA, Hegde R, Hosmani MM (2000) Vesicular arbuscular mycorrhizae and phosphate solubilizers in robusta coffee and cardamom mixed cropping systems. Karnataka J Agric Sci 13(2):489–499

Krishnamurthy Rao W (1991) Relationship of application of fertilizer in-puts in coffee plantations. Planters Cronicle 86:649–651

Krishnamurthy Rao W, Anand Alwar RP (1984) Sulphur status of soils cropped to coffee. J Coffee Res 14(2):57–60

Krishnan P, Venugopal KR, Sehagal J (1996) Soils of Kerala for optimizing land use, NBSS Publ. 48b, Nagapur, p 54

Krishnappa Naik CS, Haricharan BR, Jayarama, Surendra Swamy HB (1988) Coffee soils of Chikamagalur district. Indian Coffee 32(10): 297–298

Kulandaivelu V (2016) Impact of organic and conventional systems of coffee farming on soil properties and culturable microbial diversity. Scientifica:1–9. https://doi.org/10.1155/2016/3604026

Landon JR (1984) Booker tropical soil manual. Longman Inc., New York, p 441

Leandro P, de Sousa G-FO, Mondego JMC (2022) The rhizosphere microbiomes of five species of coffee trees. Microbiol Spec 10(2):e00444–e00522. https://doi.org/10.1128/spectrum.00444-22

Li R, Cheng J, Liu X, Wang Z, Li H, Guo J, Wang H, Nm C, Zhao L (2023) Optimizing drip fertigation at different periods to improve yield, volatile compounds and cup quality of Arabica coffee. Front Plant Sci 14:1148616. https://10.3389/fpls.2023.1148616

Liliana LC, Ramon ZR, Bernardo MA, Pablo PR, Danis M, Verdecia A, Luis G, Hernandez M (2021) Biodiversity of am fungi in coffee cultivated on eroded soil. Agronomy 11(567):2–10. https://doi.org/10.3390/agronomy11030567

Lin YM, Liu JW, Xiang P, Lin P, Ding ZH, Sternberg LDSL (2007) Tannins and nitrogen dynamics in mangrove leaves at different age and decay stages (Jiulong River Estuary, China). Hydrobiol 583:285–295

Mahdhi M, Tounekti T, Al-Turki TA, Khemira H (2017) Composition of the root mycorrhizal community associated with *Coffea arabica* in Fifa mountains (Jazan region, Saudi Arabia). J Basic Microbiol 57:691–698. https://doi.org/10.1002/jobm.201700075

Mariani S, Sabrina T, Mukhlis M (2020) Phosphate solubilizing microbes and coffee skin compost to increase robusta coffee plant growth in Andisol of Mount Sinabung Area. Bulgarian J Agr Sci 26(4):766–771

Mohan Kumar BM (1996) Tree crops in our agroforestry practices (article in Malayalam). KALPADHENU Jan-Mar 19(3):6–7

Motta ACV, Nick JA, Yorinori GT, Monte SB (2006) Horizontal and vertical distribution in soil fertility and roots in a coffee (*Coffea arabica* L.) Catuaí cultivar. Acta Scientiarum Agron 28:455–463

Muleta D, Assefa F, Börjesson E, Granhall U (2013) Phosphatesolubilising rhizobacteria associated with *Coffea arabica* L. in natural coffffee forests of southwestern Ethiopia. J Saudi Soc Agric Sci 12:73–84. https://doi.org/10.1016/j.jssas.2012.07.002

Muller LE (1966) Coffee nutrition. In: Childers NF (ed) Nutrition of fruit crops. Horticultural Publications, New Jersey, USA

Nadaf SA, Bora AR (2021) Nutrients status in arabica coffee (*Coffea arabica* L) soils of non-traditional area (NTA). Int J Plant Soil Sci 33(13):17–22. https://doi.org/10.9734/ijpss/2021/v33i1330490

Nadaf SA, Bora AR, Sunil Babu D (2019) Soil health management strategies for coffee estates of non - traditional areas. Indian Coffee LXXXIII(9):4–7

Nadaf SA, Chandrashekar N, Bora AR, Hariyappa N, Chandrappa K (2021) Secondary and micronutrients status in coffee (*Coffea arabica* L.) soils under different land use (mixed and mono shade) system of Andhra Pradesh and Odisha. Front Crop Improv 9:220–225

Nagaraja JS (2003) Development of diagnosis and recommendation integrated system (DRIS) norms for Arabica Coffee. Ph.D. Thesis submitted to University of Agricultural Sciences, Dharwad, Karnataka, India

Nair KM, Anil Kumar KS, Srinivas S, Nagaraj JS, Violet D'Souza M, Raghuramulu Y, Hegde R, Singh SK (2017) Soil quality monitoring sites for traditional coffee-growing areas of India. National Bureau of Soil Survey and Land Use Planning, Nagpur, India

Nascimento LG, de Assis GA, Fernandes MIS, Pires PS, Carvalho FJ, de Araújo NO (2020) Mulching in coffee growing: effects on productivity, maturation, grain shape and beverage quality. Res Soc Dev 9(9):e765997727. https://doi.org/10.33448/rsd-v9i9.7727

Natarajan A, Reddy PSA, Sehagal J, Velayutham M (1997) Soils of Tamil Nadu for land use planning, vol 46b. NBSS Publ, p 88

NBSS LUP (2017) Soil quality monitoring sites for traditional coffee-growing areas of India. NBSS Publ.No.1095, ICAR-National Bureau of soil survey and land use planning (NBSS LUP). Nagpur, India, p 5

Neto JF, Severiano EC, Costa KAP, Junior WSG, Goncalves WG, Andrarde R (2015) Biological soil loosening by grasses from genus Brachiaria in crop-livestock integration. Acta Sci Agron 37(3):375–383. https://doi.org/10.4025/actasciagron.v37i3.19392

Nzeyimana I, Hartemink AE, Graaff J (2013) Coffee farming and soil management in Rwanda. Outlook on Agric 42(1):47–52. https://doi.org/10.5367/oa.2013.0118

Nzeyimana I, Hartemink AE, Ritsema C, Mbonigaba JJM, Geissen V (2020) Mulching effects on soil nutrient levels and yield in coffee farming systems in Rwanda. Soil Use Manag 36:58–70. https://doi.org/10.1111/sum.12534

Ovalle-Rivera O, Laderach P, Bunn C, Obersteiner M, Schroth G (2015) Projected shifts in *Coffea arabica* suitability among major global producing regions due to climate change. PLoS One 10(4):e0124155. https://doi.org/10.1371/journal.pone.0124155

Pangle LA, Gregg JW, McDonnell JJ (2014) Rainfall seasonality and an ecohydrological feedback offset the potential impact of climate warming on evapotranspiration and groundwater recharge. Water Resour Res 50:1308–1321

Patil PL, Dasog GS (1999) Genesis and classification of ferruginous soil in Western Ghats and coastal regions of North Karnataka. Agropedol 9:1–15

Patterson E, Sim A (2013) Soil-specific response functions of organic matter mineralization to the availability of labile carbon. Glob Chang Biol 19:1562–1571

Paudel N, Kang WH (2018) Establishment of algae as bio-fertilizer for coffee plant. Int J Sci Rep 4(6):153–157. https://doi.org/10.18203/issn.2454-156.IntJSciRep20182204

Pillai KM (1984) A text book of plantation crops. Vikas Publishing House, New Delhi, p 168

Prastowo E, Mukaromah L (2020) Short-term microclimate variation and leaf litter derived soil organic carbon formation under different coffee and cocoa cropping systems. Pelita Perkebunan 36(3):212–225. https://doi.org/10.22302/iccri.jur.pelitaperkebunan.v36i3.416

Prates P Jr, Moreira BC, Da Silva M, De CS, Veloso TGR, Stürmer SL (2019) Agroecological coffee management increases arbuscular mycorrhizal fungi diversity. PLoS One 14:e0209093. https://doi.org/10.1371/journal.pone.0209093

Premkumari MS, Balasubramanian A (1993) Effect of combined application of VAM and Azospirillum on growth and nutrient uptake by coffee seedlings. Indian Coffee 57:5–11

Pumarino L, Sileshi WG, Gripenberg S, Kaartinen R, Barrios E, Muchane MN, Midega C, Jonsson M (2015) Effects of agroforestry on pest, disease and weed control: a meta-analysis. Basic and Applied Ecol 16(7):573–582. https://doi.org/10.1016/j.baae.2015.08.006

Radhakrishnan S, Raju R, Krishnamurthy Rao W (1993) Distribution, stability and trend of rainfall in south Indian coffee tracts. Indian Coffee LVII(6):5–11

Raju T (1978) Influence of continuous use of inorganic fertilizers on nutrient status of oxisols cropped to coffee (*Coffea arabica*). In: Nelliat EV, Ranganathan V, Vishveshwara S, Potti SN, Krishnamurthy Rao W, Zachariah PK (eds) In proc first National Symp Plantation Crops (PLACROSYM I). Indian Society for Plantation Crops, pp 29–38

Raju T (1985) Influence of major nutrients on availability of minor nutrients. Proceedings of second workshop on coffee, CCRI, Chikkmagaluru u Dist, Karnataka, J Coffee Res 15:21

Raju T, Deshpande PB (1985) Influence of long term application of NPK to coffee on some micronutrient contents. J Coffee Res 15(3&4):103–112

Raju T, Deshpande PB (1986) Lime - zinc interactions in coffee seedlings. J Coffee Res 16(3&4):79–88

Raju T, Deshpande PB (1987) Copper–Zinc interactions in coffee seedlings. J Coffee Res 17(1):50–58

Raju T, Krishnamurthy Rao W (1978) Studies on liming Oxisols cropped to coffee. II effect of liming to varying pH levels on chemical characteristics of soil. J Coffee Res 8(1):20–25

Raju T, Thomas MR, Ganesh KA (1982) Yield pattern of Robusta in relation to soil fertility and fertilizing practices in Wayanad district of Kerala. Indian Coffee 46(8):192–196

Ramos BZ, Toledo JPVF, Lima JM, Serafim ME, Bastos ARR, Guimaraes PTG, Coscinoe AR (2013) Gypsum doses in coffee trees: influence on calcium, magnesium, potassium and pH contents in the solution of a dystrophic red latosol. R Bras Ci Solo 37:1018–1026

Rangeshewaran RW, Krishnamurthy Rao W, Ramaiah PK (1990) Vasicular arbuscualr mycorrhiza in coffee. J Coffee Res 20(1):55–68

Ribeiro AC, Guimarães PTG, Alvarez VVH (1999) Recomendações para o uso de corretivos e fertilizantes em Minas Gerais -5ª Aproximação. SBCS, Viçosa, p 359

Robinson JBD, Hosegood PA (1963) Effect of organic mulch on fertility of a latosolic coffee soil in Kenya. Exp Agric 1:67–80

Rosolem CA, Calonego JC, Foloni JSS (2003) Lixiviação de potássio da palha de coberturas de solo em função da quantidade de chuva recebida. Rev Bras Cienc Solo 27:355–362

Salakinakop SR, Shivaprasad P, Raghuramulu Y, Paneer Selvam P (2003) Bio-nutrition for improving the vigour of coffee seedlings 6th international PGPR workshop, Calicut, India, pp 224–228

Salakinakop SR, Shivaprasad P, Raghuramulu Y (2004) Relative performance of decomposing micro-organisms for composting of coffee farm wastes. ASIC 2004:130. (PA 210)

Salakinakop SR, Shivaprasad P, Raghuramulu Y (2005) Manurial value and economics of composting of coffee (*Coffea canephora* L.) wastes. Indian J Agric Sci 75(12):814–816

Salamanca JA, Timothy AD, Horwath WR (2017) Nitrogen use efficiency of coffee at the vegetative stage as influenced by fertilizer application method. Front Plant Sci 8:223. https://10.3389/fpls.2017.00223

Schroeder PE, Winjum JK (1995) Assessing Brazil's carbon budget: II. Biotic fluxes and net carbon balance. For Ecol Manag 75(1–3):87–99

Serafim ME, Oliveira GC, Oliveira AS, Lima JM, Guimaraes PTG, Costa JC (2011) Intensive coffee cultivation management in the physiographic region of the upper San Francisco river, MG: a case study. Biosci J 27(6):964–977

Serafim ME, Oliveira GC, Curi N, Lima JM, Guimarães PTG, Lima CMP (2013) Use potential and limitations of latosols and cambisols under a conservation coffee plantation system. Biosci J 29:1640–1652

Shivaprasad P, Salakinakop SR, Pannerselvam P, Raghuramulu Y, Naidu R, Anwarulla RK (2003) Composting of coffee pulp. J Coffee Res 31(2):153–160

Shivaprasad P, Biradar IB, Salakinakop SR, Raghuramulu Y, D'souza MV, Hariyappa N, Hareesh SB, Murthy MA, Jayarama (2005) Influence of soil cultivation methods in young coffee on soil moisture, weed suppression and organic matter. J Coffee Res 33(1 & 2):1–14

Shivaprasad P, Salakinakop SR, Muralidhara HR, Raghuramulu Y, D'souza MV, Hariyappa N, Hareesh SB, Pradeep R, Anantha Murthy M, Manjunathswamy BK, Dharmapala BS, Jayarama (2007) Effect of cultivation practices and green manure crops on soil moisture conservation and growth of coffee. Indian J Soil Conserv 35(3):242–245

Shivaprasad P, Nagaraj JS, Nadaf SA, D'Souza MV (2017) Buildup of soil phosphorus in coffee plantations of Karnataka over three decades. J Plant Crops 45(1):66–71

Shivaprasad P, Hariyappa N, Chandrashekar N, Manonmani GK, Hareesh SB, Nadaf SA, Jagadeesan M, Venkatesh DH, Sujatha K (2018) Studies on soil fertility status of coffee growing regions in Wayanad district J Plantn. Crops 46(3):180–189. https://doi.org/10.25081/jpc.2018.v46.i3.2087

Shivaprasad P, Hariyappa N, Hareesh SB, Chandrashekar N, Prafulla Kumari D, Pati R, Nadaf SA, D'souza MV, Suryaprakash Rao N, Veeraswamy S, Ramesh K (2020) Studies on coffee soil nutrient status in Karnataka, Kerala and Tamil Nadu state. Int J Agric Sci 12(24):10491–10496

Siles P, Harmand J, Vaast P (2010) Effects of *Inga densiflora* on the microclimate of coffee (*Coffea arabica* L.) and overall biomass under optimal growing conditions in Costa Rica. Agrofor Syst 78(3):269–286. https://doi.org/10.1007/s10457-009-9241-y

Silva EA (2017) Propriedades físicohídricas do solo e desenvolvimento radicular do cafeeiro [dissertation]. Lavras: Universidade Federal de Lavras, p 111

Silva CA, Guimaraes PTG (2016) Coffee: vectors for increasing the productivity. In: Casarin V (ed) Agronomic information number 155. IPNI, Piracicaba, pp 13–16

Silva EA, Oliveira GC, Carducci CE, Silva BM, Oliveira LM, Costa JC (2013) Increasing doses of agricultural gypsum, aggregate stability and organic carbon in Cerrado latosol under coffee crop. Amazonian J 56:25–32

Silva EA, Oliveira GC, Silva BM, Carducci CE, Avanzi JC, Serafirm ME (2014) Aggregate stability by the high energy moisture characteristic method in an oxisol under differentiated management. R Bras C Solo 38:1633–1642

Silva EA, Oliveira GC, Carducci CE, Lima JM, Melo LBB, Benevenute PAN (2016a) Stability of soil aggregates in latosols and cambisols via standard method and sonification. African J Agric Res 11:3894–3903

Silva EA, Oliveira GC, Carducci CE, Silva BM, Serafim ME (2016b) Aggregates morphometry of a Inceptisol under conservationist system. Semina: Ciencias Agrarias 37(3):1165–1176. https://doi.org/10.5433/1679-0359.2016v37n3p1165

Silva EA, Silva SHG, Oliveira GC, Carducci CE (2016c) Root spatial distribution in coffee plants of different ages under conservation management system. African J Agric Res 11:4970–4978. https://10.5897/AJAR2016.11356

Silva BM, Geraldo CO, Milson ES, Carla EC, Erika A da Silva, Samara MB, Laura BB de Melo, Walbert JRS, Thiago HPR, Cesar HCO, Paulo TG Guimaraes (2019a) Book chapter coffee - production and research soil management and water-use efficiency in Brazilian coffee, pp 1–20. https://doi.org/10.5772/intechopen.89558

Silva BM, Oliveira GC, Serafim ME, Silva EA, Guimarães PTG, Melo LBB, Norton LD, Curi N (2019b) Soil moisture associated with least limiting water range, leaf water potential, initial growth and yield of coffee as affected by soil management system. Soil Tillage Res 189:36–43. https://doi.org/10.1016/j.still.2018.12.016

Singh TP, Varalakshmi V, Ahluwalia SK (2000) Carbon sequestration through farm forestry: case from India. Indian Forester 126:1257–1264

Souza HN, Cardoso IM, Fernandes JM, Garcia FCP, Bonfim VR, Santos AC, Carvalho AF, Mendonca ES (2010) Selection of native trees for intercropping with coffee in the Atlantic rainforest biome. Agrofor Syst 80(1):1–16. https://doi.org/10.1007/s10457-010-9340-9

Souza LA, Silva EA, Oliveira GC, Barbosa SM, Silva BM (2018) Qualitative and quantitative analysis of soil aggregates under plastic mulching associated with the application of chemical and organomineral fertilizers in coffee area. Sci Agrár 19(2):142–153

Sudha M, Balakrishna G (2004) Effect of native vesicular arbuscular mycorrhizal (VAM) fungi on the growth of coffee seedlings in the nursery. ASIC 20:130

Swarupa Glory S, Basavaraj Naik T, Reddy AGS, Raju T (1997) Effect of biofertilizers on germination of coffee seed. Indian Coffee 61(3):3–5

Sys C, Van Ranst E, Debaveye J (1991) Land evaluation parts I, II & III. Re-edited volumes of publication no. 7 of the General Administration of Cooperation Development, Brussels, Belgium, pp 273–274

Sys C, Van Ranst E, Debaveye J, Beernaert F (1993) Land evaluation Part III Crop requirements. Agri. Publ., No.7, Brussels, Belgium

Unruh JD, Houghton RA, Lefebvre PA (1993) Carbon storage in agroforestry: an estimate for sub-Saharan. Africa Clim Res 3:39–52

Vaast P, Raghuramulu Y (2012) Coffee quality in coffee agro-forestry systems of Kodagu, Western Ghats of India. Indian Coffee: The Coffee Mag 76:4–7

Vaast P, Van Kanten R, Siles P, Dzib B, Franck N, Harmand JM (2005) Shade: a key factor for coffee sustainability and quality. In: Colloque Scientifique International Sur Le Café (Nairobi: Association for Science and Information on Coffee (ASIC)). Available online at: https://agritrop.cirad.fr/529372/

Vaast P, Bertrand B, Perriot JJ, Guyot B, Génard M (2006) Fruit thinning and shade improve bean characteristics and beverage quality of coffee (*Coffea arabica* L.) under optimal conditions. J Sci Food Agr 86:197–204. https://10.1002/jsfa.2338

van Asten PJA, Wairegi LWI, Mukasa D, Uringi NO (2011) Agronomic and economic benefits of coffee–banana intercropping in Uganda's smallholder farming systems. Agric Syst 104(4):326–334. https://doi.org/10.1016/j.agsy.2010.12.004

van Asten PJA, Ochola D, Wairegi L, Nibasumba A, Jassogne L, Mukasa D (2015) Coffee-banana intercropping: implementation guidance for policymakers and investors. In practice brief | climate-smart agriculture, p 10 van der Meulen ES, Nol L, Cammeraat LH (2006) Effects of irrigation and plastic mulch on soil properties on semiarid abandoned fields. Soil Sci Soc Am J 70(3):930–939. https://doi.org/10.2136/sssaj2005.0167

Van Elzakker B (2001) Organic coffee. In: Basker PS (ed) Coffee futures: a source book of some critical issues confronting the coffee industry. CABI-FEDERACAFE Commodities Press. CABI Bioscience, Egham, UK, pp 74–81

Venkataramaiah G, Hand MB, Singh D (1974) Accumulation of copper in coffee soils. J Coffee Res 4(1):14–15

Venkataramanan D, Govindappa DA (1987) Shade requirement and productivity of coffee. J Coffee Res 17(2):16–39

Venkatasubaiah P, Safeeulla KM (1984) Rhizosphere and rhizoplane micro flora of coffee seedlings. J Coffee Res 14(1):14–18

Vijay Singh K, Sharmila C, Janardan L, Dhurvaprasad G (2018) Isolation and characterization of phosphate solubilizing bacteria from rhizosphere of coffee plant and evaluating their effects on growth and development of coffee seedlings. BioTechnol Indian J 14(5):1–10

Wallis JAN (1963) Water use by irrigated Arabica coffee in Kenya. J Agric Res 60:381–388

Wang Z, Bao X, Li X, Jin X, Zhao J, Sun J, Christie P, Li L (2015) Intercropping maintains soil fertility in terms of chemical properties and enzyme activities on a timescale of one decade. Plant Soil 391(12):265–282. https://doi.org/10.1007/s11104-015-2428-2

Wellman FL (1961) Coffee-botany, Cultivation and utilization. Interscience Publishers Inc, New York, p 438

Wilson KC (1985) Climate and soil. In: Clifford MN, Wilson KC (eds) . The AVC Publishing Co., INC., West Port, Connecticut, US

Winston E, Marsh JOLT, Lempke H, Chapman K (2005) Plant nutrition and fertiliser management, Arabica coffee manual for Lao-PDR, Edward Winston (ed) National Library, Bangkok cataloguing-in-publication. https://www.fao.org/3/ae939e/ae939e06.htm

Yamel DCPR, Rosa MA, Rosario MO, Dora TA, Gabriela H, Yakelin RY (2019) Effects of native arbuscular mycorrhizal and phosphate-solubilizing fungi on coffee plants. Agrofores Systems 93(3):961–972. https://doi.org/10.1007/s10457-018-0190-1

Young A (1975) Tropical soils and soil survey. Cambridge University Press, Cambridge, p 382

Zuviria MD, Valenzuela CR (1994) Mapping Land suitability for coffee with ILWIS. ITC J 3:301–307

Dr. S A. Nadaf is a PhD holder in Soil Science and currently serves at the Coffee Resarch Sub Station, Chettalli, Coffee Board of India, Kodagu District, Karnataka, India as a Research Assistant. His fields of specialization in research work are pedology, geospatial tools, Soil environment, and soil fertility in coffee. e-mail: drsanadaf@gmail.com

Dr. P. Shivaprasad is the Head of the Agricultural Chemistry Division at the Central Coffee Research Institute, Coffee Board of India, Chikkamagaluru District, Karnataka, India. He has completed his Doctoral degree program in Chemistry. His area of specialization includes conjunctive use of coffee effluent and pond water on the performance of Arabica and Robusta coffee and soil properties as well as studies on integrated nutrient management. shivaprasadccri@gmail.com

Dr. C. Babou has completed his M. Sc and Doctoral programme in Agronomy and is the Head of Division of Agronomy at the Central Coffee Research Institute, Coffee Board of India, Karnataka, India. He carries out on-field research programs on coffee agronomy, organic farming, farm mechanization, plant architecture management, and coffee-based crop diversification. kcbabou@gmail.com

N. Hariyappa is a postgraduate in Agriculture and serves as a Subject Matter Specialist at the Central Coffee Research Institute, Coffee Board of India, Chikkamagaluru District, Karnataka, India. His area of specialization is Soil Science and Agricultural Chemistry. hariyappan62@gmail.com

N. Chandrashekar has a postgraduate degree in Agriculture and currently works as a Research Assistant at thc Central Coffee Research Institute, Coffee Board of India, Chikkamagaluru District, Karnataka, India. He has a specialization in Soil Science and Agricultural Chemistry, chandra.gudumba@gmail.com

Prafulla Kumari is a postgraduate in Microbiology and works as a Research Assistant at the Technology Evaluation Central, Coffee Board of India, Gonikoppal Kodagu District, Karnataka, India. She specializes in the enumeration of microorganisms in coffee soils. praffulla-kumari@gmail.com

P.R. Sowmya is a Research Assistant at the Central Coffee Research Institute, Coffee Board of India, Chikkamagaluru District, Karnataka, India. With her postgraduate degree in Biochemistry, specializes in chemistry/biochemistry, soil chemistry, nutrition of coffee. soumyapr85@gmail.com

Dr. S.B. Hareesh is a Research Assistant at the Central Coffee Research Institute, Coffee Board of India, Chikkamagaluru District, Karnataka, India. He completed his M.Sc and PhD program in Chemistry and specializes in chemistry/biochemistry, soil chemistry, nutrition of coffee. hareeshsb@gmail.com

Dr. N. Rajib Pati is a Subject Matter Specialist at the Regional Coffee Research Station Coffee Board of India, Alluri Sitharamaraju District, Andhra Pradesh, India. He has postgraduate and Doctoral degree in Soil Science and Agricultural Chemistry. rajibpati@gmail.com

Dr. J.S. Nagaraja functions as the Joint Director of Research at the Central Coffee Research Institute, Coffee Board of India, Chikkamagaluru District, Karnataka, India. He has postgraduate and Doctoral degree in Soil Science and Agricultural Chemistry. His doctoral thesis is on the development of DRIS norms for coffee. jdrcoffeeboard@gmail.com

Dr. K. Chandrappa with his postgraduate degree in Soil Science and Agricultural Chemistry serves as the Deputy Director of Research at the Coffee Research Sub Station, Chettalli, Coffee Board of India, Kodagu District, Karnataka, India. His field of specialization is carbon sequestration in coffee. kurahattichandru@yahoo.co.in

Dr. M. Senthil Kumar is the Director of Research at the Central Coffee Research Institute, Coffee Board of India, Chikkamagaluru District, Karnataka. He has PhD degree in Horticulture and specializes in the research field of horticultural crops. drccri2022@gmail.com

Tea (*Camellia sinensis* (L.) O. Kuntze)

V. Krishnakumar, T. Raj Kumar, and P. Murugesan

Abstract

Tea, an important cash crop of many developing countries, has considerable socio-economic importance for rural development and poverty alleviation. More than three billion people in 160 countries and regions drink tea. Major tea producing countries of the world are China, India, Kenya, and Sri Lanka and they cultivate tea under different geo-physical environments existing in their countries. Long-term intensive tea cultivation deteriorates soil quality status and degrade land sustainability. Prolonged exploitation of tea growing soils through perennial monoculturing often causes significant changes, with acidification being the most important problem. Since tea is cultivated exclusively for leaves as the economic part, it requires heavy fertilizer application, especially of N, which often leads to high acidification and other related problems. Therefore, development and application of management strategies that enable to overcome various problems is essential to reduce the risks and ensure sustainable development of tea.

This chapter provides an extensive overview on soil acidification, its causes, and amelioration methods; role of major, secondary and micronutrients in tea growth, yield and quality; fertilizer application; chances of heavy metal contamination in the soil and their effect on tea quality; agroecological and organic management of tea plantations; best agro-management practices for sustained soil health and tea production; role of microbes in soil fertility; impact of climate change and tea growth; and the way forward for efficient soil health management of tea plantations aimed at sustained production of quality produce.

V. Krishnakumar (✉)
ICAR-Central Plantation Crops Research Institute, Regional Station, Kayamkulam, Kerala, India

T. R. Kumar · P. Murugesan
UPASI Tea Research Foundation, Regional Centre – Coonoor, The Nilgiris, Tamil Nadu, India

G. V. Thomas, V. Krishnakumar (eds.), *Soil Health Management for Plantation Crops*, https://doi.org/10.1007/978-981-97-0092-9_10

Keywords

Acid rain · Aluminum tolerance · Heavy metals · Nitrogen foot print · Nutrient budgeting · Organic tea · Quality of tea · Soil acidification · Soil ameliorant · Soil metabolomics

Abbreviations

AMF	Arbuscular mycorrhizal fungi
BFBF	Biofilm biofertilizer
BMP	Best management practice
CEC	Cation exchange capacity
CG	Catechin gallate
CRF	Controlled release fertilizer
DCD	Dicyandiamide
ECG	Epicatechin-3-gallate
EGC	Epigallocatechin
EGCG	Eepigallocatechin-3-gallate
FAA	Free amino acid
GC	Gallocatechol
GCG	Gallocatechin gallate
GHG	Green house gases
GS	Glutamine synthase
GWP	Global warming potential
IAA	Indole acetic acid
INM	Integrated nutrient management
KSB	Potassium solubilizing bacteria
MBC	Microbial biomass carbon
MOP	Muriate of potash
NES	Nutrient expert system
NI	Nitrification inhibitor
NO	Nitric oxide
NRA	Nitrate reductase activity
NUE	Nitrogen use efficiency
PGPR	Plant growth promoting rhizobacteria
SCS	Soil carbon sequestration
SOD	Superoxide dismutase
SOP	Sulfate of potash
SRF	Slow release fertilizer
TAFS	Tea agroforestry system
TAG	Triacylglycerol
TNAA	Thermal neutron activation analysis
YTD	Young tea dose

1 Introduction

Tea bush (*Camellia sinensis* (L.) O. Kuntze) is believed to have originated in China, which has a lengthy history of tea plantation (Ping et al. 2014) dating back over 3000 years (Eden 1965), with South West China serving as the world's original tea producing regions. Among the three common beverages (coffee, tea, and cocoa) consumed worldwide, tea, as an intensively managed perennial evergreen broad-leaved cash crop, is grown commercially in many countries, the northernmost being Georgia in the former Soviet Union, and the southernmost being South Africa and Argentina (Li et al. 2011). According to Falla et al. (2021), the primary locations of tea origin were Southeast Asia, south and southwest China, and the Indian subcontinent. Later on, it moved to Europe and Russia, and at present, there are more than 60 countries that cultivate tea in tropical and subtropical regions of five continents, the majority being in Asia and Africa (Pan et al. 2022). China and India occupy the first and second position among the tea-producing countries in the world and account for 36.28% and 22.61% of the global tea production, respectively (Pan et al. 2022). Although India, China, and Sri Lanka are the three major tea producers, significant quantities are grown in Java, Sumatra, Iran, Japan, and in some parts of Africa.

Globally, more than 30 lakhs cups of tea are being consumed everyday as a beverage (Williams et al. 2020) and according to Pan et al. (2022), there are about three billion tea drinkers in the world. As per the Global tea market report of 2021, the worldmarket is expected to move upward to around 72 billion US$ during 2026 from around 47 billion US$ (2020) at the compound annual growth rate (CAGR) of 7.15%.

Tea is manufactured from the apical bud and the associated two leaves immediately below it—commonly known as two young leaves and a bud (Fig. 1a, b). The bud, the young leaves, and the corresponding stems constitute the young shoot.

Fig. 1 Tea plants- (**a**) ready for plucking, (**b**) plucked shoot of two young leaves and a bud. (Photos: T Rajkumar)

Tea nutrition is very complex due to the perennial nature of the plants and the many variables including varieties, plant densities, harvesting and pruning practices followed, and climatic conditions (rainfall, temperature, day length and humidity) involved in determining crop yields.

2 The Crop

Tea is one of the widely consumed beverages in the world (Karak et al. 2011) due to its aroma, medicinal values with health effects (Khokhar and Magnusdottir 2002) derived from its rich organic and inorganic constituent elements, nutrients, and pharmacodynamic composition (Cao et al. 2018; Shen et al. 2013).

Tea is basically a woody, perennial crop with a lifespan of more than 100 years, but its peak production period is between the 30th and 50th year. Though the tea plants can grow as an evergreen bush under normal conditions and can attain a maximum height of 20–30 m (Hajiboland 2017), for the ease of cultivation, the plants are maintained at a maximum height of 60–100 cm in the vegetative stage as bushes by adopting cultural practices like centering, pruning, and plucking (de Costa et al. 2007). The plants are harvested generally from the second year onward at regular intervals of 7–10 days. In the tropics, harvest of the apical bud and the young leaves is continued throughout the year, but in temperate environments, only seasonal harvestings are carried out.

3 Climatic and Soil Requirements

3.1 Climatic Requirements

Climate is an important factor which determines where the tea plant can be cultivated and the quality of tea grown in a particular area. Although tea is native to the humid tropics and subtropical regions, it shows wide adaptability and, therefore, can be successfully grown in a wide range of climatic conditions and soils (Hajiboland 2017). Tea is also grown across a range of altitudes, from sea level up to about 2200 m above mean sea level. The tea plant requires a minimum rainfall of 1200 mm year^{-1}; however, 2500–3000 mm year^{-1} shall be the optimum. Temperature also plays a critical role in the growth, yield and quality of tea. Tea requires cool to warm temperatures with at least 5 h of sunlight per day. The mean annual temperature during the growing season is to be between 18 °C and 21 °C, and it must not fall below 14 °C.

3.2 Soil Requirements

Tea grows on a wide range of soil types ranging from the lightest of sand to heavily silt loam or even silt clay loam types (Nath 2014). High organic matter content is also important. According to Harler (1966), medium or light textured soils, which are of acidic character, are also suitable for the best growth of tea irrespective of countries. Soils with depth of 1.5–2.0 m is ideal, but in some regions (e.g., Vietnam), tea grows well on soils that are only 60–100 cm deep (Le et al. 2021). Tea is a calcifuge and does not tolerate alkaline (basic) soil, and therefore, it is generally cultivated in strongly acidic soils with pH values as low as 3.5 (Fung and Wong 2002). pH contributes significantly to changes in the soil microbial community following N application (Ma et al. 2021; Shen et al. 2013). The optimum pH for tea is 4.5–5.5.

4 Soil Physical Properties

Tea is a perennial monoculture plant and its long-term cultivation induces various changes in the soil properties. The soil in tea plantations is likely to have high BD over the years due to intensive leaching and trampling (Phukan and Baruah 2015). Conversion of abandoned land into tea plantation caused the soil aggregates to be dominated by the >5 and < 0.25 mm fractions, which were the main pools of C and N (Wang et al. (2018a). Long-term tea planting resulted in disintegration of the large macro-aggregates and degradation of bacterial functions (Wang and Ye 2020). Wang et al. (2020a) reported the mean weight diameter to be the highest in a 17-year-old tea plantation. Mishra and Francaviglia (2021) showed that soil organic carbon (SOC) was higher when soil texture was medium and fine. Therefore, soil characteristics along with the proper nutrient management practices play a good role in better tea production and maintenance of good soil health.

5 Soil Chemical Properties

5.1 Soil pH

Tea plant is an acidophilic crop. World over, tea cultivated soils are acidic in nature, as optimal pH status for the sustainable plant growth and production ranges from 4.5 to 5.8 (Malakar et al. 2022). Tea growth is inhibited when the soil pH is <4.0, and both the quality and quantity of the tea produced are negatively affected (Fung et al. 2008). Yan et al. (2020a) reported that in China, the average soil pH of tea plantation was 4.68 nationally and ranged from 3.96 to 5.48 in different provinces. Either an increase or decrease in pH (6.5–7.0 or < 4.0) affects the tea growth and quality (Chen et al. 2021b; Lin et al. 2022a; Li et al. 2023a; Wang et al. 2022a). Decreasing of soil pH causes concentration of heavy metals (Hu et al. 2017), decreases in soil microbial diversity, soil N conversion efficiency as well as N absorption capacity of plants (Lin et al. 2022b). An increase in pH improves the yield and quality attributes of tea (Lin et al. 2022a) and provides better condition in the rhizosphere with high content of NH_4-N and the ratio of NH_4-N to NO_3-N (Wang et al. 2022a).

5.2 Soil Organic Matter

Long-term tea plantation was beneficial for the accumulation of OM and nutrient fractions in soil (Li et al. 2015a; Wang et al. 2016a, 2017) including humic substance, classified into by humic acid, fulvic acid and humin. A high-yield tea plantation requires a SOM above 2.0% and higher available N and P content in the soil (Han et al. 2002). He et al. (2021) found the quality of soil humus to be better and the effect of carbon sink stronger in a 25-year-old tea plantation.

5.3 Soil Organic Carbon

Soil organic carbon (SOC), a key element of the agro-ecosystem, is also important for an optimum chemical composition and biological productivity, including fertility and nutrient holding capacity of a plantation. Successive tea planting of tea resulted in increase in pools of OC and nutrient in whole soils, except for the available K pool that showed an inverse trend (Wang et al. 2016a). Tea yield had a significant positive correlation with SOC. The land productivity was strongly related to SOC levels but not to cation exchange capacity (CEC) or pH of soils. The productivity and SOC level declined significantly with age of tea plantation (Dandeniya et al. (2022).

5.4 Soil Enzymes

An understanding of the possible roles of various soil enzymes is vital to maintain the soil health and fertility management in agricultural ecosystems. Long-term mono-cropping of tea could potentially harm the soil enzyme mechanism, which results in a significant decrease in the soil enzymatic activity (Sherene 2017). Chen et al. (2021a) opined that the relationship between bacterial composition and biologically-driven ecological processes can be considered as indicators of sustainability for the tea plantation.

5.5 Soil Quality

Soil quality is a most important factor in developing sustainable agricultural practices. Continued exploitation of soils in tea agrocenoses (perennial monoculture) deteriorates soil quality status and degrade land sustainability (Baruah et al. 2017, 2020; Wang et al. 2010). Saygın et al. (2023) reported the soil quality index of soils in a micro-catchment used for long-term intensive tea cultivation.

6 Soil Acidification

Soil acidification is a natural process that takes place at a slow rate during the weathering of soils (Bolan et al. 2023), but could be aggravated due to several reasons. Excessive use of chemical fertilizers, especially nitrogenous (Tian and Niu 2015; Ni et al. 2019; Yang et al. 2018a); nitrification process (Zhou et al. 2014); proton secretion into rhizospheres during the uptake of NH_4+ ions by roots (Wan et al. 2012a); plant uptake and the removal of base cations from the soil (Tarkalson et al. 2016); and acid deposition (Cecchini et al. 2019) are some of the reasons for acidification in tea plantation. There will also leaching of cationic nutrients (Yan et al. (2018) and potential increases in greenhouse gases (GHG) fluxes (Dai et al. 2017; Yan et al. 2018; Zheng et al. 2022). Tea leaf litterfall and/or applications of tea prunings to fields promote microbially mediated catabolism of tea polyphenols into organic acids and accelerates acidification (Arafat et al. 2020).

Tea plants prefer to absorb ammonium, which releases of more protons from their roots, and cause acidification of tea soil (Wan et al. 2013). Acidification can occur with the release of protons during nitrification (Xu 2015) for by active Al in the root system (Wan et al. 2012a, b). The degree of acidification positively correlates with years of tea planting (Su et al. 2013; Yan et al. 2018). A decrease in the base saturation percentage and increase in exchangeable Al_3+ were the main mechanisms of acidification at tea plantations (Liu et al. 2023a).

Deterioration of physico-chemical properties and the accumulation of autotoxic substances are associated with monoculture (Arafat et al. 2019). There will be lower release of H+ into the rhizosphere at a lower planting density (Li et al. 2016a).

Soil acidification causes imbalance of soil microecosystem; reduces soil nutrient cycling ability as well as absorption and utilization ability by tea plants (Han et al. 2022; Lin et al. 2022b; Ye et al. 2023b); enhances the activity of some acidophilic denitrifying bacteria and increase soil N_2O emissions (Huang et al. 2015); reduces microbial biomass and their diversity (Ma et al. 2021; Yang et al. 2019; Wang et al. 2020b); and changes enzyme activity, gene abundance, and microorganism involved in various N cycle processes leading to imbalance of soil N form ratio (Lin et al. 2022a). Therefore, it is necessary to explore soil remedial measures in acidified tea plantations for the sustainable development of tea industry (Ye et al. 2022a).

6.1 Management of Acidification

Application of organic manures to maintain sustainability in tea plantations and thus mitigate soil acidification in tea soils is suggested (Yan et al. 2020b; Yang et al. 2018a; Ye et al. 2022a; Ye et al. 2023b). Yang (2021) advised application of 300 $kg{\cdot}hm^{-2}$ dolomite powder in two consecutive years.

7 Soil Amendments and Soil Health

7.1 Humic Substances

A range of soil amendments have been used in tea plantations to improve soil health, including enhancing N availability and mitigating GHG emissions. Application of humic substances @1.2 kg ha^{-1} resulted in 18% increase in tea productivity due to enhanced N availability and uptake, increased populations of nitrifying, ammonifying, and denitrifying bacteria (Pramanik et al. 2018a).

7.2 Liming of Tea Soils

Soil acidification of tea plantations causes soil structural deterioration, reduces fertility, and increases heavy metal content, and these variations affect the growth and development of tea plants and reduce the quality (Fan et al. 2020a). Liming of acid soils is recommended to make the soil pH at the optimum level, produce higher yield without toxicity of any nutrients, and to make healthier drink. The pH of tea soil could be amended once or twice in a pruning cycle by applying dolomitic lime (Dinesh et al. 2016). As the liming materials do not move laterally, they should be uniformly broadcasted over the soil (Auler et al. 2017). Application of dolomite and biochar amendments increased soil pH (Oo et al. 2018). The use of Calciprill @40% of the recommended dose of dolomitic lime either alone or in combination with NPK was found to maintain the pH of the treated soil at or above 5.1 even after 2 years. Combined application can save 25% of the inorganic fertilizer requirement (Raj kumar 2020).

Though liming the soil to correct the acidity is beneficial, several researchers have reported growth inhibition and a decline in the quality of tea production due to over-liming (Chen and Hu 2014). Wang et al. (2018b) opined that nitrification may be optimized at around pH 4.40, and if any pH-raising practices to reduce acidification are adopted, the soil pH should be enhanced to more than the optimum pH range (around 5.10) to avoid stimulating soil nitrification. Over-liming increased soil pH and Ca levels, and the high soil and leaf Ca levels inhibited uptake of K and Al, negatively affecting the growth of tea plants (Yan et al. (2021a).

8 Soil Property and Quality of Tea

Many reports on the variations in tea responses, particularly yields due to general agro ecological conditions, and farm-specific soil characteristics are available from China (Han et al. 2007), India (Venkatesan et al. 2004), and Uganda (Carter 2007); however, soil characteristics did not affect productivity within Kericho in Kenya (Kamau 2008; Kamau et al. 2008), which could be as result of uniformity of soils characteristics within a small region. However, even within some small regions, where tea has been grown for a long time, soil quality deterioration could be seen, especially when the soil nutrients content become depleted (Dang 2002). The physical and chemical properties of the soil are important factors affecting the yield and quality of the tea (Fu et al. 2021; Liu et al. 2023b; Yang et al. 2022b). Changes in soil properties resulting from severe erosion and declining fertility have disturbed many tea plantations, leading to falling output and quality (Liu et al. 2022b). Soil pH, EC, exchangeable Ca and Mg, total concentration of Mn, and total concentration of Cu were shown as significant impact factors on free amino acid (FAA) content (Tseng and Lai 2022).

9 Tea Quality in Relation to Nutrients

The sweet and umami tastes are usually well-accepted by the consumers. The bitter and astringent tastes are usually undesired, but they are important for providing the complex sensory perceptions of the teas (Zhang et al. 2020a). The astringency of green tea is thought to be determined by polyphenols, whereas freshness and sweetness are provided by FAAs and bitterness by caffeine (Yang et al. 2022a; Zhang et al. 2020a). With adequate supply of N, many sugars that are produced by photosynthesis could be used to synthesize protein in the tea plant, resulting in an accumulation of amino acids and a decrease in polyphenols (Dong et al. 2019; Ruan et al. 2010). A deficiency of P could decrease the concentrations of total polyphenols, flavonoids, and total FAAs (Lin et al. 2012a; Santosh et al. 2018). Optimum K improves tolerance of drought stress and help the tea plant accumulate more catechins, amino acids, and caffeine (Ruan et al. 2013). Mg fertilization influences amino acid and sugar accumulation in oolong tea, and results in higher concentrations of total FAAs and a lower ratio of tea polyphenols to FAAs (Xu et al. 2021). Application of high amounts of P and K causes reallocation of photosynthates and carbohydrates to the catechin pathway through promoting malic acid metabolism in young tea shoots, which affects tea quality (Wei et al. 2022).

10 Nutrient Requirement of Tea

The content and availability of nutrients, and the degree of accessibility in soil are very dynamic due to various inorganic and biochemical processes taking place in the soils (Roy et al. 2002) and other biotic and abiotic factors consisting of temperature, water content, soil reaction, and nutrient uptake, input, and their losses (Virk et al. 2013). Nutrient requirements for commercial tea production are particularly high as the harvestable portions of tea contain the largest percentage of nutrients in the plant (Dang 2005), N being the most abundant nutrient followed by K, Ca, P, S, Mg, and Zn (Sultana et al. 2014). N plays a paramount role in establishing proteins and chlorophyll synthesis (Tongsiri et al. 2020). Long-term harvesting and pruning leads to significant deficiency of various nutrients (Zhang et al. 2020b), causing decline in plant vigor, tea yield, and quality (Sun et al. 2023). Using the quantitative evaluation of the fertility of tropical soils model, Tang et al.(2020a) worked out the nutrient requirement for producing 1 t tea (one bud and two young expanding leaves) for different seasons. This model enables estimating N, P, and K uptake of tea plants at a precise targeted yield, which helps to avoid excessive fertilization and reduce its effect on the environment and improve tea yield. Huang et al. (2022c) suggested the use of CsATG101 gene as a candidate gene to increase the yield of fresh leaves under different N levels. The work of Wang et al. (2022b) represents a comprehensive transcriptomic analysis of plant responses to NH_4+ and reveals many potential genes regulated by NH_4+ in tea plants. However, there was also a concentration range of NH_4+ nitrogen; otherwise, tea plants will be stressed under high NH_4+ conditions. Such findings may lead to improvements in N efficiency of tea plants.

11 Major Nutrients

11.1 Nitrogen

N plays a key role in improving tea yield and quality, and therefore, the plants require a large amount of N, which accounts for about 4.5% of the dry weight (Xiao et al. 2021). The amino acid content that dominates the quality of tea is mainly determined by the efficiency of N absorption by the roots of the tea plant (Huang et al. 2023). In tea plants, most amino acids are synthesized in the roots and then transported to the leaves (Ruan et al. 2019a). A reasonable N application rate exerts a positive impact on the tea root growth (Manzoor et al. 2022) and help in increasing the content of amino acid, caffeine, and catechin (Wang et al. 2016b). N fertilizer also plays an important role on the photosynthetic capacity, productivity, and nutrition status of tea plants (Wang et al. 2018c). Wang et al. (2020d) found that the effect of N addition on the soil pH was dependent on the type of N added, with ammonium bicarbonate and urea acidifying the soil while ammonium sulfate fertilizers did not affect the soil pH.

Soil acidification impacts the soil microbial community, with consequences for fundamental ecosystem processes. Long-term addition of N fertilizers led to the reduction in microbial biomass, inhibited respiration, and reduced the microbial diversity (Yang et al. 2019). Destabilization of microbial community and reduction in the complexity of bacterial networks are also reported (Ma et al. 2021). Bacterial diversity and community responses to N addition were more apparent than fungal diversity and community responses (Tang et al. 2022a).

11.1.1 Nitrogen Deficiency

N becomes deficient in the tea plant when the leaf N content is <3%, mildly deficient when it is 3.0–3.5%, and sufficiently supplied when it is >3.5% (Owuor 1997). Deficiency of N in a tea bush first shows as a lighter than normal green color in the young flush leaves, which progressively becomes quite yellow leading to the corresponding decrease of dry matter accumulation, shortening of the internodes, and leaves becoming more thin than normal (Lin et al. 2020; TRFK 2002). Low N supply decreases the contents of total FAAs, theanine, glutamic acid, and some volatile aroma compounds, but increases the contents of polyphenols and catechins which might result in the inferior sensory properties of green tea (Lin et al. (2020). N deficiency induced chlorosis of tea leaves, impaired biomass accumulation, decreased the leaf chlorophyll content and N absorption (Lin et al. 2021), and reduced the activities of N metabolism related enzymes in leaves and roots causing retardation of tea growth (Lin et al. 2023). Rebello et al. (2022) opined that factors such as harvesting of tea leaves, leaching of nutrients, gaseous emissions, and immobilization of soil microbial communities etc., will reduce the availability of N for plant growth. Therefore, maintaining yields requires soil N additions through fertilizer applications.

11.1.2 Nitrification and Nitrification Inhibitors in Tea Soils

Nitrification, which plays a critical role of the N cycle with great agricultural and environmental importance, is the soil microbial process responsible for the transformation of ammonium (NH_4+) into nitrite and then to nitrate (NO_3 -) (Han et al. 2012). In spite of the low pH, high nitrate accumulation occurs in tea soils, resulting in a great potential for diffuse pollution. The effect of different NH_4^+-based N on soil nitrification depended on the impact of the fertilizers on soil pH, and nitrification played an important role in NO emissions, but not in N_2O emissions (Wang et al. 2020c).

In recent years, nitrification inhibitors (NIs) have received extensive attention as a potential way to improve N fertilizer use (Qiao et al. 2015). NIs increases N retention and uptake since NH_4 + –N is less prone to leaching than NO_3 –N (Kumar et al. 2015a). Dicyandiamide (DCD) performed as an effective nitrification inhibitor and reduced the ratio of nitrate to total mineral N in a highly acidic tea soil (Han et al. 2008). In order to maintain the soil mineral N as NH_4+ for longer and alleviate soil acidification, Wang et al. (2020e) recommended that when fertilizing with urea, urease and nitrification inhibitors are to be applied to limit N losses. Application of 3,4-Dimethylpyrazole phosphate increased soil pH when urea was applied @ 500 kg N ha^{-1} and above, due to reduced nitrification and increased plant uptake of NO_3– (Qiao et al. 2021). Yang et al. (2023a) observed considerable increase in the denitrification potential and the absolute abundance of denitrifying genes, such as narG, nirK, norB, and nosZ.

11.1.3 N_2O Emission in Tea Plantation

The excessive application of chemical fertilizers in tea gardens in China is a major problem and > 30% of gardens are over-sprayed, and > 50% have high P and K nutrient input, with an unreasonable proportion of nutrients (Ni et al. 2019). Any amounts of N not taken up by plants will be lost to adjacent systems, causing serious environmental pollution problems (Yang et al. 2018a). Lu et al. (2019) reported that deep application of fertilizer increased N_2O emissions with high rainfall. Zhang et al. (2020c) reported that N_2O emissions are influenced mostly by soil chemical properties.

The average N_2O emission rate and emission intensity of Chinese tea gardens was 16.6 kg N_2O ha^{-1} $year^{-1}$ and 21.3 g N_2O kg^{-1} tea production, which is significantly higher than that of other crop fields (Li et al. 2016b). In Japan, the total N_2O emissions from the entire tea fields were 10.6 and 14.8 kg ha^{-1} $year^{-1}$ (Hirono and Nonaka 2012). The emission of GHG from tea plantations was similar or higher as compared to the rice-fallow system (Yao et al. 2018). On a global scale, Han et al. (2021) found fertilization to increase the mean N_2O and NO emissions from tea gardens by 44.5 Gg N $year^{-1}$ and 34.3 Gg N $year^{-1}$, respectively. Wang et al. (2020e) reported that the annual N_2O emissions from soils of global tea plantations on average was 17.1 kg N ha^{-1}which is considerably higher than those reported for cereal crop lands. Globally, the overall increase in N_2O and NO emissions by fertilization was 584% and 790%, respectively, the stimulating effect mainly due to high N application rates, climatic conditions, and soil properties (Yu et al. 2023).

Converting pruning residue to biochar and its addition to soil has the potential to mitigate soil N_2O emissions (Oo et al. 2018). Combined amendment of chicken manure and biochar was more effective in reducing N_2O emissions than organic manure alone (He et al. 2019). Yu et al. (2020) found that sorghum intercropping reduced N_2O emissions by 51.2%, while application of big urea prills reduced it by 34.7% and increase tea yield. Use of inhibitors reduced N_2O emissions by 32.2 Gg N $year^{-1}$ and biochar reduced NO emissions by 23.6 Gg N $year^{-1}$ (Han et al. 2021).

11.1.4 Nitrate Reductase Activity

Nitrate reductase (NR) enzyme catalyzes the first step in nitrate assimilatory pathway by reducing nitrate to nitrite, which is the rate limiting step that regulates the inflow of inorganic N to organic form in plants (Beevers and Hageman 1969). Venkatesan and Ganapathy (2004) found a significant correlation between green leaf yield and nitrate reductase activity (NRA) during the first and second harvestings after fertilizer application. The NRA was high in the first leaf followed by third leaf (in tea shoot of three leaves and a bud) and feeder roots. The increase in NRA was due to the increase in N or K fertilizer doses. Most of the cultivars of Assam type had high NRA followed by Cambod and China types (Venkatesan 2005). The NRA and amino acid contents were positively correlated. N or K fertilizer doses increased NRA (de Costa et al. 2007). Certain growth regulating chemicals also affected NR enzymes in tea leaf, and any alteration in NRA was found to have a direct influence on caffeine (Pandey 2010).

11.1.5 Nitrogen Leaching/Ammonia Volatilization

Considerable leaching of N and ammonia volatilization occurs in many tea plantations due to heavy application of N fertilizers (Liyanage et al. 2014), which could be reduced by split applications. A schematic diagram about the N flows into tea plantations, focusing on organic management, is presented in Fig. 2.

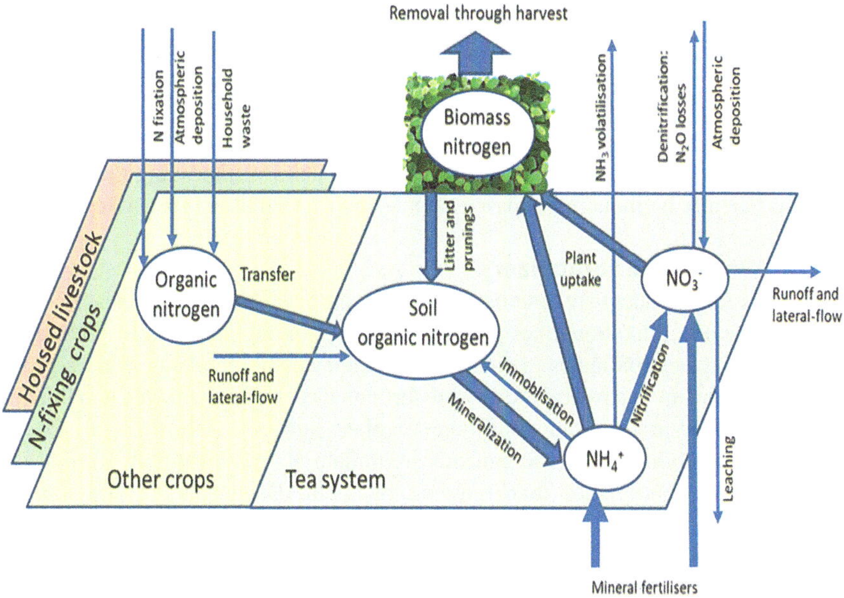

Fig. 2 N flows into tea plantations, focusing on organic management. Inorganic fertilizer additions could also be used to increase the NO_3^{-1} and NH_4 pools (with permission from Rebello et al. (2022), pp. 43–57, https://doi.org/10.3390/nitrogen3010003)

11.2 Phosphorus

Phosphorus (P) is required for the normal growth and development of higher plants, and plays very crucial role in their cellular structure and energy metabolism (Hawkesford et al. 2012). Plant roots acquire P as phosphate (Pi), primarily in the form of H_2PO_4—, from the soil solution. Depending on the soil pH, the plant will uptake either $H_2PO_4^-$ or HPO_4^{2-}(Hajiboland 2018). P also plays an important role in the energy metabolism, and is the constituent element for some biomacromolecules such as nucleic acids, phospholipids, and sugar phosphates. Lin et al. (2009) found P to promote the formation of flavonoids and increase the content of polyphenols, amino acids, and caffeine in tea.

11.2.1 Phosphorus Deficiency

P deficiency in tea manifests itself as the absence of brightness, particularly in mature leaves, and dieback of young and old woody stems (TRFK 2002). The affected tea leaves appear dull. Under gross deficiency, the smaller branches show excessive die back from the ends (Hajiboland 2018) that have been cut when pruning. It can easily be confused with "sun-scorch." Many reports on the P deficiency affecting quality parameters of tea are available (Ding et al. 2017; Li et al. 2015b; Santosh et al. 2018). An extremely high tolerance of tea plants to P deficiency is mainly due to high internal use efficiency; thus, tea is a highly P deficiency-tolerant species (Hajiboland and Salehi 2014). According to Lin et al. (2011), the high tolerance of tea plants to P-deficiency might be through (a) increasing the availability of P by inducing root release of organic acid anions; and (b) improving the ability to use P efficiently by inducing bypass enzymes involved in tissue P economy.

11.2.2 Phosphate Solubilizers

Tea plants require adequate amount of P for the proper yield of tea leaves. Because of acidic nature, P fixation takes place in tea soils. Among the various fractions of organic P, phytate dominates in soils, which is unavailable to plants. Only a group of microorganisms can hydrolyze phytate through their phytase enzyme and liberate phosphate, that in turn can contribute to plant nutrition (Ghosh et al. 2021). Phosphate solubilizing microorganisms assimilate P from insoluble forms and release soluble phosphates than required by them (Bhattacharyya and Sarmah 2018). Vyas et al. (2007) recorded *Eupenicillium parvum* for the first time during isolation of phosphate solubilizing microorganisms from the tea rhizosphere. Phukan et al. (2016) reported *Bacillus subtilis* and *Aspergillus niger* as the predominant phosphate solubilizers in P deficient acidic soils. Panda et al. (2017) found that the mineral P solubilization activity of various strains was associated with a release of organic acids and a reduction in the pH. The results also highlighted the synergistic effect of using phosphate solubilizers in combination with rock phosphate. Ghosh et al. (2021) identified two indigenous phytate solubilizing microorganisms (*Bacillus safensis* and *B. siamensis*) having the ability to increase plant available P in soil.

11.3 Potassium

In tea, when K is inadequate (< 100 μmol l^{-1}), the biomass of roots, stems, and leaves decreases, while the root–shoot ratio improves (Gong et al. 2017; Zhong et al. 2017; Zhong et al. 2018). When K is adequate, there is an increase in overall biomass. After the K concentration reaches a certain extent (800 μmol l^{-1}), the dry weights of roots, stems, and leaves plateau (Zhong et al. 2018). Ruan et al. (2013) found that when the exchangeable K content is <80 mg kg^{-1}, there is a significant increase in yield with the application of sulfate of potash (SOP) or muriate of potash (MOP). Gong et al. (2017) obtained similar results, where with an increase in K concentration, the biomass of whole plants, and other parts increased linearly, reached a peak when the K concentration was 682–865 μmol l^{-1}.

11.3.1 Potassium Deficiency

In tea plants, one of the most common signs of K deficiency is the yellow scorching or firing (chlorosis) along leaf tip margin of the dorsal surface. In severe case of K deficiency leads to the fired margin of the leaf and may fall out. There will be depletion of the starch reserve in the roots and accumulation of nitrates, degeneration of feeder roots and dieback occur (TRFK 2002; Jessy 2010). There was decrease in biomass, content of K and chlorophyll in K deficient leaves (Lin et al. 2013). When soils are in severe deficiency, young leaves gradually turn from green to light yellow (Sindhu et al. 2014) and become smaller and thinner; dormant buds increase and the internodes shorten. K deficiency in mature tea plants caused decrease in quality parameters (Zhong et al. 2018). Tea produced from K deficient soil lacks the intoxicating flavors of tea that is seen from the plants grown in well-balanced soils.

At present, >59% of China's tea gardens are deficient in Kaffecting tea quality and yield. K deficiency is widespread in Alfisols on which tea is grown (Li et al. 2021b). Li et al. (2019) reported genotype 1511 as low-K tolerant and 1601 as low-K sensitive genotype. They found the root system of 1511 to be more developed than in 1601. Huang et al. (2022) presented a detailed review on the effects of K deficiency on tea plant growth and stress response, and provided strategies for optimizing K levels in soils.

11.3.2 Potassium Solubilizers

K-solubilizing bacteria releases K from insoluble minerals by discharging organic acids (Sheng et al. 2002) and these bacteria are widely used in biological K fertilizers. Hence, inoculation of K-solubilizing microorganisms along with application of rock-K to soil has recently gained much attention. Application of K solubilizing bacteria (KSB) as biofertilizers can minimize the use of chemical fertilizers and support eco-friendly crop production. Bhattacharyya et al. (2016) suggested to choose microbial fertilizers that can improve the mineral status of K in tea soil. Bagyalakshmi et al. (2017) found the population density of KSB to differ based on location of tea plantation and cultivar. Application of KSB and mica waste during the first split instead of MOP recorded higher exchangeable K content in soil (Pramanik et al. 2021). Zhang et al. (2022a) recorded the maximum K solubilization colony with *Burkholderia* sp. and its inoculation in laterite soil improved tea plant height and total phenol content.

12 Secondary Nutrients

12.1 Calcium

Calcium (Ca) is a secondary nutrient that is essential for plant growth because it is a ubiquitous second messenger and an important component of both the cell wall and membranes (Liang et al. 2021). The proper growth and development of tea plants requires moderately acidic soils and relatively low Ca levels, and excessive Ca at high pH can damage tea plant roots (Liang et al. 2021). Ca deficiency is characterized by brittleness in older leaves and discolored laminar edges of young leaves which become dark brown in later stage. On the other hand, excess quantity of Ca in soil slows down the plant growth, young leaves become yellow, and its edges curl inwards and tips turning black. Ca can increase tea growth after application, although some research proposed that tea is a calcifuge and acidophilic plant (Karak et al. 2017), and excess Ca in soil can inhibit uptake of Zn and K (Duan et al. 2012; Yan et al. 2021a), and Mg (Liang et al. 2021).

Excessive Ca in tea leaves destroys the membrane structure of the photosynthetic system, which obstructs the electron transport chain and reduces the light energy utilization efficiency of tea leaves, thereby affecting tea growth and pigment content (Malyukova et al. 2021). Ca application resulted in higher pH of cell sap; higher accumulation of tyrosine, methionine, and valine than control plants (Malyukova et al. 2022). However, the complete removal of Ca may adversely affect the tea plant, and hence, a lesser amount of Ca content (<3 cmol kg^{-1}) is required in the tea plantation to avoid soil acidification and to protect nitrifying bacteria (Liang et al. 2021). Also, the adequate amount of Ca nutrient can enhance the tolerance of tea plant to abiotic stress (Liang et al. 2021). Application of Ca was reported to effectively lower the F content in leaves (Fung and Wong 2004). A synergistic interaction between Ca and F influenced various quality parameters in tea (Luo et al. 2019).

12.2 Magnesium

Magnesium (Mg) plays very important role in various physiological and biochemical characters and is an essential structural scaffold for many nucleotides (Hajiboland 2018; Senthurpandian et al. 2009; Xu et al. 2021). Mg, the "quality element," is an activating factor for many enzymes and has a close relationship with tea quality (Guo 2014; Ruan et al. 2012; Zhang and Ruan 2016). It is required for chlorophyll production which is critical for the black appearance of made tea and functions as a catalyst for precursors contributing to quality (Rajadurai 2021).

The availability of Mg also depends on the activity or proportion of Mg relative to soluble and exchangeable amounts of K, Ca, Na, Al, and Mn. The effects of Mg on tea quality were limited by N nutrition (Ruan et al. 2012). Ruan and Gerendas (2015) reported that the status of nutrients such as K or N is closely related to the biological effects of Mg application. Excessive application of high rates of K+ and NH+ 4 fertilizers antagonistically interferes with plant Mg uptake, thus enhancing the risk of Mg deficiency. Many results are available on the yield responses to Mg fertilization and improvement of certain quality traits (Cakmak 2013; Ruan et al. 2012; Wang et al. 2020f). According to Palani (2019), the right rate of Mg application is site-specific and its requirement depends on yield target, availability status in the soil, and the synergistic (Mg x P) and antagonistic (Mg x K) relationship of Mg with other nutrients.

Pongrac et al. (2020) concluded that old leaves of tea plants cultivated under conditions of low Al availability will have the largest concentrations of Mg, Ca, and Mn and may represent most acceptable ingredient for the preparation of tea. Liu et al. (2022a) observed an increase in Mg content in tea leaves after pruning, and the yield in the second year after pruning was higher than that of unpruned tea plants. An overall average 8.5% increase of crop yield with application of Mg fertilizers under various field conditions across the world was reported (Wang et al. 2020f).

Mg improves the photosynthetic capacity of tea plants, enhances the accumulation of primary metabolites and thus increases yield (Li et al. 2021b), but was not conducive for synthesizing secondary metabolites and the accumulation of main quality indicators of tea leaves. Therefore, the usage of Mg should be controlled to ensure the yield and quality of tea leaves (Zhang et al. 2023a). Zhang et al. (2021a) observed the Mg fertilizer use efficiency in tea plant to vary from 6.08 to 29.56%. Mg can make a difference to the quality of tea by affecting the content of specialized metabolites such as tea polyphenols, FAAs, caffeine, and aroma component (Li et al. 2021c). Mg applications increased tea yield in different seasons by an average of 14.1, 18.4, and 15.8% for spring tea, autumn tea, and annual total tea, respectively (He et al. 2023). The latest research shows that magnesium can promote tea plant growth via enhanced glutamine synthetase-mediated nitrogen assimilation (Zhang et al. 2023a).

Li et al. (2023b) identified ten different genes in tea plants, among which CsMGT5 mediates the uptake and transport of Mg^{2+} under Mg limitation. Recently, Tang et al. (2021a) isolated four *CsMGTs*, of which only CsMGT1 and CsMGT2 possessed typical Mg^{2+} transport properties. Later on, it was demonstrated by Tang et al. (2023b) that CsMGT10 mainly acts as Mg^{2+} transporter in chloroplast envelope of leaf veins, which may play a key role in leaf vein greening of tea plants. Zhang et al. (2023a) found that Mg affected the gene expression of tea leaves.

12.2.1 Magnesium Deficiency

Tea plantations are generally located in regions with acidic soil, and Mg deficiency is an important limiting factor affecting tea production, in terms of both yield and quality (Zhang et al. 2021a). Application of NPK fertilization devoid of Mg, removal of considerable Mg through harvesting of leaves, as well as high concentrations of cations (e.g., H^+, NH_4^+, K^+) in the soil solution hinder Mg absorption and enhance the risk of Mg deficiency (Gransee and Fuhrs 2013). Interveinal chlorosis along with hard and brittle nature of the mature tea leaf is the primary symptoms of Mg deficiency (Fig. 3). The symptoms of Mg deficiency appear on the lower leaves which are bright yellow with a conspicuous inverted dark green 'V' down the midrib, and sometimes extending along individual veins.

Application of high rates of K reduces Mg uptake due to the antagonism between K and Mg and results in a Mg deficiency in the crop. High application of K can increase the leaching of Mg by displacing it from cation exchange sites, leading to lower Mg availability in the root zone. Jayaganesh et al. (2011) reported severe deficiencies of Mg in many tea gardens of southern India, especially in the high-yielding tea plantations. Although Mg removal through leaf harvesting is six to eight times lesser than that of K, the higher K input reduces Mg uptake and results in the appearance of Mg deficiency in tea leaves.

Short-term Mg deficiency partially elevated the concentrations of polyphenols, FAAs, and caffeine but decreased the contents of chlorophyll and Mg, whereas a long duration (30 days) deficiency decreased these constituents (Li et al. 2021c). Mg deficiency also increased the index of catechins' bitter taste and the ratio of total polyphenols to total FAAs; thus, long-term Mg-deficient tea may taste more bitter,

Fig. 3 Magnesium deficiency symptoms (Photo: T Rajkumar)

more astringent, and less umami. Soil or foliar application of Mg fertilizers improved yield and quality (Hajiboland 2018; Xu et al. 2021). High Mg and low N had positive effects on tea quality (Zhang et al. 2023b), as under Mg deficiency, there was accumulation of N and amino acids in the roots, while metabolism of N enhanced by Mg supplementation, especially under a low N fertilizer regime.

12.3 Sulfur

The importance of S in tea nutrition was recognized long ago; however, it received little attention compared to other nutrients, mostly due to the fact that estimations of S in soil and plant are tedious. S is needed for the synthesis of S-containing essential amino acids like cysteine, methionine (Noji and Saito 2003) and for the manufacture of chlorophyll pigments. When S is in short supply, chloroplasts break down and with it, the normal dark green color of healthy tea foliage will be lost. In general, S needs were met through low analysis fertilizers like ammonium sulfate, single superphosphate, and some organic sources. With increase in productivity over time, removal of S increased considerably. Besides, increased use of high analysis fertilizers viz. urea, di-ammonium phosphate, and triple superphosphate etc. which are devoid of S further hastened its depletion from the soil (Rao and Sharma 2001). With the need for sulfur recognized in the fertilizer industry, they have designed and developed S fertilizer based on pure S being formulated as pastilles and prills (https://issuu.com/alaincharles/docs/). The S is degraded in the soil by *Thiobacillus bacteria* to form soluble sulfate (Sabagh et al. 2014) that is absorbed by the root system of the tea bush. As a component of the S bearing biochemical compounds like cysteine and methionine, S plays key role for the synthesis of chlorophyll and thiamine (Ahmed et al. 2011; Hajiboland 2018). S contributes in the many process like photosynthesis and catalyzing the carbohydrates of tea plant, which is associated to the quality of end product. Yi et al. (2016) showed that S was mainly distributed in the active parts of tea plant, such as new shoots and feeding roots> mature leaves> trunk and main roots were among the lowest. Gohain et al. (2000) found that application of gypsum recorded the highest yield, followed by ammonium sulfate.

12.3.1 Sulfur Deficiency

Tea plant is prone to S deficiency, which results in reduced plant growth rate with affected shoots showing deficiency symptoms in tea bushes, commonly called "sulfur yellows" or "yellow disease," or "tea yellows" Fig. 4) (Ahmed et al. 2011). S deficiency of tea was first seen in Malawi tea plantations during 1930s. When S is in short supply, chloroplasts break down and with it, the normal dark-green color of healthy tea foliage. Young leaves suffer much due to S deficiency.

The newly formed shoots are smaller and internodes (distance between leaves) shorter due to a slowdown in growth, which may be followed by general shoot necrosis when S deficiency persists. As a consequence, yield and quality of tea are

Fig. 4 Sulfur deficiency symptoms (Photo: T Rajkumar)

severely affected (Ananthacumaraswamy et al. 2011). Any organic or inorganic fertilizer containing S in available form can overcome the problem of "tea yellows."

13 Micronutrients

The concentration of micronutrients in made tea and tea infusion is governed by various factors such as the quality of soil where tea plants are cultivated (Yemane et al. 2008), the origin of soil (Karak and Bhagat, 2010; Szymczycha-Madeja et al. 2012), cultivation practices adopted (Jin et al. 2008), pruning operations (Kumar et al. 2015b), as well as climatic conditions of the location (Nath 2013a). Rengel (2015) indicated that the availability of micronutrients is affected by factors, particularly soil and rhizosphere pH. High soil pH results in the retention of micronutrients in the soil, thus limiting their uptake by tea plants.

13.1 Zinc

Zinc (Zn) is required for plant growth relatively to a smaller amount but is essential for normal, healthy growth, and reproduction of plants (Rudani et al. 2018; Saleem et al. 2022). Zn is required as a structural component of a large number of proteins, such as transcription factors and metalloenzymes (Figueiredo et al. 2012). Zn has important role in photosynthesis and mobilization of assimilates and has been shown to mobilize photosynthates towards pluckable shoots in tea. Leaf chlorophyll content, stomatal conductance, and net photosynthesis are adversely affected by inadequate supply of Zn (Barbora et al. 1993). Upadhyaya et al. (2013) demonstrated that $ZnSO_4$ treatment minimized the drought-induced decrease in relative water content, dry mass of leaf, and antioxidants in tea clones (TV-1, TV-17, and TV-29).

13.1.1 Zinc Deficiency/Toxicity

The first report of Zn deficiency in tea was from Sri Lanka (Tolhurst 1962). Since then, more and more tea growing countries in widely separated parts of the world have recognized the importance of Zn in increasing crop productivity. Under deficiency conditions, the youngest tissues in tea fail to develop normally (Fig. 5). Severe Zn deficiency is characterized by root apex necrosis (dieback), while sub lethal Zn deficiency induces spatially heterogeneous or interveinal chlorosis (mottle leaf), the development of reddish-brown or bronze tints (bronzing), and a range of auxin deficiency like responses such as rosetting, epinasty, "goblet" leaves and little leaf (Marschner 1995). Zn deficient plants show small and chlorotic leaves, short internodes, and longer period of dormancy (Nelson 2006).

Under Zn stress, concentrations of Zn affected the ion uptake in roots and the nutrient element transportation to leaves, leading to the different distribution of P, S, Al, Ca, Fe, and Cu in the tea leaves or roots (Zhang et al. 2017a). Venkatesan et al. (2006a) observed that Zn had an antagonism relationship with Fe in the tea leaves, promoted the translocation of Mg to the tea leaves, and suppressed the uptake of Mg

Fig. 5 Zinc deficiency symptoms (Photo: T Rajkumar)

in the tea roots. Zn deficiency can be corrected by foliar application (Raguraj et al. 2016), while application to the soil is not effective. Zn from $ZnSO_4$ (22–26% Zn) is quickly available, while that from zinc oxide (70–80% Zn) is only slowly available (Hajiboland 2018),

Zn solubilizing microorganisms are important drivers in the solubilization of insoluble Zn compounds and play an important role in soil fertility and nutrient allocation. Bhattacharya et al. (2022) isolated, characterized, and qualified the Zn-solubilizing microbes from the tea rhizosphere and the most common fungus genera were *Aspergillus* spp. and *Trichoderma* spp. Gram-positive cocci were dominant among bacteria.

13.2 Boron

Boron (B) is required by plants for proper development and differentiation of tissues, particularly the growing tips as well as for other physiological functioning of higher plants (Seeda et al. 2021). Being a weakly held anion, B can be easily leached out from soil, making acid sandy soil to be more prone to B deficiency. B is an important micronutrient for tea as its deficiency adversely affect the economic part, i.e., growing tip (Singh et al. 2014).

Fig. 6 Boron deficiency symptoms (Photo: T Rajkumar)

The first sign of B deficiency in tea plants is the restriction in growth of the terminal bud, which becomes dormant (Fig. 6). The leaves become dark green, thick, and leathery, and they are frequently misshapen and crinkled. The growing point ultimately dies and, as a result of the loss of apical dominance, many axillary buds try to grow but these also die back if sufficient B is not available. With progression of deficiency, excess cork develops first on the upper side of the petiole, but later extending to the main and lateral veins outlining them with a corky streak on both upper and lower surfaces. The veins crack as the cork develops. (https://agriculture.borax.com/crop-guides/beverage-crops/tea).

Tea plants deprived of B positively reacted to soil and foliar fertilization, and the B concentration of the leaves under the shoots increased and reached above the critical deficiency limit (>30 mg kg^{-1}). The most successful application was soil + foliar @ 4 kg B ha^{-1}, and 400 mg B l^{-1} (Balci and Taban 2022).

13.3 Copper

Copper (Cu) is essential for plant growth but increase in the concentration of both the essential and non-essential metals will result in inhibition of growth. Both shoot and root weight decreased progressively with increasing Cu concentration (Dey et al. 2014a). Dey et al. (2014b) found negative linear relationships between total chlorophyll contents and Cu concentration. Dey et al. (2015) reported that the tea clone TS-491 was more sensitive to the higher concentration of Cu compared to S3A3, and a Cu concentration beyond 400 μM was toxic to the tea plant, as it leads to the decline in the growth by retarding many biological functions. Mazumder (2015) reported that Cu accumulation in the different plant parts had a positive correlation with the Cu stress.

Tea plants deficient in copper have slightly darker foliage than normal, but in most cases, it is difficult to detect the symptoms in the field. However, severe deficiency may inhibit "fermentation" or "auto oxidation" (Willson 1969) and such leaf develops a bright orange color during fermentation and changes color very slowly through dark green to dark brown. Such cases of Cu deficiency are rare in Kenya (Anon 2012).

13.4 Manganese

Manganese (Mn) is an essential micronutrient for plants and acts as a catalyst in several important enzymatic and physiological reactions. It plays a crucial role in the soil–plant–microbial interactions. The significance of Mn as an essential micronutrient needed by tea was not given much attention until Mn deficiency was observed in many tea estates in Maskeliya area in Sri Lanka (Raguraj et al. 2019). Interveinal chlorosis and reddish brown necrotic spots on leaf blade are the typical Mn deficiency symptoms in tea. Application of ammonium sulfate (270 kg N ha^{-1} $year^{-1}$) along with $MnSO_4$ (4%) either through soil or foliar application was recommended as a remedial measure for Mn deficiency. Peyang et al. (2020) found that Cu and Mn concentrations in soil were higher in summer followed by monsoon and winter, whereas, the Zn concentration was higher in summer and winter but lower in monsoon. They also found that the concentrations of Mn were generally high in tea growing soils with respect to Cu and Zn. Tea plants with excess Mn, often due to very acid soils, appear to be normal. However, mature leaves are brittle and when crushed in the hands, crack easily with a rustling noise and the leaf surface may also develop a cracked appearance. Soil acidification enhances the solubility of metals such as Mn and Al, resulting in phytotoxicity including suppression in growth and yield (Venkatesan et al. 2007).

13.5 Fluorine

The F content of tea plants is 10–100 times higher than other plants, but tea plants can still grow normally (Xie et al. 2001) without any toxicity symptoms. This suggests that there is a built-in mechanism for F detoxification in tea plant (Peng et al. 2020; Yang et al. (2020a). The hyperaccumulation of F affects the growth of the tea plant and the biochemical composition of tea leaves and thus, tea quality (Cai et al. 2016a; Yang et al. 2015). F concentration in tea leaves depends on varieties, seasons, and maturity of leaves (Cai et al. 2016b). Under high F stress, F affects the normal growth of plants by destroying the cell structure and inhibiting enzyme activities (Anqi et al. 2022). Long et al. (2021) suggested that the key to reducing the F content in tea is to alleviate soil acidification and reduce F–Al complex content in soil. Peng et al. (2021) presented an extensive review of F absorption, transportation, F tolerance mechanisms in tea plants, as well as F bioavailability and exposure levels in tea infusions.

13.6 Iron

Iron (Fe) is an element required of the tea plant to maintain normal metabolism and promote the composition of chloroplasts and the activities of enzyme in the cells (Li et al. 2012). The perennial plants like tea have to experience differences in Fe availability in the soil due to their immobility and either deficiency or excess of this element believed to generate oxidative stress (Abdul-Kader 2007) and is responsible for several nutritional disorders affecting the physiology of plants (Becker and Asch 2005). Intermittent yellowing of leaf and coppery red discoloration are the symptoms of Fe toxicity (Hemalatha et al. 2006). Though addition of excessive Fe did not alter the translocation of P, it interfered with translocation of Mn and suppressed K uptake as well. Addition of Fe increased the soil Mg content but decreased its content in stem and root. The Ca accumulation or its translocation was not affected by Fe toxicity.

13.7 Selenium

Selenium (Se), an essential trace element with significant health benefits for humans, mainly serves as a cofactor of numerous Se-depended enzymes like glutathione peroxidase and several other antioxidant enzymes participate in protecting cells against oxidative damage (Gu et al. 2020; Wang et al. 2015). Selenite and selenate, the predominant forms of inorganic Se in soil, can be transformed into organic forms through a sequence of bio-conversion processes and absorbed by plant leaves (Fan et al. 2022a). Zhang et al. (2020d) confirmed the correlation between the total Se contents in teas and soils.

Tea plant is supposed to be a Se-enriched species (Zhao et al. 2016). Studies of Se in soil-plant cycling revealed that organic Se was synthesized in green tea (Jia et al. 2020). Se-enriched green tea (Se-GT), a product processed from the young leaves of tea plants grown in seleniferous areas, has become increasingly popular in recent years due to its more pronounced health-promoting properties (e.g., antioxidant, prebiotic, and anticarcinogenic activities) compared to common tea (Haojie et al. 2020; Ye et al. 2022b). Cao et al. (2018) reported the metabolisms of Se, glutathione, and selenocompound as well as plant hormone signal transduction to respond to selenite in tea plants. Li et al. (2021a) found that nano-selenium enhanced the defense response of tea, the nutrients, and secondary metabolites protein such as catechins contents, soluble sugar, carotenoid, and tea polyphenols. Se sources play a positive role on its content and quality of tea by increasing the antioxidant capacity of leaves (Sun et al. 2022c).

Se deficiency has been implicated with several chronic diseases, such as coronary heart disease, cancer, and Keshan (Ye et al. 2022b). Se has significant effects on the accumulation of minerals and polyphenols in tea leaves (Ye et al. 2023c). Their results elucidated the elemental and chemical basis of Se-GT and provided a rationale for the quality evaluation of Se-GT, which is of crucial importance for tea industry. Xiang et al. (2022) have provided a detailed account of the results of previous research on the effects of exogenous Se treatment on the growth and quality of tea plants.

13.8 Silicon

The benefits of Silicon (Si) nutrient are well documented in different crops when they are exposed to abiotic and biotic stresses (Debona et al. 2017). Despite Si abundance in tea soil, it is not in the available form for plant uptake because it exists mostly as an insoluble or poorly soluble form like SiO_2.The form of Si that is available to plants as a nutrient is called monosilicic acid (H_4SiO_4). While the importance of Si as a plant nutrient is still being debated, many reports are available on increasing productivity with the addition of Si nutrients through foliar and soil applications from sources such as calcium silicate (Ca_2SiO_4), potassium silicate (K_2SiO_3), silicic acid, etc. (Adrees et al. 2015). Soil application of Agrisilica powder (with 26% plant accessible Si) @150–450 kg ha^{-1} together with 100% of the recommended dose of NPK fertilizers resulted in higher tea yield (Balasubramaniam 2019).

13.9 Aluminum

Tea, known as Al-hyper-accumulator, can have a high concentration of Al in mature leaves without any symptom of toxicity, even improve growth of roots and nutrient uptake (Fan et al. 2019; Ruan and Wong 2004). Sun et al. (2020a) confirmed the beneficial effect of Al in promoting new root growth in different tea varieties. Under the optimum pH, growth and Al accumulation were markedly stimulated by Al treatment (Yamashita et al. 2020).

Accumulation of Al in tea plants is affected by plant age, cultivar type (Shu et al. 2003); application of N and P fertilizers (Fung et al. 2009); soil pH and the level of exchangeable Al (Ruan et al. 2006); leaf maturity (Carr et al. 2003) etc. Yang et al. (2022c) noticed significant differences in the distributions of Al fractions among various soils, which were attributed to the changes in soil aggregate compositions and the aggregate-associated Al fractions.

Large amounts of Al can accumulate throughout the plant, especially in mature leaves, but stimulates rather than inhibits growth (Morita et al. 2008; Sun et al. 2020a). Most of the accumulated Al is immobilized in cell walls and the rest only is stored in vacuoles (Huang et al. 2021). Mukhopadyay et al. (2012) reported that a higher Al concentration resulted in oxidative damage leading to various biochemical changes, and rise in its content along with the maturity of the leaf as well as stem tissues. Tolra et al. (2020) found that low pH and enhanced Al availability to tea plants had a strong influence on the polyphenolic pattern of tea leaves and therefore may alter both the leaves' antioxidant properties and their ability to bind Al and Fe in non-toxic form. The increased absorption and accumulation of Al by tea plant may cause soil acidification in tea plantations, leading to reduced nutrient uptake and, consequently, lower tea yield and quality (Yan et al. 2020b). Therefore, it is necessary to limit the uptake and accumulation of Al by tea plant.

13.9.1 Mechanism of Aluminum Tolerance

External exclusion and internal detoxification are the two physiological and biochemical mechanisms of Al tolerance in tea plants (Li et al. 2017; Morita et al. 2011). Al is mainly accumulated in old and mature leaves, which are less sensitive to Al stress. At the subcellular level, Al is preferentially fixed in cell walls, and once it enters into the cytoplasm, it is chelated as nontoxic forms by organic acids or phenolic compounds and later on sequestered into the vacuole of tea plants (Gao et al. 2014; Morita et al. 2008). Devi et al. (2020) indicated that differential tolerance to Al_3+ toxicity in tea varieties is linked to their oxidative stress response and, antioxidant enzymes, as a crucial constituent of the defense response. Fu et al. (2020) reported that tea plants showed high Al tolerance with plenty of monomeric catechins in leaves, and epigallocatechin gallate and polymeric proanthocyanidins in its roots. These polyphenols could bind to Al and form rPA-Al and EGCG-Al complexes in tea roots and old leaves, respectively, thus reducing Al mobility and toxicity. Fu et al. (2022) revealed that Al accumulation in tea plants was significantly and positively correlated with their flavonol content. Huang et al. (2021) found that the genes involved in phenolics biosynthesis and decomposition were very much upregulated in the leaves treated with either 0 and 4 mM Al stress, indicating they may play an important role in Al tolerance. Ray et al. (2022) suggested that Al tolerance in tea plants can be achieved through Al exclusion mechanisms from roots or by forming Al-organic acids complexes in the symplasm. Recently, Zhang et al. (2023c) presented an extensive review on the advances in the uptake, transport, accumulation of Al in tea plant, and the genetic and environmental factors that affect these processes. They also highlighted the influence of Al in tea plant based on omics approaches, including transcriptomics, proteomics, metabolomics, ionomics, and microbiomics.

14 Heavy Metals and Tea Soil Health

Soil contamination by heavy metals presents many problems for soil functions, the environment, agriculture production, food chains, or even human health. The heavy metals in tea plants are generally influenced by soil conditions and soil with high heavy metal contents appear to adversely affect the productivity (Aksuner et al. 2012: Seenivasan et al. 2016) and many studies have addressed heavy metal level in tea garden soil and compared the results with others of the world (Melucci et al. 2013; Zheng et al. 2014; Rashid et al. 2016). High levels of heavy metals can easily be transported to tea leaves through the roots of tea plants (Rashid et al. 2016). The growth media, nutrients, and soil and agrochemical inputs contribute the contamination (Dambiec et al. 2013). Traces of undesirable and toxic metals, i.e., Cd, Cr, Cu, Fe, Hg, Ni, Mn, Pb, and Zn can easily contaminate tea plants (Szymczycha-Madeja et al. 2015). Due to acidophilic nature of tea plant, it has inherent tendency to uptake metals/metalloids including the toxic ones from the soil which is of great concern worldwide (Barman et al. 2020).

Pollution of heavy metals has occurred in some tea garden soils as a result of high soil background values, the application of pesticides and chemical fertilizers containing heavy metals, and industrial activities (Li et al. 2015c; Parviz et al. 2015). In recent years, heavy metals (As, Cd, Cr, and Pb) in some tea plantation soils have exceeded the risk screening values for the soil contamination of agricultural land (GB15618-2018) in China, and therefore, excessive heavy metals in tea leaves may not be safe for human consumption (Sun et al. 2019). Barman et al. (2020) found the average concentration of Cr and As in the tested black tea samples was 10.33 and 0.11 $\mu g\ g^{-1}$, respectively. Several reports on the high content of various heavy metals in tea plantations are available (Alam et al. 2022; Li et al. 2021c; Liu et al. 2021a; Wen et al. 2022; Xu et al. 2022; Yang et al. 2022d; Zhunag et al. 2022). Yao et al. (2023) found that the roots of the "Qianmei 419" tea variety had higher Al, Cd, and Cu than in "Qianfu 4." Only the content of Cu in the leaves of the former tea variety was higher than that in the roots of the latter tea variety. They also observed that the content of various metals in the soil around the root of "Qianfu 4" to be high than in the soil around the root of the other variety.

15 Fertilizers in Tea Cultivation

Nutrient management of plantation crops has greater importance for a sustainable growth, yield and quality as well as for improvement in soil health. In recent years, agriculture has become a concern due to the pressures of the increasing fertilizer costs and increased focus on environmental protection. There is greater attention being paid to the efficient use of external inputs, including fertilizers.

The quality of tea is largely dependent on various biochemical constitutents of leaves such as polyphenols, theanine, and caffeine in the newest shoots or leaves (Li et al. 2015d; Ruan et al. 2010), which are in turn related to amino-acid uptake from the soil, especially in organic plantations. The flavor of tea depends on the ratio of amino acids to tea polyphenols. FAAs play important roles in determining the flavor of tea infusion, especially theanine, the unique amino acid in tea plants. Harvesting of tea is a continuous process and large quantities of plant nutrients are removed from the plantation, which are to be replenished through regular fertilizer application (Hoang et al. 2021; Illukpitiya et al. 2004).

Fertilization directly affects tea growth by supplying available nutrients (Ma et al. 2021; Tang et al. 2021b) and indirectly by affecting tea yield and quality through the microbial community in the soil which is critical for regulating nutrient cycling and soil multiplication (Tang et al. 2023a). Individual application of N, P, or K fertilizer cannot improve tea growth and tea plants achieve the highest yield and quality only when all of the nutrient demands of the plants are met (Dufitumukiza et al. 2020; Tang et al. 2022b).

Several research studies have been conducted across tea growing countries on the response of tea plants to fertilizer application (Amali et al. 2022; Hajiboland 2017; Kozlova and Malyukova 2021; Long et al. 2022; Ni et al. 2019; Ruan et al. 2022; Tang et al. 2022b; Wu et al. 2020a; Zhang et al. 2020a).

Tang et al. (2021c) found the nutrient expert system (NES) for tea significantly increased tea yield, quality, economic benefits, and N use efficiency, as well as decreased GHG emissions. Balanced fertilization based on the optimal nutrient requirements of tea plants improves tea yield and quality (Tang et al. 2022b). Use of slow release fertilizer will improve fertilizer use efficiency (Zhang et al. 2022b). Ruan et al. (2023) observed that LJ43 clone of tea has preferential development of secondary roots, which efficiently absorb nutrients, promote nutrient diffusion and migration in the soil, as well as promote efficient rapid nutrient absorption. It will be ideal to develop such clones for efficient utilization of applied nutrients.

15.1 Nitrogenous Fertilizers

The tea plant, being a leaf-harvested crop, requires adequate nutrients, particularly macronutrients such as N, P, and K, to fulfil the demands of sprouting leaves and metabolite synthesis (Sun et al. 2019). Tea plants require high N supply, with the fertilization rate ranging from 450 to 1200 kg N ha^{-1} $year^{-1}$ (Tang et al. 2020b), much more than that used in other ecosystems. Yang et al. (2020b) found that glutamate-derived amino acids are the most dynamic in response to various forms of N and N deficiency. The contents of FAA (e.g., theanine and glutamate) involved in N metabolism were higher under high N than under low N, while a high N supply reduced soluble sugars (e.g., glucose) in young shoots, mature leaves, and roots (Liu et al. 2021b). Ma and Ruan (2021) recommended deep side dressing of N fertilization as an ideal method for tea planting to increase N use efficiency and reduce environmental losses in forms of runoff and volatilization. Ma et al. (2021) found that N fertilization quadratically increased the yield and AA content of tea, but quadratically decreased TP content. The cup quality, aroma, liquor color, and dry leaf appearances of green tea were positively correlated and the overall quality of green tea improved as the N application rate increased (Benti et al. 2022).

Agronomic efficiency is an important indicator of tea yield, quality, and its environmental impact. Maximizing tea quality and economic benefits while minimizing its harmful environmental impact is an important strategy for sustainable tea production. Reducing N fertilizer rate helps to increase nitrogen use efficiency (NUE) and further mitigates global warming potential (GWP) and improves economic benefits and FAA content of tea (Tang et al. 2023c).

15.1.1 Nitrogen Foot Print

The N footprint is the total amount of N released into the environment as a result of consumption, and virtual N factors (VNFs) are calculated by dividing the total N loss by the amount of consumed N for a product, indicating how much N is lost to the environment with consumption (Leach et al. 2012). The N footprint provides information to consumers on their contribution to N cycle perturbation and environmental impact (Galloway et al. 2014), which are expected to lead to reduced N creation and loss (Shibata et al. 2017). Adopting N footprint concept, Hirono et al. (2021) evaluated the changes in N flows concerning green tea production and consumption in Japan from 1965 to 2016 by calculating NUE, and VNF. The VNF for green tea in Japan decreased from 54.5 in 1991 to 30.8 in 2016, such decrease resulted in a reduction in N footprint from green tea consumption. The major reasons attributed by them for this decrease were (1) increased NUE in plantations and (2) increased consumption of ready-to-drink tea, matcha, and powdered tea, indicating an increase in the efficiency of N intake from green tea by consumers.

15.2 Phosphatic Fertilizers

The availability of P becomes limited in highly weathered acid soils of the tropics and subtropics, in which free Fe and Al oxides bind native and applied P into forms unavailable to plants. The acidic conditions of the soil required for tea growth favor the formation of insoluble phosphate complexes through binding of phosphate anions with metal cations. Consequently, a large portion of the inorganic phosphates applied as fertilizers is rendered inaccessible to the plant due to rapid binding into complex forms (Lin et al. 2012a; Venkatesan and Murugesan 2004). Under P deficiency, the concentrations of total polyphenols, flavonoids, total FAAs, theanine, and aspartate in green tea decrease (Ye et al. 2021). Although P content in tea leaves is much lower than N and K contents, P fertilizer is also highly associated with improved tea growth (Ding et al. 2017). P is necessary to promote the contents of flavone (Li and Xia 2005), for the formation of tea polyphenols, particularly EGCG (Sun et al. 2019), as well as to affect the metabolism of metabolite products and minerals (Ding et al. 2017).

15.3 Potassic Fertilizers

K has been reported to play a key role in tea yield and the synthesis of amino acids, caffeine and it can greatly enhance water-extractable matter. As an enzyme activator, K can improve tea quality through increased protein concentration in leaves. In turn, protein concentration can affect the content of amino acids and caffeine available in tea. Higher amino acid content improves quality as it affects taste or organoleptic evaluation (Ruan et al. 2013).

The available K content is an index of tea yield. When the K content of soil is low, the yield of tea declines (Li et al. 2021b). It is widely acknowledged that 80 mg kg^{-1} is the minimum available K concentration required to maintain high-producing tea gardens (Zhang et al. 2004). In well-balanced tea gardens, 1 kg of soil contains approximately 10 ~ 20 g total K, which includes 101 ~ 820 mg slow-release K (average, 276 mg kg^{-1}) (Ruan et al. 2013). K supplied as MOP improved the liquor components such as theaflavin, thearubigin, their ratio at higher levels (Venkatesan et al. 2006b).

15.3.1 Potassium and Quality of Tea

As an important nutrient element for metabolism processes in tea, K greatly affects the quality, yield, and adverse circumstances resistance of tea plants (Ruan et al. 2013; Sitienei et al. 2013). Tea produced from K deficient soil lacks the intoxicating flavors of tea grown in well-balanced soils (Zhong et al. 2017). With the use of K fertilizers, the concentration of total FAA and extract from tea infusion rises considerably, while the concentration of polyphenols in autumn tea increases correspondingly (Ruan et al. 2013). Singh and Pathak (2018) opined that K offers abiotic and biotic stress toleration in tea plants and also improves biochemical parameters and organoleptic quality of tea. In young tea plants, K has little direct effect on the content of FAAs and caffeine, but has a positive effect on the content of water extract, catechins, and tea polyphenols (Lei and Yuan 2008; Zhang et al. 2022a). Application of required quantities of K improves the yield and quality of tea. K accelerates metabolism, promotes synthesis of catechins, and strengthens biotic and abiotic resistance by activating and regulating different enzymes (Huang et al. 2022a).

Drought stress limits the movement of soluble ions in the rhizosphere and restricts K+ diffusion in the soil towards plant roots, limiting K+ availability to plants. Retention of K is a key factor associated with alleviating drought-induced damage in tea plants. Zhang et al. (2018b) found that drought stress induced a massive mesophyll K+ efflux and caused a significantly lower K concentration than in non-stressed plants. Zhang et al. (2020f) reported that higher reactive oxygen species scavenging ability and plasma membrane H + -ATPase activity were major factors contributing to the higher K retention in drought tolerant tea plants. K application rate of 90–120 kg ha^{-1} maximized the tea yield and quality (Xi et al. 2023).

In order to develop high-quality tea plantations and address the contradiction between low tea yield and high tea consumption, efforts should be made to enhance the soil-available K content by optimizing the amount of K applied on tea plantations.

15.3.2 Potassium and Magnesium Fertilizers

Balanced nutrition including the nutrients K and Mg is an important measure to improve the productivity of tea. N uptake, NRA, and the accumulation of FAA especially theanine of tea plants were improved by increasing the supply of K or Mg. Mg itself positively influences the levels of amino acids and caffeine in tea leaves, but using fertilizers with both Mg and K considerably raises their contents (Zhang et al. 2020e). It is appropriate to stabilize the ratio of K_2O to MgO at 1:0.15 ~ 1:0.25 (Ruan 2007).

16 Long-Term Use of Fertilizers

Long-term use of mineral fertilizers in tea cultivation can result in problematic soil acidification, cause nutrient imbalance, and increase heavy metal levels. There could be impact on various quality parameters also due to long-term fertilizer application. Chen et al. (2021c) reported that long-term application of N increased the total FFAs and the main amino acids in fresh tea leaves, but caused decrease in the catechins, thereby reducing the phenol–ammonia ratio, which is ideal for the formation of tea liquor freshness (Fig. 7). However, there was decrease in the benzyl alcohol and 2-phenylethanol, and also downregulated the expression of *CsNES* and *CsTSB*, due to long-term N usage, which affected the formation of floral and fruity aroma compounds in fresh tea leaves. Considering the effects on taste, aroma compounds, growth, and yield, the optimal amount of N fertilizer to balance tea yield and quality was found to be 225 kg hm^{-2}.

Jahan et al. (2022) observed that continuous application of traditional fertilizers promoted the deficiency of P and mineral-based cations, such as K^{+}, Ca_2+, and Mg_2+, but increased Mn_2+ and Al_3+ toxicity in soils, which reduced levels of total flavonoids, polyphenols, soluble solids, various vitamins, and thereby the overall quality of tea. Besides, there was increase in free-radical scavenging activity (DPPH), caffeine, and tannin concentration in tea leaves, which indicated that tea plants were potentially being stressed. The schematic representation of the effects of long-term application of traditional fertilization on tea garden soil and tea leaf quality is presented in Fig. 8.

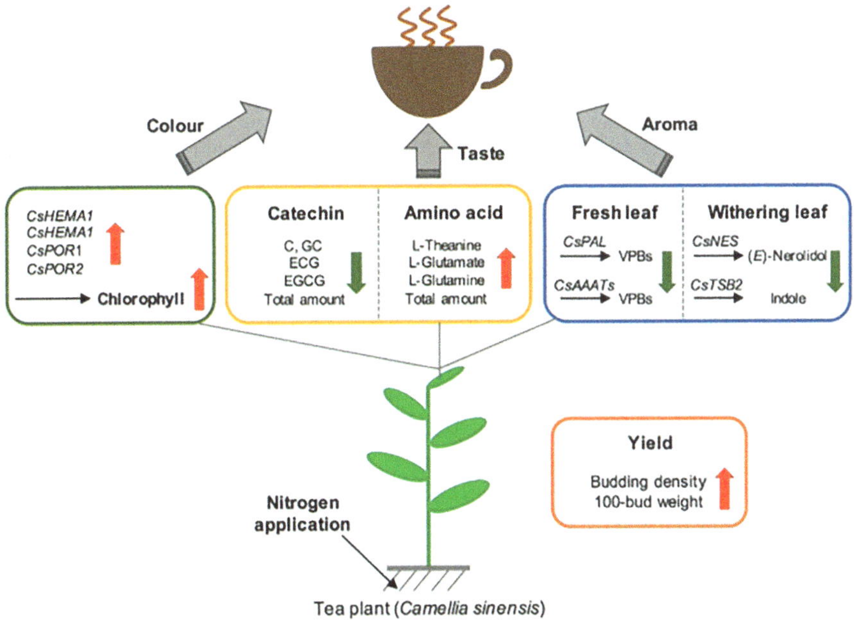

Fig. 7 Effect of long-term *N* application on tea yield and quality. Red arrows represent the increasing tendency with the increase of N application, while green arrows represent the opposite tendency. Black arrows represent the N application (with permission from Chen et al. 2021c), pp. 11, 146. https://doi.org/10.3390/metabo11030146). *VPBs* Volatile phenyl-propanoids/benzoids, *EGCG* epigallocatechin gallate, *ECG* epicatechin gallate, *GC* catechin gallate, *C* catechin, *HEMA* glutamyl-tRNA reductase, *POR* protochlorophyllide oxidoreductase, *PAL* phenylalanine lyase, *AAAT* aromatic amino acid aminotransferase, *NES* (E)-nerolidol synthase, *TSB* tryptophan synthase β-subunit

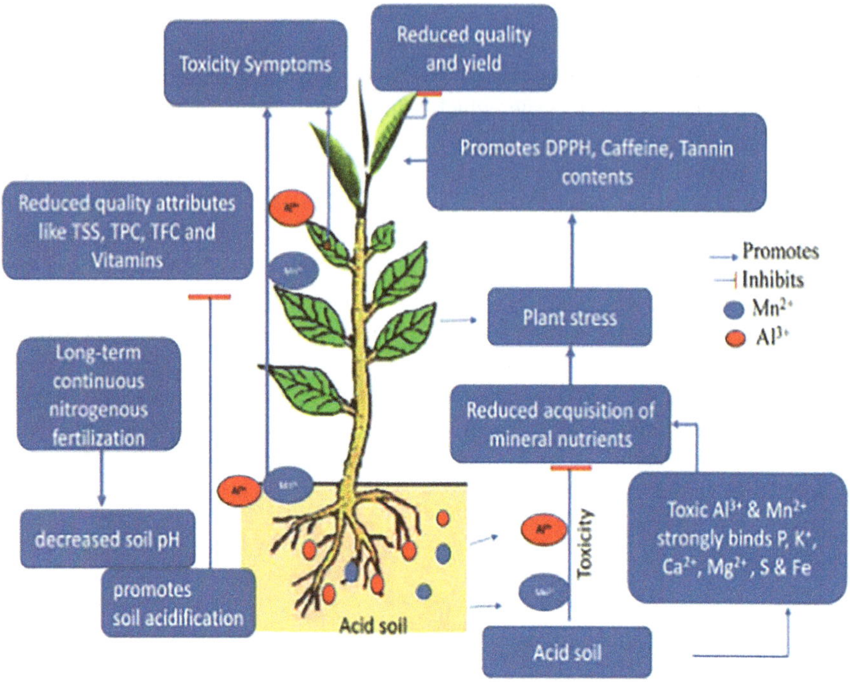

Fig. 8 Schematic representation of the effects of long-term application of traditional fertilization on tea garden soil and tea leaf quality (with permission from Jahan et al. (2022), p. 2128. https://doi.org/10.3390/agronomy12092128) (from Open access journal)

Non-supply of nutrients through fertilizer application for a long period can severely decrease yield, and affect tea quality (Zhou et al. 2022). There was down-regulation of protochlorophyllide oxidoreductase-encoding gene expression, which adversely affected chlorophyll biosynthesis, ultimately leading to leaf etiolation. Various quality attributes (taste and many aroma compounds such as green leaf volatile, linalool and its oxides) are also affected.

17 Nutrient Budgeting and Soil Health

It could be seen that, in many cases, there is excess application of chemical fertilizers for increasing tea yield, but such an approach tells up on the soil health and tea quality. Therefore, nutrient budgeting considering the soil nutrient status, release pattern, plant uptake, targeted yield, etc. should be attempted for sustainable tea production.

Chen et al. (2016) found that approximately 30% of the N was lost through storm runoff, 52% was stored in the soil mass, and only 18% of applied N were utilized by tea plants. In case of P, the respective figures were 5%, 90%, and 5%. Thus, based on nutrient budgeting, they recommended to reduce the current rate of fertilizers into half for improving their use efficiency and controlling N pollution. ul Haq and Boz (2019) suggested that the tea farmers in Rize Province Turkey can reduce their inputs use by 43% without compromising yield level by applying fertilizer in the root zone or mixed with soil, terracing of farmland, and performing of soil tests. The farmers could reduce their average fertilizer use from 115.45 to 66.45 kg (ul Haq et al. 2020) by adopting nutrient budget system.

The NES relies on understanding the relationship between yield response (the difference between the yield in the treatment receiving NPK and that with no fertilizer) and agronomic efficiency (the yield increase per unit of N, P, and K applied). The analysis by Tang et al. (2021b) indicated that the use of the NE system could contribute to a 14% increase in yields compared with conventional practices, as well as increase tea quality (increased total amino acid concentration). They also estimated that the NE system was able to reduce N_2O emissions by approximately 28% compared with conventional practices. Thus, fertilizer recommendations based on the NE system can increase economic benefits and reduce environmental risks by avoiding excessive fertilizer application for maintaining sustainable agriculture. Xie et al. (2022) opined that substituting chemical fertilizers with around 50%of organic manures can be the best strategy to balance tea yield, tea quality, and environmental protection and as a measure with the higher potential for reduction in chemical fertilizer usage.

18 Integrated Nutrient Management and Soil Health

The world has been shifting to environmentally safe and economically viable alternatives for crop production. Due to the wide-ranging adverse environmental impacts of usage of chemical fertilizers in crop production, there is a growing need to identify and apply alternative nutrient management practices, along with changing the application rates of conventional fertilizers. An appropriate combination of mineral fertilizers, organic manures, crop residues and compost of N-fixing crops, which depends on the system of land use and ecological, social, and economic conditions etc., is necessary (Gebrewold 2018). Integrated application of inorganic fertilizers and organic manures with micronutrients will help in increasing the availability of nutrients, crop yield, sustaining and restoring soil fertility in its available nutrients, and maintaining major physico-chemical characteristics of the soil.

Though the use of chemical fertilizers containing N, P, and K has increased the output of agricultural products, the strategy for improving agricultural production in developing countries should also consider supplementing the nutrients through various other processes. These options include using organic or control release fertilizers, a combination of organic and inorganic fertilizers, intercropping with legumes, adding nitrification inhibitors, biofertilizers, nutrient solubilizers, soil amendments, and changing farm-scale management practices (Rebello et al. 2022) (Fig. 9).

As in the case of any crop plant, there is a need for supply of fertilizer nutrients to plantation crops for their growth and nutrition management. Since the volume of green leaves of tea is high, the demand of and supply of total nutrients for tea should

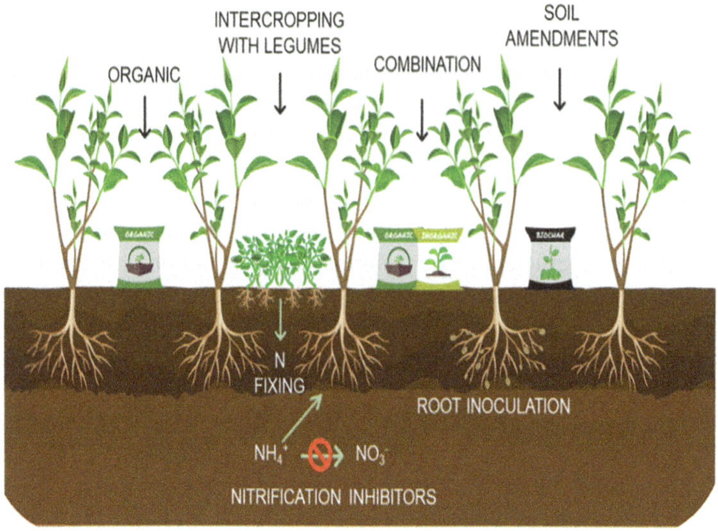

Fig. 9 Integrating alternative N management practices for tea plantations (with permission from Rebello et al. (2022), pp. 43–57, https://doi.org/10.3390/nitrogen3010003)

also be high. At present, tea plantation has mainly been dependent on chemical fertilizers, which has resulted in a decline in SOM content and soil-borne beneficial microorganisms, affecting the fertility status and soil health. Besides the problems of acidification, leaching loss, run off etc., the chemical residues that remain in the plant and soil are added threats to the soil and plant health as well as environment. These adverse conditions could be managed without compromising economic yield through adoption of alternative strategies of growth and nutrition management with biological components.

The agronomic efficiency, apparent nutrient recovery, and partial factor productivity of N, P, and K was improved with the application of Rutuba (an organic fertilizer with major, secondary and micronutrients) and NPK each @ 625 kgha^{-1}. Increase in plant growth, yields of fresh green leaf as well as made tea was also observed (Bitange et al. 2018). A combination of chicken manure and legume straw with chemical fertilizers increased total organic C and total N storage in all aggregate fractions relative to that of control (Wang et al. 2018d). Application of vermicompost (1 kg plant^{-1}) and NPK (10: 5: 10) @ 200 kg ha^{-1} year^{-1} produced the maximum tea yield (Anjaneyulu and Bhattacharjee 2019). Use of young tea dose (YTD) (a mixture of NPK recommended for young tea@ 2: 1: 3, vermicompost, pig litter, azolla biomass, and urea containing N @ 1.70, 1.10, 2.78 and 46.00%, respectively) gave better results (Baruah et al. 2019). Adoption of integrated nutrient management (INM) strategy, though increased the average yield only by 9%, in contrast there was 42% reduction of average nutrient input. The quality of tea was also maintained or improved as indicated by slightly increased contents of amino acids and tea polyphenols (Ruan et al. 2020). Results on INM practices for tea production using various chemical and organic manures such as biogas slurry (Zhang et al. 2020h); colza cake or livestock waste compost (Huang et al. 2022b); OM70 (an organic manure preparation) (Li et al. 2022); rapeseed cake (Ma et al. 2022a; Manzoor et al. 2022); cow dung (Saha et al. 2022), wheat straw compost (Fu et al. 2023) etc., are available.

19 Forms of Fertilizers in Tea

19.1 Controlled Release Fertilizers

Excessive application of fertilizers, very often,leads to the loss of a high amount of nutrients and thereby low fertilizer utilization, which severely restricts crop productivity. Establishing better fertilizer usage practices can mitigate these adverse effects of excessive fertilizer use in agricultural practices. Controlled Release Fertilizers (CRFs) which are sometimes referred to as slow-release fertilizers (SRFs) have been developed as an approach towards solving the issues that arise from the application of conventional fertilizers (Ni et al. 2010). CRFs are designed with the aim to release the active ingredient at a controlled rate and to support by providing adequate nutrients for crop production (Ni et al. 2011; Raguraj et al. 2020). SRF in tea fields could reduce N and P losses through runoff while maintaining tea yields (Liu et al. 2012). Wu et al. (2018) reported high N use efficiency with CRFs compared with the conventional urea application when expressed as a function on tea yield. By using urea–hydroxyapatite nanohybrid fertilizer, Raguraj et al. (2020) realized an increase in yield by 10–17% compared with standard fertilizer treatments. Nanohybrid fertilizer also reduced the number of fertilizer applications and the amount used by 50%.Its effect was more pronounced during unfavorable climatic conditions.

19.2 Blended Fertilizers

Tea fertilizers is available commercially in many physical and chemical forms, and each has its own uses and limitations, which provide the basis for selecting the best fertilizer for specific crops or location (Sitienei et al. 2019). The compound granular fertilizers contain all the plant nutrients specified in each granule, and the blended granular fertilizer which is a mixture of dry fertilizer granules or pills or chips, have no chemical reaction (Barnes and Fortune 2014). The range of NPK formulae already familiar to the Kenyan tea industry (25:5:5) contain N in two forms, ammonium and nitrate, often in the ratio 2:1 (Sitienei et al. 2019). However, they cannot be easily manipulated for specific soils and tea clones, and hence supplementing the NPK fertilizers with Ca and Mg and micronutrients resulted in improved soil quality, stabilized soil acidity, and better nutrients uptake hence environmental quality (Sitienei et al. 2018a). The blended fertilizer of NPKS 25:5:5:4 + 9Ca + 2.62 Mg +

trace elements was the most efficient and consistent fertilizer in production and economic returns of tea cultivation (Sitienei et al. 2018b). Use of such blended fertilizer had higher N content in the two leaves and a bud (Sitienei et al. 2018c). Okemwa and Kirwa (2020) noticed the highest percentage accumulation of elements with application of blended fertilizer (NPK 29:5:5 formulation) @75 kg N ha^{-1} $year^{-1}$ compared to standard compounded fertilizer (NPK 25:5:5 formulation), which required to be applied @ 225 kg N ha^{-1} $year^{-1}$.

19.3 Nitrogen Form of Fertilizers

N can be applied in different inorganic forms, a few examples are urea and calcium superphosphate (Liu et al. 2012), urea (Ma et al. 2021), and NPK fertilizers (Ji et al. 2018). When the consideration is for more yield and higher quality tea, improper fertilizer management, an excessive fertilization of different fertilizers decreases tea quality (Qiao et al. 2018). Tea is widely considered to be acid-tolerant and prefers NH_4+ – rich environments, rather than NO_3 – -rich ones (Yang et al. 2018a), as tea roots show a high capacity for NH4 + assimilation, can increase key enzyme activities, and improve carbohydrate status (Ruan et al. 2007b). When NH_4+ – based fertilizer is applied to the soil, it is rapidly nitrated, and the utilization rate of crops for NH_4+ -N decreases. Nitrification will also lead to soil acidification. The proper pH for tea growth is about 4.2 (Yamashita et al. 2020), and, when the pH is less than 3.8, or higher than 5.0, it is not conducive for tea plant growth. NH_4 + –based fertilizers, such as ammonium bicarbonate and urea, can help prevent rapid nitrification, and improve the N use efficiency and yield of tea.

The translocation of ^{15}N from mature leaves to new shoots was regulated by the genes involved in various processes such as autophagy, protein degradation, amino acid transformation, and transport. Based on physiological effects of NH_4+ and NO_3– on tea, Fan et al. (2020b) have confirmed that tea plants prefer NH_4+ as the dominant N source. Tang et al. (2020b) found the absorption of ^{15}N from $^{15}NH_4$ + to be more rapid than that from $^{15}NO_3$–, as there was a more efficient expression pattern of NH_4+ transporters compared with that of NO_3– transporters. Yang et al. (2018b) demonstrated the preference of tea plants for NH_4^+ at the gene level. The characteristics of the FAAs differed under the different N forms (Ruan et al. 2019a) and the ranking was NH_4+ > split NH_4+/split NO_3– > NO_3–. They concluded that NH_4+ improved the accumulations of FAAs in tea leaves, which might be attributed to the root development. Wang et al.(2021a) suggested that NO_3– as the N source to promote biosynthesis of catechins which enriched the C pool, whereas NH_4+ supply was more conducive to N metabolism, indicating that different N sources could affect the C/N balance. Wang et al. (2022b) opined that NH_4+ could be regarded as a signaling molecule with the response of catechins being higher than that of amino acids. The overall results of the study may lead to improvements in N efficiency of tea plants.

Zhang et al. (2018c) reported that *CsAMT1.2* might play an important role in ammonium uptake from soils in tea plant roots. Zhang et al. (2020b) suggested that *CsAMT1.2* was a key gene controlling N uptake in tea plants and might play a vital role in promoting NH_4+ uptake from the environment in tea roots, while Zhang et al. (2021b) inferred that *CsNRT2.4* was responsible. Zhang et al. (2022c) identified five *CsAMT2* subfamily genes in tea plant genomes and concluded that *CsAMT2.2* and *CsAMT2.3* might play roles in controlling the NH_4+ uptake from the soil to the roots. Recently Xie et al. (2023) presented an extensive review on the novel aspects of regulation of N responses in the tea plants.

19.4 Potassium Form of Fertilizers

The positive response of tea in terms of both yield and quality to K application has been reported in many studies, especially in soils containing small amounts of exchangeable K (Ruan et al. 2013). Lei (2003) opined that K fertilizers can not only balance different forms of K soils, but also work with N and P to produce high quantity of good tea. Although there are two commonly available forms of K-fertilizer (SOP and MOP) (Singh and Pathak 2018), the responses of tea yield and quality to MOP have been variable (Venkatesan et al. 2005). The ideal N:K ratios for maximum productivity and biochemical parameters would be 1:0.83 or 1:0.62 if the source of K is MOP, while it is 1:0.21 or 1:0.42 for SOP. However, MOP tentatively depressed NRA and, hence, the accumulation of FAAs, while SOP increased the contents of these quality promoting constituents of tea leaves. Similar detrimental effects of Cl from MOP on N uptake and concentration of theanine in young shoots were observed by Ruan et al. (2007a). However, within the safe dosage range, either MOP or SOP can effectively increase yield and quality of tea (Ruan et al. 2013) and the latter can increase the proportion of quality tea (Zhang et al. 2004). Venkatesan et al. (2005) reported the possibility of using lower dose of SOP to improve the quality parameters of tea. In addition, soil S availability was able to contribute to optimize tea production (Karak et al. 2015).

19.5 Polyhalite Form of Fertilizers

Polyhalite ($K_2SO_4 \cdot MgSO_4 \cdot 2CaSO_4 \cdot 2H_2O$) is a marine evaporite mineral produced by successive marine evaporation events throughout history (Kemp et al. 2016), and shows huge potential as K fertilizer due to its very low quantities of Cl but rich in K, Ca, and Mg (Ogorodova et al. 2016; Albadarin et al. 2017). POLY4 is polyhalite powder having a composition of 14% K_2O, 6% MgO, 17% CaO, and 19% S. Application of SOP and POLY4 increased soil available K, P, and S contents, exchangeable Ca and Mg as well as tea yields and economic benefits. POLY4 improved the yield by 15.1% SOP during the third year (Zhou et al. 2019).

20 Organic Manuring in Tea Plantation

20.1 Use of Organic Manures

To obtain high yield in tea cultivation, application of chemical fertilizers is generally being practiced. However, a large quantity of chemical fertilizer application in a long-term monoculture tea plantation will inevitably lead to soil acidification and a decline in fertility, leading to poor growth and yield of tea (Xie et al. 2020a). Therefore, it is necessary to restore soil fertility and sustain the development of tea planting by resorting to organic ways (Singh et al. 2014). Organic fertilizer substitution can both reduce nutrient losses and improve nutrient utilization rates and crop quality (Dimkpa et al. 2020; Marcinczyk and Oleszczuk 2022). Partial substitution of chemical fertilizers by organic amendments is essential for improving the soil quality without yield loss. The use of organic manures helps in alleviate soil acidification, and result in increased plant yields (Li et al. 2018), shape the microbial composition and recruit beneficial bacteria into the rhizosphere of tea, leading to improved tea quality and reduced heavy metals content in rhizosphere soil and tea leaves (Lin et al. 2019).

Various kinds of organic manures are being used in tea plantation. Some of them include poultry manure (Khan et al. 2013); farmyard manure (Negi and Bisht 2017); soya bean cake and livestock manures (He et al. 2019); medium rapeseed cake along with green manure (Fu et al. 2021); cow manure (Sun et al. 2021); potash derived from molasses (Raj kumar 2021a, b); and wood ash (Raj kumar et al. (2023). The wood ash treatment should be limited to a maximum of 500 kg ha^{-1} $year^{-1}$ since it might cause soil pH and electrical conductivity to rise over acceptable values, which can be harmful to soil and tea plant health.

20.2 Organic Manure and Soil Properties

Organic manure amendment is a key factor in determining the soil properties and tea productivity. Application of organic manures improves the diversity of soil bacterial community, SOM, as well as various soil nutrients (Qiu et al. 2014); considerably reduces total N losses and leaching into waters (Liu et al. 2016); elevation of soil pH, available K, SOC, total N, and microbial biomass C and N (Ji et al. 2018); mitigates soil acidification and nutrient deficiency in tea plantations, increases SOM and NH_4-N contents as well as reduces concentrations of NH_4 + and NO_3 − N in runoff compared with traditional fertilizers (Xie et al. 2019); short-term disturbance on pH, EC, and SOM (Chen et al. 2021d); increases N_2O emissions by an average of 17% with lower NO (Han et al. 2021); and increases pH values, total N, available P, and K as well as the concentration of total and exchangeable Ca_2+ and Mg_2+ in soils (Li et al. 2022). Replacement of chemical fertilizer with either 25 or 50% enhanced soil total N, P, and K (Chen et al. 2022a). The increase in the soil available nutrients, tea yield, water extracts, tea polyphenols, amino acids, and caffeine was with substitution with 50%.

Various types of organic manures exert different influence on properties of tea soils. Use of rapeseed cake and alkaline slag reduced soil exchangeable acidity and Al saturation and increased base saturation (Lei et al. 2013); both urea and oil cake stimulated annual N_2O and NO emissions (Yao et al. 2015); combined application of wood ash and refuse tea increased soil N, K, and Ca contents and tea yield (Gunarathne and Arachchi 2016); organic rapeseed cake increased the pH, total SOM, NH_4–N and NO_3–N in soils (Xie et al. 2020a; Xie et al. 2021); low dosage of sheep manures increased tea yield, quality indicators, and activity of various enzymes (Ye et al. 2023b).

20.3 Use of Organic Manures and Soil Microbes

Use of organic manures is beneficial for various microorganisms in tea plantation: organic manures made from pig and chicken to increase in the microbial communities (Wang et al. 2014d); improve the soil fungal to bacterial ratio as well as soil enzyme activity (Sun et al. 2017); increase bacterial community and diversity, accumulation of soil carbon, and nutrients (Ji et al. 2018); improve the ecosystem functions (Gu et al. 2019); reduce heavy metal content in rhizosphere soil and tea leaves (Lin et al. 2019); as well as increase the amino acids of tea (Deepshikha 2020). According to Ji et al. (2020b), the soil fungal network complexity could be a new indicator for soil quality assessment with organic substitution for soil quality improvement.

20.4 Organic Manures and Tea Quality

The quality characteristics of tea include taste, which is mainly related to amino acids, tea polyphenols, and caffeine (Hemmati et al. 2021). The content of quality-related metabolites in tea is affected by plant growth and development, which depend on various nutrient elements in the soil (Zhou et al. 2022). Tea cultivated with bio-organic manures has superior color and taste compared to tea applied with chemical fertilizers (Lin et al. 2010). Proper replacement with organic manures can effectively promote the yield and quality of teas (Gu et al. 2019; Liu et al. 2017). With use of organic manure, Xie et al. (2019) reported reduction in the ratio of polyphenols to amino acids, broadly indicative of a superior aroma and taste in green tea. An increase in caffeine levels was also reported.

20.5 Use of Biochar in Tea Plantation

Low-temperature biochar of crop residues have potential as amendment to ameliorate acidic tea garden soils. Many benefits with the use of biochar prepared using different sources in tea plantations are reported: increase in soil pH from 3.33 to 3.63 using rice husk biochar (Wang et al. 2014a); increase in exchangeable cations and reduced Al saturation of tea soils (Wang et al. (2014b); increase in soil pH, total C and C/N (Oo et al. 2018); reduction in soil N_2O emission (Ji et al. 2020a); reduction in NH_4^+-N, NO_3^--N and total N leaching in tea soils using bamboo biochar (Chen et al. (2021e); lowering of soil N_2O emissions (Han et al. 2021); reduce yield-scale N_2O emissions and increase the annual yield (Wu et al. 2021a); increase soil pH, SOM and available nutrients, and change in bacterial and fungal diversity and community composition (Yang et al. 2021); improve tea growth, increase the plant P, K, and Mg concentrations, which had positive relationships with the above ground biomass, leaf area, and rate of photosynthesis (Yan et al. 2021b); increase in tea production using biochar prepared from tea processing waste, mainly the twigs (Sipayung et al. 2022); reduce concentration of As, Cd, and Cr in made tea with tea pruning litter biochar (Borgohain et al. 2020); enhance chlorophyll content, photosynthetic rate, stomatal conductance, transpiration rate, and water use efficiency (Borgohain et al. 2023); increase soil pH, total N and total C contents, improve tea quality parameters and diversity of soil bacterial and fungal communities (Guo et al. 2023); favored N availability and alter microbial community structure with rice straw biochar along with pruned tea plant litter (Luo et al. 2023); gradual increase in total content of various nutrients (Sarmah et al. 2023); reduce water-soluble F contents in tea leaves, and contribute to improving tea polyphenols and caffeine (Wang et al. 2023a); increase soil pH, total carbon and various major and secondary soil nutrients (Zou et al. 2023). Among the various biochar application methods, homogenized application was found to be optimal for enhancing both tea plant growth and soil fertility.

21 Organic Cultivation of Tea

Tea is an important beverage consumed worldwide. It is a source of important secondary metabolites like catechins monoterpenoids, carotenoids etc. Catechins are responsible for the beneficial health effects of tea. The global organic tea market size was USD 820.8 million in 2019 and it is projected to reach USD 1918.7 million by 2027, exhibiting a CAGR of 11.2% in the near future. Among the benefits of organic technologies in tea production are higher SOM, and N, lower fossil energy inputs, yields similar to those of conventional systems, and conservation of soil moisture and water resources (especially advantageous under drought conditions). Conventional agriculture can be made more sustainable and ecologically sound by adopting some traditional organic farming technologies.

Organic tea cultivation was first started in 1983 in Sri Lanka, and thus the first organic tea production was realized. In the developing countries, consumption of organic tea rich in melatonin is on the rise owing to the increase in lifestyle related diseases. The area allocated to tea cultivation in the world is 2.9 million ha, 4.8 million tons of product is obtained from these areas, while the area allocated for organic tea cultivation is 5000 ha and the production amounts to around 4–5 thousand tons (Caykur 2016). Organic tea production has increased very rapidly in China in the recent years. Further, the conversion of conventional synthetic N fertilization to organic fertilizer practice in tea plantations has been identified as a feasible measure in the aspects of promoting soil carbon sequestration and ameliorating soil pH (Han et al. 2018a).

Organically managed tea agroecosystems can enhance soil carbon sequestration through increasing SOM levels. Organic tea soils also have higher levels of soil pH, total N content and soil microbial biomass, C, N, P and their ratios to total organic C, N and P, respectively (Han et al. 2018a). This translates into better plant nutrient content, increased water retention capacity and better soil structure and thus leads to higher yields and greater soil resilience (FAO 2009). Organically managed tea systems are further associated with higher levels of biodiversity, including natural predators for pest control (Li et al. 2014; Liu et al. 2015). Plants under organic treatments produced comparable yield as that of inorganic treatment (Saikia et al. 2014). Though pest and disease damages and the imbalance of nutrients, especially shortages of N, decreases tea yields by 20–30%, organic tea fetches 20–50% higher price compared to conventional tea, which help compensating for the loss in yield. Promoting organic tea production is, therefore, not only beneficial to farmers' incomes, but offers significant scope for climate change adaptation.

Though tea yields reported in organically managed production systems are typically 10–20% lower than in traditional agricultural ecosystems (Bisen and Singh 2012; Das et al. 2016; Han et al. 2018a), in addition to yield, the quality of tea is equally important for consumer preferences (Ahmed et al. 2014). Hallmann et al. (2007) and Kazimierczak et al. (2013) found organic teas to have higher concentration of polyphenolic flavonol and tannin than conventional teas. Darjeeling teas are produced in the 87 tea gardens covering 18,000 ha of land. According to Singh et al. (2016), >60% tea gardens have been converted to organic. However, Darjeeling tea contributes only 1% of the total tea produced in India with the annual production of 10.09 million kg of orthodox tea both black and green tea including a small amount of white tea.

21.1 Soil Property Under Organic Tea Cultivation

Organic farming system enables the sustainability of soil health through its influence on physico-chemical and biological properties. Some of the beneficial effects of organic tea cultivation due to organic cultivation are increase in pH and plant growth (Chong et al. 2008); reduction in the contents of Cd, Pb, and As in tea leaves and increase in amino acids content of tea (Lin et al. 2019); high contents of organic C and total N contents (Han et al. 2013); promote soil C sequestration and microbial biomass size and activities (Li et al. 2014); plants becoming ready for harvest at an early growth stage (Chauhan et al. 2016; Islam et al. 2017); improvement in soil nutrient status (Adhikary et al. 2019); increase in soil microbial biomass C and N (Gui et al. 2021); and the organic system to be less acidic (Bhuyan et al. 2018, 2022).

21.2 Quality of Organic Tea

The high antioxidant activity of organic tea makes it a healthy drink. Since the aroma and taste of tea products are key factors determining the quality grade of tea and its market price (Qin et al. 2013; Su et al. 2021), a significant increase in these quality indicators as a result of organic tea management practices would enhance economic benefits for the adopters. Bagchi et al. (2015) reported higher contents of secondary phenolics compounds as well as antioxidant content of tea extract from organically grown tea. Das et al. (2016) observed that organic tea fields showed higher values for leaf total catechin, EGCG, and ECG, contents although there was up to 22% reduction in tea yield. Bisen et al. (2016) found the biosynthesis of polyphenols was greatly influenced under the organic cultivation of tea. Sumi and Kabir (2018) reported that the taste, natural content, and the nutrient value of organic tea make it a popular choice for health-conscious customers. Han et al. (2018a) reported water extracts of organic tea had higher concentrations of key catechins linked to tea quality. Piyasena and Hettiarachchi (2023) also observed the organic tea to be richer in catechin, epigallocatechin, epigallocatechin-3-gallate, and other derivatives. The weights of tea bud were high in vermicompost applied fields (Shiwakoti et al. 2023). There was increase in tea yield (> 600 kg ha^{-1}) and tea quality parameters (total polyphenols, total chlorophyll, and carotenoids) with the application of optimized dose of organic inputs in Darjeeling tea production (Singh et al. 2023).

22 Management of Tea Plantations

22.1 Mono Cropping and Soil Health

Soil degradation and frequent yield decline occur after long-term continuous ratoon cropping of tea plants, an event regarded as a continuous cropping obstacle. The worsening of physicochemical properties, the accumulation of autotoxic substances, and the imbalance of microbial flora are considered the main factors contributing to this phenomenon (Zhou and Wu 2012). The abundance of bacteria and fungi decreased with the years of continuous cropping; moreover, *Acidobacteria, Actinobacteria, Chloroflexi,* and *Ascomycota* were more predominant in younger tea gardens, whereas *Proteobacteria* and *Bacteroidetes* were less abundant in the older ones (Wang 2014). Li et al. (2016c) and Zhao et al. (2012) also reported change in the bacterial community structure, diversity, and decrease in potentially beneficial plant-associated bacteria due to continuous cropping.

22.2 Best Management Practices and Soil Health

Increasing fertilizer inputs in tea fields, especially in hilly areas, poses serious potential risks of surface water contamination. Best management practices (BMPs) have been extensively implemented to control nutrient loads from non-point sources. However, little research has been conducted to evaluate the effects of dynamic

BMPs on N load reductions in tea fields with different slope gradients. Generally, tea planted on steep-sloping land leads to great losses of N, and tea fields on slopes exceeding 15° should be prioritized for BMP implementation as tea fields planted here tend to have the highest N losses. Wang et al. (2018e) conducted a study in subtropical tea fields with three BMP scenarios, viz., avoiding fertilizing before rain, contour planting and applying slow-release fertilizers, simulated to model the N load reduction using the Soil and Water Assessment Tool (SWAT). All the BMP scenarios significantly impacted N losses reduction in tea fields and that their effects vary among different slopes and months. Compared with the initial situation, the N losses rate could be reduced by approximately 24%, 28%, and 66% for three scenarios, respectively.

22.3 Impact of Residue Recycling in Tea Plantation

In conventional tea plantations, a large quantity of pruned materials is returned to the soil surface, which leads to release of high quantity of polyphenols into the soil. The accumulation of such active allelochemicals in the tea rhizosphere and subsequent shift in beneficial microbes may be the cause of acidification, soil sickness, and regeneration problem. These in turn affects plant growth, development, and result in low yield in long-term monoculture tea plantation (Arafat et al. 2020). Therefore, they suggested the removal of pruned branches from tea gardens to prevent soil sickness and for sustainable tea production. The pruning of tea plants produces large amounts of litter, the degradation of which requires the involvement of a large number of enzymes. Reports of changes in rhizosphere soil enzymes and soil microbial communities of tea plants affected by physical stress from pruning are available (Jiang et al. 2023; Pramanik et al. 2018b). Fan et al. (2017) found that the decomposition pattern of tea leaf litter followed a two-phase decomposition model, with the polyphenol/N ratio effectively regulating the degradation process. Most of the catechins decreased within 2 months. Liu et al. (2022b) suggested *Betula luminifera* as a good choice for inclusion in ecological tea gardens to increase their nutrient return capacity, maintain fertility, and generally promote the ecological development of tea gardens.

22.4 Mulching Tea Fields and Soil Health

Mulching can modify the soil microenvironment and maintain the crop yield. According to the type of materials used, it could be categorized as inorganic and organic mulching. Inorganic mulching was widely used as a low-cost and water-saving measure in agriculture areas that were susceptible to drought. Organic mulch materials are mainly obtained from plant residues, which are beneficial for soil health. The application of organic mulch on soils could not only inhibit weed germination, but also increase soil fertility and provide mineral elements for plants. Besides, it could also increase the biodiversity of soil microecosystem. Different mulches have variable effects on soil physico-chemical characteristics, bacterial and fungal communities and ecosystem functions. Zhang et al. (2020h) found that peanut hull mulching increased the diversity of fungal and bacterial communities than under polyethylene film mulching. Covering the inter row space in tea plantation with *Litsea cubeba* modified the soil microbial structure and diversity (Hao et al. 2022).

22.5 Intercropping and Soil Health in Tea Plantation

Long-term monoculture tea cultivation results in drastic decline in yield, soil degradation, and reduction of beneficial soil bacteria, which are some of the major concerns of many soil ecologists and agriculturists (Arafat et al. 2017; Gu et al. 2019; Li et al. 2016c). The monoculture mode in tea plantations also poses more environmental concerns towards soil erosions (Zhang et al. 2003). Intercropping, an eco-friendly agricultural practice of growing two or more crops together in tea plantations not only alleviates soil acidification and improves soil fertility, enabling improved crop yield and quality (Wang et al. 2014c; Zhang et al. 2017b), but also substantially optimizes and diversifies cropping systems. Tea garden intercropping is a planting mode in which tea is the main crop, and one or more plants and other crops are planted in the same tea garden using the gaps between tea rows to form an artificial vertical composite ecological tea garden. It makes full use of different spaces and soil layers on the same land, so that different plants can be intercropped with tea (Zhang et al. 2014).

Intercropping could change the composition of soil microbial communities and increase the activities of enzymes that promotes nutrients cycling in the soil (Curtright and Tiemann 2021; Farooq et al. 2021); improve the above- and below-ground environment, which is beneficial for tea plant growth, increase yield, and produce high tea quality (Lei et al. 2022); improve the abundance of C cycling related functional genes mediated by soil properties, and increase resistance to drying-rewetting disturbance (Wang et al. 2022c); and improve the soil-ecosystem multifunctionality by altering the soil environment to benefit nutrient absorption (Wang et al. 2023b). As one of the main measures of habitat management, reasonable intercropping of functional plants in tea garden can shade tea bushes and keep them warm, conserve water and soil, increase fertility and promote growth, as well as maintain micro-habitat stability (Fan et al. 2022b). If the tea has an incomplete canopy cover, intercrops can help to reduce soil erosion and run-off (Duan et al. 2019). Feng and Sunderland (2023) opined that the benefits of intercropping tea plantations exceeded those of monocropping tea plantations in terms of soil ecosystem service functions, such as water retention capacity, mineral contents, effects on energy transformation, and regulating environmental conditions.

Leguminous plants are the most popular intercropping plants used in tea gardens, as they can fix atmospheric N (Farooq et al. 2021). Intercropping of tea plants with legumes has been applied in many plantations, reducing inter-specific competition by enhancing complementarity/facilitation processes, thereby increasing tea yield (Sedaghathoor and Janatpoor 2012), changing physicochemical composition of tea leaves (Duan et al. 2021), and improving nutrient availability and enzyme activity in the soil (Brooker et al. 2015). Soybean-tea intercropping could regulate the amino acid metabolism and secondary metabolism of tea leaves and improve the total N in the soil of a tea garden (Duan et al. 2019, 2021; Huang et al. 2022; Sun et al. 2022a). Peanut (*Arachis hypogea*) could improve SOC and soil nutrient availability and affect soil enzymatic responses at different depths, especially increasing the activity of acid phosphatase, protease, and sucrose (Farooq et al. 2021).

22.5.1 Soil Nutrients in Relation to Intercropping

Most intercropping patterns improve microclimate, and soil physical properties, in turn would improve crop yield and quality. Intercropping could also change the above- and below-ground environment in which crops grow, so they can affect crop growth, metabolism, and development (Duan et al. 2019; Wu et al. 2021b; Zhang et al. 2021c). Intercropping could change the composition of soil microbial communities and increase the activities of some enzymes that promotes nutrient cycling in the soil (Curtright and Tiemann 2021; Farooq et al. 2021). The above-ground part mainly represents the microclimate, such as temperature, humidity, and light intensity. The belowground part, i.e., soil environment, defines the soil nutrients, heavy metals, microorganisms, and enzyme activities (Ma et al. 2017; Wen et al. 2020b; Zhang et al. 2020g).

Intercropping improves soil fertility and alleviates soil acidification; thus, it could reduce the input of fertilizers (Ma et al. 2017). Intercropping could improve soil nutrients in tea plantations by increasing topsoil and subsoil organic matter, as well as promoting the release and recycling of soil nutrients (Ma et al. 2022b). Many available nutrients in soil such as total and available N, available K, total and available P, and SOM are positively affected by intercropping (Ma et al. 2017). Intercropping young tea plants with soybean improved the soil NH_4+ during the flowering period of the soybean (Duan et al. 2019). Intercropping changed soil microbial abundance and enzyme activities (Bai et al. 2022; Farooq et al. 2021).

Available P and K, as well as SOM increased under Osmanthus-michelia-tea, peanut-tea, walnut-tea and *Vulpia myuros*-tea intercropping patterns (Farooq et al. 2021; Zhang et al. 2020c; Zhang et al. 2021d). However, contents of total N, P, and K didn't show much difference. The magnitude of soil nutrient changes in most intercropping tea plantations increased with soil depth (Farooq et al. 2021; Ma et al. 2017). Groundnut–tea intercropping enhanced the status of soil nutrients and positively impacted soil conservation, and was also beneficial for increasing biological diversity and ecological stability. There was positive impact on SOC, soil nutrient availability, and soil enzymatic responses at different soil depths (Farooq et al. (2021). Sorghum-tea intercropping reduced N_2O emissions and increased the available N content in soil (Lu et al. 2019; Zhang et al. 2020c). The highest available N, P, and K contents followed the order: tea- albizzia plantation>tea-arecanut plantation>tea plantation> fallow land (Alom et al. 2020). The CEC, soil electrical conductivity, organic C content, and soil microbial population were also found to be the highest in tea-albizzia cropping system indicating its superiority over other cropping systems. Inclusion of soybean as an intercrop in tea system increased SOM, total N, tea quality indices, and expression of the Camellia sinensis glutamine synthetase gene (*CsGS*) (Huang et al. 2022c). Though nutrient competition for P and K in soil existed between soybeans and tea plants, intercropped soybean boosted the uptake of N in tea plants and enhanced the growth of young tea shoots (Sun et al. 2022a).

22.5.2 Intercropping and Soil Enzyme Activities

Soil enzymes are closely related to soil microorganisms and nutrients cycling and soil health, so they could be indicators of soil health in tea plantation intercropping patterns (Farooq et al. 2021). Intercropping increases the activity of various enzymes, and in the chestnut-tea intercropping tea plantation, the activities of catalase, urease, dehydrogenase, invertase, and polyphenol oxidase were improved (Ma et al. 2017). Intercropping peanuts or walnut in tea plantation also increased the activities of acid phosphatase and protease in soil (Bai et al. 2022; Farooq et al. 2021). Invertase, urease, acid phosphatase and peroxidase promote C, P and N cycling in soil. The soil enzyme activities are different in response to different intercropping patterns. The activities of protease, sucrase, and acid phosphatase increased in peanut-tea intercropping patterns, while the catalase and urease activities were increased in chestnut-tea intercropping patterns (Farooq et al. 2021). Therefore, compared with monoculture, soil N, P, and K content gets increased in peanut-tea, chestnut-tea, and walnut-tea intercropped plantations (Bai et al. 2022; Farooq et al. 2021; Wu et al. 2021b). More importantly, soil enzyme activities were also strongly affected by soil temperature, they are generally most active in spring and early summer (Farooq et al. 2021).

The soil microbial community produces extracellular enzymes that are responsible for degrading plant residues and maintaining nutrient cycles in the soil. Intercropping increased enzyme activities by an average of 13% and the effect of intercropping varies by enzyme category, the type of plant used as the main crop or intercrop, and other experimental and environmental factors (Curtright and Tiemann 2021). The peanut–tea intercropping considerably impacted the soil enzymatic responses at different soil depths (Farooq et al. 2021). Moreover, the soil enzymatic activity was significantly correlated with SOC and total N content across all soil depths and cropping systems. An increase in the soil enzymatic activity in the intercropping system improved the reaction rate at which SOM gets decomposed and nutrients are released into the soil environment.

22.5.3 Soil Microorganisms in Relation to Intercropping

Planting patterns significantly change soil microbial abundance, composition, and diversity. Many studies have shown that the α-diversity of soil bacterial and fungal communities and the β-diversity and abundance of bacterial communities in intercropping patterns are higher compared to monoculture (Shen and Lin 2021a, b; Wang et al. 2022c; Zou et al. 2022). The soil nutrients could change microbial community diversity and abundance in soil, and vice versa. Therefore, soil microorganisms affected by many factors could change soil nutrients.

Tea plantation soil contains a variety of microorganisms, among which bacteria and fungi play an important role in maintaining soil ecological functions. Soil microorganisms could accelerate soil nutrient cycling, litter degradation, and decrease carbon emissions (Wang et al. 2022c). However, soil microbial community structure is extremely fragile, and easily affected by cultivation patterns, soil properties, and plant type (Baquerizo et al. 2016; Zhang et al. 2021d). The most common bacterial taxa at the phylum-level in tea plantation soil are *Acidobacteria, Actinobacteria, Chlorophyte,* and *Proteobacteria* (Shen and Lin 2021a, b), and *Ascomycota* and *Basidiomycota* are the most common fungal taxa at the phylum-level (Wang et al. 2017). Intercropping increases the abundance of soil beneficial microbes such as *Proteobacteria, Firmicutes, Ascomycota,* and *Mortierellomycota* and these beneficial microorganisms promote soil nutrient cycling or increase plant resistance (Ma et al. 2022b; Shen and Lin 2021a, b). With intercropping, soil microorganisms got shifted from oligotrophy to copiotrophy in soybean-tea intercropping patterns (Shen and Lin 2021a, b). Soil nutrients (SOM, available P and K) were positively correlated with beneficial bacterial communities including Proteobacteria, Firmicutes, and Bacteroidetes (Zhang et al. 2021d). In Chinese chestnut- tea intercropping system, the rhizospheric soil was enriched with several bacterial taxa, which permitted the tea plants to obtain sufficient amounts of nutrients from the soil (Wu et al. 2021b).

The interaction in the walnut-tea intercropping increased the bacterial and fungal diversity when compared to monoculture tea cultivation (Bai et al. 2022). Intercropping tea with soybean in summer and smooth vetch (*Vicia villosa*) in winter affected the bacterial communities, decreasing the relative abundance of *Bacillus* but increasing its absolute abundance (Huang et al. 2022d). Sun et al. (2022b) reported increase in the relative abundance of beneficial bacteria with tea-soybean intercropping system, while the relative abundance of denitrifying bacteria was markedly decreased. Intercropping in tea plantation with mung bean (*Vigna radiata*) and adzuki bean (*Vigna ungularis*) increased the relative abundance of *Proteobacteria buta,* but decreased relative abundance of A*ctinobacteria* (Wang et al. 2023c).

22.5.4 Agroecological Tea Management

Soil health is the capacity of soil to provide various ecosystem services. It is assessed by considering all the attributes including soil physico-chemical and biological properties (Ippolito et al. 2021; Williams et al. 2020). Different agricultural management practices can cause short- and lead long-term effects on various soil health properties (Bai et al. 2018). In conventional agriculture, use of intensive agrochemical inputs has been widely known to negatively impact soil health in comparison with conservation and organic farming (Le et al. 2021). Long-term intensive application of agrochemicals under conventional tea cultivation results in a range of serious issues, such as soil health and environmental degradation, reduced tea quality, and human health concerns (Le et al. 2021; Van Ho et al. 2019). Of late, the role of agroecology in restoring soil health, providing sustainable food production and environmental benefits, has been increasingly recognized worldwide (Dumont et al. 2021; FAO 2020; Nicholls and Altieri 2018).

According to FAO (2016), agroecology is the application of natural ecological system processes and concepts for optimizing the interactions between humans, plants, animals, and the environment. It also considers the social aspects to ensure a fair and sustainable food system This system can play vital roles in supporting food production, nutrition improvement, and food security and at the same time restore the ecosystem services and biodiversity, which are key components of sustainable agriculture (Chappell and LaValle 2011). Besides, it provides a practical way for maintaining soil quality quite often depleted by conventional management practices (Altieri et al. 2020). Agroecological practices aim at optimizing agroecological processes, environmental and public health while minimizing social-ecological costs from agricultural activities (Kerr et al. 2021).

For tea farming, numerous studies done in different tea growing countries have indicated the positive impacts of agroecological practices on soil health properties and tea quality indicators, such as the application of organic fertilizers (Gu et al. 2019; Han et al. 2021; Lin et al. 2012b) and organic mulching (Zhang et al. 2020g). Similar positive outcomes have also been recorded from other agroecological practices such as intercropping (Zhang et al. 2017b), agroforestry (Tian et al. 2013), and integrated pest/disease management (Mamun and Ahmed 2011; Shrestha and Thapa 2015). Le et al. (2023) examined the impacts of such management practices on soil health indicators, tea yield and quality, and net income of tea farmers in Northern Vietnam. They found significant enhancement of SOM by 0.8% and soil pH by 0.5 units on average. There was an increase in soil total N by 0.15–0.2% with conventional management based on chemical fertilizer applications. Soils under agroecological management favored higher soil biological parameters.

Agroecological and organic tea farming systems rely on non-chemical inputs such as organic materials and biofertilizers for maintaining crop productivity along with restoration of soil health and mitigating environmental pollution (Gui et al. 2021; Le et al. 2021). However, there is likelihood that these resources may not provide sufficient macro and micronutrients, such as N and P for crops to grow and obtain as high yields as a conventional method that employs the intensive application of synthetic fertilizers, especially during the transition period (the first 3–5 years since the conversion from conventional to organic farming management) (Doanh

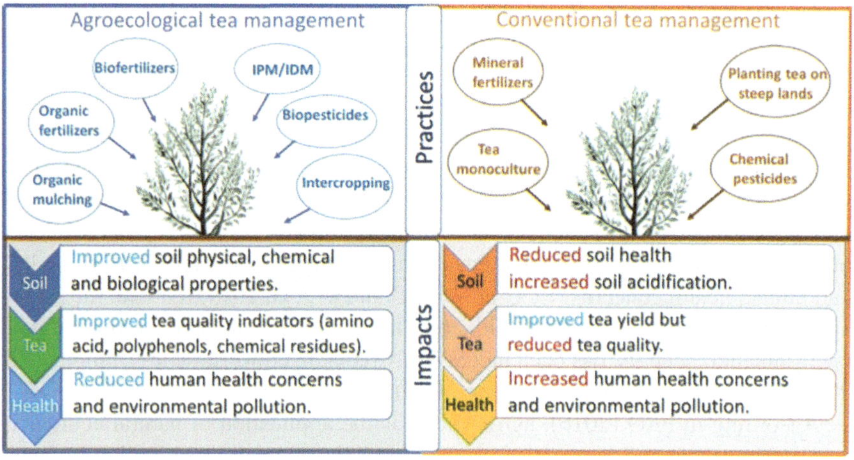

Fig. 10 Comparison of tea plantation through agroecological and conventional management practices (with permission from Le et al. (2021), pp. 589–604)

et al. 2018; Han et al. 2018a). A graphical representation of comparison of tea plantation through agroecological and conventional management practices is given in Fig. 10.

Reports on the tea harvest yield gaps between conventional and organic tea farming systems are available (Deka and Goswami 2021; Doanh et al. 2018; Qiao et al. 2016; Seufert et al. 2012). The yield gap between conventional and organic farming systems could be up to 34%, depending on conditions such as site characteristics, crop types and level of intensification. Han et al.(2018a) also concluded that tea yields in organically managed agroecosystems are generally 8–20% lower compared to those in conventional systems.

23 Microbes in Tea Cultivation

Tea soil microbiology was initially explored in 1901. Long-term tea cultivation alters the composition and structure of soil bacterial community, lowers soil microbial diversity, and leads to the decrease in beneficial bacteria. However, different soil types, agricultural management, plant species, and development stages of the same plant species can largely affect the composition of the microbial community. Different microbes including rhizobia, arbuscular mycorrhizal fungi (AMF), actinomycetes and diazotrophic bacteria play an imperative role in improving physiochemical and biological properties; nutrient cycling and their availability in tea fields, thereby improve plant growth, yield, and quality (Bag et al. 2022).

Changes in soil microbial community characteristics, including decreased diversity, weakened community function and lower community stability due to increase in N application were noticed (Ma et al. 2021). The use of a microbial consortium promotes tea development and regulates the native bacterial populations in the rhizosphere (Shang and Liu 2021). Wang et al. (2021b) indicated that the soil components in the rhizosphere of the tea plant might be altered by changing the variety of rhizobacteria. Moisture content, pH, organic carbon, and total N were the key environmental elements that had the greatest correlation with the bacterial diversity in the rhizosphere soil samples (Wang et al. 2021a). Han et al. (2022) noticed that organic fertilizer substitution and biochar amendments improved soil microbial multifunctionality by altering the abundance of keystone species. The incorporation of leguminous plants can alter the soil environment of tea plantations, which in turn influence the microbial community composition and functions, and subsequent nutrient transformations of either newly added or native organic matter (Huang et al. 2022d; Wang et al. 2022c).

Yang et al. (2022a) found the quality parameters of tea to have significant correlation with soil fungal diversity indices, indicating the possibility for use of soil fungal diversity as a biological indicator of tea quality. The ecological management of tea plantations enhanced soil fertility, microbial diversity, and microbial network complexity (Yang et al. 2022b; Yang et al. 2023b). The root system of tea plants releases root secretions into the soil and cause changes in the number and types of soil microorganisms, thereby changing soil nutrient cycling (Shen and Lin 2021a, b; Tan et al. 2022; Wang et al. 2021a).

Ghosh et al. (2021) identified two phytate solubilizing microorganisms such as *Bacillus safensis* and *B.siamensis* from tea rhizosphere, the latter showing more ability to break down phytate as well as increase the available P in soil. Gui et al. (2021) found that the structure of soil bacterial communities in tea plantations to be strongly shaped with management type as organic vs. conventional > geographical origins > duration of organic management. Intercropping soybean in a deteriorated tea plantation resulted in a shift from oligotrophy to copiotrophy (Zhao et al. 2021). Abundance of various bacterial species in tea ecosystem was reported by many researchers (Shang and Liu 2021; Shen and Lin 2021a, b; Sun et al. 2022b; Wang et al. 2021c; Xiaohu et al. 2022).

In recent years, in-depth studies have been conducted in tea rhizosphere, notably in terms of plant-microbial associations. As tea plants mature, their rhizosphere is colonized by a smaller number of microorganisms, showing that selectivity increases with age. The abundance and varieties of endophytes in the tea are influenced by its geographical surroundings (Xie et al. 2020b).

Soil fungi are key drivers regulating processes between ecosystem fertility and plant growth; however, an understanding of the responses of soil fungi community composition and diversity in deeper soil layers to the plantation and fertilization is needed. Fungi play an important role in soil quality as they decompose organic matter and cycle nutrients in the soil. Long-term tea cropping not only decreased some beneficial fungal species, but also promoted potentially pathogenic fungal species (Arafat et al. 2019; Li et al. 2020). The organic substitution altered the soil fungal community composition and soil pH, SOC, microbial biomass carbon (MBC), and available K were the important characteristics that were significantly correlated with the variation of soil fungal community (Ji et al. 2020b). Jiao et al. (2021) found community compositions to show stronger correlations with multinutrient cycling than α-diversity. Abundance of various fungal species in tea ecosystem was narrated by many researchers (Bai et al. 2022; Qiu et al. 2021; Tang et al. 2022a; Yang et al. 2022e).

23.1 Bio-Fertilizers and Tea Soil Health

Microbial biofertilizers contain live microorganisms and are environmentally safe substances. Their use reduces chemical input costs, prevents depletion of SOM, and increases crop yields. Many reports on the potential of indigenous plant growth promoting rhizobacteria (PGPR) strains for use as microbial biofertilizers to promote growth in tea plantation are available (Chakraborty et al. 2013; Dutta et al. 2015; Pandey et al. 2013; Xu et al. 2017).

Biofilms are a collection of microbial communities of rhizobacteria attached to a fungal surface including N_2 fixers, supplying N_2 through biological fixation for the root system of the plant. In turn, plant roots may provide root exudate C sources to the fungal component of the biofilm biofertilizer, which is present on root surface and rhizosphere (Seneviratne et al. 2008, 2009). Beneficial biofilms can be used as biofertilizers (BFBF), bioremediation, and nutrient mobilization when applied at high cell densities (Ghiasian 2020). BFBFs can break the dormancy of microbial forms in the soil, thereby enhance biodiversity, nutrient cycling, plant immunity, and crop production (Chandralal et al. 2020). Use of BFBF in tea cultivation provides several benefits: 50% reduction in fertilizer use as well as increase in soil microbial biomass and N_2 fixation, and decrease of soil NO_3- and pest infestation (Seneviratne et al. 2011); restoration of SOC due to increase of soil microbes, specifically N_2 fixers (Meepegamage et al. 2019); increase in soil carbon sequestration (Jayasekara et al. 2021); show an increasing trend in endophytic diazotrophs (Chandralal et al. 2020).

Table 1 PGPR and their beneficial role in tea production system

Type of PGPR	Beneficial role	Reference
Bacillus spp	Appreciable ability to produce spores and resilience to different climate conditions	Arafat et al. (2020)
Pseudomonas spp.	Better colonization capacity and their ability to suppress the fungal diseases	
Azospirillum sp.	50% reduction of N using N_2 fixing capacity	Tennakoon et al. (2021)
Brevibacterium sediminis	PGPR and antifungal properties.	Chopra et al. (2020a)
Pseudomonas aeruginosa RTE4	Indole acetic acid (IAA) production, solubilization of inorganic phosphate, production of hydrolytic enzymes, it also shows strong antagonistic activity against two selected foliar pathogenic fungi of tea	Chopra et al. (2020b)
Burkholderia pyrrocinia strain P10	ACC deaminase activity of 33.01-μmol h^{-1} mg^{-1} protein from tea rhizosphere	Han et al. (2021)
Bacillus cereus, Bacillus flexus RN 11, *Pseudomonas* sp. and *P. rhodesiae*	Ability to promote root, stem and leaf growth of tea plants	Nath et al. (2018)
Pseudomonas sp	The highest IAA and GA3 producer	Nath et al. (2022)
Bacillus cereus	The highest P solubilizer	
Acinetobacter sp	The most efficient K solubilizer	

23.2 Role of Plant Growth Promoting Rhizobacteria in Soil Health

The rhizosphere is the narrow dynamic zone of soil influenced by plant roots where intense plant-microbe interaction is found. These microorganisms can have beneficial effects on the plant health like plant growth and nutrition in agro-ecosystems. Details of some of the PGPR identified for use in tea production system and their beneficial role are presented in Table 1.

Table 2 Effect AMF on growth, yield, and quality of tea

AMF species	Effects ofAMF inoculation	Reference
Glomus intraradices, G. versiforme, and *G. etunicatum*	Improve tea plant nutrition by increasing concentrations of Fe, Zn, and Cu in plant tissues	Dash et al. (2019)
G. mosseae, Glomus intraradices	Increased callusing rate, chlorophyll content, and root resistance against pathogens	Chelangat et al. (2021)
Glomus mosseae	Alleviation of the salinity stress and improved the quality of tea leaves	Guo et al. (2021)
Claroideoglomus etunicatum	Increased plant growth parameters (height, root and shoot biomass, total leaf area, root volume) and leaf nutrient contents (N, P, K, Ca, Mg, Mn and Zn)	Shao et al. (2018)
C. etunicatum	Increased plant growth parameters (growth in root length, root and shoot biomass, plant height). Increased phosphatase activity	Shao et al. (2021)
Acaulospora scrobiculata, Glomus macrocarpum, Rhizophagus intraradices	Improvement in plant growth parameters (leaves per plant, leaf area, plant height, shoot and root length), as well as fresh and dry weight of roots and shoots.	Sharma and Kayang (2017)
Acaulospora spinosa, Glomus aggregatum, Glomus ambisporum, Glomus clavisporum, Glomus geosporum	Increased plant growth parameters (root/shoot length, dry weight; root /shoot ratio), quality parameters (caffeine contents, sugar contents, polyphenols contents)	Singh et al. (2010)
Clariodeoglomus etunicatum	Improvement of antioxidant enzyme(SOD, catalase, guaiacol peroxidase, and ascorbate peroxidase) activities during drought stress conditions in tea plant. Enhancement in plant defense mechanism which ultimately and reduced the oxidative damage of drought stress	Liu et al. (2020)
C. etunicatum	Enhanced plant growth parameters (height, root and shoots biomass, root length and volume), nutrients uptake except Fe, and stress resistance under shaded condition	Sun et al. (2020b)
C. etunicatum	Increase in shoot and root biomass production. AMF promoted tea plant growth and alleviated negative effects of drought stress by promoting gas exchange, regulating the water status of leaves, and regulating photosynthetic parameters	Dai et al. (2022)
C. etunicatum	Superior adaptability of root morphology under drought stress to promote nutrient acquisition in tea plants.	Liu et al. (2023c)

23.3 Arbuscular Mycorrhizal Fungi

Mycorrhizal fungi are responsible in improving growth of host plant species due to increased nutrient uptake, production of growth promoting substances, tolerance to drought, salinity, and synergistic interactions with other beneficial microorganisms (Sreenivasa and Bagyaraj 1989).

Details of some of the AMF and their beneficial role on tea growth, yield, and quality of tea are given in Table 2.

Apart from the above, inoculation of AMF presents many other benefits in tea production system: alleviate salt stress through the decrease of lignin and cellulose accumulation (Guo et al. 2022); promote P acquisition capacity of tea plants, especially in roots by altering root shape and modifying the expression of the transporter

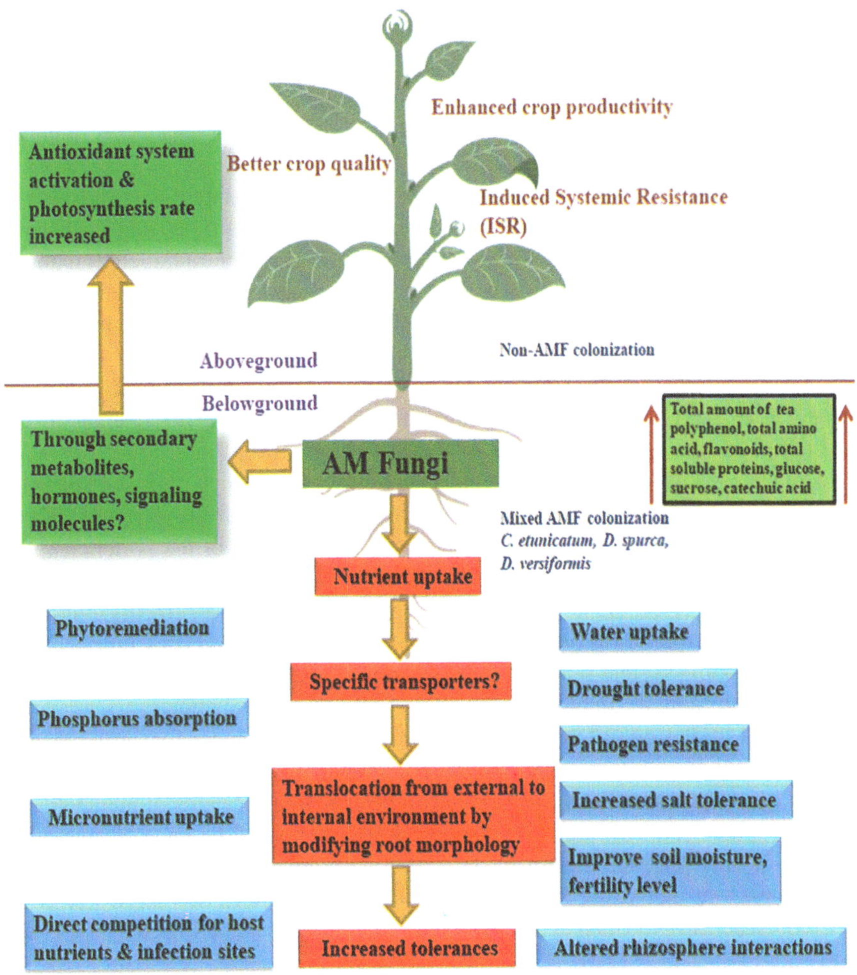

Fig. 11 Possible roles of AM fungi on the growth and development of tea plants (with permission from Bag et al. (2022), p. 126890, https://doi.org/10.1016/j.micres.2021.126890)

gene, and enhancing soil acid phosphatase activity and root CsPT1 expression (Shao et al. 2021); improving callusing rate and the chlorophyll content (Chelangat et al. 2021). Converting conventional tea production to agroecological management practices increased tea root AMF intensity by up to 24% (Le et al. 2023). Wu et al. (2020b) indicated that organic tea management significantly increased tea soil AMF contents. The possible roles of AM fungi on the growth and development of tea plants are depicted in Fig. 11.

24 Soil Metabolomics

Soil metabolomics is a large-scale study of low-molecular weight organic compounds in soil, and it provides an important means to reveal complex molecular networks and metabolic pathways in soil microbial communities and to evaluate soil function (Withers et al. 2020). In recent years, many researchers have used soil metabolomics to analyze the effects of environmental factors on soil microbial diversity and soil function, and a series of important results have been obtained (Brown et al. 2021; Sun et al. 2022b).

Metabolomics approaches allow a comprehensive understanding of various key variables, such as the chemical composition of different tea types and further the processing influence on tea chemical composition in an untargeted manner. Metabolomics is an essential aspect of systems biology increasingly being recognized as an effective tool for the assessment of the quality of tea products (Frag et al. 2023).

Metabolomics analysis revealed the responses of tea plants to excessive Ca (Liang et al. 2021) with high Ca concentrations aggravating the detrimental effects of high pH to tea roots. Lin et al. (2021) found alteration of metabolic profiles in N-deficient leaves and roots of tea plants, with N deficiency showing different effects on the primary and secondary metabolism in tea leaves and roots. These results will be useful to have comprehensive understanding of the N-deficient tea plants and for the optimized N management for sustainable development of tea plantations. Based on the non-targeted metabolomics, (Wu et al. 2021c) found 100 differential metabolites to be upregulated in the tea leaves from tea-Chinese chestnut intercropping system relative to monoculture system. The upregulated metabolites could promote the quality of tea and its health-beneficial effects.

Detailed studies on soil metabolomics are of great significance for further revealing the metabolism of soil microorganisms in tea rhizosphere and soil function caused by acidification. With increasing soil pH, the total contents of metabolite of tea rhizosphere soil increased. SOM, CEC, MBC, microbial biomass N, microbial respiration intensity, number of bacteria and actinomyces in the rhizosphere soil also showed an increasing trend, whereas the fungi number decreased (Ye et al. 2023a). According to Chen et al. (2022b), the production of various tea types such as green, white, yellow, dark, oolong, and black teas is influenced by agro management practices such as soil health management, shading, plucking, pruning, and soil composition, together with different processing methods.

25 Metagenomic Approach

Soil is, probably, the most complex and heterogeneous microbial habitat on earth, and the great diversity of soil microbial communities exceeds that of other environments. Soil metagenomics is a cultivation-independent molecular approach that explores and exploits the enormous diversity of soil microbial communities (Daniel 2005). This provides an ecological methodology for investigating potential heavy metal tolerant or resistant bacteria which are capable of coping with Cd-polluted soil (Feng et al. 2018).

SOC is an important C pool of the global ecosystem and is affected by various agricultural practices including fertilizer application. Quite often, excessive N application is an important field management measure in tea plantation systems. The mechanism underlying the impact of N fertilization on SOC, especially the microscopic mechanism, is to be studied in detail. Yang et al. (2023a) using a metagenomic approach in a tea plantation under long-term N fertilization with different rates found change in the abundance of C cycling genes, SOC degrading enzymes, especially those associated with labile and recalcitrant C degradation. There was significant difference of β-glucosidase and chitinase-expressing microbial communities as well.

26 Microbiome

The rhizosphere is considered as the "heart of the soil" where direct interactions take place between plant roots and several active microbes. Plant roots help to deposit photosynthates which influence microbial colonies into the rhizosphere and in turn microorganisms facilitate plant growth and development (Olanrewaju et al. 2019). Plant-associated microbiomes offer many advantages to the host plant, including growth promotion, nutrient uptake, stress tolerance, and resistance to pathogens (Trivedi et al. 2020). The tea microbiome has now become an important topic of research for microbiologists in the recent years as it can help the plant for soil nutrient acquisition as well as stress management. The organization of the plant microbiome is described as cotrolled by the host, especially in the rhizosphere region, where various rhizodeposits influence the populaion of microbes (Bag et al. 2022). Microorganisms can also affect tea plant gene expressions, and therefore influence their surrounding environment and plant fitness. Tea plants can modulate microbial gene expression to their benefit by controlling the composition and

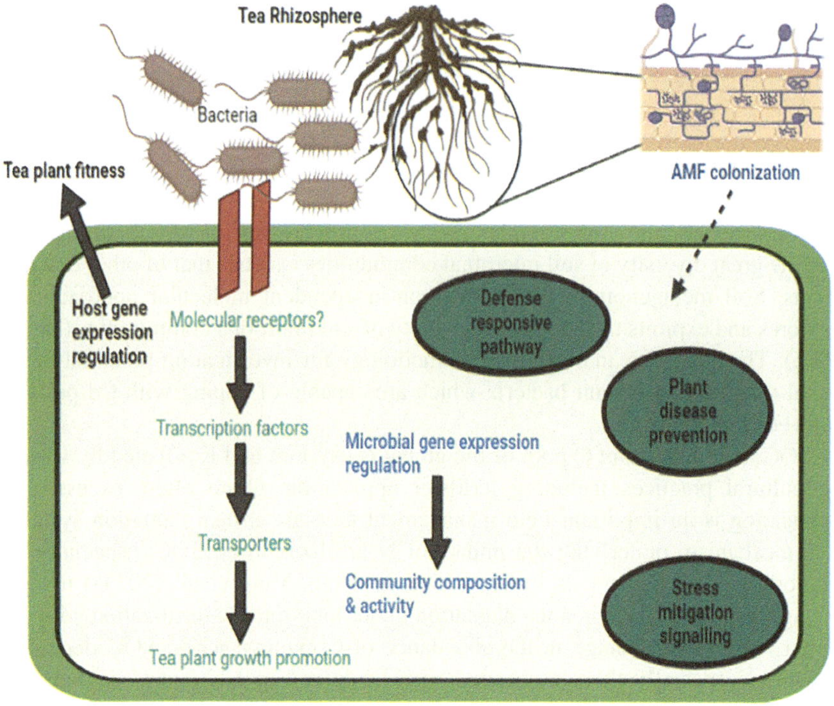

Fig. 12 A proposed model of tea plant-microbiome interactions in the rhizosphere region (with permission from Bag et al. (2022), p. 126890, https://doi.org/10.1016/j.micres.2021.126890)

activities of microbial communities. Bag et al. (2022) proposed a model of tea plant-microbiome interactions in the rhizosphere region (Fig. 12).

27 Climate Changes and Tea Production

Indiscriminate industrial growth, fossil fuel burning, and number of anthropogenic activities continuously emit tones of GHG into the atmosphere. Change in global climate variables has markedly affected the growth and productivity of plants. Excessive CO_2 concentrations, heat stress, and fluctuating precipitation pattern have altered the plant nutrient interactions (Nisa et al. 2023). The enrichment of CO_2 and other air pollutants such as SO_2 and NO_2 will cause strong acid precipitation that further increases soil acidity. It should be noted that tea soils in China are already too acidic, one-third being below pH 4.5 (Han et al. 2002). Further acidification will, therefore, degrade soil quality and tea production.

Tea is one of the plant species most affected by changes in bio-climatic suitability in Yunnan province, China (Ranjitkar et al. 2016). Han et al. (2016) showed that the annual mean temperature and extreme lowest temperature have increased by

1.0–1.6 °C and 2.1–3.8 °C, respectively, over the last 50 years. The main tea-growing areas of northeast India, Assam and West Bengal, showed a steady increase in mean temperature with the average minimum temperature having increased by about 1.3 °C over the last 100 years (Han et al. 2018b). According to these authors, climate change brings both advantages and disadvantages for the growth and development of tea and will ultimately have a considerable impact on production.

Lou et al. (2020) showed that the influence of high temperature and drought on tea production have increased significantly. Temperature was the main factor that affected the yield and climate change caused earliness in the start date of tea buds and leaves plucking period during spring. Yan et al. (2021c) found that both heat and cold extremes were associated with significant reduction of tea yields in China. Zakir (2018) pointed out that the dilution of phytochemicals as a consequence of more frequent extreme rains seems to explain the changes of the tea taste that tea farmers in Yunnan (South China) are experiencing. Using climate model projections, Jayasinghe and Kumar (2019) projected a decline of approximately 10.5%, 17%, and 8% in total "optimal," "medium," and "marginal" suitability areas respectively, implying that climate would have a negative effect on the habitat suitability of tea in Sri Lanka by 2050 and 2070. By adopting MaxEnt modelling, Jayasinghe and Kumar (2020) pointed out that the precipitation seasonality and annual precipitation were the most influential variables in Sri Lanka and India, respectively, whereas, annual mean temperature was the most effective contributor for determining the suitability of area for tea cultivation in Kenya and China.

28 Conclusion and Way Forward

Tea is widely grown in more than 52 countries of tropical and subtropical regions of the world. Since long, tea has been consumed as a beverage in over 80% of the populations worldwide, especially in Asian and African countries. Apart from its culinary usage as a popular beverage around the world, tea provides a natural source of chemicals with potential therapeutic benefits: antioxidant, anticancer, and hypolipidemic properties, to name a few.

As a leaf crop, tea plants are harvested at regular intervals. This makes it necessary to provide adequate nutrients to fulfil the demands of sprouting leaves and metabolite synthesis, particularly macronutrients such as N, P, and K. However, tea plantations are frequently over-fertilized to achieve enhanced growth and yield, which ultimately influences the soil acidification and compaction, resulting in low quality and yield. The physico-chemical properties of the soil after tea plants being grown as long-term monoculture are gradually transformed, and the unique eco-environment in the soil will be generated because of the unique biological features of tea plants (He et al. 2021; Ruan et al. 2019b).

Prolonged utilization of chemical fertilizers has a long-term negative impact on soil nutritional stabilization and rhizosphere diversification. Moreover, it will also lead to serious environmental problems, including gaseous N emissions, the leaching of potential nitrate, soil acidification, and alteration in tea quality. Therefore, the improvement of nutrient use efficiency in tea plant may be the key solution to these problems. Moreover, it is very essential to understand the N response mechanisms of tea plants and select N efficient tea varieties that can diminish the fertilizer consumption, and reduce environmental pollution.

Long-term emphasis on N fertilizer will further aggravate soil acidification, and hence, P and K become the crucial nutrient factors in soils that determine the tea plant features. LJ43 is tea variety that efficiently absorb nutrients from the soil, promote nutrient diffusion and migration in the soil, provide an opportunity for the plant to adapt to the nutrient spatial heterogeneity, and promote efficient nutrient absorption, and performs excellently in alleviating such soil acidification. Reports are available on the theoretical foundation for understanding the efficient utilization of soil nutrients by tea plant varieties (Ruan et al. 2023). Future research should focus on developing more of high-nutrient-efficient varieties of tea plant for reducing both nutrient heterogeneity and soil acidification, aimed to greatly improve the soil degradation in tea gardens.

Currently many studies have been conducted on the absorption and transfer of N in tea plants, as well as how its levels and forms affect the growth and development of tea plants. However, results from in-depth studies on how it regulates the physiological metabolism of tea plants as a whole and how it affects the genetic changes of tea plants are lacking. Therefore, the use of omics technology to understand the effects of N on the whole tea plant (such as metabolomics, proteomics, epigenomics, long non-coding RNA analysis, etc.) is an important direction for tea plant scientific research (Xie et al. 2023). Harvesting of tea is mainly based on new shoots, and N is the most needed nutrient element for tea plants, but its utilization efficiency is considerably low. In the future, it will be of interest to study the N absorption and utilization of tea plants to unravel the related genes responsible and screen tea varieties with high N utilization efficiency.

The amount of K fertilizer applied and the soil-available K content also influence the tea quality, the variation in tea yield results mainly from the tea types. Therefore, to develop high-quality tea plantations and address the gap between low tea yield and high tea consumption, efforts should be made to enhance the soil-available K content by optimizing the amount of K applied on tea plantations. Balanced application of fertilizers and efficient K usage in combination with other nutrients not only contribute to sustainable tea growth, yield and quality, but also influence plant health and reduce the environmental risks. However, our understanding about the role of K in whole-plant stress response mechanisms is very limited. Therefore, it is suggested to have detailed investigations about the molecular mechanisms of K in plant stress resistance, examine the influence of K on plant resistance to biotic and abiotic stresses in differentiated cells, tissues and organs, as well as identify common or specific response of K to distinct stress and the role of K on long-term plant responses under multiple stress conditions in nature (Wang et al. 2013).

The use of microorganisms for sustainable agriculture has now been accepted worldwide as microbes effectively colonize different plant parts and enhance crop productivity through several plant growth regulatory functions as well as reduction of plant pathogens. Efficient utilization of plant unavailable pool of soil K by the use of potassium solubilizing microbes can decrease the fertilizer requirement. The utilization of such K solubilizers is considered to be a sound strategy in improving the productivity of tea growing lands. This new technique is also claimed to have the ability to restore the productivity of degraded, marginally productive and unproductive tea fields.

Design and development of more K-efficient tea genotypes by conventional breeding or targeted biotechnological strategies is necessary. Optimized K fertilizer supply in K limited soils is important in order to increase plant response especially to drought stress via improving adaptive/resistance mechanisms of tea plants. Studies have already been conducted to understand the differential transcriptomic changes in low-potassium sensitive and low-potassium tolerant tea plant genotypes under potassium deprivation. Two tea plant genotypes, 1511 and 1601, which are tolerant and sensitive to low-K conditions, respectively were obtained from the F1 tea population of a crossbreeding program and the molecular mechanisms underlying tolerance of K^+ starvation has been elucidated (Li et al. 2019a), which can serve as a foundation for future research in selecting low potassium tolerant genotypes.

Despite nitrogen being the most important nutrient in tea cultivation, its use as fertilizer is associated with heavy losses. Such losses pollute the environment and increase greenhouse gas production and other environmental events associated with high ammonia volatilization and nitrous oxide emission. There are other problems like nitrate leaching and run-off that pollute surface and underground waters, which have human health implications. All these result in reduced N uptake and crop productivity finally affecting the efficacy and profitability of fertilizers. To address these problems, various forms of fertilizers like slow release fertilizers, controlled release fertilizers, and enhanced efficiency fertilizers are continuously being developed to regulate the release of N from fertilizers, allow improved uptake and utilization by plants, thereby lowering losses and increasing crop productivity per unit of fertilizer. More research in this direction for large scale adoption by small farmers as well as at the big plantation level are necessary. It is especially critical to develop multi-nutrient fertilizers that provide balanced nutrition to plants and humans, while improving N use efficiency and mitigating N-fertilizer effects on human and environmental health.

Many of the studies on nutrient usage have mainly been conducted in artificial climate chambers with hydroponic or pot cultivation, while there is a lack of long-term field experiments to understand the realistic response to applied N fertilizer along with other factors. For perennial tea plants, a considerable proportion of N is stored in the roots, stems, and mature leaves, and remobilized N in each organ is the main source for absorption and accumulation in the young shoots (Ma et al. 2019). Hence, the effect of fertilizer will not be clearly understood in short-term field experiments because of the large amounts of nutrients stored in tea plants, especially in mature plants (10–30 years). Besides, fertilizer use efficiency of tea plants is governed by climate, soil properties (texture, forms of available N, and pH), N-demand characteristics of different cultivars, picking season, and fertilization time. Therefore, it is necessary to comprehensively investigate the effect of long-term N application on the formation of tea quality-related metabolites in field experiments.

Varying results have been recorded on yield and quality responses to NPK application. For P and K, evaluations are necessary in different regions because where there are no beneficial effects, it could be possible to reduce application for their efficient use and to decrease costs. High rates of N reduce black tea quality and do not increase yields. Nitrogen fertilizers need to be applied at rates that are a compromise between yields and black tea quality. Such rates vary with regions and genotypes. To reach high production and quality, region- and genotype-specific fertilizer rates are necessary. Differences in quality and chemical composition have been reported for the same genotype grown in different regions and processed under identical conditions, which is mainly due to variations in the leaf biochemical constituents composition caused by the environment where plants are grown. Different regions must, therefore, optimize their production conditions to realize high quality and maintain soil health for sustained production.

The demand for sustainable tea production without compromising higher quality is a driving force for increasing conversion from conventional tea farming practices to organic-based methods; where synthetic fertilizers and plant protection chemicals are avoided. This transformation in tea management practices has been driven by the growing interest in high quality and economic efficiency of tea production, and an increased awareness of the harmful effects of agrochemicals on human health and the environment. The beneficial impacts of agroecological management practices such as using organic fertilizers, plant growth promoting rhizobacteria, biofertilizers and nutrient solubilizers, mulching, intercropping, biopesticides, and integrated pest and disease management strategies are felt and all of these practices can result in soil health improvement (biological, chemical and physical properties), reduce agrochemical input and chemical residues in soil and on tea leaves. Ultimately, agroecological tea production practices can mitigate the negative effects of chemical uses on the environment while maintaining tea productivity and quality. The research on these lines should receive greater attention.

Tea rhizosphere has been studied in-depth, notably in terms of plant-microbial associations. Continuous monocropping will lead to deterioration of soil health and quality of tea produced from such areas. The occurrence of a "negative rhizosphere impact" in mature (rather than young) tea plants is gaining attention and identified to be an interesting feature of the tea rhizosphere. Other significant and related characteristics include tea rhizosphere microbial colonization, soil pH etc. These findings have opened up new options for further research in the domain of rhizosphere engineering (Bag et al. 2022). A deeper understanding of the relationship between key microbial organisms (e.g., hub species) and their potential functions, as revealed by the network analysis, is needed. The temporal response of the microbial-driven ecological processes to fertilizer, especially of organic nature, and the relationship between the microbial community and microbial-driven ecological processes in tea crop soil are to be well understood. Therefore, future research programs should concentrate to further explore the underlying mechanisms of rhizospheric microbial interactions mediated by root exudates and to identify the contributing factors to soil sickness due to continuous tea monocropping.

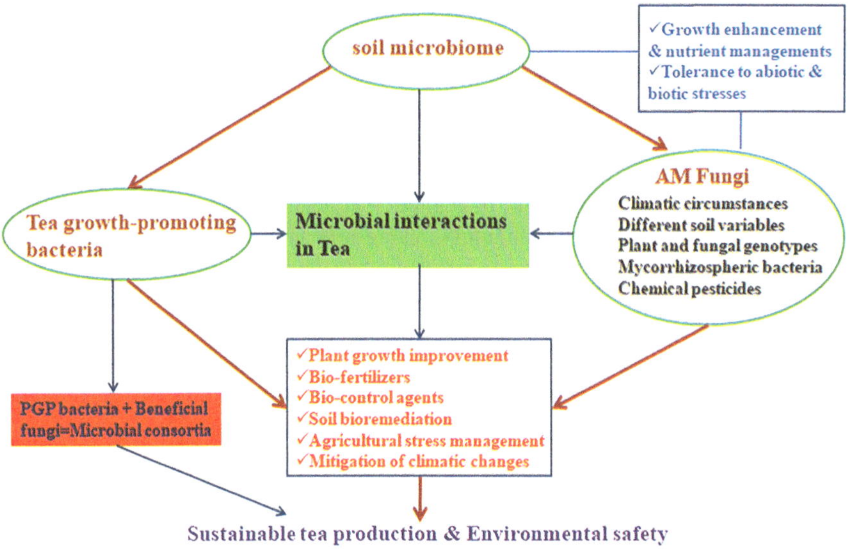

Fig. 13 Diagrammatic representation depicting the functions of AMF, and microbiome, tea growth promoting bacteria, beneficial fungi, microbial interactions for sustainable tea production, and environmental safety (with permission from Bag et al. (2022), p. 126890, https://doi.org/10.1016/j.micres.2021.126890)

A diagrammatic representation of the functions of AMF, and microbiome, tea growth promoting bacteria, beneficial fungi, microbial interactions for sustainable tea production, and environmental safety is given in Fig. 13.

The quality of tea growing soil is determined not only by the physiochemical properties of the soil but also by its microbial profile. Thus, the diversity and structure of the microbial communities within the rhizosphere have a profound impact on the growth and yield of tea plants. In a tea plantation, interactions occur between the tea plants, the growing environment, as well as soil microorganisms, which constitute the ecosystem of tea plantations. Due to the influence of environmental factors and human interventions, various changes take place in soil microbial community, which affect the growth, yield, and quality of tea plants. Only a few studies have been conducted using soil metabolomics to analyze the effects of acidification on microbial diversity and function in tea rhizosphere soil. Such studies are of great significance for further revealing the metabolism of soil microorganisms in tea rhizosphere and soil function caused by acidification. Furthermore, metagenomic methods are to be used to comprehensively analyze the differences of various functional genes in soil N transformation process and microbial communities of tea plantation with different pH values. Studying plant–microbe interactions will not only provide a better understanding of the relationship between soil metabolites and microbiomes, but it will provide an opportunity to make use of such relationships for sustainable agricultural production. The tea microbiome has now become an important topic of research for microbiologists in the recent years as it can help the plant for soil nutrient acquisition as well as stress management (Bag et al. 2022).

As soil is the largest terrestrial carbon pool, as well as a significant contributor of greenhouse gases, much progress can be made towards overcoming the adverse impact of climate change by sustainable soil management practices. One identified option is to increase soil organic carbon levels, especially with recalcitrant forms of carbon (e.g., biochar application). In general, soil health is primarily determined by the actions of the farming community. Therefore, information management and knowledge sharing are necessary to improve the sustainable behavior of practitioners and end-users.

Since the concept for "soil quality" was introduced in early 1990s, numerous revisions have been proposed (Bunemann et al. 2018; Karlen et al. 1997) and the term "soil health" is now more frequently used than "soil quality" (Bonfante et al. 2019). Soil heath is the capability of a soil to provide ecosystem services (Williams et al. 2020), and from an agricultural perspective, soil health refers to the capacity of soil to support crop productivity (Mursec 2011). Soil health is now acknowledged to consider all the soil feature indicators including soil physical, chemical, and biological properties (Allen et al. 2011; Arias et al. 2005; Cardoso et al. 2013).

One of the greatest achievements of the soil health concept is the inclusion of a biological perspective to soil management, which facilitates the addressing of long-term sustainability challenges for crop production. Historically, soil assessments were focusing on crop production, but, today, soil health also includes the role of soil in water quality, climate change, and human health. However, quantifying soil health is still dominated by chemical indicators, despite growing appreciation of the importance of soil biodiversity, mainly because of limited functional knowledge and lack of effective methods. Scientists should embrace soil health as an overarching principle that contributes to sustainability goals, rather than only a property to measure (Costantini and Mocali 2022).

In recent decades, agricultural soil has been seriously degraded by human interventions. Conventional agricultural practices such as application of agrochemicals and mechanical tillage have led to significant soil heath degradation, including reduced soil biodiversity (Alori et al. 2020). For tea farming industry, long-term conventional tea cultivation has been considered as the main driver of severe soil acidification, soil nutrient imbalance, and nutrient leaching in tea plantations (Li et al. 2016c; Yan et al. 2020a). By determining soil health, it is possible to understand those indicators that can be used to evaluate the soil's current status and, hence, to develop sustainable agricultural systems. In this regard, significant progress has been made over the last few years in the development of specific biomarkers and macromolecular probes, enabling rapid and reliable measurements of soil microbial communities. More research in these directions will make the tea production sector more environmentally friendly and sustainable.

References

Abdul-Kader DZEA (2007) Role of nitric oxide on iron homeostasis, chlorophyll biosynthesis and antioxidants system in two wheat cultivars. Am J Plant Physiol 2(4):237–250. https://doi.org/10.3923/ajpp.2007.237.250

Adhikary S, Khan M, Arobe S, Dey S, Billah S (2019) Soil chemical analysis of Kazi and Kazi organic tea garden and compared to ordinary tea gardens of Bangladesh. Open J Soil Sci 9:91–102. https://doi.org/10.4236/ojss.2019.96006

Adrees M, Ali S, Rizwan M, Zia-ur-Rehman M, Ibrahim M, Abbas F, Farid M, Qayyum MF, Irshad MK (2015) Mechanisms of silicon-mediated alleviation of heavy metal toxicity in plants: a review. Ecotoxicol Environ Safety 119:186–197. https://doi.org/10.1016/j.ecoenv.2015.05.011

Ahmed M, Hoque AR, Mamun MS (2011) Efficacy of different Sulphur formulations against red spider mite (*Oligonychus coffeae* Nietner) of tea in Bangladesh. Int J Sustain Agric Technol 7:29–33

Ahmed S, Stepp JR, Orians C, Griffin T, Matyas C, Robbat A, Cash S, Xue D, Long C, Unachukwu U, Buckley S, Small D, Kennelly E (2014) Effects of extreme climate events on tea (*Camellia sinensis*) functional quality validate indigenous farmer knowledge and sensory preferences in tropical China. PLoS One 9:e109126. https://doi.org/10.1371/journal.pone.0109126

Aksuner N, Henden E, Aker Z, Engin E, Satik S (2012) Determination of essential and non-essential elements in various tea leaves and tea infusions consumed in Turkey. Food Addit Contam 5(2):126–132. https://doi.org/10.1080/19393210.2012.675592

Alam MS, Akter S, Shemul S, Chowdhury TR, Ara PS (2022) Assessing the quality and heavy metal contamination of soil in tea gardens around Magurchara gas blowout in Bangladesh using multivariate and soil quality index methods. J Hazardous Materials Adv. https://doi.org/10.1016/j.hazadv.2022.100127. 11 pages

Albadarin AB, Lewis TD, Walker GM (2017) Granulated polyhalite fertilizer caking propensity. Powder Technol 308:193–199. https://doi.org/10.1016/j.powtec.2016.12.004

Allen DE, Singh BP, Dalal RC (2011) Soil health indicators under climate change: a review of current knowledge. In: Yin K, Singh BP, Cowie AL, Chan (eds) Soil health and climate change. Springer, New York, pp 25–45

Alom S, Das R, Ahmed CS, Baruah U, Das S (2020) Evaluation of soil quality in tea plantation under present climatic condition. Plant Archives 20(2):7352–7358

Alori ET, Adekiya AO, Adegbite KA (2020) Impact of agricultural practices on soil health. In: Giri B, Varma A (eds) Soil health, soil biology, vol 59. Springer, Cham, pp 89–98

Altieri M, Rosset P, Thrupp L (2020) The potential of agroecology to combat hunger in the developing world. A 2020 vision for food. International Food Policy Research Institute, Washington. http://agroeco.org/wp-content/uploads/2011/02/altieriIFPRI.pdf

Amali RKA, Abeyweera NHAS, Gunaratne OGKA, Gunaratne G, Kottawa-Ar LD (2022) Performance evaluation of newly developed tea (*Camellia sinensis* L.) cultivars in response to different levels of nitrogen fertilizer. Int J Agric Res 17:201–209. https://doi.org/10.3923/ijar.2022.201.209

Ananthacumaraswamy S, Hettiarachchi LSK, Dissanayake SM (2011) Soil and foliar sulfur status in some tea plantations of Sri Lanka. Comm Soil Sci Plant Analysis 34(11–12):1481–1497. https://doi.org/10.1081/CSS-120021291

Anjaneyulu A, Bhattacharjee H (2019) Yield and quality of young tea (*Camellia sinensis* (L.) O. Kuntze) with application of organic and inorganic fertilizers in arecanut garden. Res J Pharmacognosy Phytochem 8(1):949–951

Anonymous (2012) Fertilizer user guide manual for tea (*Camellia sinensis* (L.) O. Kuntze). Tea Research Foundation of Kenya (TRFK) 1st edition: March 2012. p. 35

Anqi X, Zichen W, Xiaohan X, Yi S, Genmei W, Yuhua W (2022) Research advances of fluoride accumulation mechanisms in tea plants (*Camellia sinensis*). J Tea Sci 42(3):301–315. https://doi.org/10.13305/j.cnki.jts.20220416.003

Arafat Y, Wei X, Jiang Y, Chen T, Saqib HAS, Lin S, Lin W (2017) Spatial distribution patterns of root-associated bacterial communities mediated by root exudates in different aged ratooning tea monoculture systems. Int J Mol Sci 18:1727. https://doi.org/10.3390/ijms18081727. 16 pages

Arafat Y, Tayyab M, Khan MU, Chen T, Amjad H, Awais S, Lin X, Lin W, Lin S (2019) Long-term monoculture negatively regulates fungal community composition and abundance of tea orchards. Agronomy 9(8):466. https://doi.org/10.3390/agronomy9080466

Arafat Y, Din IU, Tayyab M, Jiang Y, Chen T, Cai Z, Zhao H, Lin X, Lin W, Lin S (2020) Soil sickness in aged tea plantation is associated with a shift in microbial communities as a result of plant polyphenol accumulation in the tea gardens. Front Plant Sci 11:601. https://doi.org/10.3389/fpls.2020.00601. 13 pages

Arias ME, González-Pérez JA, González-Vila FJ, Ball AS (2005) Soil health: a new challenge for microbiologists and chemists. Int Microbiol 8(1):13–21

Auler AC, Pires LF, Santos JAB, Caires EF, Borges JAR, Giarola NFB (2017) Effects of surface-applied and soil-incorporated lime on some physical attributes of a Dystrudept soil. Soil Use Manag 33:129. https://doi.org/10.1111/sum.12330

Bag S, Mondal A, Banik A (2022) Exploring tea (*Camellia sinensis*) microbiome: insights into the functional characteristics and their impact on tea growth promotion. Microbiol Res 254(2):126890. https://doi.org/10.1016/j.micres.2021.126890. 14 pages

Bagchi A, Ghosh BC, Swain DK, Bera N (2015) Organic farming practice for quality improvement of tea and its anti-parkinsonism effect on health defense. J Phys Chem Biophys 5(2). https://doi.org/10.4172/2161-0398.1000178. 5 pages

Bagyalakshmi B, Ponmurugan P, Balamurugan A (2017) Potassium solubilization, plant growth promoting substances by potassium solubilizing bacteria (KSB) from southern Indian tea plantation soil. Biocatal Agric Biotechnol 12:116–124

Bai Z, Caspari T, Gonzalez MR, Batjes NH, Mäder P, Bünemann EK, de Goede R, Brussaard L, Xu M, Ferreira CSS, Reintam E, Fan H, Mihelič R, Glavan M, Tóth Z (2018) Effects of agricultural management practices on soil quality: a review of long-term experiments for Europe and China. Agric Ecosyst Environ 265:1–7. https://doi.org/10.1016/j.agee.2018.05.028

Bai YC, Li BX, Xu CY, Raza M, Wang Q, Wang QZ, Fu YN, Hu JY, Imoulan A, Hussain M, Xu YJ (2022) Intercropping walnut and tea: effects on soil nutrients, enzyme activity, and microbial communities. Front Microbiol 13:852342. https://doi.org/10.3389/fmicb.2022.852342. 15 pages

Balasubramaniam K (2019) Evaluation of Agripower (silicon-based fertilizer) in tea. UPASI Tea Res Foundation Ann Rep 2019:26–27

Balci M, Taban S (2022) Effect of boron applications on boron concentration of the leaves under the harvest base of tea plant. J Plant Nut 46(3):1–17. https://doi.org/10.1080/01904167.2022.2067047

Baquerizo MD, Maestre FT, Reich PB, Jeffries TC, Gaitan JJ, Encinar D, Berdugo M, Campbell CD, Singh BK (2016) Microbial diversity drives multifunctionality in terrestrial ecosystems. Nat Commun 7:10541. https://doi.org/10.1038/ncomms10541. 8 pages

Barbora BC, Rahman F, Barman TS (1993) Zinc in the nutrition of the tea plant. Two Bud 40(1):7–12

Barman T, Barooah AK, Goswami BC, Sharma N, Panja S, Khare P, Karak T (2020) Contents of chromium and arsenic in tea (*Camellia sinensis* L.): extent of transfer into tea infusion and health consequence. Biol Trace Elem Res 196:318–329. https://doi.org/10.1007/s12011-019-01889-y

Barnes B, Fortune T (2014) Blending and spreading fertilizer- Physical properties. 4 pages

Baruah N, Medhi BK, Borah N, Baruah AM, Saikia GK (2017) Effect of long term tea cultivation in deep, fine loamy, well-drained soils of Jorhat district. Assam J Soil Water Cons 16(4):347–355. https://doi.org/10.5958/2455-7145.2017.00051.0

Baruah RKG, Tamuli AK, Teron R (2019) Effect of integrated use of organic and inorganic nitrogen sources on growth, development and yield of young clonal tea (*Camellia sinensis*(L) O. Kuntze). Int Adv Sci Res Manage 4(3):95–99

Baruah N, Medhi BK, Swami S, Thakuria RK, Borkotoki B (2020) Soil quality index as affected by temporal long term tea cultivation in Jorhat district of Assam. J Environ Bot 41:1685–1695. https://doi.org/10.22438/jeb/41/6/SI-252

Becker M, Asch F (2005) Iron toxicity in rice—conditions and management concepts. J Plant Nutr Soil Sci 168:558–573. https://doi.org/10.1002/jpln.200520504

Beevers L, Hageman RH (1969) Nitrate reduction in higher plants. Ann Rev Plant Physiol 20:495–522

Benti T, Debela A, Bekele Y, Suleman S (2022) Influence of clone and nitrogen application level on quality of green tea in some selected tea (*Camellia sinensis (*L.)O. Kuntze) in Southwest Ethiopia. Heliyon. https://doi.org/10.1016/j.heliyon.2022.e10179

Bhattacharya PN, Sarmah SR, Sarma B, Madhab M, Tanti A, Borgohain K (2022) Zinc-solubilization potential of putative microorganisms isolated from tea [*Camellia sinensis* (L.) O. Kuntze] rhizosphere. Res J Microbiol 17(1):1–13. https://doi.org/10.17311/rjm.2022.1.13. 14 pages

Bhattacharyya PN, Sarmah SR (2018) The role of tea soil microflora in tea cultivation. In: Sharma VS, Gunasekare K (eds) Global tea science. Burleigh Dodds Science Publishing, Sawston, pp 135–167. https://doi.org/10.19103/AS.2017.0036.24m

Bhattacharyya PN, Dutta P, Madhab M, Phukan IK, Sarmah SR, Pathak SK (2016) Isolation of potash mobilizing microorganisms in tea soil and evaluation of their efficiency in potash nutrition in tea: a novel approach. Two Bud 63(1):8–12

Bhuyan S, Kandali GG, Deka K, Karmakar RM, Choudhury M, Borah K (2018) Impact of organic and conventional tea cultivation practices on soil health. Int J Chem Stud 6(3):502–508

Bhuyan S, Kandali GG, Choudhari B, Rajkhowa J, Pattanaik I (2022) Effect of management practices on soil acidity component in tea soil of Assam. The Pharma Innovation J 11(3):223–228

Bisen JS, Singh AK (2012) Impact of inorganic to organic cultivation practices on yield of tea in Darjeeling hills-a case study. Indian J Hort 69(2):288–291

Bisen JS, Singh AK, Chauhan RK, Singh M, Bera B (2016) Improving productivity of ageing Chinary tea plantation through rejuvenation pruning and sustaining soil fertility by use of organic inputs. Int J Agric Sci Res 6(3):433–444

Bitange N, Cheminingwa GN, Ambuko J, Nayankanga R (2018) Effect of organic and inorganic fertilizers on growth, yield and nutrient use efficiency of clonal tea (*Camellia sinensis*). Cell Bio Dev 2(1):15–26. https://doi.org/10.13057/cellbioldev/v020103

Bolan S, Padhye LP, Mulligan CN, Alonso RE, Saint-Fort R (2023) Surfactant-enhanced mobilization of persistent organic pollutants: potential for soil and sediment remediation and unintended consequences. J Hazard Mater 443:130189. https://doi.org/10.1016/j.jhazmat.2022.130189

Bonfante A, Terribile F, Bouma J (2019) Refining physical aspects of soil quality and soil health when exploring the effects of soil degradation and climate change on biomass production: An Italian case study. Soil 5(1):1. https://doi.org/10.5194/soil-5-1-2019

Borgohain A, Konwar K, Buragohain D, Varghese S, Dutta AK, Paul RK, Khare P, Karak T (2020) Temperature effect on biochar produced from tea (*Camellia sinensis* L.) pruning litters: a comprehensive treatise on physico-chemical and statistical approaches. Bioresour Technol 318:124023. https://doi.org/10.1016/j.biortech.2020.12402

Borgohain A, Sarmah M, Gogoi BB, Konwar K, Handique JG, Paul RK, Yeasin M, Pandey V, Yadav R, Malakar H, Saikia J, Deka D, Rahman FH, Panja S, Khare P, Karak T (2023) Can tea pruning litter biochar be a friend or foe for tea (*Camellia sinensis* L.) plants' growth and growth regulators? Feasible or fumes of fancy. Ind Crop Prod 195:116394. https://doi.org/10.1016/j.indcrop.2023.116394

Brooker RW, Bennett AE, Cong WF, Daniell TJ, George TS, Hallett PD, Hawes C, Iannetta PP, Jones HG, Karley AJ (2015) Improving intercropping: a synthesis of research in agronomy, plant physiology and ecology. New Phytol 206:107–117. https://doi.org/10.1111/nph.13132

Brown RW, Chadwick DR, Zang H, Jones DL (2021) Use of metabolomics to quantify changes in soil microbial function in response to fertiliser nitrogen supply and extreme drought. Soil Biol Biochem 160:108351. https://doi.org/10.1016/j.soilbio.2021.108351

Bunemann EK, Bongiorno G, Bai Z, Creamer RE, De Deyn G, de Goede R, Fleskens L, Geissen V, Kuyper TW, Mäder P (2018) Soil quality: a critical review. Soil Biol Biochem 120:105–125. https://doi.org/10.1016/j.soilbio.2018.01.030

Cai HM, Dong YY, Li YY, Li DX, Peng CY, Zhang ZZ, Wan XC (2016a) Physiological and cellular responses to fluoride stress in tea (*Camellia sinensis*) leaves. Acta Physiol Plant 38:144. https://doi.org/10.1007/S11738-016-2156-0

Cai H, Zhu X, Peng C, Xu W, Li D, Wang Y, Fang S, Li Y, Hu S, Wan X (2016b) Critical factors determining fluoride concentration in tea leaves produced from Anhui province, China. Ecotox Environ Saf 131:14–21. https://doi.org/10.1016/j.ecoenv.2016.04.023

Cakmak I (2013) Magnesium in crop production, food quality and human health. Plant Soil 368:1–4. https://doi.org/10.1007/s11104-013-1781-2

Cao D, Liu Y, Ma L, Jin X, Guo G, Tan R, Liu Z, Zheng L, Ye F, Liu W (2018) Transcriptome analysis of differentially expressed genes involved in selenium accumulation in tea plant (*Camellia sinensis*). PLoS One 13(6):e0197506. https://doi.org/10.1371/journal.pone.0197506

Cardoso EJBN, Vasconcellos RLF, Bini D, Miyauchi MYH, Santos CAD, Alves PRL, Paula AMD, Nakatani AS, Pereira JDM, Nogueira MA (2013) Soil health: looking for suitable indicators. What should be considered to assess the effects of use and management on soil health? Sci Agric 70(4):274–289. https://doi.org/10.1590/S0103-90162013000400009

Carr H, Lombi E, Küpper H, Mcgrath S, Wong M (2003) Accumulation and distribution of aluminium and other elements in tea (*Camellia sinensis*) leaves. Agronomie 23:705–710

Carter R (2007) Water management and smallholder fairtrade tea producers in South Western Uganda. MSc thesis, School of Applied Sciences, Cranfield University, Bedford

Caykur (2016) Caykur: Statistics Bulletin 2016. https://www.caykur.gov.tr

Cecchini G, Andreetta A, Marchetto A, Carnicelli S (2019) Atmospheric deposition control of soil acidification in Central Italy. Catena 182:104102. https://doi.org/10.1016/j.catena.2019.104102

Chakraborty U, Chakraborty BN, Chakraborty AP, Sunar K, Dey PL (2013) Plant growth promoting rhizobacteria mediated improvement of health status of tea plants. Indian J Biotechnol 12:20–31

Chandralal RPVT, Jayasekara APDA, Seneviratne G, Herath HMSK, Silva PDPM, Abeysinghe DC, Premarathna M (2020) Effect of biofilm biofertilizer on tea cultivation. Proceedings of the International Research Conference of Uva Wellassa University, July 29–30, 2020 Abstract p. 68

Chappell MJ, LaValle LA (2011) Food security and biodiversity: can we have both? An agroecological analysis. Agric Hum Values 28(1):3–26. https://doi.org/10.1007/s10460-009-9251-4

Chauhan RK, Singh AK, Bisen JS, Singh M, Bera B (2016) Establishment and bringing up of replanted tea: An organic approach to improve productivity and fertility of soil. Int J Agric Sci 7(1):43–51

Chelangat A, Gweyi-Onyango JP, Korir NK, Mwangi M (2021) Influence of arbuscular mycorrhizae on callusing and root colonization of tea (*Camellia sinensis*) clones in Kenya. Asian Soil Res J 5:21–26. https://doi.org/10.9734/ASRJ/2021/v5i130098

Chen J, Hu J (2014) Variation of biochemical components in different tea varieties under excessive calcium treatment. Shandong Agric Sci 46:74–76

Chen ZC, Yamaji N, Fujii-Kashino M, Ma JF (2016) A cation-chloride cotransporter gene is required for cell elongation and osmoregulation in rice. Plant Physiol 171(1):494–507. https://doi.org/10.1104/pp.16.00017

Chen YP, Tsai CF, Rekha PD, Ghate SD, Huang HY, Hsu YH, Liaw LL, Young CC (2021a) Agricultural management practices influence the soil enzyme activity and bacterial community structure in tea plantations. Bot Stud 62:8. https://doi.org/10.1186/s40529-021-00314-9

Chen X, Wang Y, Lin L, Zhang Q, Ding L, Guo W, Wang H (2021b) Effects of soil acidity on tea yield and contents of leave quality components. Chinese J Tropical Crops 42(1):260–266. https://doi.org/10.3969/j.issn.1000-2561.2021.01.035

Chen Y, Wang F, Wu Z, Jiang F, Yu W, Yang J, Chen J, Jian G, You Z, Zeng L (2021c) Effects of long-term nitrogen fertilization on the formation of metabolites related to tea quality in subtropical China. Meta 11:146. https://doi.org/10.3390/metabo11030146

Chen WC, Ko CH, Su YS, Lai WA, Shen FT (2021d) Metabolic potential and community structure of bacteria in an organic tea plantation. Appl Soil Ecol 157:103762. https://doi.org/10.1016/j.apsoil.2020.103762

Chen P, Liu Y, Mo C, Jiang Z, Yang J, Lin J (2021e) Microbial mechanism of biochar addition on nitrogen leaching and retention in tea soils from different plantation ages. Sci Total Environ 757:143817. https://doi.org/10.1016/j.scitotenv.2020.143817

Chen L, Zhou W, Zhang S, Deng L, Deng O, Luo L, Xu X, Yang C, Yu C (2022a) Improvement of tea yield and quality by chicken manure and wine lees (CMWL) substitution for chemical fertilizers in the hilly region of Western Sichuan, China. J Soil Sci Plant Nutr 22:4415–4426. https://doi.org/10.1007/s42729-022-01041-1. 12 pages

Chen D, Sun Z, Gao J, Peng J, Wang Z, Zhao Y, Lin Z, Dai W (2022b) Metabolomics combined with proteomics provides a novel interpretation of the compound differences among Chinese tea cultivars (*Camellia sinensis* var. *sinensis*) with different manufacturing suitabilities. Food Chem 377:131976. https://doi.org/10.1016/j.foodchem.2021.131976

Chong KP, Ho TY, Jalloh MB (2008) Soil nitrogen, phosphorus and tea leaf growth in organic and conventional farming of selected fields at Sabah tea plantation slope. J Sustain Dev 1(3):117–122. https://doi.org/10.5539/jsd.v1n3p117

Chopra A, Vandana UK, Rahi P, Satpute S, Mazumder PB (2020a) Plant growth promoting potential of *Brevibacterium sediminis* A6 isolated from the tea rhizosphere of Assam. India Biocat Agric Microbiol 27:101610. https://doi.org/10.1016/j.bcab.2020.101610

Chopra A, Bobate S, Rahi P, Banpurkar A, Mazumder PB, Satpute S (2020b) *Pseudomonas aeruginosa* RTE4: a tea rhizobacterium with potential for plant growth promotion and biosurfactant production. Front Bioeng Biotechnol 8:861. https://doi.org/10.3389/fbioe.2020.00861

Costantini EAC, Mocali S (2022) Soil health, soil genetic horizons and biodiversity. J Plant Nutr Soil Sci 185(1):24–34. https://doi.org/10.1002/jpln.202100437

Curtright AJ, Tiemann LK (2021) Intercropping increases soil extracellular enzyme activity: a meta-analysis. Agric Ecosyst Environ 319:107498. https://doi.org/10.1016/j.agee.2021.107489

Dai Z, Zhang X, Tang C, Muhammad N, Wu J, Brookes PC, Xu J (2017) Potential role of biochars in decreasing soil acidification— a critical review. Sci Total Environ 581:601–611. https://doi.org/10.1016/j.scitotenv.2016.12.169

Dai F, Rong Z, Wu Q, Adb Allah EF, Liu C, Liu S (2022) Mycorrhiza improves plant growth and photosynthetic characteristics of tea plants in response to drought stress. Biocell 46(5):1339–1346. https://doi.org/10.32604/biocell.2022.018909

Dambiec M, Polechonska L, Klink A (2013) Levels of essential and non-essential elements in black teas commercialized in Poland and their transfer to tea infusion. J Food Composition Analysis 31:62–66. https://doi.org/10.1016/j.jfca.2013.03.006

Dandeniya WS, Withana PA, Suriyagoda LBD, Dissanayake RJ, Ranasinghe CN, Thalagoda S (2022) Importance of soil organic carbon levels for increasing land productivity of tea smallholdings in Kandy and Kegalle districts in Sri Lanka. J Soil Sci Soc Sri Lanka 26:1–8

Dang MV (2002) Effects of tea cultivation on soil quality in the northern mountainous zone, Vietnam. Thesis Submitted to the College of Graduate Studies and Research in partial fulfilment of the requirements for the degree of Doctor of Philosophy in the Department of Soil Science, University of Saskatchewan Saskatoon-Canada. p. 175

Dang MV (2005) Soil–plant nutrient balance of tea crops in the northern mountainous region. Vietnam Agric Ecosys Environ 105:413–418. https://doi.org/10.1016/j.agee.2004.05.004

Daniel R (2005) The metagenomics of soil. Nat Rev Microbiol 3:470–478. https://doi.org/10.1038/nrmicro1160

Das S, Borua PK, Bhagat RM (2016) Soil nitrogen and tea leaf properties in organic and conventional farming systems under humid sub-tropical conditions. Organic Agric 6:119–132. https://doi.org/10.1007/s13165-015-0116-4

Dash B, Soni R, Kumar V, Suyal DC, Dash D, Goel R (2019) Mycorrhizosphere: microbial interactions for sustainable agricultural production. In: Varma A, Choudhary D (eds) Mycorrhizosphere and pedogenesis. Springer, Singapore, pp 321–338. https://doi.org/10.1007/978-981-13-6480-8_18

de Costa J, Mohotti AJ, Wijeratne MA (2007) Ecophysiology of tea. Braz J Plant Physiol 19(4):299–332. https://doi.org/10.1590/S1677-04202007000400005

Debona D, Rodrigues FA, Datnoff LE (2017) Silicon's role in abiotic and biotic plant stresses. Annu Rev Phytopathol 55:85–107. https://doi.org/10.1146/annurev-phyto-080516-035312

Deepshikha (2020) The effects of fertilizer on soil for tea production. Plant Archives 20:382–388

Deka N, Goswami K (2021) Economic sustainability of organic cultivation of Assam tea produced by small-scale growers. Sustain Prod Consump 26:111–125. https://doi.org/10.1016/j.spc.2020.09.020

Devi SS, Saha B, Awasthi JP, Regon P, Panda SK (2020) Redox status and oxalate exudation determines the differential tolerance of two contrasting varieties of 'Assam tea' [*Camelia sinensis* (L.) O. Kuntz] in response to aluminum toxicity. Hortic Environ Biotechnol 61:485–499. https://doi.org/10.1007/s13580-020-00241-x

Dey S, Mazumder PB, Paul SB (2014a) Accumulation of cu at different concentration in tea plant. IOSR JAgric Vet Sci 7(4):39–43

Dey S, Mazumder PB, Paul SB (2014b) Effect of copper on growth and chlorophyll content in tea plants (*Camellia sinensis* (L.) O. Kuntze). Int J Res Appl Nat Social Sci 2(5):223–230

Dey S, Mazumder PB, Paul SB (2015) Copper-induced changes in growth and antioxidative mechanisms of tea plant (*Camellia sinensis* (L.) O. Kuntze). Afr J Bot 14(7):582–592. https://doi.org/10.5897/AJB2014.14279

Dimkpa CO, Fugice J, Singh U, Lewis TD (2020) Development of fertilizers for enhanced nitrogen use efficiency trends and perspectives. Sci Total Environ 731:139113. https://doi.org/10.1016/j.scitotenv.2020.139113

Dinesh D, Rajan K, Thilagam VK, Khola OPS (2016) Manuring needs of tea (*Camellia sinensis*) in southern India. Rashtriya Krishi 11(1):12–14

Ding Z, Jia S, Wang Y, Xiao J, Zhang Y (2017) Phosphate stresses affect ionome and metabolome in tea plants. Plant Physiol Biochem 120:30–39. https://doi.org/10.1016/j.plaphy.2017.09.007

Doanh N, Thuong N, Heo Y (2018) Impact of conversion to organic tea cultivation on household income in the mountainous areas of northern Vietnam. Sustainability 10(12):4475. https://doi.org/10.3390/su10124475

Dong F, Hu JH, Shi YZ, Liu MY, Zhang QF, Ruan JY (2019) Effects of nitrogen supply on flavonol glycoside biosynthesis and accumulation in tea leaves (*Camellia sinensis*). Plant Physiol Biochem 138:48–57. https://doi.org/10.1016/j.plaphy.2019.02.017

Duan XH, Hu XF, Deng ZY, Chen FS (2012) Manipulation effects of calcium and aluminum on the contents of main chemical qualities of tea leaves and distribution of some microelements in different modules of tea plant. Acta Bot Boreali-Occid Sin 32:0988–0994. https://doi.org/10.3969/j.issn.1000-4025.2012.05.022

Duan Y, Shen J, Zhang X, Wen B, Ma Y, Wang Y, Fang W, Zhu X (2019) Effects of soybean–tea intercropping on soil-available nutrients and tea quality. Acta Physiol Plant 41:140. https://doi.org/10.1007/s11738-019-2932-8

Duan Y, Shang X, Liu G, Zou Z, Zhu X, Ma Y, Li F, Fang W (2021) The effects of tea plants-soybean intercropping on the secondary metabolites of tea plants by metabolomics analysis. BMC Plant Biol 21:482. https://doi.org/10.1186/s12870-021-03258-1

Dufitumukiza W, Owuor PO, Bigirimana J, Akingeneye AR, Murenzi B (2020) Influence of nitrogenous fertilizer rates and plucking intervals on tea in peatland and highland Ultisol soils of Rwanda: 1 tea yields. Int J Tea Sci 15(1):46–51

Dumont AM, Wartenberg AC, Baret PV (2021) Bridging the gap between the agroecological ideal and its implementation into practice: a review. Agronr Sustain Dev 41(3):1–17. https://doi.org/10.1007/s13593-021-00666-3

Dutta J, Handique PJ, Thakur D (2015) Assessment of culturable tea rhizobacteria isolated from tea estates of Assam, India for growth promotion in commercial tea cultivars. Front Microbiol 6:1252. https://doi.org/10.3389/fmicb.2015.01252. 13 pages

Eden T (1965) Tea. Longmans, Harlow, p 205p

Falla NM, Demasi S, Caser M, Scariot V (2021) Phytochemical profile and antioxidant properties of Italian green tea, a new high quality niche product. Horticulturae 7(5):91, 12 pages. https://doi.org/10.3390/horticulturae7050091

Fan DM, Fan K, Yu CP, Lu YT, Wang XC (2017) Tea polyphenols dominate the short-term tea (Camellia sinensis) leaf litter decomposition. J Zhejiang Univ Sci B18:99–108. https://doi.org/10.1631/jzus.B1600044

Fan K, Wang M, Gao Y, Ning Q, Shi Y (2019) Transcriptomic and ionomic analysis provides new insight into the beneficial effect of Al on tea roots' growth and nutrient uptake. Plant Cell Rep 38(6):715–729. https://doi.org/10.1007/s00299-019-02401-5

Fan ZH, Tang XJ, Zheng D, Yang Q, Chen GN, Li XW, Sun J (2020a) Study and prospect of soil acidification causes and improvement measures in tea plantation. J Tea Sci 40(1):15–25. https://doi.org/10.13305/j.cnki.jts.20200117.006

Fan K, Zhang Q, Tang D, Shi Y, Ma L, Liu M, Ruan J (2020b) Dynamics of nitrogen translocation from mature leaves to new shoots and related gene expression during spring shoots development in tea plants (*Camellia sinensis* L.). J Plant Nutr Soil Sci 183(2):180–191. https://doi.org/10.1002/jpln.201900268

Fan Z, Jia W, Du A, Xia Z, Kang J, Xue L, Sun Y, Shi L (2022a) Discovery of Se-containing flavone in Se-enriched green tea and the potential application value in the immune regulation. Food Chem 394:133468. https://doi.org/10.1016/j.foodchem.2022.133468

Fan S, Hongjing H, Yanting C, Lilin C (2022b) Effects of intercropping functional plants on the ecosystem functions and services in tea garden. J Tea Sci 42(2):151–168

FAO (2009) Food security and agricultural mitigation in developing countries: options for capturing synergies. Food and Agriculture Organization of the United Nations, Rome. 79 p

FAO (2016) Report of the working group on climate change of the FAO intergovernmental group on tea. FAO, Rome, p 84p

Farooq TH, Kumar U, Mo J, Shakoor A, Wang J, Rashid MHU, Tufail MA, Chen X, Yan W (2021) Intercropping of peanut-tea enhances soil enzymatic actitivty and soil nutrient status at different soil profiles in subtropical southern China. Plan Theory 10:881. https://doi.org/10.3390/plants10050881

Feng Y, Sunderland T (2023) Feasibility of tea/tree intercropping plantations on soil ecological service function in China. Agronomy 13:1548. https://doi.org/10.3390/agronomy13061548. 19 pages

Feng G, Xie T, Wang X, Bai J, Tang L, Zhao H, Wei W, Wang M, Zhao Y (2018) Metagenomic analysis of microbial community and function involved in cd-contaminated soil. BMC Microbiol 18:11. https://doi.org/10.1186/s12866-018-1152-5

Figueiredo DD, Barros PM, Cordeiro AM, Serra TS, Lourenço T, Chander S, Oliveira MM, Saibo NJ (2012) Seven zinc-finger transcription factors are novel regulators of the stress responsive gene OsDREB1B. J Exp Bot 63:3643–3656

Frag MA, Elmetwally F, Elghanam R, Kamal N, Hellal K, Hamezah HS, Zhao C, Mediani A (2023) Metabolomics in tea products; a compile of applications for enhancing agricultural traits and quality control analysis of *Camellia sinensis*. Food Chem 404:134628. https://doi.org/10.1016/j.foodchem.2022.134628

Fu Z, Jiang X, Li WW, Shi Y, Lai S, Zhiang J, Yao S, Liu Y, Hu J, Gao L, Xia T (2020) Proanthocyanidin-aluminum complexes improve aluminum resistance and detoxification of *Camellia sinensis*. J Agric Food Chem 68(30):7861–7869. https://doi.org/10.1021/acs.jafc.0c01689

Fu H, Li H, Yin P, Mei H, Li J, Zhou P, Wang Y, Ma Q, Jeyaraj A, Thangaraj K, Chen X, Li X, Guo G (2021) Integrated application of rapeseed cake and green manure enhances soil nutrients and microbial communities in tea garden soil. Sustainability 13(5):2967. https://doi.org/10.3390/su13052967

Fu H, Song D, Wang K, Fang F, Han S, Yang F, Ding S (2023) Application of wheat straw compost mixed with chemical fertilizer regulates soil bacterial community diversity in tea (*Camellia sinensis*) plantation. Diversity 5:580. https://doi.org/10.3390/d15040580

Fung KF, Wong MH (2002) Effects of soil pH on the uptake of Al, F and other elements by tea plants. J Sci Food Agric 82(1):146–152. https://doi.org/10.1002/jsfa.1007

Fung KF, Wong MH (2004) Application of different forms of calcium to tea soil to prevent aluminum and fluorine accumulation. J Sci Food Agric 84(12):1469–1477. https://doi.org/10.1002/jsfa.1842

Fung KF, Carr HP, Zhang J, Wong MH (2008) Growth and nutrient uptake of tea under different aluminium concentrations. J Sci Food Agric 88(9):1582–1591. https://doi.org/10.1002/jsfa.3254

Fung KF, Carr HP, Poon BHT, Wong MH (2009) A comparison of aluminum levels in tea products from Hong Kong markets and in varieties of tea plants from Hong Kong and India. Chemosphere 75(7):955–962. https://doi.org/10.1016/j.chemosphere.2009.01.003

Galloway JN, Winiwarter W, Leip A, Leach AM, Bleeker A, Erisman JW (2014) Nitrogen footprints: past, present and future. Environ Res Lett 9:115003. https://doi.org/10.1088/1748-9326/9/11/115003

Gao HJ, Zhao Q, Zhang XC, Wan XC, Mao JD (2014) Localization of fluoride and aluminum in subcellular fractions of tea leaves and roots. J Agric Food Chem 62:2313–2319. https://doi.org/10.1021/jf4038437

Gebrewold AZ (2018) Review on integrated nutrient management of tea (*Camellia sinensis* L.). Cogent Food Agric 4:1,1543536. https://doi.org/10.1080/23311932.2018.1543536. 12 pages

Ghiasian M (2020) Microbial biofilms: beneficial applications for sustainable agriculture. In: New and future developments in microbial biotechnology and bioengineering. Elsevier, Amsterdam, pp 145–155. https://doi.org/10.1016/B978-0-12-820526-6.00009-9

Ghosh S, Goswami A, Ghosh GK, Pramanik P (2021) Characterization of potent phytate solubilizing bacterial strains of tea garden soils as futuristic potent bioinoculant. Int J Curr Microbiol App Sci 10(4):470–484. https://doi.org/10.20546/ijcmas.2021.1004.049

Gohain T, Barbora AC, Deka A, Ahmed KZ (2000) Effect of different sources of Sulphur on yield and quality of tea (*Camellia sinensis* L. (O) Kuntez). J Inter Des 4(3):370–375

Gong X, Luo F, Tang X, Wang X, Li C, Wang Y, Wang Y, Du X (2017) Model construction of potassium accumulation and utilization in tea seedling. China. J Appl Ecol 28:2597–2604. https://doi.org/10.13287/j.1001-9332.201708.016

Gransee A, Fuhrs H (2013) Magnesium mobility in soils as a challenge for soil and plant analysis, magnesium fertilization and root uptake under adverse growth conditions. Plant Soil 368:5–21. https://doi.org/10.1007/s11104-012-1567-y

Gu S, Hu Q, Cheng Y, Bai L, Liu Z, Xiao W, Gong Z, Wu Y, Feng K, Deng Y, Tan L (2019) Application of organic fertilizer improves microbial community diversity and alters microbial network structure in tea (*Camellia sinensis*) plantation soils. Soil Tillage Res 195:104356. https://doi.org/10.1016/j.still.2019.104356

Gu Y, Qiu Y, Wei X, Li Z, Hu Z, Gu Y, Zhao Y, Wang Y, Yue T, Yuan Y (2020) Characterization of selenium containing polysaccharides isolated from selenium-enriched tea and its bioactivities. Food Chem 316:126371. https://doi.org/10.1016/j.foodchem.2020.126371

Gui H, Fan L, Wang D, Yan P, Li X, Zhang L, Han W (2021) Organic management practices shape the structure and associations of soil bacterial communities in tea plantations. Appl Soil Ecol 163:103975. https://doi.org/10.1016/j.apsoil.2021.103975

Gunarathne GP, Arachchi LP (2016) Manurial effect of wood ash and refuse tea on nutrient status and yield of tea (*Camellia sinensis* L.). Procedia Food Sci 6:261–266

Guo L (2014) The effect of different levels of potassium and magnesium on the change of low-molecule terpenoid and catechin amount in tea leaves. Anhui Agricultural University, Hefei

Guo S, Wang Q, Tang L, Zhang T, Li J, Xiao Y, Gao Y, Bai J, Xiao B, Gong C (2021) Inoculation with arbuscular mycorrhizal fungi reinforces tea plant's tolerance to salinity. J Plant Growth Regulation 1-20. https://doi.org/10.1007/s00344-021-10529-6

Guo S, Wang Q, Tang L, Zhang T, Li J, Xiao Y, Gao Y, Bai J, Xiao B, Gong C (2022) Inoculation with arbuscular mycorrhizal fungi reinforces tea plant's tolerance to salinity. J Plant Growth Regul 41:3498–3517. https://doi.org/10.1007/s00344-021-10529-6

Guo H, Chen L, Wang Y, Li Q, Yi Z (2023) Combined application of biochar and magnesium fertilizer effectively improved the soil environment and the tea quality in southern strongly acidic tea garden. J Soils Sediments 23:2798–2815. https://doi.org/10.1007/s11368-023-03495-x

Hajiboland R (2017) Environmental and nutritional requirements for tea cultivation. Folia Hort 29(2):199–220. https://doi.org/10.1515/fhort-2017-0019

Hajiboland R (2018) Nutrient deficiency and abundance in tea plants: metabolism to productivity. In: Han WY, Li X, Ahammed G (eds) Stress physiology of tea in the face of climate change. Springer, Singapore, pp 173–215. https://doi.org/10.1007/978-981-13-2140-5_9

Hajiboland R, Salehi SY (2014) Remobilization of phosphorus in tea plants. J Plant Nutr 37:1522–1533. https://doi.org/10.1080/01904167.2014.888743

Hallmann E, Rusaczonek A, Rembiałkowska E (2007) The phenolic compounds content and antioxidant properties of the organic and conventional teas. Proceedings of the ICPH 2007 3rd International Conference on Polyphenols and Health. Kyoto, Japan

Han W, Ruan J, Lin Z (2002) The major nutrient limiting factors in tea soils and development of series tea specialty fertilizers. J Tea Sci Res 22:70–74. https://doi.org/10.1016/j.scienta.2021.110078

Han GY, Deng XB, Jiang C, Gu XR (2007) The research progress on the cause and control of aluminum toxicity to plants. J Fujian Forestry Sci Technol 34(2):174–179

Han WY, Ma LF, Shi YZ, Ruan JY, Kemmitt SJ (2008) Nitrogen release dynamics and transformation of slow release fertiliser products and their effects on tea yield and quality. J Sci Food Agr 88(5):839–846. https://doi.org/10.1002/jsfa.3160

Han WY, Xu JM, Yi XY, Lin YD (2012) Net and gross nitrification in tea soils of varying productivity and their adjacent forest and vegetable soils. Soil Sci Plant Nutr 58(2):173–182. https://doi.org/10.1080/00380768.2012.664783

Han WY, Xu JM, Wei K, Shi RZ, Ma LF (2013) Soil carbon sequestration, plant nutrients and biological activities affected by organic farming system in tea (*Camellia sinensis* (L.) O. Kuntze) fields. Soil Sci Plant Nutr 59(5):727–739. https://doi.org/10.1080/00380768.2013.833857

Han WY, Li X, Yan P, Ahammed GJ (2016) Impact of climate change on tea economy and adaptation strategies in China. Report of the working group on climate change of the FAO intergovernmental group on tea. Food and Agriculture Organization of the United Nations, Rome, pp 61–77

Han WY, Wang DH, Fu SW, Ahmed S (2018a) Tea from organic production has higher functional quality characteristics compared with tea from conventional management systems in China. Bioll Agric Hort:120–131. https://doi.org/10.1080/01448765.2017.1396497

Han W, Li X, Zhang L, Ahammed GJ (2018b) Tea cultivation under changing climatic conditions. In: Sharma VS, Gunasekare MTK (eds) Global tea science: current status and future needs. Burleigh Dodds Science Publishing, Cambridge. https://doi.org/10.19103/AS.2017.0036.19

Han Z, Wang J, Xu P, Li Z, Liu S, Zou J (2021) Differential responses of soil nitrogen-oxide emissions to organic substitution for synthetic fertilizer and biochar amendment in a subtropical tea plantation. GCB Bioenergy 13(8):1260–1274. https://doi.org/10.1111/gcbb.12842

Han Z, Xu P, Li Z, Lin H, Zhu C, Wang J, Zou J (2022) Microbial diversity and the abundance of keystone species drive the response of soil multifunctionality to organic substitution and biochar amendment in a tea plantation. GCB Bioenergy 14:481–495. https://doi.org/10.1111/gcbb.12926

Hao HP, Xia F, Bai HT, Yan XY, Cui HX, Xie XM, Shi L (2022) Impact of *Litsea cubeba* inter-row cover on the structure of bacterial community in the tea plantation. Annals Microbiol 72:37. https://doi.org/10.1186/s13213-022-01696-2

Haojie Z, Xinyuan H, Chao Z, Lu W, Xinchao W, Yajun Y, Jianming Z, Juhui D, Jun X, Hong L, Chaoyang W, Xiangui Z, Tao L (2020) Correlation analysis between selenium accumulation in tea leaves and soil factors in selenium-rich areas. J Tea Sci 40(4):465–477. https://doi.org/10.13305/j.cnki.jts.2020.04.003

Harler CR (1966) Tea soils. In: Tea growing, 3rd edn. Oxford University Press, London, pp 34–40

Hawkesford M, Horst W, Kichey T, Lambers H, Schjoerring J, Møller SI, White P (2012) Functions of macronutrients. In: Marschner P (ed) Marschner's mineral nutrition of higher plants. Academic Press, Cambridge, pp 135–189

He T, Yuan J, Luo J, Wang W, Fan J, Liu D, Ding W (2019) Organic fertilizers have divergent effects on soil N2O emissions. Biol Fertil Soils 55:685–699. https://doi.org/10.1007/s00374-019-01385-4

He SQ, Zheng ZC, Zhu RH (2021) Long-term tea plantation effects on composition and stabilization of soil organic matter in Southwest China. Catena 199(1–2):105132. https://doi.org/10.1016/j.catena.2020.105132

He D, Chen X, Zhang Y, Huang Z, Yin J, Weng X, Yang W, Wu H, Zhang F, Wu L (2023) Magnesium is a nutritional tool for the yield and quality of oolong tea (*Camellia sinensis* L.) and reduces reactive nitrogen loss. Scientia Hort 308:111590. https://doi.org/10.1016/j.scienta.2022.111590

Hemalatha K, Venkatesan S, Jayaganesh S (2006) Characterization of iron toxicity and its influence on nutrient uptake in tea. J Plant Crops 34(3):268–272

Hemmati V, Garavand F, Khorshidian N, Cacciotti I, Goudarzi M, Chaichi M, Tiwari BK (2021) Impact of cold atmospheric plasma on microbial safety, total phenolic and favonoid contents, antioxidant activity, volatile compounds, surface morphology, and sensory quality of green tea powder. Food Biosci 44:101348. https://doi.org/10.1016/j.fbio.2021.101348

Hirono Y, Nonaka K (2012) Nitrous oxide emissions from green tea fields in Japan: contribution of emissions from soil between rows and soil under the canopy of tea plants. Soil Sci Plant Nutr 58(3):384–392. https://doi.org/10.1080/00380768.2012.686434

Hirono Y, Sano T, Eguchi S (2021) Changes in the nitrogen footprint of green tea consumption in Japan from 1965 to 2016. Environ Sci Pollut Res 28:44936–44948. https://doi.org/10.1007/s11356-021-13767-z

Hoang TX, Thang VN, Van Thu DV, Binh NN, Toan NV, Hoang DT (2021) Effects of mineral fertilizer doses and ratios on tea yield and quality. Vietnam J Agric Sci 4(2):997–1006. https://doi.org/10.31817/vjas.2021.4.2.01

Hu XF, Chen FS, Wine ML, Fang XM (2017) Increasing acidity of rain in subtropical tea plantation alters aluminum and nutrient distributions at the root-soil interface and in plant tissues. Plant Soil 417:261–274. https://doi.org/10.1007/s11104-017-3256-3

Huang Y, Long X, Chapman SJ, Yao H (2015) Acidophilic denitrifiers dominate the N_2O production in a 100-year-old tea orchard soil. Environ Sci Pollut Res 22(6):4173–4182

Huang D, Gong Z, Chen X, Wang H, Tan R, Mao Y (2021) Transcriptomic responses to aluminum stress in tea plant leaves. Sci Rep 11(1):5800. https://doi.org/10.1038/s41598-021-85393-1

Huang W, Lin M, Liao J, Li A, Tsewang W, Chen X, Sun B, Liu S, Zheng P (2022a) Effects of potassium deficiency on the growth of tea (*Camelia sinensis*) and strategies for optimizing potassium levels in soil: a critical review. Horticulture 8(7):660. https://doi.org/10.3390/horticulturae8070660. 12 pages

Huang D, Wang Y, Chen X, Wu J, Wang H, Tan R, Jiao L, Mao Y (2022b) Application of tea-specific fertilizer combined with organic fertilizer improves aroma of green tea. Horticulture 8:950. https://doi.org/10.3390/horticulturae8100950

Huang W, Ma D, Hao X, Li J, Xia L, Zhang E, Wang P, Wang M, Guo F, Wang Y, Ni D, Zhao H (2022c) CsATG101 delays growth and accelerates senescence response to low nitrogen stress in *Arabidopsis thaliana*. Front Plant Sci 13:880095. https://doi.org/10.3389/fpls.2022.880095

Huang Z, Cui C, Cao Y, Dai J, Cheng X, Hua S, Wang W, Duan Y, Petropoulos E, Wang H, Zhou L, Fang W, Zhing Z (2022d) Tea plant–legume intercropping simultaneously improves soil fertility and tea quality by changing *bacillus* species composition. Hort Res 9:uhac046. https://doi.org/10.1093/hr/uhac046

Huang Z, Wang F, Li B, Pang Y, Du Z (2023) Appropriate nitrogen form and application rate can improve yield and quality of autumn tea with drip irrigation. Agronomy 13:1303. https://doi.org/10.3390/agronomy13051303

Illukpitiya P, Shanmugaratnam N, Kjosavik DJ (2004) Tea agroecosystems in the Uva highlands of Sri Lanka. Mountain Res Dev 24(1):52–59

Ippolito JA, Ducey TF, Diaz K, Barbarick KA (2021) Long-term biosolids land application influences soil health. Sci Total Environ 791:148344. https://doi.org/10.1016/j.scitotenv.2021.148344
Islam S, Hamid FS, Amin K, Sumreen S, Zamane Q, Khan N, Khan A, Sha BH (2017) Effect of organic fertilizer on the growth of tea (*Camellia sinensis* L.). Int J Sci Basic Appl Res 36:1–9
Jahan I, Shopan J, Rahman MM, Sarkar A, Baset MA, Zhang Z, Li X, Ahammed GJ, Hasan MK (2022) Long-term traditional fertilization alters tea garden soil properties and tea leaf quality in Bangladesh. Agron 12(9):2128. https://doi.org/10.3390/agronomy12092128
Jayaganesh S, Venkatesan S, Senthurpandian VK (2011) Impact of different sources and doses of magnesium fertilizer on biochemical constituents and quality parameters of black tea. Asian J Biochem 6(3):273–281. https://doi.org/10.3923/ajb.2011.273.281
Jayasekara APDA, Premarathna M, Abeysinghe DC, Seneviratne G (2021) Biofilm biofertilizer increases soil carbon sequestration in tea and paddy cultivations. Abstract of Proceedings of Wayamba University Research Congress 2021, Senate Research and Higher Degrees Committee. pp. 7–8
Jayasinghe SL, Kumar L (2019) Modeling the climate suitability of tea [*Camellia sinensis* (L.) O. Kuntze] in Sri Lanka in response to current and future climate change scenarios. Agric Forest Meteor 272-273:102–117. https://doi.org/10.1016/j.agrformet.2019.03.025
Jayasinghe SL, Kumar L (2020) Climate change may imperil tea production in the four major tea producers according to climate prediction models. Agronomy 10:1536. https://doi.org/10.3390/agronomy10101536. 21 pages
Jessy MD (2010) Potassium management in plantation crops with special reference to tea, coffee and rubber. Karnataka J Agric Sci 24:67–74
Ji L, Wu Z, You Z, Yi X, Ni K, Guo S, Ruan J (2018) Effects of organic substitution for synthetic N fertilizer on soil bacterial diversity and community composition: a 10-year field trial in a tea plantation. Agric Ecosyst Environ 268:124–132. https://doi.org/10.1016/j.agee.2018.09.008
Ji C, Li S, Geng Y, Yuan Y, Zhi J, Yu K, Han Z, Wu S, Liu S, Zou J (2020a) Decreased N_2O and NO emissions associated with stimulated denitrification following biochar amendment in subtropical tea plantations. Geoderma 365:114223. https://doi.org/10.1016/j.geoderma.2020.114223
Ji L, Ni K, Wu Z, Zhang J, Yi X, Yang X, Ling N, You Z, Guo S, Ruan J (2020b) Effect of organic substitution rates on soil quality and fungal community composition in a tea plantation with long-term fertilization. Biol Fertil Soils 56:633–646. https://doi.org/10.1007/s00374-020-01439-y
Jia W, Liu Y, Shi L, Chu X (2020) Investigation of differentially expressed proteins induced by alteration of natural se uptake with ultrahigh-performance liquid chromatography quadrupole orbitrap uncovers the potential nutritional value in se-enriched green tea. J Agric Food Chem 68(23):6316–6332. https://doi.org/10.1021/acs.jafc.0c02130
Jiang Y, Lin X, Khan MU, Jiang W, Xu Y, Li Z, Lin W (2023) Tea pruning for the umbrella-shaped canopy can alleviate rhizosphere soil degradation and improve the ecosystem functioning of tea orchards. Catena 222:106885. https://doi.org/10.1016/j.catena.2022.106885
Jiao S, Peng Z, Qi J, Gao J, Wei G (2021) Linking bacterial-fungal relationships to microbial diversity and soil nutrient cycling. Systems 6(2):e01052–e01020. https://doi.org/10.1128/mSystems.01052-20
Jin CW, Du ST, Zhang K, Lin XY (2008) Factors determining copper concentration in tea leaves produced at Yuyao County, China. Food Chem 46(6):2054–2061. https://doi.org/10.1016/j.fct.2008.01.046
Kamau DM (2008) Productivity and resource use in ageing tea plantations. Ph.D. thesis Wageningen University, Wageningen. 140 p
Kamau DM, Wanyoko JK, Owuor PO, Ng'etich WK (2008) NPK(S) fertilizer use in commercially cultivated clone BBK 35 in different tea growing regions of Kenya: II. Final soil chemical properties. Tea 29(2):8–18
Karak T, Bhagat RM (2010) Trace elements in tea leaves, made tea and tea infusion: a review. Food Res Int 43(9):2234–2252. https://doi.org/10.1016/j.foodres.2010.08.010
Karak T, Abollino O, Bhattacharyya P, Das KK, Paul RK (2011) Fractionation and speciation of arsenic in three tea gardens soil profiles and distribution of as in different parts of

tea plant (*Camellia sinensis* L.). Chemosphere 85:948–960. https://doi.org/10.1016/j.chemosphere.2011.06.061

Karak T, Kutu FR, Nath JR, Sonar I, Paul RK, Boruah RK, Sanyal S, Sabhapondit S, Dutta AK (2015) Micronutrients (B, co, cu, Fe, Mn, Mo and Zn) content in made tea (*Camellia sinensis* L.) and tea infusion with health prospect: a critical review. Critical Rev Food Sci Nutr 57(14):2996–3034. https://doi.org/10.1080/10408398.2015.1083534

Karak T, Bora K, Paul RK, Das S, Khare P, Dutta AK, Boruah RK (2017) Paradigm shift of contamination risk of six heavy metals in tea (*Camellia sinensis* L.) growing soil: a new approach influenced by inorganic and organic amendments. J Hazardous Materials 338:250–264. https://doi.org/10.1016/j.jhazmat.2017.05.036

Karlen D, Mausbach MJ, Doran J, Cline R, Harris R, Schuman G (1997) Soil quality: a concept, definition, and framework for evaluation (a guest editorial). Soil Sci Soc Am J 61(1):4–10

Kazimierczak R, Hallmann E, Rusaczonek A, Rembiałkowska E (2013) Polyphenols, tannins and caffeine content and antioxidant activity of green teas coming from organic and non-organic production. Renewable Agric Food Systems 30(3):263–269. https://doi.org/10.1017/S1742170513000513. 7 pages

Kemp SJ, Smith FW, Wagner D, Mounteney I, Bell MCJ, Gowing CJB, Pottas TL (2016) An improved approach to characterize potash-bearing evaporite deposits, evidenced in North Yorkshire, United Kingdom. Econ Geol 111:719–742. https://doi.org/10.2113/econgeo.111.3.719

Kerr RB, Madsen S, Stüber M, Liebert J, Enloe S, Borghino N, Parros P, Muyambai DM, Prudhon M, Wezel A (2021) Can agroecology improve food security and nutrition? A review. Global Food Security 29:100540. https://doi.org/10.1016/j.gfs.2021.100540

Khan AQ, Biswas A, Zohora KFT, Saha AK (2013) Effect of poultry manure on the yield of mature tea Vis-à-Vis soil properties. Tea J Bangladesh 42:37–43

Khokhar S, Magnusdottir SGM (2002) Total phenol, catechin, and caffeine contents of teas commonly consumed in the United Kingdom. J Agric Food Chem 50:565–570. https://doi.org/10.1021/jf010153l

Kozlova NV, Malyukova LS (2021) Change in the soils' fertility level of tea agrocenoses in the transition to cultivation without mineral fertilizers in the humid subtropical zone of Russia. E3S Web of Conferences 254:05009. https://doi.org/10.1051/e3sconf/202125405009

Kumar R, Parmar BS, Walia S, Saha S (2015a) Nitrification inhibitors: classes and its use in nitrification management. In: Rakshit A, Singh HB, Sen A (eds) Nutrient use efficiency: from basics to advances. Springer, New Delhi, pp 103–122. https://doi.org/10.1007/978-81-322-2169-2_8

Kumar R, Bisen JS, Singh M, Bera B (2015b) Effect of pruning and skiffing on growth and productivity of Darjeeling tea (*Camellia sinensis* L.). Int J Tech Res Appl 3(3):28–34

Le VS, Lesueur D, Herrmann L, Hudek L, Quyen LN, Brau L (2021) Sustainable tea production through agroecological management practices in Vietnam: a review. Environ Sustain 4:589–604. https://doi.org/10.1007/s42398-021-00182-w

Le VS, Herrmann L, Brau L, Lesueur D (2023) Sustainable green tea production through agroecological management and land conversion practices for restoring soil health, crop productivity and economic efficiency: evidence from northern Vietnam. Soil Use Manag 00:1–20. https://doi.org/10.1111/sum.12885

Leach AM, Galloway JN, Bleeker A, Erisman JW, Kohn R, Kitzes J (2012) A nitrogen footprint model to help consumers understand their role in nitrogen losses to the environment. Environ Dev 1:40–66. https://doi.org/10.1016/j.envdev.2011.12.005

Lei Q (2003) Studies on potassium equilibrium in tea-grown soils and yields and qualities of tea. Southwest University, Xi'an

Lei Q, Yuan L (2008) Effect of potassium fertilizer on the absorption of mineral nutrients and yield and quality of young tea. China Soil Fert 25:41–44

Lei W, Lun XY, Rachel K, Yu W, De-Li T, Mao YE, Jiang X (2013) Combined use of alkaline slag and rapeseed cake to ameliorate soil acidity in an acid tea garden soil. Pedosphere 23:177–184. https://doi.org/10.1016/S1002-0160(13)60005-3

Lei X, Wang T, Yang B, Duan Y, Zhou L, Zou Z, Ma Y, Zhu X, Fang W (2022) Progress and perspective on intercropping patterns in tea plantations. Beverage Plant Res 2:18, 10 pages. https://doi.org/10.48130/BPR-2022-0018

Li J, Xia J (2005) Summary on nitrogen (N), phosphorus (P), potassium (K) and tea quality. Chinese Agric Sci Bull 21(1):62–65

Li CL, Xu HM, Xu J, Chun XY, Ni DJ (2011) Effects of aluminium on ultrastructure and antioxidant activity in leaves of tea plant. Acta Physiol Plant 33(3):973–978. https://doi.org/10.1007/s11738-010-0629-0

Li H, Ma Y, Liu W, Liu W (2012) Soil changes induced by rubber and tea plantation establishment: comparison with tropical rain forest soil in Xishuangbanna, SW China. Environ Manage 50(5):837–848. https://doi.org/10.1007/s00267-012-9942-2

Li XY, Liu QZ, Liu ZL, Shi WP, Yang DW, Tarasco E (2014) Effects of organic and other management practices on soil nematode communities in tea plantation: a case study in southern China. J Plant Nutr Soil Sci 177(4):604–612. https://doi.org/10.1002/jpln.201300610

Li W, Zheng Z, Li T, Zhang X, Wang Y, Yu H, He S, Liu T (2015a) Effect of tea plantation age on the distribution of soil organic carbon fractions within water-stable aggregates in the hilly region of Western Sichuan, China. Catena 133:198–205. https://doi.org/10.1016/j.catena.2015.05.017

Li F, Chung T, Pennington JG, Federico ML, Kaeppler HF, Kaeppler SM, Otegui MS, Vierstra RD (2015b) Autophagic recycling plays a central role in maize nitrogen remobilization. Plant Cell 27(5):1389–1408. https://doi.org/10.1105/tpc.15.00158

Li LH, Fu QL, Achal V, Liu YL (2015c) A comparison of the potential health risk of aluminum and heavy metals in tea leaves and tea infusion of commercially available green tea in Jiangxi, China. Environ Monit Asses 187:228. https://doi.org/10.1007/s10661-015-4445-2

Li CF, Zhu Y, Yu Y, Zhao QY, Wang SJ, Wang XC, Yao MZ, Luo D, Li X, Chen L, Yang YY (2015d) Global transcriptome and gene regulation network for secondary metabolite biosynthesis of tea plant (*Camellia sinensis*). BMC Genomics 16:560. https://doi.org/10.1186/s12864-015-1773-0. 21 pages

Li S, Li H, Yang C, Wang Y, Xue H, Niu Y (2016a) Rates of soil acidification in tea plantations and possible causes. Agric Ecosyst Environ 233:60–66. https://doi.org/10.1016/j.agee.2016.08.036

Li Y, Zheng X, Fu X, Wu Y (2016b) Is green tea still 'green'? Geo: Geogr Environ 3:e00021. https://doi.org/10.1002/geo2.21. 6 pages

Li Y, Li Z, Li ZW, Jiang YH, Weng BQ, Lin WX (2016c) Variations of rhizosphere bacterial communities in tea (*Camellia sinensis* L.) continuous cropping soil by high-throughput pyrosequencing approach. J Appl Microbiol 121:787–799. https://doi.org/10.1111/jam.13225

Li D, Shu Z, Ye X, Zhu J, Pan J, Wang W, Chang P, Cui C, Shen J, Fang Q, Zhu X, Wang Y (2017) Cell wall pectin methyl-esterification and organic acids of root tips involve in aluminum tolerance in *Camellia sinensis*. Plant Physiol Biochem 119:265–274. https://doi.org/10.1016/j.plaphy.2017.09.002

Li Q, Chai S, Li Y, Huang J, Luo Y, Xiao L, Liu Z (2018) Biochemical components associated with microbial community shift during the pile-fermentation of primary dark tea. Front Microbiol 9:1509. https://doi.org/10.3389/fmicb.2018.01509

Li Y, Wang W, Wei K, Ruan L, Wang L, Cheng H, Zhang F, Wu L, Bai P (2019) Differential transcriptomic changes in low-potassium sensitive and low-potassium tolerant tea plant (*Camellia sinensis*) genotypes under potassium deprivation. Scientia Hort 256:108570. https://doi.org/10.1016/j.scienta.2019.108570

Li YC, Li ZW, Arafat Y, Lin WX (2020) Studies on fungal communities and functional guilds shift in tea continuous cropping soils by high-throughput sequencing. Ann Microbiol 70:1–12. https://doi.org/10.1186/s13213-020-01555-y

Li D, Zhou C, Zou N, Wu Y, Zhang J, An Q, Li J, Pan C (2021a) Nanoselenium foliar application enhances biosynthesis of tea leaves in metabolic cycles and associated responsive pathways. Environ Pollution 273:116503. https://doi.org/10.1016/j.envpol.2021.116503

Li J, Li QH, Zhang XY, Zhang LY, Zhao PL, Wen T, Zhang JQ, Xu WL, Guo F, Zhao H, Wang Y, Wang P, Ni DJ, Wang ML (2021b) Exploring the effects of magnesium deficiency on the quality constituents of hydroponic-cultivated tea (*Camellia sinensis* L.) leaves. J Agric Food Chem 69(47):14278–14286. https://doi.org/10.1021/acs.jafc.1c05141

Li W, Cheng H, Mu Y, Xu A, Ma B, Wang F, Xu P (2021c) Occurrence, accumulation, and risk assessment of trace metals in tea (*Camellia sinensis*): a national reconnaissance. Sci Total Environ 792:148354. https://doi.org/10.1016/j.scitotenv.2021.148354

Li H, Hu Z, Wan Q, Mu B, Li G, Yang Y (2022) Integrated application of inorganic and organic fertilizer enhances soil organo-mineral associations and nutrients in tea garden soil. Agronomy 12(6):1330. https://doi.org/10.3390/agronomy12061330

Li CL, Xu J, Xu H-M, Liu J, Zhang LX, Wang ZK (2023a) Analysis of growth resistance mechanisms and causes in tea plants (*Camellia sinensis*) in high-pH regions of northern China. Front Nutr 10:1131380. https://doi.org/10.3389/fnut.2023.1131380

Li J, Hu X, Zhang R, Li Q, Xu W, Zhang L, Guo F, Zhao H, Wang P, Wang Y, Ni D, Wang M (2023b) The plasma membrane magnesium transporter CsMGT5 mediates magnesium uptake and translocation under magnesium limitation in tea plants (*Camellia sinensis* L.). Scientia Hort 310:111711. https://doi.org/10.1016/j.scienta.2022.111711

Liang LL, Song YK, Qian WJ, Ruan JY, Ding ZT, Zhang QF (2021) Metabolomics analysis reveals the responses of tea plants to excessive calcium. J Sci Food Agr 101(13):5678–5687. https://doi.org/10.1002/jsfa.11222

Lin ZH, Chen LS, Chen RB, Zhang FZ, Jiang HX, Tang N (2009) CO_2 assimilation, ribulose-1,5-bisphosphate carboxylase/oxygenase, carbohydrates and photosynthetic electron transport probed by the JIP-test, of tea leaves in response to phosphorus supply. BMC Plant Biol 9:43. https://doi.org/10.1186/1471-2229-9-43. 12 pages

Lin Q, Ge C, Chen Y (2010) Contents and quality safety assessment of heavy metals in tea garden soil of Shekou township, Fu'an City. Tea Sci Technol 4:8–10

Lin ZH, Chen LS, Chen RB, Zhang FZ, Jiang HX, Tang N, Smith BR (2011) Root release and metabolism of organic acids in tea plants in response to phosphorus supply. J Plant Physiol 168(7):644–652. https://doi.org/10.1016/j.jplph.2010.09.017

Lin ZH, Qi YP, Chen RB, Zhang FZ, Chen LS (2012a) Effects of phosphorus supply on the quality of green tea. Food Chem 130(4):908–914. https://doi.org/10.1016/j.foodchem.2011.08.008

Lin WX, Chen T, Zhou MM (2012b) New dimensions in agroecology. China J Eco-Agric 20:253–264

Lin ZH, Zhong QS, Chen CS, Chen ZH, You XM (2013) Effects of different potassium level on leaf photosynthesis of tea seedling. J Tea Sci 33(3):261–267

Lin W, Lin M, Zhou H, Wu H, Li Z, Lin W (2019) The effects of chemical and organic fertilizer usage on rhizosphere soil in tea orchards. PLoS One 14(5):e0217018. https://doi.org/10.1371/journal.pone.0217018

Lin ZH, Zhong QS, Chen CS, Ruan QC, Chen ZH, You XM (2020) Effects of nitrogen supply on the biochemical attributes of green tea. Int J Agric Biol 23:573–581. https://doi.org/10.17957/IJAB/15.1326

Lin ZH, Chen CS, Zhong QS, Ruan QC, Chen ZH, You XM, Shan RY, Li XL (2021) The GC-TOF/MS-based metabolomic analysis reveals altered metabolic profiles in nitrogen deficient leaves and roots of tea plants (*Camellia sinensis*). BMC Plant Biol 21:506. https://doi.org/10.1186/s12870-021-03285-y

Lin S, Liu Z, Wang Y, Li J, Wang G, Zhang W, Wang H, He H (2022a) Soil acidification associated with changes in inorganic forms of N reduces the yield of tea (*Camellia sinensis*). Arch Agron Soil Sci 69(9):1660–1673. https://doi.org/10.1080/03650340.2022.2104452

Lin S, Liu Z, Wang Y, Li J, Wang G, Ye J, Wang H, He H (2022b) Soil metagenomic analysis on changes of functional genes and microorganisms involved in nitrogen-cycle processes of acidified tea soils. Front Plant Sci 13:998178. https://doi.org/10.3389/fpls.2022.998178

Lin ZH, Chen CS, Zhao SQ, Liu Y, Zhong QS, Ruan QC, Chen ZH, You XM, Shan RY, Li XL, Zhang YZ (2023) Molecular and physiological mechanisms of tea (*Camellia sinensis* (L.) O. Kuntze) leaf and root in response to nitrogen deficiency. BMC Genomics 24(1):27. https://doi.org/10.1186/s12864-023-09112-y. 16 pages

Liu Z, Yang J, Yang Z, Zou J (2012) Effects of rainfall and fertilizer types on nitrogen and phosphorus concentrations in surface runoff from subtropical tea fields in Zhejiang, China. Nutr Cycl Agroecosyst 93(3):297–307. https://doi.org/10.1007/s10705-012-9517-x

Liu S, Li Z, Sui Y, Schaefer DA, Alele PO, Chen J, Yang X (2015) Spider foraging strategies dominate pest suppression in organic tea plantations. BioControl 60(6):839–847. https://doi.org/10.1007/s10526-015-9691-2

Liu R, Kang Y, Wan S, Pei L (2016) Evaluation of methods of nutrient and water management on tea performance and nutrient loss in the Danjiangkou reservoir area, China. Arch Agron Soil Sci 62:1123–1135. https://doi.org/10.1080/03650340.2015.1115017

Liu MY, Burgos A, Ma LF, Zhang QF, Tang DD, Ruan JY (2017) Lipidomics analysis unravels the effect of nitrogen fertilization on lipid metabolism in tea plant (*Camellia sinensis* L.). BMC Plant Biol 17:165. https://doi.org/10.1186/s12870-017-1111-6

Liu CY, Wang YJ, Wu QS, Yang TY, Kuca K (2020) Arbuscular mycorrhizal fungi improve the antioxidant capacity of tea (*Camellia sinensis*) seedlings under drought stress. Notulae Botanicae Horti Agrobotanici Cluj-Napoca 48(4):1993–2005. https://doi.org/10.15835/48412066

Liu J, Lu W, Zhang N, Su D, Zeer, Du H, Hu K (2021a) Collaborative assessment and health risk of heavy metals in soils and tea leaves in the southwest region of China. Int J Environ Res Public Health 18:10151. https://doi.org/10.3390/ijerph18191015

Liu J, Liu M, Fang H, Zhang Q, Ruan J (2021b) Accumulation of amino acids and flavonoids in young tea shoots is highly correlated with carbon and nitrogen metabolism in roots and mature leaves. Front Plant Sci 12:756433. https://doi.org/10.3389/fpls.2021.756433

Liu Y, Tian J, Liu B, Zhuo Z, Shi C, Xu R, Xu M, Liu B, Ye J, Sun L, Liao H (2022a) Effects of pruning on mineral nutrients and untargeted metabolites in fresh leaves of *Camellia sinensis* cv. Shuixian. Front Plant Sci 13:1016511. https://doi.org/10.3389/fpls.2022.1016511

Liu S, Yang R, Peng X, Hou C, Ma J, Guo J (2022b) Contributions of plant litter decomposition to soil nutrients in ecological tea gardens. Agric 12(7):957. https://doi.org/10.3390/agriculture12070957

Liu Y, Zhang M, Li Y, Zhang Y, Huang X, Yang Y, Zhu H, Xiong H, Jiang T (2023a) Influence of nitrogen fertilizer application on soil acidification characteristics of tea plantations in karst areas of Southwest China. Agriculture 13:849. https://doi.org/10.3390/agriculture13040849

Liu Z, Yang D, Zhang G, Zheng L, Chen C, Sun X, Yu F (2023b) Effects of soil physical and chemical properties on the quality of Nanjing 'Yuhua' tea, a type of famous green tea. Horticulturae 9:189. https://doi.org/10.3390/horticulturae9020189. 10 pages

Liu CY, Hao Y, Wu XL, Dai FJ, Abd Allah EF, Wu QS, Liu SR (2023c) Arbuscular mycorrhizal fungi improve drought tolerance of tea plants via modulating root architecture and hormones. Plant Growth Regul. https://doi.org/10.1007/s10725-023-00972-8

Liyanage LRMC, Jayakody AN, Gunaratne GP (2014) Ammonia volatilization from frequently applied fertilizers for the low-country tea growing soils of Sri Lanka. Trop Agric Res 26(1):48–61. https://doi.org/10.4038/tar.v26i1.8071

Long H, Jiang Y, Li C, Liao S, Shi S, Huang C, Zhao S, Li X, Liao Y (2021) Effect of urea feeding on transforming and migrating soil fluorine in a tea garden of hilly region. Environ Geochem Health 43:5087–5098. https://doi.org/10.1007/s10653-021-00949-4

Long L, Shi Y, Ma L, Ruan J (2022) Characterization of young shoot population, yield, and nitrogen demands of tea (*Camellia sinensis* L.) harvested under different standards. Horticulturae 8:275, 15 pages. https://doi.org/10.3390/horticulturae8040275

Lou W, Sun K, Zhao Y, Deng S, Zhou Z (2020) Impact of climate change on inter-annual variation in tea plant output in Zhejiang, China. Int J Climatol 41(S1):E479–E490. https://doi.org/10.1002/joc.6700

Lu TX, Wu YZ, Shen JL, Chen D, Jiang WQ, Wang J, Li Y, Wu JS (2019) N_2O emissions from a tea field with deep application of nitrogen fertilizer and intercropping with white clover. Huanjing Kexue 40(9):4221–4229. https://doi.org/10.13227/j.hjkx.201901254

Luo J, Ni D, He C, Zhang S, Liu S, Du Y, Chen Y (2019) Influence of exogenous calcium on the physiological, biochemical, phytochemical and ionic homeostasis of tea plants (*Camellia sinensis* (L.) O. Kuntze) subjected to fluorine stress. Plant Growth Regul 87(4):455–465. https://doi.org/10.1007/s10725-019-00478-2

Luo Y, Zhang Y, Wang Y, Sun Y, Xia X, Su Y, Liao W (2023) Combined application of biochar and pruned tea plant litter benefits nitrogen availability for tea and alters microbial community structure. Agronomy 13:1465. https://doi.org/10.3390/agronomy13061465. 17 pages

Ma L, Ruan J (2021) Deep placement of nitrogen fertilizer in autumn improves N utilization by spring tea (*Camellia sinensis* L.). J Plant Nutr Soil Sci 184(4):430–438. https://doi.org/10.1002/jpln.202000140

Ma YH, Fu S, Zhang X, Zhao K, Chen HYH (2017) Intercropping improves soil nutrient availability, soil enzyme activity and tea quantity and quality. Appl Soil Ecol 119:171–178. https://doi.org/10.1016/j.apsoil.2017.06.028

Ma L, Shi Y, Ruan J (2019) Nitrogen absorption by field-grown tea plants (*Camellia sinensis*) in winter dormancy and utilization in spring shoots. Plant Soil 442:127–140. https://doi.org/10.1007/s11104-019-04182-y

Ma L, Yang X, Shi Y, Yi X, Ji L, Cheng Y, Ni K, Ruan J (2021) Response of tea yield, quality and soil bacterial characteristics to long-term nitrogen fertilization in an eleven-year field experiment. Appl Soil Ecol 166:103976. https://doi.org/10.1016/j.apsoil.2021.103976

Ma L, Zhu Y, Geng S, Ruan J (2022a) Response of nutritional status and tea quality to the rate and substitution of chemical fertilizers with organic manure. Horticulture 8:1198. https://doi.org/10.3390/horticulturae8121198

Ma Q, Song L, Niu Z, Qiu Z, Sun H, Ren Z, Wu H, Wang Y, Mei H, Li X, Ding Z (2022b) Pea-tea intercropping improves tea quality through regulating amino acid metabolism and flavonoid biosynthesis. Foods 11:3746. https://doi.org/10.3390/foods11223746. 13 pages

Malakar H, Timsina G, Dutta J, Borgohain A, Deka D, Babu A, Paul RK, Yeasin M, Rahman FH, Panja S, Karak T (2022) Sick or rich: assessing the selected soil properties and fertility status across the tea-growing region of Dooars, West Bengal, India. Front Plant Sci 13:1017145. https://doi.org/10.3389/fpls.2022.1017145

Malyukova LS, Koninskaya NG, Orlov YL, Samarina LS (2022) Effects of exogenous calcium on the drought response of the tea plant (*Camellia sinensis* (L.) Kuntze). Peer J 10:e13997. https://doi.org/10.7717/peerj.13997

Malyukova LS, Pritula ZV, Kozlova NV, Velikiy AV, Rogozhina EV, Kerimzade VV, Samarina LS (2021) Effects of calcium-containing natural fertilizer on *Camellia sinensis* (L.) Kunize. Bangaladesh J Bot 50:179–187. https://doi.org/10.3329/bjb.v50i1.52686

Mamun M, Ahmed M (2011) Integrated pest management in tea: prospects and future strategies in Bangladesh. The J Plant Protect Sci 3(2):1–13

Manzoor ML, Ni K, Ruan J (2022) Effect of integrated use of rapeseed cake, biochar and chemical fertilizers on root growth, nutrients use efficiency and productivity of tea. Agronomy 12(8):1823. https://doi.org/10.3390/agronomy12081823

Marcinczyk M, Oleszczuk P (2022) Biochar and engineered biochar as slow- and controlled-release fertilizers. J Clean Prod 339:130685. https://doi.org/10.1016/j.jclepro.2022.130685

Marschner H (1995) Mineral nutrition of higher plants, 2nd edn. Academic Press, London, p 889

Mazumder PB (2015) Copper induced changes in growth and antioxidative mechanisms of tea plant (*Camellia sinensis* (L.) O. Kuntze). Afr J Biotech 14(7):582–592. https://doi.org/10.5897/AJB2014.1427

Meepegamage SW, Seneviratne G, Rajapaske RGSC (2019) Biofilm biofertilizer application can increase soil carbon content in tea plantation. In: Proceedings of the ISSM-2019. Soil: underpinning life and environment, pp 55–56

Melucci D, Locatelli M, Locatelli C (2013) Trace level voltammetric determination of heavy metals and total mercury in tea matrices (*Camellia sinensis*). Food Chem Toxicol 62:901–907. https://doi.org/10.1016/j.fct.2013.10.029

Mishra G, Francaviglia R (2021) Land uses, altitude and texture effects on soil parameters. A comparative study in two districts of Nagaland, Northeast India. Agriculture 11(2):171, 13 pages. https://doi.org/10.3390/agriculture11020171

Morita A, Yanagisawa O, Takatsu S, Maeda S, Hiradate S (2008) Mechanism for the detoxification of aluminum in roots of tea plant (*Camellia sinensis* (L.) Kuntze). Phytochemistry 69:147–153. https://doi.org/10.1016/j.phytochem.2007.06.007

Morita A, Yanagisawa O, Maeda S, Takatsu S, Ikka T (2011) Tea plant (*Camellia sinensis* L.) roots secrete oxalic acid and caffeine into medium containing aluminum. Soil Sci Plant Nutr 57(6):796–802. https://doi.org/10.1080/00380768.2011.629176

Mukhopadyay M, Bantawa P, Das A, Sarkar B, Bera B, Ghosh P, Mondal TK (2012) Changes of growth, photosynthesis and alteration of leaf antioxidative defence system of tea [*Camellia sinensis* (L.) O. Kuntze] seedlings under aluminum stress. Biometals 25(6):1141–1154. https://doi.org/10.1007/s10534-012-9576-0

Mursec M (2011) Agricultural practices impact on soil quality and health: case study of Slovenian irrigated or organic orchards. Earth Sciences. Universitéde Bourgogne, 2011. Doctoral thesis To obtain the grade of Doctor of University of Burgundy and University of Maribor 200p

Nath TN (2013a) The status of micronutrients (Mn, Fe, Cu, Zn) in tea plantations in Dibrugarh district of Assam, India. Int Res J Environ Sci 2(6):25–30

Nath TN (2013b) Status of macronutrients in some selected tea growing soils in Dibrugarh Sivasagar districts of Assam, India. Int J Sci Res 2:273–275

Nath TN (2014) Soil texture and total organic matter content and its influences on soil water holding capacity of some selected tea growing soils in Sivasagar district of Assam, India. Int J Chem Sci 12(4):1419–1429

Nath R, Sharma GD, Borooah M (2018) Evaluation of endophytic bacillus spp. of tea (*Camellia sinensis* L.) for their plant growth promoting activities. Int J Life Sci Res 6(2):106–118

Nath R, Dulumoni T, Dutta SG, Madhumita B (2022) Isolation, characterization and evaluation of plant growth promoting activities of bacterial endophytes of tea (*Camellia sinensis* (L.) O. Kuntze) res. J Biotech 17(7):98–110

Negi GCS, Bisht V (2017) Promoting organic tea farming in mid-hills of north-west Himalaya, India. Tea 38(2):57–67

Nelson S (2006) Zinc deficiency in tea (*Camellia sinensis*). University of Hawaii, Honolulu (HI). 1 p. (Plant Disease; PD-34), http://hdl.handle.net/10125/12430

Ni B, Liu M, Lü S, Xie WY (2010) Multifunctional slow-release organic–inorganic compound fertilizer. J Agric Food Chem 58:12373–12378. https://doi.org/10.1021/jf1029306

Ni B, Liu M, Lü S, Xie L, Wang Y (2011) Environmentally friendly slow-release nitrogen fertilizer. J Agric Food Chem 59(18):10169–10175. https://doi.org/10.1021/jf202131z

Ni K, Liao W, Yi X, Niu S, Ma L, Shi Y, Zhang Q, Liu M, Ruan J (2019) Fertilization status and reduction potential in tea gardens of China. J Plant Nutr Fertil 25:421–432. https://doi.org/10.11674/zwyf.18078

Nicholls CI, Altieri MA (2018) Pathways for the amplification of agroecology. Aroecoly Sustain Food Sys 42(10):1170–1193. https://doi.org/10.1080/21683565.2018.1499578

Nisa KU, Tarfeen N, Nisa Q, Wani S (2023) Climate change and plant nutrient availability: challenges and assessment strategies. Sustainable Plant Nutrition- Molecular Interventions and Advancements for Crop Improvement. pp. 71–86, https://doi.org/10.1016/B978-0-443-18675-2.00015

Noji M, Saito K (2003) Sulphur amino acids: biosynthesis of cysteine and methionine. In: Abrol YP, Ahmad A (eds) Sulphur in plants. Springer, Dordrecht, pp 135–144. https://doi.org/10.1007/978-94-017-0289-8_7

Ogorodova LP, Vigasina MF, Mel'chakova LV, Bryzgalov IA, Ksenofontov DA (2016) Enthalpy of formation of natural polyhalite. Geochem Int 54(7):645–649. https://doi.org/10.1134/S0016702916070077

Okemwa EK, Kirwa JN (2020) Optimization of *Camellia sinensis* crop productivity by use of blended fertilizers. World J Appl Chem 5(2):20–25. https://doi.org/10.11648/j.wjac.20200502.12

Olanrewaju OS, Ayangbenro AS, Glick BR, Babalola OO (2019) Plant health: feedback effect of root exudates-rhizobiome interactions. Appl Microbiol Biotechnol 103(3):1155–1166. https://doi.org/10.1007/s00253-018-9556-6

Oo AZ, Sudo S, Win KT, Shibata A, Sano T, Hirono Y (2018) Returning tea pruning residue and its biochar had a contrasting effect on soil N_2O and CO_2 emissions from tea plantation soil. Atmos 9(3):109. https://doi.org/10.3390/atmos9030109

Owuor PO (1997) Fertilizer use in tea: the case of nitrogen. Tea 18:132–143

Palani N (2019) Magnesium fertilization is essential for tea gardens of southern India. Better Crops-South Asia:15–19

Pan SY, Nie Q, Tai HC, Song XL, Tong YF, ZhangLJF WXW, Lin ZH, Zhang YY, Ye DY, Yi Zhang Y, Wang XY, Zhu PL, Chu ZS, Yu ZL, Liang C (2022) Tea and tea drinking: China's outstanding contributions to the mankind. Chin Med 17:27. https://doi.org/10.1186/s13020-022-00571-1. 40 pages

Panda P, Choudhury A, Chakraborty S, Ray DB, Deb S, Patra PS, Mahato B, Paramanik B, Singh AK, Chauhan RK (2017) Phosphorus solubilizing bacteria from tea soils and their phosphate solubilizing abilities. Int J Biores Sci 4(2):113–125. https://doi.org/10.5958/2454-9541.2017.00018.4

Pandey R (2010) Influence of certain chemicals on nitrate reductase activity and its correlation with caffeine in tea [(*Camellia sinensis* (L) O. Kuntze)] leaves. J Plant Crops 38(2):132–137

Pandey A, Singh S, Palni LMS (2013) Microbial inoculants to support tea industry in India. Indian J Biotechnol 12:13–19

Parviz M, Eshghi N, Asadi S, Teimoory H, Rezaei M (2015) Investigation of heavy metal contents in infusion tea samples of Iran. Toxin Rev 34(3):157–160. https://doi.org/10.3109/15569543.2015.1072562

Peng CY, Liu HF, Qiao HH, Luo J, Liu XM, Hou RY, Wan X, Cai H (2020) Evaluation of the feasibility of short-term electrodialysis for separating naturally occurring fluoride from instant brick tea infusion. J Sci Food Agric 100(1):168–176. https://doi.org/10.1002/jsfa.10011

Peng CY, Xu XF, Zhu HY, Ren YF, Niu HL, Hou RY, Wan XC, Cai HM (2021) Metabolics and ionomics responses of tea leaves (*Camellia sinensis* (L.) O. Kuntze) to fluoride stress. Plant Physiol Biochem 158:65–75. https://doi.org/10.1016/j.plaphy.2020.11.024

Peyang B, Samanta R, Karak T (2020) Seasonal variation of copper, manganese and zinc in tea (*Camellia sinensis* L.) growing soils of Jorhat district of Assam. Indian J Appl Res 10(4):56–58. https://doi.org/10.36106/ijar

Phukan IK, Baruah A (2015) Studies on soil physical, chemical and microbiological properties under compacted and non-compacted tea soils of south bank. Int J Sci Envt Technol 4(1):253–263

Phukan IK, Safique S, Jahan A, Dutta J, Phukan I (2016) Effect of phosphate solubilizing microorganisms on soil available phosphate and growth of young tea. Two Bud 63(1):4–7

Ping X, Liyun Y, Moucheng L, Fei P (2014) Soil characteristics and nutrients in different tea garden types in Fujian Province, China. J Res Econ 5(4):356–363. https://doi.org/10.5814/j.issn.1674-764x.2014.04.011

Piyasena KGNP, Hettiarachchi LSK (2023) Comparison of tea quality parameters of conventionally and organically grown tea, and effects of fertilizer on tea quality: a mini-review. Food Chem Adv 3:10399. https://doi.org/10.1016/j.focha.2023.100399

Pongrac P, Tolrà R, Hajiboland R, Vogel-Mikus K, Kelemen M, Vavpetic P, Pelicon P, Barcelo J, Regvar M, Poschenrieder C (2020) Contrasting allocation of magnesium, calcium and manganese in leaves of tea (*Camellia sinensis* (L.) Kuntze) plants may explain their different extraction efficiency into tea. Food Chem Toxicol 135:110974. https://doi.org/10.1016/j.fct.2019.110974

Pramanik P, Safique S, Jahan A (2018a) Humic substrates facilitated nitrogen uptake in tea bushes by influencing biochemical and microbiological properties in soil of Northeast India. J Plant Nutr 41:2376–2385. https://doi.org/10.1080/01904167.2018.1510510

Pramanik P, Phukan M, Ghosh S, Goswami AJ (2018b) Pruned tea bushes secrete more root exudates to influence microbiological properties in soil. Arch Agron Soil Sci 64:1172–1180. https://doi.org/10.1080/03650340.2017.1414945

Pramanik P, Kalita C, Borah K, Kalita P (2021) Combined application of mica waste and *bacillus pseudomycoides* as a potassium solubilizing bio-fertilizer reduced the dose of potassium fertilizer in tea-growing soil. Agroecol Sust Food 45(5):732–744. https://doi.org/10.1080/21683565.2020.1847232

Qiao C, Liu L, Hu S, Compton JE, Greaver TL, Li Q (2015) How inhibiting nitrification affects nitrogen cycle and reduces environmental impacts of anthropogenic nitrogen input. Glob Chang Biol 21:1249–1257. https://doi.org/10.1111/gcb.12802

Qiao Y, Halberg N, Vaheesan S, Scott S (2016) Assessing the social and economic benefits of organic and fair trade tea production for small-scale farmers in Asia: a comparative case study of China and Sri Lanka. Renewable Agric Food Syst 31(3):246–257. https://doi.org/10.1017/S1742170515000162

Qiao C, Xu B, Han Y, Wang J, Wang X, Liu L, Liu W, Wan S, Tan H, Liu Y, Zhao X (2018) Synthetic nitrogen fertilizers alter the soil chemistry, production and quality of tea. a meta-analysis. Agron Sustain Dev 38(1):10. https://doi.org/10.1007/s13593-017-0485-z

Qiao C, Mia S, Wang Y, Hou J, Xu B (2021) Assessing the effects of nitrification inhibitor dmpp on acidification and inorganic N leaching loss from tea (*Camellia sinensis* L.) cultivated soils with increasing urea–N rates. Sustainability 13:994, 8 pages. https://doi.org/10.3390/su13020994

Qin Z, Pang X, Chen D, Cheng H, Hu X, Wu J (2013) Evaluation of Chinese tea by the electronic nose and gas chromatography–mass spectrometry: correlation with sensory properties and classification according to grade level. Food Res Int 53(2):864–874. https://doi.org/10.1016/j.foodres.2013.02.005

Qiu S, Wang LM, Huang DF, Lin XJ (2014) Effects of fertilization regimes on tea yields, soil fertility, and soil microbial diversity. Chile J Agr Res 74(3):333–339. https://doi.org/10.4067/S0718-58392014000300012

Qiu Q, Fan D, Wang Y, Huang D, Wang Y, Ma J, Wang X (2021) Effect of calcium cyanamide on soil fungal community in successive tea-cuttings nursery. Agriculture 11:716. https://doi.org/10.3390/agriculture11080716

Raguraj S, Dharmakeerthi RS, Gunaratne OGKA, Wijayathunge WMS (2016) Evaluating the suitability of foliar application of micronutrient combinations for mature tea (*Camellia sinensis* (L.) O.Kuntze) in Uva region of Sri Lanka. Sri Lanka J Tea Sci 81(1–2):41–51

Raguraj S, Gunaratne GP, Wijethunga WMS (2019) Remediation of manganese deficiency in tea (*Camellia sinensis* (L.) O. Kuntze) plantation of Maskeliya soil series of Sri Lanka. In: 35th Tocklai Conference Sustainable Tea-Challenges and the way forward. Feb 22–23 2019, Jorhart, Assam, India

Raguraj S, Wijayathunga WMS, Gunaratne GP, Amali RKA, Priyadarshana G, Sandaruwan C, Karunratne V, Hettiarachchi LSK, Kottegod N (2020) Urea–hydroxyapatite nanohybrid as an efficient nutrient source in *Camellia sinensis* (L.) Kuntze (tea). J Plant Nutr 43:2383–2394. https://doi.org/10.1080/01904167.2020.1771576

Raj Kumar T (2020) Efficacy of calciprill as an alternate to dolomite lime. UPASI Tea Res Foundation Ann Rep 2020, pp. 27–30

Raj Kumar T (2021a) Potash derived from molasses as an alternate to MOP. Newsletter UPASI Tea Res Foundation 31(1):3–4

Raj Kumar T (2021b) Evaluation of Parrys potash from molasses in conventional and organic tea fields. UPASI Tea Res Foundation Ann Rep 2021, pp. 33–35

Raj Kumar T, Nivetha B, Revanthraaj Mohan AM (2023) A note on wood ash for application in tea fields. UPASI Planters' Chronicle 119(4 & 5):4–6

Rajadurai R (2021) Balanced plant nutrition critical for sustainable tea cultivation (https://www.historyofceylontea.com/ceylon-publications/ceylon-tea-articles

Ranjitkar S, Sujakhu NM, Lu Y, Wang Q, Wang M, He J, Mortimer PE, Xu J, Kindt R, Zomer RJ (2016) Climate modelling for agroforestry species selection in Yunnan Province, China. Environ Modelling Software 75:263–272. https://doi.org/10.1016/j.envsoft.2015.10.027

Rao TN, Sharma PK (2001) Sulphur nutrition of tea - global review. Int J Tea Sci 1(1):20–27

Rashid MH, Fardous Z, Chowdhury MAZ, Alam MK, Bari ML, Moniruzzaman M, Gan SH (2016) Determination of heavy metals in the soils of tea plantations and in fresh and processed tea leaves: an evaluation of six digestion methods. Chem Central J 10:7. https://doi.org/10.1186/s13065-016-0154-3. 13 pages

Ray D, Baruah PM, Agarwala N (2022) Aluminum in tea plants: phytotoxicity, tolerance and mitigation. In: Naeem M, Ansari AA, Macovei A, Aftab T, Gill SS (eds) Hazardous and Trace Materials in Soil and Plants: sources, effects and management, pp 217–229. https://doi.org/10.1016/B978-0-323-91632-5.00018-5

Rebello R, Burgess PJ, Girkin NT (2022) Identifying sustainable nitrogen management practices for tea plantations. Nitrogen 3:43–57. https://doi.org/10.3390/nitrogen3010003

Rengel Z (2015) Availability of Mn, Zn and Fe in the rhizosphere. J Soil Sci Plant Nutr 15(2):397–409. https://doi.org/10.4067/s0718-95162015005000036

Roy AK, Kumari R, Chakraborty BN, Chakraborty U (2002) VA mycorrhizae in relation to growth of different tea varieties. Mycorrhiza News 14:9–11

Ruan JY (2007) Potassium and magnesium nutrition of tea plants and management technology. Sci Agric Sinica 40(1):256–266

Ruan JY, Gerendas J (2015) Absorption of foliar-applied urea-N-15 and the impact of low nitrogen, potassium, magnesium and sulfur nutritional status in tea (*Camellia sinensis* L.) plants. J Soil Sci Plant Nut 61:653–663. https://doi.org/10.1080/00380768.2015.1027134

Ruan J, Wong MH (2004) Aluminium absorption by intact roots of the Al-accumulating plant *Camellia sinensis* L. Agronomie 24(3):137–142. https://doi.org/10.1051/agro:2004012

Ruan J, Ma L, Shi Y (2006) Aluminium in tea plantations: mobility in soils and plants, and the influence of nitrogen fertilization. Environ Geochem Health 28(6):519–528. https://doi.org/10.1007/s10653-006-9047-z

Ruan JY, Gerendas J, Hardter R, Sattelmacher B (2007a) Effect of nitrogen form and root-zone pH on growth and nitrogen uptake of tea (*Camellia sinensis*) plants. Ann Bot 99:301–310. https://doi.org/10.1093/aob/mcl258

Ruan JY, Gerendás J, Härdter R, Sattelmacher B (2007b) Effect of root zone pH and form and concentration of nitrogen on accumulation of quality-related components in green tea. J Sci Food Agric 87(8):1505–1516. https://doi.org/10.1002/jsfa.2875

Ruan JY, Haerdter R, Gerendas J (2010) Impact of nitrogen supply on carbon/nitrogen allocation: a case study on amino acids and catechins in green tea [*Camellia sinensis* (L.) O. Kuntze] plants. Plant Biol (Stuttg) 12:724–734. https://doi.org/10.1111/j.1438-8677.2009.00288.x

Ruan JY, Ma LF, Yang YJ (2012) Magnesium nutrition on accumulation and transport of amino acids in tea plants. J Sci Food Agric 92(7):1375–1383. https://doi.org/10.1002/jsfa.4709

Ruan JY, Ma L, Shi Y (2013) Potassium management in tea plantations: its uptake by field plants, status in soils, and efficacy on yields and quality of teas in China. J Plant Nutr Soil Sci 176(3):450–459. https://doi.org/10.1002/jpln.201200175

Ruan L, Wei K, Wang L, Cheng H, Wu L, Li H (2019a) Characteristics of free amino acids (the quality chemical components of tea) under spatial heterogeneity of different nitrogen forms in tea (*Camellia sinensis*) plants. Molecules 24:415, 9 pages. https://doi.org/10.3390/molecules24030415

Ruan L, Wang LY, Wei K, Cheng H, Li HL, Shao S (2019b) Comparative analysis of nitrogen spatial heterogeneity responses in low nitrogen susceptible and tolerant tea plants (*Camellia sinensis*). Sci Hortic 246:182–189. https://doi.org/10.1016/j.scienta.2018.10.063

Ruan JY, Ma LF, Yi XY, Shi YZ, Ni K, Liu MY, Zhang QF (2020) Integrated nutrient management in tea plantation to reduce chemical fertilizer and increase nutrient use efficiency. J Tea Sci 40(1):85–95. https://doi.org/10.13305/j.cnki.jts.20200117.003

Ruan L, Cheng H, Ludewig U, Li J, Chang SX (2022) Root foraging strategy improves the adaptability of tea plants (*Camellia sinensis* L.) to soil potassium heterogeneity. Int J Mol Sci 23:8585. https://doi.org/10.3390/ijms23158585

Ruan L, Li X, Song Y, Li J, Palansooriya KN (2023) Effects of tea plant varieties with high- and low-nutrient efficiency on nutrients in degraded soil. Plan Theory 12:905. https://doi.org/10.3390/plants12040905

Rudani K, Pate V, Prajapati K (2018) The importance of zinc in plant growth -a review the importance of zinc in plant growth -a review. Int Res J Nat Appl Sci 5(2):38–48

Sabagh H, Khoramivafa M, Honarmand SJ, Al-Agha AB (2014) Effect of thiobacillus bacteria, Sulphur and manure on the nutrient and pH of soil in garlic (*Allium sativum*). Int J Biosci 5(4):186–193

Saha AK, Biswas A, Zohora KFT, Ali NM, Chowdhury FJ (2022) Response of organic and inorganic sources of manure and fertilizer on soil properties and yield of tea. Tea J Bangladesh 48:1–13

Saikia GK, Barbora AC, Deka MK (2014) Pest, disease and weed incidence and crop yield as influenced by organic culture in tea [*Camellia sinensis* (L.) O. Kuntze]. Agric Sci Digest 34(2):119–122. https://doi.org/10.5958/0976-0547.2014.00028.7

Saleem HM, Usman K, Rizwan M, Al Jabri H, Alsafran M (2022) Functions and strategies for enhancing zinc availability in plants for sustainable agriculture. Front Plant Sci 13:1033092. https://doi.org/10.3389/fpls.2022.1033092

Santosh KC, Liu M, Zhang Q, Fan K, Shi Y, Ruan J (2018) Metabolic changes of amino acids and flavonoids in tea plants in response to inorganic phosphate limitation. Int J Mol Sci 19:3683, 11 pages. https://doi.org/10.3390/ijms19113683

Sarmah M, Borgohain A, Gogoi BB, Yeasin M, Paul RK, Malakr H, Handique JG, Saikia J, Deka D, Khare P, Karak T (2023) Insights into the effects of tea pruning litter biochar on major micronutrients (cu, Mn, and Zn) pathway from soil to tea plant: An environmental Armour. J Hazard Mater 442:129970. https://doi.org/10.1016/j.jhazmat.2022.129970

Saygın F, Şavşatlı Y, Dengiz O, Yazıcı K, Namlı A, Karataş A, Şenol ND, Akça MO, Pacci S, Karapıçak B, Ay A, Demirkaya S (2023) Soil quality assessment based on hybrid computational approach with spatial multi-criteria analysis and geographical information system for sustainable tea cultivation. J Agric Sci 161(2):187–204. https://doi.org/10.1017/S0021859623000138

Sedaghathoor S, Janatpoor G (2012) Study on effect of soybean and tea intercropping on yield and yield components of soybean and tea. J Agric Biol Sci 7(9):664–671

Seeda AMA, El-Nour AEAA, Yassen AA, Hammad SA (2021) Boron, structure, functions and its interaction with nutrients in plant physiology. A review. Middle East. J Agric Res 10(1):117–179

Seenivasan S, Anderson TA, Muraleedharan N (2016) Heavy metal content in tea soils and their distribution in different parts of tea plants, *Camellia sinensis* (L). O Kuntze. Environ Monit Assess 188(7):428. https://doi.org/10.1007/s10661-016-5440-y

Seneviratne G, Kecskes ML, Kenendy IR (2008) Biofilmed biofertilizers: novel inoculants for efficient nutrient use in plants. ACIAR Proc 130:126–130

Seneviratne G, Thilakaratne R, Jayasekara A, Seneviratne K, Padmathilake K, De Silva M (2009) Developing beneficial microbial biofilms on roots of non-legumes: a novel biofertilizing technique. In: Khan M, Zaidi A, Musarrat J (eds) Microbial strategies for crop improvement. Springer, Berlin. https://doi.org/10.1007/978-3-642-01979-1_3

Seneviratne G, Jayasekara APDA, de Silva MSDL, Abeysekara UP (2011) Developed microbial biofilms can restore deteriorated conventional agricultural soils. Biokhimiya/Biochemistry 43(5):1059–1062. https://doi.org/10.1016/j.soilbio.2011.01.026

Senthurpandian VK, Venkatesan S, Jayaganesh (2009) Calcium and magnesium releasing capacity of Alfisols under tea in South India. Geoderma 152(3–5):239–242. https://doi.org/10.1016/j.geoderma.2009.06.005

Seufert V, Ramankutty N, Foley JA (2012) Comparing the yields of organic and conventional agriculture. Nature 485(7397):229–232. https://doi.org/10.1038/nature11069

Shang J, Liu B (2021) Application of a microbial consortium improves the growth of *Camellia sinensis* and influences the indigenous rhizosphere bacterial communities. J Appl Microbiol 130(6):2029–2040. https://doi.org/10.1111/jam.14927

Shao YD, Zhang DJ, Hu XC, Wu QS, Jiang CJ, Xi TJ, Gaos B, Kuca K (2018) Mycorrhiza-induced changes in root growth and nutrient absorption of tea plants. Plant Soil Environ 64(6):283–289. https://doi.org/10.17221/126/2018-PSE

Shao YD, Hua X, Wu QS, Yang TY, Srivastavad AK, Zhang DJ, Gaoe XB, Kuca K (2021) Mycorrhizas promote P acquisition of tea plants through changes in root morphology and P transporter gene expression. South Afr J Bot 137:455–462. https://doi.org/10.1016/j.sajb.2020.11.028. 8 pages

Sharma D, Kayang H (2017) Effects of arbuscular mycorrhizal fungi (AMF) on *Camellia sinensis* (L.) O. Kuntze under greenhouse conditions. J Exp Biol 5:235–241. https://doi.org/10.18006/2017.5(2).235.241

Shen FT, Lin SJ (2021a) Shifts in bacterial community associated with green manure soybean intercropping and edaphic properties in a tea plantation. Sustainability 13(20):11478. https://doi.org/10.3390/su132011478. 16 pages

Shen FT, Lin SH (2021b) Priming effects of cover cropping on bacterial community in a tea plantation. Sustainability 13(8):4345. https://doi.org/10.3390/su13084345

Shen CL, Chyu MC, Wang JS (2013) Tea and bone health: steps forward in translational nutrition. Am J Clinical Nutr 98:1694S–1699S. https://doi.org/10.3945/ajcn.113.058255

Sheng XF, He LY, Huang WY (2002) The conditions of releasing potassium by a silicate dissolving bacterial strain NBT. Agric Sci China 1:662–666

Sherene T (2017) Role of soil enzymes in nutrient transformation: a review. Bio Bull 3(1):109–131

Shibata H, Galloway JN, Leach AM, Cattaneo LR, Cattell Noll L, Erisman JW, Gu B, Liang X, Hayashi K, Ma L, Dalgaard T, Graversgaard M, Chen D, Nansai K, Shindo J, Matsubae K, Oita A, Su MC, Mishima SI, Bleeker A (2017) Nitrogen footprints: regional realities and options to reduce nitrogen loss to the environment. Ambio 46:129–142. https://doi.org/10.1007/s13280-016-0815-4

Shiwakoti LD, Chalise K, Dahal P, Shiwakoti R, Katuwal N, Yadav KC (2023) Effect of fertilizer application on tea plant productivity and phytochemicals in prepared black tea. Cogent Food Agric 9(1):2184013. https://doi.org/10.1080/23311932.2023.2184013

Shrestha G, Thapa RB (2015) Tea pests and pesticide problems and integrated management. J Agric Environ 16:188–200. https://doi.org/10.3126/aej.v16i0.19852

Shu WS, Zhang ZQ, Lan CY, Wong MH (2003) Fluoride and aluminium concentrations of tea plants and tea products from Sichuan Province, PR China. Chemosphere 52:1475–1482. https://doi.org/10.1016/S0045-6535(03)00485-5

Sindhu SS, Parmar P, Phour M (2014) Nutrient cycling: potassium solubilization by microorganisms and improvement of crop growth. In: Parmar N, Singh A (eds) Geomicrobiology and biogeochemistry. Springer, Berlin, pp 175–198

Singh AK, Pathak SK (2018) Potassium in tea (*Camellia sinensis* (L) O. Kuntze) cultivation from soil to cup quality- a review. Agric Rev (Karnal) 39(1):40–47. https://doi.org/10.18805/AG.R-1731

Singh S, Pandey A, Kumar B, Palni LMS (2010) Enhancement in growth and quality parameters of tea [*Camellia sinensis* (L.) O. Kuntze] through inoculation with arbuscular mycorrhizal fungi in an acid soil. Biol Fertil Soils 46:427–433. https://doi.org/10.1007/s00374-010-0448-x

Singh AK, Chauhan RK, Bisen JS (2014) Role of soil organic matter in soil health sustainability. Int J Agric Sci 5(2):219–227

Singh AK, Bisen JS, Chauhan RK, Choubey M, Kumar R, Kumar N (2016) Tea research for Darjeeling tea industry - various aspects. In: Bag N, Bag A, Palni LMS (eds) Tea: technological initiatives. New India Publishing Agency, New Delhi, pp 195–239

Singh AK, Chauan R, Sharma CK, Bera B (2023) Sustaining Darjeeling organic tea (*Camellia sinensis* L) production and tea quality through organic inputs in the Himalayan foothills. J Sci Plant Nutr 23:1774–1796. https://doi.org/10.1007/s42729-023-01137-2

Sipayung HN, Wu KT, Liu DY, Chen CT (2022) Growth enhancement of *Camellia sinensis* with biochar. PRO 10(2):199. https://doi.org/10.3390/pr10020199. 14 pages

Sitienei K, Home PG, Kamau DM, Wanyoko JK (2013) Nitrogen and potassium dynamics in tea cultivation as influenced by fertilizer type and application rates. Am J Plant Sci 4(1):59–65. https://doi.org/10.4236/ajps.2013.41010

Sitienei K, Kamiri HW, Nduru GM, Kamau DM, Nyabundid WK, Morogoe M (2018a) Effects of blended fertilizers on yields of mature tea clones TRFK 6/8 and BBK 35 grown in Kenyan highlands. Indonesian J Agric Sci 19(1):17–24. https://doi.org/10.21082/ijas.v19n1.2018

Sitienei K, Kamiri HW, Nduru GM, Kamau DM (2018b) Nutrient budget and economic assessment of blended fertilizer use in Kenya tea industry. Appl Environ Soil Sci 2018:2563293. https://doi.org/10.1155/2018/2563293. 11 pages

Sitienei K, Kamiri HW, Nduru GM, Kamau DM (2018c) Effects of blended fertilizers on leaf nutrients content of mature clonal tea in Kenya. J Exp Res 6(2):43–51

Sitienei K, Kamiri HW, Nduru GM, Kamau DM, Nyabundi WK, Morogo M (2019) Effects of blended fertilizers on yields of mature clonal tea grown in Kenyan highlands. Int J Tea Sci 14(1):36–43. https://doi.org/10.20425/ijts1415

Sreenivasa MN, Bagyaraj DJ (1989) Use of pesticides for mass production of vesicular arbuscular mycorrhizal inoculum. Plant Soil 119:127–132

Su YJ, Liao WY, Wang YJ, Zhang YL, Wu XR, Hu SG, Sun L (2013) Influences of soil pH and cultivation years on active aluminum species distribution from tea soils in southern Anhui. China. J Agro Environ Sci 32:721–728

Su TC, Yang MJ, Huang HH, Kuo CC, Chen LY (2021) Using sensory wheels to characterize consumers' perception for authentication of Taiwan specialty teas. Food Rev 10(4):836. https://doi.org/10.3390/foods10040836

Sultana J, Siddique MNA, Kamaruzzaman M, Halim MA (2014) Conventional to ecological: tea plantation soil management in Panchagarh district of Bangladesh. J Sci Technol Environ Inform 1(1):27–35. https://doi.org/10.18801/jstei.010114.03

Sumi RS, Kabir G (2018) Factors affecting the buying intention of organic tea consumers of Bangladesh. J Open Innov Technol Market Complexity 4(3):24. https://doi.org/10.3390/joitmc4030024

Sun H, Wang QX, Liu N, Li L, Zhang CG, Liu ZB, Zhang YY (2017) Effects of different leaf litters on the physicochemical properties and bacterial communities in Panax ginseng-growing soil. Appl Soil Ecol 111:17–24. https://doi.org/10.1016/j.apsoil.2016.11.008

Sun LL, Liu Y, Wu LQ, Liao H (2019) Comprehensive analysis revealed the close relationship between N/P/K status and secondary metabolites in tea leaves. ACS Omega 4(1):176–184. https://doi.org/10.1021/acsomega.8b02611

Sun L, Zhang M, Liu X, Mao Q, Shi C, Kochian LV, Liao H (2020a) Aluminium is essential for root growth and development of tea plants (*Camellia sinensis*). J Integrative Plant Biol 62(7):984–997. https://doi.org/10.1111/jipb.12942

Sun M, Yuan D, Hu X, Zhang D, Li Y (2020b) Effects of mycorrhizal fungi on plant growth, nutrient absorption and phytohormones levels in tea under shading condition. Not Bot Horti Agrobot Cluj 48(4):2006–2020. https://doi.org/10.15835/nbha48412082

Sun L, Fan K, Wang L, Ma D, Wang Y, Kong X, Li H, Ren Y, Ding Z (2021) Correlation among metabolic changes in tea plant *Camellia sinensis* (L.) shoots, green tea quality and the application of cow manure to tea plantation soils. Molecules 26(20):6180. https://doi.org/10.3390/molecules26206180

Sun L, Dong X, Wang Y, Maker G, Agarwal M, Ding Z (2022a) Tea-soybean intercropping improves tea quality and nutrition uptake by inducing changes of rhizosphere bacterial communities. Microorganisms 10:2149. https://doi.org/10.3390/microorganisms10112149

Sun L, Wang Y, Ma D, Wang L, Zhang X, Ding Y, Fan K, Xu Z, Yuan C, Jia H, Ren Y, Ding Z (2022b) Differential responses of the rhizosphere microbiome structure and soil metabolites in tea (*Camellia sinensis*) upon application of cow manure. BMC Microbiol 22:55. https://doi.org/10.1186/s12866-022-02470-9

Sun MF, Wang JJ, Liu W, Yin P, Guo GY, Tong CL, Chang YL (2022c) Effect and mechanism of exogenous selenium on selenium content andquality of tea leaves. Notulae Botanicae Horti Agrobotanici Cluj-Napoca 50(3):12814. https://doi.org/10.15835/nbha50312814

Sun L, Zhang Z, Li Y, Ruan J, Karak T, Yang T (2023) Editorial: mineral nutrients on tea yield and quality formation. Front Plant Sci 14:1192432. https://doi.org/10.3389/fpls.2023.1192432

Szymczycha-Madeja A, Welna M, Pohl P (2012) Elemental analysis of teas and their infusions by spectrometric methods. Trends Analytical Chem 35:169–171. https://doi.org/10.1016/j.trac.2011.12.005

Szymczycha-Madeja A, Welna M, Pohl P (2015) Determination of essential and non-essential elements in green and black teas by FAAS and ICP OES simplified – multivariate classification of different tea products. Microchem J 121:122–129. https://doi.org/10.1016/j.microc.2015.02.009

Tan X, Xie H, Yu J, Wang Y, Xu J, Xu P, Ma B (2022) Host genetic determinants drive compartment-specific assembly of tea plant microbiomes. Plant Biotechnol J 20(11):2174–2186. https://doi.org/10.1111/pbi.13897

Tang S, Liu Y, Zheng N, Li Y, Ma Q, Xiao H, Zhou X, Xu X, Jiang T, He P, Wu L (2020a) Temporal variation in nutrient requirements of tea (*Camellia sinensis*) in China based on QUEFTS analysis. Sci Rep 10:1745, 10 pages. https://doi.org/10.1038/s41598-020-57809-x

Tang D, Liu MY, Zhang Q, Ma L, Shi Y, Ruan J (2020b) Preferential assimilation of NH_4^+ over NO_3^- in tea plant associated with genes involved in nitrogen transportation, utilization and catechins biosynthesis. Plant Sci 291:110369. https://doi.org/10.1016/j.plantsci.2019.110369

Tang L, Xiao L, Huang Y, Xiao B, Gong C (2021a) Cloning and magnesium transport function analysis of CsMGTs genes in tea plants (*Camellia sinensis*). J Tea Sci 416:761–776

Tang S, Zheng N, Ma Q, Zhou J, Sun T, Zhang X, Wu L (2021b) Applying nutrient expert system for rational fertilisation to tea (*Camellia sinensis*) reduces environmental risks and increases economic benefits. J Cleaner Production 305:127197. https://doi.org/10.1016/j.jclepro.2021.127197

Tang S, Zhou J, Pan W, Tang R, Ma Q, Xu M, Qi T, Ma Z, Fu H, Wu L (2022a) Impact of N application rate on tea (*Camellia sinensis*) growth and soil bacterial and fungi communities. Plant Soil 475:343–359. https://doi.org/10.1007/s11104-022-05372-x

Tang S, Pan W, Tang R, Ma Q, Zhou J, Zheng N, Wang J, Sun T, Wu L (2022b) Effects of balanced and unbalanced fertilisation on tea quality, yield, and soil bacterial community. Appl Soil Ecol 175:104442. https://doi.org/10.1016/j.apsoil.2022.104442

Tang S, Zhou J, Pan W, Sun T, Liu M, Tang R, Li A, Ma Q, Wu L (2023a) Effects of combined application of nitrogen, phosphorus, and potassium fertilizers on tea (*Camellia sinensis*) growth and fungal community. Appl Soil Ecol 181:104661. https://doi.org/10.1016/j.apsoil.2022.104661. 10 pages

Tang L, Xiao L, Cgen E, Lei X, Ren J, Yang Y, Xiao B, Gong C (2023b) Magnesium transporter CsMGT10 of tea plants plays a key role in chlorosis leaf vein greening. Plant Physoil Biochem 201:107842. https://doi.org/10.1016/j.plaphy.2023.107842

Tang S, Fu H, Pan W, Zhou J, Xu M, Han K, Chen K, Ma Q, Wu L (2023c) Improving tea (*Camellia sinensis*) quality, economic income, and environmental benefits by optimizing agronomic nitrogen efficiency: a synergistic strategy. Eur J Agron 142:126673. https://doi.org/10.1016/j.eja.2022.126673

Tarkalson DD, Payero JO, Hergert GW, Cassman KG (2016) Acidification of soil in a dry land winter wheat-sorghum/corn-fallow rotation in the semiarid U.S. Great Plains. Plant Soil 283:367–379. https://doi.org/10.1007/s11104-006-0027-y

Tennakoon PLK, Rajapaksha RMCP, Hettiarachchi LSK (2021) Indigenous plant growth-promoting rhizobacterial consortia greatly reduces fertilizer need for tea nurseries: characterization and evaluation. Sri Lanka J Food Agric 7(1):41–52. https://doi.org/10.4038/sljfa.v7i1.93

Tian D, Niu S (2015) A global analysis of soil acidification caused by nitrogen addition. Environ Res Lett 10(2):024019. https://doi.org/10.1088/1748-9326/10/2/024019

Tian Y, Cao F, Wang G (2013) Soil microbiological properties and enzyme activity in Ginkgo–tea agroforestry compared with monoculture. Agrofor Syst 87:1201–1210. https://doi.org/10.1007/s10457-013-9630-0

Tolhurst JAH (1962) Zn deficiency in tea in Ceylon. Tea Q 33:134–137

Tolra R, Martos S, Hajiboland R, Poschenrieder C (2020) Aluminium alters mineral composition and polyphenol metabolism in leaves of tea plants (*Camellia sinensis*). J Inorg Biochem 204:110956. https://doi.org/10.1016/j.jinorgbio.2019.110956

Tongsiri P, Tseng WY, Shen Y, Lai HY (2020) Comparison of soil properties and organic components in infusions according to different aerial appearances of tea plantations in Central Taiwan. Sustainability 12(11):4384. https://doi.org/10.3390/su12114384

TRFK (2002) (tea research Foundation of Kenya), tea growers' handbook, 5th edn. Kijabe Printing Press, Kijabe, p 261p

Trivedi P, Leach JE, Tringe SG, Sa T, Singh BK (2020) Plant–microbiome interactions: from community assembly to plant health. Nat Rev Microbiol 18:607–621. https://doi.org/10.1038/s41579-020-0412-1

Tseng WY, Lai HY (2022) Comprehensive analysis revealed the specific soil properties and foliar elements respond to the quality composition levels of tea (*Camellia sinensis* L.). Agronomy 12(3):670. https://doi.org/10.3390/agronomy12030670

ul Haq S, Boz I (2019) Estimating the efficiency level of different tea farming systems in Rize Province Turkey. Ciência Rural 49(12):e20181052. https://doi.org/10.1590/0103-8478cr20181052

ul Haq S, Boz I, Shahbaz P, Yildirim C (2020) Evaluating eco-efficiency and optimal levels of fertilizer use based on the social cost and social benefits in tea production. Environ Sci Pollut Res 27:33008–33019. https://doi.org/10.1007/s11356-020-09533-2

Upadhyaya H, Dutta BK, Panda SK (2013) Zinc modulates drought-induced biochemical damages in tea [*Camellia sinensis* (L) O Kuntze]. J Agric Food Chem 61:6660–6670. https://doi.org/10.1021/jf304254z

Van Ho B, Nanseki T, Chomei Y (2019) Profit efficiency of tea farmers: case study of safe and conventional farms in northern Vietnam. Environ Dev Sustain 21(4):1695–1713. https://doi.org/10.1007/s10668-017-0073-z

Venkatesan S (2005) Impact of genotype and micronutrient applications on nitrate reductase activity of tea leaves. J Sci Food Agric 85(3):513–516. https://doi.org/10.1002/jsfa.1986

Venkatesan S, Ganapathy MNK (2004) Nitrate reductase activity in tea as influenced by various levels of nitrogen and potassium fertilizers. Commun Soil Sci Plant Anal 35:1283–1291. https://doi.org/10.1081/CSS-120037546

Venkatesan S, Murugesan S (2004) Phosphorus fixing capacity of tea soils of Anamallais and Nilgiris. J Plantain Crops 32(2):65–69

Venkatesan S, Murugesan S, Senthur Pandian VK, Ganapathy MNK (2005) Impact of sources and doses of potassium on biochemical and green leaf parameters of tea. Food Chem 90(4):535–539. https://doi.org/10.1016/j.foodchem.2004.05.014

Venkatesan S, Hemalatha KV, Jayaganhesh S (2006a) Zinc toxicity and its influence on nutrient uptake in tea. Am J Plant Physiol 1(2):185–192. https://doi.org/10.3923/ajpp.2006.185.192

Venkatesan S, Senthurpandian VK, Murugesan S, Maibuam W, Ganapathy MNK (2006b) Quality standards of CTC black teas as influenced by sources of potassium fertiliser. J Sci Food Agric 86(5):799–803. https://doi.org/10.1002/jsfa.2418

Venkatesan S, Hemalatha KV, Jayaganesh S (2007) Characterization of manganese toxicity and its influence on nutrient uptake, antioxidant enzymes and biochemical parameters in tea. Res J Phytochem 1(2):52–60

Venkatesan S, Murugesan S, Ganapathy MNK, Verma DP (2004) Long-term impact of nitrogen and potassium fertilizers on yield, soil nutrients and biochemical parameters of tea. J Sci Food Agric 84(14):1939–1944. https://doi.org/10.1002/jsfs.1897

Virk SS, Mullenix DK, Sharda A, Hall JB, Wood CW, Fasina OO, McDonald TP, Pate GL, Fulton JP (2013) Case study: distribution uniformity of a blended fertilizer applied using a variable-rate spinner-disc spreader. Appl Eng Agric 29(5):627–636. https://doi.org/10.13031/aea.29.9774

Vyas P, Rahi P, Chauhan A, Gulati A (2007) Phosphate solubilization potential and stress tolerance of *Eupenicillium parvum* from tea soil. Mycol Res 111:931–938. https://doi.org/10.1016/j.mycres.2007.06.003

Wan Q, Xu RK, Li XH (2012a) Proton release by tea plant (*Camellia sinensis* L.) roots as affected by nutrient solution concentration and pH. Plant Soil Environ 58(9):429–434. https://doi.org/10.17221/326/2012-PSE

Wan Q, Xu RK, Li XH (2012b) Proton release from tea plant (*Camellia sinensis* L.) roots induced by Al (III) under hydroponic conditions. Soil Res 50(6):482–488. https://doi.org/10.1071/SR12099

Wan Q, Xu RK, Li XH (2013) Effect of forms of nitrogen on proton release from tea plant roots under hydroponic condition. Acta Pedol Sin 50(4):720–725

Wang H (2014) The response of soil microbial community structure to cultivating age, fertilization and high temperature in tea orchard. Zhejiang University, China, Dissertation

Wang S, Ye S (2020) Dynamics of bacterial function characteristics in soil aggregates during the tea-planting process. Geoderma 377:114609. https://doi.org/10.1016/j.geoderma.2020.114609

Wang H, Xu RK, Wang N, Li XH (2010) Soil acidification of alfisols as influenced by tea cultivation in eastern China. Pedosphere 20(6):799–806. https://doi.org/10.1016/S1002-0160(10)60070-7

Wang M, Zheng QS, Shen QR, Guo SW (2013) The critical role of potassium in plant stress response. Int J Mol Sci 14(4):7370–7390. https://doi.org/10.3390/ijms14047370

Wang Y, Yin R, Liu R (2014a) Characterization of biochar from fast pyrolysis and its effect on chemical properties of the tea garden soil. J Anal Appl Pyrol 110:375–381. https://doi.org/10.1016/j.jaap.2014.10.006

Wang L, Butterly CR, Wang Y, Herath HMSK, Xi YG, Xiao XJ (2014b) Effect of crop residue biochar on soil acidity amelioration in strongly acidic tea garden soils. Soil Use Manag 30(1):119–128. https://doi.org/10.1111/sum.12096

Wang ZG, Jin X, Bao XG, Li XF, Zhao JH, Sun JH, Christie P, Li L (2014c) Intercropping enhances productivity and maintains the most soil fertility properties relative to sole cropping. PLoS One 9(12):e113984. https://doi.org/10.1371/journal.pone.0113984

Wang H, Yang SH, Yang JP, Lv YM, Zhao X, Pang JL (2014d) Temporal changes in soil bacterial and archaeal communities with different fertilizers in tea orchards. J Zhejiang Univ Sci B 15(11):953–965. https://doi.org/10.1631/jzus.B1400114

Wang Y, Li Y, Liu Y, Chen X, Wei X (2015) Extraction, characterization and antioxidant activities of se-enriched tea polysaccharides. Int J Biol Macromol 77:76–84. https://doi.org/10.1016/j.ijbiomac.2015.02.052

Wang S, Li T, Zheng T (2016a) Effect of tea plantation age on the distribution of soil organic carbon and nutrient within micro-aggregates in the hilly region of western Sichuan, China. Ecol Eng 90:113–119. https://doi.org/10.1016/j.ecoleng.2016.01.046

Wang Y, Fu D, Pan SL, Ding Z (2016b) The coupling effect of water and fertilizer on the growth of tea plants [*Camellia sinensis* (L.) O. Kuntz]. J Plant Nutr 39:620–627. https://doi.org/10.1080/01904167.2015.1047521

Wang S, Li T, Zheng Z (2017) Distribution of microbial biomass and activity within soil aggregates as affected by tea plantation age. Catena 153:1–8. https://doi.org/10.1016/j.catena.2017.01.029

Wang S, Li T, Zheng Z (2018a) Tea plantation age effects on soil aggregate-associated carbon and nitrogen in the hilly region of western Sichuan, China. Soil Till Res 180:91–98. https://doi.org/10.1016/j.still.2018.02.016

Wang J, Zhang B, Tian Y, Zhang H, Cheng Y, Zhang J (2018b) A soil management strategy for ameliorating soil acidification and reducing nitrification in tea plantations. Eur J Soil Biol 88:36–40. https://doi.org/10.1016/j.ejsobi.2018.06.001

Wang Y, Hu X, Hou Z, Ning J, Zhang Z (2018c) Discrimination of nitrogen fertilizer levels of tea plant (*Camellia sinensis*) based on hyperspectral imaging. J Sci Food Agric 98(12):4659–4664. https://doi.org/10.1002/jsfa.8996

Wang L, Huang D, Li Q, He C, Zhang H, Liu C, Li F, Huang Y (2018d) Organic carbon and total nitrogen distribution in aggregates from yellow-red soils under tea plantations with different fertilizer managements. J Tea Sci 38(4):342–352

Wang W, Xie Y, Bi M, Wang X, Lu Y, Fan Z (2018e) Effects of best management practices on nitrogen load reduction in tea fields with different slope gradients using the SWAT model. Appl Geogr 90:200–213. https://doi.org/10.1016/j.apgeog.2017.08.020

Wang S, Yao X, Zhang Z, He X, Ye S (2020a) Soil aggregation and aggregate-related exchangeable base cations under different aged tea (*Camellia sinensis* L.) plantations in the hilly regions of southern Guangxi, China. Soil Sci Plant Nutr 66(4):636–644. https://doi.org/10.1080/00380768.2020.1780471

Wang HB, Zhang QX, Chen XT, Wang YH, Lin LW, Ye JH, Zhang Q, Jia XL, He HB (2020b) Analysis of microbial diversity of tea tree (*Camellia sinensis* (L.) O. Kuntze) sick rhizospheric soil using soil metaproteomic technology. Allelopath J 51(2):147–155. https://doi.org/10.26651/allelo.j/2020-51-2-1296

Wang J, Tu X, Zhang H, Cui J, Ni K, Chen J, Cheng Y, Zhang J, Chang SX (2020c) Effects of ammonium-based nitrogen addition on soil nitrification and nitrogen gas emissions depend on fertilizer-induced changes in pH in a tea plantation soil. Sci Total Environ 10(747):141340. https://doi.org/10.1016/j.scitotenv.2020.141340

Wang HB, Chen XT, Wang YH, Ye JH, Jia XL, Zhang Q, He HB (2020d) Effects of tea garden soil on aroma components and related gene expression in tea leaves. J Appl Bot Food Qual 93:105–111. https://doi.org/10.5073/JABFQ.2020.093.013

Wang Y, Yao Z, Pan Z, Wang R, Yan G, Liu C, Su Y, Zheng X, Bahl KB (2020e) Tea-planted soils as global hotspots for N_2O emissions from croplands. Environ Res Lett 15:104018. https://doi.org/10.1088/1748-9326/aba5b2. 12 pages

Wang Z, Hassan MU, Nadeem F, Wu L, Zhang F, Li X (2020f) Magnesium fertilization improves crop yield in most production systems: a meta-analysis. Front Plant Sci 10:1727. https://doi.org/10.3389/fpls.2019.01727

Wang Y, Wang YM, Lu YT, Qiu QL, Fan DM, Wang XC, Zheng XQ (2021a) Influence of different nitrogen sources on carbon and nitrogen metabolism and gene expression in tea plants (*Camellia sinensis* L.). Plant Physio Biochem 167:561–566. https://doi.org/10.1016/j.plaphy.2021.08.034

Wang M, Sun H, Xu L, Xu Z (2021b) Bacterial diversity in tea plant (*Camellia sinensis*) rhizosphere soil from Qinling mountains and its relationship with environmental elements. Plant Soil 460(1):403–415. https://doi.org/10.1007/s11104-020-04822-8

Wang Y, Cheng X, Yang T, Su Y, Lin S, Zhang S, Zhang Z (2021c) Nitrogen-regulated theanine and flavonoid biosynthesis in tea plant roots: protein-level regulation revealed by multiomics analyses. J Agric Food Chem 69(34):10002–10016. https://doi.org/10.1021/acs.jafc.1c02589

Wang YH, Hong L, Wang YC, Yang YW, Lin LW, Ye JH, Jia XL, Wang HB (2022a) Effects of soil nitrogen and pH in tea plantation soil on yield and quality of tea leaves. Allelopath J 55(1):51–61. https://doi.org/10.26651/allelo.j/2022-55-1-1370

Wang Y, Ouyang JX, Fan DM, Wang SM, Xuan YM, Wang XC, Zheng XQ (2022b) Transcriptome analysis of tea (*Camellia sinensis*) leaves in response to ammonium starvation and recovery. Front Plant Sci 13:963269. https://doi.org/10.3389/fpls.2022.963269

Wang T, Duan Y, Liu G, Shang X, Liu L, Zhang K, Li J, Zou Z, Zhu X, Fang W (2022c) Tea plantation intercropping green manure enhances soil functional microbial abundance and multifunctionality resistance to drying-rewetting cycles. Sci Total Environ 810:151282. https://doi.org/10.1016/j.scitotenv.2021.151282

Wang H, Wang L, Shen X, Zhou LZ (2023a) Biochar addition to tea garden soils: effects on tea fluoride uptake and accumulation. Biochar 5:37. https://doi.org/10.1007/s42773-023-00220-2. 16 pages

Wang T, Duan Y, Lei X, Cao Y, Liu L, Shang X, Wang M, Lv C, Ma Y, Fang W, Zhu X (2023b) Tea plantation intercropping legume improves soil ecosystem multifunctionality and tea quality by regulating rare bacterial taxa. Agronomy 13:1110. https://doi.org/10.3390/agronomy13041110. 18 pages

Wang S, Zhang X, Li X, Shen J, Sun L, Zaman S, Wang Y, Ding Z (2023c) Different changes of bacterial diversity and soil metabolites in tea plants-legume intercropping systems. Front Plant Sci 14:1110623. https://doi.org/10.3389/fpls.2023.1110623. 10 pages

Wei K, Liu M, Shi Y, Zhang H, Ruan J, Zhang Q, Cao M (2022) Metabolomics reveal that the high application of phosphorus and potassium in tea plantation inhibited amino-acid accumulation but promoted metabolism of flavonoid. Agron 12:1086. https://doi.org/10.3390/agronomy12051086. 14 pages

Wen B, Zhang X, Ren S, Duan Y, Zhang Y, Zhu X, Wang Y, Ma Y, Fang W (2020) Charactristics of soil nutrients, heavy metals and tea quality in different intercropping patterns. Agrofor Syst 94(3):963–974. https://doi.org/10.1007/s10457-019-00463-8

Wen X, Zhang Z, Huang X (2022) Heavy metals in karst tea garden soils under different ecological environments in southwestern China. Trop Ecol 63:495–505. https://doi.org/10.1007/s42965-022-00218-0

Willson KC (1969) The mineral nutrition of tea. Potash Review:1–18

Williams H, Colombi T, Keller T (2020) The influence of soil management on soil health: an on-farm study insouthern Swededn. Geoderma 360:111401. https://doi.org/10.1016/j.geoderma.2019.114010

Withers E, Hill PW, Chadwick DR, Jones DL (2020) Use of untargeted metabolomics for assessing soil quality and microbial function. Soil Biol Biochem 143:107758. https://doi.org/10.1016/j.soilbio.2020.107758

Wu W, Zheng Z, Li T, He S, Zhang X, Wang Y, Liu T (2018) Distribution of inorganic phosphorus fractions in water-stable aggregates of soil from tea plantations converted from farmland in the hilly region of western Sichuan, China. J Soils Sediments 18:906–916. https://doi.org/10.1007/s11368-017-1834-x

Wu Z, Jiang F, Zhang L, You Z (2020a) Study on soil nitrogen status of Tieguanyin tea gardens in Anxi county of Fujian Province. Soil 1:16–24

Wu T, Liu W, Wang D, Zou Y, Lin R, Yang Q, Gbokie T, Bughio MA, Li Q, Jn W (2020b) Organic management improves soil phosphorus availability and microbial properties in a tea plantation after land conversion from longan (*Dimocarpus longan*). Appl Soil Ecol 154:103642. https://doi.org/10.1016/j.apsoil.2020.103642

Wu Y, Li Y, Wang H, Wang Z, Fu X, Shen J, Wang Y, Liu X, Meng L, Wu J (2021a) Response of N_2O emissions to biochar amendment on a tea field soil in subtropical Central China: a three-year field experiment. Agric Ecosyst Environ 318:107473. https://doi.org/10.1016/j.agee.2021.107473

Wu T, Qin Y, Li M (2021b) Intercropping of tea (*Camellia sinensis* L.) and Chinese chestnut: variation in the structure of rhizosphere bacterial communities. J Soil Sci Plant Nutr 21:2178–2190. https://doi.org/10.1007/s42729-021-00513-0

Wu T, Zou R, Pu D, Lan Z, Zhao B (2021c) Non-targeted and targeted metabolomics profiling of tea plants (*Camellia sinensis*) in response to its intercropping with Chinese chestnut. BMC Plant Biol 21:55. https://doi.org/10.1186/s12870-021-02841-w

Xi S, Chu H, Zhou Z, Li T, Zhang S, Xu X, Pu Y, Wang G, Jia Y, Liu Z (2023) Effect of potassium fertilizer on tea yield and quality: a meta-analysis. Eur J Agron 144:126677. https://doi.org/10.1016/j.eja.2023.126767

Xiang J, Rao S, Chen Q, Zhang W, Cheng S, Cong X, Zhang Y, Yang X, Xu F (2022) Research progress on the effects of selenium on the growth and quality of tea plants. Plan Theory 11:2491. https://doi.org/10.3390/plants11192491

Xiao Y, Li M, Liu Y, Xu S, Zhong K, Wu Y, Gao H (2021) The effect of *Eurotium cristatum* (MF800948) fermentation on the quality of autumn green tea. Food Chem 358:129848. https://doi.org/10.1016/j.foodchem.2021.129848

Xiaohu W, Liujian Y, Shengbo W, Shuang H, Qixia Z, Liqin Z (2022) Comparison of soil microbial communities between artificial tea plantation and wild tea plantation in China. Open Acc J Biomed Sci 4(4):2002–2010. https://doi.org/10.38125/OAJBS.000480

Xie S, Yang F, Feng H, Yu Z, Liu C, Wei C, Liang T (2021) Organic fertilizer reduced carbon and nitrogen in runoff and buffered soil acidification in tea plantations: evidence in nutrient contents and isotope fractionations. Sci Total Envt 762:143059. https://doi.org/10.1016/j.scitotenv.2020.143059

Xie ZM, Ye ZH, Wong MH (2001) Distribution characteristics of fluoride and aluminum in soil profiles of an abandoned tea plantation and their uptake by six woody species. Environ Int 26(5–6):341–346. https://doi.org/10.1016/S0160-4120(01)00010-1

Xie S, Feng H, Yang F, Zhao Z, Hu X, Wei C, Liang T, Li H, Geng Y (2019) Does dual reduction in chemical fertilizer and pesticides improve nutrient loss and tea yield and quality? A pilot study in a green tea garden in Shaoxing, Zhejiang Province, China. Environ Sci Pollut Res 26:2464–2476. https://doi.org/10.1007/s11356-018-3732-1

Xie S, Yang F, Feng H, Yu Z, Liu C, Wei C, Liang T (2020a) Organic fertilizer reduced carbon and nitrogen in runoff and buffered soil acidification in tea plantations: evidence in nutrient contents and isotope fractionations. Sci Total Environ 762:143059. https://doi.org/10.1016/j.scitotenv.2020.143059

Xie H, Feng X, Wang M, Wang Y, Awasthi KM, Xu P (2020b) Implications of endophytic microbiota in *Camellia sinensis*: a review on current understanding and future insights. Bioengineered 11(1):1001–1015. https://doi.org/10.1080/21655979.2020.1816788
Xie S, Yang F, Feng H, Yu Z, Wei X, Liu C, Wei C (2022) Potential to reduce chemical fertilizer application in tea plantations at various spatial scales. Int J Environ Res Public Health 19:5243. https://doi.org/10.3390/ijerph19095243,10.3390/ijerph19095243
Xie X, Sun Z, Zhang X, Han X (2023) Novel aspects of regulation of nitrogen responses in the tea plant (*Camellia sinensis* L.). Agronomy 13:144. https://doi.org/10.3390/agronomy13010144
Xu RK (2015) Research progresses in soil acidification and its control. Soil 47(2):238–244
Xu S, Zhou S, Ma S, Jiang C, Wu S, Bai Z, Zhuang G, Zhuang X (2017) Manipulation of nitrogen leaching from tea field soil using a *Trichoderma viride* biofertilizer. Environ Sci Pollut Res 24(5):1–10. https://doi.org/10.1007/s11356-017-0355-x
Xu J, Wu L, Tong B, Yin J, Huang Z, Li W, Li X (2021) Magnesium supplementation alters leaf metabolic pathways for higher flavor quality of oolong tea. Agriculture 11:120. https://doi.org/10.3390/agriculture11020120
Xu J, Wang S, Yao T, She X, Gan Z (2022) Vertical distributions and bio availabilities of heavy metals in soil in An-tea plantations in Qimen County, China. Processes 10:664. https://doi.org/10.3390/pr10040664
Yamashita H, Fukuda Y, Yonezawa S, Morita A, Ikka T (2020) Tissue ionome response to rhizosphere pH and aluminum in tea plants (*Camellia sinensis* L.), a species adapted to acidic soils. Plant-Environ Interactions 1:152–164. https://doi.org/10.1002/pei3.10028
Yan P, Shen C, Fan L, Li X, Zhang L, Zhang L, Han W (2018) Tea planting affects soil acidification and nitrogen and phosphorus distribution in soil. Agric Ecosyst Environ 254:20–25. https://doi.org/10.1016/j.agee.2017.11.015
Yan P, Wu L, Wang D, Fu J, Shen C, Li X, Zhang L, Zhang L, Fan L, Wenyan H (2020a) Soil acidification in Chinese tea plantations. Sci Environ 715:136963. https://doi.org/10.1016/j.scitotenv.2020.136963
Yan P, Han WY, Li X, Zhang LP, Zhang L (2020b) Present situation and analysis of soil acidification in Chinese tea garden. Scientia Agric Sinica 53(4):795–813. https://doi.org/10.3864/j.issn.0578-1752.2020.04.011
Yan P, Zou Z, Zhang J, Yuan L, Shen C, Ni K, Sun Y, Li X, Zhang L, Zhang L, Fu J, Han W (2021a) Crop growth inhibited by over-liming in tea plantations. Bever Plant Res 1:9. https://doi.org/10.48130/BPR-2021-0009. 7 pages
Yan P, Shen C, Zou Z, Fu J, Li X, Zhang L, Zhang L, Han W, Fan L (2021b) Biochar stimulates tea growth by improving nutrients in acidic soil. Scientia Hort 283:110078. https://doi.org/10.1016/j.scienta.2021.110078
Yan Y, Jeong S, Park CE, Mueller ND, Piao S, Park H, Joo J, Chen K, Wang X, Liu J, Zheng C (2021c) Effects of extreme temperature on China's tea production. Environ Res Lett 16:04404. https://doi.org/10.1088/1748-9326/abede6
Yang X, Yu Z, Zhang B, Huang J, Zhang Y, Fang F, Li C, Zhu H, Chen Y (2015) Effect of fluoride on the biosynthesis of catechins in tea [*Camellia sinensis* (L.) O. Kuntze] leaves. Scientia Hort 184:78–84. https://doi.org/10.1016/j.scienta.2014.12.031
Yang X, Ni K, Shi Y, YiX ZQ, Fang L, Ma L, Ruan J (2018a) Effects of long-term nitrogen application on soil acidification and solution chemistry of a tea plantation in China. Agric Ecosyst Environ 252:74–82. https://doi.org/10.1016/j.agee.2017.10.004
Yang Y, Wang F, Wan Q, Ruan J (2018b) Transcriptome analysis using RNA-Seq revealed the effects of nitrogen form on major secondary metabolite biosynthesis in tea (Camellia sinensis) plants. Acta Physiol Plant 40:127. https://doi.org/10.1007/s11738-018-2701-0. 17 pages
Yang XD, Ma LF, Ji LF, Shi YZ, Yi XY, Yang QL, Ni K, Ruan JY (2019) Long term nitrogen fertilization indirectly affects soil fungi community structure by changing soil and pruned litter in a subtropical tea (*Camellia sinensis* L.) plantation in China. Plant Soil 444(1–2):409–426. https://doi.org/10.1007/s11104-019-04291-8

Yang P, Liu Z, Zhao Y, Cheng Y, Li J, Ning J, Yang Y, Huang J (2020a) Comparative study of vegetative and reproductive growth of different tea varieties response to different fluoride concentrations stress. Plant Physiol Biochem 154:419–428. https://doi.org/10.1016/j.plaphy.2020.05.038

Yang TY, Li HP, Tai YL, Dong CX, Cheng XM, Xia E, Chen ZP, Li F, Wan XC, Zhang ZL (2020b) Transcriptional regulation of amino acid metabolism in response to nitrogen deficiency and nitrogen forms in tea plant root (*Camellia sinensis* L.). Sci Rep 10:6868

Yang W, Li C, Wang S, Zhou B, Mao Y, Rensing C, Xing S (2021) Influence of biochar and biochar-based fertilizer on yield, quality of tea and microbial community in an acid tea orchard soil. Appl Soil Ecol 166:104005. https://doi.org/10.1016/j.apsoil.2021.104005

Yang G, Zhou D, Wan R, Wang C, Xie J, Ma C, Li Y (2022a) HPLC and high-throughput sequencing revealed higher tea-leaves quality, soil fertility and microbial community diversity in ancient tea plantations: compared with modern tea plantations. BMC Plant Biol 22:239. https://doi.org/10.1186/s12870-022-03633-6

Yang X, Leng Y, Zhou Z, Shang H, Ni K, Ma L, Yi X, Cai Y, Ji L, Ruan J, Shi Y (2022b) Ecological management model for the improvement of soil fertility through the regulation of rare microbial taxa in tea (*Camellia sinensis* l.) plantation soils. J Environ Manag 308:114595. https://doi.org/10.1016/j.jenvman.2022.114595

Yang J, Zheng Z, Li T, Ye D, Wang Y, Huang H, Yu H, Liu T, Zhang X (2022c) Variations in aluminum fractions within soils associated with different tea (*Camellia sinensis* L.) varieties: insights at the aggregate scale. Plant Soil 480:121–133. https://doi.org/10.1007/s11104-022-05563-6

Yang B, Ren S, Zhang K, Li S, Zou Z, Zhao X, Ma Y, Zhu X, Fang W (2022d) Distribution of trace metals in a soil–tea leaves–tea infusion system: characteristics, translocation and health risk assessment. Environ Geochem Health 44:4631–4645. https://doi.org/10.1007/s10653-021-01190-9

Yang Y, Kim J, Chung JO, Cho D, Roh JH, Hong YD, Kim WG, Kang H (2022e) Variations in the composition of tea leaves and soil microbial community. Biol Fertil Soils 58:167–179. https://doi.org/10.1007/s00374-021-01615-8

Yang X, Tang S, Ni K, Shi Y, Yi X, Ma Q, Cai Y, Ma L, Ruan J (2023a) Long-term nitrogen addition increases denitrification potential and functional gene abundance and changes denitrifying communities in acidic tea plantation soil. Envt Res 216:114679. https://doi.org/10.1016/j.envres.2022.114679

Yang X, Ni K, Shi Y, Yi X, Ji L, Wei S, Jiang Y, Zhang Y, Cai Y, Ma Q, Tang S, Ma L, Ruan J (2023b) Metagenomics reveals N-induced changes in carbon-degrading genes and microbial communities of tea (*Camellia sinensis* L.) plantation soil under long-term fertilization. Sci Environ 856(2):159231. https://doi.org/10.1016/j.scitotenv.2022.159231

Yao Z, Wei Y, Liu C, Zheng X, Xie B (2015) Organically fertilized tea plantation stimulates N_2O emissions and lowers NO fluxes in subtropical China. Biogeosciences 12(20):5915–5928. https://doi.org/10.5194/bg-12-5915-2015

Yao Z, Zheng X, Liu C, Wang R, Xie B, Bahl KB (2018) Stand age amplifies greenhouse gas and NO releases following conversion of rice paddy to tea plantations in subtropical China. Agric For Meteorol 248:386–396

Yao X, Chen H, Zhang B, Lu L (2023) Transcriptome analysis reveals differentially expressed genes involved in aluminum, copper and cadmium accumulation in tea 'Qianmei 419' and 'Qianfu 4'. Plan Theory 12:2580. https://doi.org/10.3390/plants12132580

Ye HM, Yuan XY, He H (2021) The distribution of phosphorus forms in Wuyi rock region and its effect on tea quality-related constituents in tea garden soil. Pol J Environ Stud 30:4331–4341. https://doi.org/10.15244/pjoes/132815

Ye J, Wang Y, Wang Y, Hong L, Jia X, Kang J, Lin S, Wu Z, Wang H (2022a) Improvement of soil acidification in tea plantations by long-term use of organic fertilizers and its effect on tea yield and quality. Front Plant Sci 13. https://doi.org/10.3389/fpls.2022.1055900. 13 pages

Ye Y, He J, He Z, Zhang N, Liu X, Zhou J, Cheng S, Cai J (2022b) Evaluation of the brewing characteristics, digestion profiles, and neuroprotective effects of two typical se enriched green teas. Foods 11(14):2159. https://doi.org/10.3390/foods11142159

Ye J, Wang Y, Lin S, Wang Y, Chen P, Hong L, Jia X, Kang J, Wu Z, Wang H (2023a) Metabolomics analysis of the effect of acidification on rhizosphere soil microecosystem of tea tree. Front Plant Sci 14:1137465. https://doi.org/10.3389/fpls.2023.1137465. 12 pages

Ye J, Wang Y, Kang J, Chen Y, Hong L, Li M, Jia Y, Wang Y, Jia X, Wu Z, Wang H (2023b) Effects of long-term use of organic fertilizer with different dosages on soil improvement, nitrogen transformation, tea yield and quality in acidified tea plantations. Plan Theory 12:122. https://doi.org/10.3390/plants12010122. 25 pages

Ye Y, Yan W, Peng L, He J, Zhang N, Zhou J, Cheng S, Jie Cai J (2023c) Minerals and bioactive components profiling in se-enriched green tea and the Pearson correlation with se. LWT 175:114470. https://doi.org/10.1016/j.lwt.2023.114470. 8 pages

Yemane M, Chandravanshi BS, Wondimu T (2008) Levels of essential and non-essential metals in leaves of the tea plant (*Camellia sinensis* L.) and soil of Wushwush farms, Ethiopia. Food Chem 107(3):1236–1243. https://doi.org/10.1016/j.foodchem.2007.09.058

Yi PJ, Ting HS, Jiong ZY, Yuelin Z, You LW, Yan HW (2016) Characteristics of Sulphur distribution in tea plant and the effect of Sulphur application on tea yield and quality. J Tea Sci 36(6):575–586

Yu J, Lin S, Shaaban M, Ju W, Fang L (2020) Nitrous oxide emissions from tea garden soil following the addition of urea and rapeseed cake. J Soils Sediments 20:3330–3339. https://doi.org/10.1007/s11368-020-02641-z

Yu H, Han X, Zhang X, Meng X, Yue Z, Liu X, Zhang N, Li Y, Yu Y, Yao H (2023) Fertilizer-induced N_2O and NO emissions in tea gardens and the main controlling factors: a recent three-decade data synthesis. Sci Environ 871:162054. https://doi.org/10.1016/j.scitotenv.2023.162054

Zakir M (2018) Review on impacts of climate change on tea yield, yield components and quality. Int J Res Stud Agric Sci 4(5):24–37. https://doi.org/10.20431/2454-6224.0405005

Zhang QF, Ruan J (2016) Tea: analysis and tasting. In: Caballero B, Finglas PM, Toldrá F (eds) Encyclopedia of food and health. Academic Press, Oxford, pp 256–267

Zhang YH, Hang H, Peng BZ, Yang H (2003) Soil erosion and its impacts on environment in Yixing tea plantation of Jiangsu Province. Chinese Geograph Sci 13(2):142–148. https://doi.org/10.1007/s11769-003-0008-5

Zhang Y, Zeng Y, Chang S, Wang X (2004) Available potassium content of red soil tea plantation-and efficiency of spreading potassium. J Tea Commun 46:11–13

Zhang X, Chen J, Liang Y (2014) Advances in the effects of intercropping on ecological factors, growth and economic benefits of young tea garden. Guizhou Agric Sci 42:67–71

Zhang Y, Wang Y, Ding Z, Wang H, Song L, Jia S, Ma D (2017a) Zinc stress affects ionome and metabolome in tea plants. Plant Physiol Biochem 111:318–328. https://doi.org/10.1016/j.plaphy.2016.12.014

Zhang Z, Zhou C, Xu Y, Huang X, Zhang L, Mu W (2017b) Effects of intercropping tea with aromatic plants on population dynamics of arthropods in Chinese tea plantations. J Pest Sci 90:227–237. https://doi.org/10.1007/s10340-016-0783-2

Zhang X, Wu H, Chen L, Liu L, Wan X (2018b) Maintenance of mesophyll potassium and regulation of plasma membrane H~+– ATPase are associated with physiological responses of tea plants to drought and subsequent rehydration. Crop J 6(6):611–620. https://doi.org/10.1016/j.cj.2018.06.001

Zhang F, Liu Y, Wang L, Bai P, Ruan L, Zhang C, Wei K, Cheng H (2018c) Molecular cloning and expression analysis of ammonium transporters in tea plants (*Camellia sinensis* (L.) O. Kuntze) under different nitrogen treatments. Gene 658:136–145. https://doi.org/10.1016/j.gene.2018.03.024

Zhang L, Cao QQ, Granato D, Xu YQ, Ho CT (2020a) Association between chemistry and taste of tea: a review. Trends Food Sci Technol 101:139–149. https://doi.org/10.1016/j.tifs.2020.05.015

Zhang F, Wang L, Bai P, Wei K, Zhang Y, Ruan L, Wu L, Cheng H (2020b) Identification of regulatory networks and hub genes controlling nitrogen uptake in tea plants [*Camellia sinensis* (L.) O. Kuntze]. J Agric Food Chem 68(8):2445–2456. https://doi.org/10.1021/acs.jafc.9b06427

Zhang KB, Liu XL, Kang M, Wang Y, Shen JL, Li Y (2020c) N_2O emissions from tea plantations with sorghum intercropping and application of big urea pills. Environ Monitoring Assess 41(5):2434–2444. https://doi.org/10.13227/j.hjkx.201911090

Zhang H, Hao X, Zhou C, Wang L, Wang X, Yang Y, Zeng J, Sun L, Dai J, Xiang J, Luo H, Wang C, Zhang X, Liu T (2020d) Correlation analysis between selenium accumulation in tea leaves and soil factors in selenium-rich areas. J Tea Sci 40(4:465–477

Zhang X, Wu H, Chen J, Chen L, Chang N, Ge G, Wan X (2020e) Higher ROS scavenging ability and plasma membrane H+-ATPase activity are associated with potassium retention in drought tolerant tea plants. J Plant Nutr Soil Sci 183(4):406–415. https://doi.org/10.1002/jpln.202000007

Zhang X, Luo JL, Zhang C, Chen SX (2020f) The effects of three fertilization treatments on soil fertility and yield and quality of fresh leaves in tea gardens. Materials Sci Forum 984:153–159

Zhang S, Wnag Y, Sun L, Qiu C, Ding Y, Gu H, Wang L, Wang Z, Ding Z (2020g) Organic mulching positively regulates the soil microbial communities and ecosystem functions in tea plantation. BMC Microbiol 20:103. https://doi.org/10.1186/s12866-020-01794-8

Zhang X, Jiang H, Wan X, Li Y (2020h) The effects of different types of mulch on soil properties and tea production and quality. J Sci Food Agric 100:5292–5300. https://doi.org/10.1002/jsfa.10580

Zhang Q, Tang D, Yang X, Geng S, He Y, Chen Y, Yi X, Ni K, Liu M, Ruan J (2021a) Plant availability of magnesium in typical tea plantation soils. Front Plant Sci 12:641501. https://doi.org/10.3389/fpls.2021.641501. 9 pages

Zhang F, He W, Yuan Q, Wei K, Ruan L, Wang L, Cheng H (2021b) Transcriptome analysis identifies CsNRT genes involved in nitrogen uptake in tea plants, with a major role of CsNRT2.4. Plant Physiol Biochem 167:970–979. https://doi.org/10.1016/j.plaphy.2021.09.024

Zhang G, Chu X, Zhu H, Zou D, Li L, Du L (2021c) The response of soil nutrients and microbial community structures in long-term tea plantations and diverse agroforestry intercropping systems. Sustainability 13(14):7799. https://doi.org/10.3390/su13147799

Zhang X, Wang N, Hou M, Wu H, Jiang H, Zhou Z, Chang N, Wang Q, Wan X, Jiang J, Shen Z, Li Y (2022a) Contribution of K solubilising bacteria (*Burkholderia* sp.) promotes tea plant growth (*Camellia sinensis*) and leaf polyphenols content by improving soil available K level. Funct Plant Biol 49(3):283–294. https://doi.org/10.1071/FP21193

Zhang G, Chu X, Zhu H, Zu D, Li L, Du L (2021) The response of soil nutrients and microbial community structures in long-term tea plantations and diverse agroforestry intercropping systems. Sustain For 13(14):7799. https://doi.org/10.3390/su13147799

Zhang H, Dong C, Meng L, Reng L, Li J, Zhang N, Bao L (2022b) Effect of soil fertilizer application on soil nutrient migration and tea quality of plateau-puerh tea. J Geoscience Environ Protection 10:273–286. https://doi.org/10.4236/gep.2022.104017

Zhang W, Lin L, Wang T, Chen M, Song B, Sun W (2022c) Genome-wide identification of amt2-type ammonium transporters reveal that CsAMT2.2 and CsAMT2.3 potentially regulate nh4 + absorption among three different cultivars of *Camellia sinensis*. Int J Mol Sci 23:15661. https://doi.org/10.3390/ijms232415661

Zhang Y, Zhang Q, Wang Y, Lin S, Cheng P, Wang Y, Du M, Jia X, Wang H, Ye J (2023a) Effects of magnesium on transcriptome and physicochemical index of tea leaves. Plan Theory 12:1810. https://doi.org/10.3390/plants12091810

Zhang Q, Shi Y, Hu H, Shi Y, Tang D, Ruan J, Fernie AR, Liu MY (2023b) Magnesium promotes tea plant growth via enhanced glutamine synthetase-mediated nitrogen assimilation. Plant Physiol 192(2):1321–1337. https://doi.org/10.1093/plphys/kiad143

Zhang X, Liu L, Luo S, Ye X, Wen W (2023c) Research advances in aluminum tolerance and accumulation of tea plant (*Camellia sinensis*). Beverage Plant Res 18. https://doi.org/10.48130/BPR-2023-0018. 11 pages

Zhao J, Wu X, Nie C, Wu T, Dai W, Liu H, Yang R (2012) Analysis of unculturable bacterial communities in tea orchard soils based on nested PCR-DGGE. World J Microbiol Biotechnol 28:1967–1979. https://doi.org/10.1007/s11274-011-0999-5

Zhao H, Huang J, Li Y, Song X, Luo J, Yu Z, Ni D (2016) Natural variation of selenium concentration in diverse tea plant (*Camellia sinensis*) accessions at seedling stage. Scientia Hortic 198:163–169. https://doi.org/10.1016/j.scienta.2015.11.026

Zhao C, He X, Dan X, Zhao J, Huang X, Cai Z, Meng H, Zhang J (2021) Specific dissolved organic matter components drive the assembly of a core microbial community in acidic soil of ammonium-preferring plants. Catena 207:105584. https://doi.org/10.1016/j.catena.2021.105584

Zheng H, Li JL, Li HH, Hu GC, Li HS (2014) Analysis of trace metals and perfluorinated compounds in 43 representative tea products from South China. J Food Sci 79(6):1123–1129. https://doi.org/10.1111/1750-3841.12470

Zheng Q, Ding J, Lin W, Yao Z, Li Q, Xu C, Zhuang S, Kou X, Li Y (2022) The influence of soil acidification on N_2O emissions derived from fungal and bacterial denitrification using dual isotopocule mapping and acetylene inhibition. Environ Pollut 303:119076. https://doi.org/10.1016/j.envpol.2022.119076

Zhong Q, Lin Z, Zhang H, Chen Z, You X, Shan R, Chen C (2017) Effects of different potassium levels on main biochemical components of fresh leaves of tea seedlings. J Tea Sci 37(1):49–59

Zhong Q, Lin Z, Chen C, Chen Z, You X, Shan R (2018) Effects of varied potassium supply on growth and antioxidant enzyme activities in leaves of tea seedlings. Acta Tea Sin 59:12–18

Zhou X, Wu F (2012) Dynamics of the diversity of fungal and fusarium communities during continuous cropping of cucumber in the greenhouse. FEMS Microbiol Ecol 80(2):469–478. https://doi.org/10.1111/j.1574-6941.2012.01312.x

Zhou J, Xia F, Liu X, He Y, Xu J, Brookes PC (2014) Effects of nitrogen fertilizer on the acidification of two typical acid soils in South China. J Soils Sediments 14(2):415–422. https://doi.org/10.1007/s11368-013-0695-1

Zhou Z, Chen K, Yu H, Chen Q, Wu F, Zeng X, Tu S, Qin Y, Meakin R, Fan X (2019) Changes in tea performance and soil properties after three years of polyhalite application. Agron J 111(4):1967–1976. https://doi.org/10.2134/agronj2018.06.0393

Zhou B, Chen Y, Zheng L, Cui Y, Li J, Tang H, Liu J, Tang J (2022) Soil nutrient deficiency decreases the postharvest quality-related metabolite contents of tea (*Camellia sinensis* (L.) Kuntze) leaves. Food Chem 377:132003. https://doi.org/10.1016/j.foodchem.2021.132003

Zhunag Z, Mi Z, Kong L, Wang Q, Schweiger AH, Wan Y, Li H (2022) Accumulation of potentially toxic elements in Chinese tea (*Camellia sinensis*): towards source apportionment and health risk assessment. Sci Total Environ 851(1). https://doi.org/10.1016/j.scitotenv.2022.158018

Zou Y, Shen F, Zhong Y, Lv C, Pokharel SS, Fang W, Chen F (2022) Impacts of intercropped maize ecological shading on tea foliar and functional components, insect pest diversity and soil microbes. Plan Theory 11:1883. https://doi.org/10.3390/plants11141883

Zou Z, Mi W, Li X, Hu Q, Zhang L, Zhang L, Fu J, Li Z, Han W, Yan P (2023) Biochar application method influences root growth of tea (*Camellia sinensis* L.) by altering soil biochemical properties. Scientia Hort 315:111960. https://doi.org/10.1016/j.scienta.2023.111960

Dr V Krishnakumar superannuated as Principal Scientist (Agronomy) and Head, ICAR-Central Plantation Crops Research Institute, Regional Station, Kayamkulam, Kerala, India. As an agronomist, he has undertaken various field research programs, farmer-community oriented technology dissemination activities, administration, and manpower management programs. He is a recipient of the ICAR Award for Outstanding Interdisciplinary Team Research in Agriculture and Allied Sciences for the biennium 2005–06.In his career of 37 years, he had worked as the Principal Investigator, Co-investigator of various National and International Projects, published more than 100 research articles both in the National and International journals; edited ten books and contributed chapters; and edited and published technical bulletins, training manuals, and other compilations. He served as Editor, Journal of Plantation Crops and currently is a member of the Editorial Board of Current Science. dr.krishnavkumar@gmail.com

Dr T Rajkumar is currently serving as Soil Chemist in the Chemistry Division, UPASI Tea Research Foundation, Coonoor, Tamil Nadu, India. He started his career as a Soil and Sugar Chemist (Asst. Manager) in R&D Center of E.I.D Parry India Ltd. He has completed his Ph. D in the area of nanomaterials and their applications.He has worked as a National Post-Doctoral Fellow under the DST-SERB, New Delhi scheme at the Indian Institute of Science Education and Research (IISER), Pune. He published more than 13 research articles in reputed peer-reviewed international journals. Currently, he is focusing on the field trials in organic/conventional tea gardens with soil amendments and foliar nutrients for the sustainable tea cultivation. He is also covering the research works in the field of nano-fertilizers and their impacts in tea fields. rajvinayagam@gmail.com

Dr. P. Murugesan is serving as the Quality Manager and Advisory Officer at the UPASI Tea Research Foundation (TRF), Regional Center in Coonoor, Tamil Nadu, India. He completed his Ph. D in Plant Chemistry at the UPASI Tea Research Institute in 2020, specializing in Tea Brew Quality. He also held the pivotal position of Technical Manager, overseeing National Accreditation Board for Testing and Calibration Laboratories (NABL) responsibilities until March 2020.With a portfolio of more than 13 articles published in esteemed peer-reviewed international journals, his current focus revolves around analytical and advisory services, as well as field and factory trials related to nutritional and quality advancements at the Regional Center in Coonoor. Since April 2020, he has assumed the crucial role of Quality Manager for NABL.mpmsun@gmail.com

New Paradigms in Soil Health Management for Sustainable Production of Plantation Crops

George V. Thomas, V. Krishnakumar, and S. R. Prabhu

Abstract

It is crucial to achieve economic and ecological sustainability in the production of plantation crops in view of their importance to support the livelihood of millions of farmers and agro-industrial entrepreneurs by supplying food, fiber, and industrial raw materials. Plantations are developed by clearing forests, and these perennial crops are confined to the same piece of land for decades. Due to the practice of unsustainable agronomic practices and enhanced influence of climate change on crops in recent years, the production system of plantation crops is confronted with several serious challenges. As soils influence agroecosystem productivity, new directions in soil health management based on ecological principles are required to attain sustainable intensification in plantations. The new paradigms in soil health management and crop nutrition in plantation crops need to be based on remediation, maintenance, and enhancement of soil health. Recent research efforts did result in the development of several low cost, simple, and viable technologies to enhance soil health and plantation ecosystem properties capable of achieving sustainability in the plantations. Management strategies for soil carbon sequestration by enhancing above-ground diversity and proper recycling of lignin-rich biomass have been shown to enhance below-ground biodiver-

G. V. Thomas (✉)
ICAR—Central Plantation Crops Research Institute, Kasaragod, Kerala, India

V. Krishnakumar
ICAR—Central Plantation Crops Research Institute, Regional Station, Kayamkulam, Kerala, India

S. R. Prabhu
TerraBioGen Technologies, Burnaby, BC, Canada

G. V. Thomas, V. Krishnakumar (eds.), *Soil Health Management for Plantation Crops*, https://doi.org/10.1007/978-981-97-0092-9_11

sity and together, these practices will contribute to achieve resilience in plantation farming systems. A holistic approach in soil health management in an agroecosystem mode, using integrated nutrient sources, harnessing the potential of biodiversity and carbon sequestration by recycling lignocellulosic plantation biomass will not only help to restore the degraded soils but also will enable to achieve sustainability in plantation crops by enhancing resilience of farming systems to extreme climatic events.

Keywords

Plantation crops · Plantation sustainability · Soil health · Ligno-cellulosic biomass · Soil biodiversity · Biomass recycling · Carbon sequestration

Abbreviations

ABCS	Arecanut-based cropping system
ACC	1-Aminocyclopropane-1-carboxylic acid
AM	Arbuscular mycorrhizae
AMF	Arbuscular mycorrhizal fungi
C/N	Carbon/Nitrogen
CO_2 e	Carbon dioxide equivalent
CPCRI	Central Plantation Crops Research Institute
DNA	Deoxyribonucleic acid
EFB	Empty fruit bunches
g	Gram(s)
GAP	Good agricultural practices
GIS	Geographical information system
Gt	Gigaton(s)
ha	Hectare(s)
HCN	Hydrogen cyanide
IAA	Indole-3-acetic acid
INM	Integrated nutrient management
ISRO	Indian Space Research Organization
K_2O	Potassium oxide
NGS	Next-generation sequencing
PGP	Plant growth promoting
PGPR	Plant growth-promoting rhizobacteria
RS	Remote sensing
SOC	Soil organic carbon
SOM	Soil organic matter

1 Introduction

Ever-increasing global population and resulting resource consumption are placing high demands on agriculture and natural resources (Foley et al. 2011) so far not experienced by humanity. Agricultural intensification driven by the anthropogenic inputs of fertilizers, pesticides, and irrigation, though enabled to enhance global food production, has led to strikingly homogenous agricultural landscapes, reduced habitats for plants, animals and microbes, loss of biodiversity, reduced ecological regulations, enhanced water stress, soil erosions, climate change, and more pest and disease incidences (Duru et al. 2015). Despite many technological advancements with respect to the development of high-yielding varieties and chemical-intensive agriculture, environmental issues are increasing, and resources are becoming scarce, affecting the sustainability of agricultural production. According to the renowned soil scientist Lal (2015), the major constraint in achieving an increase in agricultural production is due to the on-going soil degradation process which results in a decline in soil quality, which in turn decreases the ecosystem goods and services provided by soil. Several factors such as soil erosion, loss of soil organic matter (SOM), acidification, salinization, loss of soil biodiversity, loss of soil fertility, and nutrient imbalance contribute to soil degradation and decline in soil quality (Lal 2015). An urgent need was felt to revive damaged ecosystems on a global scale, resulting in the declaration of the "Decade of Ecosystem Restoration" for the decade 2021–2030 by the United Nations to create awareness to stop and reverse ecosystem degradation (Dessureault-Rompré 2022). Based on recent estimates, food systems contribute to roughly a third of global greenhouse gas emissions. In 2019, food systems emitted 16.5 Gt CO_2e globally and the largest contributors were agriculture, land use, and land-use change activities and the remaining emissions coming from other downstream and upstream activities such as transport, industrial processes, waste management, and packaging (Costa et al. 2022), clearly indicating that there is an urgent need to change the ways we produce and consume food to reduce ecosystem degradation.

Anthropogenic activities related to farming and on-going climate change result in land degradation, which is supported by reduced land productivity, decline in biodiversity, and reduced economic returns. This is a major threat to sustainable agricultural production. The Indian Space Research Organization published in 2021 a report entitled "*Desertification and Land degradation Atlas of India*" which confirms that almost 30% of India's geographical area is under degradation. Maximum degradation takes place in unirrigated agriculture land (38%), followed by forest land (22%), land with scrub (14%), agriculture-irrigated land (8%), and others. Water erosion has been reported to be the major cause of erosion (37%) followed by vegetation degradation (31%) and wind erosion (18%). Almost all Indian states reported enhanced degradation in the past 1.5 decades, and the most rapid increase was reported in the biodiversity-rich northeastern states (ISRO 2021; Pandey and Sengupta 2021).

Global agricultural production is largely dependent on monocropping, chemically synthesized fertilizers, chemical pesticides, irrigation, etc. This industrial agricultural approach based on external inputs has led to several environmental problems such as groundwater pollution, eutrophication of water bodies caused by soil erosion and more availability of soluble nutrients, and leaching loss of nutrients (Tilman et al. 2001, 2002). There is an urgent need to implement sustainable and efficient agricultural practices to overcome the adverse effects of agricultural expansion (Tilman 1999). It is possible to make significant progress in this direction by reducing expansion of agriculture, "yield gap" reduction on low productive lands, enhancing cropping efficiency, and reducing resource waste. Taken together, these strategies can help to significantly increase food production and at the same time should help to reduce adverse ecological impacts of agriculture (Foley et al. 2011). Several management strategies have been suggested for mitigation of soil degradation and these include implementation of techniques to reduce soil erosion, ensuring positive SOC and N budgets, enhancing the biodiversity and activities of soil biota (micro, meso, and macro), and improving soil structural stability and pore geometry (Lal 2015). As a dynamic and living natural system, soil has significant role in functioning of terrestrial ecosystems. Soil health is the foundation of healthy agroecosystems. Doran et al. (1996) defined soil health as "the continued capacity of soil to function as a living system, within the ecosystem and land use boundaries, to sustain biological productivity, quality of air and water environment, promote plant, animal and human health." The living organisms, both prokaryotic and eukaryotic residing in soil, are critically important for the ecosystem services provided by soils, and soil biological productivity will support sustainable agriculture (Hatfield and Walthall 2015). More than 25% of the earth's total biodiversity is supported by soils, which in turn supports life on land and water through continuous nutrient cycling and retention, supporting agricultural production, remediation of various pollutants, and climate regulation (Bach et al. 2020). Soil biota supports several ecosystem services such as the regulation of biogeochemical cycles, regulation of atmospheric trace gases, retention and delivery of required nutrients to plants, maintenance of soil structure and soil fertility, control of soil erosion, provision of clean drinking water, mitigation of floods and droughts, and pest and disease control (Lal 2015).

On a global scale, plantation crops occupy significant acreage: Oil palm 19 million ha (Murphy et al. 2021), coconut 12.3 million ha (Nayar 2017), rubber 11 million ha (Singh et al. 2021), cacao 14–15 million ha (Amponsah-Doku et al. 2022), coffee 10 million ha (Jha et al. 2014), and tea 5 million ha (Wang et al. 2020). When we analyze the data on plantation crops, it becomes clear that these perennial crops occupy only less than 2% of the total cultivated area in India, however, contribute 75% of export earnings from agricultural products or 12.72% of the total export earnings. Six major plantation crops (coconut, arecanut, cashew, tea, coffee, and rubber) occupy around 5.12 million ha in India. Other plantation crops such as oil palm, cocoa, small cardamom, black pepper, and nutmeg also showed steady increase in area over the last five decades (Anandaraj 2015). It is also important to mention that the plantation sector provides livelihood for about 23 million workers and more than 1.5 million growers (Abraham and Madhavan 2020). However, due to the adoption of unsustainable agroecosystem management practices, and ongoing climate change, the production system of plantation crops is facing several serious challenges. These challenges are primarily characterized by a degraded soil resource base with multiple nutrient deficiencies, reduced organic carbon content and biodiversity, moisture stress, and increased pest and disease incidence. All these factors affect soil and crop productivity and economic viability, making these crops more vulnerable to climate change.

To address these challenges and to achieve sustainability, it is crucially important to enhance the capacity of plantations to function as a complex biological system capable of sustaining biological productivity, environmental quality, and to promote soil, plant, animal, and human health. Crop diversification and intensification along with animal husbandry and provision of integrated crop nutrition and crop protection have proved to be one of the best models for sustainably optimizing system productivity. Closed nutrient cycles leading to considerably reduced external nutrient input supply can be achieved by effectively recycling lignocellulosic biomass from plantations through employment of local earthworms and microbial resources. Despite the growing importance of soil biology for maintaining and improving soil health (Tarafdar 2022), soil biological investigations attracted only limited attention in R&D programs of even fast-expanding plantation crops, which result in significant changes in the way land is used. However, in the recent past plantation soil health and soil biology received significant importance on a global basis. This positive change should help in the coming years to direct the management of plantations sustainably. It might be possible to engineer the rhizospheres (Ryan et al. 2009) of plantation crops through various agronomic interventions, resulting in several beneficial below-ground characteristics of plants such as improved root architecture, nutrient uptake, and nitrogen fixation, thus helping the plantation owners to usher into very much needed new green revolution (Herder et al. 2010). Utilization of efficient strains of microbial resources as bio-fertilizer formulations and use of biostimulants would help in bio-nutrition, plant growth promotion, and abiotic and biotic stress management in plantation crops.

The need to produce more food, feed, and fiber to support growing population during this climate change scenario provides arguments for an increased intensification of agriculture. Under these circumstances, it has been suggested to try a promising option based on eco-functional intensification of farming, such as organic farming, an approach based on using and enhancing locally available and natural resources and processes to improve soil productivity, while minimizing negative ecological impacts. This concept is based on the active roles played by soil organisms, which regulate several soil-based ecosystem services such as cycling of nutrients, soil erosion control, and pest and disease regulation. There are many studies on positive effects of organic farming on soil health and quality including microbial community traits (Lori et al. 2017). Complex interactions between different components have been reported in these bio-intensive systems and fertility management in organic farming relies on a long-term approach rather than the more short-term solutions common in conventional agriculture (Watson et al. 2006). By applying various agroecological principles, it has been shown that it is possible to manage many plantation crops through organic farming technologies (Chattopadhyay et al. 2015; Nampoothiri 2001; Thomas 2010).

2 Plantation Crops: Soil Constraints Limiting Sustainable Productivity

Plantation crops, viz., coconut palm (*Cocos nucifera*), arecanut palm (*Areca catechu*), oil palm (*Elaeis guineensis*), cashew (*Anacardium occidentale*), cocoa (*Theobroma cacao*), coffee (*Coffea arabica*), rubber (*Hevea brasiliensis*) and spices, are cultivated to meet the needs of food, oil, beverage, industrial raw materials, and fiber and evidently drive the rural economy in the tropical and subtropical regions of India. These crops play an important role in elevating the socio-economic status of millions who depend on these groups of crops for cultivation, processing, value addition, marketing, and export. Being perennial crops with long lifespan, they are also important in view of their role in ecosystem services and their impact on environment. However, plantation agriculture creates several environmental issues (Hartemink 2005b), as most of the plantations are created by clearing natural forests. As these crops are dominantly cultivated in small and marginal holdings, the sustainability of plantation agriculture is confronted with several challenges related to climate change and deterioration of production base. The rapid expansion of plantations is a major driver of deforestation, soil degradation, CO_2 emissions, and decline and loss of biodiversity in Southeast Asia (Panklang et al. 2021; Peerawat et al. 2018; Qaim et al. 2020; Singh et al. 2021; Warren-Thomas et al. 2018). It has been reported that the predominance of small holder cocoa and oil palm cultivation significantly increases deforestation (Kroeger et al. 2017). The oil palm crop system faces several challenges such as a general lack of the ability to withstand higher temperature and erratic rainfall and increasing incidence of pests/diseases (Murphy et al. 2021). Indian Cardamom hills support intensive cardamom cultivation, and this has resulted in serious ecological and environmental problems, which in turn can adversely affect long-term sustainability of cardamom cultivation. The adverse effects of intensive cardamom cultivation on its forest environment had been significant, and increase in surface air temperature was observed in the cardamom hills during the period 1990–2020 (Murugan et al. 2022). Monoculture of plantation crops is one of the major causes for deterioration of these agroecosystems. There are several reports from plantation crop-grown soils of different states indicating soil constraints as an important factor affecting the productivity of these crops. Malhotra et al. (2017) reviewed various reports on diagnosis and management of various soil fertility constraints in coconut plantations and reported that coconut palms grown in laterite, red, and sandy soils faced multiple problems such as poor nutrient levels, soil acidity, and low cation exchange capacity. Reddy and Upadhyay (2002) found that coconut productivity in coastal sandy soils is adversely affected by high soil temperature, poor water holding capacity, low nutrient content, poor soil aggregate stability, etc. Enhancement of the content of soil organic matter (SOM) is a pre-requisite for the successful management of coconut palms grown in such problem soils. Shivaprasad et al. (2018) reported the results of a study on the fertility status of coffee plantation soils in Wayanad district in Kerala, India, which revealed that most (98%) soils were acidic, and deficiencies of calcium, magnesium, and boron were common, thus adversely affecting the productivity of coffee plants.

Remote sensing (RS) and geographic information system (GIS)-based studies in Tamil Nadu, India proved that 31.5% of total coconut planted area had soil constraints, particularly related to soil depth, soil texture drainage, and soil reaction (Selvamani and Duraisami 2018). Deterioration of soil condition with respect to very low organic carbon content, multi-nutrient deficiencies, and low carbon sequestration has been reported in intensively cultivated cashew plots when compared to that in forest and scrub lands (Srinivasan et al. 2019). A study centered around arecanut and coconut palm plantation soils in Karnataka indicated that soils are low to medium in available N and K in surface and sub-surface. Due to high P fixing capacity and removal than replenishment, the soils were low in available phosphorus. Arecanut soils were generally more fertile and higher available N and K_2O were observed in arecanut soils than in coconut soils (Vasundhara et al. 2021). These reports indicated that soil degradation is a major constraint influencing the sustainable productivity of many plantation crops, and it necessitates technological interventions to mitigate the soil constraints with appropriate soil health improvement strategies to achieve sustainability in production.

Soil and leaf nutrient status of cashew plants growing along the coastal belt in India revealed several nutrient constraints causing lower productivity. The cashew plantation soils were deficient in nitrogen, potassium, calcium, magnesium, zinc, and copper. This study also proved the ill effects of neglecting fertilizer and manure application in cashew on SOC, nitrogen, and potassium, which adversely affected the uptake of nutrients by cashew plants as clearly shown from the leaf nutrient status. The use of soil ameliorants to correct soil acidity has been recommended in cashew soils to make calcium, magnesium, and micronutrients such as zinc, copper, and manganese available for the crop (Mangalassery et al. 2021). Dinesh et al. (2014) reported that soils of Kerala state, India, accumulate significant quantities of phosphorus. However, most of these major nutrients will be in unavailable form, necessitating adoption of various biological strategies and agronomic interventions like liberal use of organic manures and use of phosphorus solubilizing microbial inoculants (Simpson et al. 2011) to reduce fertilizer usage by releasing fixed phosphorus.

Cocoa plantations are usually created by clearing natural forests. After the initial conversion of forests to cocoa agroforests, sustaining production is difficult as trees age, together with increase in pest and disease pressure. To sustain the production, various conservation measures can be combined with cocoa production by using opportunities to combine biodiversity conservation, sustainability, and carbon storage (Clough et al. 2009). It has been shown that crop intensification may reduce ecological resilience of cocoa production systems (Vaast and Somarriba 2014).

3 Technological Advances in Diversified and Multifunctional Land Use Systems

3.1 Complex Cropping Systems-Based Management of Plantation Crops

In industrial agriculture and plantations, the emphasis is on monocultures, and this has resulted in very serious ecosystem imbalances. Perennial monocultures often decline in productivity over time due to the pressure from pests and diseases. This is especially true in the case of woody perennial tree crops where crop rotation is impossible (Vukicevich et al. 2016). Long-term monocultures of coffee can have adverse effects on soil health and soil microbial activities (Zhao et al. 2018). Practical and sustainable solutions for this kind of basic issue in perennial monocultures are possible through agrotechniques based on ecological principles (Vukicevich et al. 2016). The benefits of the time-tested practice of growing several plant species together in managed systems with trees and crops in various tiers deserve serious attention, as it has been proved that crop biodiversity increases ecosystem productivity and resistance to climate extremes (Isbell et al. 2015). Higher plant diversity can enhance microbial diversity, reduce pests, and diseases and thus reduce the decline of productivity of perennial crops (Vukicevich et al. 2016). Plants and their living roots function as the main engines in soil restoration efforts (Dessureault-Rompré 2022) because plant roots mediate significant carbon sequestration (Obersteiner and Klein 2022). Intercropping can stimulate the release of soluble carbon compounds in soil (Wang et al. 2021), which can enhance soil microbial activities.

Perennial tree-based ecosystem services are the best features of more sustainable agroecosystems (Barrios et al. 2018) and agroforestry systems can play a leading role in biodiversity conservation (Bhagwat et al. 2008). Rubber-based agroforestry systems have been shown to support biodiversity without adversely affecting latex yields (Warren-Thomas et al. 2019). Coconut palm-based multispecies systems in tropical home gardens and shaded perennial systems are typical examples of such systems. The ecological principle of these complex systems is niche complementarity, which means that the cropping systems that are structurally and functionally more complex than monocultures of trees or crops result in efficient capture and utilization of resources such as nutrients, light, and water. Unexplored or under-exploited areas of benefits of multiple tree and crop systems include their ecosystem services such as biodiversity conservation, climate regulation, and carbon storage. These multi-species cropping systems represent an agroecological wonder, and the basic laws governing these systems could be utilized to design more sustainable and productive agroecosystems (Nair 2017; Renard and Tilman 2021). When cassava fields were converted to a young rubber plantation, significant changes on microbial biomass and activities and fungal composition were observed; however, this conversion did not impact the bacterial and macrofaunal diversity (Peerawat et al. 2018). This unique soil biological effect could be due to the cultivation of pineapple in the inter-row of the young rubber plantation (Peerawat et al. 2018). Cocoa agroforestry systems were shown to improve the soil quality in degraded pastures (Suárez et al.

2021). When a perennial crop like coconut is cultivated as a monocrop, the resource utilization and farm income are much lower than the actual potential, and the plantations are exposed to the harmful effects of natural forces, resulting in poor soil health and depletion of biodiversity. The development of diversified and multifunctional land use systems has been shown to provide several advantages such as increased on-farm income, better environmental quality, and conservation of natural resources. The research efforts on enhancing above-ground diversity by way of developing appropriate diversified cropping systems in plantation crops such as coconut and arecanut are driven by the fact that when monocropped, only a small portion of the available natural resources such as solar radiation, water, nutrients, and land space is utilized by these crops (Bavappa et al. 1986; Nelliat et al. 1974).

Diversified cropping systems with complex crop arrangements are usually created on a small scale rely on local resources and are reasonably stable and productive, resulting in a high income per unit of labor and energy. These complex polycultures and agroforestry systems in the tropics adopted by small farmers mimic the structure and function of natural communities, therefore, have several benefits such as high biodiversity, closed nutrient cycling, and resistance to pest invasion (Altieri and Nicholls 2004). This way, now it is possible to design several cropping systems by combining several concepts and approaches of ecology, agriculture, and genetics (Gaba et al. 2015). To effectively utilize natural resources, meet farmers' needs, and make farming economically viable, many cropping systems have been designed for plantation crops by integrating annuals, biennials, and perennials in the same piece of land (Nair 1977; Thomas et al. 2010).

A wide diversity of crops such as tropical tuber crops, rhizomatous spice crops, cereals, vegetables, pulses, oil seeds, fruit crops, flower crops, and medicinal and aromatic crops were found to be suitable for growing in the interspaces of coconut palms. Perennial crops such as cocoa, clove, nutmeg, coffee, black pepper, cinnamon, cardamom, mango, and papaya can also be successfully grown as mixed crops in the coconut gardens. High-density multispecies cropping system or multistoried cropping system involves integration of coconut or arecanut palms with annual and perennial crops of varying plant heights, rooting, and canopy patterns in the same piece of land so as to maximize utilization of solar radiation, nutrients, and moisture (Bavappa et al. 1986). Mixed farming systems involve integration of subsidiary enterprises such as livestock, poultry, rabbitry, pisci-culture, sericulture, and goat rearing along with the cultivation of fodder and pasture in the interspaces of coconut palms (Subramanian et al. 2014). It was suggested to integrate livestock components in arecanut palm ecosystem to help the system to reduce ecological imbalances, adapt to climate change and provide various ecosystem services arising due to continuous cultivation of areca palm (Sujatha and Bhat 2015). Suja et al. (2023) demonstrated that organic management of coconut-tuber crop systems offered resilience with higher system productivity and profitability. The coconut yield under intercropping with tuber crops in organic mode was high by 7–13% in comparison to monocrop of coconut maintained by the farmers. By combining annual and perennial plants with farm animals in a dynamic manner, coconut farming with agroforestry concepts has proved to create a sustainable ecosystem capable of landscape restoration. These complex systems reduce crop failure, generate extra income, and balance the functioning of ecosystem by increasing biodiversity, enhancing soil quality, and encouraging carbon sequestration pathways (Dissanayaka et al. 2023). In an experiment involving a coffee plantation intercropped brachiaria in coffee rows, the positive impact on soil enzyme activities was mainly due to the growth of the grass (Rodrigues et al. 2021). Agroforestry system developed by integrating coconut palms with compatible fruit trees and multipurpose tree species is a good model for additional farm benefits and potential for carbon sequestration (CPCRI 1989). In coconut-based homestead farming system, diverse species of agricultural, horticultural, and miscellaneous plants are grown and along with the integration of animal husbandry enterprises, meet the requirements of ecological and economic sustainability (John 2014).

In cocoa agroforestry, continuous litter fall, and the resulting litter layer support higher abundance and diversity of soil fauna (da Silva Moço et al. 2009) and in such systems, it is possible to achieve high biodiversity and high crop yields (Clough et al. 2011). Based on the nutrient stocks in the litter, cocoa agroforests have the potential productive capacity to sustain production in different harvest cycles (Fontes et al. 2014). Rubber-based agroforestry systems characterized by soil surface cover resulting from constant leaf litter fall and extensive root systems lead to increased SOC and nitrogen, improved soil aggregation, reduced soil erosion, decreased carbon, and nitrogen loss, and finally improved rates of accumulation of carbon and nitrogen (Chen et al. 2017). The conversion of rubber monocultures to rubber-based agroforestry has become a common trend in the recent past. The introduction of a legume shrub (*Flemingia macrophylla*) in rubber plantations can change soil carbon and nitrogen fractions and improve soil labile organic carbon and nitrogen (Liu et al. 2018). Rubber–*F. macrophylla* systems have been shown to slow down the release of cations and mitigate soil acidification relative to rubber monoculture (Liu et al. 2019). Essentially, these complex cropping systems work based on the principle of "niche complementary system," resulting in the utilization of resources in a sustainable way. Sustainable intensification of agriculture through intercropping practices has been recently promoted as the basis for a new green revolution (Martin-Guay et al. 2018) and is particularly true with plantation crops, which provide ample opportunities for intercropping during the long-life cycle.

3.2 Biodiversity and Conservation-Based Plantation Agriculture

The integration of diversified crops including perennials, biennials, and annuals into monocrop plantations is an innovative technological approach to enhance the productivity of the whole system and to support environmental sustainability. In complex cropping systems with diverse crop species, significantly higher number of niches are formed above and belowground, and these support the colonization of an array of natural organisms. These organisms create and maintain the ecological equilibrium between species through interacting life cycles and food webs and enhance the functions of nutrient, carbon, and water cycles. With the introduction of diverse crops of variable canopy and root architecture as spatio-temporal assemblages of annuals and/or perennials, system sustainability is achieved through functional complementarities, beneficial biological interactions, and synergism between plant species in the agroecosystem. In biodiversity-based agriculture, enhanced biological processes such as organic matter decomposition, cycling of nutrients, minimization of nutrient losses, and management of pests are important in the development of sustainable systems (Duru et al. 2015).

In mixed tree systems, "wood wide web," the mycorrhizal networks that connect different tree species and transfer carbon, has ecological and functional significance. Simard (2018) in a controlled experiment, birch saplings exposed to radioactive CO_2 gas, converted the radioactive gas to radioactive sugar by photosynthesis. Traces of radioactive sugar appeared in the leaves of fir sapling growing nearby in two hours. The exchange is mainly through mycelia of fungi and may extend through the whole forest. The intricate and vast system of fungal connections developed under tree crop ecosystems transfer carbon between tree species and make systems sustainable.

Scientific studies suggest that modification of management strategies can result in better system productivity through harnessing of ecosystem services to a greater extent. In cropping systems, soil management practices should be designed to increase SOM content and soil health by creating habitat diversity to better support soil biological communities. Agronomic practices, such as conservation tillage, use of cover crops, diversified crops, incorporation of perennial species, use of organic or biofertilizers, retention of crop residue, and weed control by mulching/cultural practices, could greatly help to prevent soil degradation and result in enhanced soil health (Lehman et al. 2015). Sustainable agriculture seeks to ameliorate soil health, sequester carbon, store water, and build healthier farm ecosystems. Protection of soil surface is an important aspect of protecting soil quality. Manorama et al. (2021) noted that in a plantation with oil palm monocrop, just mulching was able to enhance SOC content by 52%.

"Mimicking the Nature" has been suggested as a sustainable path to achieve ecological agriculture (Blok and Gremmen 2016). By "doing it the nature's way," agro-technology and scientific innovation can be combined to achieve harmony with the natural ecosystems of mother earth. Cropping systems can be designed from nature (Malézieux 2012) and by designing systems capable of mimicking nature's functions, various agroforestry systems can remain productive and support several ecosystem services (Rodriguez et al. 2021; Wilson and Lovell 2016). The coconut-based farming systems in Kerala, India, are living examples of nature-based solutions in agriculture (Mohan Kumar and Kunhamu 2022).

3.3 Augmented Below-Ground Biodiversity and Biological Activities in Diversified Systems

Higher microbial activity is associated with higher plant density (Vukicevich et al. 2016). Accordingly, in the perennial crop-based cropping/farming systems, plant-beneficial microbial communities have been shown to be enhanced in a significant manner in rhizospheres and endorhizospheres of crops when compared to the monocrop stands (Bopaiah and Shetty 1991a, b). Several function-specific microbes, namely, biological nitrogen-fixing bacteria, phosphate solubilizing bacteria and fungi, and plant growth-promoting bacteria (PGPR) were seen in greater levels in various cropping systems when compared to that under monocropping systems (Thomas et al. 1991). The coconut palms in diversified farming systems harbored unique endophytic diazotrophs viz., *Azoarcus* spp., *Burkholderia* spp., *Herbaspirillum frisingense,* and *Arthrobactor* spp. with nitrogen fixation and plant growth promotion properties (Thomas and Prabhu 2003). Plant beneficial bacteria possessing plant growth promoting (PGP) traits such as production of IAA, ACC deaminase, HCN, siderophore, chitinases, antibiotics, growth on free media, and phosphate solubilization were present in high numbers in plantation crop-based cropping systems (Priya et al. 2012; Litty et al. 2010). Several AM species belonging to ten genera viz. *Acaulospora, Claroideoglomus, Dentiscutata, Diversispora, Funneliformis, Gigaspora, Glomus, Redeckera, Scutellospora,* and *Septoglomus* were found, indicating high richness of AM fungi in coconut palm rhizosphere under intercropping (Rajesh et al. 2015). Higher incidence of vesicular-arbuscular mycorrhizal symbiosis was observed in intercropped and mix-cropped coconut plantations (Thomas 1988). Earthworm communities in cropping systems, as evidenced by density and biomass, were high in diversified systems and were absent in monocropped plots under conventional farming. In alley-cropping system involving oil palm intercropped with pineapple, bamboo, black pepper, cacao, and bactris palms, soil faunal biodiversity and related ecosystem services were significantly enhanced by alley cropping (Ashraf et al. 2018).

Soil enzymes such as dehydrogenase, urease, and phosphatase that are indicative of soil biological activity were significantly higher in diversified cropping systems when compared to a monocrop. It was observed that mineralization of carbon and nitrogen was high in diversified cropping systems when compared to a monocrop (Bopaiah and Shetty 1991a). Various positive effects of organic farming practiced in coconut-based integrated farming systems on soil health included improved soil physical, chemical, and biological properties when compared to the inorganic nutrient management practices (Palaniswami et al. 2008).

3.4 Interrelationships Between Below-Ground Diversity and Above-Ground Diversity

There are strong relations between plant diversity and soil microbes in terrestrial ecosystem, which suggests the relevance of plant diversity in the functioning of belowground ecosystems (Chen et al. 2019). Barrios et al. (2018) provided evidence for this relationship in coffee and tea-based cropping systems. Loss of native earthworm populations can result in 76% lower soil macroporosity when shade trees are absent in coffee-based cropping system. Increased tree cover in tea estates can contribute 53% increased tea yield. Averill et al. (2022) analyzed 80 experiments to prove that the restoration of native soil microbiomes can help to accelerate plant biomass production to the extent of 64% across various ecosystems. These studies indicate that there is enormous potential to enhance soil productivity within managed landscapes, as agriculture and forestry are the dominant uses of land on Earth. In addition to improved crop productivity, stimulating higher soil microbial biodiversity in managed landscapes is a critical, but under appreciated opportunity. Reduced abundance of certain microbial groups such as firmicutes has been reported to result in increased plant disease incidence (Lee et al. 2021), opening varied possibilities for managing plantation crop diseases through manipulation of soil microbial community structure. Several field experiments conducted over many decades did generate lots of data supporting better efficiency of ecosystem functioning in cropping systems with perennial plantation crops as evidenced by higher yields of the main crop, overall higher productivity of the cropping system, and enhanced resource use efficiency (Thomas et al. 2022).

3.5 Planned Biodiversity, Agroecological Function, and System Productivity

It has been established that agronomic practices influence ecosystem services, and agricultural sustainability depends on the ecosystem services rendered by the production systems. An understanding of the contribution of various agricultural practices on ecosystem services will enable us to identify the management practices that enable to develop sustainable production systems. Studies conducted on arecanut palm-based cropping systems (ABCS) in west coast of India provided significant insights on ecosystem services and energy use efficiency in ABCS under organic farming and integrated nutrient management (INM) (Paramesh et al. 2019). The benefits of organic farming include its effectiveness in improving cropping system profitability, soil quality, energy efficiency, and environmental quality. Based on these indications, it is pertinent to undertake further studies to elucidate the relationship between planned biodiversity, agroecological function, and total productivity.

4 Lignocellulosic Biomass from Plantation as a Resource for Sustaining Plantation Productivity

4.1 Utilization of Lignocellulosic Biomass from Plantation Crops for Improving Soil Health and Plant Productivity

Plantation crops generate significant quantities of lignocellulosic organic matter, which can be effectively used for improving and maintaining plantation soil health. Agroforestry systems generate 6.5–7.4 Mg ha^{-1} of annual mean biomass for recycling in soil (Rodríguez et al. 2023; Schroeder 1993). There is an urgent need for practical solutions for implementing the conversion processes related to valorization of plantation biomass to soil amendments and fertilizers. Disposal of oil palm empty fruit bunch (EFB) solid wastes is becoming a challenge with an increased level of production of palm oil. Even though this is a challenge, it could be turned into an opportunity by utilizing the biomass as a resource capable of increasing the crop productivity. The use of appropriate conversion technology for EFB to use as soil ameliorants can enhance crop yield by improving soil properties and reducing environmental pollution (Anyaoha et al. 2018). Plantation wastes being rich in lignocellulosic complexes biodegrade in soil slowly and the benefits in terms of improved soil health and fertility will be visible after a lag time. It will take 18–24 months after application of EFB waste from oil palm plantations to see an improvement in soil fertility and soil biology parameters such as earthworm counts (Carron et al. 2015). It is critically important to implement simple, low cost and viable physical, chemical, and biological technologies for managing plantation wastes. To effectively implement biological resource-based technologies, it is pragmatic to construct organic waste treatment and bioorganic fertilizer production facilities in the plantations, which will solve the problem of waste transport or sanitary hazards (Chojnacka et al. 2020).

Plantation crops generate significant quantities of ligno-cellulosic biomass, which is not fully utilized to improve soil health. Nutrient recycling from waste to crop (Hidalgo et al. 2021) by converting biomass to bio-stimulants and bio-fertilizers (Puglia et al. 2021) has become a major theme in sustainable management of soils. Plantation waste biomass can be effectively recycled to meet a part of the nutrient needs of plantations (Embrandiri et al. 2015; Fidelis and Rao 2017; Gandahi and Hanafi 2014; Sadasivuni et al. 2015; Thomas 2010). Coffee and tea wastes can be converted to fertilizers and plant growth-promoting formulations (Ebid et al. 2008; Tarashkar et al. 2023; Vitale et al. 2023). In addition, plantation wastes such as tea leaf waste have been reported to be of value in bioremediation (Debnath et al. 2022) and would help to remove or reduce toxic chemicals from plantation soils. About 80% of cocoa fruit biomass comprising of pod husks, bean shells, and cocoa sweatings is discarded as waste biomass (Vásquez et al. 2019), which when properly planned can be used as a resource for various biotechnological applications and as a plant nutrient source. Every year large quantities of nutrients in cocoa ecosystems are utilized for plant growth and cycling of plant nutrients through recycling of cocoa ecosystem is very essential for maintaining cocoa productivity (Hartemink 2005a). Technology for production of potash from cocoa pod husks has been developed and commercialized (Kone et al. 2020; Simpson et al. 1985). Most of the phosphorus used in farming comes from mining, and this supply is not sustainable. The current situation of less availability and high price of P fertilizers should trigger concerted actions globally to help face the global phosphorus challenge. Phosphorus waste reduction and efficient phosphorus recycling can reduce imported phosphorus (Brownlie et al. 2023). A significant portion of phosphorus needs of plantation crops can be met through efficient recycling of waste biomass generated in the system. In India, rubber plantations produce over 7 ton of litter ha^{-1} and experiments indicated that litter-derived P can meet phosphorus demands of a mature rubber plantation (Jessy et al. 2009). Pectin is known to enhance the population of phosphate-solubilizing bacteria (Mise et al. 2020), and there are possibilities for enhancing the solubilization of phosphates in plantation soils using pectin-rich wastes from plantations in the form of soil amendments. Cocoa pod husk is known to be a rich source of pectin (Barrios-Rodríguez et al. 2022) and research work in the direction of pectin-induced phosphate solubilization is hoped to yield rewarding results.

Under natural conditions, lignocellulosic plantation waste biomass undergoes slow decomposition. In oil palm plantations, leaflets and fronds decompose faster as compared to empty fruit bunches (EFB) and rachis. The decomposition of biomass and nutrient release rates were faster during the initial stages and leaflets released the highest total amount of nutrients other than K (Moradi et al. 2013). Lignocellulosic biomass from plantations needs to undergo certain level of pretreatment to make it amenable for easy composting (Ma and Nichols 2004; Mosier et al. 2005). Lime pretreatment is a simple, safe, and easily adoptable technique to facilitate faster microbial degradation (Kaar and Holtzapple 2000; Ma and Nichols 2004; Prabhu and Thomas 2002; Sierra et al. 2009).

Waste lignin has unique properties to allow controlled slow release of nitrogen fertilizers (Behin and Nader 2016; Chen et al. 2020; Fischer and Schiene 2002; Gu et al. 2019; Jiao et al. 2018, 2019; Lu et al. 2022). Additionally, phosphorus fertilizer use efficiency can be enhanced when applied along with lignin (Rotondo et al. 2018). Lignin and lignin derivatives from empty fruit bunch of oil palm have been used as an additive substance in urea fertilizer to reduce the solubility of urea nitrogen (Arianie 2010). Lignin can also reduce nitrification process (Huang et al. 2003). In addition to lignin, many plantation wastes are rich in polyphenols. Polyphenols can influence the pools and fluxes of inorganic and organic soil nutrients (Hättenschwiler and Vitousek 2000). Humified phenol-rich plant residues from waste tea leaves can reduce soil urease activity (Sivapalan et al. 1983), which has the potential to enhance fertilizer use efficiency in tea estates. As recycling of plantation biomass rich in lignin results in build-up of humus or humic substances in soil, it might be interesting to monitor humus build-up through various sustainable soil and crop management technologies. Elucidation of supramolecular nature has recently permitted soil scientists to develop techniques for fractionation of soil humus known as Humeomics, which enables in-depth studies on the chemical nature of soil humus fractions (Piccolo et al. 2019). It would be valuable to adopt humeomics technologies in the plantation industry. Each biomass has specific lignin chemistry and coconut lignin has been shown to be unique (Rencoret et al. 2013). Characterizing humic substances produced from the biomass recycled from various plantation crops will be quite useful.

4.2 Organic Biomass-Based Low-Cost Technologies for Sustaining Plantation Productivity

4.2.1 Leguminous Cover Crops and Tress in the Plantations

Several bioresources-based low-cost and simple technologies have been developed that are important for producing farm inputs that can substitute, supplement chemical fertilizers enable to grow the plantation crops in the organic mode, and reduce crop decline in perennial systems. In perennial monocultures, maintenance of non-crop vegetation such as leguminous cover crops is an important agrotechnique, which can support higher soil microbial diversity and help to reduce the negative effects of cultivating monocultures of perennial crops for a long time on the same land (Vukicevich et al. 2016). Legume crops can directly alter bulk soil chemistry capable of altering soil microbial communities, finally leading to improved productivity of degraded lands (Zhou et al. 2019). Cover crops have also been proved to support sustainable ecological intensification (Wittwer et al. 2017) and adjust and adapt to the negative influences of climate change (Kaye and Quemada 2017). Given these benefits, legume cover crops are common in many plantations. A simple and easily adoptable agrotechnique involving cultivation of leguminous nitrogen-fixing cover crops, *Pueraria phaseoloides* and *Calopogonium mucunoides* in basins of coconut palms yield 15–20 kg of biomass and 150–200 g of biological nitrogen per basin and the incorporation of this biomass bring about significant improvement in soil health properties and enabled substitution of 50% fertilizer nitrogen (Thomas and Shantaram 1984; Thomas et al. 2001). In oil palm plantations, leguminous cover crops have been shown to add 150 kg N $ha^{-1}year^{-1}$ through the process of biological nitrogen fixation (Agamuthu and Broughton 1985). In rubber plantations, *P. phaseoloides* cover crop can deliver 250 kg N $ha^{-1}year^{-1}$ and of this 39–46% of the biologically fixed nitrogen can be transferred from the legume to the trees (Clermont-Dauphin et al. 2016). Cover crops can increase soil microbial population, biological activity, and microbial diversity (Kim et al. 2020) and reduce soil-borne diseases (Larkin 2015). Dinesh et al. (2006) reported various long-term effects of leguminous cover crops on biochemical and biological properties of coconut plantation soils and the benefits included improved soil microbial biomass and soil enzyme activities. Buyer et al. (2017) found that soil microbial community structure was significantly altered by cover crops maintained in cocoa plantations. Liu et al. (2021) reported that rhizosphere P cycling was improved when legume cover crop was present in young rubber plantations, and biomass of cover crop contributed to the regulation of rhizosphere phosphorus cycling and availability. As of now, there are several knowledge gaps on the impacts of cover crops on soil resources in tree cropping systems (Rodriguez-Ramos et al. 2022), and further work in this direction would add to the knowledge to keep plantation soils biologically active.

Fixation of significant levels of atmospheric nitrogen by *Gliricidia sepium* trees intercropped in cocoa plantation has been established by employing ^{15}N isotope dilution technique. *Rhizobium tropici* and *Rhizobium etli* associated with root nodules of *Gliricidia* produced nitrogen-rich biomass capable of theoretically supplying 31.4–38.0 kg N ha^{-1} $year^{-1}$ (Kaba et al. 2019). Tea plants intercropped with soybeans and smooth vetch improved tea leaf quality and soil fertility attributes by changing the composition of *Bacill*us sp in the plantation soil (Huang et al. 2022).

4.2.2 Earthworm-Based Technologies

Several bioresource technologies for managing plantation waste and to produce soil-improving organic products are available now. The technology for bioconversion of lignin-rich coconut biomass using *Eudrilus* sp., a local epiegic earthworm gave new directions for biological soil fertility management in coconut (Prabhu et al. 1998). Earthworms can stimulate soil microbial activity (Binet et al. 1998) and the vermicompost from coconut biomass increased soil fertility through the addition of plant nutrients, growth hormones, soil enzymes, and plant beneficial microbes such as free-living nitrogen fixers, phosphate solubilizers, fluorescent pseudomonads, and silicate solubilizers (Gopal et al. 2009). Earthworms have also been used for composting coir pith combined with cow dung and *Gliricidia* leaves (Jayakumar et al. 2022), coconut husks mixed with cattle dung (Quadar et al. 2022), cocoa and arecanut biomass (Chowdappa et al. 1999), oil palm mill wastes (Rupani et al. 2017; Singh et al. 2011), oil palm empty fruit bunches (Sabrina et al. 2009), coffee pulp (Orozco et al. 1996; Raphael and Velmourougane 2011), spent coffee grounds (González-Moreno et al. 2020; Hanc et al. 2021), tea wastes (Badhwar et al. 2020), cashew litter (Parthasarathi et al. 2016), and rubber leaves (Chaudhuri et al. 2003). Ashraf et al. (2017) reported that vermicompost from oil palm biomass can enhance the growth of oil palm seedlings. It would be of interest to see if the application of plantation waste-based vermicompost can improve the yield of bioactive compounds such as oleoresins in spice crops as proved in the case of chicory (Gholami et al. 2018).

Earthworms are ecosystem engineers and have developed intimate relationships with several microbes and other microbial communities in soil. Microorganisms, which are mostly in dormant stage in soil, are activated in earthworm gut through the action of intestinal mucus (Lavelle 2011). This microbial activation phenomenon in vermicasts produced from coconut wastes was clearly demonstrated (Gopal et al. 2009, 2017). Soil-dwelling earthworms can be used to enhance microbial activities and improve degraded soils. In-soil earthworm technology was shown to restore degraded soils and improve soil fertility and crop yields in tea plantations (Senapati et al. 1994, 1999). Earthworm-based organic fertilization technology has been shown to increase the number and activity of soil microorganisms in a significant way and enhance soil enzyme activities, resulting in restoration of the damaged soil ecosystem of tea garden (Tang et al. 2016).

4.2.3 Other Bioresource Technologies for Managing Plantation Wastes

Lignocellulosic biomass from plantation crops undergo biodegradation slowly, causing pollution problems as in the case of coir pith. India and Sri Lanka together produce over 0.7 million tons of coir pith (Van Doren et al. 2017) and as reviewed by Prabhu and Thomas (2002), coir pith has enviable physical and chemical characteristics suitable for enhancing soil quality once properly composted products are used in crop production. White rot fungi such as *Pleurotus* species can break down lignin (Camarero et al. 1996) and can be employed for enhancing the biodegradation process of plantation wastes. Ghosh et al. (2007) developed a fast and large-scale process for composting coir pith using *Pleurotus* inoculum. Shashirekha and Rajarathnam (2007) noticed that coir pith gets better colonized by *Pleurotus* when mixed with straw, highlighting the importance of blending various organic wastes with plantation wastes to produce fast and good quality composts. A bioresource technology developed for co-composting of coir pith in conjunction with poultry manure, lime, and rock phosphate resulted in the production of quality organic manure with physico-chemical and biological characteristics for use as an important crop production input in plantations (Thomas et al. 2013). Coir pith can also be co-composed with cow manure (Tripetchkul et al. 2012). Mghaiouini et al. (2022) developed a composting technology for co-composting coffee marc in combination with cattle manure. Vico et al. (2018) valorized date palm pruning biomass by co-composting with urban and agri-food sludge. Jiang et al. (2023) reported the use of coffee husk compost for improving soil quality and photosynthesis in arabica coffee plants.

Thomas et al. (1998) developed a technology for biodegradation of coconut wastes and to produce edible mushroom using *Pleurotus sajor-caju*. Similar technologies are available for converting oil palm wastes to edible mushrooms (Aubrey et al. 2022). The organic wastes left after mushroom production is a valuable organic soil amendment and also finds application as a plant-growing substrate (Medina et al. 2009), which is possibly due to the plant growth-promoting substances produced by *Pleurotus* sp. (Pham et al. 2019), in addition to the humic substances produced during lignocellulosic biodegradation. Spent mushroom wastes from plantation crop biomass can be gainfully used for raising nursery seedling of various plantation crops, thus contributing to the circular economy of plantations. Globally, there is a need to find replacements for peat for sustainable soil-less cultivation (Atzori et al. 2021), and plantation wastes can find a place in this niche market. In addition, spent mushroom waste can help to degrade and eliminate various aromatic toxins in plantation soils thanks to the ability of laccase and peroxidase enzymes produced by *Pleurotus* sp. (Rodrıguez et al. 2004).

New composting technologies for utilizing lignocellulosic wastes by employing cellulolytic nitrogen-fixing *Klebsiella* species are now available (Harindintwali et al. 2020). Other well-known diazotrophs such as *Rhizobium* spp. (Jackson et al. 2017) and *Azospirillum* spp. (Kupryashina et al. 2015) can degrade lignin and could be valuable for developing technologies for composting plantation wastes and adding value as a biofertilizer. Lignocellulosic biomass from oil palm can also be broken down using plant growth-promoting and disease-suppressing *Trichoderma* spp. (Siddiquee et al. 2017). Given the opportunities available for value-added biomass products with biofertilizer and biopesticide attributes, research work in this direction would pay rich dividends. Another area of interest is conversion of lignin to humic acids, as humic acids have emerged as crop biostimulants (Jindo et al. 2020). Various humic acids have been proved to have beneficial effects on morphological and physiological parameters when applied on plants (Nardi et al. 2000; Trevisan et al. 2010). There are technologies available for converting lignin to humic substances (Lee et al. 2019). Muniraj et al. (2021) reported that tyrosinase secreted by *Bacillus aryabhattai* can produce humic substances from coir pith wastes.

5 Utilization of Crop Microbiota Diversity for Sustainable Plantation Management

5.1 Beneficial Microorganisms Form an Integral Part of Healthy Plantations

The soil microbiome is the most important and complex biological component of soil ecosystems, as it harbors millions of microbes belonging to innumerable taxonomic groups. Soil microbes together with environmental factors contribute to shaping the soil microbiome in space and time dimensions (Islam et al. 2020). Beneficial interactions with root zone microorganisms have been suggested to reduce the reliance on external crop inputs, increase disease resistance, and reduce abiotic crop stresses (Schmidt et al. 2022a). Multiple sustainability goals can be simultaneously addressed when soil biota is involved in the assessment of the quality of land management practices because soil organisms are closely tied to multiple processes that ecosystems rely on (Bach et al. 2020). In the case of coffee plant, the soil microbiota diversity at farming and processing levels is characterized by the ability to promote plant growth, control pests, improve coffee quality, and manage organic wastes sustainably (Duong et al. 2020). "Core microbiota" capable of colonizing the plant system in coffee has been identified (Duong et al. 2020), and similar studies need to be conducted in various plantation crops for biotechnological applications. To fully utilize the capabilities of soil microorganisms in agriculture, it is important to integrate soil biodiversity with other disciplines (Venturini et al. 2023).

Each plantation crop selects and enriches a unique set of microorganisms in root zones and plant tissues. Bacterial species associated with the coffee rhizosphere were shown to be more diverse and compositionally different from their respective control soil samples (Caldwell et al. 2015). Tea plants grown in India are associated with plant growth-promoting and pathogen-suppressive bacterial strains (Dutta and Thakur 2017). Plants usually drive the process of selection of microbes from the bulk soil for association with roots (Hartmann et al. 2008). Interestingly, rhizospheres of established tea bushes in India have been shown to be dominated by *Bacillus species* (Pandey and Palni 1997). Regardless of the edaphic conditions and geographic location, date palm (*Phoenix dactylifera*) invariably selects similar Gamma proteobacteria- and Alpha proteo-bacteria-dominated bacterial communities (Mosqueira Santillán et al. 2019). If there is a specific association between a plantation crop and microbial species, microbial inoculant product development should be based on the associated species. Additionally, it would make optimization of agronomic management practices easier, as the target microbial group to be favored through management practices is known. It is expected that intensive research and development activities in this direction will happen in the ensuing years. The possibility of manipulating and agronomically managing microbiome to enhance soil health and soil fertility (Chaparro et al. 2012) opens several practical opportunities for sustainably managing plantations. As suggested by Pulleman et al. (2022), targeted stimulation of native beneficial soil biota deserves attention as a foundation for the development of cost-effective sustainable management strategies. This strategy would be of great value in enhancing microbial activities and soil quality in plantations. Hartman et al. (2018) reported that various cropping practices can manipulate the relative abundance of soil and root-associated microbes sensitive to various agronomic practices. Similar studies in plantation soils are predicted to enable scientists and growers to attempt to manipulate microbiome capable of enhancing soil health, as suggested by Qiu et al. (2019) for in situ microbiome engineering. Understanding the role of endophytic communities is of relevance for the improvement of plantation soil and plant health. Coffee and rubber plants are actively associated with a variety of endophytic microorganisms (Araujo et al. 2020; Asad et al. 2023; Chaverri et al. 2010; De Souza et al. 2008; Gazis and Chaverri 2010; Nascimento Brito et al. 2023; Rodrigues et al. 2021). The unique ecosystem offered by the endorhizosphere of coffee plants favored a variety of community of fungal endophytes and the novel endophytic *Trichoderma* sp. from healthy plants were found to be effective for controlling coffee tracheomycosis caused by *Fusarium* spp. (Mulaw et al. 2010; Mulaw et al. 2013). Abraham et al. (2013) reported endophytic *Alcaligenes* sp. and *Pseudomonas aeruginosa* in rubber trees, and these endophytes were shown to be antagonistic against the fungal pathogen *Phytophthora*. A better understanding of cocoa root-associated microbiome and efforts to optimize the same can help to efficiently deliver various ecosystem services by cocoa plantations (Schmidt et al. 2022a). To analyze data from a cocoa agroforestry system, it was revealed that there exists a strong relationship between microbial diversity and economic yields from the plantation and disease incidence rates. The shade trees in the plantation were shown to favor increased availability of N and P in the rhizospheres of neighboring cocoa trees without reducing cocoa yields. In a long term, this positive influence of the agroforestry system on soil fertility will likely enhance cocoa yields, soil productivity, and soil microbial composition (Schmidt et al. 2022b). Aravind et al. (2012) developed stem and root bacterization technique for pre-plant inoculation of endophytic bacteria in black pepper, which in addition to protecting rooted cuttings from infection enhanced growth. An endophytic biofilm-forming *Bacillus* spp. (Azri et al. 2018) and Arbuscular Mycorrhizal (AM) Fungi (Sundram et al. 2022) were shown to enhance the growth of oil palm. Falcäo et al. (2014) isolated growth-promoting and antagonistic *Bacillus subtilis* from the internal tissues of cocoa plants, which would be of practical interest in raising healthy cocoa seedlings in commercial nurseries. Such technologies would be welcome in all plantation crops, as nursery inoculation of beneficial microbes would be viable and easily adoptable by growers.

Recent studies revealed an interesting and unexpected association between plantation crops such as oil palm and rubber with endophytic basidiomycetous fungi (Martin et al. 2015; Pinruan et al. 2010). There is only very limited knowledge on the diversity and ecological roles of these plant tissue colonizing symbiotic fungi associated with these plantation crops. There is some evidence that some strains of these basidiomycetous fungi can inhibit the growth of the deleterious oil palm pathogen *Ganoderma* (Naidu et al. 2016). Further investigations in other plantation crops should reveal any specific patterns of occurrence and specific roles played by these endophytic fungi in the growth and health of crops. Further studies are needed to develop efficient microbiota-based alternatives to improve crop growth as well as quality.

Plantation crops are known to harbor a number of novel plant-beneficial diazotrophic bacteria such as *Herbaspirillum* sp. in coconut palm (Thomas and Prabhu 2003), *Gluconacetobacter* in coffee (Fuentes-Ramírez et al. 2001; Jimenez-Salgado et al. 1997) and *Phytobacter palmae* in oil palm (Madhaiyan et al. 2020). There is evidence that diazotrophic bacteria are involved in fixing atmospheric nitrogen in association with oil palm (Reis et al. 2000; Zakry et al. 2012).

Research efforts centered around coffee plantations in Brazil revealed genetic signatures of diverse population of unreported plant growth-promoting bacteria, implying that new groups of plant-beneficial bacteria may exist in association with coffee plants and further studies on these novel groups of potential plant growth-promoting bacteria can be used to improve coffee production (Caldwell et al. 2015). Clearly, microbial diversity in plantation soils is poorly understood as of now and what is known is just the tip of iceberg. There exists microbial dark matter, which usually constitutes over 99% of the total soil microbial population. Most of the prokaryotes in the environment cannot be cultured in the laboratory using conventional microbiological techniques, making it hard to study and utilize these microbes (Schloss and Handelsman 2005). Culturomics, a technique for high-throughput culturing, has enabled us to discover and describe several new taxa by testing several culture conditions and media (Lagier et al. 2015). Application of culturomics approaches also will enable us to better understand the interaction between microorganisms and plantation crops to utilize the associations for sustainable agriculture.

5.2 Metagenomic Approach to Study Microbiota Diversity

Recently, it has become clear that soil microbiome has significant potential for promoting plant growth, increasing crop productivity, and alleviating biotic and abiotic stresses faced by crops (Zhang et al. 2021). In the recent years, culture-independent strategies have been increasingly used for studying soil microbiome, and the adoption of next-generation sequencing (NGS) techniques permits meta-barcoding analysis to identify a whole microbial community present in an environmental DNA sample (Santos et al. 2020). These innovative technologies provide a clear insight into crop-associated microbial communities. Such a study on coffee plant-associated microbes revealed the presence of significantly high microbial diversity encompassing 607 genera and 923 species belonging to the bacterial and fungal kingdoms (Duong et al. 2020). Plant microbiome can be engineered through new plant breeding strategies, and this innovation can be adopted to sustain crop productivity (Arif et al. 2020). Also, further research is needed to elucidate the dynamic interaction between crop plants, the rhizosphere microbiome, and the environment to harness microbiome to increase crop yield and quality. Rhizo-microbiome has emerged as a new and fascinating factor contributing to hybrid vigor. It was evident from a study conducted by Wagner et al. (2020) with the parents and the hybrid developed in maize, the hybrid vigor was not observed when grown in sterile soil. When the rebuilding of soil was done with seven species of bacteria, they could observe the hybrid vigor as well as parent hybrid relationship, indicating the role of rhizo-microbiome in contributing to the heterosis.

5.3 Microbial Biotechnologies for Sustainable Plantation Management

Industrially produced microbial inoculants are usually not that effective and don't provide consistent positive effects given the fact that the inoculated strains fail to survive in alien soils (Kaminsky et al. 2019). It is an opportunity for plantation agriculture to mass culture appropriate plant-beneficial microbes by utilizing plantation waste and the microbes adapted to plantation waste chemistry are more likely to survive in plantation soils, leading to better crop response to inoculation. Plantation crop and processing wastes can be used for effectively mass culturing beneficial microorganisms to produce microbial fertilizers (Anith 2009; ChandraMohanan et al. 2013; Das et al. 2021; Gopal et al. 1999; Pandey et al. 2000; Sriram et al. 2010). Organic soil amendments, leguminous cover crops, and microbial products when used appropriately can help to manage several serious diseases of plantation crops (Bouchon and Hoopen 2022; Thomas et al. 1993a) and abiotic crop stress (Cherif et al. 2015; Hazzouri et al. 2020; Thomas et al. 1993b).

There are convincing reports on the beneficial symbiotic association between nitrogen-fixing bacteria in plantation crops (de Carvalho et al. 2008; Reis et al. 2000; Thomas and Prabhu 2003; Varmazyari and Cakmakci 2018). Recently, microbial inoculation of perennial tree roots with plant growth-promoting bacteria has been promoted as a promising technology for promoting growth of trees (Chanway 1997). Carrier-based biofertilizer formulations have been developed with efficient strains of PGPR, arbuscular mycorrhizal fungi (AMF), phosphate solubilizers, and nitrogen-fixing bacteria for application to seedlings and adult palms as low cost and eco-friendly inputs. A talk-based formulation of PGPR, *Pseudomonas* sp. and a soil-based AM formulation with AMF, *Claroideoglomus etunicatum* have been commercialized as effective bio-inoculants for coconut seedlings. It has been shown that inoculation of oil palm seedlings with nitrogen-fixing *Bacillus sphaericus* can significantly increase the content of nitrogen and dry matter weight of leaflets and rachis (Zakry et al. 2012).

6 Future Strategies for Sustainable Agriculture Through Soil Health Management

6.1 Strategies to Enhance Carbon Sequestration by Terrestrial Woody Biomass and Below-Ground Systems

Soil stores the largest pool of carbon on this globe (2157–2293 Pg), which is twice the quantity of carbon in the atmosphere. Given this fact, small changes in soil C status may have a serious effect on the global carbon balance and on climate change (Arias et al. 2005). The capacity of soils to "store" stabilized carbon in managed systems is adversely affected by agronomic practices that enhance the loss of SOM or limit its formation and retention. In total, 50–75% of decline in total SOM content of soils in agricultural systems (i.e., ~60 Mg C ha^{-1}) is common. It is a priority to build lost soil carbon to mitigate climate change and to support crop health and ecosystem services derived from SOM (Richardson et al. 2014). Management of SOM is the critical aspect of sustaining soil productivity given the fact that it influences several soil properties. Even minimal changes in SOM content can have large impacts on key soil quality parameters. Agronomic practices capable of encouraging the maintenance and improvement of SOM are critically important for sustaining soil functions (Chenu et al. 2019; Powlson et al. 2011b).

In plantations, sequestration of carbon happens in live plant biomass and soil organic matter generated from litter fall. Increasing plant density per unit area through the adoption of high-density multispecies cropping systems and efficient recycling of plantation waste biomass would favor significant carbon sequestration in the plantations. Jessy et al. (2013) reported that 590–1758 kg ha^{-1} of carbon is recycled per year through fine roots produced by rubber trees. Lignocellulosic biomass from plantation crops can be converted to biochars as shown in the case of coconut biomass (Gopal et al. 2020), which when recycled in plantation soils helps to build soil health. Ecosystem services provided by trees are increasingly recognized as important features of sustainable agroecosystems (Barrios et al. 2018; Reed et al. 2017). Accordingly, agroforestry systems have gained importance as a carbon sequestration strategy because of the potential of the system for carbon storage in multiple plant species and soil systems. This potential benefit has been calculated to be substantial; however, it has not been fully exploited (Leakey 2014; Montagnini and Nair 2004). The potential for carbon sequestration in agroforestry systems has been estimated to be between 12 and 228 Mg ha^{-1} (Albrecht and Kandji 2003). Perennial plantation crops such as coconut, arecanut, oil palm, and rubber provide agronomic opportunities for intercropping with several crops. In addition, cocoa, coffee, black pepper, ginger. are normally grown under the shade of multipurpose trees in various agroforestry systems. Total carbon stock in a shaded coffee system has been calculated to be 81 Mg ha^{-1} (Dossa et al. 2008). Joseph et al. (2022) reported that rubber-based agroforestry systems favor the accumulation of organic carbon in soil when compared to the monocropped rubber plantations. Various plantation crop-based cropping systems can mediate significant carbon sequestration through the production of massive live biomass and continuous accumulation of soil carbon aided by continuous litter fall. Clearly, plantation crops enhance plant diversity per unit area and add varied quality biomass to soil, supporting the efforts of carbon sequestration. Several studies have reported that CO_2 sequestration was significantly greater under perennial crops as compared to annual crops (Ganeshamurthy et al. 2019). Sequestration of atmospheric carbon into terrestrial woody biomass and soil carbon pool is vital strategy for mitigation of climate change. Integrating trees into farming systems has been reported to be the stand-alone agronomic technology to enhance the environmental sustainability as well as productivity of farming systems. Plantation crops, particularly plantation-based agroforestry systems with several diverse tree species that form components of the system, have great potential to sequester significant quantities of carbon in the woody biomass. Diversified land management systems with vegetable crops as intercrops in coconut plantation resulted in increased levels of above- and below-ground carbon sequestration (Kumar and Maheswarappa 2019). The same study reported an incremental increase in carbon sequestration by palms after two years to the tune of 3.01 t ha^{-1} under intercropping when compared to 2.31 t ha^{-1} in coconut monocrop. Inclusion of organic and bio-nutrient sources such as vermicompost, vermiwash, coir pith compost, farm yard manure, and biofertilizer also brought out significant improvement in the carbon stocks in soil. A 50-year coconut plantation under mono-cropping can fix 98.2 C t ha^{-1} through above- and below-ground carbon sequestration (Bhagya et al. 2017). Average total carbon in a cocoa agroforestry was calculated to be 117 ± 47 Mg ha^{-1}, out of which 42% was contributed by fresh above-ground biomass (Somarriba et al. 2013). Cocoa cultivation under diverse shade tree cover can permit high carbon storage and this will not result in yield losses (Rajab et al. 2016).

Investigations on carbon stocks in cashew-growing soils of coastal Karnataka, India, revealed higher levels of carbon sequestration in natural cashew plantation management than in intensively managed plantations (Srinivasan et al. 2019). Agroforestry systems with perennial stands of horticultural and forestry tree crops can become very efficient carbon sinks by virtue of their carbon sequestration potential. Research attention should be devoted to identifying compatible multipurpose tree species that could be grown with plantation crops, without competing with the main crop for nutrients, water, and space, and to harness complementary benefits and the carbon sequestration potential in woody biomass as well as in the soil. Fast expansion of natural rubber plantations in Asia and other regions has resulted in changes in soil carbon stocks with potential impacts on climate change. Carbon sequestration occurred after the conversion of arable land to natural rubber plantations, while carbon losses prevailed if forest land was converted to rubber (Blagodatsky et al. 2016). The rubber-leguminous shrub systems greatly improved SOC sequestration rate, and rubber-leguminous shrub systems could mitigate local climate warming by reducing greenhouse gas emissions and improving soil organic sequestration rate (Rao et al. 2021). When compared to SOM and microbial biomass, crop residues have a very wide ratio of nutrients to carbon and this fact indicates various complex reactions involved in the conversion of carbon in crop residues into SOM and microbial biomass. SOM is essentially derived from microbial detritus, rather than via direct deposition of recalcitrant plant materials (Richardson et al. 2014). Evidently, the efficient recycling of plantation biomass and implementation of agronomic management practices capable of enhancing soil microbial activities would certainly help to build soil carbon stocks. As plantation biomass has a wide C/N ratio (Prabhu and Thomas 2002), it is important to use plant litter with varied qualities and nutrient compositions to favor fast microbial action leading to the formation of humus. It is also of critical importance to prevent loss of nitrogen from soil, thus making the nutrient available for humus formation. Biomass with a high active polyphenol content can reduce faster nitrogen release, thus preventing its loss (Handayanto et al. 1997) and nitrification as proved in the case of tea wastes (Tang et al. 2021). Plantation wastes have high polyphenol content, and it is hypothesized that efficient recycling of biomass would help to conserve nitrogen required for microbial activities and subsequent humus formation.

6.2 Use of Biostimulants in Plantation Crops

As defined by du Jardin (2015), a plant biostimulant is any substance or microorganism applied to plants with the aim to enhance nutrition efficiency, abiotic stress tolerance, and/or crop quality traits, regardless of its nutrient content. The use of biostimulants such as humic acids, protein hydrolysates, and microbial products is becoming common in horticulture to enhance crop growth and protect from biotic and abiotic stresses (Colla and Rouphael 2015; Jindo et al. 2020). Crop biostimulants usually function through their action as prebiotics capable of enhancing microbial activities (Vassileva et al. 2020), and many products can be developed from agro-industrial by-products (Xu and Geelen 2018), including plantation wastes.

Plants can utilize proteins as a source of nitrogen without getting broken down through microbial action (Paungfoo-Lonhienne et al. 2008), and amino acids and peptides have significant plant growth and health-promoting and abiotic stress-relieving properties (Colla et al. 2017). Protein wastes from coconut and oil palm kernels and seafood industry wastes can positively impact growth and productivity of plantation crops, as evidenced by historic uses of coconut oil cakes and sea food wastes in coconut and arecanut cultivation. Raguraj et al. (2022a) used chicken feathers to produce protein hydrolysates, which were used for boosting the growth of tea seedlings in the nursery. Spent coffee grounds and proteins extracted from it can be converted into valuable crop biostimulants through proper processing techniques. These biostimulants have been shown to possess auxin-like activities (Massaya et al. 2022). The recent interest in the use of humic substances in agriculture and horticulture did result in increased demand and cost of humic and fulvic acid formulations separated from mined humic materials, the supply of which is not sustainable. Hence, there is scientific and commercial interest to produce humus – like organic materials from lignocellulosic waste materials that resemble humic substances. Raguraj et al. (2022b) extracted two humic fractions from tea wastes rich in lignocellulosic compounds discarded from tea manufacturing factories. Tea waste derived humic-like substances when applied on tea seedlings raised in the nursery at a rate of 80 mg L^{-1} water concentration promoted seedling growth, proving that tea waste can be used for manufacturing a potential plant bio-stimulant, which can be used instead of mined humic products. Various lignins and lignin derivatives have recently been shown to have plant bio-stimulant actions (Savy and Cozzolino 2022; Savy et al. 2016). As plantation wastes are rich in lignin compounds, research and development activities in plantation waste lignin-based biostimulants would help to reduce fertilizer inputs and manage crops sustainably. Agro-industrial wastes and extracts of various plants can be used for producing nanomaterials with biostimulant action (Tolisano and Del Buono 2023), and lignin nanoparticles have been proved to stimulate plant growth and biochemical traits in plants (Tolisano et al. 2023). These studies open many opportunities for manufacturing a variety of biostimulants from lignocellulosic wastes from plantation crops.

More research-based data is required to expand the use of this practice in the commercial plantation industry. This new area of research assumes more importance given the fact that crop biostimulants can help to increase crop nutrient uptake (Halpern et al. 2015), mitigate adverse effects of climate change on crops (Del Buono 2021; Li et al. 2020), and enhance plant metabolites of commercial interest (Anaya et al. 2022; Ertani et al. 2014), which would be of interest to enhance oleoresin yield from spice crops.

7 Take Home Message for Improving Plantation Soil Health and Sustainability

Plantation agriculture even though increased farm income and reduced poverty has contributed to deforestation and losses in ecosystem functions (Qaim et al. 2020) and the best efforts to reduce negative environmental impacts would require the stakeholders to concentrate the efforts on improving soil health. Soil health continues to be a challenge for microbiologists and chemists (Arias et al. 2005), so also is the task of practically using the information we have on soil microbiome to enhance soil health and improve plant nutrition in agroecosystems (Mitter et al. 2021). Soils are under constant change and function as the channel for the flow of energy produced by photosynthesis, thus function as an ecosystem (Janzen 2015) and will remain the central aspect of the concept of One Health which states that the health of human beings is not isolated, instead closely connected with the health of plants, animals, and environments (Banerjee and van der Heijden 2023). As there is undisputable data on microbial involvement on soil carbon formation (Bradford et al. 2013; Kallenbach et al. 2016), clearly, any attempt to sustain plantations will require improvements in organic inputs and soil biological activities. Application of appropriate technologies for efficient utilization of organic biomass generated in the plantations would be the top priority in the upcoming years to overcome the shortcomings resulted from the expansion of plantation acreage. Just this one developmental activity would greatly contribute to the circular economy of plantations, as suggested by Chojnacka et al. (2020). Despite varied challenges in improving soil health of tropical plantations, based on the accumulated scientific knowledge, many practical directions can be suggested.

Through the 4 per 1000 strategy (increasing SOC stocks in the topsoil by 0.4% year^{-1}) soil degradation and soil erosion can be potentially limited, resulting in preservation of mineral nutrients required for crop growth and the buildup of SOM, which would help to reduce fertilizer use (Soussana et al. 2017). We need to let soils work for us. Appropriate management of soil microbial populations can potentially reduce nutrient leakage from root zones and enhance nutrient use efficiency and agroecosystem productivity. Soil management practices can have a significant influence on microbial composition (Jurburg et al. 2020). By intelligently managing tillage and residue, it could be possible to manipulate microbial habitats and create congenial conditions for soil microbes to thrive. As soil ecology is very complex, in soil ecological research involving soil organisms and plant growth, all possible factors involved should be taken into consideration (Elliott and Coleman 1988). The concept of soil health needs to be approached holistically by involving multiple disciplines that integrate physico-chemical and biological approaches for soil management to achieve soil-supporting sustainable agriculture (Arias et al. 2005).

Agricultural operations capable of increasing SOC content are believed to benefit soil properties and functioning. Even such agronomic practices that increase SOC, at the same time, with no or minimal benefits for climate change mitigation are likely to be beneficial for soils. Hence our efforts to enhance SOC accumulation should continue without regrets (Lehmann and Kleber 2015; Powlson et al. 2011a). Fundamental principles of agroecological farming include biomass recycling, maintenance and enhancement of functionally diverse species and genetic resources in the agroecosystem, limiting the loss of nutrients, improving soil conditions for better plant growth and implementation of management practices capable of supporting beneficial biological interactions and synergies (Migliorini and Wezel 2017). Implementation of even limited activities related to these principles will invariably help to make soil organically rich and biologically active. Plantations incapable of receiving certified organic status will also benefit from undertaking agronomic practices based on some or all these agroecological principles and economically benefit from long-term soil health. Powlson et al. (2011b) emphasized the views published in the paper entitled "The soil carbon dilemma: shall we hoard it or use it?" by Janzen (2006) that it is the biological turnover of SOM rather than just the quantity of SOM accumulated is the critically important factor for soil functioning. A portion of SOC should breakdown to support the functioning of soil microbial activities, as soil microbial activities are carbon limited (Soong et al. 2020), and biologically active soils will have good carbon turn over.

There is practical evidence in support for the proposed direction. Many plantation owners have benefited from the implementation some of the agroecological principles while using some fertilizers as well. Studies in tea plantations revealed that small changes in management practices can enhance biodiversity (Chowdhury et al. 2021). In spite of lower tea leaf yields, agroecological tea adopters earned more than USD 8400 ha year^{-1} than the farmers practicing conventional tea estate management. This study proved that it is economically and environmentally more sustainable to produce organic tea than conventional tea (Le et al. 2023). Using *Urochloa decumbens* + coffee husk/organic compound + controlled-release fertilizers resulted in higher growth and yield of the coffee plants, supporting the view that the adoption of agroecological principles in plantations will support plant nutrition and ensure higher yields (Voltolini et al. 2023).

8 Conclusions

The urgent need to alter our current agricultural production systems has been well recognized in view of its impact on the deterioration of health of the planet and its people. Soil degradation has emerged as a major constraint limiting the productivity of plantation crops, and technological interventions to mitigate the soil constraints with appropriate soil health improvement strategies assume great significance in achieving sustainability in production agriculture. The improvement of soil health through the implementation of novel and innovative agronomic approaches is possible (Mortimer et al. 2015; Shahane and Shivay 2021) and several advances made, as described in this review would be helpful in attaining sustainability in plantation crops. It has been proved that plantation agriculture can be managed to achieve agronomic and economic sustainability (Khatun et al. 2017; Luke et al. 2020; Van Der Vossen 2005) through the adoption of appropriate agronomic and post-harvest processing technologies. Advancements made in development of diversified and multifunctional land use systems with high above-ground diversity have revealed empowerment of below-ground biodiversity and soil health improvement. Several bioresources-based low-cost technologies have been developed, which proved beneficial to minimize or avoid chemical fertilizer inputs and enabled the production of plantation crops by improving soil health in a sustainable way. As suggested by De Corato (2020), concerted efforts should be made for on-farm composting and stabilization of plantation wastes using a variety of available bioresource technologies, thus supporting the evolution of circular plantation economy.

Future strategies for sustainable agriculture should be based on biodiversity and conservation-based approaches to enhance soil health, system productivity, and ecosystem services. As mitigation of climate change is most vital to achieve sustainability, strategies to promote sequestration of atmospheric carbon into terrestrial woody biomass and soil carbon pool need greater attention. The tremendous potential available to promote plant growth by harnessing soil microbiome holds promise as a sustainable strategy to increase crop production and to alleviate biotic and abiotic stresses faced by the crops. Bae et al. (2009) reported that beneficial endophytic fungi are capable of enhancing plant growth and protecting cocoa from drought stress. Beneficial soil microbes also help plantation crops to survive and yield under limited nutrient conditions prevailing in natural conditions (Gopal et al. 2022). As plant–microbe networks are weakened by agrochemical use (Huang et al. 2019), microbial technologies capable of reducing and eliminating chemicals are of significant relevance in the current scenario. Sufficiently funded research programs to elucidate such beneficial microbial mechanisms are urgently required to make available these technologies for use in plantation agriculture through the involvement of commercially motivated science-based companies. The emerging opportunities for adding significant value to plantation wastes through the production of a variety of lignin-based biostimulants and products for enhancing fertilizer use efficiency need to be explored seriously. Lignin nanoparticles have been predicted to have an important role in supporting sustainable agriculture (Santo Pereira et al. 2022). To facilitate a holistic approach in soil health management in plantation crops, promotion of cross-sectorial collaboration and resource exchange in various plantation sectors, as proposed by Gontard et al. (2018), would help to overcome the challenges in achieving sustainability. For example, oil palm leaf compost can be used as a media amendment for raising rubber seedlings (Stevanus et al. 2016) and coir pith can be used as a low-cost seed germination media for rubber (Joseph and Jessy 2015). As suggested for coffee (Philpott et al. 2007), strategic work plans for triple certification (organic, fair trade, and shade coffee) at the farm level may enhance biodiversity, increase farmers' benefits, and this would ultimately lead to successful conservation strategies in other plantation crops as well. As opined by van Noordwijk and Brussaard (2014), enforcement of Good Agricultural Practices (GAP) in world trade would pave the way for incremental improvement in locally sustainable food systems that are efficient at the global level and through this, it should be possible to achieve sustainable intensification in plantation crops.

References

Abraham V, Madhavan M (2020) Performance of the plantation sector during the COVID-19 pandemic. Indian Econ J 68:438–456. https://doi.org/10.1177/0019466220988064

Abraham A, Philip S, Jacob KC, Jayachandran K (2013) Novel bacterial endophytes from *Hevea brasiliensis* as biocontrol agent against *Phytophthora* leaf fall disease. BioControl 58:675–684. https://doi.org/10.1007/s10526-013-9516-0

Agamuthu P, Broughton WJ (1985) Nutrient cycling within the developing oil palm legume ecosystem. Agric Ecosyst Environ 13:111–123. https://doi.org/10.1016/0167-8809(85)90054-4

Albrecht A, Kandji ST (2003) Carbon sequestration in tropical agroforestry systems. Agric Ecosyst Environ 99:15–27. https://doi.org/10.1016/S0167-8809(03)00138-5

Altieri MA, Nicholls CI (2004) An agroecological basis for designing diversified cropping systems in the tropics. J Crop Improv 11:81–103. https://doi.org/10.1300/J411v11n01_05

Amponsah-Doku B, Daymond A, Robinson S, Atuah L, Sizmur T (2022) Improving soil health and closing the yield gap of cocoa production in Ghana—a review. Sci Afr 15:e01075. https://doi.org/10.1016/j.sciaf.2021.e01075

Anandaraj M (2015) Status of plantation crops sector in India. J Plant Crops 43:1–8

Anaya JA, Álvarez I, García MJ, Lizama V (2022) Application of green tea extract and catechin on the polyphenolic and volatile composition of Monastrell red wines. Int J Food Sci Technol 57(9):6097–6111. https://doi.org/10.1111/ijfs.15970

Anith KN (2009) Mature coconut as a bio-fermentor for multiplication of plant growth promoting rhizobacteria. Curr Sci 97:1647–1653

Anyaoha KE, Sakrabani R, Patchigolla K, Mouazen AM (2018) Critical evaluation of oil palm fresh fruit bunch solid wastes as soil amendments: prospects and challenges. Resour Conserv Recycl 136:399–409. https://doi.org/10.1016/j.resconrec.2018.04.02

Araujo KS, Brito VN, Veloso TG, de Leite TS, Alves JL, da Hora Junior BT, Moreno HL, Pereira OL, Mizubuti ES, de Queiroz MV (2020) Diversity and distribution of endophytic fungi in different tissues of *Hevea brasiliensis* native to the Brazilian Amazon forest. Mycol Prog 19:1057–1068. https://doi.org/10.1007/s11557-020-01613-4

Aravind R, Kumar A, Eapen SJ (2012) Pre-plant bacterisation: a strategy for delivery of beneficial endophytic bacteria and production of disease-free plantlets of black pepper (*Piper nigrum* L.). Arch Phytopathol Plant Protect 45(9):1115–1126. https://doi.org/10.1080/03235408.2012.657894

Arianie L (2010) Potential of lignin, lignin sulfonate and lignin acetate from palm empty bunch as an additive substance in urea fertilizer as an effort to reduce the solubility of urea nitrogen.

In: Advanced materials research, vol 93. Trans Tech Publications Ltd, pp 409–412. https://doi.org/10.4028/www.scientific.net/AMR.93-94.409

Arias ME, González-Pérez JA, González-Vila FJ, Ball AS (2005) Soil health: a new challenge for microbiologists and chemists. Int Microbiol 8(1):13–21

Arif I, Batool M, Schenk PM (2020) Plant microbiome engineering: expected benefits for improved crop growth and resilience. Trends Biotechnol 38:1385–1396. https://doi.org/10.1016/j.tibtech.2020.04.015

Asad S, Priyashantha AKH, Tibpromma S, Luo Y, Zhang J, Fan Z, Zhao L, Shen K, Niu C, Lu L, Promputtha I, Karunarathna SC (2023) Coffee-associated endophytes: plant growth promotion and crop protection. Biology 12(7):911. https://doi.org/10.3390/biology12070911

Ashraf AN, Zulkefly S, Adekunle SM, Samad MY (2017) Growth and biomass yield of oil palm (*Elaeis guineensis*) seedlings as influenced by different rates of vermicompost. Eur J Eng Technol Res 2(8):17–21. https://doi.org/10.24018/ejeng.2017.2.8.405

Ashraf M, Zulkifli R, Sanusi RA, Tohiran KA, Terhem R, Moslim R, Norhisham AR, Ashton-Butt A, Azhar B (2018) Alley-cropping system can boost arthropod biodiversity and ecosystem functions in oil palm plantations. Agric Ecosyst Environ 260:19–26. https://doi.org/10.1016/j.agee.2018.03.017

Atzori G, Pane C, Zaccardelli M, Cacini S, Massa D (2021) The role of peat-free organic substrates in the sustainable management of soilless cultivations. Agronomy 11(6):123. https://doi.org/10.3390/agronomy11061236

Aubrey MLL, Chin CFS, Seelan JSS, Chye FY, Lee HH, Rakib MRM (2022) Conversion of oil palm by-products into value-added products through oyster mushroom (*Pleurotus ostreatus*) cultivation. Horticulturae 8(11):1040. https://doi.org/10.3390/horticulturae8111040

Averill C, Anthony MA, Baldrian P, Finkbeiner F, Hoogen J, Kiers T, Kohout P, Hirt E, Smith GR, Crowther TW (2022) Defending Earth's terrestrial microbiome. Nat Microbiol 7(11):1717–1725. https://doi.org/10.1038/s41564-022-01228-3

Azri MH, Ismail S, Abdullah R (2018) An endophytic *Bacillus* strain promotes growth of oil palm seedling by fine root biofilm formation. Rhizosphere 5:1–7. https://doi.org/10.1016/j.rhisph.2017.10.003

Bach EM, Ramirez KS, Fraser TD, Wall DH (2020) Soil biodiversity integrates solutions for a sustainable future. Sustainability 12(7):2662. https://doi.org/10.3390/su12072662

Badhwar VK, Singh S, Singh B (2020) Biotransformation of paper mill sludge and tea waste with cow dung using vermicomposting. Bioresour Technol 318:124097. https://doi.org/10.1016/j.biortech.2020.124097

Bae H, Sicher RC, Kim MS, Kim SH, Strem MD, Melnick RL, Bailey BA (2009) The beneficial endophyte *Trichoderma hamatum* isolate DIS 219b promotes growth and delays the onset of the drought response in *Theobroma cacao*. J Exp Bot 60(11):3279–3295. https://doi.org/10.1093/jxb/erp165

Banerjee S, van der Heijden MG (2023) Soil microbiomes and one health. Nat Rev Microbiol 21(1):6–20. https://doi.org/10.1038/s41579-022-00779-w

Barrios E, Valencia V, Jonsson M, Brauman A, Hairiah K, Mortimer PE, Okubo S (2018) Contribution of trees to the conservation of biodiversity and ecosystem services in agricultural landscapes. Int J Biodivers Sci Ecosyst Serv Manage 14:1–16. https://doi.org/10.1080/21513732.2017.1399167

Barrios-Rodríguez YF, Salas-Calderón KT, Orozco-Blanco DA, Gentile P, Girón-Hernández J (2022) Cocoa Pod husk: a high-pectin source with applications in the food and biomedical fields. ChemBioEng Rev 9(5):462–474. https://doi.org/10.1002/cben.202100061

Bavappa KVA, Khader KBA, Biddappa CC, Khan HH, Kasturi Bai KV, Ramadasan A, Sundararaju P, Bopaiah BM, Thomas GV, Misra LP, Balasimha D, Bhat NT, Bhat SK (1986) Coconut and arecanut based high density multispecies cropping system. J Plant Crops 14(2):74–87

Behin J, Nader S (2016) Utilization of waste lignin to prepare controlled slow-release urea. Int J Recycl Org Waste Agric 5:289–299. https://doi.org/10.1007/s40093-016-0139-1

Bhagwat SA, Willis KJ, Birks HJB, Whittaker RJ (2008) Agroforestry: a refuge for tropical biodiversity? Trends Ecol Evol 23:261–267. https://doi.org/10.1016/j.tree.2008.01.005

Bhagya HP, Maheswarappa HP, Surekha, Bhat R (2017) Carbon sequestration potential in coconut-based cropping systems. Indian J Hort 74(1):1–5. https://doi.org/10.5958/0974-0112.2017.00004.4

Binet F, Fayolle L, Pussard M (1998) Significance of earthworms in stimulating soil microbial activity. Biol Fertil Soils 27:79–84. https://doi.org/10.1007/s003740050403

Blagodatsky S, Xu J, Cadisch G (2016) Carbon balance of rubber (*Hevea brasiliensis*) plantations: a review of uncertainties at plot, landscape and production level. Agric Ecosyst Environ 221:8–19. https://doi.org/10.1016/j.agee.2016.01.025

Blok V, Gremmen B (2016) Ecological innovation: biomimicry as a new way of thinking and acting ecologically. J Agric Environ Ethics 29(2):203–217. https://doi.org/10.1007/s10806-015-9596-1

Bopaiah BM, Shetty HS (1991a) Soil microflora and biological activities in the rhizosphere and root regions of coconut based multistoried cropping and monocropping systems. Soil Biol Biochem 23:89–94. https://doi.org/10.1016/0038-0717(91)90167-I

Bopaiah BM, Shetty HS (1991b) Microbiology and fertility in coconut- based mixed farming and coconut monocropping systems. Trop Agric 68(2):135–138

Bouchon AS, Hoopen GMT (2022) Integrated management of verticillium wilt of cacao. Front Agron 4:911969. https://doi.org/10.3389/fagro.2022.911969

Bradford MA, Keiser AD, Davies CA, Mersmann CA, Strickland MS (2013) Empirical evidence that soil carbon formation from plant inputs is positively related to microbial growth. Biogeochemistry 113:271–281. https://doi.org/10.1007/s10533-012-9822-0

Brownlie WJ, Sutton MA, Cordell D, Reay DS, Heal KV, Withers PJA, Vanderbeck I, Spears BM (2023) Phosphorus price spikes: a wake-up call for phosphorus resilience. Front Sust Food Syst 7:1088776. https://doi.org/10.3389/fsufs.2023.1088776

Buyer JS, Baligar VC, He Z, Arévalo-Gardini E (2017) Soil microbial communities under cacao agroforestry and cover crop systems in Peru. Appl Soil Ecol 120:273–280. https://doi.org/10.1016/j.apsoil.2017.09.009

Caldwell AC, Silva LC, da Silva CC, Ouverney CC (2015) Prokaryotic diversity in the rhizosphere of organic, intensive, and transitional coffee farms in Brazil. PLoS One 10(6):e0106355. https://doi.org/10.1371/journal.pone.0106355

Camarero S, Bockle B, Martinez MJ, Martinez AT (1996) Manganese-mediated lignin degradation by *Pleurotus pulmonarius*. Appl Environ Microbiol 62(3):1070–1072. https://doi.org/10.1128/aem.62.3.1070-1072.1996

Carron MP, Pierrat M, Snoeck D et al (2015) Temporal variability in soil quality after organic residue application in mature oil palm plantations. Soil Res 53(2):205–215. https://doi.org/10.1071/SR14249

ChandraMohanan R, Peter PK, Sharadraj KM (2013) Production technology of coir pith cake formulation of *Trichoderma harzianum*. J Plant Crops 41(2):214–218

Chanway CP (1997) Inoculation of tree roots with plant growth promoting soil bacteria: an emerging technology for reforestation. For Sci 43:99–112. https://doi.org/10.1093/forestscience/43.1.99

Chaparro JM, Sheflin AM, Manter DK, Vivanco JM (2012) Manipulating the soil microbiome to increase soil health and plant fertility. Biol Fertil Soils 48:489–499. https://doi.org/10.1007/s00374-012-0691-4

Chattopadhyay N, Hore JK, Bhattacharya A (2015) Prospect of organic cashew production in India. Acta Hortic 1080:337–341. https://doi.org/10.17660/ActaHortic2015.1080.45

Chaudhuri PS, Pal TK, Bhattacharjee G, Dey SK (2003) Rubber leaf litters (*Hevea brasiliensis* var. RRM 600) as vermiculture substrate for epigeic earthworm, *Perionyx excavatus, Eudrilus eugeniae* and *Eisenia fetida*. Pedobiologia 47:796–800. https://doi.org/10.1078/0031-4056-00261

Chaverri P, Gazis RO, Samuels GJ (2010) *Trichoderma amazonicum*, a new endophytic species on *Hevea brasiliensis* and *H. guianensis* from the Amazon basin. Mycologia 103:139–151. https://doi.org/10.3852/10-078

Chen C, Liu W, Jiang X, Wu J (2017) Effects of rubber-based agroforestry systems on soil aggregation and associated soil organic carbon: implications for land use. Geoderma 299:13–24. https://doi.org/10.1016/j.geoderma.2017.03.021

Chen C, Chen HYH, Chen X, Huang Z (2019) Meta-analysis shows positive effects of plant diversity on microbial biomass and respiration. Nat Commun 10:1332. https://doi.org/10.1038/s41467-019-09258-y

Chen J, Fan X, Zhang L, Chen X, Sun S, Sun RC (2020) Research progress in lignin-based slow/controlled release fertilizer. Chem Sus Chem 13:4356–4366. https://doi.org/10.1002/cssc.202000455

Chenu C, Angers DA, Barré P, Derrien D, Arrouays D, Balesdent J (2019) Increasing organic stocks in agricultural soils: knowledge gaps and potential innovations. Soil Tillage Res 188:41–52. https://doi.org/10.1016/j.still.2018.04.011

Cherif H, Marasco R, Rolli E, Ferjani R, Marco F, Asma S, Francesca M, Ikram B, Sara B, Abdellatif B, Ameur C, Daniele D, Hadda O (2015) Oasis desert farming selects environment-specific date palm root endophytic communities and cultivable bacteria that promote resistance to drought. Environ Microbiol Rep 7:668–678. https://doi.org/10.1111/1758-2229.12304

Chojnacka K, Moustakas K, Witek-Krowiak A (2020) Bio-based fertilizers: a practical approach towards circular economy. Bioresour Technol 295:122223. https://doi.org/10.1016/j.biortech.2019.122223

Chowdappa P, Biddappa CC, Sujatha S (1999) Efficient recycling of organic wastes in arecanut (*Areca catechu* L.) and cocoa (*Theobroma cacao* L.) plantations through vermicomposting. Indian J Agric Sci 69:563–566

Chowdhury A, Samrat A, Devy MS (2021) Can tea support biodiversity with a few "nudges" in management: evidence from tea growing landscapes around the world. Glob Ecol Conserv 31:e01801. https://doi.org/10.1016/j.gecco.2021.e01801

Clermont-Dauphin C, Suvannang N, Pongwichian P, Chevlan V, Hammecker C, Harmand JM (2016) Dinitrogen fixation by the legume cover crop *Pueraria phaseoloides* and transfer of fixed N to *Hevea brasiliensis*—impact on tree growth and vulnerability to drought. Agric Ecosyst Environ 217:79–88. https://doi.org/10.1016/j.agee.2015.11.002

Clough Y, Faust H, Tscharntke T (2009) Cacao boom and bust: sustainability of agroforests and opportunities for biodiversity conservation. Conserv Lett 2:197–205. https://doi.org/10.1111/j.1755-263X.2009.00072.x

Clough Y, Barkmann J, Juhrbandt J (2011) Combining high biodiversity with high yields in tropical agroforests. Proc Natl Acad Sci U S A 108:8311–8316. https://doi.org/10.1073/pnas.1016799108

Colla G, Rouphael Y (2015) Biostimulants in horticulture. Sci Hort 196:1–2. https://doi.org/10.1016/j.scienta.2015.10.044

Colla G, Hoagland L, Ruzzi M, Cardarelli M, Bonini P, Canaguier R, Rouphael Y (2017) Biostimulant action of protein hydrolysates: unraveling their effects on plant physiology and microbiome. Front Plant Sci 8:2202. https://doi.org/10.3389/fpls.2017.02202

Costa C, Wollenberg E, Benitez M et al (2022) Roadmap for achieving net-zero emissions in global food systems by 2050. Sci Rep 12:15064. https://doi.org/10.1038/s41598-022-18601-1

CPCRI (1989) Annual report 1988. Central Plantation Crops Research Institute, Kasaragod, p 37

da Silva Moço MK, da Gama-Rodrigues EF, da Gama-Rodrigues AC, Machado RCR, Baligar VC (2009) Soil and litter fauna of cacao agroforestry systems in Bahia, Brazil. Agrofor Syst 76:127–138. https://doi.org/10.1007/s10457-008-9178-6

Das MM, Aguilar CN, Haridas M, Sabu A (2021) Production of bio-fungicide, *Trichoderma harzianum* CH1 under solid-state fermentation using coffee husk. Bioresour Technol Rep 15:100708. https://doi.org/10.1016/j.biteb.2021.100708

de Carvalho AL, Alves BJ, Baldani VL, Reis VM (2008) Application of 15N natural abundance technique for evaluating biological nitrogen fixation in oil palm ecotypes at nursery stage in pot experiments and at mature plantation sites. Plant Soil 302(1):71–78. https://doi.org/10.1007/s11104-007-9456-5

De Corato U (2020) Agricultural waste recycling in horticultural intensive farming systems by on-farm composting and compost-based tea application improves soil quality and plant health: a review under the perspective of a circular economy. Sci Total Environ 738:139840. https://doi.org/10.1016/j.scitotenv.2020.139840

De Souza JT, Bailey BA, Pomella AWV, Erbe EF, Murphy CA, Bae H et al (2008) Colonization of cacao seedlings by *Trichoderma stromaticum*, a mycoparasite of the witches' broom pathogen, and its influence on plant growth and resistance. Biol Control 46:36–45. https://doi.org/10.1016/j.biocontrol.2008.01.010

Debnath B, Haldar D, Purkait MK (2022) Environmental remediation by tea waste and its derivative products: a review on present status and technological advancements. Chemosphere 300:134480. https://doi.org/10.1016/j.chemosphere.2022.134480

Del Buono D (2021) Can biostimulants be used to mitigate the effect of anthropogenic climate change on agriculture? It is time to respond. Sci Total Environ 751:141763. https://doi.org/10.1016/j.scitotenv.2020.141763

Dessureault-Rompré J (2022) Restoring soil functions and agroecosystem services through phytotechnologies. Front Soil Sci 2:927148. https://doi.org/10.3389/fsoil.2022.927148

Dinesh R, Suryanarayana MA, Chaudhuri SG, Sheeja TE, Shiva KN (2006) Long-term effects of leguminous cover crops on biochemical and biological properties in the organic and mineral layers of soils of a coconut plantation. Eur J Soil Biol 42:147–157. https://doi.org/10.1016/j.ejsobi.2005.12.004

Dinesh R, Srinivasan V, Hamza S, Anandaraj M (2014) Massive phosphorus accumulation in soils: Kerala's continuing conundrum. Curr Sci 106:343–344

Dissanayaka DMNS, Dissanayake DKRPL, Udumann SS, Nuwarapaksha TD, Atapattu AJ (2023) Agroforestry—a key tool in the climate-smart agriculture context: a review on coconut cultivation in Sri Lanka. Front Agron 5:116275. https://doi.org/10.3389/fagro.2023.1162750

Doran JW, Sarrantonio M, Liebig MA (1996) Soil health and sustainability. Adv Agron 56:1–54

Dossa EL, Fernandes ECM, Reid WS, Ezui K (2008) Above- and belowground biomass, nutrient and carbon stocks contrasting an open-grown and a shaded coffee plantation. Agrofor Syst 72:103–115. https://doi.org/10.1007/s10457-007-9075-4

du Jardin P (2015) Plant biostimulants: definition, concept, main categories and regulation. Sci Hortic 196:3–14. https://doi.org/10.1016/j.scienta.2015.09.021

Duong B, Marraccini P, Maeght JL, Vaast P, Lebrun M, Duponnois R (2020) Coffee microbiota and its potential use in sustainable crop management. A review. Front Sustain Food Syst 4:607935. https://doi.org/10.3389/fsufs.2020.607935

Duru M, Therond O, Martin G, Martin-Clouaire R, Magne MA, Justes E, Journet EP, Aubertot JN, Savary S, Bergez JE, Sarthou JP (2015) How to implement biodiversity-based agriculture to enhance ecosystem services: a review. Agron Sustain Dev 35:1259–1281. https://doi.org/10.1007/s13593-015-0306-1

Dutta J, Thakur D (2017) Evaluation of multifarious plant growth promoting traits, antagonistic potential and phylogenetic affiliation of rhizobacteria associated with commercial tea plants grown in Darjeeling, India. PLoS One 12(8):e0182302. https://doi.org/10.1371/journal.pone.0182302

Ebid A, Ueno H, Ghoneim A, Asagi N (2008) Nitrogen uptake by radish, spinach and "Chingensai" from composted tea leaves, coffee waste and kitchen garbage. Compost Sci Util 16(3):152–158. https://doi.org/10.1080/1065657X.2008.10702373

Elliott ET, Coleman DC (1988) Let the soil work for us. Ecol Bull 39:23–32

Embrandiri A, Quaik S, Rupani P, Srivastava V, Singh P (2015) Sustainable utilization of oil palm wastes: opportunities and challenges. In: Singh RP, Sarkar A (eds) Waste management. Nova Science Publishers, New York, pp 231–246

Ertani A, Pizzeghello D, Francioso O, Sambo P, Sanchez-Cortes S, Nardi S (2014) *Capsicum chinensis* L. growth and nutraceutical properties are enhanced by biostimulants in a long-term period: chemical and metabolomic approaches. Front Plant Sci 5:375. https://doi.org/10.3389/fpls.2014.00375

Falcão LL, Werneck SJO, Vilarinho BR, da Silva JP, Pomella AWV, Marcellino LH (2014) Antimicrobial and plant growth-promoting properties of the cacao endophyte *Bacillus subtilis* ALB629. J Appl Microbiol 116(6):1584–1592. https://doi.org/10.1111/jam.12485

Fidelis C, Rao RBK (2017) Enriched cocoa pod composts and their fertilizing effects on hybrid cocoa seedlings. Int J Recycl Org Waste Agric 6:99–106. https://doi.org/10.1007/s40093-017-0156-8

Fischer K, Schiene R (2002) Nitrogenous fertilizers from lignins. In: Hu T (ed) Chemical modification, properties and usage of lignin. Kluwer Academics/Plenum Publishers, New York, pp 167–198

Foley JA, Ramankutty N, Brauman KA, Cassidy ES, Gerber JS, Johnston M, Mueller ND, O'Connell C, Ray DK, West PC, Balzer C, Bennett EM, Carpenter SR, Hill J, Monfreda C, Polasky S, Rockstrom J, Sheehan J, Siebert S, Tilman D, Zaks DPM (2011) Solutions for a cultivated planet. Nature 478:337–342. https://doi.org/10.1038/nature10452

Fontes AG, Gama-Rodrigues AC, Gama-Rodrigues EF, Sales MVS, Costa MG, Machado RCR (2014) Nutrient stocks in litterfall and litter in cocoa agroforests in Brazil. Plant Soil 383:313–335. https://doi.org/10.1007/s11104-014-2175-9

Fuentes-Ramírez LE, Cristales BR, Hernández TA, Salgado TJ, Wang ET, Romero EM, Mellado JC (2001) Novel nitrogen-fixing acetic acid bacteria, *Gluconacetobacter johannae* sp. nov. and *Gluconacetobacter azotocaptans* sp. nov., associated with coffee plants. Int J Syst Evol Microbiol 51(4):1305–1314. https://doi.org/10.1099/00207713-51-4-1305

Gaba S, Lescourret F, Boudsocq S, Enjalbert J, Hinsinger P, Journet EP, Navas ML, Wery J, Louarn G, Malézieux E, Pelzer E, Prudent M, Ozier-Lafontaine H (2015) Multiple cropping systems as drivers for providing multiple ecosystem services: from concepts to design. Agron Sustain Dev 35:607–623. https://doi.org/10.1007/s13593-014-0272-z

Gandahi AW, Hanafi MM (2014) Bio-composting oil palm waste for improvement of soil fertility. In: Maheshwari DK (ed) Composting for sustainable agriculture. Springer International Publishing Switzerland, pp 209-244. https://doi.org/10.1007/978-3-319-08004-8_11

Ganeshamurthy AN, Kalaivanan D, Rajendiran S (2019) Carbon sequestration potential of perennial horticultural crops in India tropics. In: Ghosh PK, Mahanta SK, Mandal D, Mandal B, Ramakrishnan S (eds) Carbon management in tropical and subtropical terrestrial systems. Springer, Singapore, pp 333–348. https://doi.org/10.1007/978-981-13-9628-1_20

Gazis R, Chaverri P (2010) Diversity of fungal endophytes in leaves and stems of wild rubber trees (*Hevea brasiliensis*) in Peru. Fungal Ecol 3:240–254. https://doi.org/10.1016/j.funeco.2009.12.001

Gholami H, Saharkhiz MJ, Fard FR, Ghani A, Nadaf F (2018) Humic acid and vermicompost increased bioactive components, antioxidant activity and herb yield of Chicory (*Cichorium intybus L.*). Biocatal Agric Biotechnol 14:286–292. https://doi.org/10.1016/j.bcab.2018.03.021

Ghosh PK, Sarma US, Ravindranath AD, Radhakrishnan S, Ghosh P (2007) A novel method for accelerated composting of coir pith. Energy Fuels 21(2):822–827. https://doi.org/10.1021/ef060513c

Gontard N, Sonesson U, Birkved M, Majone M, Bolzonella D, Celli A, Angellier-Coussy H, Jang GW, Verniquet A, Broeze J, Schaer B, Batista AP, Sebok A (2018) A research challenge vision regarding management of agricultural waste in a circular bio-based economy. Crit Rev Environ Sci Technol 48:614–654. https://doi.org/10.1080/10643389.2018.1471957

González-Moreno MA, García Gracianteparaluceta B, Marcelino Sádaba S, Zaratiegui Urdin J, Robles Domínguez E, Pérez Ezcurdia MA, Seco Meneses A (2020) Feasibility of vermicomposting of spent coffee grounds and silverskin from coffee industries: a laboratory study. Agronomy 10(8):1125. https://doi.org/10.3390/agronomy10081125

Gopal KV, Anandaraj M, Sarma Y (1999) Evaluation of substrates for mass multiplication of fungal biocontrol agents *Trichoderma harzianum* and *T. virens*. J Spices Aromat Crops 8(2):207–210

Gopal M, Gupta A, Sunil E, Thomas GV (2009) Amplification of plant beneficial microbial communities during conversion of coconut leaf substrate to vermicompost by *Eudrilus* sp. Curr Microbiol 59:15–20. https://doi.org/10.1007/s00284-009-9388-9

Gopal M, Bhute SS, Gupta A, Prabhu SR, Thomas GV, Whitman WB, Jangid K (2017) Changes in structure and function of bacterial communities during coconut leaf vermicomposting. Antonie Van Leeuwenhoek 110:1339–1355. https://doi.org/10.1007/s10482-017-0894-7

Gopal M, Gupta A, Hameed SK, Sathyaseelan N, Rajeela THK, Thomas GV (2020) Biochars produced from coconut palm biomass residues can aid regenerative agriculture by improving soil properties and plant yield in humid tropics. Biochar 2:211–226. https://doi.org/10.1007/s42773-020-00043-5

Gopal M, Gupta A, Arunachalam V et al (2022) Autochthonous nutrient recycling driven by soil microbiota could be sustaining high coconut productivity in Lakshadweep islands sans external fertilizer application. World J Microbiol Biotechnol 38:213. https://doi.org/10.1007/s11274-022-03373-7

Gu H, Zhang Y, Li X, Li W, Huang S (2019) Lignin improves release behavior of slow-release fertilizers with high content of urea. J Appl Polym Sci 136:48238. https://doi.org/10.1002/app.48238

Halpern M, Bar-Tal A, Ofek M, Minz D, Muller T, Yermiyahu U (2015) The use of biostimulants for enhancing nutrient uptake. Adv Agron 130:141–174. https://doi.org/10.1016/bs.agron.2014.10.001

Hanc A, Hrebeckova T, Grasserova A, Cajthaml T (2021) Conversion of spent coffee grounds into vermicompost. Bioresour Technol 341:125925. https://doi.org/10.1016/j.biortech.2021.125925

Handayanto E, Cadisch G, Giller KE (1997) Regulating N mineralization from plant residues by manipulation of quality. In: Cadisch G, Giller KE (eds) Driven by nature: plant litter quality and decomposition. CAB International, Wallingford, pp 175–185

Harindintwali JD, Zhou J, Yu X (2020) Lignocellulosic crop residue composting by cellulolytic nitrogen-fixing bacteria: a novel tool for environmental sustainability. Sci Total Environ 715:136912. https://doi.org/10.1016/j.scitotenv.2020.136912

Hartemink AE (2005a) Nutrient stocks, nutrient cycling, and soil changes in cocoa ecosystems: a review. Adv Agron 86:227–253. https://doi.org/10.1016/S0065-2113(05)86005-5

Hartemink AE (2005b) Plantation agriculture in the tropics: environmental issues. Outlook Agric 34:11–21. https://doi.org/10.5367/0000000053295150

Hartman K, van der Heijden MGA, Wittwer RA, Banerjee S, Walser JC, Schlaeppi K (2018) Cropping practices manipulate abundance patterns of root and soil microbiome members paving the way to smart farming. Microbiome 6:1–14. https://doi.org/10.1186/s40168-017-0389-9

Hartmann A, Schmid M, van Tuinen D, Berg G (2008) Plant-driven selection of microbes. Plant Soil 321:235–257. https://doi.org/10.1007/s11104-008-9814-y

Hatfield JL, Walthall CL (2015) Soil biological fertility: foundation for the next revolution in agriculture? Commun Soil Sci Plant Anal 46:753–762. https://doi.org/10.1080/00103624.2015.1005227

Hättenschwiler S, Vitousek PM (2000) The role of polyphenols in terrestrial ecosystem nutrient cycling. Trends Ecol Evol 15:238–243. https://doi.org/10.1016/S0169-5347(00)01861-9

Hazzouri KM, Flowers JM, Nelson D, Lemansour A, Masmoudi K, Amiri KMA (2020) Prospects for the study and improvement of abiotic stress tolerance in date palms in the post-genomics era. Front Plant Sci 11:293. https://doi.org/10.3389/fpls.2020.00293

Herder GD, Isterdael GV, Beeckman T, Smet ID (2010) The roots of a new green revolution. Trends Plant Sci 15:600–607. https://doi.org/10.1016/j.tplants.2010.08.009

Hidalgo D, Corona F, Martín-Marroquín JM (2021) Nutrient recycling: from waste to crop. Biomass Convers Biorefin 11:207–217. https://doi.org/10.1007/s13399-019-00590-3

Huang YZ, Feng ZW, Zhang FZ, Liu SQ (2003) Effects of lignin on nitrification in soil. J Environ Sci 15(3):363–366

Huang R, Mcgrath S, Hirsch P, Clark I, Storkey J, Wu L et al (2019) Plant–microbe networks in soil are weakened by century-long use of inorganic fertilizers. Microb Biotechnol 12:1464–1475. https://doi.org/10.1111/1751-7915.13487

Huang Z, Cui C, Cao Y, Dai J, Cheng X, Hua S, Wang W, Duan Y, Petropoulos E, Wang H, Zhou L, Fang W, Zhing Z (2022) Tea plant–legume intercropping simultaneously improves soil fertility and tea quality by changing *Bacillus* species composition. Hortic Res 9:uhac046. https://doi.org/10.1093/hr/uhac046

Isbell F, Craven D, Connolly J (2015) Biodiversity increases the resistance of ecosystem productivity to climate extremes. Nature 526:574–577. https://doi.org/10.1038/nature15374

Islam W, Noman A, Naveed H, Huang Z, Chen HY (2020) Role of environmental factors in shaping the soil microbiome. Environ Sci Pollut Res 27:41225–41247. https://doi.org/10.1007/s11356-020-10471-2

ISRO (2021) Indian Space Research Organization 2021. Desertification and land degradation atlas of India. 2021

Jackson CA, Couger MB, Prabhakaran M, Ramachandriya KD, Canaan P, Fathepure BZ (2017) Isolation and characterization of Rhizobium sp. strain YS-1r that degrades lignin in plant biomass. J Appl Microbiol 122(4):940–952. https://doi.org/10.1111/jam.13401

Janzen HH (2006) The soil carbon dilemma: shall we hoard it or use it? Soil Biol Biochem 38:419–424. https://doi.org/10.1016/j.soilbio.2005.10.008

Janzen HH (2015) Beyond carbon sequestration: soil as conduit of solar energy. Eur J Soil Sci 66(1):19–32. https://doi.org/10.1111/ejss.12194

Jayakumar M, Emana AN, Ramasamy S, Mohanadoss P, Kumar KA, Muthusamy G, Kim W, Karmegam N (2022) Detoxification of coir pith through refined vermicomposting engaging *Eudrilus eugeniae*. Chemosphere 291:132675. https://doi.org/10.1016/j.chemosphere.2021.132675

Jessy MD, Nair AS, Bai MM, Rajendran P, Punnoose KI (2009) Self–sustainability of phosphorous cycle in rubber (*Hevea brasiliensis*) plantations: annual recycling through litter and removal through latex. J Plant Crops 37(3):177–184

Jessy MD, Prasannakumari P, Abraham J (2013) Carbon and nutrient cycling through fine roots in rubber (*Hevea brasiliensis*) plantations in India. Exp Agric 49(4):556–573. https://doi.org/10.1017/S0014479713000203

Jha S, Bacon CM, Philpott SM, Ernesto Méndez V, Läderach P, Rice RA (2014) Shade coffee: update on a disappearing refuge for biodiversity. BioScience 64(5):416–428. https://doi.org/10.1093/biosci/biu038

Jiang Z, Lou Y, Liu X, Sun W, Wang H, Liang J, Guo J, Li N, Yang Q (2023) Combined application of coffee husk compost and inorganic fertilizer to improve the soil ecological environment and photosynthetic characteristics of arabica coffee. Agronomy 13(5):1212. https://doi.org/10.3390/agronomy13051212

Jiao GJ, Xu Q, Cao SL, Peng P, She D (2018) Controlled-release fertilizer with lignin used to trap urea/hydroxymethyl urea/urea-formaldehyde polymers. Bioresources 13(1):1711–1728

Jiao GJ, Peng P, Sun SL, Geng ZC, She D (2019) Amination of biorefinery technical lignin by Mannich reaction for preparing highly efficient nitrogen fertilizer. Int J Biol Macromol 127:544–554. https://doi.org/10.1016/j.ijbiomac.2019.01.076

Jimenez-Salgado T, Fuentes-Ramirez LE, Tapia-Hernandez A, Mascarua-Esparza MA, Martinez-Romero E, Caballero-Mellado J (1997) *Coffea arabica* L., a new host plant for Acetobacter diazotrophicus, and isolation of other nitrogen-fixing acetobacteria. Appl Environ Microbiol 63:3676–3683. https://doi.org/10.1128/aem.63.9.3676-3683.1997

Jindo K, Olivares FL, Malcher DJDP, Monedero MAS, Kempenaar C, Canellas LP (2020) From lab to field: role of humic substances under open-field and greenhouse conditions as biostimulant and biocontrol agent. Front Plant Sci 11:426. https://doi.org/10.3389/fpls.2020.00426

John J (2014) Homestead farming in Kerala: a multi-faceted land-use system. Rev Agrarian Stud 4(1):80–94

Joseph P, Jessy MD (2015) Coir pith: a cost-effective alternative germination medium for rubber seed (*Hevea brasiliensis*). Rubber Sci 28(3):237–246

Joseph P, Jessy MD, Mohan M (2022) Soil carbon pools under rubber (*Hevea brasiliensis*) based agroforestry systems in South India. Agrofor Syst 96:1121–1133. https://doi.org/10.1007/s10457-022-00770-7

Jurburg SD, Shek KL, McGuire K (2020) Soil microbial composition varies in response to coffee agroecosystem management. FEMS Microbiol Ecol 96(9):fiaa164. https://doi.org/10.1093/femsec/fiaa164

Kaar WE, Holtzapple MT (2000) Using lime pretreatment to facilitate the enzyme hydrolysis of corn stover. Biomass Bioenerg 18:189–199. https://doi.org/10.1016/S0961-9534(99)00091-4

Kaba JS, Zerbe S, Agnolucci M, Scandellari F, Abunyewa AA, Giovannetti M, Tagliavini M (2019) Atmospheric nitrogen fixation by gliricidia trees (*Gliricidia sepium* (Jacq.) Kunth ex Walp.) intercropped with cocoa (*Theobroma cacao* L.). Plant Soil 435:323–336. https://doi.org/10.1007/s11104-018-3897-x

Kallenbach CM, Frey SD, Grandy AS (2016) Direct evidence for microbial-derived soil organic matter formation and its ecophysiological controls. Nat Commun 7:13630. https://doi.org/10.1038/ncomms13630

Kaminsky LM, Trexler RV, Malik RJ, Hockett KL, Bell TH (2019) The inherent conflicts in developing soil microbial inoculants. Trends Biotechnol 37:140–151. https://doi.org/10.1016/J.TIBTECH.2018.11.011

Kaye JP, Quemada M (2017) Using cover crops to mitigate and adapt to climate change. A review. Agron Sustain Dev 37:4. https://doi.org/10.1007/s13593-016-0410-x

Khatun R, Reza MIH, Moniruzzaman M, Yaakob Z (2017) Sustainable oil palm industry: the possibilities. Renew Sustain Energy Rev 76:608–619. https://doi.org/10.1016/j.rser.2017.03.077

Kim N, Zabaloy MC, Guan K, Villamil MB (2020) Do cover crops benefit soil microbiome? A meta-analysis of current research. Soil Biol Biochem 142:107701. https://doi.org/10.1016/j.soilbio.2019.107701

Kone K, Akueson K, Norval G (2020) On the production of potassium carbonate from cocoa pod husks. Recycling 5:23. https://doi.org/10.3390/recycling5030023

Kroeger A, Bakhtary H, Haupt F, Streck C (2017) Eliminating deforestation from the cocoa supply chain. Climate Focus North America Inc., Washington. 61p

Kumar KSN, Maheswarappa HP (2019) Carbon sequestration potential of coconut-based cropping systems under integrated nutrient management practices. J Plant Crops 47(2):107–114. https://doi.org/10.25081/jpc.2019.v47.i2.5776

Kupryashina MA, Petrov SV, Ponomareva EG, Nikitina VE (2015) Ligninolytic activity of bacteria of the genera *Azospirillum* and *Niveispirillum*. Microbiology 84:791–795. https://doi.org/10.1134/S0026261715060041

Lagier JC, Hugon P, Khelaifia S, Fournier PE, La Scola B, Raoult D (2015) The rebirth of culture in microbiology through the example of culturomics to study human gut microbiota. Clin Microbiol Rev 28:237–264. https://doi.org/10.1128/CMR.00014-14

Lal R (2015) Restoring soil quality to mitigate soil degradation. Sustainability 7:5875–5895. https://doi.org/10.3390/su7055875

Larkin RP (2015) Soil health paradigms and implications for disease management. Annu Rev Phytopathol 53:199–221. https://doi.org/10.1146/annurev-phyto-080614-120357

Lavelle P (2011) Earthworms as ecosystem engineers. In: Glinski J, Horabik J, Lipiec J (eds) Encyclopedia of agrophysics. Springer New York, New York, pp 233–235

Le VS, Herrmann L, Bräu L, Lesueur D (2023) Sustainable green tea production through agroecological management and land conversion practices for restoring soil health, crop productivity and economic efficiency: evidence from northern Vietnam. Soil Use Manag 39:1185. https://doi.org/10.1111/sum.12885

Leakey RR (2014) The role of trees in agroecology and sustainable agriculture in the tropics. Annu Rev Phytopathol 52:113–133. https://doi.org/10.1146/annurev-phyto-102313-045838

Lee JG, Yoon HY, Cha JY, Kim WY, Kim PJ, Jeon JR (2019) Artificial humification of lignin architecture: top-down and bottom-up approaches. Biotechnol Adv 37:107416. https://doi.org/10.1016/j.biotechadv.2019.107416

Lee SM, Kong HG, Song GC, Ryu CM (2021) Disruption of Firmicutes and Actinobacteria abundance in tomato rhizosphere causes the incidence of bacterial wilt disease. ISME J 15:330–347. https://doi.org/10.1038/s41396-020-00785-x

Lehman RM, Cambardella CA, Stott DE, Martinez VA, Manter DK, Buyer JS, Maul JE, Smith JL, Collins HP, Halvorson JJ, Kremer RJ, Lundgren JG, Ducey TF, Jin VL, Karlen DL (2015) Understanding and enhancing soil biological health: the solution for reversing soil degradation. Sustainability 7:988–1027. https://doi.org/10.3390/su7010988

Lehmann J, Kleber M (2015) The contentious nature of soil organic matter. Nature 528(7580):60–68. https://doi.org/10.1038/nature16069

Li Y, Zhang Q, Ou L, Ji D, Liu T, Lan R, Jin L (2020) Response to the cold stress signaling of the tea plant (*Camellia sinensis*) elicited by chitosan oligosaccharide. Agronomy 10(6):915. https://doi.org/10.3390/agronomy10060915

Litty T, Gupta A, Gopal M, George P, Thomas GV (2010) Plant growth promoting potential of *Bacillus* spp. isolated from rhizosphere of cocoa (*Theobroma cacao* L.). J Plant Crops 38(2):97–104

Liu CA, Nie Y, Zhang YM, Tang JW, Siddique KHM (2018) Introduction of a leguminous shrub to a rubber plantation changed the soil carbon and nitrogen fractions and ameliorated soil environments. Sci Rep 8(1):1–11. https://doi.org/10.1038/s41598-018-35762-0

Liu CA, Nie Y, Rao X, Tang JW, Siddique KHM (2019) The effects of introducing *Flemingia macrophylla* to rubber plantations on soil water content and exchangeable cations. Catena 172:480–487. https://doi.org/10.1016/j.catena.2018.08.038

Liu C, Wang QW, Jin Y, Tang J, Lin F, Olatunji OA (2021) Perennial cover crop biomass contributes to regulating soil P availability more than rhizosphere P-mobilizing capacity in rubber-based agroforestry systems. Geoderma 401:115218. https://doi.org/10.1016/j.geoderma.2021.115218

Lori M, Symnaczik S, Mader P, Deyn GD, Gattinger A (2017) Organic farming enhances soil microbial abundance and activity-a meta-analysis and meta-regression. PLoS One 12(7):e0180442. https://doi.org/10.1371/journal.pone.0180442

Lu J, Cheng M, Zhao C, Li B, Peng H, Zhang Y, Shao Q, Hassan M (2022) Application of lignin in preparation of slow-release fertilizer: current status and future perspectives. Ind Crops Prod 176(983–986):114267. https://doi.org/10.1016/j.indcrop.2021.114267

Luke SH, Advento AD, Aryawan AA (2020) Managing oil palm plantations more sustainably: large-scale experiments within the biodiversity and ecosystem function in tropical agriculture (BEFTA) program. Front For Global Change 2:75. https://doi.org/10.3389/ffgc.2019.00075

Ma YB, Nichols DG (2004) Phytotoxicity and detoxification of fresh coir dust and coconut shell. Commun Soil Sci Plant Anal 35:205–218. https://doi.org/10.1081/CSS-120027644

Madhaiyan M, Saravanan VS, Blom J (2020) *Phytobacter palmae* sp. nov., a novel endophytic, N2 fixing, plant growth promoting Gamma-proteobacterium isolated from oil palm (*Elaeis guineensis* Jacq.). Int J Syst Evol Microbiol 70:841–884. https://doi.org/10.1099/ijsem.0.003834

Malézieux E (2012) Designing cropping systems from nature. Agron Sustain Dev 32(1):1529. https://doi.org/10.1007/s13593-011-0027-z

Malhotra S, Maheswarappa HP, Selvamani V, Chowdappa C (2017) Diagnosis and management of soil fertility constraints in coconut (*Cocos nucifera*): a review. Indian J Agric Sci 87(6):711–726

Mangalassery S, Nayak MG, Susan PP, Rupa TR, Behera SK, Srinivasan V (2021) Delineating the nutrient constraints and developing nutrient norms for cashew (*Anacardium occidentale* L.) in coastal India. J Plant Nutr 44:2627–2639. https://doi.org/10.1080/01904167.2021.1921198

Manorama K, Behera SK, Suresh K, Prasad MV, Mathur RK, Harinarayana P (2021) Mulching and technological interventions avoid land degradation in an intensive oil palm (*Elaeis guineensis* Jacq.) production system. Land Degrad Dev 32:3785–3797. https://doi.org/10.1002/ldr.3886

Martin R, Gazis R, Skaltsas D, Chaverri P, Hibbett D (2015) Unexpected diversity of basidiomycetous endophytes in sapwood and leaves of Hevea. Mycologia 107(2):284–297. https://doi.org/10.3852/14-206

Martin-Guay MO, Paquette A, Dupras J, Rivest D (2018) The new green revolution: sustainable intensification of agriculture by intercropping. Sci Total Environ 615:767–772. https://doi.org/10.1016/j.scitotenv.2017.10.024

Massaya J, Mills-Lamptey B, Chuck CJ (2022) Soil amendments and biostimulants from the hydrothermal processing of spent coffee grounds. Waste Biomass Valor 13:2889–2904. https://doi.org/10.1007/s12649-022-01697-x

Medina E, Paredes C, Pérez-Murcia MD, Bustamante MA (2009) Spent mushroom substrates as component of growing media for germination and growth of horticultural plants. Bioresour Technol 100(18):4227–4232. https://doi.org/10.1016/j.biortech.2009.03.055

Mghaiouini R, Benzibiria N, Monkade M et al (2022) Formulation of new biostimulant of plant and soil correction based on humic acids extracted by magnetized water from compost from the waste of coffee marc and cattle manure. Waste Biomass Valor 13:453–465. https://doi.org/10.1007/s12649-021-01535-6

Migliorini P, Wezel A (2017) Converging and diverging principles and practices of organic agriculture regulations and agroecology. A review. Agron Sustain Dev 37:1–8. https://doi.org/10.1007/s13593-017-0472-4

Mise K, Koyama Y, Matsumoto A, Fujita K, Kunito T, Senoo K, Otsuka S (2020) Pectin drives microbial phosphorus solubilization in soil: evidence from isolation-based and community-scale approaches. Eur J Soil Biol 97:103169. https://doi.org/10.1016/j.ejsobi.2020.103169

Mitter EK, Tosi M, Obregon D, Dunfield KE, Germida JJ (2021) Rethinking crop nutrition in times of modern microbiology: innovative biofertilizer technologies. Front Sustain Food Syst 5:1–2. https://doi.org/10.3389/fsufs.2021.606815

Mohan Kumar B, Kunhamu TK (2022) Nature-based solutions in agriculture: a review of the coconut (*Cocos nucifera* L.)-based farming systems in Kerala, "the Land of Coconut Trees". Nature-Based Solutions 2:100012. https://doi.org/10.1016/j.nbsj.2022.100012

Montagnini F, Nair P (2004) Carbon sequestration: an underexploited environmental benefit of agroforestry systems. Agrofor Syst 61:281–295. https://doi.org/10.1023/B:AGFO.0000029005.92691.79

Moradi A, Teh CBS, Goh KJ, Husni MHA, Ishak CF (2013) Decomposition and nutrient release temporal pattern of oil palm residues. Ann Appl Biol 164(2):208–219. https://doi.org/10.1111/aab.12094

Mortimer PE, Gui H, Xu J, Zhang C, Barrios E, Hyde KD (2015) Alder trees enhance crop productivity and soil microbial biomass in tea plantations. Appl Soil Ecol 96:25–32. https://doi.org/10.1016/j.apsoil.2015.05.012

Mosier NS, Wyman CE, Dale BE, Elander RT, Lee YY, Holtzapple MT, Michael R (2005) Features of promising technologies for pretreatment of lignocellulosic biomass. Bioresour Technol 96:673–686. https://doi.org/10.1016/j.biortech.2004.06.025

Mosqueira Santillán MJ, Marasco R, Fusi M (2019) Consistent bacterial selection by date palm root system across heterogeneous desert oasis agroecosystems. Sci Rep 9(1):4033. https://doi.org/10.1038/s41598-019-40551-4

Mulaw TB, Kubicek CP, Druzhinina I (2010) The rhizosphere of Coffea arabica in its native highland forests of Ethiopia provides a niche for a distinguished diversity of Trichoderma. Diversity 2:527–549. https://doi.org/10.3390/d2040527

Mulaw TB, Druzhinina IS, Christian P, Kubicek CP, Lea A (2013) Novel endophytic *Trichoderma* spp. isolated from healthy *Coffea arabica* roots are capable of controlling coffee *Tracheomycosis*. Diversity 5:750–766. https://doi.org/10.3390/d5040750

Muniraj I, Shameer S, Ramachandran P, Uthandi S (2021) *Bacillus aryabhattai* TFG5-mediated synthesis of humic substances from coir pith wastes. Microb Cell Fact 20(1):1–9. https://doi.org/10.1186/s12934-021-01538-x

Murphy DJ, Goggin K, Paterson RRM (2021) Oil palm in the 2020s and beyond: challenges and solutions. CABI Agric Biosci 2(1):1–22. https://doi.org/10.1186/s43170-021-00058-3

Murugan M, Ashokkumar K, Alagupalamuthirsolai M, Anandhi A, Ravi R, Dhanya MK, Sathyan T (2022) Understanding the effects of cardamom cultivation on its local environment using novel systems thinking approach-the case of Indian cardamom hills. Front Sustain Food Syst 6:728651. https://doi.org/10.3389/fsufs.2022.728651

Naidu Y, Idris AS, Madihah AZ, Kamarudin N (2016) *In vitro* antagonistic interactions between endophytic basidiomycetes of oil palm (*Elaeis guineensis*) and *Ganoderma boninense*. J Phytopathol 164(10):779–790. https://doi.org/10.1111/jph.12498

Nair PKR (1977) Multispecies crop combinations with tree crops for enhanced productivity in the tropics. Gartenbauwissenschqft 42:145–150

Nair PKR (2017) Managed multi-strata tree+ crop systems: an agroecological marvel. Front Environ Sci 5:88. https://doi.org/10.3389/fenvs.2017.00088

Nampoothiri KUK (2001) Organic farming-its relevance to plantation crops. J Plant Crops 29(1):1–9

Nardi S, Pizzeghello D, Gessa C, Ferrarese L, Trainotti L, Casadoro G (2000) A low molecular weight humic fraction on nitrate uptake and protein synthesis in maize seedlings. Soil Biol Biochem 32:415–419. https://doi.org/10.1016/S0038-0717(99)00168-6

Nascimento Brito V, Lana Alves J, Sírio Araújo K, de Souza LT, Borges de Queiroz C, Liparini Pereira O, de Queiroz MV (2023) Endophytic *Trichoderma* species from rubber trees native to the Brazilian Amazon, including four new species. Front Microbiol 14:1095199. https://doi.org/10.3389/fmicb.2023.1095199

Nayar NM (2017) The coconut: phylogeny, origins, and spread. Elsevier Academic Press, 224 p. https://doi.org/10.1016/B978-0-12-809778-6.00001-2

Nelliat EV, Bavappa KVA, Nair PKR (1974) Multi-storied cropping—a new dimension in multiple cropping for plantations. World Crops 26(6):262–266

Obersteiner S, Klein T (2022) Closing in on the last frontier: C allocation in the rhizosphere. Global Change Biol 28(23):6835–6837. https://onlinelibrary.wiley.com/doi/full/10.1111/gcb.16388.

Orozco FH, Cegarra J, Trujillo LM, Roig A (1996) Vermicomposting of coffee pulp using the earthworm *Eisenia fetida*: effects on C and N contents and the availability of nutrients. Biol Fertil Soils 22:162–166. https://doi.org/10.1007/BF00384449

Palaniswami C, Thomas GV, Subramaniam P, Dhanapal R, Selvamani V (2008) Comparative evaluation of soil health and fertility under organic, inorganic and integrated nutrient management in coconut based mixed farming system. In: Singh HP, Thomas GV (eds) Organic horticulture-principles, practices, and technologies. Westvile Publishing House, New Delhi, pp 216–222

Pandey A, Palni LMS (1997) Bacillus species: the dominant bacteria of the rhizosphere of established tea bushes. Microbiol Res 152(4):359–365. https://doi.org/10.1016/S0944-5013(97)80052-3

Pandey K, Sengupta R (2021) Dry as dust. Down to Earth, pp 20–22

Pandey A, Soccol CR, Nigam P, Brand D, Mohan R, Rousso S (2000) Biotechnological potential of coffee pulp and coffee husk for bioprocesses. Biochem Eng J 6(2):153–162. https://doi.org/10.1016/S1369-703X(00)00084-X

Panklang P, Thoumazeau A, Chiarawipa R, Sdoodee S, Sebag D, Gay F, Thaler P, Brauman A (2021) Rubber, rubber and rubber: how 75 years of successive rubber plantation rotations affect topsoil quality? Land Degrad Dev 33(8):1159–1169. https://doi.org/10.1002/ldr.4171

Paramesh V, Arunachalam V, Nath AJ (2019) Enhancing ecosystem services and energy use efficiency under organic and conventional nutrient management system to a sustainable arecanut based cropping system. Energy 187:1–10. https://doi.org/10.1016/j.energy.2019.115902

Parthasarathi K, Balamurugan M, Prashija KV, Jayanthi L, Ameer Basha S (2016) Potential of *Perionyx excavatus* (Perrier) in lignocellulosic solid waste management and quality vermifertilizer production for soil health. Int J Recycl Org Waste Agric 5:65–86. https://doi.org/10.1007/s40093-016-0118-6

Paungfoo-Lonhienne C, Lonhienne TG, Rentsch D, Robinson N, Christie M, Webb RI, Gamage HK, Carroll BJ, Schenk PM, Schmidt S (2008) Plants can use protein as a nitrogen source without assistance from other organisms. Proc Natl Acad Sci U S A 105(11):4524–4452. https://doi.org/10.1073/pnas.0712078105

Peerawat M, Blaud A, Trap J, Chevallier T, Alonso P, Gay F, Thaler P, Spor A, Sebag D, Chutinan C, Suvannang N, Sajjaphan K, Brauman A (2018) Rubber plantation ageing controls soil biodiversity after land conversion from cassava. Agric Ecosyst Environ 257:92–102. https://doi.org/10.1016/j.agee.2018.01.034

Pham MT, Huang CM, Kirschner R (2019) The plant growth-promoting potential of the mesophilic wood-rot mushroom *Pleurotus pulmonarius*. J Appl Microbiol 1127(4):1157–1171. https://doi.org/10.1111/jam.14375

Philpott SM, Bichier P, Rice R, Greenberg R (2007) Field testing ecological and economic benefits of coffee certification programs. Conserv Biol 21(4):975–985. https://doi.org/10.1111/j.1523-1739.2007.00728.x

Piccolo A, Spaccini R, Savy D, Drosos M, Cozzolino V (2019) The soil humeome: chemical structure, functions and technological perspectives. In: Vaz S Jr (ed) Sustainable agrochemistry—a compendium of technologies. Springer, Heidelberg, pp 183–222. https://doi.org/10.1007/978-3-030-17891-8_7

Pinruan U, Rungjindamai N, Choeyklin R, Lumyong S, Hyde KD, Jones EBG (2010) Occurrence and diversity of basidiomycetous endophytes from the oil palm, *Elaeis guineensis* in Thailand. Fungal Divers 41:71–88. https://doi.org/10.1007/s13225-010-0029-1

Powlson DS, Whitmore AP, Goulding KW (2011a) Soil carbon sequestration to mitigate climate change: a critical re-examination to identify the true and the false. Eur J Soil Sci 62(1):42–55. https://doi.org/10.1111/j.1365-2389.2010.01342.x

Powlson DS, Gregory PJ, Whalley WR, Quinton JN, Hopkins DW, Whitmore AP, Hirsch PR, Goulding KWT (2011b) Soil management in relation to sustainable agriculture and ecosystem services. Food Policy 36(1):S72–S87. https://doi.org/10.1016/j.foodpol.2010.11.025

Prabhu SR, Thomas GV (2002) Biological conversion of coir pith into a value-added organic resource and its application in agri-horticulture: current status, prospect and perspective. J Plant Crops 30:1–17

Prabhu SR, Subramanian P, Bidappa CC, Bopaiah BM (1998) Prospects and improving coconut productivity through vermiculture technology. Indian Coconut J 29:79–84

Priya G, Gupta A, Gopal M, Thomas L, Thomas GV (2012) Screening and *in vitro* evaluation of phosphate solubilizing bacteria from rhizosphere and roots of coconut palms (*Cocos nucifera* L.) growing in different states of India. J Plant Crops 40(1):61–64

Puglia D, Pezzolla D, Gigliotti G, Torre L, Bartucca ML, Buono DD (2021) The opportunity of valorizing agricultural waste, through its conversion into biostimulants, biofertilizers, and biopolymers. Sustainability 13(5):2710. https://doi.org/10.3390/su13052710

Pulleman MM, de Boer W, Giller KE, Kuyper TW (2022) Soil biodiversity and nature-mimicry in agriculture; the power of metaphor? Outlook Agric 51(1):75–90. https://doi.org/10.1177/00307270221080180

Qaim M, Sibhatu KT, Siregar H, Grass I (2020) Environmental, economic, and social consequences of the oil palm boom. Annu Rev Resour Econ 12:321–344. https://doi.org/10.1146/annurev-resource-110119-024922

Qiu Z, Egidi E, Liu H, Kaur S, Singh BK (2019) New frontiers in agricultural productivity: optimized microbial inoculants and in situ microbiome engineering. Biotechnol Adv 37:107371. https://doi.org/10.1016/j.biotechadv.2019.03.010

Quadar J, Chowdhary AB, Dutta R, Angmo D, Rashid F, Singh S, Singh J, Vig AP (2022) Characterization of vermicompost of coconut husk mixed with cattle dung: physicochemical properties, SEM, and FT-IR analysis. Environ Sci Pollut Res 29(58):87790–87801. https://doi.org/10.1007/s11356-022-21899-z

Raguraj S, Kasim S, Jaafar NM, Nazli MH (2022a) Growth of tea nursery plants as influenced by different rates of protein hydrolysate derived from chicken feathers. Agronomy 12(2):299. https://doi.org/10.3390/agronomy12020299

Raguraj S, Kasim S, Jaafar NM, Nazli MH, Amali RKA (2022b) A comparative study of tea waste derived humic-like substances with lignite-derived humic substances on chemical composition, spectroscopic properties and biological activity. Environ Sci Pollut Res 29:60631–60640. https://doi.org/10.1007/s11356-022-20060-0

Rajab AY, Leuschner C, Barus H, Tjoa A, Hertel D (2016) Cacao cultivation under diverse shade tree cover allows high carbon storage and sequestration without yield losses. PLoS One 11(2):e0149949. https://doi.org/10.1371/journal.pone.0149949

Rajesh KPP, Thomas GV, Gupta A, Gopal M (2015) Diversity, richness and degree of colonization of arbuscular mycorrhizal fungi in coconut cultivated along with intercrops in high productive zone of Kerala, India. Symbiosis 65:125–141. https://doi.org/10.1007/s13199-015-0326-2

Rao X, Liu CA, Tang JW, Nie Y, Lioang MY, Shen WJ, Siddique KHM (2021) Rubber-leguminous shrub systems stimulate soil N2O but reduce CO2 and CH4 emissions. For Ecol Manag 480:118665. https://doi.org/10.1016/j.foreco.2020.118665

Raphael K, Velmourougane K (2011) Chemical and microbiological changes during vermicomposting of coffee pulp using exotic (*Eudrilus eugeniae*) and native earthworm (*Perionyx ceylanesis*) species. Biodegradation 22:497–507. https://doi.org/10.1007/s10532-010-9422-4

Reddy DVS, Upadhyay AK (2002) Impact of integrated nutrient management of the mineral nutrition and yield of WCT coconut in littoral sandy soil at Kasaragod. In: Sreedharan K, Vinod Kumar PK, Jayarama, Chulaki BM (eds) Proc. PLACROSYM-XV, Mysore, Karnataka, December 10–13, 2002, pp 274–282

Reed J, van Vianen J, Foli S, Clendenning J, Yang K, MacDonald M, Petrokofsky G, Padoch C, Sunderland T (2017) Trees for life: the ecosystem service contribution of trees to food production and livelihoods in the tropics. For Policy Econ 84:62–71. https://doi.org/10.1016/j.forpol.2017.01.012

Reis VM, Baldani JI, Baldani VLD, Dobereiner J (2000) Biological dinitrogen fixation in gramineae and palm trees. Crit Rev Plant Sci 19:227–247. https://doi.org/10.1080/07352680091139213

Renard D, Tilman D (2021) Cultivate biodiversity to harvest food security and sustainability. Curr Biol 31(19):R1154–R1158. https://doi.org/10.1016/j.cub.2021.06.082

Rencoret J, Marques G, Gutiérrez A, Jiménez-Barbero J, Martínez AT, del Río JC (2013) Structural characterization of lignin isolated from coconut (*Cocos nucifera*) coir fibers. J Agric Food Chem 61(10):34–2445. https://doi.org/10.1021/jf304686x

Richardson AE, Kirkby CA, Banerjee S, Kirkegaard JA (2014) The inorganic nutrient cost of building soil carbon. Carbon Manage 5(3):265–268. https://doi.org/10.1080/17583004.2014.923226

Rodrigues RN, Reis Junior FB, Lopes AA, Rocha OC, Guerra AF, Veiga AD, Mendes IC (2021) Soil enzymatic activity under coffee cultivation with different water regimes associated to liming and intercropped brachiaria. Ciência Rural 52(3):e20200532. https://doi.org/10.1590/0103-8478cr20200532

Rodrıguez E, Nuero O, Guillén F, Martınez AT, Martınez MJ (2004) Degradation of phenolic and non-phenolic aromatic pollutants by four *Pleurotus* species: the role of laccase and versatile peroxidase. Soil Biol Biochem 36(6):909–916. https://doi.org/10.1016/j.soilbio.2004.02.005

Rodriguez MCH, Evans HC, de Abreu LM, Macedo DM, Ndacnou MK, Bekele BK et al (2021) New species and records of *Trichoderma* isolated as mycoparasites and endophytes from cultivated and wild coffee in Africa. Sci Rep 11:5671–5630. https://doi.org/10.1038/s41598-021-84111-1

Rodríguez W, Suárez JC, Casanoves F (2023) Total litterfall and leaf-litter decomposition of *Theobroma grandiflorum* under different agroforestry systems in the western Colombian Amazon. Agrofor Syst 97:1541. https://doi.org/10.1007/s10457-023-00876-6

Rodriguez-Ramos JC, Scott N, Marty J, Kaiser D, Hale L (2022) Cover crops enhance resource availability for soil microorganisms in a pecan orchard. Agric Ecosyst Environ 337:108049. https://doi.org/10.1016/j.agee.2022.108049

Rotondo F, Coniglio R, Cantera L, Pascua ID, Clavijo L, Dieste A (2018) Lignin-based coatings for controlled P-release fertilizer consisting of granulated simple superphosphate. Holzforschung 72(8):637–643. https://doi.org/10.1515/hf-2017-0176

Rupani PF, Embrandiri A, Ibrahim MH, Shahadat M, Hansen SB, Mansor NN (2017) Bioremediation of palm industry wastes using vermicomposting technology: its environmental application as green fertilizer. 3 Biotech 7:1–8. https://doi.org/10.1007/s13205-017-0770-1

Ryan PR, Dessaux Y, Thomashow LS, Weller DM (2009) Rhizosphere engineering and management for sustainable agriculture. Plant Soil 321:363–383. https://doi.org/10.1007/s11104-009-0001-6

Sabrina DT, Hanafi MM, Mahmud TM, Azwady AN (2009) Vermicomposting of oil palm empty fruit bunch and its potential in supplying of nutrients for crop growth. Compost Sci Util 17(1):61–67. https://doi.org/10.1080/1065657X.2009.10702401

Sadasivuni S, Bhat R, Pallem C (2015) Recycling potential of organic wastes of arecanut and cocoa in India: a short review. Environ Technol Rev 4(1):91–102. https://doi.org/10.1080/09593330.2015.1077897

Santo Pereira AD, de Oliveira JL, Savassa SM, Rogério CB, de Medeiros GA, Fraceto LF (2022) Lignin nanoparticles: new insights for a sustainable agriculture. J Clean Prod 345:131145. https://doi.org/10.1016/j.jclepro.2022.131145

Santos A, van Aerle R, Barrientos L, Martinez-Urtaza J (2020) Computational methods for 16S metabarcoding studies using nanopore sequencing data. Comput Struct Biotechnol J 18:296–305. https://doi.org/10.1016/j.csbj.2020.01.005

Savy D, Cozzolino V (2022) Novel fertilising products from lignin and its derivatives to enhance plant development and increase the sustainability of crop production. J Clean Prod 366:132832. https://doi.org/10.1016/j.jclepro.2022.132832

Savy D, Cozzolino V, Nebbioso A, Drosos M, Nuzzo A, Mazzei P, Piccolo A (2016) Humic-like bioactivity on emergence and early growth of maize (*Zea mays* L.) of water-soluble lignins isolated from biomass for energy. Plant Soil 402:221–233. https://doi.org/10.1007/s11104-015-2780-2

Schloss PD, Handelsman J (2005) Metagenomics for studying unculturable microorganisms: cutting the Gordian knot. Genome Biol 6:229. https://doi.org/10.1186/gb-2005-6-8-229

Schmidt JE, DuVal A, Isaac ME, Hohmann P (2022a) At the roots of chocolate: understanding and optimizing the cacao root-associated microbiome for ecosystem services: a review. Agron Sustain Dev 42(2):1–19. https://doi.org/10.1007/s13593-021-00748-2

Schmidt JE, Firl A, Hamran H, Imaniar NI, Crow TM, Forbes SJ (2022b) Impacts of shade trees on the adjacent cacao rhizosphere in a young, diversified agroforestry system. Agronomy 12(1):195. https://doi.org/10.3390/agronomy12010195

Schroeder P (1993) Organic matter cycling by tropical agroforestry systems: a review. J Trop For Sci 7:462–474

Selvamani V, Duraisami VP (2018) Mapping soil constraints using Rs and GIS for the major coconut growing regions of Tamil Nadu, India. J Plant Crops 46(3):190–195

Senapati BK, Panigrahi PK, Lavelle P (1994) Macrofaunal status and restoration strategy in degraded soil under intensive tea cultivation in India. In: Transactions of the 15th world congress of soil science, vol 4A. ISSS, Acapulco, pp 64–75

Senapati BK, Lavelle P, Giri S, Pashanasi B, Alegre J, Decaens T, Jimenez JJ, Albrecht A, Blanchart E, Mahieux M, Rousseaux L, Thomas R, Panigrahi PK, Venkatachalam M (1999) In-soil technologies for tropical ecosystems. In: Lavelle P, Brussaard L, Hendrix PF (eds) Earthworm management in tropical agroecosystems. CAB International, Wallingford, pp 199–237

Shahane AA, Shivay YS (2021) Soil health and its improvement through novel agronomic and innovative approaches. Front Agron 69:680456. https://doi.org/10.3389/fagro.2021.680456

Shashirekha MN, Rajarathnam S (2007) Bioconversion and biotransformation of coir pith for economic production of *Pleurotus florida*: chemical and biochemical changes in coir pith during the mushroom growth and fructification. World J Microbiol Biotechnol 23:1107–1114. https://doi.org/10.1007/s11274-006-9340-0

Shivaprasad P, Hanyappa N, Chandrasekhar N, Manonmani GK, Hareesh SB, Nadaf SA, Jagadeesan M, Venkatesh DH, Sujatha K (2018) Studies on soil fertility status of coffee growing districts of Wayanad district. J Plant Crops 46(3):180–189. https://doi.org/10.25081/jpc.2018.v46.i3.2087

Siddiquee S, Shafawati SN, Naher L (2017) Effective composting of empty fruit bunches using potential *Trichoderma* strains. Biotechnol Rep 13:1–7. https://doi.org/10.1016/j.btre.2016.11.001

Sierra R, Granda B, Holtzapple MT (2009) Lime pretreatment. In: Mielenz JR (ed) Biofuels. Methods in molecular biology, vol 581. Humana Press, Totowa, pp 115–124. https://doi.org/10.1007/978-1-60761-214-8_9

Simard SW (2018) Mycorrhizal networks facilitate tree communication, learning, and memory. In: Baluska F, Gagliano M, Witzany G (eds) Memory and learning in plants. Springer, Cham, pp 191–213. https://doi.org/10.1007/978-3-319-75596-0_10

Simpson BK, Oldham JH, Martin AM (1985) Extraction of potash from cocoa pod husks. Agric Wastes 13(1):69–73. https://doi.org/10.1016/0141-4607(85)90013-7

Simpson RJ, Oberson A, Culvenor RA, Ryab MH, Veneklaas EJ, Lambers H, Lunch JP, Ryab PR, Delhaize E, Smith SE, Harvey PR, Richarson AE (2011) Strategies and agronomic interventions to improve the phosphorus-use efficiency of farming systems. Plant Soil 349:89–120. https://doi.org/10.1007/s11104-011-0880-1

Singh RP, Embrandiri A, Ibrahim MH, Esa N (2011) Management of biomass residues generated from palm oil mill: vermicomposting a sustainable option. Resour Conserv Recycl 55:423–434. https://doi.org/10.1016/j.resconrec.2010.11.005

Singh AK, Liu W, Zakari S, Wu J, Yang B, Jiang XJ, Zhu X, Zou X, Zhang W, Chen C, Singh R (2021) A global review of rubber plantations: impacts on ecosystem functions, mitigations, future directions, and policies for sustainable cultivation. Sci Total Environ 796:148948. https://doi.org/10.1016/j.scitotenv.2021.148948

Sivapalan K, Fernando V, Thenabadu MW (1983) Humified phenol-rich plant residues and soil urease activity. Plant Soil 170:143–146. https://doi.org/10.1007/BF02374757

Somarriba E, Cerda R, Orozcoa L, Cifuentes M, Dávila H, Espin T, Mavisoy H, Ávila G, Alvarado E, Poved V, Astorga C, Say E, Deheuvels O (2013) Carbon stocks and cocoa yields in agroforestry systems of Central America. Agric Ecosyst Environ 173:46–57. https://doi.org/10.1016/j.agee.2013.04.013

Soong JL, Fuchslueger L, Marañon-Jimenez S, Torn MS, Janssens IA, Penuelas J, Richter A (2020) Microbial carbon limitation: the need for integrating microorganisms into our understanding of ecosystem carbon cycling. Global Change Biol 26(4):1953–1961. https://doi.org/10.1111/gcb.14962

Soussana JF, Lutfalla S, Smith P, Lal R, Chenu C, Ciais P (2017) Answer to the viewpoint "sequestering soil organic carbon: a nitrogen dilemma". Environ Sci Technol 51(20):11502. https://doi.org/10.1021/acs.est.7b03932

Srinivasan R, Natarajan A, Kumar KA, Lalitha M (2019) Carbon stocks in major cashew growing soils of coastal Karnataka, India. J Plant Crops 47(1):55–61. https://doi.org/10.25081/jpc.2019.v47.i1.5536

Sriram S, Savitha MJ, Ramanujam B (2010) *Trichoderma*-enriched coco-peat for the management of Phytophthora and Fusarium diseases of chilli and tomato in nurseries. J Biol Control 13:311–316

Stevanus CT, Bintarti AF, Setiawan N (2016) The effect of oil palm frond-based compost as growing media amendment for rubber (*Hevea brasiliensis*, Müll. Arg.) planting material. Malays J Soil Sci 20:95–209

Suárez LR, Salazar JCS, Casanoves F, Being MAN (2021) Cacao agroforestry systems improve soil fertility: comparison of soil properties between forest, cacao agroforestry systems, and pasture in the Colombian Amazon. Agric Ecosyst Environ 314:107349. https://doi.org/10.1016/j.agee.2021.107349

Subramanian P, Dhanapal R., Thomas GV, Maheswarappa HP, Thamban C (2014) Coconut based integrated farming system. Tech. Bull. No.78, ICAR—Central Plantation Crops Research Institute, Kasaragod, 32 p

Suja G, Jaganathan D, Byju G (2023) Organic management of tuberous intercrops for resilience, higher yield and profit from coconut plantations: insights from validation experiments in Kerala, India. J Plant Crops 51(1):1–8. https://doi.org/10.25081/jpc.2023.v51.i1.8466

Sujatha S, Bhat R (2015) Resource use and benefits of mixed farming approach in areca-nut ecosystem in India. Agric Syst 141:126–137. https://doi.org/10.1016/j.agsy.2015.10.005

Sundram S, Othman R, Idris AS, Angel LP, Meon S (2022) Improved growth performance of *Elaeis guineensis* Jacq. through the applications of arbuscular mycorrhizal (AM) fungi and endophytic bacteria. Curr Microbiol 79:155. https://doi.org/10.1007/s00284-022-02842-4

Tang J, Zhou B, Li J, Tang H, Cao J (2016) Effects of earthworm bio-organic fertilization technology on soil microbial characteristics and enzyme activities of tea plants. J Tea Sci 36(1):45–51

Tang S, Ma Q, Luo J, Xie Y, Hashmi MLUR, Pan W, Zheng N, Liu M, Wu L (2021) The inhibition effect of tea polyphenols on soil nitrification is greater than denitrification in tea garden soil. Sci Total Environ 778:146328. https://doi.org/10.1016/j.scitotenv.2021.146328

Tarafdar JC (2022) Role of soil biology on soil health for sustainable agricultural production. In: Giri B, Kapoor R, Wu QS, Varma A (eds) Structure and functions of pedosphere. Springer Nature, pp 67–81. https://doi.org/10.1007/978-981-16-8770-9_3

Tarashkar M, Matloobi M, Qureshi S, Rahimi A (2023) Assessing the growth-stimulating effect of tea waste compost in urban agriculture while identifying the benefits of household waste carbon dioxide. Ecol Indic 151:110292. https://doi.org/10.1016/j.ecolind.2023.110292

Thomas GV (1988) Vesicular-arbuscular mycorrizhal symbiosis in coconut (*Cocos nucifera*) in relation to the root (wilt) disease and intercropping or mixed cropping. Indian J Agric Sci 57(2):145–147

Thomas GV (2010) Technological advances in organic farming of plantation crops. In: Singh HP, Thomas GV (eds) Organic horticulture, principles, practices and technologies. Westville Publishing House, New Delhi, pp 32–47

Thomas GV, Prabhu SR (2003) Association of diazotrophic and plant growth promoting rhizobacteria with coconut palm (*Cocos nucifera* L.). In: Proceedings of the 6th international PGPR workshop, IISR, Calicut, 5–10th October 2003, pp 20–26

Thomas GV, Shantaram MV (1984) *In situ* cultivation and incorporation of green manure legumes in coconut basins; an approach to improve soil fertility and microbial activity. Plant Soil 80:373–380

Thomas GV, Iyer R, Bopaiah BM (1991) Beneficial microbes in the nutrition of coconut. J Plant Crops 19(2):127–138

Thomas GV, George M, Shantaram MV (1993a) Effect of growing green manure legumes in the basins of root (wilt) affected coconut gardens. In: Nair MK, Khan HH, Gopalasundaram P, Rao EVV (eds) Advances in coconut research and development. New Delhi, Oxford and IBH Publishing, pp 421–442

Thomas GV, Rajagopal V, Bopaiah BM (1993b) VA-mycorrhizal association in relation to drought tolerance in coconut. J Plant Crops 21:98–103

Thomas GV, Prabhu SR, Reeny MZ, Bopaiah BM (1998) Evaluation of lignocellulosic biomass from coconut palm as substrate for cultivation of *Pleurotus sajor-caju* (Fr.) Singer. World J Microbiol Biotechnol 14:879–882. https://doi.org/10.1023/A:1008881124903

Thomas GV, Biddappa CC, Prabhu SR (2001) Evaluation of N2 fixing cover legumes as green manures for nitrogen substitution in coconut palm. Trop Agric 78(1):13–18

Thomas GV, Krishnakumar V, Maheswarappa HP, Palaniswami C (2010) Coconut based cropping/farming systems. Central Plantation Crops Research Institute, Kasaragod, 231p

Thomas GV, Palaniswami C, Prabhu SR, Gopal M, Gupta A (2013) Co-composting of coconut coir pith with solid poultry manure. Curr Sci 104:245–250

Thomas GV, Subramanian P, Gopal M, Gupta A, Prabhu SR (2022) Unraveling the potential of belowground and aboveground biodiversity for sustainable management of the health of plantation crop soils in coastal agroecosystem. In: Lama T, Burman D, Mandal UK, Sarangi SK, Sen H (eds) Transforming coastal zone for sustainable food and income security. Springer, pp 633–643. https://doi.org/10.1007/978-3-030-95618-9_48

Tilman D (1999) Global environmental impacts of agricultural expansion: the need for sustainable and efficient practices. Proc Natl Acad Sci U S A 96:5995–6000. https://doi.org/10.1073/pnas.96.11.599

Tilman D, Fargione J, Wolff B, D'antonio C, Dobson A, Howarth R, Schindler D, Schlesinger WMH, Simberloff D, Swackhamer D (2001) Forecasting agriculturally driven global environmental change. Science 292:281–284. https://doi.org/10.1126/science.1057544

Tilman D, Cassman KG, Matson PA, Naylor R, Polasky S (2002) Agricultural sustainability and intensive production practices. Nature 418:671–677. https://doi.org/10.1038/nature0101

Tolisano C, Del Buono D (2023) Biobased: biostimulants and biogenic nanoparticles enter the scene, vol 885. Sci Total Environ, p 163912. https://doi.org/10.1016/j.scitotenv.2023.163912

Tolisano C, Luzi F, Regni L, Proietti P, Puglia D, Gigliotti G, Michele AD, Priolo D, Buono DD (2023) A way to valorize pomace from olive oil production: lignin nanoparticles to biostimulate maize plants. Environ Technol Innov 31:103216. https://doi.org/10.1016/j.eti.2023.103216

Trevisan S, Pizzeghello D, Ruperti B, Francioso O, Sassi A, Palme K, Quaggiotti S, Nardi S (2010) Humic substances induce lateral root formation and expression of the early auxin-responsive IAA19 gene and DR5 synthetic element in Arabidopsis. Plant Biol 12(4):604–614. https://doi.org/10.1111/j.1438-8677.2009.00248.x

Tripetchkul S, Pundee K, Koonsrisuk S, Akeprathumchai S (2012) Co-composting of coir pith and cow manure: initial C/N ratio vs physico-chemical changes. Int J Recycl Organic Waste Agric 1(1):1–8. https://doi.org/10.1186/2251-7715-1-15

Vaast P, Somarriba E (2014) Trade-offs between crop intensification and ecosystem services: the role of agroforestry in cocoa cultivation. Agrofor Syst 88:947–956. https://doi.org/10.1007/s10457-014-9762-x

Van Der Vossen HAM (2005) A critical analysis of the agronomic and economic sustainability of organic coffee production. Exp Agric 41(4):449–473. https://doi.org/10.1017/S0014479705002863

Van Doren J, Van den Hurk K, Broekmeulen D (2017) Can coir pith replace peatmoss for horticultural purposes in the future? Acta Hortic 1266:43–48. https://doi.org/10.17660/ActaHortic.2019.1266

van Noordwijk M, Brussaard L (2014) Minimizing the ecological footprint of food: closing yield and efficiency gaps simultaneously? Curr Opin Environ Sustain 8:62–67. https://doi.org/10.1016/j.cosust.2014.08.008

Varmazyari A, Çakmakçı R (2018) Diversity and metabolic potential of the dominant culturable N2-fixing and P-solubilising bacteria from tea (*Camellia sinensis* L.) rhizosphere. Front Environ Microbiol 4(2):45–54. https://doi.org/10.11648/j.fem.20180402.12

Vásquez ZS, de Carvalho Neto DP, Pereira GV, Vandenberghe LP, de Oliveira PZ, Tiburcio PB, Rogez HL, Neto AG, Soccol CR (2019) Biotechnological approaches for cocoa waste management: a review. Waste Manag 90:72–83. https://doi.org/10.1016/j.wasman.2019.04.030

Vassileva M, Flor-Peregrin E, Malusá E, Vassilev N (2020) Towards better understanding of the interactions and efficient application of plant beneficial prebiotics, probiotics, postbiotics and synbiotics. Front Plant Sci 11:1068. https://doi.org/10.3389/fpls.2020.01068

Vasundhara R, Prakash NB, Kumar KA, Hegde R, Dharumarajan S (2021) Soil fertility status of coconut and arecanut growing soils. J Plant Crops 49(2):83–93. https://doi.org/10.25081/jpc.2021.v49.i2.7254

Venturini AM, Gontijo JB, Mandro JA, Berenguer E, Peay KG, Tsai SM, Bohannan BJM (2023) Soil microbes under threat in the Amazon rainforest. Trends Ecol Evol 38(8):693–696. https://doi.org/10.1016/j.tree.2023.04.014

Vico A, Pérez-Murcia MD, Bustamante MA, Agullo E, Marhuenda-Egea FC, Saez JA, Parfedes C, Perez-Espinosa A, Moral R (2018) Valorization of date palm (*Phoenix dactylifera* L.) pruning biomass by co-composting with urban and agri-food sludge. J Environ Manag 226:408–415. https://doi.org/10.1016/j.jenvman.2018.08.035

Vitale E, Motta CM, Avallone B, Amoresano A, Fontanarosa C, Battaglia G, Spinelli M, Fogliano C, Paradiso R, Arena C (2023) Sustainable reuse of expresso coffee by-products as a natural fertilizer to improve growth and photosynthesis in cucumber (*Cucumis sativus* L.) plants. Waste Biomass Valor 8:1–7. https://doi.org/10.1007/s12649-023-02143-2

Voltolini GB, Vilela MS, Alecrim AO, da Silva LC, Guimaraes RJ, Netto PM, Rezende TT (2023) Techniques for sustainable coffee crop management: impacts on growth and yield. Arch Agron Soil Sci 69(12):2413–2242. https://doi.org/10.1080/03650340.2022.2157406

Vukicevich E, Lowery T, Bowen P, Úrbez-Torres JR, Hart M (2016) Cover crops to increase soil microbial diversity and mitigate decline in perennial agriculture. A review. Agron Sustain Dev 36(3):1–14. https://doi.org/10.1007/s13593-016-0385-7

Wagner MR, Roberts JH, Balint-Kurti P, Holland JB (2020) Heterosis of leaf and rhizosphere microbiomes in field grown maize. New Phytol 228:1055–1069. https://doi.org/10.1111/nph.16730

Wang Y, Yao Z, Pan Z, Wang R, Yan G, Liu C, Su Y, Zheng X, Butterbach-Bahl K (2020) Tea-planted soils as global hotspots for N2O emissions from croplands. Environ Res Lett 15(10):104018. https://doi.org/10.1088/1748-9326/aba5b2

Wang D, Yi W, Zhou Y, He S, Tang L, Yin X, Zhao P, Long G (2021) Intercropping and N application enhance soil dissolved organic carbon concentration with complicated chemical composition. Soil Tillage Res 210:104979. https://doi.org/10.1016/j.still.2021.104979

Warren-Thomas EM, Edwards DP, Bebber DP, Chhang P, Diment AN, Lambrick FH, Maxwell JF, Nut M, O'Kelly HJ, Theilade I, Dolman PM (2018) Protecting tropical forests from the rapid expansion of rubber using carbon payments. Nat Commun 9:1–12. https://doi.org/10.1038/s41467-018-03287-9

Warren-Thomas E, Nelson L, Juthong W, Bumrungsri S, Brattström O, Stroesser L, Chambon B, Penot É, Tongkaemkaew U, Edwards DP, Dolman PM (2019) Rubber agroforestry in Thailand provides some biodiversity benefits without reducing yields. J Appl Ecol 57(1):17–30. https://doi.org/10.1111/1365-2664.13530

Watson CA, Atkinson D, Gosling P, Jackson LR, Rayns FW (2006) Managing soil fertility in organic farming systems. Soil Use Manag 18(1):239–247. https://doi.org/10.1111/j.1475-2743.2002.tb00265.x

Wilson MH, Lovell ST (2016) Agroforestry—the next step in sustainable and resilient agriculture. Sustainability 8:574. https://doi.org/10.3390/su8060574

Wittwer RA, Dorn B, Jossi W, van der Heijden MGA (2017) Cover crops support ecological intensification of arable cropping systems. Sci Rep 7:41911. https://doi.org/10.1038/srep41911

Xu L, Geelen D (2018) Developing biostimulants from agro-food and industrial by-products. Front Plant Sci 9:1567. https://doi.org/10.3389/fpls.2018.01567

Zakry FA, Shamsuddin ZH, Rahim KA, Zakaria ZZ, Rahim AA (2012) Inoculation of *Bacillus sphaericus* UPMB-10 to young oil palm and measurement of its uptake of fixed nitrogen using the ^{1}7N isotope dilution technique. Microbes Environ 27(3):257–262. https://doi.org/10.1264/jsme2.me11309

Zhang J, Cook J, Nearing JT, Zhang J, Raudonis R, Glick BR, Langille MGI, Cheng Z (2021) Harnessing the plant microbiome to promote the growth of agricultural crops. Microbiol Res 245:1266–1290. https://doi.org/10.1016/j.micres.2020.126690

Zhao Q, Xiong W, Xing Y, Sun Y, Lin X, Dong Y (2018) Long-term coffee monoculture alters soil chemical properties and microbial communities. Sci Rep 8(1):6116. https://doi.org/10.1038/s41598-018-24537-2

Zhou J, Zhang F, Huo Y, Wilson GWT, Cobb AB, Xu X, Xiong X, Liu L, Zhang Y (2019) Following legume establishment, microbial and chemical associations facilitate improved productivity in degraded grasslands. Plant Soil 443:273–292. https://doi.org/10.1007/s11104-019-04169-9

Dr. George V. Thomas is a Former Director of the Central Plantation Crops Research Institute, Kasaragod of the Indian Council for Agricultural Research, Government of India. He has a research and research management career spanning 36 years in plantation crops, particularly on coconut. He made significant contributions to soil microbiology, soil fertility, plant–microbe interactions, biomass recycling and bio-formulations development in plantation crops. He was chairman, International Coconut Genetic Resources Network; President, Indian Society for Plantation Crops, Editor, *Journal of Plantation Crops* and Fellow of four Professional Societies. He is a recipient of ICAR-Outstanding Team Research Award in Agriculture and Allied Sectors, as Team Leader for 2005–2006. He also served as Emeritus Scientist of Kerala State Council for Science, Technology and Environment and Director of Council for Food Research and Development, Konni, Kerala, India. Dr. Thomas has over 200 research publications and edited 10 books.

Dr. V. Krishnakumar completed his M. Sc (Agri) in Agronomy from the College of Agriculture, Vellayani, Thiruvananthapuram of the KAU and Ph. D in Agronomy from the Tamil Nadu Agricultural University, Coimbatore, India and served various state and central government research institutions. He had worked as the Principal Investigator, Co-investigator of various National and International Projects. He superannuated as Principal Scientist (Agronomy) and Head, ICAR-Central Plantation Crops Research Institute, Regional Station, Kayamkulam, Kerala, India. As an agronomist, he has undertaken various field research programs, farmer-community-oriented technology dissemination activities, administration, and manpower management programs. He is a recipient of the ICAR Award for Outstanding Interdisciplinary Team Research in Agriculture and Allied Sciences for the biennium 2005–2006. In his career of 37 years, he has published more than 100 research articles both in the National and International journals; edited 10 books and contributed chapters; edited and published technical bulletins, training manuals, and other compilations. He served as Editor, *Journal of Plantation Crops* and currently is a member of the Editorial Board of Current Science.

S. R. Prabhu is a soil microbiologist currently serving as the Chief Scientific Officer at TerraBioGen Technologies, Burnaby, BC, Canada. He served for 10 years at CPCRI, Kasaragod and focused on recycling of ligno-cellulosic biomass using earthworms and root-associated and endophytic plant growth promotion. His current interests are development of novel crop biostimulants and culturing hard-to-culture microbes of agricultural importance.

SPRINGER NATURE

GPSR Compliance

The European Union's (EU) General Product Safety Regulation (GPSR) is a set of rules that requires consumer products to be safe and our obligations to ensure this.

If you have any concerns about our products, you can contact us on ProductSafety@springernature.com

In case Publisher is established outside the EU, the EU authorized representative is:

Springer Nature Customer Service Center GmbH
Europaplatz 3
69115 Heidelberg, Germany

uthorised representative in the EU is Springer
vice Centre GmbH, Europaplatz 3, 69115 Heidelberg,
any concerns regarding our products, please
@springernature.com

I Group (UK) Ltd, Croydon, CR0 4YY
30/09/2026
225318-0019